Example 3. A new dorm is to be built on campus. Because of expenses anc dorm room will have an 18-meter perimeter, including windows and doors. Th… much floor space (area) as possible. Which rectangle should they choose?

Using a Table

Length	*Width*	*Perimeter*	*Area*
1	8	18	8
2	7	18	14
3	6	18	18
4	5	18	20
5	4	18	20
6	3	18	18
7	2	18	14
8	1	18	8

Using a Graph

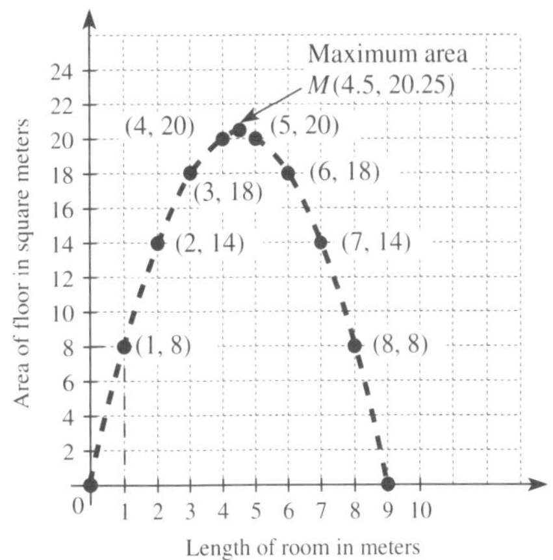

Using an Equation

Perimeter $= 2(\text{Length} + \text{Width}) = 18$

Area $=$ Length $\times$ Width, or

$$l + w = 9 \quad \text{or} \quad w = 9 - l$$

$$A = l(9 - l) = 9l - l^2$$

QUADRATIC FUNCTION

Example 4. Investigate the relationship between rate (speed) and time it takes for a car to travel 500 miles.

Using a Table

Rate (mph)	*Time (hours)*	*Distance (miles)*
10	50	500
20	25	500
30	16.7	500
40	12.5	500
50	10	500
60	8.3	500
70	7.1	500
80	6.25	500

Using a Graph

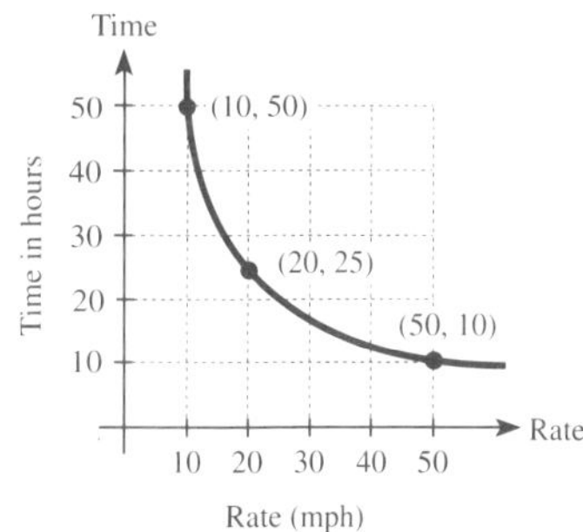

Using an Equation

Time equals distance divided by rate, or

$$T = \frac{500}{R}$$

RATIONAL FUNCTION

Intermediate Algebra
Applications and Problem Solving

SECOND EDITION

Elizabeth Difanis Phillips
Michigan State University

Thomas Butts
University of Texas at Dallas

J. Michael Shaughnessy
Portland State University

HarperCollins*CollegePublishers*

Sponsoring Editor: Karin E. Wagner/Anne Kelly
Senior Development Editor: Kathy Richmond
Cover Design: Lesiak/Crampton Design Inc.: Cynthia Crampton
Photo Researcher: Sandy Schneider/Karen Koblik
Project Coordination and Text Design: Elm Street Publishing Services, Inc.
Compositor: Weimer Graphics, Inc.
Printer and Binder: R. R. Donnelley & Sons Company
Art Studio: Rolin Graphics Inc.
Cover, Painting, *Saginaw Hill,* by William Conger, 1992, oil on canvas, 50″ x 60″, Courtesy Roy Boyd Gallery, Chicago, IL.

Intermediate Algebra: Applications and Problem Solving, Second Edition

Library of Congress Cataloging-in-Publication Data

Phillips, Elizabeth Difanis, 1937–
Intermediate algebra: applications and problem solving / Elizabeth Difanis Phillips, Thomas Butts, J. Michael Shaughnessy.—2nd ed.
p. cm.
Includes index.
ISBN 0-06-045220-X
1. Algebra. I. Butts, Thomas, 1943– . II. Shaughnessy, J. Michael, 1946– . III. Title.
QA154.2.P53 1993
512′.9—dc20 93-4030

93 94 95 96 9 8 7 6 5 4 3 2 1

Contents

CHAPTER

CHAPTER 6

Quadratic Equations and Quadratic Functions 334

Rational Exponents and Exponential Equations 432

CHAPTER 8

CHAPTER 9

Preface

The Course and Its Audience

Intermediate Algebra: Applications and Problem Solving, second edition, is designed for students who will be going on to study college algebra, finite mathematics, or statistics, or for students who are liberal arts or elementary education majors, or for students who want to brush up on their algebra skills. This book starts with a review of the arithmetic of whole numbers, so it is appropriate for a student who has not had any algebra.

Approach

As in the first edition, we strongly believe that algebra can be taught in a way that is both stimulating and interesting to the student by continually posing relevant problems and applications. Recent position statements in the *Curriculum and Evaluation Standards* of the National Council of Teachers of Mathematics and in Mathematics Association of America publications echo our sentiments. As students pursue the solution to a particular problem, they can observe any inherent underlying patterns. They can conjecture, test, discuss, verbalize, and generalize these patterns. Students can discover the salient features of the pattern, construct understandings of concepts and relationships, and develop a language to talk about the pattern. Throughout this process students encounter mathematical concepts and procedures in context. In this way symbols can take on meaning for the students and students can develop an intuitive understanding of the concepts and procedures while at the same time developing problem-solving and critical-thinking skills.

Posing problems and applications early and promoting the use of graphing calculators provide natural opportunities to introduce functions. When relationships between quantities are studied, knowledge about important mathematical relationships and functions emerges. Functions are used constantly throughout the book in an informal way. Functional relationships are used continually throughout the book and are identified as they naturally occur. These relationships are first discussed in tables and graphical representations and then in symbolic form. These representations provide the basis for a major theme of this book: Algebra is a language to describe patterns and relationships. Chapter 8 is the capstone, in which the concept of a function is revisited and is studied in depth, building on examples that have been developed throughout previous chapters.

Distinguishing Features of the Book

Multiple Representations

Students need to see many different representations of mathematical ideas rather than just the symbolic form. For example, a problem can be represented verbally, through visual models, as data in a table, as a graph, or in symbols. Each representation reveals some aspect of the problem that may not be so evident in the other representations. With the advent of graphing calculators, graphical representations have become a more powerful tool to understand and solve problems. This book places a heavy emphasis on graphical approaches to algebraic concepts and on the connections between various representations of algebraic concepts. Throughout the text, tables, graphs, and symbols are used simultaneously to represent ideas and to solve problems. Students are asked to apply various representations and to translate information from one representation to another.

Equivalent Expressions

The notions of equivalent algebraic expressions and equivalent equations (same solution set) are introduced throughout the text. This approach helps to avoid the ambiguous and varied meaning of the word *simplify,* which often confuses students. We want students to realize that many expressions are equivalent to a given algebraic expression and that the form used in a particular situation depends heavily on the context and the information provided.

Related Examples

Often examples are grouped together to point out the salient features of a concept or a procedure providing variations on a theme. Such grouped sets include building complex procedures from simple ones, identifying what are and are not examples of a concept, and making connections between multiple representations of a concept or connecting several closely related concepts. Pertinent commentary appears next to the solution steps for each example to point out subtleties or difficult steps, to recall needed facts, and to provide a rationale for the solution process. In addition, common errors are discussed. Students are also asked to find errors in given solutions.

Strategies and Concepts

Throughout the book applications, patterns, and examples are used to demonstrate important concepts and strategies. After this motivational discussion, the concepts and strategies are then highlighted in colored boxes. These boxes are preceded by narrative outlining the transition between worked examples and the boxes themselves.

Problem Sets

The problem sets are partitioned into four parts:

I. *Procedures* help students to apply algebraic skills and algorithmic processes.

II. *Concepts* enable students to investigate and compare, to explore and make connections between several representations of a concept, and to extend and enrich their understanding of concepts.

III. *Applications* allow students to use concepts and procedures to solve problems from a variety of settings including science, business, agriculture, forestry, engineering, medicine, and social science.

IV. *Extensions* challenge students to investigate more challenging problems related to the current material or foreshadowing material to come. These problems include puzzles, historical anecdotes, interesting patterns, and further examination of concepts and applications already introduced.

Writing

In each chapter there are problems in Parts II, III, and IV that encourage students to write about their understanding of concepts. For instance, students are asked to write an explanation, including examples, connecting the graph of the quadratic equation $y = ax^2 + bx + c$ to the solutions of the equation $0 = ax^2 + bx + c$. In another case, students are asked to describe and compare strategies for solving a particular type of equation.

Group Work

We have found that many of the problems in Parts II, III, and IV are especially suitable for small-group problem solving. For instance, Exercises 57–64 in Section 2.2 are adaptable to small-group discussion. These problems contain a set of statements, and students must decide whether they are always true, sometimes true, or never true. For the statements that are always true or never true, students must give reasons why this is so. If the statement is sometimes true, students must give an example showing when it is true and one showing when it is false. Another sample is Exercise 55 in Section 2.2, which makes for a good group discussion on the various uses of the "–" symbol. Students must include possible misconceptions or errors in their discussion.

Focus on Technology

Calculator use is integrated naturally into the examples and problem sets. We strongly encourage that students have access to a graphing calculator (or computer lab with graphing software). Throughout the book, we introduce the use of graphing calculators to emphasize the connections between graphical solutions and symbolic solutions. Students should be able to recognize and sketch basic groups in a family of functions such as linear, quadratic, and exponential functions. While it is possible to draw all of the graphs with paper and pencil, graphing calculators are preferred because they allow for greater exploration of functional relationships.

Content Highlights

The first section of Chapter 1, "What Is Algebra?" gives an overview of the book: Algebra is a *brief* and *general language* used to describe mathematical patterns and patterns in the world. The rest of the chapter provides a review of arithmetic and sets the tone and format for the book.

The first example in Chapter 2 demonstrates a need for exponents and illustrates an exponential function. This pattern is represented in a table, as a graph, and then with symbols. After many examples of how exponents are defined using both numbers and symbols and illustrations of exponential patterns, the properties of exponents are introduced.

The first example of Chapter 3 uses the cost of renting cars to illustrate a linear function. Three methods for modeling a set of data are illustrated: table, graph, and equation. Proportions (including percents) are defined and solved as linear equations. The concepts of conditional equation, identity, and contradiction are illustrated using linear equations. Finally, the car rental problem is used to introduce linear inequalities.

The advantages of and the need for graphs that arose briefly in the three previous chapters are explored in more detail in Chapter 4. Slope, intercepts, parallel and perpendicular lines, and 2×2 linear systems are examined in appropriate applications.

In the first example in Chapter 5, a glider traveling in the wind is used to illustrate rational equations. Equivalent rational expressions and applications are presented to explain operations with rational expressions and how to solve rational equations. The emphasis is on finding equivalent expressions or equivalent equations.

Applications continue to play an important role in Chapter 6, where they are used to introduce properties of quadratic functions. An important focus in this chapter is on the relationship between $y = ax^2 + bx + c$ and $0 = ax^2 + bx + c$.

The basic properties of integral exponents are reviewed in Chapter 7 and extended to rational exponents. Graphing and applications continue to promote an intuitive understanding of radicals and rational exponents and to help students make sense of the symbols. This chapter culminates by solving exponential and radical equations.

Chapter 8 studies the function concept more formally and reviews material in previous chapters by recasting tables, graphs, and general patterns in function notation. Transformational relationships between functions are investigated by looking at translations and reflections of graphs.

In Chapter 9 the emphasis is on applications of exponential and logarithmic functions using graphs and calculators. This chapter provides a springboard for students who go on to courses in basic statistics, business, chemistry, or physics.

Course Flexibility

Although Chapter 1 can be omitted, it is worth doing a few problems from Section 1.1 to set the theme of algebra as a language that uses problems, tables, graphs, or equations to study relationships and functions. The order of the chapters can be altered if appropriate problems are omitted that use concepts from a previous chapter. Other possible orders are as follows.

- Chapters 1, 3, 4, 2, 5, 6, 7, 8, 9
- Chapters 1, 2, 3, 4, 6, 5, 7, 8, 9

Supplement Package to Accompany This Text

For the Instructor

The *Instructor's Manual: Tests and Solutions* contains complete, worked-out solutions to all of the even-numbered section exercises. Also included are printed tests and answers to all the test problems.

The *HarperCollins Test Generator/Editor for Mathematics with QuizMaster* is available in IBM and Macintosh versions and is fully networkable. The test generator enables instructors to select questions by objective, section, or chapter or to use a ready-made test for each chapter. The editor enables instructors to edit any preexisting data or to create their own questions. The software is algorithm-driven and allows the instructor to regenerate constants while maintaining problem type so that a large number of test or quiz items are available in both multiple-choice or open-response formats. The system features printed graphics and accurate mathematics symbols. QuizMaster enables instructors to create tests and quizzes using the Test Generator/Editor and save them to disk so that students can take the test or quiz on a stand-alone computer or network. QuizMaster then grades the test or quiz and allows the instructor to create reports on individual students or classes.

A *videotape series* is available to accompany *Intermediate Algebra: Applications and Problem Solving,* second edition.

For the Student

The *Student's Solution Manual* includes complete, worked-out solutions to the odd-numbered problems in the text with the exception of writing and discussion exercises where more than one answer may be appropriate. To order, use ISBN 0-06-501927-X.

An *Interactive Tutorial Software with Management System* is available in IBM and Macintosh versions and is fully networkable. This software is algorithm-driven and automatically regenerates constants so that a student will not see the numbers repeat in a problem type if he or she revisits any particular section. The tutorial is self-paced and provides unlimited opportunities to review lessons and to practice problem solving. When students give a wrong answer, they can request to see the problem worked out. The problem is menu-driven for easy use, and on-screen help can be obtained at any time with a single keystroke. Students' scores are automatically recorded and can be printed for a permanent record. The optional management system lets instructors record student scores on disk and print diagnostic reports for individual students or classes. For student convenience and home use the tutorial is also available for purchase. To order, use ISBN 0-06-502193-2 for the Macintosh version, ISBN 0-06-502233-5 for IBM.

Acknowledgments

For their assistance in markedly improving the quality of this book, many people deserve our heartfelt thanks. The following reviewers supplied much useful criticism after careful readings of various versions of the manuscript:

Radwan Al-Jarrah
Ohio State University

Jerry Burkhart
Mankato State University

Edgar M. Chandler
Paradise Valley Community College

Joyce Cundiff
Principia College

Bettie A. DeGryse
Black Hawk College

Margaret M. Donlan
University of Delaware

Barbara Edwards
Portland State University

Elaine M. Hale
Georgia State University

Louise S. Hasty
Austin Community College–Rio Grande

Martha Ann Larkin
Southern Utah University

Arthur Lieberman
Cleveland State University

Margaret Barnes Morrison
San Jacinto College–Central

Stuart Moskowitz
Butte College

Ernest Palmer
Grand Valley State University

Willard Raiffeisen
Texas State College

Pamela Reisner
Wittenberg University

Donna Fields Rochon
Western Washington University

C. Donald Smith
Louisiana State University–Shreveport

Thomas J. Stillman
University of San Francisco

Shirley M. Thompson
North Lake College

Patrick Wagener
Los Medanos College

Martha Wilson
University of Delaware

Lois Yamakoshi
Los Medanos College

The following instructors completed a survey about their intermediate algebra courses:

Eugenia D. Allen
York Technical College

Robert Buck
Slippery Rock University of Pennsylvania

Charlotte M. Grossbeck
State University of New York–Cobleskill

Kay D. Haralson
Austin Peay State University

Harold N. Hauser
Mount Hood Community College

Jaclyn LeFebvre
Illinois Central College

Don H. Martin
Grayson County College

Ernest Palmer
Grand Valley State University

Donald Platte
Mercyhurst College

Willard Raiffeisen
Texas State College

Pamela Reisner
Wittenberg University

Donna Fields Rochon
Western Washington University

Larry Runyan
Shoreline Community College

Dorothy Schwellenbach
Hartnell College

R. Alvin Vaughn
Ulster County Community College

Patrick Wagener
Los Medanos College

Shelda Warren
Moorehead State University

Melvin R. Woodard
Indiana University of Pennsylvania

Class Testing

We are especially grateful to Pamela Reisner, Wittenberg University, and Penny Slingerland and Allison Warr, Mount Hood Community College, for taking on the effort and the adventure of using earlier drafts of the second edition in their classrooms.

We thank Pamela Reisner, Wittenberg University, and Dan and Sherrie Nicol, University of Wisconsin–Platteville, for their assistance in checking the mathematical accuracy of the answers.

Thanks also go to the editorial staff at HarperCollins, especially Kathy Richmond, Anne Kelly, Karin E. Wagner, and to Joyce E. Skoog, Pamela Johnson, and JoAnn Learman of Elm Street Publishing Services for their work in making this second edition a reality.

E. D. P.
T. B.
J. M. S.

CHAPTER

1 The Real Numbers and Graphs

What is algebra and why do we need to know it? What is the connection between arithmetic and algebra? This chapter seeks to answer these questions.

Source: Pete Saloutos / The Stock Market

1.1 WHAT IS ALGEBRA?

Algebra as a Language: Tables, Graphs, Equations

Mathematics is the study of patterns in the world around us. Algebra—the language of mathematics—is used to describe and analyze patterns. The use of mathematics to study a wide array of patterns that occur in science, social science, psychology, forestry, engineering, nursing, economics, and business is rapidly increasing. In this book we will study patterns from many of these areas. The purpose of this section is to provide an overview of algebra as a language and as a tool to investigate and solve problems. You are not expected at this time to completely understand all the examples. The remaining chapters will develop techniques that will help you to analyze patterns and solve problems.

As a first example, consider the following problem:

EXAMPLE 1

A new dorm is to be built on campus. Because of expenses and other considerations the architects decide that each dorm room will measure 18 meters around the edges of the room, including windows and doors. The architects ask a group of student advisors to decide on the dimensions of the room. The students decide they want the rectangle that will give them the greatest amount of floor space. Which rectangle should they choose?

Solution

To help the students make a decision, we first recognize that the measure around the edges of the room is equivalent to the perimeter of the rectangle. So our problem is to find the rectangle with a perimeter of 18 meters that has the greatest area. We know that a rectangle has two dimensions: length and width. Recall that the **perimeter** of a rectangle is twice the sum of the length and the width, or perimeter = 2(length + width), and the **area** of a rectangle is length times width. (See Figure 1.1.) We will solve this problem using three methods: a table, a graph, and an equation.

Method 1: Using a Table Let us begin by using trial and error to find some rectangles with dimensions in whole numbers. (We will use units in place of meters. Thus our solution will be valid for all units of measure—meters, feet, etc.) For example, if the length is 7 units, then the width is 2 units, since $2(7 + 2) = 18$ units. The area of this rectangle is 7 times 2, or 14 square units. Some of the rectangles that have a perimeter of 18 units

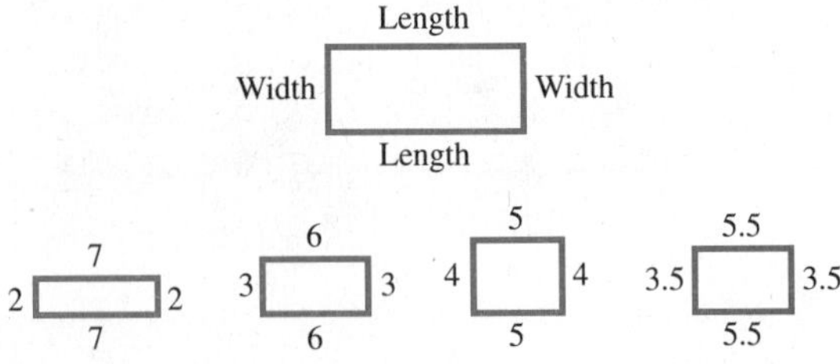

Figure 1.1

Rectangle	*Length in units*	*Width in units*	*Perimeter in units*	*Area in square units*
A	1	8	18	8
B	2	7	18	14
C	3	6	18	18
D	4	5	18	20
E	5	4	18	20
F	6	3	18	18
G	7	2	18	14
H	8	1	18	8

Table 1.1

are described in Figure 1.1. To find all the rectangles with whole number dimensions and their corresponding areas, it is convenient to systematically start with the smallest value for the length (or width) and arrange the values in a table (Table 1.1).

From Table 1.1 we begin to observe some patterns. For example, as the length increases the width decreases, but the area increases to 20 square units and then starts to decrease. The rectangle with the greatest area has a length of 4 and a width of 5 (or a length of 5 and a width of 4). For whole numbers it appears that the rectangle with a perimeter of 18 units having the greatest area is the rectangle that is most like a square.

If we let the dimensions of the rectangle be any real numbers (including fractions and decimals), is there a rectangle with a perimeter of 18 that has an area greater than 20? Judging from Table 1.1, a good guess would be a rectangle that occurs between the two rectangles with dimensions 4 by 5 and 5 by 4. Let us try a length between 4 units and 5 units, say 4.5 units. The width must also be 4.5 since the perimeter $= 18 = 2(4.5 + 4.5)$. The area is 4.5 times 4.5, or 20.25 square units. This area is slightly more than 20 square units. By trying numbers slightly greater or slightly smaller than 4.5 we can conjecture that the rectangle with dimensions of 4.5 units by 4.5 units has the greatest area for a perimeter of 18 units.

There are an infinite number of rectangles with a perimeter of 18. It would be impossible to list them all in a table. However, we conjecture that only one rectangle has the greatest area. Note that this rectangle is a square.

Method 2: Using a Graph Another way to organize all the data in Table 1.1 is with a graph. To form a graph we use two perpendicular lines, which are called the **axes.** In the graph (Figure 1.2), the numbers on the horizontal axis represent the lengths of the rectangles and the numbers on the vertical axis represent the widths of the rectangles. The points, A, B, C, D, E, F, G, and H represent the whole number lengths and widths of some rectangles with a perimeter of 18. These points correspond to the first two columns of entries in Table 1.1. For example, point A represents the rectangle that is 1 unit in length and 8 units in width. We say the **coordinates** of point A are (1, 8). The horizontal value is always given first. Point B represents the rectangle that is 2 units in length and 7 units in width. Thus point B has the coordinates (2, 7). To plot the point C that represents the rectangle that is 3 units long and 6 units wide, we start from the origin and move 3 units

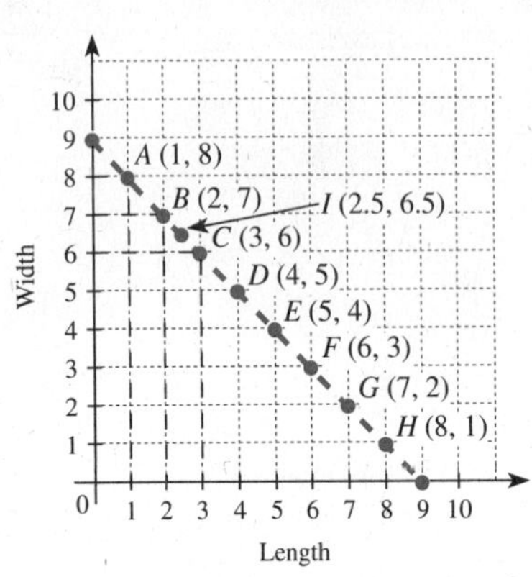

Figure 1.2

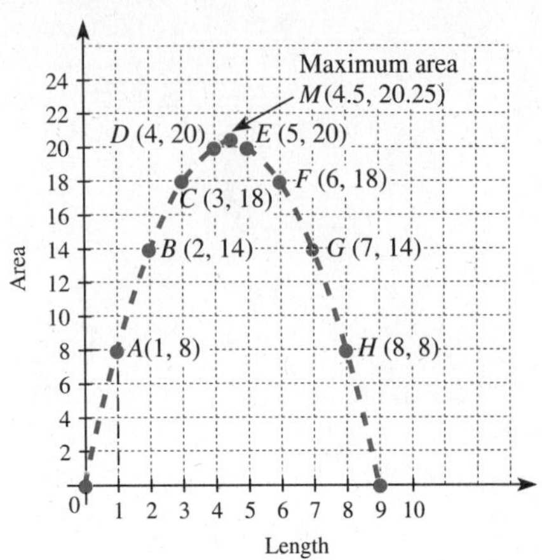

Figure 1.3

to the right to point 3 on the horizontal axis and then 6 units up the vertical line through 3. This point C has the coordinates (3, 6).

From the graph (Figure 1.2), it is easy to visualize the relationship between length and width. The points whose coordinates represent the length and width of the rectangles appear to lie on a line that has been sketched. This line represents the lengths and widths of *all* the rectangles with perimeters of 18. For example, if a rectangle's length is 2.5 units, then its width is 6.5 units. Point I (2.5, 6.5) represents this rectangle. From the graph it is also easy to observe that as the length increases, the width decreases. This relationship between the length and width is called a **linear function** since its graph is a line. We say that the *length is a function of (or depends on) the width*. Linear functions will be studied in more detail in Chapters 3 and 4.

The relationship between the lengths of all the rectangles with a perimeter of 18 and their corresponding areas can also be represented with a graph (Figure 1.3). We let the horizontal axis represent the lengths and the vertical axis represent the areas. The points A, B, C, D, E, F, G, and H represent the eight entries in the first and last columns of Table 1.1. For example, point A represents the rectangle that is 1 unit in length and 8 square units in area. Point D represents the rectangle that is 4 units in length and 20 square units in area. The coordinates of points A and D are (1, 8) and (4, 20), respectively.

From the graph (Figure 1.3), it is easy to visualize the relationship between the length and area; as the length increases, the area increases to a certain point and then starts to decrease. If we extend the lengths to include any real number, then a good guess for the maximum area is the rectangle with a length midway between 4 and 5 at 4.5. The area of this rectangle is 20.25 square units. The coordinates of the maximum point M are (4.5, 20.25).

We can draw a smooth curve through the points on the graph in Figure 1.3. (See Figure 1.4.) We can then use the graph to estimate the lengths and the corresponding areas of any rectangle with a perimeter of 18. For example, in Figure 1.4

1. If the length is 0 or 9, then the area is 0, but the perimeter is still 18 since $2(0 + 9) = 18$.

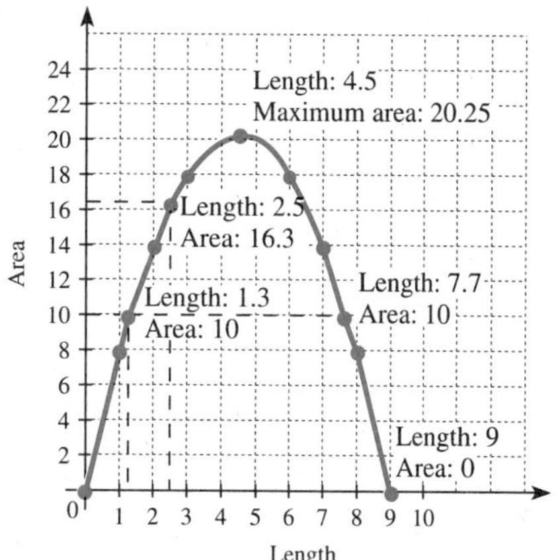

Figure 1.4

2. If the length is 2.5, then the area is approximately 16.3 square units. The perimeter is still 18 since the width is 6.5 and $2(2.5 + 6.5) = 18$.
3. If the area is 10 square units, then the length can be either approximately 1.3 units or 7.7 units.

It is also easy to see that this graph has a point representing the maximum area. A graph with this shape is called a **parabola.** This relationship between the length and area of rectangles with a fixed perimeter is called a **quadratic function.** The *area is a function of (depends on) the length.* These functions will be studied in more detail in Chapter 6.

The graph in Figure 1.4 represents all the rectangles with a perimeter of 18 units. From Figure 1.4 it is easy to estimate the area of a rectangle given a specific length; conversely, it is easy to estimate the length of a rectangle given its area. The graph is a more complete method than a table to represent all the rectangles with a given perimeter. But the values are only approximate. Let's see if we can find another way to represent the lengths, widths, and areas of rectangles with a perimeter of 18.

Method 3: Using an Equation Another way to attack this problem is to describe the four quantities of length, width, perimeter, and area using variables.

Definition

A **variable** is a symbol, usually a letter, that stands for any number in a specified set of numbers.

We will use variables to describe some of the patterns we have observed. Let L represent the length, W represent the width, and A represent the area of a rectangle with a perimeter of 18 units (Figure 1.5).

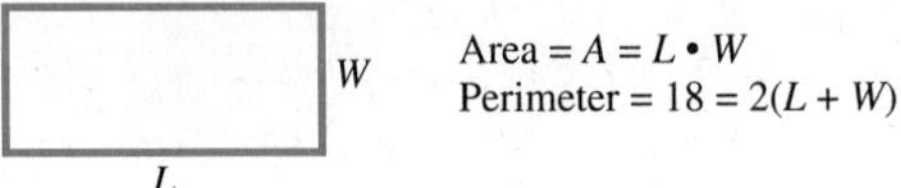

Figure 1.5

Since the perimeter is 18, we have the equation

$$18 = 2(L + W)$$

and $\quad 9 = L + W$

We divided both sides of the equation by 2. We can also observe that the sum of the length and width is equal to half the perimeter.

Therefore,

$$9 - L = W \text{ (or } 9 - W = L) \qquad \textbf{Equation (1.1)}$$

Equation 1.1 says that if we know the length of the rectangle of perimeter 18, we can determine the width by subtracting the length from 9. The expression $9 - L$ represents the width. For example, pick any real number—say 2.7—for the length; the corresponding width is $9 - 2.7$, or 6.3. If we systematically substitute values for one dimension into Equation 1.1, we can determine the corresponding value for the other dimension. Equation 1.1 represents the widths and lengths of *all* the rectangles with a perimeter of 18. The equation describes the graph in Figure 1.2.

Thus the lengths and widths of a rectangle can be represented by

- a table (see Table 1.1);
- a graph (see Figure 1.2);
- an equation (see Equation 1.1).

Each of these representations has its strengths and its weaknesses:

- The table is easy to read, but it is not complete.
- The graph is complete and the line on the graph allows us to easily visualize the relationship between length and width, but the graph represents approximate values.
- The equation is both complete and exact, but it does not allow us to easily visualize the relationships between the length and width or between the length and area.

To answer the question in Example 1 (Which rectangle has the greatest area?), we can employ all three methods. We can use a table (see Table 1.1) and then by trial and error determine that the rectangle with the greatest area is that with a length and width of 4.5 units and an area of 20.25 square units. We can also use a graph (see Figure 1.4) and estimate the greatest area by estimating the coordinates of its highest point. Finally, we can use an equation. We know that area $= A = L \cdot W$

and that $\quad W = 9 - L \qquad$ **Equation (1.1)**

Thus $\quad A = L(9 - L) \qquad$ **Equation (1.2)**

In place of the width in the equation $A = L \cdot W$, we can substitute an equivalent expression.

We notice that this relationship is not a line. It is a parabola.

Equation 1.2 represents *all* the areas of the rectangles with perimeters of 18 and lengths L. For example, if we let the length be 7, then the area is $7(9 - 7)$, or 14. If the length is 1.7, then the area is $1.7(9 - 1.7)$, or 12.41. In a later chapter we will show how Equation 1.2 can be used to quickly and exactly determine the length of the rectangle that has the greatest area.

In analyzing the questions raised in Example 1, we looked at three methods to represent the data: a **table,** a **graph,** and an **equation.** This process of representing the data to obtain a solution or information is a key part of algebra. It is the language of algebra. In particular, the equations $9 - L = W$ and $A = L(9 - L)$ represent the symbolic language of algebra. Symbolic language (algebraic equations) is both complete and exact. The advantages of symbolic language are that

1. it is brief;

2. it is general.

As we continue our study of algebra, we will find that all three representations are important in describing patterns.

The key to finding the pattern is to find the relationship between two quantities (variables). One quantity usually depends on (or is a function of) the other quantity. We use algebra to express the relationship.

In the next example we will try to analyze the pattern in the problem using the three different methods described in Example 1.

EXAMPLE 2

If the Acme Car Rental firm charges \$29.95 per day plus 12¢ a mile to rent a Ford Escort, how much would it cost to rent the Escort for one day?

Solution

On what does the cost depend? It depends on or is a function of the number of miles we drive, which is a variable in this problem. The feasible choices for the number of miles driven are the nonnegative real numbers.

Using a Table We can compute the cost for various numbers of miles and arrange them in a table in some systematic way. Let's try to compute the costs for increments of 50 miles.

On a graphing calculator, you can use either the TABLE mode or the repeat feature to efficiently perform the calculations. For example, insert $29.95 + 50 \cdot (0.12)$ and obtain the sum 35.95. Then pressing the repeat key gives $29.95 + 50 \cdot (0.12)$. Edit this line by changing 50 to 100 and then obtain the sum 41.95. Continue the process.

It is impossible to list all the possible values for the number of miles in a table. If we drive 75 miles, we can observe from Table 1.2 that the cost will be midway between

Miles driven in one day	*Total cost*	
0	$29.95	
50	$35.95	$29.95 + 50($0.12)
100	$41.95	$29.95 + 100($0.12)
150	$47.95	$29.95 + 150($0.12)
200	$53.95	$29.95 + 200($0.12)

Table 1.2

$35.95 and $41.95, or $38.95, since 75 miles is midway between 50 and 100 miles. To estimate the cost of driving 190 miles, the table is not as convenient; we can say that the cost will be less than $53 but certainly more than $47.

Using a Graph We will represent the data in Table 1.2 as points on a graph. Let the horizontal axis represent the number of miles driven and the vertial axis represent the total cost in dollars (see Figure 1.6). It appears that the points all lie on a straight line.

From the graph it is easy to estimate both the number of miles and the costs. For example, if we drive 75 miles, the cost is approximately $38. If we drive 167 miles, the cost is approximately $50. If we want to know how far we can drive for $90, we can follow a horizontal line from the point 90 on the vertical axis until it intersects the graph and then follow a vertical line down until it intersects the horizontal axis. This point of intersection occurs at approximately 500 miles (see Figure 1.6).

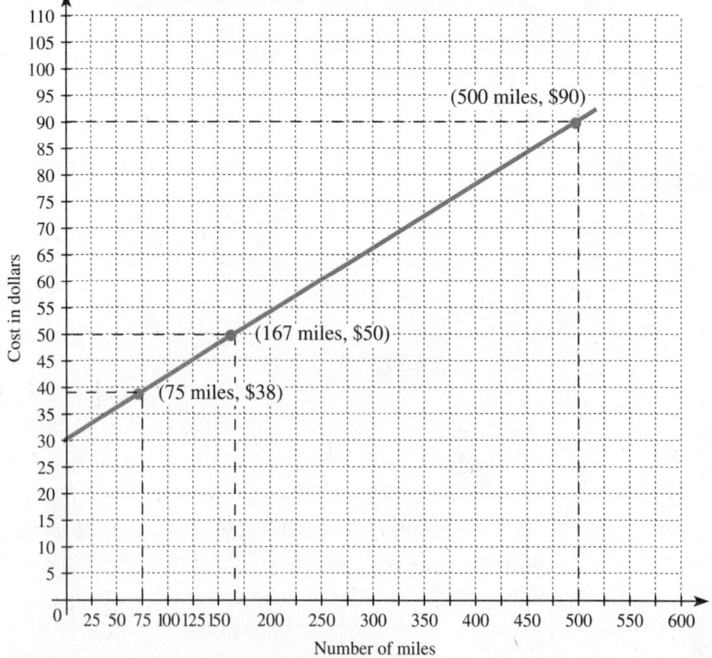

Figure 1.6

Using an Equation To help us arrive at the correct equation, let's study the process we used to obtain the costs for a specific number of miles driven. For example, if we drive 100 miles, then the

$$\begin{aligned} \text{cost} &= \$29.95 \text{ plus } 0.12 \text{ times } 100 \\ &= \$29.95 + 0.12(100) \\ &= \$41.95. \end{aligned}$$

In general, if we let m denote the number of miles driven and c denote the cost, then

$$c = \$29.95 + 0.12m.$$

NOTE: The table is very helpful in finding the relationship (or the pattern) between the two variables, which can then be written symbolically. ■

This equation allows us to compute the cost for any number of miles. We simply substitute the number of miles driven for m in the equation and compute the cost. The cost depends on the number of miles driven. This is another example of a **linear function.** It describes the relationship of the graph in Figure 1.6. ■

In the next two examples we will try to describe numerical patterns using the language of algebra.

EXAMPLES 3–4

How would you describe the integers in these sequences? (The integers are the set of numbers containing the positive and negative whole numbers and zero.)

3. $\ldots -15, -10, -5, 0, 5, 10, 15, 20, \ldots$

4. $\ldots -7, -2, 3, 8, 13, 18, 23, 28, \ldots$

Solutions

3. Each integer is a multiple of 5. If we let n represent an integer, then in algebraic language, each integer in the sequence is of the form 5 times n, or $5(n)$, or more briefly, $5n$.

If n, the variable, is replaced by any integer, $5n$ describes a member of the sequence. If n is 11, then $5n$ becomes 5 times 11, or 55. Hence 55 is a member of the sequence.

Remember that if there is no operation sign between a number and a variable, then multiplication is the implied operation, as in $5n$. But if the variable is replaced by a specific number, then for clarity, multiplication is indicated by parentheses or by a raised dot, as in $5(11)$ or $5 \cdot 11$.

4. Each member of the sequence is 2 less than a multiple of 5 *or* each member is 3 more than a multiple of 5. In algebraic language, the two most popular answers would be: "5 times n plus 3" or "5 times n minus 2." More briefly, we write $5n + 3$ or $5n - 2$. This depends on where we let $n = 0$ or where we start counting. See Figure 1.7.

Once again, for every integer n, the expression $5n + 3$ (or $5n - 2$) represents a member of the sequence. For example, if n is replaced by 6, $5n + 3$ is $5(6) + 3$, or 33, and $5n - 2$ is 28. Both 28 and 33 are in the sequence. (Notice that both sequences could also be described as "add 5 to the previous member.")

NOTE: We can also describe the patterns in Examples 3 and 4 with graphs (Figure 1.7). The points lie on a line, but not all the points on the line are part of the pattern. We call this graph a discrete graph. Both patterns are examples of linear functions.

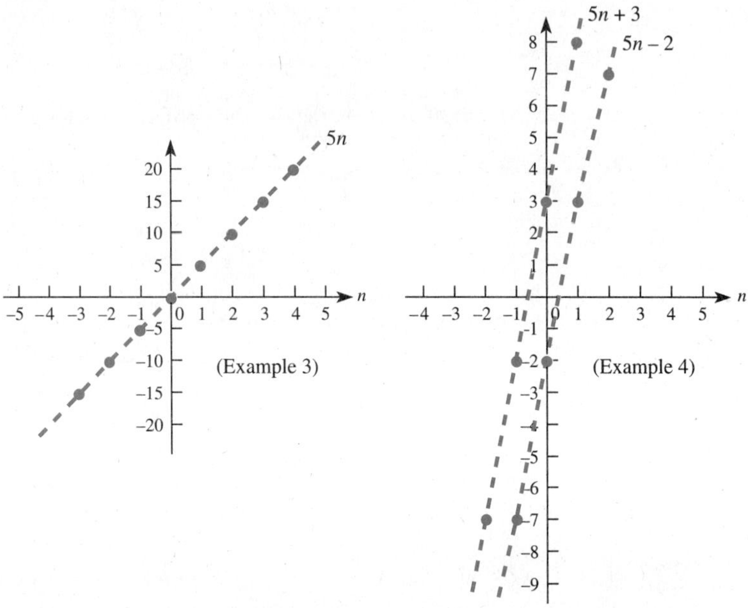

Figure 1.7

EXAMPLES 5–7

Each of the following cases reveals a pattern. Use an algebraic statement to describe the pattern. Do you think the algebraic statement is always true? That is, does the pattern apply to all numbers?

5. $3 + 5 = 5 + 3$

$10.1 + 9.2 = 9.2 + 10.1$

$1{,}876 + 2{,}017 = 2{,}017 + 1{,}876$

6. $2 + 2 = 2 \cdot 2$

$\frac{4}{3} + 4 = \left(\frac{4}{3}\right) \cdot 4$

$6 + 1.2 = 6(1.2)$

7. $5 + (-5) = 0$

$1.2 + (-1.2) = 0$

$\frac{3}{4} + \left(-\frac{3}{4}\right) = 0$

Solutions

5. $a + b = b + a$. In other words, the order in which we add two numbers does not matter. This statement is always true. This pattern is called the **commutative property** of addition.

6. $x + y = x \cdot y$. By letting $x = 8$ and $y = 2$, we see that $8 + 2 \neq 8 \cdot 2$, so the statement is not always true.

7. $a + (-a) = 0$. An English statement describing this pattern is "if a number is added to its opposite, the result is zero." This statement is always true.

From Examples 5 and 7, we observe that algebraic language is a general and brief way to describe the laws of arithmetic. In fact, ***algebra may be thought of as a generalized form of arithmetic.***

EXAMPLE 8

The typical interest rate on a credit card purchase is 1.5% per month for the first \$400 and 1% per month for the amount exceeding \$400. Suppose we just purchased a stereo system. How much interest would we pay for the first month where interest is charged?

Solution

The amount of interest depends on the price of the stereo. (Remember that $1.5\% = 0.015$.)

If the stereo costs \$295.98, then the interest is

$$0.015(\$295.98) = \$4.44.$$

If the stereo costs \$695, the interest is

$$0.015(\$400) + 0.01(\$295) = \$6 + \$2.95 = \$8.95.$$

In general, if we denote the price of the stereo by s, then the amount of interest we would pay is

$0.015s$, if $s \leq 400$ (s is less than or equal to 400)

$0.015(400) + 0.01(s - 400)$, if $s > 400$ (s is greater than 400)

In the next example we will try to describe a geometric pattern with an algebraic expression or sentence.

EXAMPLE 9

(Optional) How many diagonals does a convex polygon have?

Solution

This question is considerably more difficult to answer than the others. First let us be sure we understand the question. A polygon is a geometric figure with sides consisting of line segments. Triangles, rectangles, pentagons, hexagons, and octagons (stop signs) are polygons. A diagonal is a line segment that joins two nonadjacent vertices of a polygon. The colored line segments in the illustrations below are diagonals (Figure 1.8).

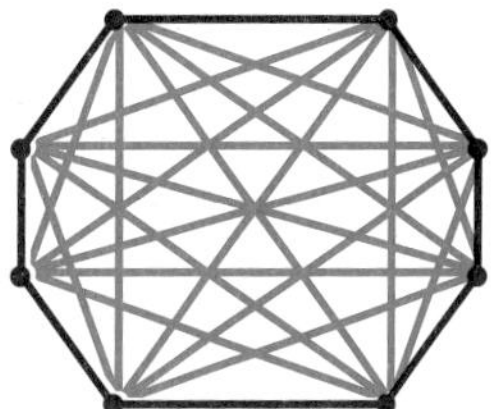

Figure 1.8

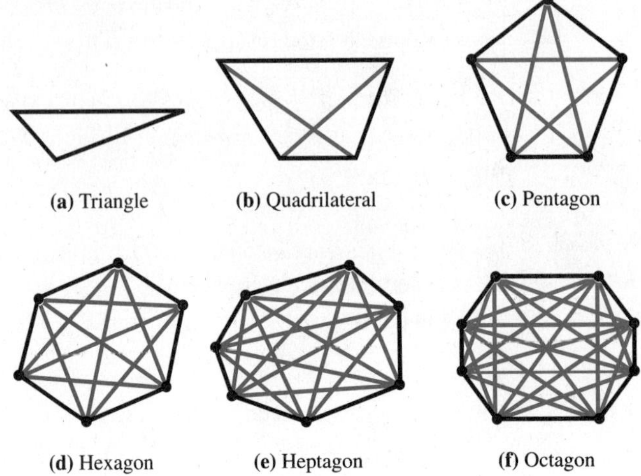

Figure 1.9

There are two ways of solving this problem:

Using a Table We first try a few examples and make a table of our results. Let n denote the number of sides of the polygon and d the number of diagonals. By examining the following polygons, we arrive at Table 1.3. (It would be instructive for you to draw your own figures and count the number of diagonals in each case.)

Polygon	n	d
a	3	0
b	4	2
c	5	5
d	6	9
e	7	14
f	8	20

Table 1.3

n	d	*Amount of increase*
3	0	
4	2	+2
5	5	+3
6	9	+4
7	14	+5
8	20	+6

Table 1.4

If $n = 3$, then $d = 0$.
If $n = 4$, then $d = 2$. $(n - 2 = 2)$
If $n = 5$, then $d = 2 + 3$. $(n - 2 = 3)$
If $n = 8$, then $d = 2 + 3 + 4 + 5 + 6$. $(n - 2 = 6)$

By examining Table 1.3 and Table 1.4, we note that in each case the increase in the number of diagonals is one more than the number of diagonals given in the previous case (see Table 1.4). Therefore, one possible formula would be $d = 2 + 3 + 4 + \cdots + (n - 2)$. The three dots mean we continue the pattern until we reach $(n - 2)$. In general, we stop at $(n - 2)$, since we stopped at 3 for a five-sided polygon, we stopped at 5 for a seven-sided polygon, and so on.

Using Algebra On the other hand, if we thought carefully about how we drew these diagonals, then we could observe the following facts (see Figure 1.10):

1. *One* diagonal is drawn from each vertex in a *four-sided* polygon (a).
2. *Two* diagonals are drawn from each vertex in a *five-sided* polygon (b).
3. *Three* diagonals are drawn from each vertex in a *six-sided* polygon (c).

Generalizing, we can say that $n - 3$ diagonals are drawn from each vertex in an *n-sided* polygon (d).

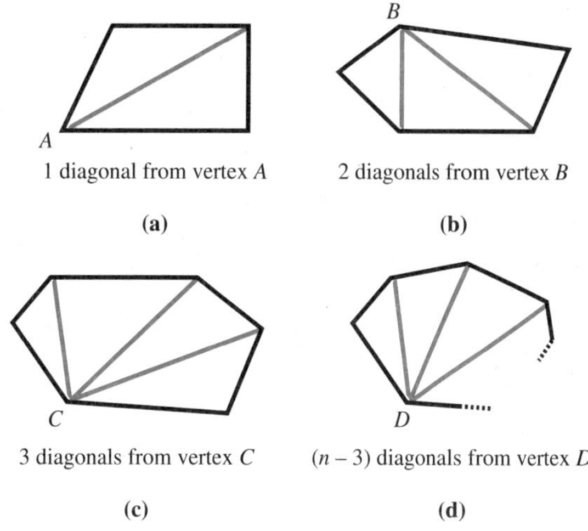

Figure 1.10

Consequently, we can come to the following conclusions:

1. There are $n - 3$ diagonals drawn from each vertex.
2. There are n vertices in an n-sided polygon.
3. Each diagonal can be drawn from two vertices, so we must divide by 2 so that each diagonal is not counted twice. Therefore,

$$d = \frac{n(n-3)}{2}$$

As a by-product of analyzing this problem in two different ways, we have also discovered the following formula:

$$2 + 3 + 4 + 5 + \cdots + (n - 2) = \frac{n(n-3)}{2}, \text{ for } n > 3$$

From this example, we see that algebraic language can be used to describe geometric patterns. ***Algebra is a powerful language used to describe patterns in arithmetic, geometry, science, and the world around us.*** In this book we will investigate how to translate these concepts into algebraic language and how to use algebra to solve problems.

1.1 EXERCISES

I. Procedures

1–5. **(a)** Write an algebraic expression or formula to describe the patterns in each sequence. The variable n in each case represents an element from the set of integers. **(b)** Sketch a graph of each pattern.

1. The members of the sequence
$\ldots -8, -4, 0, 4, 8, 12, 16, \ldots$

2. The members of the sequence
$\ldots -7, -3, 1, 5, 9, 13, 17, \ldots$

3. The members of the sequence
$\ldots -11, -1, 9, 19, 29, 39, 49, \ldots$

4. The members of the sequence
$\ldots -9, -4, -1, 0, 1, 4, 9, \ldots$

5. The members of the sequence
$\ldots -27, -8, -1, 0, 1, 8, 27, \ldots$

II. Concepts

6–10. **(a)** Write an algebraic statement to describe each of the following patterns. **(b)** Test your statement by replacing the variables with several more numbers. **(c)** Is your statement true for all numbers?

6. $2 + 6 + 7 = 7 + 6 + 2$

$13 + 5 + 8 = 8 + 5 + 13$

$1 + 2 + 3 = 3 + 2 + 1$

7. $7 \cdot 5 = 5 \cdot 7$

$(19.3) \cdot (34.2) = (34.2) \cdot (19.3)$

$(-3.1) \cdot (4.7) = (4.7) \cdot (-3.1)$

8. $3(5 + 2) = (3 \cdot 5) + (3 \cdot 2)$

$1.2(5.7 + 3.1) = [(1.2)(5.7)] + [(1.2)(3.1)]$

$\frac{3}{4}\left(\frac{1}{7} + \frac{2}{11}\right) = \left(\frac{3}{4} \cdot \frac{1}{7}\right) + \left(\frac{3}{4} \cdot \frac{2}{11}\right)$

9. $5^2 - 3^2 = (5 + 3)(5 - 3)$

$(8.14)^2 - (3.23)^2 = (8.14 + 3.23)(8.14 - 3.23)$

$\left(\frac{5}{16}\right)^2 - \left(\frac{1}{8}\right)^2 = \left(\frac{5}{16} + \frac{1}{8}\right)\left(\frac{5}{16} - \frac{1}{8}\right)$

10. Write a paragraph explaining what you think is meant by "Algebra is the language of mathematics." Give some examples in your discussion.

III. Applications

11. Repeat Example 1, but use a perimeter of 20. Be sure to do both parts of the problem: the first involving length and width; the second involving length and area.

(a) Make a table for all the whole number dimensions of the rectangle. Use the table to make a graph, and then describe the relationship with an algebraic statement.

(b) Compare the tables, graphs, and equations with those obtained in Example 1.

12. Repeat Example 2, but have the car agency charge $19.25 a day plus 20¢ a mile.

(a) Make a table for a few values, sketch a graph, and find an algebraic statement that relates cost and miles.

(b) Compare the table, graph, and equation to those obtained in Example 2.

13–31. Write an algebraic statement to describe each of the following patterns. Describe your variable(s).

13. The perimeter of a square.

14. The area of a square.

15. The area of a right triangle.

16. The area of a circle.

17. The perimeter (called the circumference) of a circle.

18. The volume of a box (rectangular solid).

19. The surface area of a cube.

20. The area of the shaded region. (The circles have the same center.)

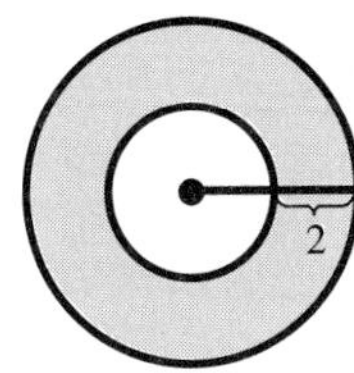

21. The area of the shaded region. (The circle and square have the same center.)

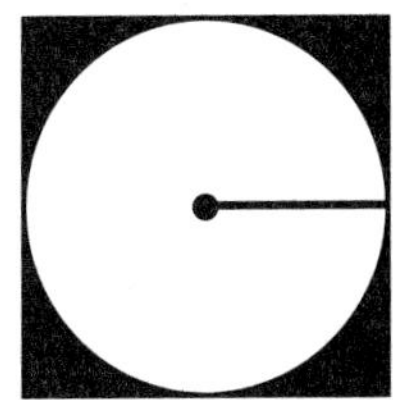

22. The distance traveled by a car in three hours.

23. The time it takes to walk 350 km.

24. The cost of a box of pens, if one pen costs 59¢.

25. The amount of money in a bag of quarters.

26. The amount of money in a bag of quarters and dimes.

27. The amount of interest earned on a savings account with an annual interest rate of 7.25%.

28. The cost of renting a car for one day at a rate of 37¢ per mile.

29. The cost of renting a car for one day for $28 plus 12¢ per mile.

30. An electrician's wages for one job, if her typical wage is $35 for the first half hour, and $20 for each additional half hour.

31. The income tax paid on a taxable income by a couple earning between $17,000 and $41,000, if the rate of taxation is $2,550 plus 28% of the amount over $17,000.

IV. Extensions

32–34. Discover an algebraic expression that will help answer each of the following questions.

32. Which integers can be written as the sum of three consecutive integers? *Note:* If x is an integer, then $x + 1$ is the next consecutive integer after x. (For example, $6 = 1 + 2 + 3$.)

33. Which integers can be written as the sum of four consecutive integers? (For example, $18 = 3 + 4 + 5 + 6$.)

34. Which integers can be written as the difference of two squares? (For example, $4 = 2^2 - 0^2$, $5 = 3^2 - 2^2$.)

35. A *triangular number* is one which can be pictured as a triangular array of dots. For example, 1, 3, 6, and 10 are the first four triangular numbers (see below).

(a) Which numbers are triangular?

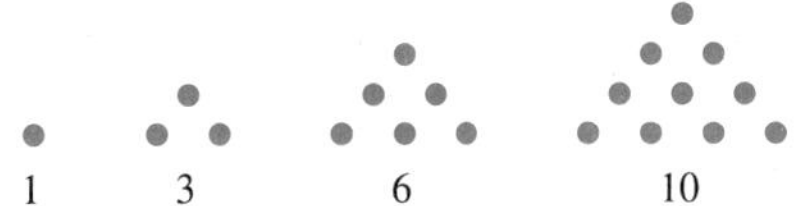

(b) Graph their relationship.

36. Two points determine one line in a plane. Three noncollinear points determine three lines in a plane.

(a) How many lines are determined by four points, five points, and n points? (Assume that no three points lie on the same line.)

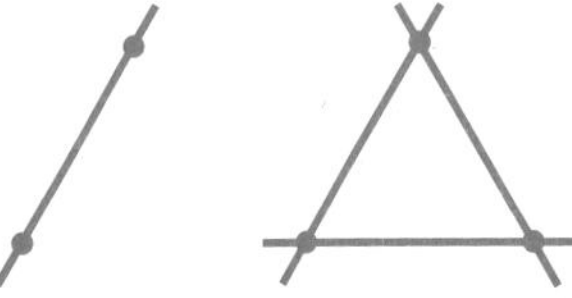

(b) Graph their relationship.

37. (a) If 6 people enter a room and they all shake hands, how many handshakes are there?

(b) How many handshakes are there if there are 10 people? n people?

1.2 REAL NUMBERS, THEIR PROPERTIES, AND GRAPHS

EXAMPLE 1

A familiar parlor trick involving the arithmetic of numbers goes as follows:

LEADER: Pick any number, add 15, multiply by 2, subtract 30, and tell me your answer.
FOLLOWER 1: My answer is 20.
LEADER: Your original number is 10.
FOLLOWER 2: My answer is 25.
LEADER: Your original number is $12\frac{1}{2}$.

Magic? No, not if we analyze the problem algebraically.

Suppose we let the original number picked be represented by n. Then adding 15 yields $n + 15$. Multiplying by 2 yields $2(n + 15) = 2n + 30$. Subtracting 30 yields $2n + 30 - 30 = 2n$. The leader simply has to divide the last answer $2n$ by 2 to obtain the original number n. Try it.

The leader represented this series of arithmetic steps in a general form by using algebra. By doing this, the leader can easily see how the final answer is related to the original number chosen. The leader can be even more impressive by requiring more complicated arithmetic computations. ■

Algebra can be thought of as a brief and general way to represent arithmetic. As we will see, it is the general language of arithmetic. Therefore, before we begin our study of algebra, we will spend some time reviewing the important properties and operations of arithmetic.

Real Numbers

In arithmetic we work with various sets of numbers.

- The numbers 1, 2, 3, . . . , are called the **natural numbers** or **counting numbers.**
- The numbers . . . -2, -1, 0, 1, 2, 3, . . . , are called the **integers.** The integers are made up of the counting numbers 1, 2, 3, . . . , (positive integers), their opposites -1, -2, . . . , (negative integers), and the number zero.
- The set of numbers containing $\frac{3}{1}$, $\frac{1}{2}$, $\frac{2}{3}$, $-\frac{4}{5}$, . . . , is called the set of **rational numbers.** The rational numbers are all numbers of the form $\frac{p}{q}$, where p is any integer and q is a positive integer. The rational numbers contain all the integers, since integers can be represented by $\frac{p}{1}$; $2 = \frac{2}{1}$, $-4 = \frac{-4}{1}$, $0 = \frac{0}{1}$, and so on. We recall that $\frac{2}{3}$ means $2 \div 3$; thus, $\frac{p}{q}$ can be thought of as $p \div q$.

The rational numbers cannot describe the solutions to all mathematical problems. An example arises in finding the length of the side of a square when you are given its area. Recall that the area of a square is the number of unit squares (squares with sides equal to the length of one unit) needed to exactly cover the original square.

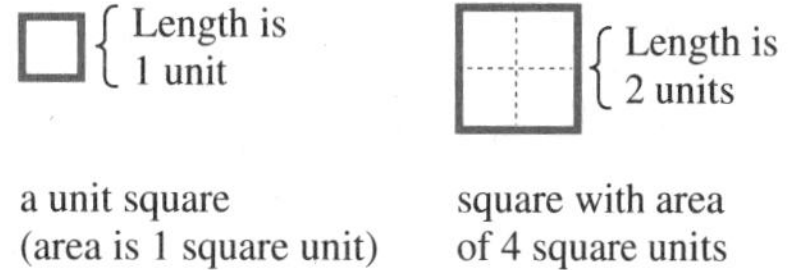

Figure 1.11 (a)

Consider a square with sides having a length of 2.5 units. Its area is 6.25 square units (Figure 1.11[b]).

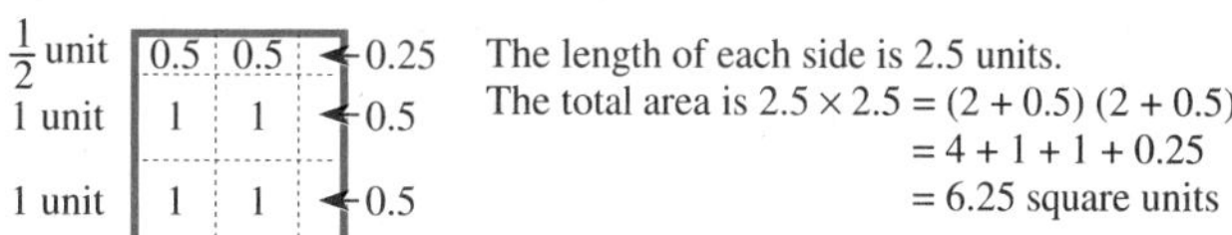

Figure 1.11 (b)

In general, if x is the length of the side of a square, then $x \cdot x = x^2$ is the area of the square.

Figure 1.11 (c)

Conversely, if the area of the square is known, then we can determine the length of one of its sides by finding a number which when multiplied by itself equals the area. For example, if the area of a square is 4 square units (Figure 1.11[d]), we seek a number which when multiplied by itself gives 4. That is, if $x^2 = 4$, then $x = 2$. This process is called finding the "square root" of the number 4, written $\sqrt{4}$. Thus $x = \sqrt{4} = 2$.

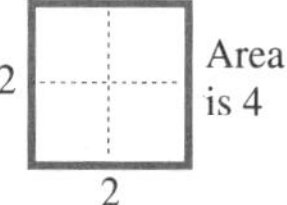

Figure 1.11 (d)

Definition

If p and q are positive numbers, then $\sqrt{p} = q$ means $q^2 = p$. $\sqrt{p}$ is read "the **square root** of p."

The symbol $\sqrt{\ }$ is called the square root; $\sqrt{9} = 3$ since $9 = 3^2$; $\sqrt{256} = 16$ since $256 = 16^2$. *Note:* Even though $(-3)^2 = 9$, $\sqrt{9} = 3$. By definition, the square root is *always* positive or zero.

We can interpret the square root of a number as the length of a square whose area is the given number.

Now consider the problem of finding the length of a side of a square whose area is 2 (Figure 1.12). Since $x^2 = 2$, then $x = \sqrt{2}$. But what number squared yields 2? Since $1^2 = 1$ and $2^2 = 4$, then $\sqrt{2}$ lies between 1 and 2. By trying several rational numbers between 1 and 2, it does not appear that $\sqrt{2} = \frac{p}{q}$ for any rational number $\frac{p}{q}$. For example, $\left(\frac{3}{2}\right)^2 = \frac{9}{4} \neq 2$. But we now know that $\sqrt{2}$ lies between 1 and $\frac{3}{2}$. Most calculators have a square root key. Using the calculator to compute $\sqrt{2}$, we obtain the number 1.4142135. But multiplying 1.4142135 by itself does not yield exactly 2. Actually, $\sqrt{2} = 1.4142135\ldots$. The dots mean the decimal part goes on and on (even though it appears to stop on the calculator). In fact, $\sqrt{2}$ *cannot* be expressed as the quotient of two whole numbers: $\sqrt{2}$ is *not* a rational number.

Area = 2
or
$x^2 = 2$
x
x

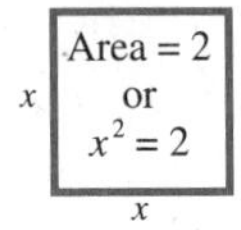

Figure 1.12

EXAMPLES 2–3

Find the length of the side of a square whose area is

2. 12.25 square units

3. 5 square units

Solutions

2. We can use a guess and check method to find a number x such that $x^2 = 12.25$. We know that $3 \times 3 = 9$ and $4 \times 4 = 16$, so x is between 3 and 4. If we choose x to be 3.5, we find that $3.5 \times 3.5 = 12.25$. So the length of the side of the square is 3.5. We can also use our calculator to compute the square root of 12.25: $\sqrt{12.25} = 3.5$ (Figure 1.13).

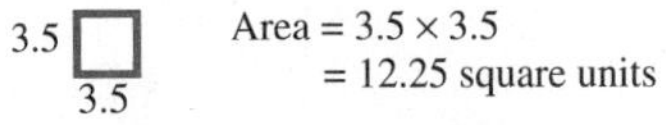

3.5
3.5
Area = 3.5×3.5
= 12.25 square units

Figure 1.13

3. We can estimate the length to be between 2 and 3 since $2 \times 2 = 4$ and $3 \times 3 = 9$. Since 5 is closer to 4 than 9, we can estimate the length to be 2.2 or 2.3. We find $2.2 \times 2.2 = 4.84$ and $2.3 \times 2.3 = 5.29$. If we continue this process we see that $\sqrt{5}$ is not a rational number and that 2.25 is only an estimate for $\sqrt{5}$. Using the calculator we find $\sqrt{5} = 2.36068$, an even closer estimate for the length of a square whose area is 5. (See Figure 1.14.) Both the method of guess and check and the method of using the square root button on the calculator give approximations for the square root of 5. The decimal expansion of $\sqrt{5}$ has an infinite number of places after the decimal point.

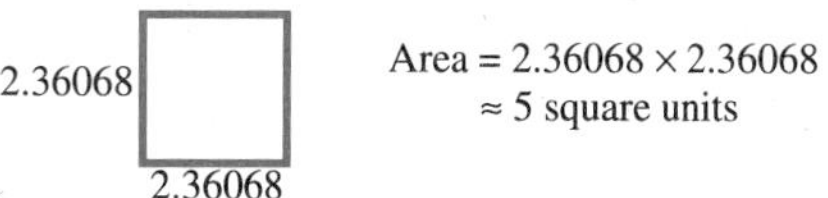

Figure 1.14

There are infinitely many examples of numbers that are not rational: $\sqrt{3}$, $\sqrt{5}$, $\sqrt{17}$, and π, to name a few. The number π is a constant that occurs in many geometric problems. For any circle with a circumference C and a diameter d, π is the ratio of circumference to diameter; $\pi = \frac{C}{d}$.

Thus,

$$C = \pi d.$$

The exact value of this number $\frac{C}{d}$ is not rational. We write $\pi \approx 3.14$ ($\approx$ means "**approximately equal to**"). These numbers, $\sqrt{2}$, $\sqrt{3}$, $\sqrt{5}$, $\sqrt{11/3}$, π, etc., are called **irrational numbers.** The set of all irrational and rational numbers is called the set of **real numbers.** The following tree summarizes the relationships among the set of real numbers. (See Figure 1.15.)

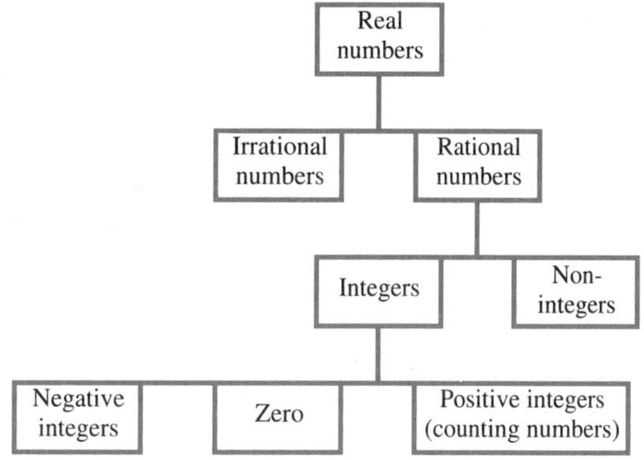

Figure 1.15

Properties of the Real Numbers

The Distributive Property

A very important property of the real numbers can be illustrated by examining the area of a rectangle in two different ways.

EXAMPLE 4

The distributive property can be illustrated geometrically by cutting the area of a rectangle into two subrectangles. Consider a rectangle with width equal to 3 and length equal to 7 that has been subdivided into squares. Find its area in two different ways.

Solution

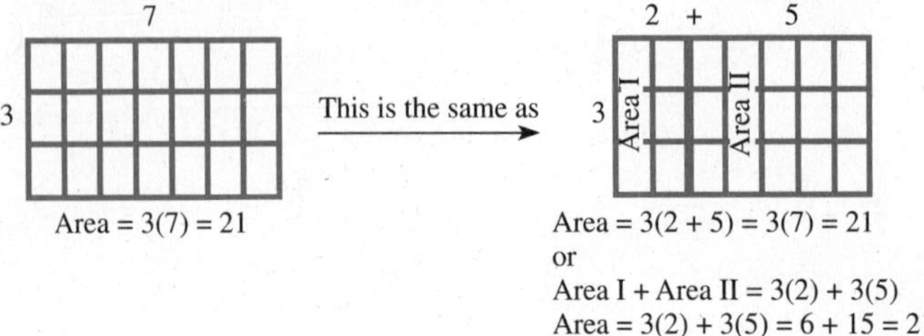

Figure 1.16

We observe that $3(2 + 5) = 3(2) + 3(5)$. That is, the area can be written as a *product of two factors,* 3 and $(2 + 5)$, or as a *sum of two terms,* $3(2)$ and $3(5)$. ■

Here is another illustration of this property connecting the operations of multiplication and addition.

EXAMPLE 5 Calculate $22(57 + 43)$.

Solution

We can add and then multiply:

$$22(57 + 43) = 22(100) = 2{,}200$$

An alternate method is to multiply and then add:

$$22(57 + 43) = 22(57) + 22(43) = 1{,}254 + 946 = 2{,}200$$

In the second method, multiplication is distributed over addition. Hence, the equivalence of these two processes is called the **distributive property.** ■

We can also distribute multiplication over subtraction.

EXAMPLE 6 Calculate $15(10 - 7)$

Solution

$$15(10 - 7) = 15(3) = 45$$

or

$$15(10 - 7) = 15(10) - 15(7) = 150 - 105 = 45.$$ ■

We can state the distributive properties in more general terms using algebra. Let a, b, and c represent any three real numbers.

The Distributive Properties

$$a(b + c) = ab + ac$$
$$a(b - c) = ab - ac$$

A geometric interpretation of the distributive property is illustrated by examining the area of a rectangle in two different ways (Figure 1.17).

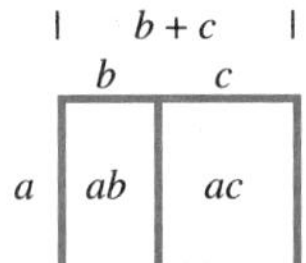

The area of the large rectangle is $a(b + c)$ (length times width) or $ab + ac$ (the sum of the area of the two smaller rectangles); thus $a(b + c) = ab + ac$.

Figure 1.17

The **distributive property** is very useful in that it connects the operations of addition/subtraction and multiplication/division.

Other Properties of the Real Numbers

Other properties of the real numbers can also be stated in general algebraic language. Let $a, b, c, \ldots$ be variables representing any real numbers:

Properties of Addition and Multiplication of Real Numbers

Addition	*Property*	*Multiplication*
$a + b = b + a$	Commutative	$ab = ba$
$(a + b) + c = a + (b + c)$	Associative	$(ab)c = a(bc)$
$a + 0 = a$	Identity element	$a \cdot 1 = a$
$a + (-a) = 0$	Inverse element	$a \cdot \frac{1}{a} = 1, a \neq 0$
	Multiplication by zero	$a \cdot 0 = 0$
If $a + b = c$, then $a = c - b$ and $b = c - a$.	Inverse operation	If $ab = c$ and $a, b \neq 0$, then $a = \frac{c}{b}$ and $b = \frac{c}{a}$.
	Zero factor	If $ab = 0$, then $a = 0$ or $b = 0$.

We have been using these properties "automatically" for a long time. In the next two sections, we will use them when we review the operations for real numbers—in particular, fractions and integers. We will continue to use these properties throughout the book to find equivalent algebraic expressions and to find solutions to algebraic equations.

EXAMPLE 7

Represent the steps of the parlor trick in Example 1 by a series of arithmetic statements to see if you can discover the trick.

Solution

	Directions	*Computation*	*The result*
Step 1	Pick a number, say	10	10
Step 2	Add 15	$10 + 15$	25
Step 3	Multiply by 2	$2(10 + 15)$	50
Step 4	Subtract 30	$2(10 + 15) - 30$	20
Step 5	What is your number?	20 is my number	
Step 6	Your original number is 10.		

Look at Step 4 and rewrite the computation using the distributive property:

$$2(10 + 15) - 30 = 2(10 + 15) - 30$$
$$= 2(10 + 15) - 2(15) \quad \text{Rewrite 30 as 2(15)}$$
$$= 2[(10 + 15) - (15)] \quad \text{Apply the distributive property; (10 + 15) is thought of as one factor in the term 2(10 + 15).}$$
$$= 2(10) \quad \text{Compute the numbers inside the brackets.}$$

This series of steps will always end in a number that is twice the original number. Thus the leader takes the last number and divides by two to identify the original number.

The Number Line

The real numbers can be represented on a **number line.** We let any point be zero, pick an arbitrary length to represent one unit, and then locate the integers. Further subdividing locates the other real numbers. Every real number corresponds to a point on the number line, and conversely, every point corresponds to a real number. (See Figure 1.18.)

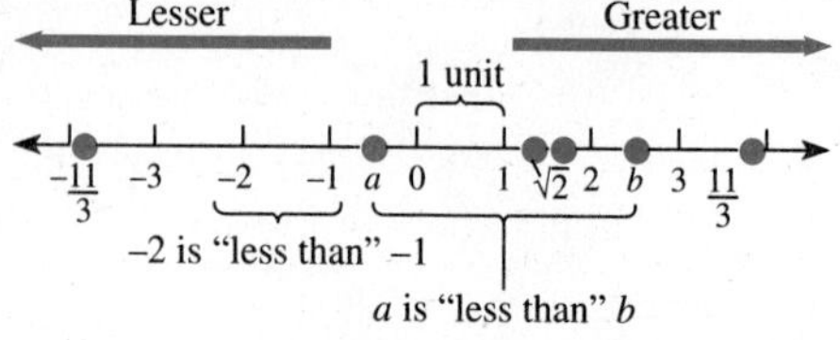

Figure 1.18

The numbers increase as we go from left to right. For example, -2 is less than -1, and -2 is greater than -3; -2 is less than all numbers to the right of -2. Similarly, -2 is greater than all numbers to the left of -2.

Definition

If a number a lies to the left of a number b on the number line, then a **is less than** b, and we write $a < b$. Similarly, if a number b is to the right of the number a, then b **is greater than** a, and we write $b > a$.

EXAMPLE 8

Locate the following numbers on a number line. Use the number line to determine the largest and smallest numbers in the set.

$$3.1, 0, -2.5, 1, \frac{3}{4}, \sqrt{2}, -0.5.$$

Solution

Draw a number line and locate the points.

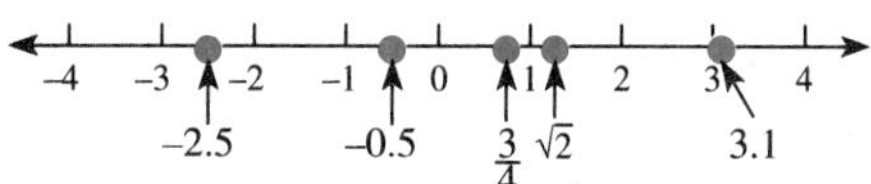

Figure 1.19

Since the number -2.5 lies to the left of the other numbers, it is the smallest number in the set. Similarly, 3.1 lies to the right of the other numbers, so it is the greatest. The numbers are arranged in increasing order on the number line from left to right.

Inequalities

Comparison of Quantities

The need to compare two quantities to determine which quantity is larger arises in many situations. Consider the following example.

EXAMPLE 9

Suppose the government established a minimum income of \$15,000 per year for every family. Under this program, any family with an income less than \$15,000 would be subsidized by the government. What does this mean?

Solution

It means, for example, if the family income is \$14,000, then the government will add \$1,000 to it. If the family income is \$14,500, then the government will add \$500, and so on. In general, if the family income is x dollars, then the government will add c dollars so that the total income is $x + c =$ \$15,000 (x is less than \$15,000).

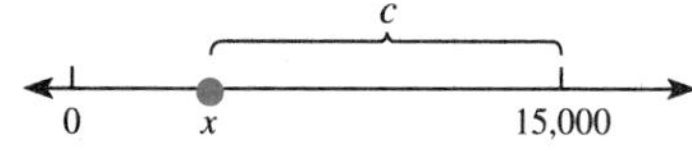

Figure 1.20

For x to be less than \$15,000 means that a quantity must be added to x to bring the total up to \$15,000. As long as the income x corresponds to a point to the left of \$15,000 on the number line, a benefit is due the family. ■

This example motivates the following general property of *less than:*

Property of Less Than

x is **less than** y, written $x < y$, if there is a positive number c such that $x + c = y$.

Adding the number c shifts the value *towards the right,* from x to $x + c = y$. This is equivalent to our earlier definition: "$x < y$ if x is to the left of y on the number line" (see Figure 1.21). Also, if x is less than y, then y is **greater than** x, written $y > x$.

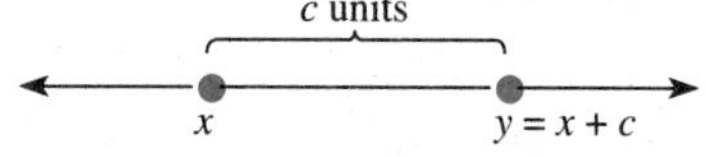

Figure 1.21

We know $4 < 9$, since there is a positive number, 5, such that $4 + 5 = 9$. This is equivalent to saying $4 - 9 = -5$ or $4 - 9 < 0$. Similarly, $9 > 4$, since $9 - 4 > 0$. We generalize this result as follows:

Test for Inequality

$$x < y \text{ is equivalent to } x - y < 0$$

and

$$x > y \text{ is equivalent to } x - y > 0$$

EXAMPLES 10–11

Determine whether each of the following inequalities is true or false.

10. $-4 > -1$

11. $3(7 - 5) < 4(9 - 6)$

Solutions

10. False: $-4 \not> -1$ since $-4 - (-1) < 0$. However, $-4 < -1$ is a true statement since the point -4 lies to the left of the point -1 on a number line. *Note:* $-4 \not> -1$ is read "-4 is not greater than -1."

11. True: $3(7 - 5) < 4(9 - 6)$. We compute each side of the inequality: $3(7 - 5) = 6$ and $4(9 - 6) = 12$.
Since $6 < 12$, we know that $3(7 - 5) < 4(9 - 6)$. ■

Graphing Inequalities on a Number Line

Analyzing the signs $<$ and $>$ in terms of "to the left" and "to the right" enables us to interpret inequalities on the number line. For example, we may need to locate all the numbers that are less than 3. There are many such numbers, and they all lie to the left of point 3. This set of numbers is described by the symbol { }, which means "**set of,**" and $\{x|x < 3\}$, which is read "the set of all real numbers x that are less than 3," or equivalently, "the set of all points on the number line to the left of the point 3." Similarly, $\{x|x \geq -4\}$ is read "the set of all real numbers x that are greater than -4 or equal to -4," or equivalently, "the set of all points on the number line that are to the right of the point -4, including the point -4."

EXAMPLES 12–13

Sketch the following sets on the number line.

12. $\{x|x < 3\}$ **13.** $\{x|x \geq 3\}$

Solutions

12. Sketch $\{x|x < 3\}$ on the number line.

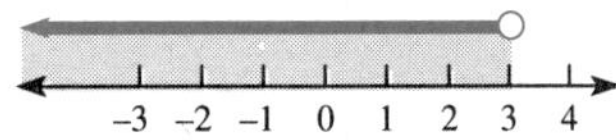

Figure 1.22

The open circle above 3 indicates that 3 *is not* included in the solution set. The shading above the number line indicates all the points to the left of 3 are included.

13. We must identify all the numbers *greater than* 3 or *equal to* 3.

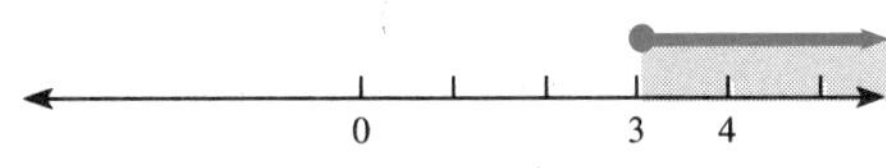

Figure 1.23

The closed circle indicates the point 3 is included in the set. ■

EXAMPLES 14–16

Sketch the following sets on the number line.

14. $\{x|x < 3 \text{ and } x > -4\}$

15. $\{x|x > 3 \text{ or } x < -4\}$

16. $\{x|-2 < x \leq 5\}$

Solutions

14. We must identify all points on the number line that are less than 3 *and* (simultaneously) greater than -4. These points are all the points to the left of 3 and (simultaneously) to the right of -4.

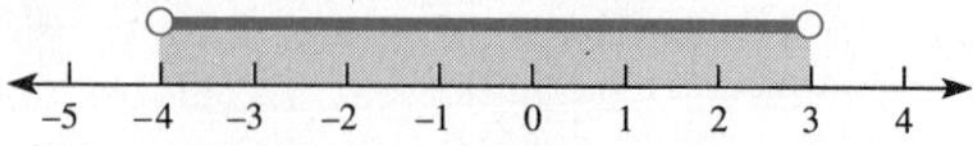

Figure 1.24

The condition "$x < 3$ and $x > -4$" is usually written as $-4 < x < 3$ and is read "x is between -4 and 3."

15. We must identify all points on the number line that are greater than 3 *or* less than -4. These are all the points to the right of 3 *or* to the left of -4.

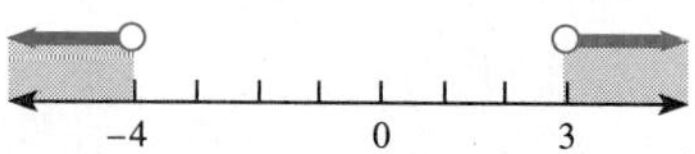

Figure 1.25

16. Sketch $\{x| -2 < x \leq 5\}$. That is, find and indicate all numbers between -2 and 5, including the point 5.

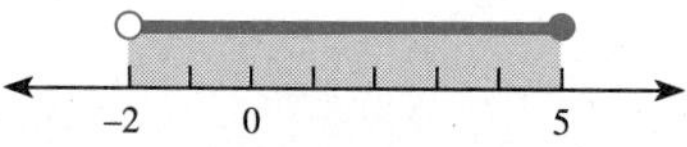

Figure 1.26

Note that this inequality is equivalent to $\{x|x > -2 \text{ and } x \leq 5\}$. ■

EXAMPLES 17–19

Express each of the following by using mathematical symbols ($<, >, \leq, \geq$). Graph each set on a number line.

17. x is less than or equal to 3.

18. x is between -3 and 3.

19. x is greater than 3, or x is less than -3.

Solutions

17. $\{x|x \leq 3\}$

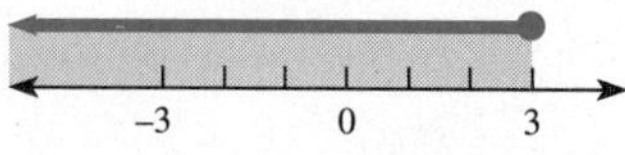

Figure 1.27

18. $\{x|-3 < x < 3\}$

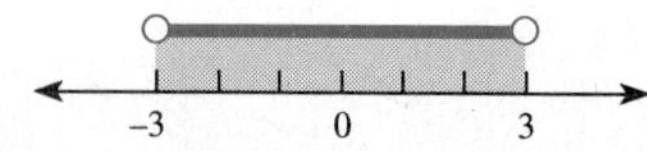

Figure 1.28

NOTE: We could also describe this set as $\{x|x > -3 \text{ and } x < 3\}$. ■

19. $\{x|x > 3 \text{ or } x < -3\}$

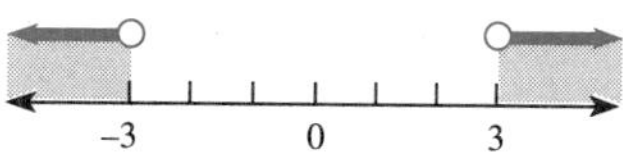

Figure 1.29

We often need to estimate costs or measurements. Usually, this means that we find a range or an interval of values for the actual cost or measurement.

EXAMPLE 20

The We Try Harder (WTH) car rental agency charges \$140 a week plus 10¢ a mile to rent a car for a week. We plan to drive between 1,600 and 1,900 miles. Use an inequality statement to describe the cost c of renting a car for a week.

Solution

If we drive the minimum number of miles, 1,600, then the

$$\text{cost} = 140 + 0.10(1{,}600) = 140 + 160 = \$300.$$

If we drive the maximum number of miles, 1,900, then the

$$\text{cost} = 140 + 0.10(1{,}900) = 140 + 190 = \$330.$$

Therefore the cost is between \$300 and \$330. We write

$$300 \le c \le 330$$ ■

EXAMPLE 21

A theorem in geometry states that "the sum of the lengths of any two sides of a triangle is greater than the length of the third side."

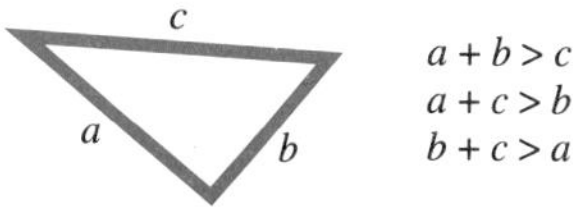

Figure 1.30

(a) Which of the following sets of numbers could possibly be the lengths of the sides of a triangle?

1. 3, 7, 9 **2.** 3, 7, 11

(b) Two sides of a triangle have lengths 3 and 7. What are some possible values for the length of the third side?

Solutions

(a) We add the lengths of each set of two sides and check to see that this sum is larger than the third length.

Figure 1.31 Figure 1.32

1. $\left.\begin{array}{l} 3+7>9 \\ 7+9>3 \\ 3+9>7 \end{array}\right\}$ **All three expressions are true. Thus 3, 7, and 9 could be the lengths of the sides of a triangle.**

2. $3+7 \not> 11$ **We need go no further. Thus 3, 7, and 11 are not the lengths of the sides of a triangle.**

(b) 8 will work since

$$3+7>8$$
$$3+8>7$$
$$7+8>3$$

$\frac{9}{2}$ will work since

$$3+7>\tfrac{9}{2}$$
$$3+\tfrac{9}{2}>7$$
$$7+\tfrac{9}{2}>3$$

In fact, any number c *between* 4 and 10 will work, $4<c<10$. ■

The Cartesian Coordinate System

We can represent expressions involving one variable on the number line (Figure 1.33). For example, if $x<5$, then all the numbers that make this a true statement are the points to the left of the point 5 on the number line.

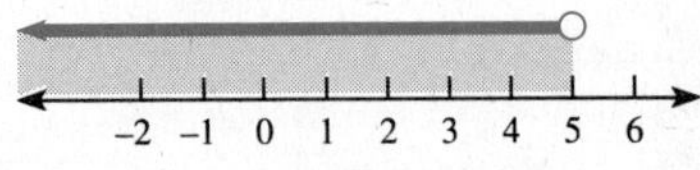

Figure 1.33

If we want to represent graphically the relationship between two variables as in Example 1 in Section 1.1, then we need to use a plane or **Cartesian coordinates** (named after its originator, René Descartes, a renowned seventeenth century French mathematician and philosopher).

The **coordinate axes** are two perpendicular number lines, with an arrow indicating the positive direction. The horizontal axis is usually called the ***x*-axis,** and the vertical axis is usually called the ***y*-axis.** They intersect at the **origin** (the zero point on both axes). The **quadrants** are numbered counterclockwise from I to IV (see Figure 1.34).

A point is located by giving its **coordinates** (a, b), where a denotes the number of units from the origin in the x-direction and b denotes the number of units from the origin in the y-direction (see Figure 1.35).

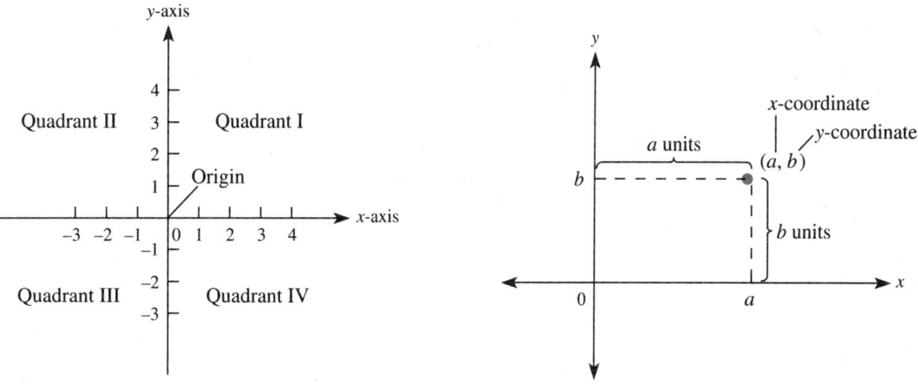

Figure 1.34 **Figure 1.35**

We know that a and b exist for each point in the plane, since every point on the x-axis and on the y-axis has a real number associated with it. By reversing this process, we see that every point in the plane can be labeled by a unique pair of real numbers. The first number in the pair is called the ***x*-coordinate,** and the second number in the pair is called the ***y*-coordinate** of the point.

EXAMPLE 22 Find the coordinates of each of the points A, B, C, D, E, F, and G.

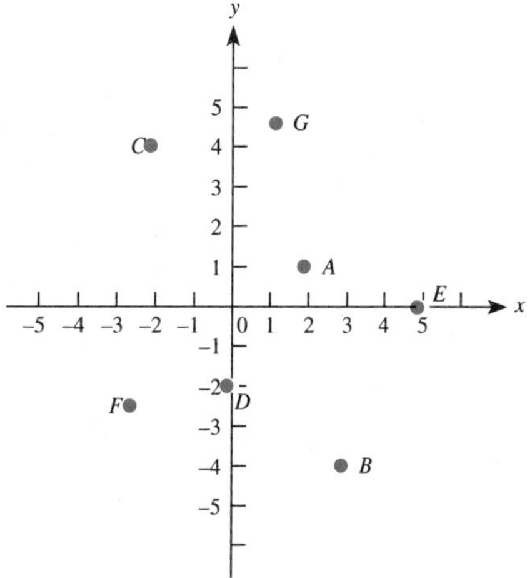

Figure 1.36

Solution

The coordinates of A are (2, 1). — **Starting at the origin, go two units in the positive x-direction and one unit up in the positive y-direction.**

The coordinates of B are (3, −4). — **Starting at the origin, go three units in the positive x-direction and four units in the negative y-direction.**

The coordinates of C are (−2, 4). — **Starting at the origin, go two units to the *left* and then four units *up*.**

The coordinates of D are (0, −2). — **Starting at the origin, go zero units to the *right* and two units *down*.**

The coordinates of E are (5, 0). — **Go five units to the *right* and zero units *up (or down)*.**

The coordinates of F are (−2.5, −2.5). — **The coordinates of F are not integers. The best we can do is to approximate them.**

The coordinates of G are (1.4, 4.7). — **We approximate them.** ■

EXAMPLE 23

Locate each of the following points on the graph: $A\ (-3, 2)$, $B\ (3, -2)$, $C\ (-3, -2)$, $D\ (3, 0)$, $E\ (0, -2)$, $F\ (\pi, 3.27)$, and $G\ (-1.5, 4.75)$.

Solution

To locate point $A\ (-3, 2)$, *we start at the origin* and *go to the left three units* and *then go two units up.* A similar procedure will locate the other points. Note that the location of point F is approximate ($\pi \approx 3.14$)

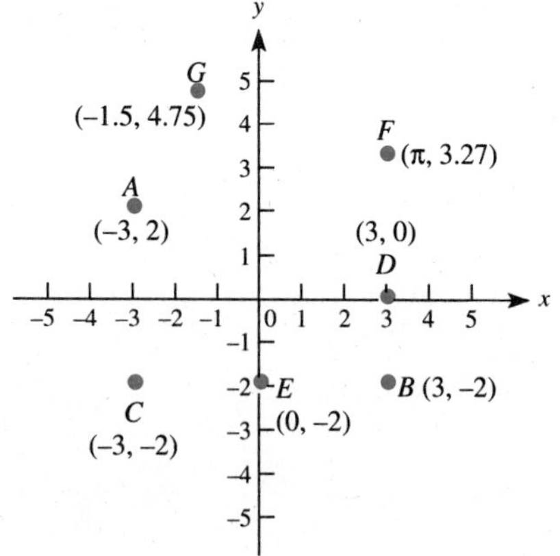

Figure 1.37 ■

We observe from Examples 22 and 23 that any point on the x-axis will have a y-coordinate of zero. In general, if P is a point on the x-axis, its coordinates are $(p, 0)$. Similarly, a point Q on the y-axis will have its x-coordinate equal to zero. We write $(0, q)$. Notice also in Example 23 that points E, B, and C have the same y-coordinate (-2). They all lie on a horizontal line that intercepts the y-axis at -2. Similarly, points A and C have the same x-coordinate (-3). They lie on a vertical line intercepting the x-axis at -3. We conclude this section with an application.

EXAMPLE 24 The You Drive It Car Rental Agency has a plan that does not include an initial cost. Some of its prices are listed in Table 1.5.

(a) Graph the entries on a Cartesian coordinate plane. Assume the pattern continues.

Miles driven	*Cost*
0	$ 0
100	$ 26
200	$ 52
300	$ 78
400	$104
500	$130

Table 1.5

(b) What is the cost to drive 270 miles in one day?

(c) How many miles can be driven in one day for $200?

(d) What is the cost per mile?

Solutions

(a)

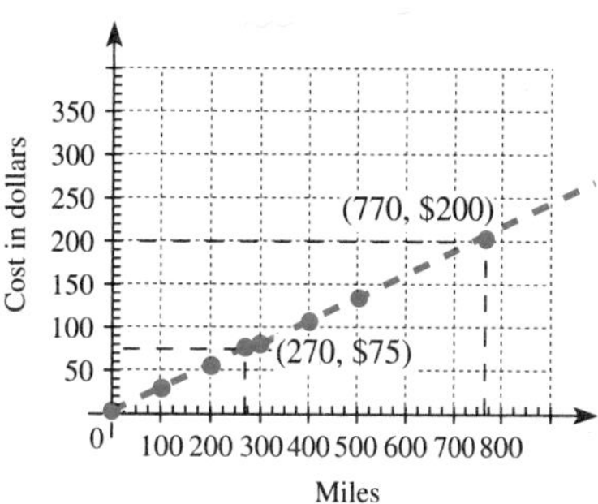

Figure 1.38

(b) From Table 1.5 we estimate the cost to drive 270 miles to be between $70 and $80. From the graph we can estimate the cost to be about $75.

(c) If we continue the table, we estimate the number of miles that can be driven for $200 to be 750; from the graph, we estimate about 770 miles.

(d) Every 100 miles cost $26. So the cost per mile is $26 ÷ 100, or $0.26 (26¢) per mile.

NOTE: On the graph (Figure 1.38), the data appear to lie on a straight line. ■

In this section we examined some important properties of real numbers—in particular, the distributive property and the inequality relationships among real numbers. We also looked at coordinate graphing in more detail.

1.2 EXERCISES

I. Procedures

1–4. Find the length of the side of a square whose area is given.

1. 25 square units **2.** 2.56 square units

3. 17 square units **4.** 199 square units

5 and 6. Express the area of rectangle *ABCD* in two different ways.

5.

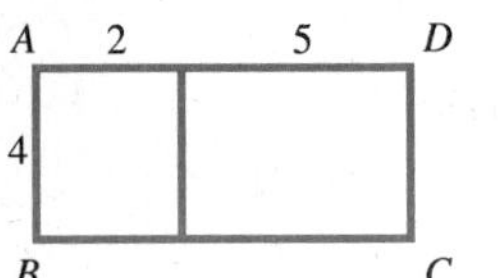

6.

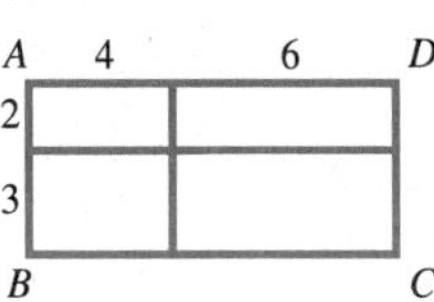

7 and 8. Use an area model to show that the following statements are true.

7. $3(7 + 2) = 3(7) + 3(2)$ **8.** $5(6) + 5(2) = 5(6 + 2)$

9 and 10. Draw a rectangle whose area is represented by the following expression.

9. $5(6 + 10)$ **10.** 120

11–13. For each of the following "number tricks," try several numbers. See if you can figure out the trick.

11. Choose a number. Multiply by 3. Add 6. Divide by 3. Subtract the original number. What is the result?

12. Choose a number. Double it. Add 6. Divide by 2. Subtract the original number. What is the result?

13. Choose a number. Add 4. Multiply by 2. Subtract 6. Subtract the original number. What is the result?

14 and 15. Locate each set of numbers as points on a number line. Give the largest and smallest number in each set.

14. $-8, \sqrt{36}, 1.5, -2.5, -3, \sqrt{3}, -\sqrt{2}$

15. $\sqrt{11}, 0, -2, 3.5, -\sqrt{8}, 2.75, -1$

16–25. Determine whether each inequality is true or false.

16. $6 < -12$ **17.** $2 \geq -5$

18. $-6 \leq -6$ **19.** $0 \geq -6$

20. $-11.6 > -11.602$ **21.** $0.999 > 0.9999$

22. $\sqrt{8} < 3$ **23.** $\sqrt{5} > 2.2$

24. $6(9 - 7) \leq 54 - 40$ **25.** $2(6 - 3) - 6 \geq 0$

26–31. **(a)** Express each of the following algebraically using the symbols $<, >, \leq, \geq$. **(b)** Sketch each set on a number line.

26. x is less than 9.

27. a is greater than or equal to -8.

28. x is between -1 and 2.

29. m is less than -2 or greater than 1.

30. y is greater than -2 and less than -1.

31. w is less than 3 and greater than 0.

32–37. **(a)** Sketch each of the following on the number line. **(b)** Write an English statement to describe each set.

32. $\{x|x > -1\}$ **33.** $\{x|x \leq 4\}$

34. $\{x|0 < x < 3\}$ **35.** $\{x|x > 3 \text{ or } x < -3\}$

36. $\{x|x > -5 \text{ and } x > -1\}$ **37.** $\{x|x \leq 5 \text{ or } x > 2\}$

38 and 39. Locate the following points on a Cartesian coordinate system.

38. $A\,(4, 2), B\,(4, -2), C\,(-4, -2), D\,(-4, 2), E\,(2, 4), F\,(-2.5, -4.2), G\,(0, -2), H\,(-4, 0), I\,(0, 0)$

39. $A\,(3, 0), B\,(3, 1), C\,(3, -1), D\,(3, 6), E\,(3, -7), F\,(3, \pi), G\,(0.5, -1.75)$

40 and 41. Write the approximate coordinates of each point.

40. **41.**

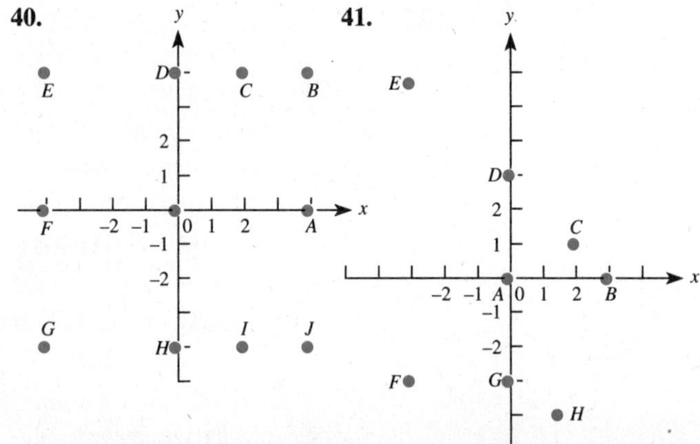

II. Concepts

42–45. Fill in the box with a number to make a true statement.

42. $7(4 + \square) = 49$

43. $8(\square - 3) = 56$

44. $\square \cdot 10 + \square \cdot 13 = 138$

45. $\square \cdot 5 - \square \cdot 2 = 9$

46–48. Without changing the order of the numbers below, insert parentheses and/or addition signs so that the computation yields the given number for the answer.

$$3 \quad 2 \quad 4 \quad 1$$

Example: The following arrangement of 3, 2, 4, 1 yields an answer of 9. (Are there other possibilities?)

$$3 + (2 + 4)1 = 9$$

46. 13 **47.** 21 **48.** 10

49. Find three whole numbers between $\sqrt{5}$ and $\sqrt{50}$.

50. Find three real numbers between $\sqrt{10}$ and 5.

51–56. The algebraic equivalent of each of the following graphs is included in the statements (a) through (h) that follow the graphs. After each graph, place the letter corresponding to the statement that matches each graph. Some of the statements do not match any of the given graphs.

51.

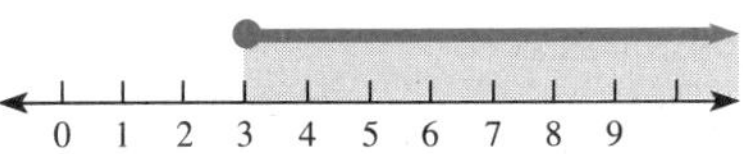

52.

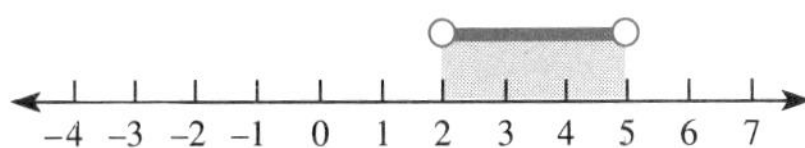

53.

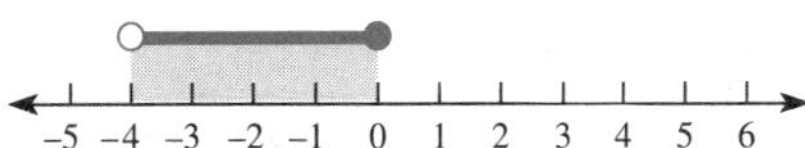

54.

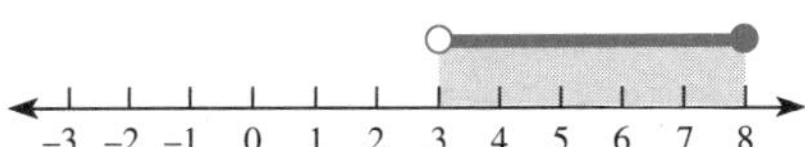

55.

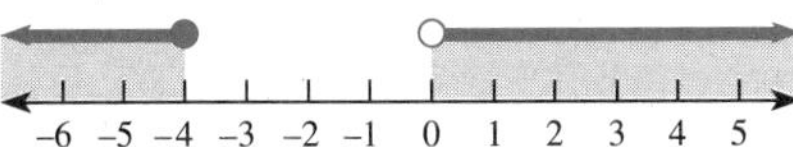

56.

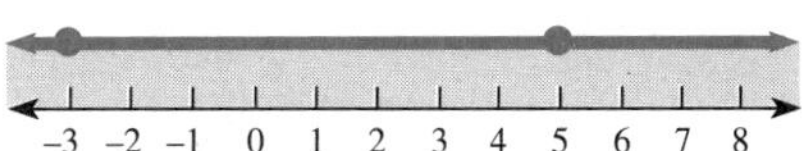

(a) $x > 0$ or $x \leq -4$ **(b)** $2 < x < 5$

(c) $3 < x < 8$ **(d)** $-4 < x \leq 0$

(e) $x \leq +5$ or $x \geq -3$ **(f)** $3 < x < 8$

(g) $x > 3$ and $x \leq 8$ **(h)** $x \geq 3$

57. Describe the process of graphing points on a coordinate system.

III. Applications

58–70. An inequality is a mathematical statement that contains one or more of the following symbols: $<$, $>$, $\leq$, or $\geq$. Write an inequality that describes each of the following situations.

58. The temperature in Flagstaff, Arizona, on July 4 was over 92°F. Let *T* be the actual temperature. Describe *T*.

59. The 1990 census gives a population of 1,006,877 for Dallas, Texas. It was estimated that there were at least 2.5 million people in the greater metropolitan area of Dallas. How many people, *P*, live in the metropolitan area but not in Dallas?

60. John's grade in math is a B. The range for a B is 80–89%. Describe the range for John's percentage grade, *S*.

61. On a 100-point test, John needs at least 78 points to obtain a B. Let *S* be John's score on the test. If John wants a B, describe the range of values for *S*.

62. The minimum speed on an interstate highway is 45 and the maximum speed is 65. Describe the speed, *m*, of a car going

(a) too slow. **(b)** too fast.

(c) within the speed limit.

63. The Glee Club plans on selling Bach T-shirts for \$6.95. Its expenses are \$25 for advertising plus \$5.25 for each T-shirt. How many T-shirts, *t*, must the club sell before it realizes a profit?

64. The We Try Harder car rental agency charges \$200 a week plus 11¢ a mile for a camper. Describe the cost for a week if you plan to drive between 2,000 and 2,400 miles.

65. Which of the following sets of numbers can be the lengths of the sides of a triangle?

(a) 7, 6, 5 **(b)** 3, 4, 8

(c) 8.3, 7.62, 15.905

66. Give two values for the length of one side, s, of a triangle if the lengths of the other two sides are

(a) 8 and 11. **(b)** 3.56 and 1.03.

67. An electrical motor is designed to run on any voltage in a range from 208 to 220 volts, plus or minus 10%. Describe the actual range of voltages, v, within which the motor will run.

68. The minimum wage in the United States is $4.25 per hour. Describe the hourly wage, w, of a worker in the United States in companies that follow the minimum wage guidelines.

69. An employee in the ABC manufacturing company pays the minimum wage to all new workers. Increases are granted regularly, with a maximum of $10.75 per hour. Describe the range for the hourly wage of a worker at the ABC company.

70. You would like at least $500 a year interest on an investment of P dollars. If the annual interest rate is 8%, describe the range of values for P.

71–73. For each of the following sets of data draw a Cartesian graph and answer the questions.

71. The You Drive It car rental agency

Miles driven	*Cost*
0	$12
50	$22
100	$32
150	$42
200	$52
250	$62

(a) What is the cost for 120 miles?

(b) How many miles can be driven for $90?

(c) Describe the pattern in words.

72. The area of a square

Length of a side	*Area*
1 cm	1 square cm
2	4
3	9
4	16
5	25
6	36

(a) What is the area of a square whose side is 3.2 cm long?

(b) What is the length of a square whose area is 50 square cm?

(c) Describe the pattern in words.

73. The population of fruit flies

Units of time	*Number of flies*
1	2
2	4
3	8
4	16
5	32
6	64

(a) How many flies will there be after 12 units of time?

(b) In how many hours will there be 2,048 flies?

(c) Describe the pattern in words.

74. An investor purchased 100 shares of HAL Computer stock on January 1. The value of the stock over the year is given by the following graph.

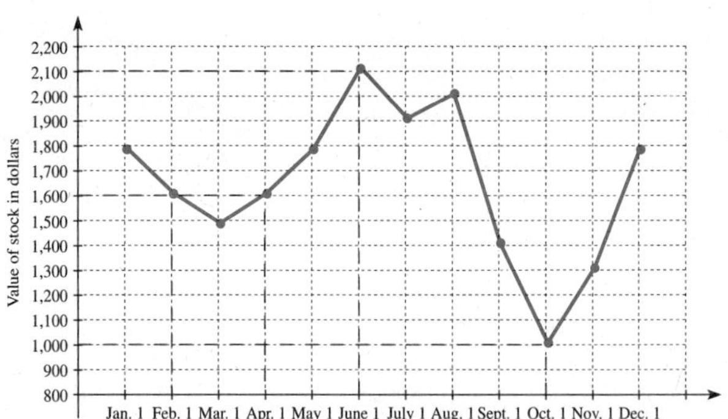

Label the points A, B, C, D on the graph and give their coordinates.

POINT A: The value of the stock on April 1.
POINT B: The day(s) that the stock's value was $1,600.
POINT C: The value of the stock that is the highest for the year.
POINT D: The day(s) that the stock is of lowest value.

IV. Extensions

75 and 76. **(a)** Find the dimensions of the rectangle $ABCD$ if the areas of three of the smaller rectangles are given. **(b)** What is the area of the shaded rectangle?

75.

A D

36	(shaded)
42	70

B C

76.

A D

60	72
144	(shaded)

B C

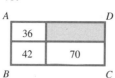

77. If $a + 7 = b + 9$, which is larger, a or b? Give a reason for your answer.

78. If $x - 7 = y - 9$, which is larger, x or y? Give a reason for your answer.

79. If $a + 5$ is a positive real number and a is an integer, what is the smallest possible value for a?

80. If $x + 5$ is a negative real number and x is an integer, describe the possible values for x.

81 and 82. For each of the following shaded regions, give the coordinates of four points that are **(a)** outside the region, **(b)** inside the region, **(c)** on the boundary of the region (other than the labeled points).

81. **82.**

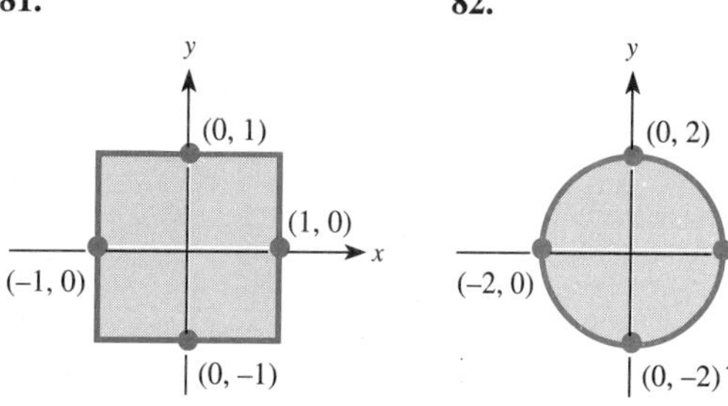

83. Triangle Inequality Experiment. Cut a straw into three pieces and see if the three pieces will form a triangle. What is the longest length one of the pieces can be and still form a triangle?

1.3 OPERATIONS ON POSITIVE REAL NUMBERS

In this section we will review operations on positive real numbers—in particular, fractions. Consider the example.

EXAMPLE 1

Seven days before the big football game between the Gold Diggers and Blue Devils, which will take place at the stadium of the Blue Devils, fans of the Gold Diggers come at night and paint half the Blue Devils' field gold. The second night they paint half of what is left. On the third night they paint half of what is left. If this continues, will they paint the entire field gold before the big game?

Solution

To answer this question, it will help to draw pictures and make a table of how much of the field has been painted. (See Figure 1.39).

After seven nights they have painted $\frac{1}{2} + \frac{1}{4} + \frac{1}{8} + \frac{1}{16} + \frac{1}{32} + \frac{1}{64} + \frac{1}{128} = \frac{127}{128}$ of the field. Thus there will be $\frac{1}{128}$ of the field unpainted. Although this is a very small amount, the entire field will not be painted. If the process is continued, will the field eventually be painted? The answer is no. Theoretically, there will always be a very small piece left, but in reality it would be difficult to see it.

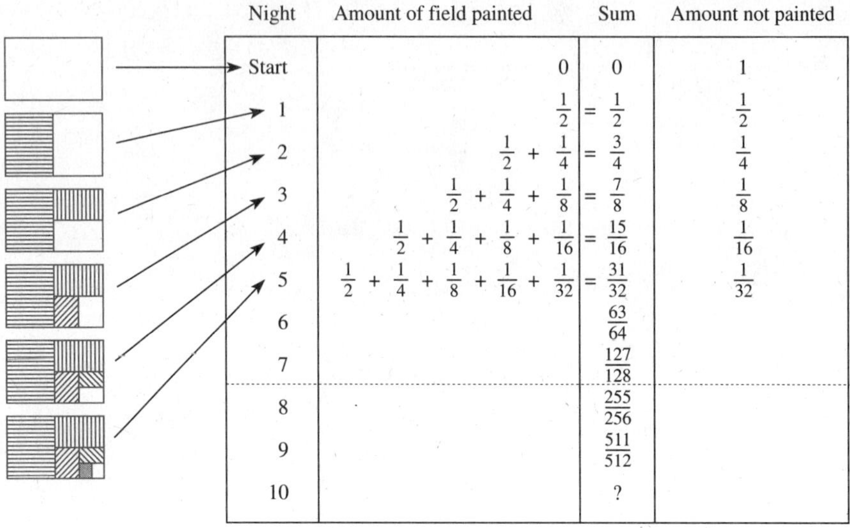

Night	Amount of field painted	Sum	Amount not painted
Start	0	0	1
1	$\frac{1}{2}$ =	$\frac{1}{2}$	$\frac{1}{2}$
2	$\frac{1}{2}+\frac{1}{4}$ =	$\frac{3}{4}$	$\frac{1}{4}$
3	$\frac{1}{2}+\frac{1}{4}+\frac{1}{8}$ =	$\frac{7}{8}$	$\frac{1}{8}$
4	$\frac{1}{2}+\frac{1}{4}+\frac{1}{8}+\frac{1}{16}$ =	$\frac{15}{16}$	$\frac{1}{16}$
5	$\frac{1}{2}+\frac{1}{4}+\frac{1}{8}+\frac{1}{16}+\frac{1}{32}$ =	$\frac{31}{32}$	$\frac{1}{32}$
6		$\frac{63}{64}$	
7		$\frac{127}{128}$	
8		$\frac{255}{256}$	
9		$\frac{511}{512}$	
10		?	

Figure 1.39

In this problem, the picture helps in visualizing how we multiply fractions. For example, on night two there is $\frac{1}{2}$ of the field left to be painted, but only $\frac{1}{2}$ of this will be painted.

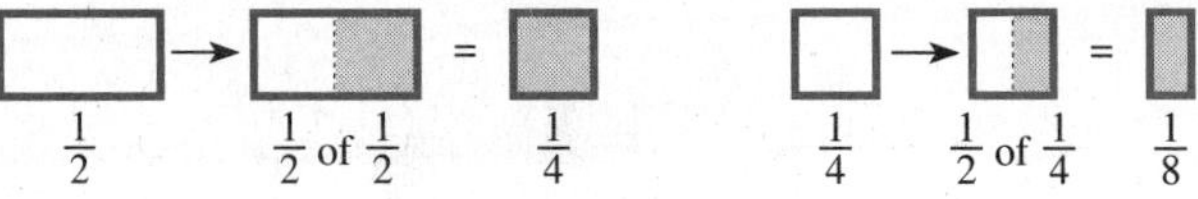

Figure 1.40

$$\text{We have} \quad \frac{1}{2} \times \frac{1}{2} = \frac{1}{4} \quad \text{and} \quad \frac{1}{2} \times \frac{1}{4} = \frac{1}{8}$$

Multiplication and Division of Fractions

To multiply two fractions, say $\frac{2}{3}$ and $\frac{4}{5}$, we can use a rectangle as we did in Example 1. First divide the area into thirds and shade two of these (use horizontal markings). Then divide the area into fifths and shade four of these (use vertical markings).

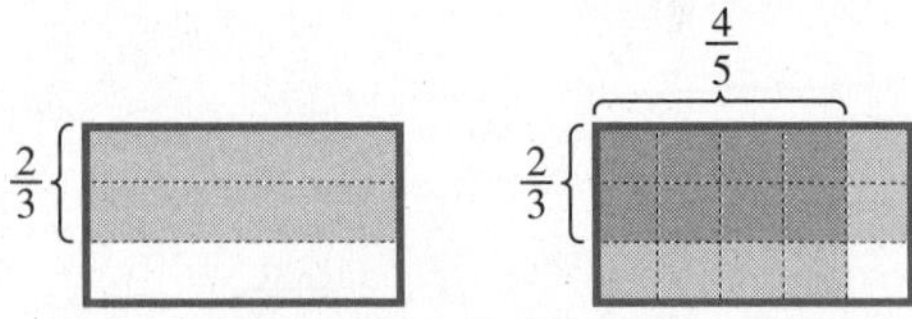

Figure 1.41

Figure 1.41 suggests that we obtain a rectangle partitioned into 15 equal parts, 8 of which overlap. Another way to multiply two fractions is to multiply the numerators and the denominators to form a new fraction:

$$\frac{2}{3} \times \frac{4}{5} = \frac{8}{15} = \frac{2 \times 4}{3 \times 5}$$

Division is related to multiplication. For example, what does it mean to divide $\frac{2}{3}$ by $\frac{5}{7}$? Suppose $\frac{2}{3} \div \frac{5}{7}$ is equal to $\frac{p}{q}$.

$$\frac{2}{3} \div \frac{5}{7} = \frac{p}{q}$$

This is equivalent to the multiplication statement

$$\frac{2}{3} = \frac{p}{q} \times \frac{5}{7}$$

$$\frac{2}{3} \times \frac{7}{5} = \frac{p}{q} \times \frac{5}{7} \times \frac{7}{5}$$

We choose to multiply both sides of the equation by $\frac{7}{5}$ because $\frac{5}{7} \times \frac{7}{5} = 1$ and multiplying both sides of an equality by the same number does not change the equality.

or

$$\frac{2}{3} \times \frac{7}{5} = \frac{p}{q} \times (1)$$

$$\frac{2}{3} \times \frac{7}{5} = \frac{p}{q}$$

$$\frac{14}{15} = \frac{p}{q}$$

therefore,

$$\frac{2}{3} \div \frac{5}{7}$$

is equivalent to

$$\frac{2}{3} \times \frac{7}{5}$$

To divide by a fraction, we have shown that we can invert the divisor and multiply:

$$\frac{2}{3} \div \frac{5}{7} = \frac{2}{3} \times \frac{7}{5} = \frac{14}{15}$$

In general, we have

Multiplication and Division of Fractions

$$\frac{p}{q} \cdot \frac{r}{s} = \frac{pr}{qs}, \text{ where } q, s \neq 0$$

$$\frac{p}{q} \div \frac{r}{s} = \frac{p}{q} \cdot \frac{s}{r} = \frac{ps}{qr}, \text{ where } q, r, s \neq 0$$

EXAMPLE 2

Examine the sequence of positive fractions below:

$$\frac{1}{2}, \frac{1}{3}, \frac{1}{4}, \frac{1}{5}, \frac{1}{6}, \ldots$$

(a) What are the next four fractions in the sequence?

(b) Describe the pattern.

(c) Sketch a graph.

Solutions

(a) The next four fractions are $\frac{1}{7}, \frac{1}{8}, \frac{1}{9}, \frac{1}{10}$.

(b) The numerator is constant, 1, and the denominator is increasing by one. The fraction is getting smaller and smaller. If n is any positive integer, then any fraction in the sequence can be represented by $\frac{1}{n}$. Also, as n gets larger and larger, $\frac{1}{n}$ gets closer and closer to zero.

(c) The graph of the sequence is on a Cartesian coordinate plane, with values of n along the horizontal axis and values of $\frac{1}{n}$ along the vertical axis (Figure 1.42).

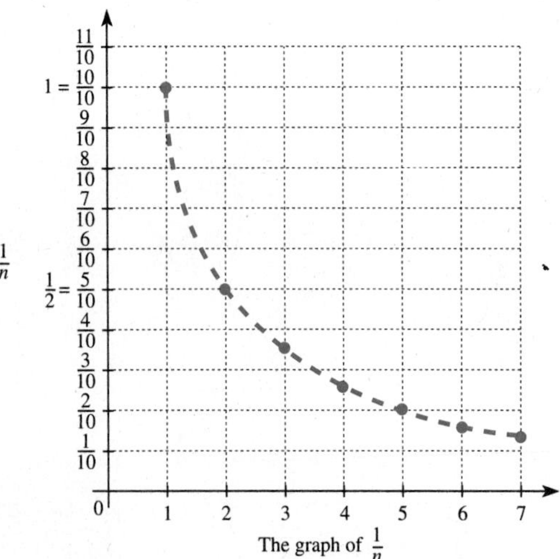

Figure 1.42

Suppose in the graph we let n be any real number. Let us examine the behavior of the graph for values of n between 0 and 1. Notice that as n gets close to zero, $\frac{1}{n}$ rapidly increases. We can say that it is "going toward infinity." The graph will never intersect the vertical axis. This suggests that we cannot divide by zero. To help see this in another way, recall that $\frac{8}{2} = 4$ because $8 = 4 \cdot 2$. What does $\frac{8}{0}$ mean? Let $\frac{8}{0} = \boxed{?}$ Thus, $8 = 0 \cdot \boxed{?}$. Since zero times any number is 0, *there is no number* that we can put in the box to make the equation hold true. This is the reason we cannot divide by zero. Thus, we emphasize that *no* denominator can be zero. **Division by 0 is undefined.** Similarly, $\frac{0}{0}$ is undefined, since infinitely many numbers n would satisfy $0 = 0 \cdot n$. To reiterate, the expression $\frac{p}{0}$ *is undefined for any number p.* ■

Equivalent Fractions

The following picture illustrates the concept of equivalent fractions.

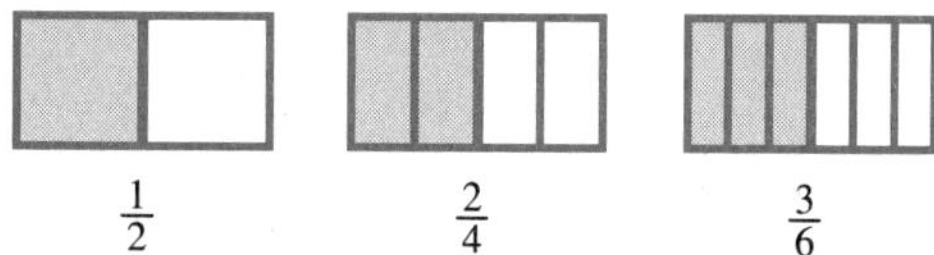

$\frac{1}{2}$ $\frac{2}{4}$ $\frac{3}{6}$

Figure 1.43

We note that $\frac{1}{2} = \frac{2}{4} = \frac{3}{6}$.

Another way to view equivalent fractions is to consider the property that any number is unchanged when multiplied by one.

Since $a \cdot 1 = a$, then $\frac{2}{3} \cdot 1 = \frac{2}{3}$. But 1 can be represented by $\frac{1}{1}, \frac{2}{2}, \frac{3}{3}$, and so on. Thus

$$\frac{2}{3} \cdot 1 = \frac{2}{3} \cdot \frac{2}{2} = \frac{2 \cdot 2}{3 \cdot 2} = \frac{4}{6}$$

Conversely, we have

$$\frac{9}{6} = \frac{3 \cdot 3}{3 \cdot 2} = \frac{3}{3} \cdot \frac{3}{2} = 1 \cdot \frac{3}{2} = \frac{3}{2}$$

These examples illustrate a very important property of fractions:

The Fundamental Property of Fractions

$$\frac{ax}{bx} = \frac{a}{b}, \text{ where } b \text{ and } x \neq 0$$

When $\frac{ax}{bx} = \frac{a}{b}$ and a and b have no factor other than 1 in common, then $\frac{a}{b}$ is called the **equivalent simplest form** of $\frac{ax}{bx}$.

The fundamental property of fractions implies two processes:

1. $\frac{a}{b} = \frac{a}{b} \cdot \frac{x}{x} = \frac{ax}{bx}$ $(x, b \neq 0)$

To go from $\frac{a}{b}$ to $\frac{ax}{bx}$, we multiply the numerator and denominator by x. That is, we multiply the fraction by $\frac{x}{x}$ or 1.

2. $\frac{ax}{bx} = \frac{a}{b} \cdot \frac{x}{x} = \frac{a}{b} \cdot 1 = \frac{a}{b}$ $(x, b \neq 0)$

To go from $\frac{ax}{bx}$ to $\frac{a}{b}$, we divide by a common factor, x, in the numerator and denominator. Both processes are a way of naming (and writing) equivalent fractions.

The fractions $\frac{1}{10}, \frac{1}{100}, \frac{1}{1000}, \ldots$ can also be written as the decimals 0.1, 0.01, 0.001, ... Conversely, decimals can be written as fractions. For example, $0.125 = \frac{125}{1000} = \frac{1}{8}$; $12.5 = 12\frac{5}{10}$ or $12\frac{1}{2}$, and $1.123 = 1\frac{123}{1000}$. Fractions can be written as decimals and decimals can be written as fractions.

EXAMPLES 3–6 Express the following fractions in simplest form.

3. $\frac{344}{76}$ 4. $\frac{1\frac{1}{3}}{2\frac{2}{3}}$ 5. $\frac{27}{10}$ 6. $\frac{8.1}{0.27}$

Solutions

3. $\frac{344}{76} = \frac{86 \cdot 4}{19 \cdot 4} = \frac{86}{19}$ **Fundamental property of fractions.**

4. $\frac{1\frac{1}{3}}{2\frac{2}{3}} = \frac{\frac{4}{3}}{\frac{8}{3}}$

$= \frac{1 \cdot \left(\frac{4}{3}\right)}{2 \cdot \left(\frac{4}{3}\right)}$

$= \frac{1}{2}$ **Fundamental property of fractions.**

Notice that we could have proceeded as follows:

$\frac{1\frac{1}{3}}{2\frac{2}{3}} = 1\frac{1}{3} \div 2\frac{2}{3} = \frac{4}{3} \div \frac{8}{3} = \frac{4}{3} \cdot \frac{3}{8} = \frac{1}{2}$ **Recall that $\frac{p}{q} = p \div q$.**

5. $\frac{27}{10}$ is in simplest form. **There are no common whole number factors in the numerator and denominator (other than 1).**

6. $\frac{8.1}{0.27} = \frac{30(0.27)}{1(0.27)}$

$= 30$ **This could also be done on a calculator.**

Addition and Subtraction of Fractions

The pattern in Example 1 also illustrates addition of fractions. For example, after two nights the amount of the field that has been painted is $\frac{1}{2} + \frac{1}{4} = \frac{3}{4}$. To arrive at this answer, we express $\frac{1}{2}$ as $\frac{2}{4}$ and combine

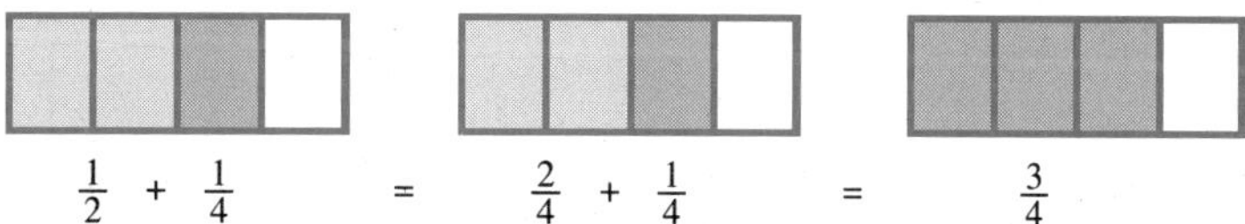

$$\frac{1}{2} + \frac{1}{4} \quad = \quad \frac{2}{4} + \frac{1}{4} \quad = \quad \frac{3}{4}$$

Figure 1.44

To add fractions, we need to rename them so each denominator represents a number of the same-sized pieces of the whole. This is called **changing each fraction to an equivalent fraction with a "common denominator."**

We can use the fundamental property of fractions to write equivalent fractions with common denominators. For example,

$$\begin{aligned}\frac{1}{2} + \frac{1}{3} &= \left(\frac{1}{2} \cdot \frac{3}{3}\right) + \left(\frac{1}{3} \cdot \frac{2}{2}\right) \qquad \text{Rename } \tfrac{1}{2} \text{ and } \tfrac{1}{3} \text{ as } \tfrac{3}{6} \text{ and } \tfrac{2}{6}.\\ &= \frac{3}{6} + \frac{2}{6}\\ &= \frac{5}{6}\end{aligned}$$

In this example, we used the common denominator of 6, but any common multiple of 2 and 3 will work. For example, $\frac{1}{2} + \frac{1}{3} = \frac{6}{12} + \frac{4}{12} = \frac{10}{12} = \frac{5}{6}$. Recall that to add or subtract fractions, we rename them to find a common denominator.

EXAMPLES 7–10

Perform the following operations and express the answer in simplest form.

7. $\frac{3}{7} - \frac{2}{7}$ **8.** $\frac{4}{9} + \frac{2}{3}$ **9.** $\frac{3}{4} + \frac{1}{6}$ **10.** $1 + \frac{3}{5} - \frac{1}{2} - \frac{1}{6}$

Solutions

7. $\frac{3}{7} - \frac{2}{7} = \frac{3 - 2}{7} = \frac{1}{7}$

8. $$\begin{aligned}\frac{4}{9} + \frac{2}{3} &= \frac{4}{9} + \frac{2 \cdot 3}{3 \cdot 3}\\ &= \frac{4 + 6}{9}\\ &= \frac{10}{9}\end{aligned}$$

Fundamental property of fractions. We note that 9 is a common multiple of 3 and 9. It is the smallest common multiple. We call it the least common multiple of 3 and 9.

The **least common denominator** of $\frac{a}{b}$ and $\frac{c}{d}$ is the *least common multiple* of b and d.

While any common multiple will work for common denominators, the least common multiple will usually produce an answer in simplest form.

9. Least common denominator:

$$\frac{3}{4}+\frac{1}{6}=\frac{9}{12}+\frac{2}{12}=\frac{11}{12}$$

Least common denominator is 12; 12 is the smallest common multiple of 4 and 6.

Any common denominator:

$$\frac{3}{4}+\frac{1}{6}=\frac{3\cdot 6}{4\cdot 6}+\frac{1\cdot 4}{6\cdot 4}=\frac{22}{24}=\frac{11}{12}$$

Any common multiple will do. The product of the denominators is a multiple of each denominator. We could have used 36, 60, etc. as denominators.

10. $1+\frac{3}{5}-\frac{1}{2}-\frac{1}{6}=\frac{30}{30}+\frac{18}{30}-\frac{15}{30}-\frac{5}{30}$

30 is the least common denominator. All fractions are renamed to have 30 as a denominator.

$$=\frac{30+18-15-5}{30}$$

$$=\frac{28}{30}$$

$$=\frac{14}{15}$$

Rewrite in simplest form. ■

Addition and Subtraction of Fractions

1. Where the fractions have the same denominator,

$$\frac{a}{c}\pm\frac{b}{c}=\frac{a\pm b}{c},\text{ where } c\neq 0$$

2. Where the fractions have different denominators,

$$\frac{a}{b}\pm\frac{c}{d}=\frac{ak}{D}\pm\frac{cl}{D}=\frac{ak\pm cl}{D},$$

where $D=bk=dl\neq 0$ is a common multiple of b and d.

EXAMPLES 11–12

Perform the following operations. Express the answers in simplest form.

11. $2+\frac{2}{3}\left(1-\frac{3}{5}\right)$

12. $\dfrac{\frac{3}{4}+\frac{12}{5}}{\frac{3}{10}}$

Solutions

> *Note:* To avoid ambiguity, we agree to *multiply and divide before adding or subtracting*. To further clarify the order of operations, parentheses are used.

11. In the expression $2 + \frac{2}{3}(1 - \frac{3}{5})$, the multiplication, $\frac{2}{3}(1 - \frac{3}{5})$, must be done first. But $\frac{2}{3}(1 - \frac{3}{5})$ can be done in two ways. We can either distribute multiplication over subtraction *or* combine the numbers inside the parentheses and then multiply.

$$2 + \frac{2}{3}\left(1 - \frac{3}{5}\right) = 2 + \frac{2}{3}(1) - \frac{2}{3}\left(\frac{3}{5}\right) \qquad \text{or} \qquad = 2 + \frac{2}{3}\left(\frac{2}{5}\right)$$

$$= 2 + \frac{2}{3} - \frac{2}{5} \qquad\qquad = 2 + \frac{4}{15}$$

$$= \frac{30}{15} + \frac{10}{15} - \frac{6}{15} \qquad\qquad = \frac{30}{15} + \frac{4}{15}$$

$$= \frac{34}{15} \qquad\qquad = \frac{34}{15}$$

12.
$$\frac{\frac{3}{4} + \frac{12}{5}}{\frac{3}{10}} = \left(\frac{3}{4} + \frac{12}{5}\right) \div \frac{3}{10}$$

$$= \left(\frac{15 + 48}{20}\right) \div \frac{3}{10}$$

$$= \frac{63}{20} \div \frac{3}{10}$$

$$= \frac{63}{20} \cdot \frac{10}{3}$$

$$= \frac{21}{2}$$

We could also perform the addition first:

$$\frac{\frac{3}{4} + \frac{12}{5}}{\frac{3}{10}} = \frac{\frac{63}{20}}{\frac{3}{10}} = \frac{63}{20} \div \frac{3}{10} = \frac{63}{20} \cdot \frac{10}{3} = \frac{21}{2}$$ ∎

EXAMPLES 13–14

Calculate. Round off answers to two decimal places.

13. $2.3(1359.15 - 987.65)$

14. $\dfrac{1.853 \times 50.59 \times 111.2}{147.6}$

Solutions

13. Before we begin, it might be wise to estimate our answer. Even with calculators, we can make mistakes. First we observe that

$$2.3(1359.15 - 987.65) \approx 2(1400 - 1000) \approx 2(400) \approx 800$$

Note: "≈" means "approximately equal to."

Using a calculator, we have

$$2.3(1359.15 - 987.65) = 854.45$$

So 800 is a reasonable estimate for 854.45.

14. Estimating, we have

$$\frac{1.853 \times 50.59 \times 111.2}{147.6} \approx \frac{2 \times 50 \times 100}{150} \approx \frac{200}{3} \approx 67$$

Using a calculator, we have

$$\frac{1.853 \times 50.59 \times 111.2}{147.6} = 70.625011 \approx 70.63$$

Recall that by convention, in the absence of a real context, the digits 0, 1, 2, 3, and 4 are *rounded down*. That is, the digit immediately preceding remains the same. The digits 5, 6, 7, 8, and 9 are *rounded up*. That is, the digit immediately preceding increases by one.

It is a good practice to estimate the answer before starting the detailed calculations. For a lengthy computation such as that in Example 14, it is *not* recommended to round off intermediary results, since this practice might cause an accumulation in errors. Carry these digits in your calculator until the final rounding steps for greater accuracy.

EXAMPLES 15–16

Perform the operations in each sequence and answer the questions.

15. $\frac{1}{2}$

$\frac{1}{2} + \frac{1}{4}$

$\frac{1}{2} + \frac{1}{4} + \frac{1}{8}$

$\frac{1}{2} + \frac{1}{4} + \frac{1}{8} + \frac{1}{16}$

16. $\frac{1}{3}$

$\frac{1}{3} + \frac{1}{9}$

$\frac{1}{3} + \frac{1}{9} + \frac{1}{27}$

$\frac{1}{3} + \frac{1}{9} + \frac{1}{27} + \frac{1}{81}$

(a) What is the sum of the fractions in the fifth row? The tenth row?

(b) Describe the pattern.

(c) If the pattern continues, what number is the sum approaching?

Solutions

15. (a) $\frac{31}{32}, \frac{1,023}{1,024}$

(b) The denominator of the sum doubles each time and the numerator is one less than the denominator.

(c) If we represent the sums as decimals (0.5, 0.75, 0.875, 0.9375, 0.9675, . . . , 0.9990239375), we see that the sum is approaching 1.

16. (a) $\frac{121}{243}, \frac{29,524}{59,049}$

(b) The denominator of the sum triples each time (is multiplied by 3 each time) and the numerator is almost half of the denominator. (Subtract one from the denominator and divide by two to get the numerator.)

(c) Represent the sums as decimals (0.333 . . . , 0.444 . . . , 0.48148148 . . . , 0.49794238 . . . , . . . , 0.49991532 . . .); we see that the sum is approaching $\frac{1}{2}$.

We conclude this section by examining an important application of fractions.

Ratio

We have often seen or heard statements such as: Canseco went 3 for 4 today. The daily rate dropped to 1 birth per 10,000 people. It consumes 10 times its weight in excess stomach acid. Buy three tires, get the fourth one free. Lettuce is 69¢ a pound. One inch represents 150 miles. The Supreme Court upheld a previous decision on child custody by a vote of 5 to 4 today.

Each of these statements involves the quantitative *comparison* of two quantities or two measures. In mathematics, such statements are called *ratios.* A ratio is a concise way to express the relative (to each other) sizes of two measures. The statements above could be shortened to: 3 hits in 4 tries; 1 birth per 10,000 people; 10 grams to every 1 gram; 3 bought for 1 free; 69 cents to 1 pound; 150 miles to 1 inch; 5 yes to 4 no votes. In fact, if we remove the dimensions, we are left with the pure ratios: 3 to 4, 1 to 10,000, 10 to 1, 3 to 1, 69 to 1, 150 to 1, 5 to 4. Of course, these ratios without dimensions are not as interesting, because the context of their meaning has been removed.

Definition

A **ratio** of a number b to a number c ($c \neq 0$) is the quotient $\frac{b}{c}$.

Although expressed as a fraction $\frac{b}{c}$, we often read a ratio as "b is to c." The ratio of b to c can also be written b:c. Let's look at several examples of ratios.

EXAMPLE 17

A baseball player got 190 hits last season, but he also struck out 112 times. What is the ratio of the number of hits to the number of strikeouts?

Solution

We are asked to find the ratio of 190 hits to 112 strikeouts. This ratio can be expressed as

$$\frac{190 \text{ hits}}{112 \text{ strikeouts}} \quad \text{or just} \quad \frac{190}{112}$$

Now, while $\frac{190}{112}$ is a correct response, we could perhaps express it in a simpler way: $\frac{190}{112} = \frac{95}{56} \approx \frac{1.7}{1}$. So, we can say the ratio of hits to strikeouts is 1.7 hits (approximately) to every strikeout. ■

EXAMPLE 18

Of 76 geese captured, 19 of the geese had red bands around their necks.

(a) What is the ratio of banded geese to the total number captured?

(b) What is the ratio of unbanded geese to banded geese?

Solutions

(a) We can write 19 banded geese to 76 captured as

$$\frac{19 \text{ banded}}{76 \text{ total geese}} \quad \text{or just} \quad \frac{19}{76} = \frac{1}{4}$$

The ratio is simply 1 to 4.

(b) We are not told how many unbanded geese there are. However, we can easily see that

$$\text{Number of unbanded geese} = \begin{matrix}\text{Total number}\\ \text{of geese}\end{matrix} - \begin{matrix}\text{Number of}\\ \text{banded geese}\end{matrix}$$

so

$$\text{Number of unbanded geese} = 76 - 19 = 57$$

The ratio of unbanded to banded geese is then

$$\frac{57 \text{ unbanded}}{19 \text{ banded}} \quad \text{or} \quad \frac{57}{19} = \frac{3}{1}$$

The ratio is 3 to 1. There are three times as many unbanded geese as banded geese in the capture. ■

Percent

A very important use of ratio is percent. **Percent** is a ratio whose denominator is 100. Percent means *parts per 100* or *parts of 100.* Percents are very useful in comparing two fractions. Consider the following example.

EXAMPLE 19

At the end of an intramural basketball tournament, Chris, Pat, and Terry were comparing their free-throw averages. Their free-throw record for the tournament is

Chris:	16 out of 23
Pat:	12 out of 18
Terry:	21 out of 31

Terry claimed to be the best because she made the most free throws; Pat said she was the best because she missed the fewest; Chris said he was the best because he had the best free-throw average. Who is correct?

Solution

In basketball, it is customary to express one's free-throw record as a percent consisting of the ratio of free throws made to total free throws attempted.

$$\text{Chris:} \quad \text{Free-throw average} = \frac{16}{23}$$

$$\text{Pat:} \quad \text{Free-throw average} = \frac{12}{18}$$

$$\text{Terry:} \quad \text{Free-throw average} = \frac{21}{31}$$

To find the best free-throw percentage, we must compare the fractions. We can convert the fractions to decimals (rounded to the nearest hundredth) and then to percents. In that way we are comparing fractions that count "parts" of equal size.

$$\text{Chris:} \quad \frac{16}{23} \approx 0.70$$

$$\text{Pat:} \quad \frac{12}{18} \approx 0.67$$

$$\text{Terry:} \quad \frac{21}{31} \approx 0.68$$

Chris has the best average. In basketball, free-throw averages are usually expressed as percentage. Thus, Chris has a $\frac{70}{100}$ or 70% free-throw average; Pat has a $\frac{67}{100}$ or 67% free-throw average; Terry has a $\frac{68}{100}$ or 68% free-throw average. ■

Other uses of percent will occur throughout this book. Some common uses are to express increases or decreases in quantities such as

sales discounts: A \$30 shirt marked down \$6 is said to have a $\frac{6}{30}$ or 20% markdown.

population rise: A city with a population of 1,000 increases to 1,050 after one year. It had a $\frac{50}{1{,}000}$ or 5% increase in population.

In this section we studied fractions and in particular we looked at special fractions—ratio and percent. The fundamental property of fractions, $\frac{ax}{bx} = \frac{a}{b}$ $(b, x \neq 0)$, is very important.

1.3 EXERCISES

I. Procedures

1–20. Compute each of the following. Express all answers in simplest form.

1. $\frac{7}{12} + \frac{19}{12}$

2. $\frac{3}{10} + \frac{11}{10} + 1$

3. $\frac{17}{12} - \frac{5}{6}$

4. $\frac{6}{7} - \frac{1}{14} + 2$

5. $\frac{3}{7} \cdot \frac{49}{75}$

6. $\frac{4}{15} \cdot \frac{45}{56}$

7. $\frac{16}{24} \div \frac{12}{64}$

8. $\frac{3}{5} \div \frac{10}{6}$

9. $\frac{(6-4)}{2} - \frac{1}{3}$

10. $\left(\frac{2+3}{6}\right) \div 15$

11. $\frac{2}{3}\left(\frac{6}{5} + \frac{3}{10}\right)$

12. $\frac{3}{2}(9 - 5)$

13. $\frac{1}{4} - \frac{1}{3}\left(\frac{3}{4} - \frac{2}{3}\right)$

14. $4 - \frac{2}{3}\left(2 - \frac{7}{8}\right)$

15. $\frac{3}{5}(7 - 2) - \frac{5}{6}(10 - 8)$

16. $\frac{5}{24} \cdot \frac{16}{30} + 3$

17. $\dfrac{\frac{15}{18}}{\frac{24}{45}}$

18. $\dfrac{\frac{3}{2} + \frac{1}{6}}{\frac{3}{4}}$

19. $\frac{23}{12} - 2 + \frac{1}{3} + \frac{7}{6}$

20. $\frac{35}{18} \div \frac{7}{12} - 1$

21–28. **(a)** Estimate each answer. **(b)** Perform the indicated operations. Round off answers to three decimal places.

21. $2.076 + 1.76 - 0.74$
22. $5.06 - 4.982$
23. $(3.27)(0.79)$
24. $0.215 + 0.04$
25. $0.06(3 + 0.05)$
26. $1.2(2.1 - 1.4)$
27. $\dfrac{1{,}000(31.5275)}{39.012 - 12.003}$
28. $\dfrac{3.61(18.075 - 9.025)}{2.01(41.111 + 60.203)}$

29. Which pairs of fractions are equivalent?

(a) $\frac{1}{2}, \frac{3}{6}$ **(b)** $\frac{3}{15}, \frac{6}{30}$

(c) $\frac{15}{25}, \frac{4}{5}$ **(d)** $\dfrac{\frac{3}{4}}{\frac{8}{9}}, \frac{27}{32}$

(e) $\frac{91}{7}, \frac{13}{1}$ **(f)** $\frac{3+2}{8+2}, \frac{3}{8}$

(g) $\frac{3}{5}, \frac{6(17-2)}{10(17-2)}$

II. Concepts

30. **(a)** Divide 3 by $\frac{3}{7}$. **(b)** Divide $\frac{3}{7}$ by 3.

31. **(a)** Divide $\frac{3}{7}$ by $\frac{4}{5}$. **(b)** Divide $\frac{4}{5}$ by $\frac{3}{7}$.

(c) Multiply $\frac{3}{7}$ by $\frac{4}{5}$. **(d)** Multiply $\frac{4}{5}$ by $\frac{3}{7}$.

32. Divide $\frac{65}{100}$ by $\frac{39}{120}$, and multiply the result by $\frac{2}{3}$.

33. Subtract $\frac{15}{8}$ from $\frac{20}{9}$, and multiply the result by $\frac{9}{5}$.

34–36. Find three fractions equivalent to each of the following fractions. Draw a picture to illustrate the equivalency.

34. $\frac{10}{36}$ 35. 2.4 36. $3\frac{2}{3}$

37–40. Draw a picture to illustrate each of the following statements.

37. $\frac{35}{10} = 3\frac{1}{2}$

38. $\frac{3}{4} \times \frac{8}{10} = \frac{6}{10}$

39. $\frac{2}{3} + \frac{5}{6} = \frac{9}{6}$

40. $\frac{3}{4} \div \frac{1}{8} = 6$

41. Find three fractions between $\frac{1}{18}$ and $\frac{1}{19}$.

42–49. Fill in the box with a number to make a true statement.

42. $\frac{5}{8} = \frac{\square}{24}$

43. $7(4 + \square) = 49$

44. $8(\square - 3) = 4$

45. $\square \cdot 5 - \square \cdot 2 = \frac{3}{2}$

46. $\frac{\square}{3} = \frac{24}{18}$

47. $\frac{9}{\square} = \frac{3}{2}$

48. $\frac{5}{6} + \square = \frac{3}{2}$

49. $\frac{1}{4} \cdot \square = \frac{3}{8}$

50–53. Make up your own ratio problem about each of the following statements. Can you make up more than one?

50. The heart rate of a marathon runner increases from 52 beats per minute to 160 beats per minute.

51. There were 17 rainy days in April.

52. Russia has 1,169 females to every 1,000 males.

53. The gross national product was about $5,465,100,000,000 in the United States, and the population was approximately 250 million.

54. Write an explanation of how multiplication and division of fractions are related and explain why division is the "inverse" of multiplication.

55. Explain how fractions, decimals, and percents are all related. Give examples.

III. Applications

56. In Example 1

(a) On what day will the Gold Diggers paint $\frac{1,023}{1,024}$ of the field gold?

(b) What part of the field is painted on day 15?

(c) Graph the data: Let the horizontal axis represent the row number and let the vertical axis represent the amount of the field painted. Describe the graph.

57 and 58. Examine the patterns and answer the questions. **(a)** What is the sum of the fractions in the fifth row? The tenth row? **(b)** Describe the pattern. **(c)** If the pattern continues, what number is the sum approaching? (Hint: express each sum as a decimal.) **(d)** Graph the data. Let the horizontal axis represent the row number and let the vertical axis represent the sum. Describe the graph.

57. $\frac{1}{5}$

$\frac{1}{5} + \frac{1}{25}$

$\frac{1}{5} + \frac{1}{25} + \frac{1}{125}$

$\frac{1}{5} + \frac{1}{25} + \frac{1}{125} + \frac{1}{625}$

58. $\frac{1}{10}$

$\frac{1}{10} + \frac{1}{100}$

$\frac{1}{10} + \frac{1}{100} + \frac{1}{1,000}$

$\frac{1}{10} + \frac{1}{100} + \frac{1}{1,000} + \frac{1}{10,000}$

59. A cricket is on the number line. She starts at the point 0 and jumps half the distance to the point 1. She lands on the point $\frac{1}{2}$. Then she jumps half of the remaining distance from the point $\frac{1}{2}$ to the point 1. If the cricket continues the process of jumping half the remaining distance to 1 each time.

(a) On what point will she be after seven jumps? After ten jumps?

(b) Will the cricket ever arrive at the point 1?

(c) Draw a picture to illustrate your answers in parts (a) and (b).

60–66. Solve each of the following problems.

60. A student drives 145 miles on 10 gallons of gas. What is the rate in miles per gallon?

61. The moon takes 27.3 days to travel 1,500,000 miles. Express this rate in miles per hour and in miles per minute.

62. Twenty-two pounds of flour cost $3.54. What is the cost in cents per pound?

63. A recipe for coconut bread calls for $4\frac{3}{4}$ cups of flour and $1\frac{2}{3}$ cups of sugar.

(a) What is the ratio of flour to sugar?

(b) What is the ratio of sugar to flour?

64. McDonald's U.S. sales in 1991 were $12,500,000,000. Assuming a population of 250,000,000, how much money per person was spent at McDonald's in 1991?

65. The following table represents the population and area of some of the world's largest cities in 1990. Calculate the population density (number of people per square mile) of each city.

(a) Which city has the highest population density?

(b) Which city has the lowest population density?

City	*Population in thousands*	*Area in square miles*
Mexico City	20,207	522
New York	14,622	1,274
Lagos, Nigeria	7,602	56
Hong Kong	5,655	23
Toronto	3,108	54
Sao Paulo	18,052	451
Singapore	2,695	78

66. Here is a list of world speed records.

Women's speed	*Distance*	*Men's speed*
10.49 seconds	100 meters	9.90 seconds
21.34 seconds	200 meters	19.72 seconds
47.60 seconds	400 meters	43.29 seconds
1:53.28 minutes	800 meters	1:41.73 minutes
3:52.47 minutes	1,500 meters	3:29.46 minutes

(a) For both the men and women calculate the ratio of meters to seconds for each distance for both men and women. (Remember to change minutes to seconds.)

(b) Does this ratio change significantly as the distance increases?

(c) Sketch a graph of the men's and women's speed and distance on the same coordinate axis. Let the horizontal axis represent distance.

(d) Use the graph to estimate the distance that each can run in 1 minute, 3 minutes, and 4 minutes.

(e) Estimate how long it would take each to run 500 meters, 1,000 meters, and 2,000 meters.

67–77. The following problems will give you practice with percents.

67 and 68. Part of each grid has been shaded. What percent of the area is shaded? Unshaded?

67.

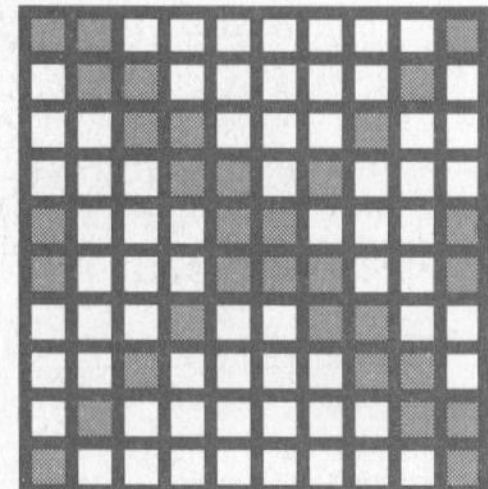

68.

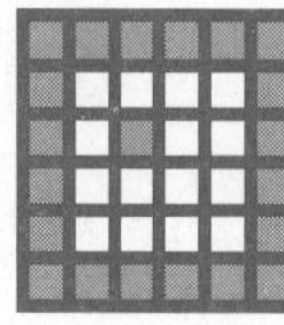

69. The free-throw records for three basketball players are given below. Find the free-throw percentage for each player. Which player has the best percentage?

John: 15 out of 29
Mary: 23 out of 30
Mickie: 17 out of 25

70–73. Calculate the percent increase or percent decrease in each of the following:

70. The price of an automobile went from \$8,575 to \$10,500.

71. A \$1,000 investment earned \$288 in interest for one year.

72. A population of 90 spotted horned owls decreased to 69 owls in one year.

73. A \$95 dress was on sale for \$60.

74. The Portland Clam Diggers lost 12 of 20 kickball games. What percent of the games did they win?

75. Ten out of every 2,000 parts produced by an assembly line are defective. What percent of 2,000 parts are defective?

76. Out of 2,874 votes cast in a local election, 1,972 were cast by women. What percent of the votes were cast by women? By men?

77. Your monthly expenditures are

Rent plus utilities: \$425	Car payment: \$250
Insurances: \$100	Food: \$280
Miscellaneous: \$300	

If your monthly take-home check is \$1,500,

(a) what percent of your monthly salary goes for each of the above?

(b) what percent of your monthly salary is available for savings?

78–83. The following problems will give you practice in computing with decimals.

78. From a piece of wire 27.635 inches long, a 9.874-inch piece was cut. What is the length of the remaining piece if 0.125 inches was wasted in making the cut?

79. Three steel plates having thicknesses of 0.726 inches, 0.508 inches, and 0.631 inches are to be riveted together.

(a) What will be their combined thickness?

(b) What is the average thickness of the plates?

80. What is the total thickness of 70 sheets of aluminum, if one sheet has a thickness of 0.13 inches?

81. If the large wheel of an antique bike is 14.36 feet in circumference, how many times (to the nearest whole number) will it turn in going one mile (5,280 feet)?

82. A good time for the 100-yard dash is 9.1 seconds. What is this speed in miles per hour to the nearest hundredth of a mile?

83. At the beginning of a trip, the odometer on a car (with a full tank) reads 43,291.5 miles. After the trip, which took six hours, the odometer read 43,480.2 miles, and the driver needed 10.5 gallons of gas to fill the gas tank.

(a) How many miles to the gallon did the car get?

(b) What was the average speed on the trip?

84. The graphs below represent the profit on sweatshirt sales for the Spirit Club. Which graph best describes the profit from the sales of sweatshirts? Explain your reasoning.

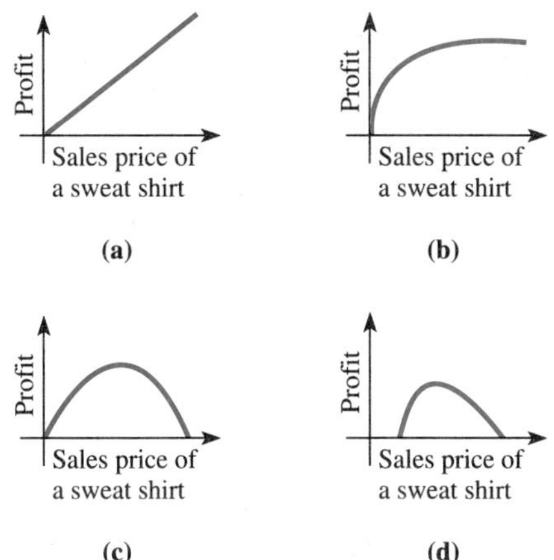

85. The graph below represents the speed of a cross country skier at certain times during a cross country trek. Describe the skier's journey at points *A, B, C, D, E, F* and *G*.

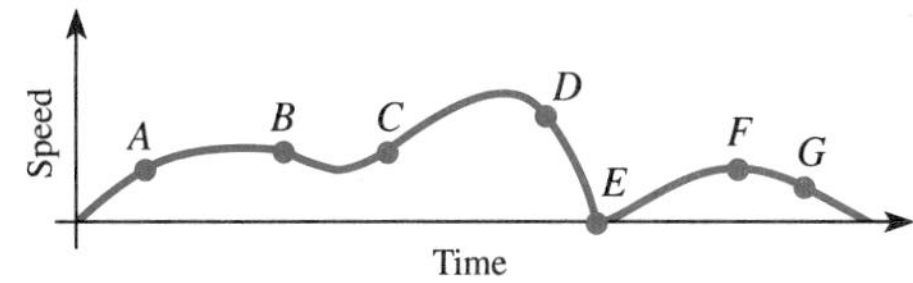

IV. Extensions

86. If the same positive number is added to the numerator and denominator of a positive fraction, is the resulting fraction equal to, less than, or greater than the original fraction? That is, if $a > 0$ and $b > 0$, compare the fraction $\frac{a}{b}$ to $\frac{(a+c)}{(b+c)}$.

87. Take any pair of fractions or integers, say 2 and 5. Add 1 to the second number and divide by the first: $\frac{(5+1)}{2} = 3$ and form a new pair 5, 3. Repeat: $\frac{(3+1)}{5} = \frac{4}{5}$; the new pair is $3, \frac{4}{5}$.

(a) Repeat until you obtain the original pair of numbers.

(b) Try the same thing with another pair of numbers.

(c) Describe the pattern.

88. *Famous Question* (Looking ahead): What should the ratio of the length (L) to the width (W) of a rectangle be if a square whose sides are equal to W is cut off from the original rectangle, and the result will be a rectangle that is the same "shape" as the original rectangle. *Note:* You may not have the tools to complete this problem now, but we want you to at least think about the statement of the problem. We will return to this problem several times. The Greeks called this ratio the "Golden Ratio." It appears in some of their architecture.

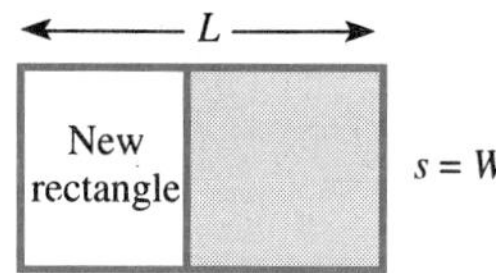

89. Examine the pattern below.

(a) Give two more examples that fit the pattern.

(b) Write a general algebraic statement for the pattern.

(c) Is it always true?

$$\frac{1}{1\cdot 2} = \frac{1}{2}$$

$$\frac{1}{1\cdot 2} + \frac{1}{2\cdot 3} = \frac{2}{3}$$

$$\frac{1}{1\cdot 2} + \frac{1}{2\cdot 3} + \frac{1}{3\cdot 4} = \frac{3}{4}$$

$$\frac{1}{1\cdot 2} + \frac{1}{2\cdot 3} + \frac{1}{3\cdot 4} + \frac{1}{4\cdot 5} = ?$$

1.4 OPERATIONS ON NEGATIVE REAL NUMBERS

The following situations are some of the many applications involving negative numbers:

- Temperatures can be above zero (+), below zero (−), or zero (0).
- Business can be in the black (gain or +), in the red (loss or −), or break even (zero) at the end of the year.
- Land elevations are given as above sea level (plus), below sea level (minus), or at sea level (zero).
- Direction can be given as east (+) or west (−) of a stationary object (zero).
- Saving accounts include deposits (positive), withdrawals (negative), or no transaction (zero).
- In football, yards are gained (+), lost (−), or no yards are gained or lost (zero).
- On television, the scoring for bowling tournaments is read "ahead by 8 pins" (+8), "behind by 11 pins" (−11), or the bowlers are even (0).
- Time counted before a spaceship launch is negative (−), time at the launch is zero (0), and time after the launch is positive (+).
- In the stock market, the Dow Jones average is either up (+), down (−), or unchanged (zero).

Before we investigate further some of the applications of negative numbers, we need to review how to operate with negative numbers.

Addition and Subtraction

The number line can be used to illustrate the addition and subtraction of real numbers. We first examine the operations on integers. For example, to illustrate $(+3) + (-3) = 0$, *we must start at zero.* We interpret $+3$ as "go to the *right three places*" and -3 as "go to the *left three places*" (see Figure 1.45). We call $+3$ and -3 **opposites,** or the **additive inverses** of each other.

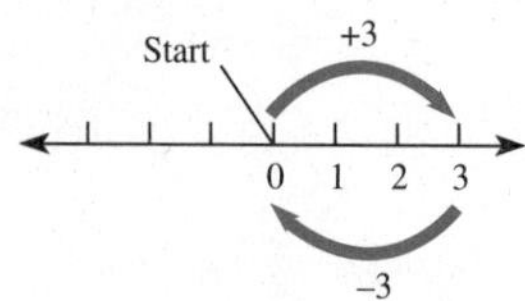

Figure 1.45

EXAMPLES 1–2

Find the sum using the number line.

1. $5 + (-8)$

2. $-3 + (-4)$

Solutions

1. $5 + (-8) = -3$

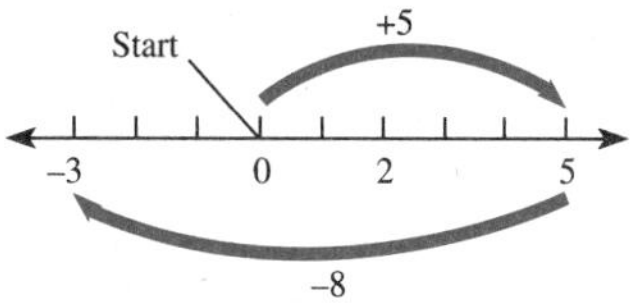

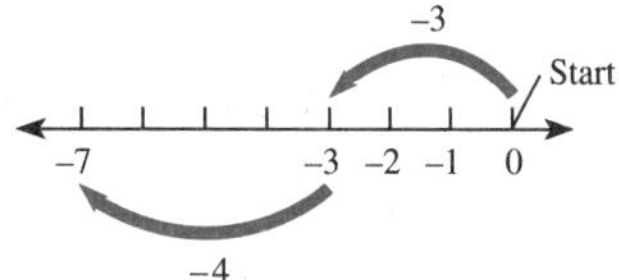

Figure 1.46

2. $-3 + (-4) = -7$

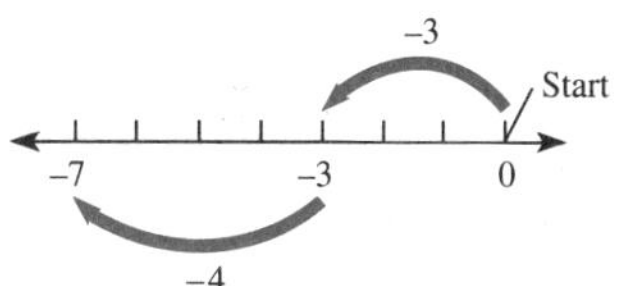

Figure 1.47

EXAMPLES 3–4

Find the difference by using the number line.

3. $-3 - 4$

4. $-7 - (-3)$

Solutions

Subtraction is the opposite of addition. Therefore, on the number line, subtraction means to reverse the direction of the subtracted number by adding its opposite.

3. $-3 - 4 = -3 + (-4) = -7$

Reverse the direction of +4

Figure 1.48

Again we observe that $-3 - 4 = -3 + (-4)$

4. $-7 - (-3) = -7 + (+3) = -4$

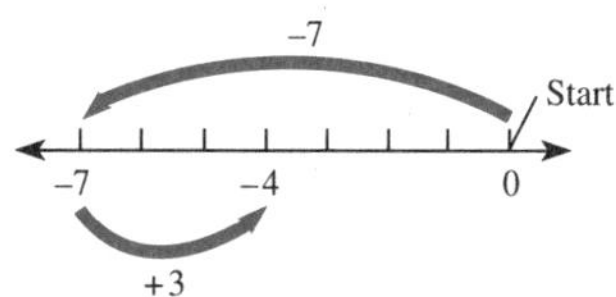

Reverse the direction of –3

Figure 1.49

We observe that $-7 - (-3) = -7 + 3 = -4.$

From the preceding examples, we observe that the symbol $-$ is used in two different ways: as an adjective to indicate the opposite of a number and as a verb to indicate the operation performed.

Definition

Subtracting a number is the same as adding its opposite:

$$a - b = a + (-b)$$

The symbol $-a$ is also called the additive inverse of a.

This relationship between addition and subtraction yields another general result that involves parentheses. Consider the following example:

EXAMPLE 5

At the beginning of the week, Sam's checking account had a balance of \$75.50. During the week, he wrote two checks. On Tuesday he wrote a check for \$15.00. On Friday he wrote a check for \$20.50. What was his balance at the end of the week?

Solution

The solution can be obtained in two ways. After the first check, Sam had $(75.50 - 15)$ dollars. At the end of Friday, he had $(75.50 - 15) - 20.50$ dollars, or 40 dollars. Alternatively, Sam withdrew a total of $(15 + 20.50)$ dollars during the week. Thus his balance was $75.50 - (15 + 20.50)$, or 40 dollars.

NOTE: $75.50 - (15 + 20.50) = 75.50 - 15 - 20.50 = 40$ ■

In general, if there is a negative sign immediately preceding a parentheses, we can remove the parentheses and add the opposite of *every term* inside the parentheses:

A Property of Parentheses

$$a - (b + c) = a + (-b) + (-c) = a - b - c$$

Thus

$$16 - (5 + 3) = 16 + (-5) + (-3) = 16 - 5 - 3 = 8$$

and

$$10 - (5 - 2) = 10 + (-5) + (+2) = 10 - 5 + 2 = 7$$

Multiplication and Division

To multiply integers, we can think of $3(-2)$ as repeated addition:

$$3(-2) = \underbrace{-2 + -2 + -2}_{\text{3 times}} = -6$$

Therefore, *a negative number times a positive number is a negative number.* Since the order in which we multiply does not matter, we have $3(-2) = (-2)(3) = -6$.

The opposite of 2(3) is $-(2)(3)$, or -6. Thus

$$3(-2) = (-2)(3) = -(2)(3) = -6$$

The following patterns suggest a way to define the multiplication of two negative integers:

Pattern 1

$$\begin{aligned} -3(4) &= -12 \\ -3(2) &= -6 \\ -3(0) &= 0 \\ -3(-2) &= +6 \\ -3(-4) &= 12 \end{aligned} \quad \text{(each step } +6\text{)}$$

We can also observe another pattern:

Pattern 2

$$3 + (-3) = 0$$

so

$$-2[3 + (-3)] = 0$$
$$(-2)(3) + (-2)(-3) = 0$$
$$-6 + ? = 0$$

To be consistent with the distributive property,

$$(-2)(-3) = 6$$

For these patterns to be consistent, we claim that $-3(-2) = +6$ and $-3(-4) = +12$. These patterns hold for *all* negative real numbers.

For division of negative integers, we need only observe that division is the inverse of multiplication:

Since $-3(4) = -12$, then $-3 = \frac{-12}{4}$, and $4 = \frac{-12}{-3}$.

Since $(-3)(-4) = 12$, then $-3 = \frac{12}{-4}$, and $-4 = \frac{12}{-3}$.

A fraction has three signs associated with it: one for the numerator, one for the denominator, and one for the entire fraction. From the above we can see that

$$\frac{-a}{b} = \frac{a}{-b} = -\frac{a}{b}$$

If two numbers are both positive (or both negative), we say they have **like signs.** If one of the numbers is positive and one of the numbers is negative, we say they have **unlike signs.**

Summary: Multiplication and Division of Real Numbers

1. If two numbers with like signs are multiplied (or divided), the result is a positive number.
2. If two numbers with unlike signs are multiplied (or divided), the result is a negative number.

It is sometimes common practice to read $-x$ as "negative x" or "minus x." "Minus x" is incorrect, since we are not necessarily subtracting; we may be finding the opposite of the number x. "Negative x" may also lead to some misunderstandings. For example,

If $x = 3$, then $-x = -(3) = -3$. So $(-x)$ is a *negative number.*
If $x = -3$, then $-x = -(-3) = +3$. So $(-x)$ is a *positive number.*
If $x = 0$, then $-x = -(0) = 0$. So $(-x)$ is *zero.*

To avoid this confusion, it is preferable to call $-x$ the **opposite** of x or the **additive inverse** of x.

Notice also that

$$(-1)x = -(1)(x) = -(1x) = -x$$

Definition

The opposite (or the additive inverse) of a number is a negative one times the number.

$$-x = (-1)x$$

EXAMPLES 6–8

Compute.

6. $(-3)(-2)(8)$ **7.** $\frac{15(-5)}{-3}$ **8.** $\frac{(3.6)(12)}{4 - 16}$

Solutions

6. $(-3)(-2)(8) = (6)(8) = 48$

7. $\frac{15(-5)}{-3} = \frac{-75}{-3} = 25,$ or $\frac{15(-5)}{-3} = (-5)(-5) = 25.$

8. $\frac{(3.6)(12)}{4 - 16} = \frac{(3.6)(12)}{-12} = -3.6$ ■

Equivalent Expressions with Parentheses

Some numerical expressions may need two or more sets of parentheses to indicate the necessary operations. However, two sets of parentheses may be confusing, so we use brackets, [ ], in place of the second, outermost set. In very complicated expressions, braces, { }, may also be used for further clarification. We agree to always work with the innermost inclusion symbol first, and then work our way out.

To avoid the possibility of having two different answers for the same problem, we also need to agree on the order of performing operations. For example, in the expression $3 + 2(4)$, we multiply first and then add:

$$3 + 2(4) = 3 + 8 = 11$$

Clearly, if we add the 3 to the 2 first and then multiply, we obtain an answer of 20, rather than 11. We do not want two different answers for the same expression. Here is a summary for the order of operations:

Order of Operations

In an expression with several operations and parentheses, we usually compute first any expression within a symbol of inclusion (parentheses, brackets, braces, fraction bars, etc.), always working from the innermost inclusion symbols. Next, multiply and divide as encountered in order from left to right. Last, add and subtract in order from left to right.

NOTE: Whether we first compute inside the parentheses depends on the situation. In any case we always multiply and divide before we add or subtract. Study the following examples. ■

EXAMPLES 9–10

Compute.

9. $5 - 2[3 - (6 - 2)]$

10. $30\left(\frac{2}{90} + \frac{7}{120}\right) + \frac{1}{12}$

Solutions

9.
$$\begin{aligned} 5 - 2[3 - (6 - 2)] &= 5 - 2[3 + (-6) + (2)] \\ &= 5 - 2[-1] \\ &= 5 + 2 \\ &= 7 \end{aligned}$$

Innermost parentheses first, or $-(6 - 2) = -6 + 2 = -4$; and then $5 - 2[3 - 4] = 5 - 2[-1]$. Multiply before adding.

10.
$$\begin{aligned} 30\left(\frac{2}{90} + \frac{7}{120}\right) + \frac{1}{12} &= 30\left(\frac{2}{90}\right) + 30\left(\frac{7}{120}\right) + \frac{1}{12} \\ &= \frac{2}{3} + \frac{7}{4} + \frac{1}{12} \end{aligned}$$

The distributive property. By inspection, we see that 30 is a factor of both denominators of the fractions inside the parentheses. So it seems better to multiply first, rather than simplify the expression inside the parentheses.

$$= \frac{30}{12} \text{ or } \frac{5}{2}$$

Both $\frac{30}{12}$ or $\frac{5}{2}$ are acceptable answers.

NOTE: We could have added the two fractions inside the parentheses first and then multiplied. The computation seems easier if we apply the distributive property first.

Absolute Value

Throughout this course, exercises involving distance will occur. Distance is a nonnegative quantity. For example, on the number line, the distance from the origin to the point 3 is 3 units, and the distance from the origin to the point -3 is also 3 units (Figure 1.50).

The **absolute value** of the number x is the distance from the origin to the point x on the number line.

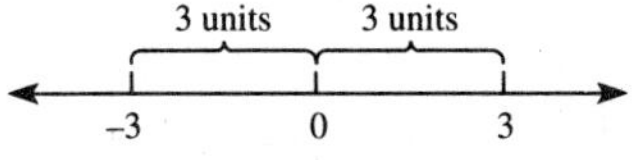

Figure 1.50

Definition of Absolute Value

The distance between a point x and the origin is called the **absolute value** of x and is denoted by $|x|$.

For example, $|3| = |-3| = 3$, and $|-8.75| = |8.75| = 8.75$. *Pairs of opposite numbers have the same absolute value.* They are the same distance from the origin. This observation leads us to an algebraic definition of absolute value:

Algebraic Definition of Absolute Value

$$|x| = \begin{cases} x, & \text{if } x \text{ is positive or } 0 \\ -x, & \text{if } x \text{ is negative} \end{cases}$$

Thus $|0| = 0$, $|4| = 4$, and $|-4| = -(-4) = 4$. Remember that distance is always positive or zero. Hence $|x|$ is always a nonnegative number. (Figure 1.51)

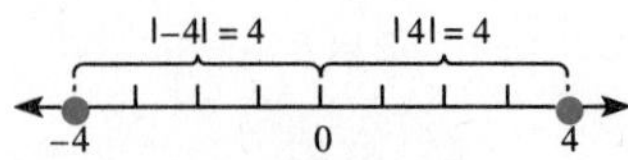

Figure 1.51

Distance on the number line can be extended to include any two points. For example, the distance between the points 7 and 3 is the same as the distance between the points 3 and 7. (See Figure 1.52.) That is,

$$|7 - 3| = |3 - 7| = 4$$

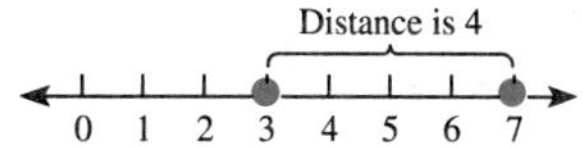

Figure 1.52

Thus to find the distance between any two points that represent the numbers a and b on the number line, we take the absolute value of the difference between a and b. *The distance between a and b is $|a - b| = |b - a|$.*

EXAMPLES 11–13

Calculate the absolute value of each of the following expressions.

11. -7 **12.** $46 - 20$ **13.** $\dfrac{6 - 3(8 - 6)}{4}$

Solutions

11. $|-7| = -(-7) = 7$

12. $|46 - 20| = |26| = 26$

13. $\left|\dfrac{6 - 3(8 - 6)}{4}\right| = \left|\dfrac{6 - 6}{4}\right| = |0| = 0$ ■

Some calculators have an absolute value key, usually $\boxed{\text{ABS}}$. Check the answers to Examples 11–13 using your calculator.

EXAMPLES 14–15

In each of the following expressions, find all real numbers x that make the statement true.

14. $|x| = 6$ **15.** $|x - 2| = 3$

Solutions

14. $|x| = 6$ means x is a point 6 units from the origin. Thus $x = 6$ or $x = -6$.

15. $|x - 2| = 3$ means the distance between the point x and the point 2 is 3 units.

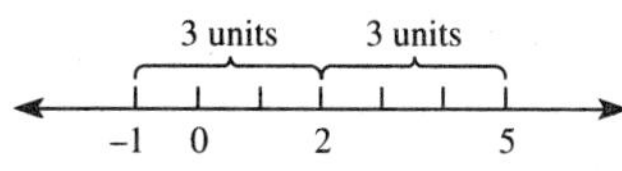

Figure 1.53

Thus $x = -1$ or $x = 5$.

Check: $|-1 - 2| = |-3| = 3$ and $|5 - 2| = |3| = 3$. ■

EXAMPLES 16–17

Express the following using absolute value.

16. x is two units from the origin. **17.** x is two units from the point 6.

Solutions

16. x is two units from the origin means $x = 2$ or $x = -2$.

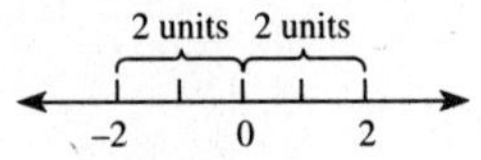

Figure 1.54

so $|x| = 2$

17. x is two units from the point 6 means $x = 4$ or $x = 8$.

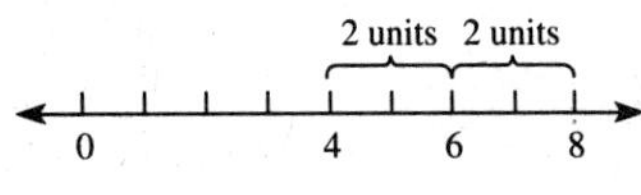

Figure 1.55

so $|x - 6| = 2$ ■

The following example illustrates how signed numbers are used in applications.

EXAMPLE 18

Four weeks ago, you decided to go on a diet. You weighed 210 pounds. During the first week, you lost 12 pounds, and during the second week, you lost $3\frac{1}{2}$ pounds more. However, your birthday occurred during the third week, and because of the celebration feast and other things, you gained 10 pounds. During the fourth week, your willpower returned, and you lost $6\frac{1}{2}$ pounds.

(a) How many pounds did you gain or lose during the four-week period?

(b) What was your weight at the end of the four-week period?

(c) What was the average weight loss per week?

Solutions

(a)
$$\underbrace{(-12)}_{\text{lost 12 pounds}} + \underbrace{\left(-3\tfrac{1}{2}\right)}_{\text{lost } 3\frac{1}{2} \text{ pounds}} + \underbrace{10}_{\text{gained 10 pounds}} + \underbrace{\left(-6\tfrac{1}{2}\right)}_{\text{lost } 6\frac{1}{2} \text{ pounds}}$$

$$= -15\tfrac{1}{2} + 10 + \left(-6\tfrac{1}{2}\right)$$

$$= -5\tfrac{1}{2} + \left(-6\tfrac{1}{2}\right) = -12 \text{ pounds}$$

(b) So you *lost* 12 pounds during the four-week period; your weight then was $210 - 12 = 198$ pounds.

(c) Total weight lost in four weeks $\div$ 4 = average weight loss per week: $12 \div 4 = 3$ pounds per week. ■

The next example is another important use of "average" in statistics.

EXAMPLE 19 The salaries of all the employees of a small company are listed below:

$81,500	$24,350	$10,500
$59,000	$24,000	$4,000
$32,000	$22,000	$3,750
$28,100	$18,000	

(a) What is the median salary of the company?

(b) What is the mean (average) salary of the company?

(c) Which gives a better description of the company's salary pattern, the mean or the median? Why?

(d) Another employee making $23,000 joins the company. Now what is the median salary?

Solutions

(a) The **median** is the middle number of a set of numbers arranged in order of size. The 11 salaries as listed are arranged in decreasing order. The median is the 6th number from the top or bottom, which is $24,000.

(b) The **mean** or average is the sum of all the numbers in a set divided by the number of members in the set. In this case, the sum of the salaries divided by the number of salaries is as follows:

$$\frac{\left(\begin{array}{c}81{,}500 + 59{,}000 + 32{,}000 + 28{,}100 + 24{,}350 + 24{,}000 \\ + 22{,}000 + 18{,}000 + 10{,}500 + 4{,}000 + 3{,}750\end{array}\right)}{11} = \$27{,}927.27$$

(c) The mean gives the impression that a typical salary is higher than it actually is. Only four salaries are higher than the mean. The median might be more representative of the salaries in this company.

(d) Since there are 12 salaries, the median is half way between the sixth and seventh salaries, namely $23,500. ■

On many calculators, it is possible to find the mean of a set of numbers by entering the numbers and then pressing the appropriate key. Most graphing calculators are capable of sorting a set of numbers into increasing order to help find the median.

In this section we reviewed the properties of the negative real numbers. In particular we studied the operations, including parentheses, on the set of integers. This allowed us to study distance (absolute value) on the number line. We concluded by examining an important application from statistics—mean and median.

1.4 EXERCISES

I. Procedures

1–10. Compute each of the following without using a calculator.

1. $6 - 3 + 2$
2. $-6 - 3 + 2$
3. $6 - (-3) - 2$
4. $6 - 3 - (-2)$
5. $6(-3)(-2)$
6. $(-6)(3)(2)$
7. $\frac{-6}{3} + 2$
8. $\frac{6}{-3} + 2$
9. $\frac{6(-3)}{-2}$
10. $\frac{6 - 3}{-2}$

11–44. Compute each of the following without using a calculator. Write any fraction that occurs as an answer in simplest form. Be careful of the order of operations.

11. $-4(-6)(-2)$
12. $-4(-6) - 2$
13. $4 - 6(5 - 3)$
14. $4 + 6(5 - 3)$
15. $(4 - 6)(5 - 3)$
16. $(4 + 6)(5 - 3)$
17. $-8 - 2(3 - 4) - (5 - 7)$
18. $-8 - 2(4 - 3) - (7 - 5)$
19. $\frac{-3(3 - 3)}{-3}$
20. $\frac{-3(-3 - 3)}{-3}$
21. $\left(\frac{-24}{15}\right)\left(\frac{-10}{40}\right)\left(\frac{5}{8}\right)$
22. $\left(\frac{1}{-2}\right)\left(\frac{-6}{5}\right)\left(\frac{-15}{9}\right)$
23. $\frac{-3}{3} \div \frac{-6}{10}$
24. $\frac{-3}{7} \div \frac{75}{49}$
25. $\frac{(-6)(-8)(2)}{12 - 8}$
26. $\frac{(-3)(15)(-1)}{(-5)(-10)}$
27. $\frac{-5 - 6}{(-2)(-3)} - \frac{3(-5)}{-1}$
28. $\frac{(-4 - 6)}{-2} + \frac{2(-3)}{-6}$
29. $-2 - 4[3 - 2(4 - 6)]$
30. $12 - [8 - (1 - 9)]$
31. $3 - 2[4 + 6(3 - 1)]$
32. $-3 + 2[(4 + 6)(3 - 1)]$
33. $\frac{1}{4} - \frac{1}{3}\left(\frac{3}{4} - \frac{2}{3}\right)$
34. $\frac{1}{4} - \frac{1}{3}\left(\frac{3}{4} - \frac{6}{8}\right)$
35. $1 - \frac{2}{3}\left(\frac{5}{6} + \frac{2}{3}\right)$
36. $3 + \frac{1}{2}(-5 + 3)$
37. $\frac{3}{5} - 2[3 - 2(4 - 3)]$
38. $3 - 2\left[\frac{1}{2} - 4\left(1 - \frac{1}{4}\right)\right]$
39. $|5 - 2|$
40. $|2 - 5|$
41. $|0 - 5|$
42. $|5| + |-5|$
43. $|6 - 3(5 - 9)|$
44. $\frac{|-15(8 - 12)|}{|36 - 30|}$

45–50. Using a calculator, compute each problem. Round off all answers to three decimal places.

45. $3.45 - 2.001 + 1.093$
46. $0.0075 - (1.003 + 8.119 - 1.34)$
47. $\frac{(0.0012)(379{,}500)}{-4{,}980}$
48. $\frac{(-211)(0.602)}{(2{,}000)(-0.005)}$
49. $\frac{1.89 - 3.21(1.01 + 3.05)}{5.25 - 6.14}$
50. $\frac{2.11(0.009) - 10.34}{13.9 - 4.89}$

51–53. Find the mean (average) and median of each set of numbers.

51. $70, 12, -14, 1, 14, 0, -13, 14$
52. $-1, 3.5, -2.4, 6.05, 1.1, -3.2, -5$
53. $\frac{1}{2}, -\frac{3}{4}, 1.5, -\frac{3}{8}, 2.25, 10.5, -\frac{12}{5}$

II. Concepts

54–57. In each of the problems below, the student has made an error in the solution. Find the error and provide the correct solution. Compute and simplify as much as possible.

54. $\frac{-3 + 2}{-3} = 2$ **(incorrect)**
55. $5 - 2(6 - 2) = 3(6 - 2) = 3(4) = 12$ **(incorrect)**
56. $5 - 2(6 - 2) = 5 - 2(6) - 2(2)$
    $= 5 - 12 - 4 = -11$ **(incorrect)**
57. $\frac{-3(-5)}{6 - 2} = \frac{-8}{6 - 2} = \frac{-8}{4} = -2$ **(incorrect)**

58–64. Put a number in the box to make the statement true.

58. $4 + 2\square = 0$

59. $-3\square = 18$

60. $\square \cdot 0 = 0$

61. $2\square - 3(-2) = 0$

62. $2(4 + \square) = 18$

63. $2(4 + \square) = -18$

64. $2(4 + \square) = 0$

65–68. Insert the signs $<$, $>$, $\leq$, $\geq$, or $=$ in the box to make each statement true.

65. $|-6| \square |-5 + 2|$

66. $|-2(4)| \square |1 + 5|$

67. $|-4(3 - 1)| \square |-6|$

68. $|3(8 - 5)| \square |4 - 6 - 7|$

69. **(a)** Complete the table. For example, if $x = 6$ and $y = 3$, then $x - y = 6 - 3 = 3$, and $y - x = -3$.

x	y	$x - y$	$y - x$	$\|x - y\|$	$\|y - x\|$
6	3	3	-3		
6	2				
6	1				
6	0				
6	-1				
6	-2				
6	-3				

(b) Give the general rule for the results of Exercise 69(a).

70. Describe the significance of absolute value. Include some examples to illustrate your discussion.

71–76. **(a)** Write an algebraic statement using absolute value for the following statements: x is a point on the number line. **(b)** Find all x that make the statements true.

71. x is 4 units from the origin.

72. x is 2.5 units from the origin.

73. x is 4 units from the point 5.

74. x is 4 units from the point 2.

75. x is 3 units from the point -4.

76. x is 3 units from the point -1.

77–82. **(a)** Write an English statement in terms of distance on the number line for each of the following. **(b)** Find all x that make the statements true. **(c)** Draw a graph of the solution on a number line.

77. $|x| = 5$

78. $|x| = \frac{1}{2}$

79. $|x - 8| = 5$

80. $|x - 2| = 5$

81. $|x - (-1)| = 3$

82. $|x - (-3)| = 3$

83. **(a)** Give two examples of a set of ten different numbers whose average (mean) is -4.

(b) How many possibilities are there for ten numbers to average -4?

84. **(a)** Give two examples of a set of ten different numbers whose average (mean) is $\frac{1}{2}$.

(b) Are there other possibilities? Why?

85–90. Mark each of the following statements AT for always true, ST for sometimes true, or NT for never true. If a statement is sometimes true (ST), give one example of when the statement is true and one example of when the statement is false.

Example: $-(-x)$ is positive.
Answer: ST. True if $x = 3$, since $-(-3) = +3$. False if $x = -2$, since $-[-(-2)] = -2$.

85. The sum of two negative numbers is a negative number.

86. The difference of two negative numbers is a negative number.

87. The product of two negative numbers is a negative number.

88. The sum of two positive numbers and one negative number is a positive number.

89. A number is either positive or negative.

90. For any number x, the expression $-x$ is negative.

III. Applications

91–105. Solve the following problems.

91. How many years elapsed from the rise of the Roman Empire in 753 B.C. to the fall of the Roman Empire in 422 A.D.?

92. Is $-47.2°C$ warmer or colder than $-23.1°C$? By how much?

93. During four days, Mercy Hospital received 12 new patients and discharged 9, received 14 and discharged 21, received 5 and discharged 12, and received 11 and discharged 10. How did the number of patients in the hospital at the end of the four-day period compare with the number of patients at the start of the four-day period?

94. The ferryboat that was the only connection between Angel Island and the mainland made three round trips. It carried 83 persons to the island on its first trip and returned 114 persons to the mainland. It then delivered 109 and returned 121 and delivered 114 and returned 98.

Compare the population of the island after all the trips to the population before the first trip.

95. Suppose the Dallas Cowboys gain 5 yards on every play. They are now on the 50-yard line.

(a) On what yard line were they four plays ago?

(b) On what yard line will they be after the next five plays?

96. The pilot of a jet traveling at 27,000 feet above sea level is ordered by ground control to descend 8,000 feet. Later, the pilot is ordered to climb 3,500 feet to a new altitude. At what altitude will the jet then be flying?

97. Jack bought four objects for \$2.15, \$3.05, \$3.40, and \$3.85. He later sold them for \$2.25, \$2.85, \$3.15, and \$2.75, respectively. What was the net financial result of the transactions?

98. On a revolving charge account, Mrs. Collins purchased \$27.50 worth of clothing and \$120.60 worth of furniture. She then made two monthly payments of \$32.00 each. If the interest charges for the period of two months were \$3.25, what did Mrs. Collins owe on the account?

99. A spelunker (cave explorer) had gone down vertically 2,340 feet into the 3,300-foot pothole cave called Gouffre Berger, the world's deepest cave, located in the Isere province of France. On her ascent, the spelunker climbed 732 feet, slipped back 25 feet, then slipped back another 60 feet, climbed 232 feet, and finally slipped back 32 feet. Use addition of signed numbers, starting with $-2{,}340$, to find her final position.

100. During a hot summer day, a cold front passed through early in the morning and caused the temperature to drop an average of 4° per hour all day long. The temperature at noon was 75°.

(a) What was the temperature at 4:00 P.M.?

(b) What was the temperature at 9:00 A.M.?

(c) Draw a graph of the temperature for each hour of the day starting at noon.

101. The elevations above sea level $(+)$ or below sea level $(-)$ of some high and low spots in the world are given at the top of next column:

The Dead Sea	-395 meters
Death Valley	-86 meters
Mt. McKinley	6,194 meters
Mt. Everest	8,848 meters

(a) How much higher is Death Valley than the Dead Sea?

(b) How much higher is Mt. Everest than the Dead Sea?

(c) How much lower is Death Valley than Mt. McKinley?

102. The tourist bureau in Bismarck, North Dakota, wrote a travel brochure for its winter season. The average temperatures in January for the past ten years are given by the chart at the bottom of the page.

(a) When the bureau writes in its brochure, "During January, you can expect an average (mean) temperature of ______ degrees," what number will go in the blank?

(b) What is the median temperature?

103. During the last football game with the Seattle Seahawks, Sam ran the ball ten times. The total yards gained or lost on each play were 10 yards, 2 yards, -5 yards, -7 yards, 8 yards, 20 yards, -3 yards, 15 yards, -1 yard, 25 yards.

(a) What was the mean (average) number of yards gained or lost per play?

(b) What was the median number of yards gained or lost per play?

(c) Which of these is a better description of Sam's performance? Why?

104. A math class with 20 students has the following test scores on a 100-point test: 100, 95, 92, 90, 88, 87, 87, 87, 82, 80, 80, 80, 79, 79, 69, 50, 41, 30, 22, 9.

(a) What are the mean and median grades?

(b) Which gives a better description of the class's performance?

105. Give two examples of a company with eight employees whose

Chart for Exercise 102

Year	1983	1984	1985	1986	1987	1988	1989	1990	1991	1992
Temperature in January	$+2°$	$-4°$	$0°$	$-6°$	$-3°$	$+1°$	$-7°$	$-4°$	$-5°$	$-2°$

(a) mean (average) salary is \$20,000. What is the median salary of this company?

(b) median salary is \$20,000. What is the mean salary of this company?

106. The following graph describes the temperature at Deep Gulch during a day in January. Describe the temperature at certain times in the day as indicated by points *A, B, C, D, E, F,* and *G.*

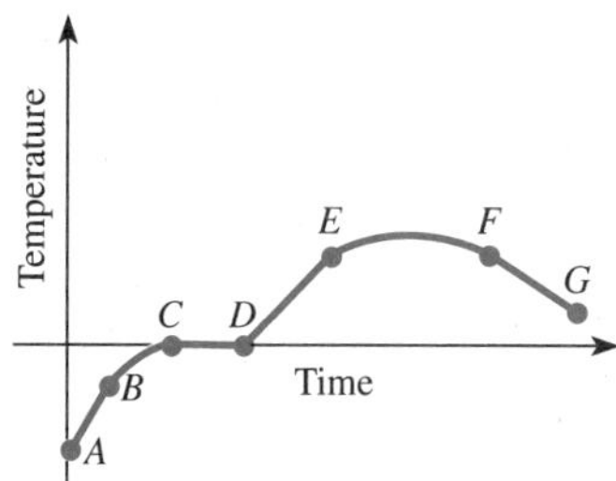

IV. Extensions

107. A cricket jumps along a number line as follows: He starts at zero, moves forward one unit, then backward two units, forward three units, backward four units, forward five units, and so on. Where will he be after

(a) 15 moves?

(b) 2,000 moves?

(c) 2,001 moves?

(d) n moves?

(e) On what move will the cricket hit the point $+100$?

(f) Will the cricket eventually hit every integer?

108–113. Insert operation signs ($+$, $-$, $\cdot$, $\div$) and parentheses in order to make the following statements true:

Example: $3 \quad 1 \quad 2 = 1$
$(3 - 1) \div 2 = 1$

108. $12 \quad 4 \quad 2 \quad 2 = 4$

109. $4 \quad 10 \quad 2 \quad 3 = 36$

110. $2 \quad 7 \quad 6 \quad 3 \quad 8 \quad 10 = -3$

111. $8 \quad 6 \quad 4 \quad 2 = 4$

112. $8 \quad 6 \quad 4 \quad 2 = 7$

113. $8 \quad 6 \quad 4 \quad 2 = 1$

114. **(a)** Make as many different integers as you can by using the integers 12, 6, and 2 (each integer used only one time and in the given order), any of the four operation signs, and parentheses.

Examples: $12 + 6 + 2 = 20$ $\quad \left(\frac{12}{6}\right) - 2 = 0$
$12 - 6 - 2 = 4$ $\quad 12 \div (6 \times 2) = 1$
$12(6 + 2) = 96$ $\quad -12(6 - 2) = -48$

(b) Do the same for the four integers in 1995.

(c) Do the same using four 4s.

115. **(a)** Find the sum of the integers $-2{,}000$ to 2,000 inclusive.

(b) Find the sum of the integers -995 to 1,000 inclusive.

116. Give a table of values and then sketch a graph for each of the following.

(a) $y = |x|$

(b) $y = |x| + 2$

(c) $y = |x + 2|$

CHAPTER 1 SUMMARY

Algebra is used to describe patterns in mathematics and in the world around us. In particular, algebra is a brief and general language used to describe much of what we learned in arithmetic. This language consists of tables, graphs, and equations. For this reason, it is important to review the properties and operations of the real numbers. All the applications in this chapter were solved by using arithmetic. Many of these applications will occur in a more general form in the rest of the book. Specifically, we will continue to study the relationship between two quantities, such as the linear function and quadratic function described in Examples in Sections 1.1 and 1.2.

Important Words and Phrases

variable	absolute value
table	less than
equation	greater than
graph	approximately equal to

square root
Cartesian coordinate plane
graphs of inequalities on a number line
equivalent fractions
set
additive inverse of a number
opposite of a number
distance on the number line
number line
ratio
integers
rational numbers
irrational numbers
counting numbers
real numbers
mean
median
percent
function

Important Properties and Procedures

- Distributive property: $a(b + c) = ab + ac$
- Fundamental property of fractions: $\frac{ax}{bx} = \frac{a}{b}$, $b \neq 0, x \neq 0$
- Properties for addition and multiplication of real numbers
- Subtraction property: $a - b = a + (-b)$
- Order of operations and finding equivalent expressions
- Multiplication by zero: $a \cdot 0 = 0$
- Division by zero is undefined
- Expressing numbers in equivalent simplest form
- Operations on real numbers
- Graphing on a number line
- Graphing on a Cartesian coordinate plane

REVIEW EXERCISES

I. Procedures

1–24. Compute the following. Express any fraction that appears in the answer in its equivalent simplest form.

1. $\frac{5}{7} + \frac{4}{3}$

2. $-\frac{3}{4} + \frac{1}{3} + \frac{5}{12}$

3. $1 + \frac{7}{4} + \frac{5}{2}$

4. $\frac{-7 - 9 - 3}{19}$

5. $\frac{-4}{15} \cdot \frac{35}{28}$

6. $\frac{16}{24} \div \frac{12}{64}$

7. $4 - (8 + 7)$

8. $11 - 3(6 + 2)$

9. $\dfrac{\frac{(2 + 3)}{6}}{15}$

10. $\dfrac{\frac{8}{9}}{3 - 5}$

11. $\frac{-5(-5 - 5)}{-5}$

12. $\frac{-(4 - 11)}{(-14)(-2)}$

13. $15 - 4[9 - (3 + 2)]$

14. $10 - 2[-3 - (9 - 4)] - 14$

15. $-2(2 - 2 + 2 - 2 + 2) + 4$

16. $\frac{2}{5} - 3[2 - 2(5 - 4)]$

17. $3 - 4[-3 - (-2)] - 3(8 - 9)$

18. $-\frac{3}{2}(4 - 11) - \frac{3}{2}$

19. $1 - \frac{2}{5}\left(\frac{10}{7} - \frac{15}{14}\right)$

20. $\frac{1}{3} - \frac{1}{4}\left(\frac{3}{4} - \frac{2}{3}\right)$

21. $-2\,|5 - 7|$

22. $3 + |6 - 3| - 5$

23. $3 - 3\,|9 - 7|$

24. $-2(-3 - |-4|) + 3\,|4 - 6|$

25 and 26. Express the area of the rectangle *ABCD* in two different ways.

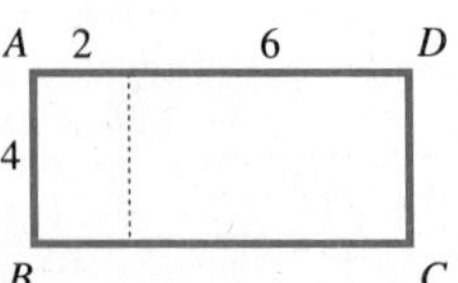

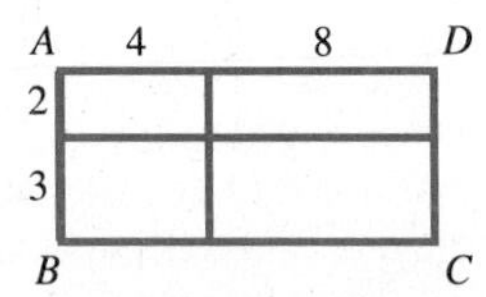

27 and 28. Draw a rectangle whose area is represented by the following expression.

27. $5(6 + 2)$

28. 36

29 and 30. Locate the following numbers on a number line. Give the greatest and smallest number.

29. $3.12, -\frac{5}{12}, -1, 2.5, -\frac{1}{2}, 2.45, \sqrt{5}$

30. $-4, \frac{3}{2}, -2.1, 0, -3, 2\frac{1}{2}, \pi$

31 and 32. **(a)** Estimate the answer to each problem. **(b)** Use a calculator to compute the answer to each problem. Round off all answers to three decimal places.

31. $\dfrac{1.69 + 4.23(2.027 + 3.03)}{6.258 - 5.146}$

32. $\dfrac{323(0.701)}{3{,}000(0.004)} \div \dfrac{15.752}{30.055}$

33–36. **(a)** Write an algebraic statement using absolute value to express each of the following. **(b)** Find all values of x in each case that make the statement true.

33. x is 3 units from the origin.

34. x is 3 units from the point 2.

35. x is 2 units from the point 3.

36. The distance between the point 4 and x is 5 units.

37–46. Insert $<$, $>$, or $=$ in the box to make each statement true.

37. $\dfrac{3.151}{-3.052} \square -1$

38. $3 - 5[-7 - (1 - 3)] \square 0$

39. $\dfrac{33}{5} - \dfrac{19}{3} \square 0$

40. $\dfrac{5}{7}\left(\dfrac{2.7}{3} - \dfrac{3.6}{4}\right) \square 0$

41. $\dfrac{53}{79} \square \dfrac{201}{300}$

42. $\sqrt{987} \square 10\pi$

43. $-2.345 + 6.789 \square 2.345 - 6.789$

44. $\dfrac{-2.345}{2.123} \square -1$

45. $\dfrac{(1.234)(-2.345)}{(-3.456)(4.567)} \square 0$

46. $\dfrac{1.234 - 2.345}{3.456 - 4.567} \square 0$

47–50. **(a)** Express each of the following statements algebraically by using the symbols $<, >, \leq, \geq$. **(b)** Sketch each set on the number line.

47. x is greater than or equal to -5.

48. x is a negative number.

49. x is between -3 and 6.

50. x is larger than 2, or x is less than 0.

51 and 52. Write an algebraic statement that describes each of the following sets.

51.

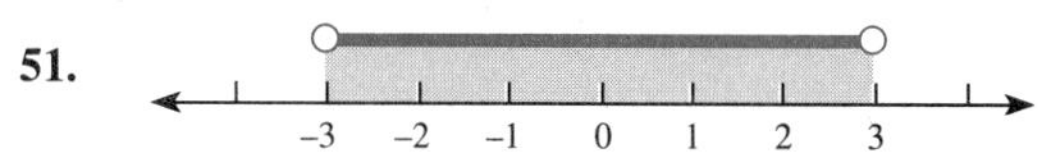

52.

-1 0 1 2 3 4

53–56. **(a)** Sketch each set on the number line. **(b)** Write an English statement to describe each set.

53. $\{x | x < 1\}$

54. $\{x | x > 2 \text{ or } x \leq 2\}$

55. $\{x | x > 1 \text{ and } x < 4\}$

56. $\{x | -1 < x \leq 3\}$

II. Concepts

57–64. Decide whether each of the following statements is true or false.

57. If $x + 2 > 5$, then $x > 5$.

58. If $x - 3 < 0$, then $x > 3$.

59. If $x + 6 > 0$, then $x < 6$.

60. $|x| \geq 0$ for all real numbers x.

61. $-x \leq 0$ for all real numbers x.

62. If $a < 0$ and $b < 0$, then $a + b < 0$.

63. If $a < 0$ and $b < 0$, then $ab < 0$.

64. The statement "x is a nonnegative number" is equivalent to saying "$x > 0$."

65 and 66. Perform the following computations.

65. Add 3 and -7. Subtract -6 from the answer.

66. Divide $\frac{3}{4}$ by $\frac{9}{8}$. Multiply the answer by $-\left(\frac{3}{2}\right)$.

67. Write a brief explanation of the importance of the distributive property. How does it relate addition and multiplication (factors and terms)?

68. What is the difference between mean and median? Give an example where the median is more useful.

69–74. Fill in the box with a number that makes the statement true.

69. $\frac{4}{5} = \frac{-20}{\square}$

70. $\frac{6}{5} \cdot \frac{\square}{21} = \frac{4}{7}$

71. $-3(\square + 2) = 15$

72. $-3(\square + 2) = -15$

73. $3 \cdot \square + 15 = 21$

74. $-\frac{49}{63} = \frac{\square}{9}$

75. If $x - 8$ is a positive real number and x is a positive integer, what is the smallest possible value for x?

76. If $x - 7$ is a negative real number and x is an integer, what is the largest value of x?

77. Find four rational numbers between $\frac{1}{5}$ and $\frac{1}{4}$.

78. Find four real numbers between $\sqrt{2}$ and $\sqrt{10}$.

79. Describe how you can decide whether one number is less than another number. Give examples using various types of numbers.

III. Applications

80–93. Solve each of the following problems.

80. A public school has an athletic budget of \$128,000, of which \$35,000 is for the women's athletic program and \$93,000 is for the men's athletic program.

(a) What is the ratio of the money spent on women's athletics to the money spent on men's athletics?

(b) What is the ratio of money for women's athletics to the total athletic budget?

(c) What is the ratio of money for men's athletics to the total athletic budget?

81. A cake recipe calls for $2\frac{1}{2}$ cups of flour and $1\frac{1}{3}$ cups of sugar. What is the ratio of flour to sugar?

82. On a state map, $\frac{1}{4}$ inch represents 50 miles.

(a) What is the ratio of inches to miles?

(b) On the map, 1 inch will represent how many miles?

83. A woman bought a new car and drove 8,000 miles in nine months.

(a) What is the ratio of miles driven to months?

(b) At this rate, how many miles will she have driven in a year?

84. The average daily temperature for the first week of January in Butte, Montana, was −2°C, −12°C, 10°C, 5°C, −3°C, 10°C, 12°C. What was the average mean temperature for the week? What was the median temperature?

85. At 5 A.M. the temperature was −17.3°C. During the next 7 hours, the temperature rose a total of 35.7°C. Assuming the temperature continued to rise at this rate, what was the temperature at 12:00 P.M.?

86. On Monday, April 1, the balance in your checking account was \$259.65. During the week of April 1–6, you wrote checks for \$52.10, \$35.75, and \$117.18, and you made a deposit of \$125. At the end of the week, on April 6, what was your balance?

87. During the first quarter of a football game, the net yardage gained on six different downs was 30, 8, −5, −20 (penalty), 15, and 22 yards. What was the average mean yardage gained for the six downs?

88. A plane has a speed of 260 mph and flies with the wind, which has a speed of 40 mph.

(a) How long will it take the plane to fly 900 miles?

(b) Assuming the wind does not change, how long will it take the plane to fly back?

89. A piece of wire is 36.568 cm in length. A piece 12.135 cm is cut from the wire. If 0.125 cm is wasted in the cut, what is the length of the remaining piece of wire?

90. **(a)** Calculate the miles per gallon a T-car gets if 20.5 gallons of gas are used for a 715-mile trip.

(b) At this rate, how many gallons of gas will be needed for a 1,000-mile trip?

91. Which of the following sets of numbers can be the lengths of the sides of a triangle?

(a) 3, 5, 8 **(b)** 31.75, 21.29, 53.045

92. If the lengths of two sides of a triangle are $\frac{11}{3}$ and $\frac{15}{4}$, describe some possible values for the third side.

93. The You Drive It car agency (UDI) charges \$22 a day plus 10¢ a mile.

(a) If you drive 1,500 miles in four days, what is the cost?

(b) Describe the range of values for the cost if you rent the car for seven days and plan to drive between 2,000 and 2,500 miles.

94 and 95. For each problem, **(a)** make a graph to correspond to the values in the chart, **(b)** answer the questions, **(c)** describe the patterns in words.

94.

Miles driven in one day	*Cost*
0	$17
50	20
100	23
150	26
200	29
250	32

(a) How much does it cost to drive 375 miles?

(b) How many miles can be driven for $50?

95.

Number of credits	*Cost*
0	$100
2	220
4	340
6	460
8	580
10	700

(a) How much would it cost to take 17 credits?

(b) How many credits can be taken for $1,000?

96. The test scores on a 100-point chemistry test were 99, 98, 90, 88, 88, 85, 83, 82, 80, 80, 78, 77, 74, 74, 70, 69, 60, 58, 51, 20, 19, 5, 0.

(a) Calculate the mean and median test scores.

(b) Which is a better description of the performance of the class?

97. Calculate the percent increase or decrease in each case.

(a) The population of Oz went from 1,500 to 820.

(b) The cost of tuition went from $150 a credit to $170 a credit.

98. Which graph models most realistically the relationship between the time it takes to run a race and the distance? Explain your reasoning. If you think a different graph is better, draw it and explain your reasoning.

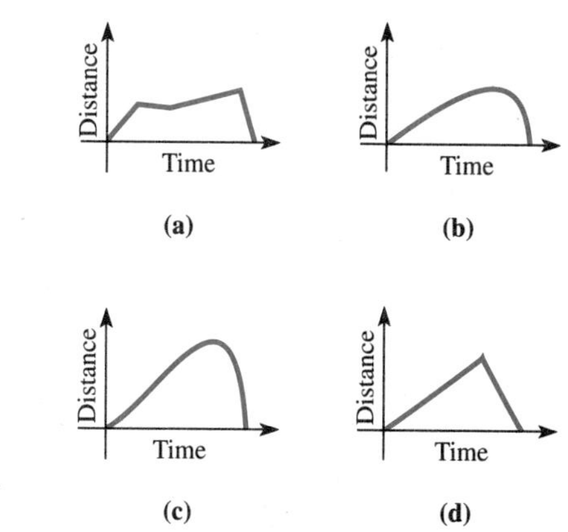

99. The graph below shows the number of cans of soft drink sold in a machine on a typical day.

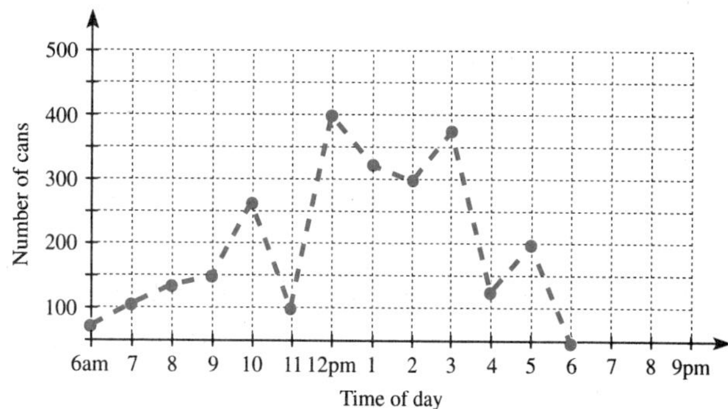

Describe in words how the number of cans of soft drink varies during the day. Explain why you think the relationship varies.

CHAPTER

2 Algebraic Expressions

The last chapter provided an overview of algebra as a language used to describe relationships between quantities. We now continue by studying concepts necessary to develop the language needed to describe patterns. In this chapter we will look at a special growth pattern—involving exponents and the exponential function and explore other uses of exponents in algebraic expressions.

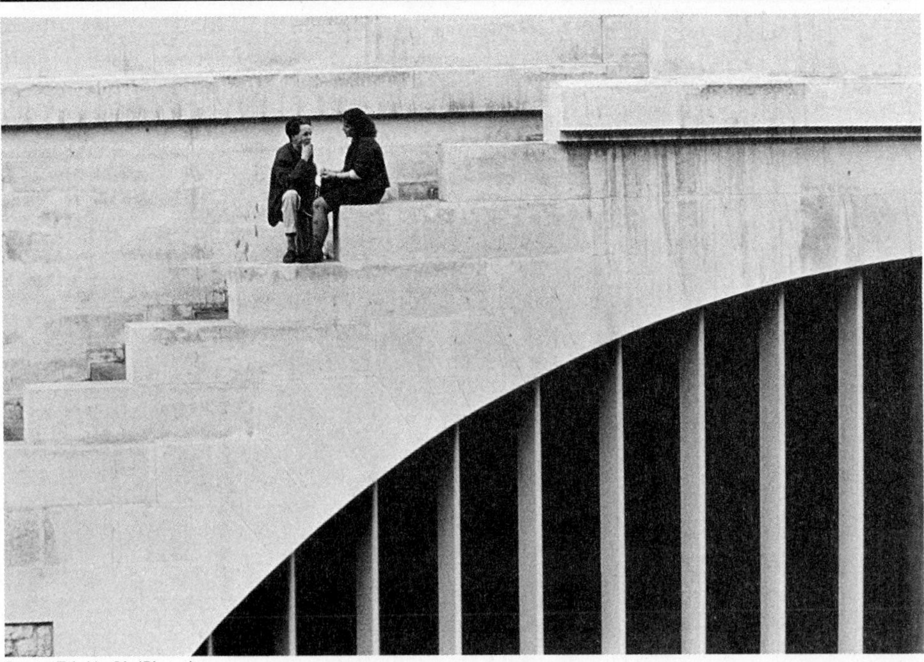

Source: Takahisa Ide / Photonica

2.1 EXPLORING EXPONENTIAL RELATIONS: TABLES, GRAPHS, EQUATIONS

The phrase "it is growing exponentially" is used to describe many things, such as the national deficit, inflation, population growth, and other rapidly increasing quantities. The following is an example involving the spread of rumors.

EXAMPLE 1

A visitor to a city of five million people starts a rumor that alien beings have invaded the city. The visitor tells two people about the aliens. The next day these two people each tell two more people about the invaders. Subsequently, each person who hears about the aliens tells two new people on the next day and then tells no one else. At this rate, how long will it take for one million people to hear about the aliens on the same day? (Make a guess!)

Solution

The first day, two people hear the rumor. The second day, each of those two people tells two more so 2×2 people learn about the aliens on day 2.

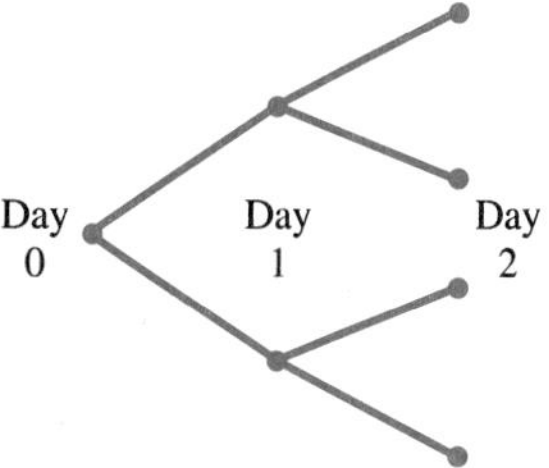

Figure 2.1

On the third day, $2 \times 2 \times 2 = 8$ people hear about the aliens.

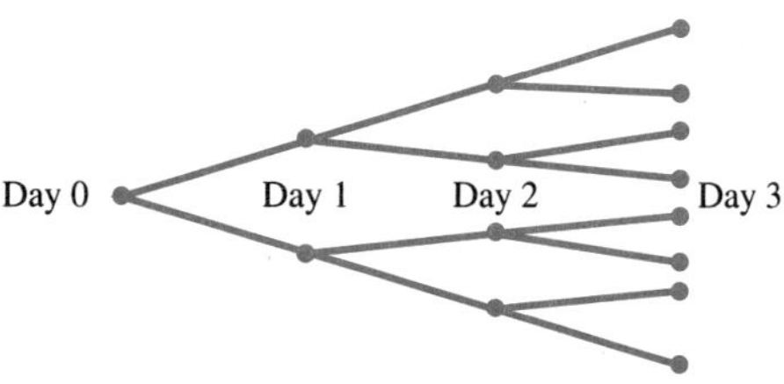

Figure 2.2

On the fourth day, $2 \times 2 \times 2 \times 2 = 16$ people hear. On the tenth day, $2 \times 2 \times 2 \times 2 \times 2 \times 2 \times 2 \times 2 \times 2 \times 2 = 1{,}024$ people hear. On the fifteenth day, $2 \times 2 \times 2 \times 2 \times 2 \times 2 \times 2 \times 2 \times 2 \times 2 \times 2 \times 2 \times 2 \times 2 \times 2 = 32{,}768$ people hear. On the 20th day, $2 \times 2 \times 2 \times 2 \times 2 \times 2 \times 2 \times 2 \times 2 \times 2 \times 2 \times 2 \times 2 \times 2 \times 2 \times 2 \times 2 \times 2 \times 2 \times 2 = 1{,}048{,}576$ people hear about the aliens. (*Note:* Such a rumor about an alien invasion actually started in the United States on Halloween, 1938, as a result of an Orson Welles radio play based on *War of the Worlds.*)

Using a Table The relationship between the number of the day and the number of people learning about the aliens on that day can be represented in a table.

d = days	N = number of people who hear about aliens on the dth day
1	2
2	4
3	8
4	16
5	32
6	64
7	128

Table 2.1

Using a Graph A graph of the data in this table shows the rapid growth of the number of people hearing the rumor as the days increase.

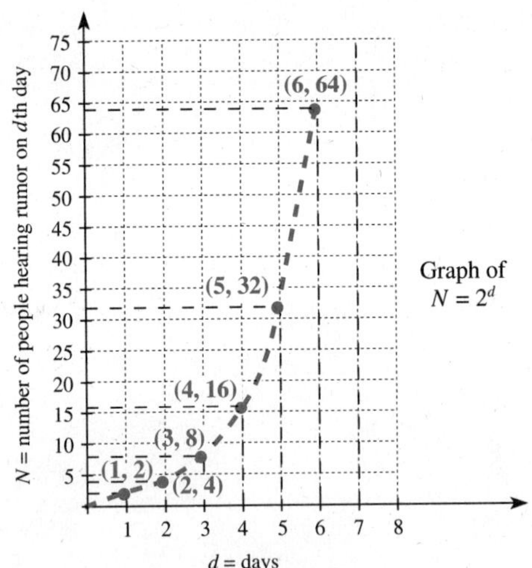

Figure 2.3

Using an Equation Observe the pattern in Table 2.1 and graph (Figure 2.3). The number of people who hear the rumor on day 10 is the product of 10 twos. We can write $2 \times 2 \times 2 \times 2 \times 2 \times 2 \times 2 \times 2 \times 2 \times 2$ as 2^{10}. This pattern can be described more generally by the equation $N = 2^d$, where d is the number of the day on which N people hear about the aliens. This situation is an example of an **exponential function.** The number of people (N) who hear about the rumor on a given day depends on (is a function of) the number of days (d). We will study such functions in more detail in Chapters 7 and 9. For now, we will look at exponents. ■

To avoid lengthy repetition as in Example 1 ($2 \times 2 \times 2 \ldots \times 2$), we write $2 = 2^1$, $2 \times 2 = 2^2$, $2 \times 2 \times 2 = 2^3, \ldots, 2 \cdot 2 \cdot 2 \cdot 2 \cdot 2 \cdot 2 \cdot 2 \cdot 2 \cdot 2 \cdot 2 = 2^{10}$. In general,

$$2^n = \underbrace{2 \cdot 2 \cdot 2 \cdot \ldots \cdot 2}_{n \text{ times}}$$

Definition

$$x^n = \underbrace{x \cdot x \cdot x \cdot \ldots \cdot x}_{n \text{ times}},$$

where n is a positive integer and x is any real number.

The expression x^n is called a **power.** It is made up of two distinct parts, x and n: x is the **base,** and n is the **exponent.** For example, 2^6 is read "2 to the sixth power"; 6 is the exponent, and 2 is the base. Likewise, b^3 is read "b to the third power" or b "cubed"; 3 is the exponent, and b is the base.

The definition

$$x^n = \underbrace{x \cdot x \cdot \ldots \cdot x}_{n \text{ times}}$$

implies that x is used as a factor n times. Thus

$$\begin{aligned} 3^5 &= 3 \cdot 3 \cdot 3 \cdot 3 \cdot 3 \\ (-6)^3 &= (-6)(-6)(-6) \\ -6^3 &= -(6)(6)(6) \\ (2x)^4 &= (2x)(2x)(2x)(2x) \end{aligned}$$

In general,

$$(\quad)^n = \underbrace{(\quad) \cdot (\quad) \cdot (\quad) \cdot \ldots \cdot (\quad)}_{n \text{ times}}$$

for any equivalent quantity in the parentheses.

By convention, the exponent applies only to the number or expression that is *directly to the left of the exponent:*

$$\begin{aligned} 7 \cdot 3^5 &= 7 \cdot 3 \cdot 3 \cdot 3 \cdot 3 \cdot 3 \\ (7 \cdot 3)^5 &= (7 \cdot 3)(7 \cdot 3)(7 \cdot 3)(7 \cdot 3)(7 \cdot 3) \\ 2x^3 &= 2 \cdot x \cdot x \cdot x \\ (2x)^3 &= (2x)(2x)(2x) \end{aligned}$$

It is especially important to keep this convention in mind when working with negative quantities:

$$-2^4 = -(2)(2)(2)(2) = -16$$
$$(-2)^4 = (-2)(-2)(-2)(-2) = +16$$
$$-x^4 = -(x)(x)(x)(x) = -x^4$$
$$(-x)^4 = (-x)(-x)(-x)(-x) = x^4$$

We agree that in -2^4, the 4 applies only to the 2. Parentheses are necessary to indicate that the negative quantity, -2, is to be raised to the fourth power. Thus $-2^4 = -16$, and $(-2)^4 = 16$.

Properties of Positive Integral Exponents

There are many instances where it is convenient to use exponential notation in doing arithmetic. Similar situations arise in algebra. The following examples illustrate how algebra can be used to generalize arithmetic patterns.

EXAMPLES 2–3

Use the definition of exponents to express each as a power of one quantity.

2. $2^5 \cdot 2^2$ **3.** $x^3 \cdot x^4$

Solutions

2. $$2^5 \cdot 2^2 = \underbrace{\underbrace{(2 \cdot 2 \cdot 2 \cdot 2 \cdot 2)}_{5\text{ times}}\underbrace{(2 \cdot 2)}_{2\text{ times}}}_{5\,+\,2\text{ times}} = 2^7$$

3. $$x^3 \cdot x^4 = \underbrace{\underbrace{(x \cdot x \cdot x)}_{3\text{ times}}\underbrace{(x \cdot x \cdot x \cdot x)}_{4\text{ times}}}_{3\,+\,4\text{ times}} = x^7$$

The pattern of these examples leads us to the first property of exponents:

Property 1

For any real number x and any positive integers m and n, $x^m \cdot x^n = x^{m+n}$.

EXAMPLES 4–9

Write each of these expressions as a power of one quantity, if possible.

4. $(-1)^5(-1)^6$ **5.** $(-4)^3(-4)^5$ **6.** $y \cdot y^7$

7. x^2y^3 **8.** $x^2 + x^3$ **9.** $x^3 \cdot x^2 \cdot x^5$

Solutions

4. $(-1)^5(-1)^6 = (-1)^{5+6}$
$= (-1)^{11}$
$= -1$

5. $(-4)^3(-4)^5 = (-4)^{3+5}$
$= (-4)^8$
$= 4^8$ — A negative number raised to an even power is positive.

6. $y \cdot y^7 = y^{1+7}$
$= y^8$ — *Note:* $y = y^1$.

7. x^2y^3 — Property 1 does not apply since *the base is not the same* for both the x^2 and y^3 factors.

8. $x^2 + x^3$ — This cannot be expressed as the power of one number. The base is the same in each term, but we are *adding* rather than multiplying. Property 1 applies only if *bases are the same and the two quantities are being multiplied*.

9.
$$x^3x^2x^5 = \underbrace{\underbrace{(x \cdot x \cdot x)}_{3 \text{ times}}\ \underbrace{(x \cdot x)}_{2 \text{ times}}\ \underbrace{(x \cdot x \cdot x \cdot x \cdot x)}_{5 \text{ times}}}_{3+2+5 \text{ times}}$$
$$= x^{3+2+5}$$
$$= x^{10}$$

The definition for exponents can also be used to analyze expressions such as $(2^2)^3$ and $(x^5)^4$. Observe the following patterns.

$$\begin{aligned}(2^2)^3 &= (2^2)(2^2)(2^2)\\ &= 2^{2+2+2}\\ &= 2^{2(3)}\\ &= 2^6\\ (x^5)^4 &= (x^5)(x^5)(x^5)(x^5)\\ &= x^{5+5+5+5}\\ &= x^{5(4)}\\ &= x^{20}\end{aligned}$$

In general, we have

$$(x^m)^n = \underbrace{x^m \cdot x^m \cdot \ldots \cdot x^m}_{n \text{ times}} = x^{\overbrace{m + m + \cdots + m}^{n \text{ times}}} = x^{mn}$$

Property 2

$(x^m)^n = x^{mn}$, where x is any real number, and m and n are positive integers.

EXAMPLES 10–13

Write each of the following expressions as a power of one quantity, if possible.

10. $(3^2)^4$ **11.** $[(-2)^3]^4$ **12.** $(x^4)^6$ **13.** $(2x^3)^2$

Solutions

10. $(3^2)^4 = 3^{2\cdot 4}$
$= 3^8$

11. $[(-2)^3]^4 = (-2)^{3\cdot 4}$
$= (-2)^{12}$
$= 2^{12}$

12. $(x^4)^6 = x^{4\cdot 6}$
$= x^{24}$

13. $(2x^3)^2 = (2x^3)(2x^3)$ **Careful, Property 2 only applies to one factor.**
$= 2\cdot 2\cdot x^3\cdot x^3$
$= 2^{1(2)}x^{3(2)}$
$= 2^2x^6$ ■

The pattern in Example 13 leads us to extend Property 2 to include expressions such as $(2x)^3$ or $(xy)^n$. In general,

$$(xy)^n = \underbrace{(xy)(xy)\ldots(xy)}_{n \text{ times}} = \underbrace{(x\cdot x\cdot\ldots\cdot x)}_{n \text{ times}}\underbrace{(y\cdot y\cdot\ldots\cdot y)}_{n \text{ times}} = x^n y^n$$

Property 3

$(xy)^n = x^n y^n$ where x and y are any real numbers and n is a positive integer.

EXAMPLES 14–17

Express without using parentheses.

14. $(x^3y^5)^4$ **15.** $(-3x^2y^3)^4$ **16.** $-3(x^2y^3)^4$ **17.** $(x^2 + y^3)^4$

Solutions

14. $(x^3y^5)^4 = x^{3(4)}y^{5(4)} = x^{12}y^{20}$

15. $(-3x^2y^3)^4 = (-3)^4x^{2(4)}y^{3(4)}$ *Note:* **It is important to include the parentheses in $(-3)^4$.**
$= 81x^8y^{12}$

16. $-3(x^2y^3)^4 = -3x^{2(4)}y^{3(4)}$ **The exponent 4 applies only to (x^2y^3).**
$= -3x^8y^{12}$

17. $(x^2 + y^3)^4$ **Property 3 applies only to a product, never a sum.**

NOTE: $(x^2 + y^3)^4 = (x^2 + y^3)(x^2 + y^3)(x^2 + y^3)(x^2 + y^3)$. ■

Equivalent Algebraic Fractions with Exponents

There are many equivalent ways to express a given fraction. Recall the fundamental property of fractions:

$$\frac{ax}{bx} = \frac{a}{b}, \quad \text{where } b \neq 0, x \neq 0.$$

The process of going from $\frac{ax}{bx}$ to $\frac{a}{b}$ removes identical factors from the numerator and denominator. If a and b have no common factor other than 1, then $\frac{a}{b}$ is said to be in its *equivalent simplest form.*

EXAMPLES 18–20

Express each of the following fractions in simplest form.

18. $\frac{2^3}{2^5}$ **19.** $\frac{(3m)^2}{9m}$ **20.** $\frac{-4x^2y^3}{2x^5y}$

Solutions

18. $\frac{2^3}{2^5} = \frac{2^3 \cdot 1}{2^3 \cdot 2^2}$ **Find the common factor, if any, in the numerator and denominator by using property 1.**

$= \frac{1}{2^2}$ **The fundamental property of fractions.**

$= \frac{1}{4}$

19. $\frac{(3m)^2}{9m} = \frac{9m^2}{9m}$ **Property 3.**

$= \frac{(9m) \cdot m}{(9m) \cdot 1}$ **Property 1.**

$= m$ **The fundamental property of fractions.**

20. $\frac{-4x^2y^3}{2x^5y} = \frac{-2 \cdot 2x^2yy^2}{2x^2x^3y}$ **Try to find the common factors in the numerator and denominator.**

$= \frac{-2y^2(2x^2y)}{x^3(2x^2y)}$ **Similarly,** $\frac{-2 \cdot 2\cancel{x}\cancel{x}\cancel{y}yy}{2xxx\cancel{x}\cancel{x}\cancel{y}} = \frac{-2yy}{xxx}$

$= \frac{-2y^2}{x^3}$ **Simplest form of the fraction.** ■

Algebraic expressions represent numbers. We often need to evaluate expressions for specific values of the variables.

EXAMPLE 21

Evaluate each of the expressions below for the given values of the variables.

x	y	x^2y	x^3y^2	$\frac{x^2y}{x^3y^2}$
1	-2			
-1	2			
-1	-2			
-100	1,000			

Table 2.2

Solution

21. See Table 2.3 on next page.

Note that the last column is a quotient of the two previous columns. With equivalent fractions, note that

$$\frac{x^2y}{x^3y^2} = \frac{x^2y}{x^2y \cdot xy} = \frac{1}{xy}.$$

It is simpler to evaluate

$$\frac{1}{xy}$$

than it is to evaluate

$$\frac{x^2y}{x^3y^2}.$$

EXAMPLE 22

Which of these expressions are equivalent to each other ($x \neq 0$, $y \neq 0$)?

(a) $\frac{x^3y^2}{x^2y}$ **(b)** $-\frac{1}{8}$ **(c)** $\frac{1}{xy}$ **(d)** $\frac{x^2y}{x^3y^2}$ **(e)** xy **(f)** $\frac{(2x^2)^3}{-(2x)^6}$ **(g)** $\frac{2}{(-2)^4}$

Solution

To see whether two expressions are equivalent, we can compare their simplest forms.

(a) $\frac{x^3y^2}{x^2y} = \frac{x \cdot x \cdot x \cdot y \cdot y}{x \cdot x \cdot y} = xy$

x	y	x^2y	x^3y^2	$\frac{x^2y}{x^3y^2}$
1	-2	$1^2 \cdot (-2)$ $= 1 \cdot (-2)$ $= -2$	$1^3 \cdot (-2)^2$ $= 1 \cdot (4)$ $= 4$	$\frac{1^2(-2)}{1^3(-2)^2}$ $= \frac{-2}{4}$ $= -\frac{1}{2}$
-1	2	$(-1)^2 \cdot 2$ $= 1 \cdot 2$ $= 2$	$(-1)^3 \cdot (2)^2$ $= -1 \cdot 4$ $= -4$	$\frac{(-1)^2 \cdot 2}{(-1)^3(2)^2}$ $= \frac{2}{-4}$ $= -\frac{1}{2}$
-1	-2	$(-1)^2(-2)$ $= 1 \cdot (-2)$ $= -2$	$(-1)^3(-2)^2$ $= -1\ (4)$ $= -4$	$\frac{(-1)^2(-2)}{(-1)^3(-2)^2}$ $= \frac{-2}{-4}$ $= \frac{1}{2}$
-100	1,000	$(-100)^2 \cdot (1{,}000)$ $= (10{,}000) \cdot (1{,}000)$ $= 10{,}000{,}000$	$(-100)^3 \cdot (1{,}000)^2$ $= (-1{,}000{,}000) \cdot 1{,}000{,}000$ $= -100{,}000{,}000{,}000$ (What is this number?)	$\frac{(-100)^2 \cdot (1{,}000)}{(-100)^3 \cdot (1{,}000)^2}$ $= \frac{(-100)^2 \cdot (1{,}000)}{(-100)^2 \cdot (-100) \cdot (1{,}000) \cdot (1{,}000)}$ $= \frac{1}{(-100)(1{,}000)}$ $= -\frac{1}{100{,}000}$

Table 2.3

(b), (c), and **(e)** are already in simplest form.

(d) $\frac{x^2y}{x^3y^2} = \frac{x^2y}{x^2 \cdot x \cdot y \cdot y} = \frac{1}{xy}$

(f) $\frac{(2x^2)^3}{-(2x)^6} = \frac{(2x^2)(2x^2)(2x^2)}{-(2x)(2x)(2x)(2x)(2x)(2x)}$ **Property 2. The exponent in the denominator does not apply to the negative sign.**

$= \frac{2^3 \cdot x^6}{-2^6 \cdot x^6}$ **Properties 1 and 3 of exponents.**

$$= \frac{2^3}{-2^6}$$ **Equivalent fractions.**

$$= \frac{2^3}{-2^3 \cdot 2^3}$$ **Property 1 of exponents.**

$$= \frac{1}{-2^3} = \frac{1}{-8} = -\frac{1}{8}$$

(g) $\dfrac{2}{(-2)^4} = \dfrac{2}{(-2)(-2)(-2)(-2)} = \dfrac{2}{16} = \dfrac{1}{8}$

Thus we see that (a) and (e) are equivalent, (c) and (d) are equivalent, and (b) and (f) are equivalent. ■

An important use of exponents is found in calculating interest on loans, savings accounts, and charge accounts. Consider the following examples.

EXAMPLE 23

We are considering putting $1,000 in a savings account that pays an effective annual interest rate of 8%. How much money would we have at the end of five years if no money is withdrawn at any time?

Solution

Recall that the interest equals the rate of interest times the amount of principal.

At the end of	*interest is*	*and the amount of money in our account is principal and interest.*
1 year	$(0.08)1000$	$1000 + (0.08)1000 = (1.08)1000$
2 years	$(0.08)[(1.08)1000]$	$(1.08)(1000) + 0.08[(1.08)1000]$ $= (1.08)(1000)(1 + 0.08)$ $= (1.08)^2(1000)$
3 years	$(0.08)(1.08)^2(1000)$	$(1.08)^2(1000) + (0.08)(1.08)^2 1000$ $= (1.08)^3 1000$
4 years	$(0.08)(1.08)^3(1000)$	$(1.08)^4 1000$
5 years	$(0.08)(1.08)^4(1000)$	$(1.08)^5 1000$

Table 2.4

Using the exponent key on a calculator, we find that we will have $\$(1.08)^5 1000$, or $1,469.33 at the end of five years. ■

In general, if P is the amount of money to be invested (the principal) and I is the *annual* interest rate, then at the end of n years, the total amount of money, A, satisfies $A = (1 + I)^n P$.

EXAMPLE 24

A certain mutual fund offers a $12\frac{1}{2}\%$ APR (annual percentage rate). How much money would you have at the end of ten years if you start with an initial investment of \$10,000?

Solution

From the preceding example, the amount of money accumulated after n years at an interest rate I on a principal investment P is $(1 + I)^nP$. Assume no money is withdrawn.

$$I = 12\tfrac{1}{2}\% = 0.125 \quad \text{To calculate, we must change interest rate to a decimal.}$$
$$n = 10 \text{ years}$$
$$P = \$10{,}000$$
$$\text{Total money} = (1 + 0.125)^{10}(10{,}000) = (1.125)^{10}(10{,}000)$$

Using your calculator, you obtain \$32,473.21. After ten years, an initial investment of \$10,000 invested at a $12\frac{1}{2}\%$ interest rate will yield \$32,473.21. You have more than tripled your initial investment. ■

EXAMPLE 25

The water in Loch Ness is very murky. (This is one reason given for not being able to find the Loch Ness monster.) Suppose that for every meter of water, only half the light filters through. How much light comes through 2 meters down, 3 meters down, 100 meters down?

Solution

One meter down, there is $\frac{1}{2}$ the original light.

Two meters down, there is $\frac{1}{2}\left(\frac{1}{2}\right) = \left(\frac{1}{2}\right)^2$ or $\frac{1}{4}$ of the original light.

Three meters down, there is $\left(\frac{1}{2}\right)\left(\frac{1}{2}\right)\left(\frac{1}{2}\right) = \left(\frac{1}{2}\right)^3$ or $\frac{1}{8}$ of the original light.

One hundred meters down, there is $\left(\frac{1}{2}\right)\left(\frac{1}{2}\right)\cdots = \left(\frac{1}{2}\right)^{100}$ of the original light (not much!).

We can express this relationship algebraically as $F = \left(\frac{1}{2}\right)^m$, where m is the water depth in meters and F is the fraction (percentage) of the original light at a depth of m meters. The graph is given below.

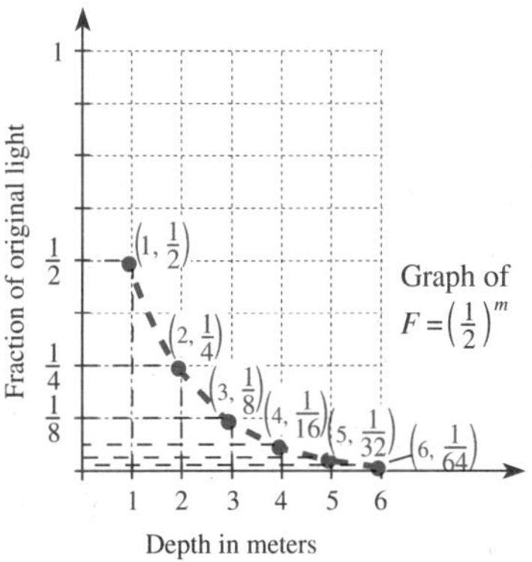

Figure 2.4

This situation is another example of an exponential function. ■

Algebra as a Language

Throughout this book we will continually return to the theme of algebra as a language and to the process of expressing problem situations in this mathematical language.

We have already examined expressions for many different quantities: area, perimeter, volume, average rate, time, distance, tax rate, and interest rate, among others. Let us review some key words that show how quantities are related to one another. For example, there are many words that signify +, the operation of addition. Some common words that suggest addition are "plus," "added to," "increased by," "more than," and "the sum of." Can you think of others? These words are replaced by + in the translation from words to mathematical symbols. Common translations for − (subtraction) include "subtract," "minus," "difference," "less than," "decreased by," and "diminished by."

EXAMPLE 26

ENGLISH PHRASE	MATHEMATICAL PHRASE
The sum of a number n and 5	$n + 5$
7 more than a number n	$n + 7$
3 less than a number p	$p - 3$
A number x increased by 15	$x + 15$
A number x decreased by 15	$x - 15$
The difference between two numbers x and y	$x - y$ or $y - x$

Such words as "times," "product," "multiplied by," and "of" (as in one-half of a number) suggests multiplication. Division is implied by words such as "quotient," "divided by," "ratio of," and "per."

EXAMPLE 27

ENGLISH PHRASE	MATHEMATICAL PHRASE
Twice a number n	$2n$
The sum of 3 times a number x and 5	$3x + 5$
The product of two numbers x and y	xy
A number n divided by 2	$\frac{n}{2}$
The ratio of two numbers x and y	$\frac{x}{y}$
Miles per hour	$\frac{\text{mi}}{\text{hr}}$
10% of a number	$0.10n$
Light intensity at a depth of m meters if $\frac{1}{2}$ the light gets through for each meter of depth	$\left(\frac{1}{2}\right)^m$

Note: **In algebra we prefer to use a fraction $\frac{x}{y}$ rather than $x \div y$.**

2.1 EXERCISES

I. Procedures

1–12. Use the properties of exponents to write an equivalent expression for each of the following expressions. Express fractions in simplest form.

1. $(-4)^2(4)$

2. $(-3)^3(-3)^2$

3. $(-1)^{100}(-1)^{9999}$

4. $(-2)^5 + 2^4$

5. $\dfrac{4^5}{4^3}$

6. $\dfrac{-4^5}{4^3}$

7. $\dfrac{(-4)^5}{(-4)^3}$

8. $\dfrac{-4^5}{-4^3}$

9. $\dfrac{4^3}{5^3}$

10. $\dfrac{3^3 \cdot 9^2}{3}$

11. $\dfrac{6^2 \cdot 3^3}{4^2}$

12. $\dfrac{12^8}{6^{10}}$

13–32. Use the properties of exponents. Express fractions in equivalent simplest form. Assume all variables represent positive numbers.

13. x^6x^7

14. x^6y^6

15. $(a^2b)^3$

16. $(ab^2)^4(ab)$

17. $(-2a^2b)^3$

18. $(3x^3y)(-2x^2y^3)$

19. $(0.02x^2)(0.01x^3)$

20. $(0.02x)^2(0.01x)^3$

21. $-xy(x^3y^2)$

22. $(3^m \cdot 3^n)^2$

23. $\dfrac{12x^6}{6x^{12}}$

24. $\dfrac{4x^2}{2^4x}$

25. $\dfrac{(3xy)^2}{bx^2y}$

26. $\dfrac{18x^3yz^5}{9xy^4z^4}$

27. $\dfrac{-1.8a^2b^3}{2a^5b}$

28. $\dfrac{(-m^2np^2)^3}{mnp}$

29. $(2x)^n(4x)^m$

30. $\dfrac{(4x)^n}{8x^m}, n > m$

31. $(2x + 3y)^5$

32. $y^{2137} + (-y)^{2137}$

33–36. Evaluate the following expressions for the values given.

33. $(2x^3)(3y)(-z^2)$, if $x = -1$, $y = 2$, and $z = -3$.

34. $\dfrac{3(a^2b)^3}{3b}$, if $a = 2.37$ and $b = -1.84$.

35. x^{1000}, if $x = 1$; if $x = -1$.

36. $\dfrac{x^{259}}{x^{255}}$, if $x = 2$; if $x = 1$.

II. Concepts

37–40. Rewrite each of the following expressions using exponents.

37. x raised to the fourth power

38. 4 raised to the xth power

39. 3 used as a factor 10 times

40. x^2 used as a factor 10 times

41–46. Insert parentheses and/or brackets to make the following statements true. For example, the solution for $3a^{10}b^2 = 3a^5b^2$ is $3a^{10}b^2 = 3(a^5b)^2$.

41. $2a^{12}b^6 = 2a^2b^6$

42. $9x^4y^2 = 3x^2y^2$

43. $-2ab^2 = 4a^2b^2$

44. $2a^2b^3c^4 = 16a^8b^{12}c^4$

45. $2a^2b^3c^4 = 4a^2b^3c^4$

46. $1 + 2^3 + 3^2 = 900$

47–49. Write three equivalent expressions for each of the given expressions.

47. $(x^3)^6$

48. $\dfrac{(x^2y^3)^2}{(2x)^2}$

49. $\dfrac{3^2 \cdot 2^5}{(4 \cdot 3)^3}$

50. Make up a sentence that suggests each of the following algebraic phrases.

(a) $x + 3$ **(b)** $\dfrac{x}{3}$ **(c)** $3x$ **(d)** $3 - x$

51–58. Change each of the following English phrases into a mathematical phrase.

51. Twice a number.

52. Sum of a number and 5.

53. Three times the sum of a number and 5.

54. A number divided by 2.

55. The cost of six books, each of which costs x dollars.

56. The distance a car travels in six hours if it goes x mph.

57. The time it takes an airplane to go 500 miles if it travels x mph.

58. The amount of sales tax on a stove costing x dollars if the tax rate is 4%.

59–68. After each numbered statement, place the algebraic expression from the list below that most closely matches it. Answers in the list below may be used more than once.

$\frac{x}{4}$

$10x + 50$

$x^2 + x$

$4x$

59. Four times a number.

60. Ten times the sum of a number and 5.

61. The sum of a number and the square of that number.

62. A number divided by 4.

63. The length of a rectangle whose length is ten times the width, if the width is $x + 5$.

64. The distance traveled by a car in four hours going x mph.

65. The average rate of a car that traveled x miles in 4 hours.

66. The perimeter of a square whose side is x inches long.

67. The total cost of x books if each book costs $4.

68. The area of a rectangle whose length is x feet and whose width is $x + 1$ feet.

69. In Example 1, did everyone in the city learn about the existence of the aliens by the end of the twentieth day? Give a reason for your answer.

III. Applications

70–79. Solve each of the following problems.

70. The expression for determining the amount of money acquired in n years by an initial investment P (principal) at an interest rate I (expressed as a decimal) *compounded annually* is given by $P(1 + I)^n$.

(a) How much money would a person have after five years if $1000 were invested at $6\frac{1}{2}\%$?

(b) How much would be acquired after ten years? Estimate your answer before you compute it.

71. The formula for calculating the volume of certain solids is given. Evaluate the volume of each solid for the values given in chart at bottom of page.

72. If you trace your family tree back 12 generations, how many sets of grandparents would be in the twelfth generation back? Assume no duplicate role for any individual. Draw a graph of the relationship between the number of generations and number of ancestors.

73. A certain amoeba (a microscopic, single-celled animal) reproduces by dividing in half every three hours. Draw a graph relating time to number of amoebae. If we start

Chart for Exercise 71

	Solid	*Formula for volume*	*Given*	*Volume*
(a)	Sphere	$\frac{4}{3}\pi r^3$	$\pi \approx 3.14$, $r = 10^2$ cm	
(b)	Rectangular prism	lwh	$l = 10$ cm, $w = 100$ cm, $h = 1000$ cm	
(c)	Cube	e^3	$e = 1.02$ cm	
(d)	Cylinder	$\pi r^2 h$	$\pi \approx 3.14$, $r = 0.05$ cm, $h = 1.5$ cm	
(e)	Cone	$\frac{1}{3}\pi r^2 h$	$\pi \approx 3.14$, $r = 1.1$ cm, $h = 2$ cm	

with one amoeba, how many amoebae will there be at the end of

(a) 3 hours? **(b)** 6 hours? **(c)** 12 hours?

(d) 24 hours? **(e)** Describe the pattern in words.

74. Assume the inflation rate is 10% a year. Draw a graph relating years to amount you would pay. If you buy a product this year for $500 and if inflation continues at this rate, how much could you expect to pay for the product

(a) two years from now? **(b)** three years from now?

(c) five years from now? **(d)** n years from now?

(e) Describe the pattern in words.

75. A man wants to dig a hole 16 meters deep in the ground. The ground is soft near the surface, but digging deeper gets harder. He obtains two estimates for doing the job:

ESTIMATE A: $1,000 per meter.

ESTIMATE B: $1 for just 1 meter, $2 for 2 meters, $4 for 3 meters, $8 for 4 meters . . . each additional meter will make the total cost double.

(a) Which estimate is cheaper?

(b) Which estimate is cheaper for a 14-meter hole? For a 15-meter hole?

(c) Draw a graph of each relationship.

(d) Describe the relationship in words.

76. If we start with a unit cube (each edge is one unit in length), what is the volume of the resulting cube

(a) if each edge is doubled?

(b) if each edge is tripled?

(c) if each edge is increased 100 times?

77. See Exercise 71 for the formulas for volume. What is the change in volume of

(a) a cone, if both the radius and height are doubled?

(b) a sphere, if the radius is doubled?

(c) a cylinder, if both the radius and height are doubled?

78. Suppose a special type of window glass, 1 inch thick, lets in only $\frac{2}{3}$ of the light. If two thicknesses of glass are used together, then only $\frac{2}{3}(\frac{2}{3})$ of the light will come through.

(a) If three thicknesses of glass are used, how much light comes through?

(b) If ten thicknesses of glass are used, how much light comes through?

(c) If n thicknesses of glass are used, how much light comes through?

(d) Draw a graph relating number of thicknesses of glass to the amount of light.

79. Suppose you receive a chain letter (which, by the way, is illegal) with five names on it. You are asked to send a dollar to the person named at the top, cross the name out, add your name to the bottom, and then send out five copies of the letter to your friends with the same instructions. If no one breaks the chain, how much money will you receive?

IV. Extensions

80. If 1,729 is the smallest number that can be represented as the sum of two cubes in two different ways, find the two ways.

81. **(a)** What is the last digit of 2^{100}?

(b) What is the last digit of 3^{101}?

82. A cricket is on the number line again. He starts at zero and would like to hop to the point labeled 1. He has a slight problem: He can only hop half the distance from where he is to the point of destination. That is, after the first hop, he will be at the point $\frac{1}{2}$. After the second hop, he will be at the point $\frac{3}{4}$ (half of the distance from $\frac{1}{2}$ to 1). After the third hop, he will be on $\frac{7}{8}$ (half of the distance from $\frac{3}{4}$ to 1), and so on.

(a) Where will he be after ten hops?

(b) Where will he be after n hops?

(c) Will he ever get to point 1?

83. Indicate which of the following are prime numbers: $2^2 - 1, 2^3 - 1, 2^4 - 1, 2^5 - 1, 2^6 - 1, 2^7 - 1, 2^8 - 1, 2^9 - 1, 2^{10} - 1, 2^{11} - 1, 2^{12} - 1, 2^{100} - 1$.
For what values of n can we say for certain that $2^n - 1$ will not be prime? A prime number is a natural number greater than 1 divisible only by itself and 1. *Hint:* In the above examples, what values of n led to a number whose final digit showed it was not prime?

84. Wanda's Hamburgers advertises "Have your hamburger 256 different ways." It offers pickles, onions, lettuce, tomato slices, catsup, mustard, mayonnaise, and cheese as toppings. Show why there are 256 ways to fix a hamburger.

85. Rotello's Pizza claims that it makes "over one million different kinds of single pizzas." In order to make this claim, how many toppings must it offer? Explain your strategy.

86. A unit cube 1 cm × 1 cm × 1 cm is to be cut into smaller cubes:

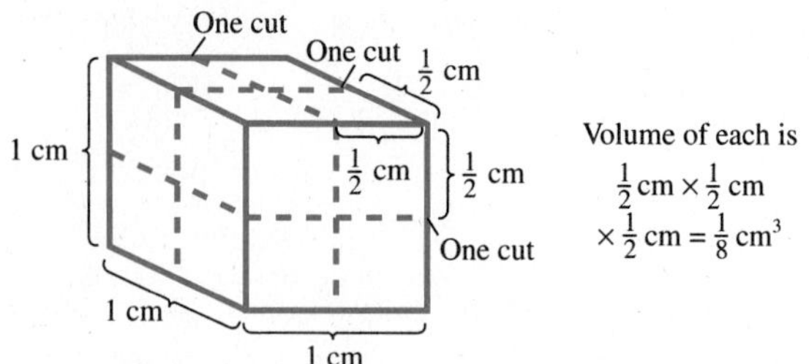

If each edge is subdivided into	*How many cuts are needed to split the block?*	*How many cubes result?*	*What is the volume of each smaller cube?*
2 equal parts	3	8	$\frac{1}{8}$ cm^3
3 equal parts			
4 equal parts			
10 equal parts			
100 equal parts			
1,000 equal parts			

2.2 INTEGRAL EXPONENTS AND SCIENTIFIC NOTATION

There is a story about a king of Persia who rewarded the inventor of chess with a prize of as much wheat as could be put on the squares of a chessboard by starting with one grain on square 1, two grains on square 2, four grains on square 3, and so on. On the last square there would be 2^{63} grains. How big is 2^{63}? If we use our calculators, we get 9.2234E18. The number displayed means 9.2234×10^{18}. A number of this form is said to be in **scientific notation.**

In the last section we discussed the amount of light at various depths of Loch Ness. At 100 meters down, we claim there is $\left(\frac{1}{2}\right)^{100}$ times the amount of light there is at the surface. If we calculate $\left(\frac{1}{2}\right)^{100}$ on our calculators, we might obtain 7.8886E−31. Again the answer is written in scientific notation; that is, 7.8886E−31 = 7.8886×10^{-31}. But what does 10^{-31} mean?

Negative Integral Exponents

To use a calculator, we need to know scientific notation. Since scientific notation involves integer exponents, we consider the meaning of negative exponents. The following pattern suggests a definition for 10^{-31}.

$$\begin{aligned}
10^5 &= 100{,}000 & &\text{divide by a factor of 10}\\
10^4 &= 10{,}000 & &\text{divide by a factor of 10}\\
10^3 &= 1{,}000 & &\text{divide by a factor of 10}\\
10^2 &= 100 & &\text{divide by a factor of 10}\\
10^1 &= 10 & &\text{divide by a factor of ?}\\
10^0 &= ? & &\text{divide by a factor of ?}\\
10^{-1} &= ? & &\text{divide by a factor of ?}\\
10^{-2} &= ?
\end{aligned}$$

To have a consistent pattern, we need to let $10^0 = 1$, $10^{-1} = \frac{1}{10}$, and $10^{-2} = \frac{1}{10^2}$. We make this general definition of a **zero exponent** and a **negative exponent:**

> **Definition**
>
> $x^0 = 1$ for all real numbers, $x \neq 0$; $x^{-n} = \frac{1}{x^n}$ for all real numbers, $x \neq 0$, and n, any positive integer.

By this definition, x^{-n} is another name for $\frac{1}{x^n}$. For example, $2^{-2} = \frac{1}{2^2}$, and $x^{-3} = \frac{1}{x^3}$.

EXAMPLES 1–3

Rewrite each of the following expressions without negative exponents.

1. 10^{-5} **2.** $\dfrac{1}{10^{-5}}$ **3.** 1.56×10^{-3}

Solutions

1. $10^{-5} = \dfrac{1}{10^5}$ Definition of negative exponents.

$= 0.00001$

2. $\dfrac{1}{10^{-5}} = \dfrac{1}{\left(\dfrac{1}{10^5}\right)}$ Use the definition of negative exponents in the denominator.

$= \dfrac{1}{\dfrac{1}{10^5}} = 1 \div \dfrac{1}{10^5}$

$= \dfrac{1}{1} \cdot \dfrac{10^5}{1} = 10^5$

Thus $\dfrac{1}{10^{-5}} = 10^5$.

3. $1.56 \times 10^{-3} = 1.56 \times \frac{1}{10^3}$

$$= \frac{1.56}{10^3} = 0.00156$$

Examples 1 and 2 are special cases of the following general statements:

$$x^{-n} = \frac{1}{x^n} \quad \text{and} \quad \frac{1}{x^{-n}} = x^n \qquad x \neq 0;\ n \text{ is an integer}$$

Since x^{-n} and $\frac{1}{x^n}$ are different names for the same thing, we would expect our properties for exponents to apply to negative exponents.

EXAMPLES 4–9

Find an equivalent expression without negative exponents.

4. $(10^{-5})^6$ **5.** $(10^{-5})^{-6}$ **6.** $(10^{-5})(10^6)$

7. $(10^{-5})(10^{-6})$ **8.** $\frac{10^5}{10^{-6}}$ **9.** $\frac{10^{-5}}{10^{-6}}$

Solutions

4. $(10^{-5})^6 = \left(\frac{1}{10^5}\right)^6$

$= \frac{1}{10^{30}}$ or 10^{-30}

Since $\frac{1}{10^{30}} = 10^{-30}$, we have $(10^{-5})^6 = 10^{(-5)6} = 10^{(-5)(6)} = 10^{-30}$.

5. $(10^{-5})^{-6} = \left(\frac{1}{10^5}\right)^{-6}$

We could have done $\frac{1}{(10^{-5})^6}$ first.

$= \frac{1}{\left(\frac{1}{10^5}\right)^6}$

We must be careful with the parentheses.

$= \frac{1}{\left(\frac{1}{10^{30}}\right)}$

$= 10^{30}$

Therefore $(10^{-5})^{-6} = 10^{(-5)(-6)} = 10^{30}$.

6. $(10^{-5})(10^6) = \frac{1}{10^5} \cdot 10^6$

$= 10$

So $10^{-5}10^6 = 10^{-5+6} = 10^1$.

7. $(10^{-5})(10^{-6}) = \frac{1}{10^5} \cdot \frac{1}{10^6}$

$= \frac{1}{10^{11}}$ or 10^{-11}

Again we note that $(10^{-5})(10^{-6}) = 10^{-5+(-6)} = 10^{-11}$.

8. $\frac{10^5}{10^{-6}} = 10^5 \cdot \frac{1}{10^{-6}}$

$$= 10^5 \cdot 10^6$$

$$= 10^{11} \qquad \text{So } \frac{10^5}{10^{-6}} = 10^{5-(-6)} = 10^{11}.$$

9. $\dfrac{10^{-5}}{10^{-6}} \qquad = \dfrac{1}{10^5} \cdot 10^6$

$$= 10 \qquad \text{So } \frac{10^{-5}}{10^{-6}} = 10^{-5-(-6)} = 10.$$

These examples suggest that the properties for positive integral exponents can be extended to include all integers: the positive integers, the negative integers, and zero. This fact is indeed true. In addition, there is another pattern involving exponents, which you may have already observed. Consider:

$$\frac{10^5}{10^2} = \frac{10^3 \cdot 10^2}{10^2} = 10^3$$

This implies that

$$\frac{10^5}{10^2} = 10^5 \cdot \frac{1}{10^2} = 10^5 \cdot 10^{-2} = 10^{5-2} = 10^3$$

Similarly,

$$\frac{x^9}{x^4} = \frac{x^5x^4}{1 \cdot x^4} = x^5$$

This implies that

$$\frac{x^9}{x^4} = x^9x^{-4} = x^{9-4} = x^5$$

In general,

$$\frac{x^m}{x^n} = x^m x^{-n} = x^{m-n}$$

for all integers m, n and all real $x \neq 0$. We can now summarize our knowledge of exponents.

Definitions

If n is a positive integer and $x \neq 0$, then

(a) $x^n = \underbrace{x \cdot \ldots \cdot x}_{n \text{ times}}$

(b) $x^{-n} = \dfrac{1}{x^n}$

(c) $x^0 = 1$

Properties of Exponents

If m and n are any integers and $x \neq 0$, then these hold:

Property 1: $x^m x^n = x^{m+n}$

Property 2: $(x^m)^n = x^{mn}$

Property 3: $(xy)^n = x^n y^n$

Property 4: $\dfrac{x^m}{x^n} = x^{m-n}$

EXAMPLES 10–15

Write each expression in equivalent simplest form without negative exponents.

10. x^3x^{-4} **11.** $(x^{-2})^3$ **12.** $(-2x^2)^{-3}$

13. $\dfrac{x^0x^2}{x}$ **14.** $\dfrac{2x^{-3}}{3x}$ **15.** $\dfrac{2(x^{-1}y)^{-1}}{3xy^{-1}}$

Solutions

10. $x^3x^{-4} = x^{3-4}$ Property 1.

$= x^{-1}$

$= \dfrac{1}{x}$ Definition.

11. $(x^{-2})^3 = x^{-2(3)}$ Property 2.

$= x^{-6}$

$= \dfrac{1}{x^6}$ Definition.

12. $(-2x^2)^{-3} = (-2)^{-3}(x^2)^{-3}$ Property 3.

$= \dfrac{1}{(-2)^3}x^{-6}$

$= \dfrac{1}{-8x^6}$ We could also have written $(-2x^2)^{-3} = \dfrac{1}{(-2x^2)^3} = \dfrac{1}{-8x^6}$.

$= -\dfrac{1}{8x^6}$

13. $\dfrac{x^0x^2}{x} = x$ $x^0 = 1$.

14. $\dfrac{2x^{-3}}{3x} = \dfrac{2x^{-3-1}}{3}$ Property 4.

$= \dfrac{2x^{-4}}{3}$

$= \dfrac{2}{3x^4}$

15. $\dfrac{2(x^{-1}y)^{-1}}{3xy^{-1}} = \dfrac{2x^{1}y^{-1}}{3xy^{-1}}$ Property 3.

$= \dfrac{2}{3}$ ■

EXAMPLE 16 Given the values of x and y, compute the value of each expression.

x	y	$x^{-1} \cdot y^{-2}$	$\dfrac{x^{-1}}{y^{-2}}$	$x^{-1} + y^{-2}$
2	3			
−2	3			
2	−3			
−100	10			

Table 2.5

Solution

x	y	$x^{-1} \cdot y^{-2}$	$\dfrac{x^{-1}}{y^{-2}}$	$x^{-1} + y^{-2}$
2	3	$2^{-1} \cdot 3^{-2} = \dfrac{1}{2} \cdot \dfrac{1}{3^2}$ $= \dfrac{1}{2} \cdot \dfrac{1}{9}$ $= \dfrac{1}{18}$	$\dfrac{2^{-1}}{3^{-2}} = \dfrac{3^2}{2^1}$ $= \dfrac{9}{2}$	$2^{-1} + 3^{-2}$ $= \dfrac{1}{2} + \dfrac{1}{9}$ $= \dfrac{18}{36} + \dfrac{4}{36}$ $= \dfrac{22}{36} = \dfrac{11}{18}$
−2	3	$(-2)^{-1} \cdot 3^{-2} = \dfrac{1}{(-2)} \cdot \dfrac{1}{3^2}$ $= -\dfrac{1}{18}$	$\dfrac{(-2)^{-1}}{3^{-2}} = \dfrac{3^2}{-2}$ $= -\dfrac{9}{2}$	$(-2)^{-1} + 3^{-2}$ $= -\dfrac{1}{2} + \dfrac{1}{9}$ $= -\dfrac{18}{36} + \dfrac{4}{36}$ $= -\dfrac{14}{36} = -\dfrac{7}{18}$

Table 2.6 **(Continued on next page)**

Table 2.6 (continued)

x	y	$x^{-1} \cdot y^{-2}$	$\dfrac{x^{-1}}{y^{-2}}$	$x^{-1} + y^{-2}$
2	-3	$2^{-1} \cdot (-3)^{-2} = \frac{1}{2} \cdot \frac{1}{(-3)^2} = \frac{1}{2} \cdot \frac{1}{9} = \frac{1}{18}$	$\frac{2^{-1}}{(-3)^{-2}} = \frac{(-3)^2}{2} = \frac{9}{2}$	$2^{-1} + (-3)^{-2} = \frac{1}{2} + \frac{1}{(-3)^2} = \frac{1}{2} + \frac{1}{9} = \frac{22}{36} = \frac{11}{18}$
-100	10	$(-100)^{-1} \cdot 10^{-2} = \frac{1}{(-100)} \cdot \frac{1}{10^2} = \left(-\frac{1}{100}\right) \cdot \frac{1}{100} = -\frac{1}{10{,}000} = -.0001$	$\frac{(-100)^{-1}}{10^{-2}} = -\frac{10^2}{100} = -\frac{100}{100} = -1$	$(-100)^{-1} + 10^{-2} = -\frac{1}{100} + \frac{1}{10^2} = -\frac{1}{100} + \frac{1}{100} = 0$

Table 2.6

Note the difference between column 3 and column 5 in Example 16. This is a frequent source of errors. ■

EXAMPLES 17–20

For each relationship between x and y, make a table of values and plot the (x, y) pairs on a coordinate system. Graphing calculators will help here.

17. $y = x$ **18.** $y = x^{-1}$ **19.** $y = x^2$ **20.** $y = x^{-2}$

Solutions

17. $y = x$

Using a Table In each case, we will put in values for x first, then calculate the y value that corresponds.

x	y
1	1
2	2
5	5
-1	-1
0	0
$-\frac{1}{2}$	$-\frac{1}{2}$
-3	-3

Table 2.7

In this case, x and y have the same value.

Using a Graph

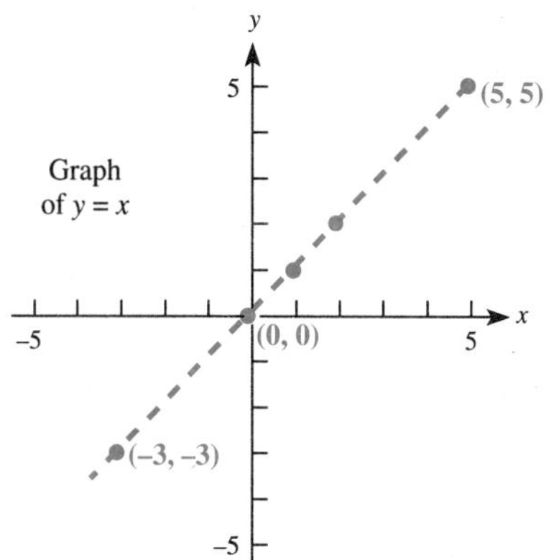

Figure 2.5

You can obtain a more complete picture on a graphing calculator. The points appear to be on a straight line. We will study straight lines in Chapter 4.

18. $y = x^{-1}$ is the same as $y = \frac{1}{x}$.

Using a Table

x	y
1	1
2	$\frac{1}{2}$
3	$\frac{1}{3}$
5	$\frac{1}{5}$
0	no value
-1	-1
-3	$-\frac{1}{3}$
$\frac{1}{2}$	2
$\frac{1}{4}$	4
$-\frac{1}{2}$	-2

Table 2.8

In this case, y is the reciprocal of x. There is no value for $y = \frac{1}{x}$ when $x = 0$, as $\frac{1}{0}$ is not defined.

Using a Graph

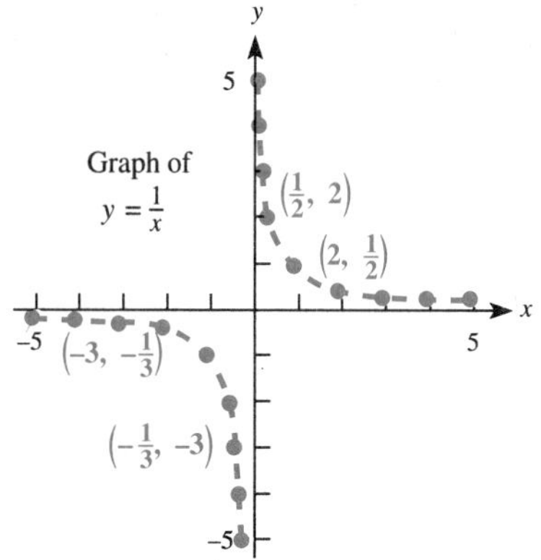

Figure 2.6

Using the points in the table, we can obtain a graph like this. A more complete graph can be achieved on a graphing calculator. There is no point on the graph when $x = 0$.

19. $y = x^2$

Using a Table

x	y
1	1
2	4
3	9
-1	1
-2	4
-3	9
0	0
$\frac{1}{2}$	$\frac{1}{4}$
$-\frac{1}{2}$	$\frac{1}{4}$
$2\frac{1}{2}$	$6\frac{1}{4}$
$-2\frac{1}{2}$	$6\frac{1}{4}$

Table 2.9

In this case, y is the square of x, and so y is always *nonnegative.*
$0^2 = 0$
$\left(2\frac{1}{2}\right)^2 = \left(\frac{5}{2}\right)^2 = \frac{25}{4} = 6\frac{1}{4}$

Using a Graph

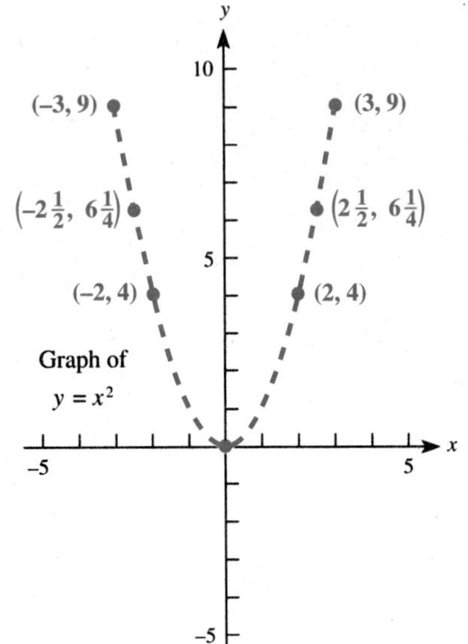

Figure 2.7

Since y is nonnegative, the graph never goes below the x-axis. A more complete graph can be achieved on a graphing calculator. We will study graphs like this in Chapter 6. These types of graphs are called parabolas and are examples of quadratic functions.

20. $y = x^{-2}$ is the same as $y = \frac{1}{x^2}$.

Using a Table

x	y
1	1
-1	1
2	$\frac{1}{4}$
-2	$\frac{1}{4}$
3	$\frac{1}{9}$
-3	$\frac{1}{9}$
$\frac{1}{2}$	4
$-\frac{1}{2}$	4
$\frac{1}{3}$	9
$-\frac{1}{3}$	9
0	no value
2.5	$\frac{1}{6.25}$
-2.5	$\frac{1}{6.25}$

Table 2.10

In this case, y is the *reciprocal of the square* of x. Thus, y is again nonnegative. Also, there is no value for $\frac{1}{x^2}$ when $x = 0$.
$\frac{1}{(2.5)^2} = \frac{1}{6.25} = 0.16$

Using a Graph

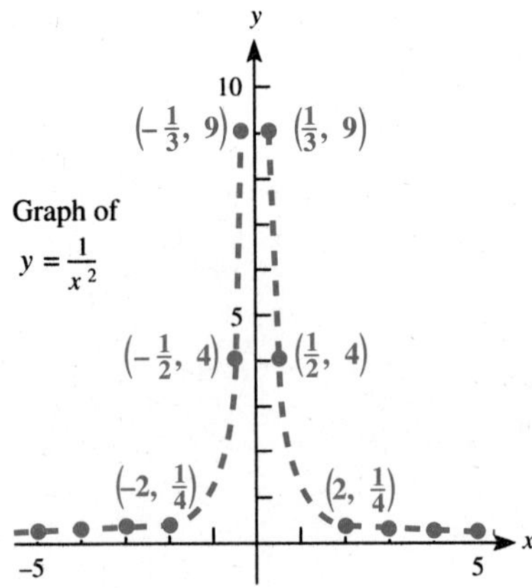

Figure 2.8

A more complete graph can be achieved on a graphing calculator. There is no point on this graph when $x = 0$. We will study these types of graphs in more detail in Chapters 5 and 8.

Scientific Notation

An important application of exponents is their use in simplifying computations involving extremely large or extremely small numbers. We now answer the questions asked in the beginning of this section about scientific notation on calculators.

Definition

A number is written in **scientific notation** if it is expressed in the form

$$m \times 10^c$$

where c is any integer and m is greater than or equal to 1 and less than 10. That is,

$$1 \leq m < 10$$

Each of the numbers below is written in scientific notation:

$$\begin{aligned} 21.33 &= 2.133 \times 10^1 \\ -4{,}765 &= -4.765 \times 10^3 \\ 0.00123 &= 1.23 \times 10^{-3} \end{aligned}$$

To change a number from standard form to scientific notation, write the number as a product of a number between 1 and 10 by *factoring* out the appropriate power of 10. For example,

$$4{,}765 = 4.765 \times 1{,}000 = 4.765 \times 10^3$$ **The decimal moves 3 places to the left.**

$$0.00123 = 1.23 \times \frac{1}{1{,}000} = 1.23 \times 10^{-3}$$ **The decimal moves 3 places to the right.**

To change a number expressed in scientific notation into standard form, we reverse the process by *multiplying* by the appropriate power of 10. For example,

$$3.207 \times 10^5 = 3.207 \times 100{,}000 = 320{,}700$$ **The decimal point moves 5 places to the right.**

$$7.001 \times 10^{-4} = 7.001 \times \frac{1}{10{,}000} = 0.0007001$$ **The decimal point moves 4 places to the left.**

We can use scientific notation to *estimate* answers before we do any calculations.

EXAMPLE 21

Estimate the following, then check your estimate using a calculator.

21. $\dfrac{(1{,}923{,}000)(0.0015)}{(0.00032)(45{,}400)}$

Solution

21. $$\frac{(1{,}923{,}000)(0.0015)}{(0.00032)(45{,}400)} \approx \frac{(2 \times 10^6)(1.5 \times 10^{-3})}{(3 \times 10^{-4})(4.5 \times 10^4)}$$
$$\approx \frac{3 \times 10^{6-3}}{13.5 \times 10^{-4+4}}$$
$$\approx \frac{1 \times 10^3}{4 \times 10^0}$$
$$\approx 0.25 \times 10^3$$
$$\approx 250$$

In this example, we estimate the magnitude of the power of 10. Note that an estimate of 2,500 would be off by a factor of 10, while an estimate of 25,000 would be off by a factor of 10^2. Use of scientific notation can help us to estimate a number accurately to the nearest power of 10.

Using a calculator, we obtain 198.5. ■

Scientific notation is useful in computing with very large or very small numbers. Consider the following example:

EXAMPLE 22

A certain computer can perform an arithmetic computation in 2.4×10^{-9} seconds. How many similar arithmetic computations can the computer perform in

(a) 1 minute? **(b)** 24 hours?

Solutions

(a) 60 seconds = 1 minute. The number of computations solved in one minute equals

$$\frac{1 \text{ computation}}{2.4 \times 10^{-9} \text{ seconds}} \times 60 \text{ seconds}$$
$$= \frac{60}{2.4 \times 10^{-9}}$$
$$= \frac{6 \times 10}{2.4 \times 10^{-9}}$$
$$= \frac{6 \times 10^{10}}{2.4}$$
$$= 2.5 \times 10^{10} = 25{,}000{,}000{,}000$$ **What number is this?**

(b) 24 hours = $60 \times 60 \times 24$ seconds. The number of computations solved in 24 hours equals

$$\frac{60 \times 60 \times 24}{2.4 \times 10^{-9}} = \frac{6 \times 6 \times 2.4 \times 10^3}{2.4 \times 10^{-9}}$$
$$= 3.6 \times 10^{13} = 36{,}000{,}000{,}000{,}000$$ **What number is this?** ■

2.2 EXERCISES

I. Procedures

1–32. Use the properties of exponents to express all answers without negative exponents in simplest form.

1. 3^{-3}

2. -3^{-3}

3. $-4(4)^{-2}$

4. $(10 \cdot 10^{-2})^2$

5. $\left(\frac{3}{4}\right)^{-3}$

6. $\left(\frac{-3}{4}\right)^{-3}$

7. $10^{-5} \cdot 10^{11}$

8. $\frac{10^{-6} \cdot 10^{12}}{10^{-4}}$

9. $10^{-1} + 10^2$

10. $(10^3 + 10^2)^{-1}$

11. $(0.02)^{-2}$

12. $(2.5)^{-3}$

13. $(4 \cdot 10^{-6})(2 \cdot 10^4)$

14. $\frac{(-2 \cdot 10^2)^2}{-2^2 \cdot 10^{-1}}$

15. $\frac{2^0 - 2^{-2}}{2 - 2(2)^{-2}}$

16. $\frac{2^3 \cdot 2^{-2} \cdot 2^4}{2^{-1} \cdot 2^0 \cdot 2^{-3}}$

17. $\left(\frac{1}{3}\right)^2 - (-3)^2$

18. $\left(\frac{1}{2}\right)^3 - (-2)^{-3}$

19. x^3x^{-3}

20. $\frac{x^3}{x^{-3}}$

21. $(x^3)^{-3}$

22. $\left(\frac{-x}{y}\right)^{-2}$

23. $\frac{x^{-100}}{x^{-105}}$

24. $\frac{x^{100}}{y^{-100}}$

25. $x^{-2}x^3x^{-4}$

26. $(x^{-2}y^2)(x^3y^0)$

27. $\frac{x^{-2}x^3}{x^{-4}x^0}$

28. $(x^4y)^{-2}$

29. $\frac{2xy^2}{4x^2y^{-2}}$

30. $\frac{(-5)^3x^3}{5(-x)^3}$

31. $\frac{-2(xy)^3}{(-xy)^{-2}}$

32. $\frac{(3x^{-2}y^3)^{-1}}{2x^5y^2}$

33. Complete the table below. Observe any patterns.

x	y	$x^{-2}y^3$	$\frac{x^{-2}}{y^3}$	$x^{-2} + y^3$
2	2			
−2	2			
2	−2			
−2	−2			
−10	10			
100	−10			

34. Write each of the following numbers in scientific notation.

(a) 2,731 **(b)** 2.731 **(c)** 0.002731
(d) 273.1 **(e)** 0.2731 **(f)** 0.0002731
(g) 27.31 **(h)** 273,100.000 **(i)** 0.0000002731

35. Write each number in expanded form (standard form).

(a) 6.93×10^{-4} **(b)** 7.4964×10^{-8}
(c) 4.234×10^8 **(d)** 3.002×10^5

36–41. Compute each of the following without a calculator. Use the properties of exponents.

36. $\frac{(3.6 \cdot 10^{-3})(3 \cdot 10^{-2})}{(1.2 \cdot 10^5)(9 \cdot 10^{-5})}$

37. $\frac{(4.2 \cdot 10^4)(3.2 \cdot 10^{-3})(3 \cdot 10^5)}{(2 \cdot 10^{-2})(4.8 \cdot 10^6)}$

38. $\frac{3{,}600(0.03)}{120{,}000(0.00009)}$

39. $\frac{(0.000096)(2{,}000)}{1{,}600(0.0012)}$

40. $\frac{(0.000075)(1{,}800)}{(2.5)(0.09)10^3}$

41. $\frac{28(0.0045)}{140(1{,}500)}$

42–45. **(a)** Estimate each of the following. **(b)** Then compute each using a calculator. Write the answer in both scientific notation and standard form.

42. $\frac{(379{,}500)(0.0012)}{4{,}980}$

43. $\frac{650}{5{,}100(0.00013)}$

44. $\frac{211(0.6)}{2{,}000(0.005)}$

45. $\frac{(0.00349)(1{,}049.5)}{(80{,}000)(79.345)}$

II. Concepts

46–49. Locate the decimal point in the numbers on the right side of each equation in order to obtain the best estimate for each exercise.

46. $\frac{145{,}215(279{,}000)}{67{,}596{,}240} \approx 605$

47. $\frac{(0.0000071426)(2{,}731{,}827)}{(0.000036)(2{,}112)} \approx 266$

48. $\dfrac{296{,}840{,}000(0.0009997)}{(0.0510706)(0.024)} \approx 241$

49. $\dfrac{(29{,}640)(3{,}100{,}000)}{(179{,}687)(6{,}410)} \approx 7{,}942$

50–53. Make a table of values and plot the x-y pairs on a coordinate axis. Compare the graphs to those in Examples 17–20.

50. $y = 2x$

51. $y = \dfrac{1}{2x}$

52. $y = 2x^2$

53. $y = \dfrac{1}{2x^2}$

54. Write a brief paragraph explaining what scientific notation is and why it is useful.

55. Write a brief paragraph explaining three different uses and meanings of the "$-$" symbol as it appears in -8, $3x - y$, and 2^{-3}.

56. Compute $(.5)^{200}$ on your calculator. What was your result? Why do you think this result was obtained?

57. Using the fact that $10^3 \approx 2^{10}$, mentally find a value of n so that

(a) $2^n > 10^{50}$ **(b)** $2^n > n^{10}$

58–64. Decide whether each statement is AT for always true, ST for sometimes true, or NT for never true. If a statement is ST, give an example illustrating a case where the statement is true and an example illustrating a case where the statement is false.

58. $(3x)^0 = 1$

59. $3x^0 = 1$

60. x^{-2} is a negative number.

61. $(-1)^n = -1$

62. If $x \neq 0$, then x^{-2} is a negative number.

63. If $x \neq 0$, then x^{-3} is a negative number.

64. If $x \neq 0$, then $\dfrac{x^{2n}}{x^n} = x^2$.

III. Applications

65–76. Solve each of the following problems.

65. The average distance of the Earth to the sun is about 9.3×10^7 miles, or 1 A.U. (astronomical unit). How far will the Earth travel in its yearly orbit around the sun?

66. The average distance of Mercury to the sun is about 3.6×10^7 miles.

(a) How far will Mercury travel in one orbit around the sun?

(b) Approximately how many orbits will Mercury make in the time it takes the Earth to make one orbit?

67. A microsecond is one millionth of a second. A nanosecond is 10^{-9} seconds $= 0.000000001$ second.

(a) How many nanoseconds make a microsecond?

(b) A micromicrosecond is 10^{-12} seconds. How many micromicroseconds make a nanosecond? A microsecond?

68. A light-year is equal to $3.1 \cdot 10^{16}$ feet. If 1 mile $= 5{,}280$ feet, then approximately how many miles are in a light-year?

69. Light travels $1.86 \cdot 10^5$ miles per second. Estimate how far it will travel in

(a) 1 minute. **(b)** 1 hour. **(c)** 1 day.

(d) 365 days.

70. A certain computer can perform 75 million multiplications a second. How many multiplications can the computer do in

(a) 1 minute? **(b)** 1 hour? **(c)** 1 day?

(d) 365 days?

71. McDonald's restaurants claim to have sold about 1.8 billion hamburgers in the United States in 1990. If there are approximately 250 million people in the United States, how many hamburgers is that per person?

72. One germ can produce 281,000,000,000,000 more germs in 24 hours. Approximately how many germs will be produced by

(a) 10 germs in 24 hours?

(b) 1,000 germs in 24 hours?

(c) 1 million germs in one week?

73. A proton weighs about 1.67×10^{-27} kg, while an electron weighs about 9.11×10^{-31} kg. How many times heavier than an electron is a proton?

74. The mass of the Earth is about 5.975×10^{24} kg, while the mass of the moon is about 7.343×10^{22} kg. How many times heavier than the moon is the Earth?

75. The cells in every type of tissue are nearly the same size, 3×10^{-4} cm in length, about half that amount in width and thickness. What is the volume of a typical cell? Assume that the cell is shaped like a rectangular prism.

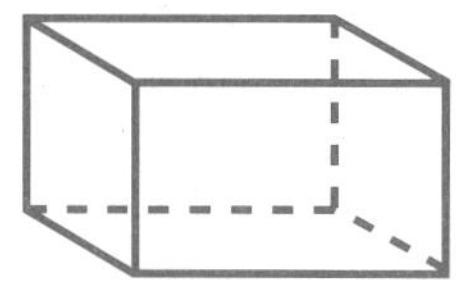

76. The diameter of the average red corpuscle cell is 8×10^{-5} cm. Assume the cell is spherical in shape, and compute the following quantities.

    (a) The volume of the red corpuscle.

    (b) The surface area of the red corpuscle.

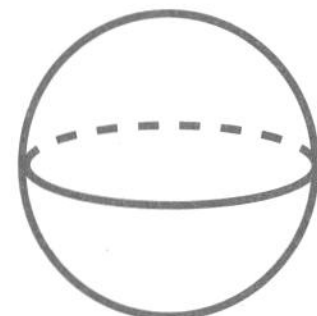

IV. Extensions

77. How many digits are in the expanded form of

    (a) 3^{20}? **(b)** 2^{100}? **(c)** $3^{100} \cdot 5^{60}$?

78. The German astronomer Johann Bode (1747–1826) formulated a law giving the distance from each planet to the sun. The law is $D = 0.4 + 0.3 \times 2^n$, where $n = 0$ for Venus, 1 for Earth, 2 for Mars, and on up to 8 for Pluto, and D is the distance in A.U. (astronomical units; see Exercise 65).

    (a) Consult a table of distances of the various planets to the sun to verify the accuracy of Bode's formula.

    (b) Why do you think the estimates given by the formula are inaccurate for Neptune and Pluto?

79. Will all of the people in the world fit in a cubic mile? Answer this question by making appropriate estimates for **(a)** The space (volume) occupied by an average person and **(b)** The world's population (5.384 billion in 1992).

80. Find the missing entries in the table.

x	y	$x^{-1}\cdot y$	$\frac{x^{-1}}{y}$	$x^{-1} + y$
	2			$\frac{7}{2}$
2				3
	2	$\frac{2}{3}$		
2			$\frac{3}{2}$	
		1	$\frac{1}{16}$	

2.3 POLYNOMIALS

Expressions, Factors, and Terms

Since algebra is a language, we need to look more closely at algebraic expressions. An **algebraic expression** contains a meaningful combination of variables, numbers, and the signs of operations. Algebraic expressions are mathematical phrases that can be combined to form mathematical sentences, called equations or inequalities. Expressions are used to describe patterns.

The algebraic expression $n + 2$ can represent infinitely many numbers. For example,

$$\text{If } n = 0, \quad \text{then} \quad n + 2 = 0 + 2 = 2.$$

$$\text{If } n = \frac{-1}{2}, \quad \text{then} \quad n + 2 = \frac{-1}{2} + 2 = \frac{3}{2}.$$

$$\text{If } n = \pi, \quad \text{then} \quad n + 2 = \pi + 2.$$

Each number is obtained by *adding* two numbers, n and 2. We call n and 2 *terms* of the expression, $n + 2$. The expression $5n - 2$ represents a number that is obtained by subtracting the number 2 from $5n$. Again, $5n$ and 2 are terms of the expression $5n - 2$.

Terms are separated from each other by addition or subtraction. The expression $5n$ contains one term, $5n$. However, this term, $5n$, is obtained by *multiplying* the two numbers 5 and n. We call 5 and n *factors* of the term $5n$. In general, if $ab = c$, then a and b are each a **factor** of c.

A **term** is an algebraic expression containing only products or quotients. An algebraic expression may have many terms, but the terms will always be separated from one another by addition or subtraction.

CAUTION Many mathematical errors can be traced to a lack of understanding of the difference between a term and a factor. For example, consider $a + 2$ and $5(a + 2)$. The expression $a + 2$ by itself is a *sum* of two terms, a and 2, while the expression $5(a + 2)$ contains only one term, $5(a + 2)$. Moreover, $5(a + 2)$ is also a *product of two factors,* 5 and $(a + 2)$. Similarly, $6a + 5(a + 2)$ is the sum of two terms $6a$ and $5(a + 2)$. The difference between factors and terms (products and sums) is very important. Consider the following example. ■

EXAMPLE 1

Without rewriting, which of the following are expressed as a product? If the expression is a product, is x a factor of the product?

(a) x **(b)** $6x^2$ **(c)** $6x + 1$

(d) $x(x + 1)$ **(e)** $(2 + x)(x + 1)$ **(f)** $2 + x(x + 1)$

Solutions

(a) x can be considered as a product: $x = x \cdot 1$. Therefore x is a factor of x. That is, every number is a factor of itself.

(b) $6x^2$ is a product: $6x^2 = 6 \cdot x \cdot x$. Therefore x is a factor of $6x^2$.

(c) $6x + 1$ *is not* a product. It is a sum of two terms, $6x$ and 1.

(d) $x(x + 1)$ is a product and x is a factor of $x(x + 1)$.

(e) $(2 + x)(x + 1)$ is a product, but x *is not* a factor.

(f) $2 + x(x + 1)$ *is not* a product. The entire expression is not expressed as a product. It is the sum of two terms, 2 and $x(x + 1)$. ■

The Distributive Property Revisited

As we have seen, the distributive property can be illustrated geometrically by cutting the area of a rectangle into two subrectangles.

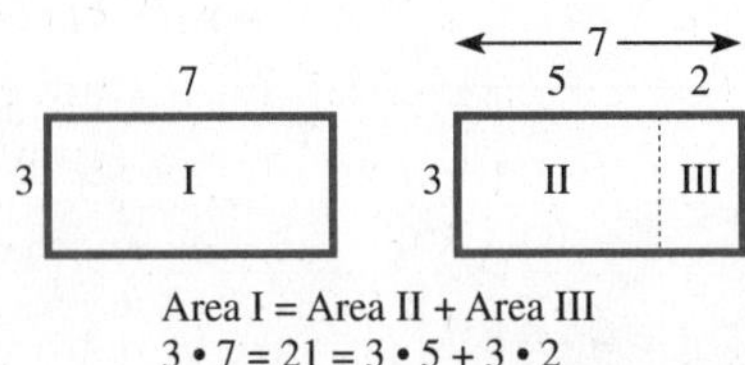

Figure 2.9

In general, consider a rectangle with width a and length $b + c$. Its area is $a(b + c)$, or $ab + ac$.

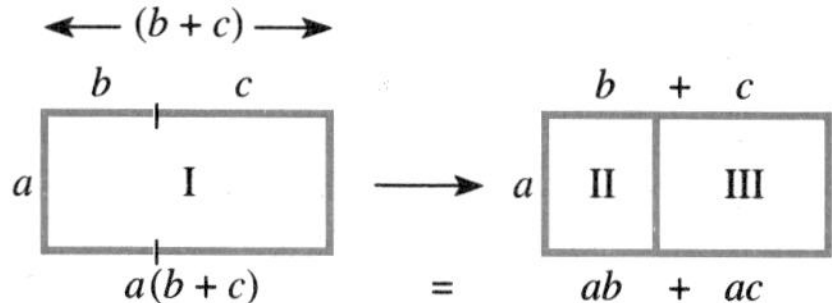

Area of rectangle I = Area of rectangle II + Area of rectangle III

Figure 2.10

The distributive property connects the operations of multiplication (products) and addition (sums).

The Distributive Property

$a(b + c)$	$=$	$ab + ac$
a product with *two factors*		a sum with *two terms*

The distributive property read in one direction is called *multiplying* and in the other direction is called *factoring.*

Multiplying	Factoring
$a(b + c) = ab + ac$	$ab + ac = a(b + c)$

Note also:

$a(b + c) = ab + ac = ba + ca = (b + c)a$ Commutative and distributive properties.

Thus, the commutative property of multiplication allows us to distribute from the left or from the right.

$a(b + c) = ab + ac$ is the *left* distributive property

$(b + c)a = ba + ca$ is the *right* distributive property

The following examples show how the distributive property is used.

EXAMPLES 2-5

Write each of the following products as a sum or as a difference of terms. That is, perform the indicated multiplication.

2. $3(x + 2)$ **3.** $(2x - 1)x$ **4.** $5(a + b + c)$ **5.** $-2(1 - x)$

Solutions

2. $3(x + 2) = 3x + 3(2)$ — **The distributive property.**

$= 3x + 6$

3. $(2x - 1)x = x(2x - 1)$ — **Multiplication is commutative: $(2x - 1)x = x(2x - 1)$.**

$= x(2x) - x(1)$

$= 2x^2 - x$

4. $5(a + b + c) = 5[(a + b) + c]$

$= 5(a + b) + 5c$

$= 5a + 5b + 5c$ — **This tells us that we can multiply over any number of terms.**

5. $-2(1 - x) = -2[1 + (-x)]$

$= (-2)(1) + (-2)(-x)$

$= -2 + 2x$ ■

Example 5 points out that algebraic expressions behave the same as arithmetic expressions.

$-(a + b) = (-a) + (-b) = -a - b$ like $-(5 + 7) = (-5) + (-7) = -12$

$-(a - b) = -[a + (-b)] = -a + b$ like $-(5 - 7) = -[5 + (-7)] = -5 + 7 = 2$

EXAMPLES 6–10

Write each of the following sums or differences as a product. This process is called **factoring.**

6. $5x + ax$ **7.** $5 + 5a$ **8.** $20x^2 - 15x^4y$

9. $6a^3b - 3b^3 + 9a^3b^2$ **10.** $3(a + 1) - b(a + 1)$

Solutions

6. $5x + ax = x(5 + a)$ — **The distributive property.**

7. $5 + 5a = 5 \cdot 1 + 5 \cdot a = 5(1 + a)$ — **$5 = 5 \cdot 1$ is always a factor of itself.**

8. $20x^2 - 15x^4y = (5x^2)(4) - (5x^2)(3x^2y)$ — **First find the largest common factor of each term. In this case, it is $5x^2$. Then use the distributive property to factor out the common factors from each term.**

$= (5x^2)(4 - 3x^2y)$

9. $6a^3b - 3b^3 + 9a^3b^2 = (3b)(2a^3) - (3b)(b^2) + (3b)(3a^3b)$ — **Look for the largest common factor. Factor out the common *factors*.**

$= 3b(2a^3 - b^2 + 3a^3b)$

10. $3(a + 1) - b(a + 1) = (3 - b)(a + 1)$ — **Here the common factor is $(a + 1)$.** ■

Terms such as $5x^2$ and $3x^2$ are called *like terms,* since they both contain the same variable to the same power. The numerical factor is called the *coefficient* of the variable. The coefficients in $5x^2$ and $3x^2$ are 5 and 3, respectively. Like terms can have more than one variable. For example, $3xy$ and $6xy$ are like terms, but $3x^2y$ and $6xy$ are *not* like terms. The exponent of each variable in like terms must be the same. **To add like terms we can add their coefficients** and then multiply the sum by the variable part, as can be seen from the distributive property. For example,

$$5x + 3x = (5 + 3)x = 8x$$
$$-5a^3b^2 + 8a^3b^2 = (-5 + 8)a^3b^2 = 3a^3b^2$$
$$2y^2 + 2y^2 + 2y^2 + 2y^2 = (2 + 2 + 2 + 2)y^2 = 8y^2$$

Definition of a Polynomial

The algebraic expressions $3x^2 + 1$, $4xy + y$, $x^5 - x^4 + 1$, and $3x^2yz^3$ are examples of **polynomials.** Polynomial is a Greek word, "poly" meaning many and "nomial" meaning term. Hence, "polynomial" means "many terms."

We classify polynomials by degree and by term. The **degree** of a polynomial in one variable is given by the highest exponent of the variable in the polynomial. Further, a polynomial with one term is called a **monomial.** A polynomial with two terms is called a **binomial.** A polynomial with three terms is called a **trinomial.** Thus

$3x^2$ is a monomial of degree 2.

$x^3 + 2$ is a binomial of degree 3.

$x^4 + x^3 + 1$ is a trinomial of degree 4.

15 is a monomial of degree 0. $15 = 15 \cdot 1 = 15 \cdot x^0$

Algebra is a general form of arithmetic. Thus we combine polynomials in a manner similar to how we combine arithmetic expressions. The properties and rules for the order of operations on algebraic expressions are the same as those for arithmetic.

Addition and Subtraction of Polynomials

EXAMPLES 11–13

Perform the indicated operations and combine like terms.

11. $3(x^2 - 2x) + 6x - x^2$

12. $x(xy - 1) - (x^2y + 2x)$

13. $2x^2y - 6[x - x(xy - y)] - xy$

Solutions

11. $3(x^2 - 2x) + 6x - x^2 = 3x^2 - 6x + 6x - x^2$ Distributive property.

$= 3x^2 - x^2 - 6x + 6x$ Commutative property of addition.

$= 2x^2$ Combine like terms.

12. $x(xy - 1) - (x^2y + 2x) = x(xy) - x(1) - (x^2y + 2x)$ Distributive property.

$= x^2y - x - x^2y - 2x$ Recall that $-(a + b) = -a - b$.

$= x^2y - x^2y - x - 2x$ Commutative property of addition.

$= -3x$ Combine like terms.

13. $2x^2y - 6[x - x(xy - y)] - xy$

$= 2x^2y - 6[x - x^2y + xy] - xy$ **Work with the innermost parentheses first.**

$= 2x^2y - 6x + 6x^2y - 6xy - xy$

$= 2x^2y + 6x^2y - 6x - 6xy - xy$

$= 8x^2y - 6x - 7xy$ ■

Algebra as a Language

Polynomials enable us to represent many relationships in the mathematical sciences. Sometimes the relationship between quantities is not explicitly stated but is implied by the physical nature of the situation or by recalling a previously learned fact, as in the next two examples.

EXAMPLE 14

Write an expression for

(a) The amount of acid in x milliliters (ml) of a 10% acid solution.

(b) The perimeter of an equilateral triangle whose sides are x cm long.

Solutions

(a) In x milliliters of solution (water mixed with acid), 10% of the mixture is acid, and 90% of the mixture is water. The implied relationship is multiplication:

$$\text{Amount of acid} = 10\%(x \text{ ml}) = 0.10(x \text{ ml})$$

It is helpful to illustrate this relationship with a few examples.

An acid solution contains:	ACID + WATER
50 ml of a 10% acid solution contains:	$10\%(50 \text{ ml}) + 90\%(50 \text{ ml})$ $= 0.10(50 \text{ ml}) + 0.90(50 \text{ ml})$ $= 5 \text{ ml} + 45 \text{ ml}$ $= 50$ ml of solution
100 ml of a 10% acid solution contains:	$10\%(100 \text{ ml}) + 90\%(100 \text{ ml})$ $= 10 \text{ ml} + 90 \text{ ml}$ $= 100$ ml of solution
x ml of a 10% acid solution contains:	$10\%(x \text{ ml}) + 90\%(x \text{ ml})$ $= 0.10x + 0.90x$ $= x$ ml of solution

Therefore the amount of acid in x liters of a 10% acid solution is equal to $0.10x$ ml.

There are similar problems that involve saline (salt) solutions, sucrose solutions (for intravenous feedings), butterfat in milk, alcohol solutions, and fertilizer that contains 50% phosphate. Such problems are commonly called mixture problems.

(b) The perimeter of a triangle is the total length of its three sides. In an equilateral triangle, all three sides are equal in length. Therefore the perimeter of an equilateral triangle with all sides x cm long is equal to $x + x + x$, or $3x$ cm. The perimeter equals $3x$.

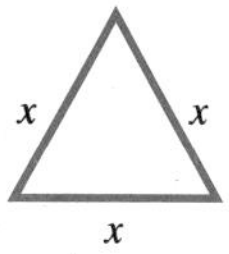

x x x

Figure 2.11

We can use the techniques for algebraic expressions we have just studied to express more complicated problems in algebraic form.

EXAMPLES 15–18

Write an algebraic statement for the following situations.

15. The sum of two numbers, where the second number is four more than six times the first number.

16. The distance traveled during an eight-hour trip if you traveled at an average rate r mph for the first six hours on the highway and at an average rate of 30 mph less than this rate for the last two hours in the city.

17. The amount of money in a bag of 1,000 coins that are dimes and quarters.

18. The cost of renting a car for one day, where the charge is \$21.95 per day plus 17¢ a mile.

Solutions

15. Let x represent the first number. The second number is

$$\underbrace{4}_{\text{four}} \quad \underbrace{+}_{\text{more than}} \quad \underbrace{6(x)}_{\text{six times the first}}$$

The sum of the two numbers is

$$\underbrace{x}_{\text{first number}} \quad \underbrace{+}_{\text{sum of}} \quad \underbrace{4 + (6x)}_{\text{second number}}$$

This sum can be written:

$$x + 4 + 6x = 4 + 7x$$

16. Recall that distance = rate × time. To analyze this problem, it would help to draw a picture in order to sort out the data (facts).

time = 6 hours	time = 2 hours
rate = r	rate = $r - 30$
distance = $6r$	distance = $2(r - 30)$

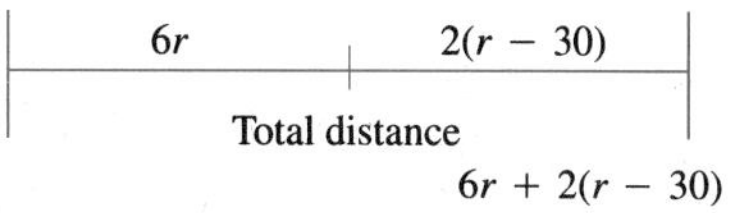

$6r$ $2(r - 30)$

Total distance

$6r + 2(r - 30)$

$$\text{Total distance} = \underbrace{6r}_{\substack{\text{distance for}\\\text{first 6 hours}\\\text{at a rate } r}} + \underbrace{2(r - 30)}_{\substack{\text{distance for}\\\text{last 2 hours}\\\text{at a rate 30 mph}\\\text{less than } r}}$$

$= 6r + 2r - 60$ The distributive property.

$= 8r - 60$ Expression for distance.

17. If d is the number of dimes, then $1{,}000 - d$ must be the number of quarters, since there is a total of 1,000 coins.

$$\begin{aligned}
\text{Amount of money in dimes} &= 10(d) \text{ cents}\\
\text{Amount of money in quarters} &= 25(1{,}000 - d) \text{ cents}\\
\text{Total money in the 1,000 coins} &= 10(d) \text{ cents} + 25(1{,}000 - d) \text{ cents}\\
&= [10d + 25(1{,}000) - 25d] \text{ cents}\\
&= (25{,}000 - 15d) \text{ cents}\\
\text{Money expressed in dollars} &= \left(\frac{25{,}000 - 15d}{100}\right) \text{dollars}\\
&= (250 - 0.15d) \text{ dollars}
\end{aligned}$$

18. Let m be the number of miles driven in one day. The cost of renting a car is $\$21.95 + 0.17m$. ■

Polynomials often are used to express relationships between two variables. In such cases we can also represent the relationship in a table of values or in a graph.

EXAMPLES 19–22

For each relationship between x and y, make a table of values and plot the x-y pairs on a coordinate axis. In each case, what happens to the value of y as the value of x increases?

19. $y = x + 2$ **20.** $y = -x + 2$ **21.** $y = x^2 + 2$ **22.** $y = -x^2 + 2$

Solutions

19. $y = x + 2$

Using a Table

x	y
0	2
1	3
2	4
−1	1
−2	0
−3	−1
$2\frac{1}{2}$	$4\frac{1}{2}$
$-2\frac{1}{2}$	$-\frac{1}{2}$

Table 2.11

These are just a few of the *many, many* possibilities for points.

Using a Graph

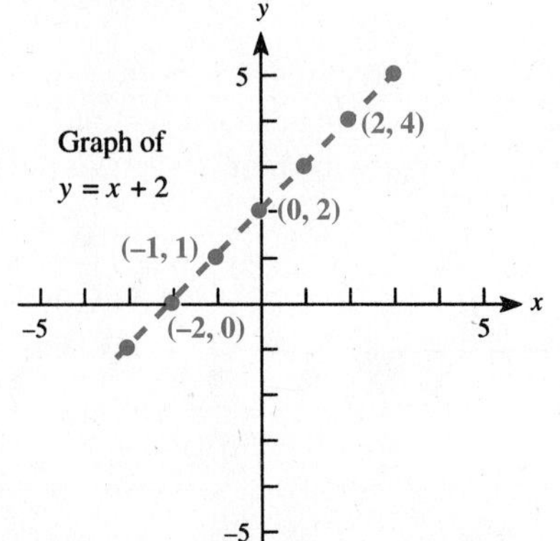

Figure 2.12

The x-y pairs appear to lie on a straight line. A more complete graph can be obtained on a graphing calculator. In this example, y increases as x increases.

20. $y = -x + 2$

Using a Table

x	y
0	2
1	1
2	0
-1	3
-2	4
-3	5
$2\frac{1}{2}$	$-\frac{1}{2}$
$-2\frac{1}{2}$	$4\frac{1}{2}$
3	-1

Table 2.12

For example, when $x = -3$, $-x + 2 = -(-3) + 2 = 5$.

Using a Graph

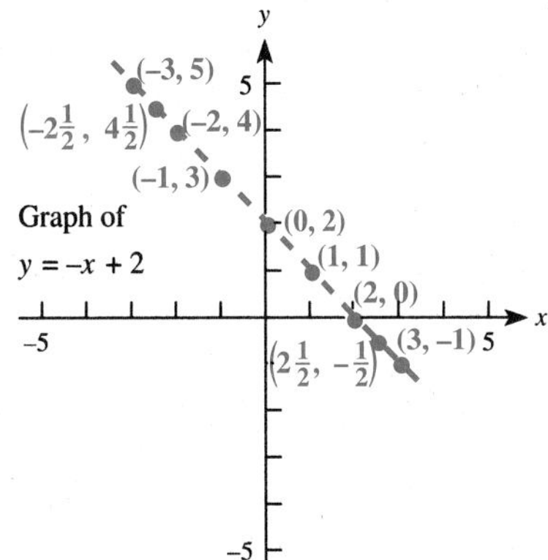

Figure 2.13

Again, the x-y pairs appear to lie on a straight line. A more complete graph can be obtained on a graphing calculator. In this example, y decreases as x increases.

21. $y = x^2 + 2$

Using a Table

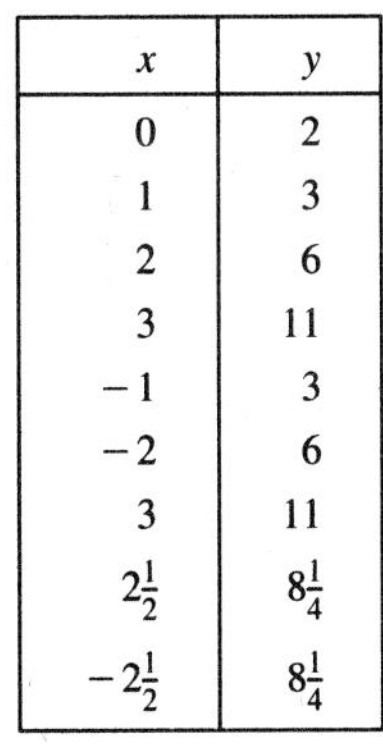

x	y
0	2
1	3
2	6
3	11
-1	3
-2	6
3	11
$2\frac{1}{2}$	$8\frac{1}{4}$
$-2\frac{1}{2}$	$8\frac{1}{4}$

Table 2.13

Notice that $x = \pm 1$ gives $y = 3$, $x = \pm 2$ gives $y = 6$, $x = \pm 3$ gives $y = 11$. When $x = 2\frac{1}{2}$, $x^2 + 2$ $= \left(\frac{5}{2}\right)^2 + 2 = \frac{25}{4} + \frac{8}{4} = \frac{33}{4}$.

Using a Graph

Figure 2.14

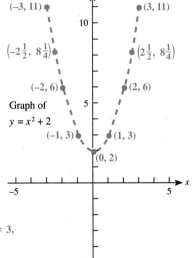

This graph always lies above the x-axis, since $y = x^2 + 2$ is always positive. Note the mirror image around the y-axis. In this example, as x increases through negative values to 0, y decreases. Then, as x increases through positive values, y increases. The minimum value of y ($y = 2$) occurs when $x = 0$.

22. $y = -x^2 + 2$

Using a Table

x	y
0	2
1	1
-1	1
2	-2
-2	-2
3	-7
-3	-7
$2\frac{1}{2}$	$-4\frac{1}{4}$
$-2\frac{1}{2}$	$-4\frac{1}{4}$

Table 2.14

Note that when $x = -2, x^2 = -(-2)^2 + 2 = -4 + 2 = -2$.
When $x = 2\frac{1}{2}, -x^2 + 2 = -\left(\frac{5}{2}\right)^2 + 2 = \frac{-25}{4} + \frac{8}{4} = \frac{-17}{4} = -4\frac{1}{4}$.

Using a Graph

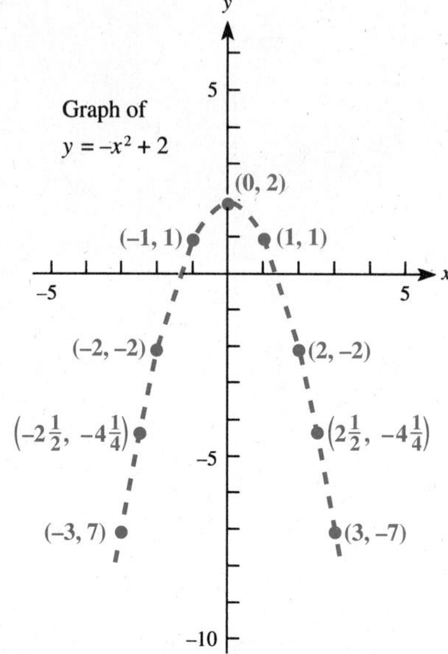

Figure 2.15

Again note that there is a mirror image in the y-axis. Once again, $x = \pm 1$ gives $y = 1$, $x = \pm 3$ gives $y = -7$. This occurs because x is squared, thus the sign of x won't change the value of y. In this example, as x increases through negative values to 0, y increases. Then, as x increases through positive values, y decreases. The maximum value of y ($y = 2$) occurs when $x = 0$. ■

Examples 19 and 20 are examples of linear functions. We will study them in detail in Chapters 3 and 4. Examples 21 and 22 are examples of quadratic functions. We will study them in detail in Chapter 6.

2.3 EXERCISES

1. Procedures

1–8. **(a)** Which of the following are expressed as a product? **(b)** If it is a product, is $x + 2$ a factor?

1. $2x$ **2.** $2 + x$

3. $3(2 + x)$ **4.** $3(2 + x) + y^2$

5. $3(2 + x)y^2$ **6.** $(x + 2)(x + 1)$

7. $(x + 2)(x + 1) + 6$ **8.** $2x^2 - x(x + 2)$

9–16. Use the distributive property to write each of the following expressions as a sum or a difference of terms.

9. $2x(3 + y)$ **10.** $7.8 + x(1.1 + 0.1x)$

11. $\frac{3}{2}(1.2x - 9y)$ **12.** $-(x - 2)5y$

13. $-5y(x + 2)$ **14.** $(a + 2)b - (a + 2)c$

15. $2x(x - 1) + 3x^3$ **16.** $17(10a - 11b - 5c)$

17–23. Use the distributive property to write the following sums or differences as the product of two or more factors.

17. $5ax - 25x$

18. $6x + 15x$

19. $3a + 6ab + 15a$

20. $(x + y)a - (x + y)b$

21. $6by + 18y + b + 3$

22. $\underbrace{3a + 3a + \cdots + 3a}_{\text{10 terms}}$

23. $\underbrace{(3 + a) + (3 + a) + \cdots + (3 + a)}_{\text{11 terms}}$

24–33. Perform the indicated operations and combine any like terms.

24. $x(x - 1) + 2x^2 + 3$

25. $5 - 2(ax^2 - 3) + a(2x^2 - 10)$

26. $3(a^2 - 5) + 2a(a - 1)$

27. $x^2y + 3x(xy - 1) - 4y(x^2 - 1)$

28. $3x - 2x(1 - x) - 6(x - 1)$

29. $(x + 4)x + 1 - 2x[3 - 2(x + 1)]$

30. $n - 2n[1 - (n + 1)] - (n + 1)$

31. $x - [2x^2y - (x - 1)] - [-y(2x^2 + y)]$

32. $(6x^4 - 6x^3 - 2x - 1) - 3x(2x^3 - x^2 + 1)$

33. $3x^{10} - 6(x^5 - x^2) - x^4(x^6 - x) + x^2$

34–39. **(a)** In the following polynomials, state the degree of the polynomial. **(b)** Decide whether they are monomial, binomial, trinomial, or none of these.

34. $2x$

35. $x + 2$

36. $3(x + 2)$

37. $x^2 + 3x + 6$

38. $5x^2 - 6x^5 - 10$

39. $x^2 + 3x(x^3 + 2)$

II. Concepts

40–43. Perform the following operations.

40. Add $2x$ to $3y - 2x$.

41. Subtract $4 - t$ from $8 - 2t$.

42. Subtract $16x - 3y$ from $18x - 5y$ and multiply the result by $\frac{1}{2}$.

43. Subtract $-(2a - b)$ from $(6a + b)$ and multiply the result by the reciprocal of $6a$.

44–46. Show how each of the figures below illustrates the distributive property by writing the area as both a product and a sum.

44.

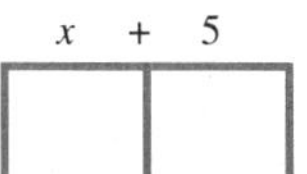

45.

y

3 + x

46.

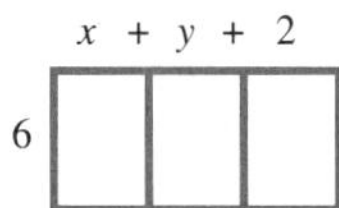

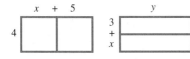

47–50. Insert parentheses on the left side of the equality in order to make each statement true. For example, $3b - 2b = 3b - 6b$ if $3(b - 2b) = 3b - 6b$.

47. $3x + 2 = 3x + 6$

48. $4b - 2b = -4b$

49. $6a + 6a + 1 = 12a + 6$

50. $ax + 1 + bx + 1 = (a + b)(x + 1)$

51. **(a)** Rewrite the number 12 as a product with two factors and as a product with three factors.

(b) Rewrite 12 as a sum of two terms and as a sum of three terms.

(c) Rewrite 12 as a product of two factors such that one factor is the sum of two terms.

52. Repeat all parts of Exercise 51 using $30x^2$.

53–56. Fill in the box with a number to make each statement true.

53. $5(3 + \square) = 45$

54. $7(4 - \square) = -28$

55. $(x + \square)\,3 = 3x + 6$

56. $9(x + 2) - \square\,(x + 1) = 5x + 14$

57. Write a brief paragraph explaining the difference between *factors* and *terms*. Give some examples.

58. Write a brief description explaining what a polynomial is. Give examples of different kinds. Include what is meant by *like terms*.

59. A student in your class asks why the distributive property is useful. Write a note back to the student discussing the merits of the distributive property.

60–63. In each case, make a table of x-y pairs for the given polynomial relationship, and then plot both equations on the same coordinate system. (Use a graphing calculator to draw a more complete graph.)

60. **(a)** $y = 2x + 3$ **(b)** $y = -2x + 3$

61. **(a)** $y = \left(\frac{1}{2}\right)x - 1$ **(b)** $y = \left(-\frac{1}{2}\right)x - 1$

62. **(a)** $y = 3x^2 + 2$ **(b)** $y = -3x^2 - 2$

63. **(a)** $y = x^2 + x$ **(b)** $y = -x^2 + x$

64. **(a)** Describe the similarities and differences between each pair of graphs in Exercises 60–63.

(b) Describe the behavior of the value of y as the values of x increase.

65 and 66. In each case, make a table of values for the two expressions given. What do you notice about the values for each expression?

65. $x - y$ and $-(y - x)$

66. $(x + y)^2$ and $x^2 + y^2$

III. Applications

67–73. Write an algebraic statement to describe each situation. Combine like terms where possible.

67. The sum of two numbers. The second number is four more than six times the first number n.

68. The perimeter of a rectangle. The length is 4 feet longer than the width w.

69. The amount of profit made on a sale of x widgets. Each widget costs \$4 to make and each is sold for \$6.

70. The amount of interest earned in one year on two savings accounts: There are x dollars in the first account with a 7.5% annual interest rate and $(1{,}000 - x)$ dollars in the second account with a 9% annual interest rate.

71. The cost of renting a car per day, when the rate is \$10 a day plus 24¢ per mile for the first 50 miles and 15¢ per mile for each mile over 50 miles.

72. The amount of money collected for ticket sales to a concert: There were x reserved seats at \$4 per seat and $(4{,}000 - x)$ unreserved seats at \$3 per seat.

73. Two trains leave a station at the same time, one traveling east and the other traveling west. What is the distance between the two trains after t hours if the rate of the first train is 60 mph and the rate of the second train is 50 mph?

74–79. In these problems, change each of the following English phrases to a mathematical phrase involving *two* variables. For example, suppose we want to know the amount of interest earned in one year on two savings accounts if there are x dollars in the first account with a 5% annual interest rate and y dollars in the second account with a 6% annual interest rate. The solution would be $0.05x + 0.06y$.

74. The sum of two numbers (call them x and y).

75. The sum of three times one number and four times another number.

76. The perimeter of a rectangle of length l and width w.

77. The total cost of meat for a picnic if you bought x pounds of hot dogs at \$1.60 per pound and y pounds of hamburger at \$1.85 per pound.

78. The total amount of money in cents if you have x quarters and y dimes.

79. The total amount of money collected from ticket sales to a concert if there were x reserved seats at \$4 per seat and y unreserved seats at \$3 per seat.

80–83. For each statement, write two expressions, one a product and one a sum of terms.

EXAMPLE: The total amount of money in x dimes and y quarters.

SOLUTION: Money in cents:

$$\underset{\text{sum}}{10x + 25y} = \underset{\text{product}}{5(2x + 5y)}$$

80. The total amount of money in cents in x dimes and y nickels.

81. The total amount of money collected by selling x tickets on Monday, y tickets on Tuesday, and z tickets on Wednesday if each ticket sold for \$2.50.

82. The area of a rectangle whose width is $x + 2$ and whose length is $\frac{x}{2}$.

83. The total number of pages typed in two hours by Mary and Sam together, if Mary types 20 pages per hour and Sam types x pages per hour.

84. In the rectangle $ABCD$, find the area of the shaded part if the areas of the smaller rectangles are $5x$, x^2, and $2x$.

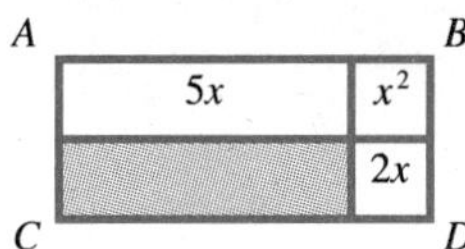

85. Find an algebraic expression for the area of the shaded part in each figure. The circle and square in each case have the same center.

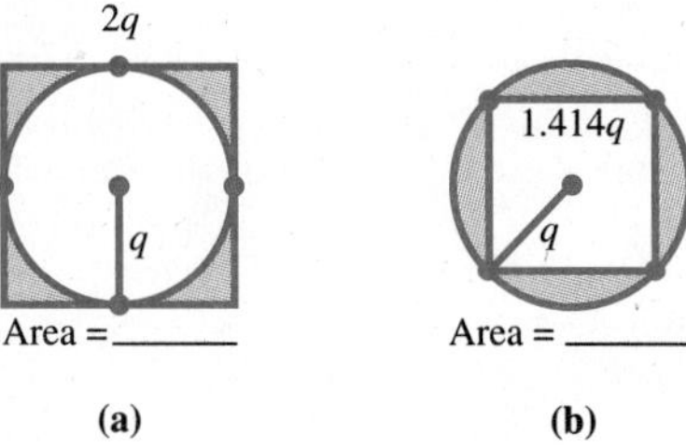

(c) Which of the shaded areas is the larger? Let $\pi \approx 3.142$.

86. **(a)** If a sphere and cylinder have the same radius, which one has the larger surface area? Why?

(b) Which one has the larger volume? Why?

(c) What happens if the radii are equal to the height?

87. A ball (sphere) is dropped into a cylinder. Water is then poured into the cylinder until it is full.

(a) How much water is needed if the radius of the ball equals the radius of the cylinder (a tight squeeze)?

(b) Calculate the volume of water in part (a) if the radius is equal to 4 cm and the height is 8 cm.

88. (a) If a cone and cylinder have the same radius, and the height of the cylinder equals the slant height of the cone, which has the larger surface area? By how much?

(b) Repeat part (a) using volumes. (See problem 71, Section 2.1.)

IV. Extensions

89. Box puzzle: Consider a 3 × 3 square.

(a) Pick four numbers, say 2, 8, 6, and 3, and enter them on the border as follows:

	6	3	Sum
2			
8			
Sum			

(b) Enter the product of the border numbers into the top four squares:

	6	3	→ Sum □
2	12	6	
8	48	24	
↓ Sum □			

(c) Add across the rows and columns.

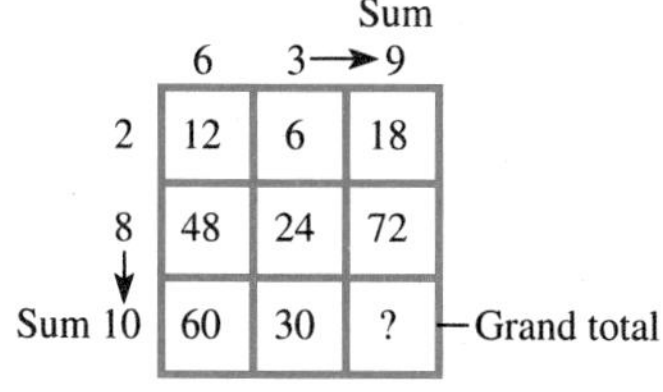

(d) What is the relationship between the number in the lower right-hand square and the border numbers?

(e) Try this puzzle again by picking another set of four numbers for the border. Does the same relationship hold for the lower right-hand square and the border numbers?

(f) Show why this relationship holds by using p, q, r, and s to represent the border numbers.

(g) Now work backward. Find the border numbers, and fill in the squares so that the sum in the lower right-hand square is 2001.

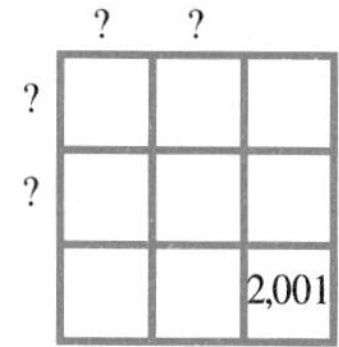

90. Use algebra to show why the following math puzzles (tricks) work.

(a) Think of a number. Double it. Add 6. Divide by 2. Subtract the number you thought of. Give the result.

(b) Think of a number. Add 4. Multiply by 2. Subtract 6. Divide by 2. Subtract the number you thought of. Give the result.

(c) Finish the steps to the following problem so that it will give the same answer no matter what number we start with. Think of a number. Add 2. Triple the result.

91. You are the winner of a contest, and you have a choice of the following two prizes: $5,000 cash or money received each day for a year (365 days) as follows: $1 the first day, $2 the second day, $3 the third day . . . $10 the tenth day . . . $365 the three hundred and sixty-fifth day. Which prize would you choose? Why?

2.4 MULTIPLICATION AND DIVISION OF POLYNOMIALS

Multiplication of Polynomials

Earlier in this chapter we multiplied monomials by monomials and polynomials by a monomial. For example,

$$(3x^2y)(2x^3y^2) = 6x^5y^3$$
$$(3x^2y)(2xy + 3y + 1) = (3x^2y)(2xy) + (3x^2y)(3y) + (3x^2y)(1)$$
$$= 6x^3y^2 + 9x^2y^2 + 3x^2y$$

In general, the distributive property can be used repeatedly to multiply polynomials with several terms. It is also possible to picture multiplication of binomials of the first degree with a rectangle model.

EXAMPLES 1–6

Multiply the following polynomials

1. $(x + 1)(x + 2)$

2. $(2x - 1)(x - 3)$

3. $(2x + 1)(x + 3)$

4. $(x + 1)(x - 2)$

5. $(x - 1)(x^2 + x + 1)$

6. $(x^2 - x - 1)(x^2 + x + 1)$

Solutions

1. If we represent x^2 as an "x-square" like this,

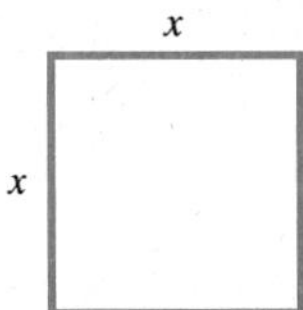

Figure 2.16

then we can represent $(x + 1)(x + 2)$ as the following rectangle.

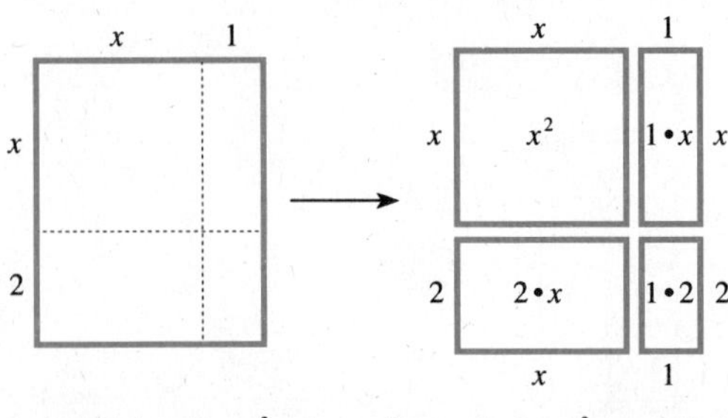

Figure 2.17

We can depict the total area of this rectangular region with four subrectangles and add the areas of the four rectangles. Notice how the distributive property comes into play here.

On the other hand, we could just apply the distributive law in a symbolic way.

$(x + 1)(x + 2) = x(x + 2) + 1(x + 2)$ **Right distributive property with $(x + 2)$ as a factor. $(a + b)c = ac + bc$.**

$= x^2 + 2x + x + 2$ **Left distributive property, with x as a factor. $a(b + c) = ab + ac$.**

$= x^2 + 3x + 2$ **Combine like terms.**

2. In this case, we need to subtract the area of the shaded regions to obtain the area of the unshaded rectangles, which have dimension $(2x - 1)$ by $(x - 3)$.

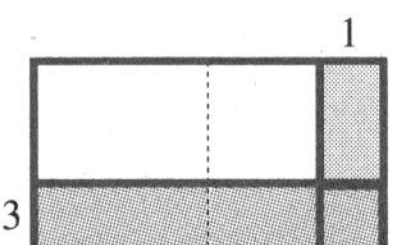

Figure 2.18

Area of two "x-squares" minus the shaded parts.

Figure 2.19

$= 2x^2 - 3 \cdot x - 3 \cdot x - 1 \cdot x + 3$. (Notice that when we subtract the second area of $3x$ and the area of $1x$, they overlap in a rectangle of area 3. Thus, *we must add this piece back on,* since we would have subtracted it off twice.)

Using the distributive property,

$$\begin{aligned}(2x - 1)(x - 3) &= 2x(x - 3) - 1(x - 3)\\ &= 2x^2 - 6x - x + 3\\ &= 2x^2 - 7x + 3\end{aligned}$$

3. $$\begin{aligned}(2x + 1)(x + 3) &= 2x(x + 3) + 1(x + 3)\\ &= 2x^2 + 6x + x + 3\\ &= 2x^2 + 7x + 3\end{aligned}$$

Can you make a rectangle model for this multiplication?

4. $$\begin{aligned}(x + 1)(x - 2) &= x(x - 2) + 1(x - 2)\\ &= x^2 - 2x + x - 2\\ &= x^2 - x - 2\end{aligned}$$

Can you make a rectangle model for this multiplication?

5. $$\begin{aligned}(x - 1)(x^2 + x + 1) &= (x - 1) \cdot x^2 + (x - 1) \cdot x + (x - 1) \cdot 1\\ &= x \cdot x^2 - 1 \cdot x^2 + x \cdot x - 1 \cdot x + x \cdot 1 - 1 \cdot 1\\ &= x^3 - x^2 + x^2 - x + x - 1 = x^3 - 1\end{aligned}$$ **Combine like terms.**

6. $(x^2 - x - 1)(x^2 + x + 1)$

$= (x^2 - x - 1) \cdot x^2 + (x^2 - x - 1) \cdot x + (x^2 - x - 1) \cdot 1$ — $(x^2 - x - 1)$ **distributes over three terms.**

$= x^4 - x^3 - x^2 + x^3 - x^2 - x + x^2 - x - 1$

$= x^4 - x^2 - 2x - 1$ — **Combine like terms.**

There is a pattern when we multiply two polynomials. Notice that each term in the first polynomial is multiplied by every term in the second polynomial. In Example 1,

$$(x + 1)(x + 2) = \underbrace{x \cdot x} + \underbrace{x \cdot 2} + \underbrace{1 \cdot x} + \underbrace{1 \cdot 2}$$

Similarly, in Example 5,

$$(x - 1)(x^2 + x + 1) = \underbrace{x \cdot x^2} + \underbrace{x \cdot x} + \underbrace{x \cdot 1} - \underbrace{1 \cdot x^2} - \underbrace{1 \cdot x} - \underbrace{1 \cdot 1}$$

We could represent the action of the product with a diagram like this:

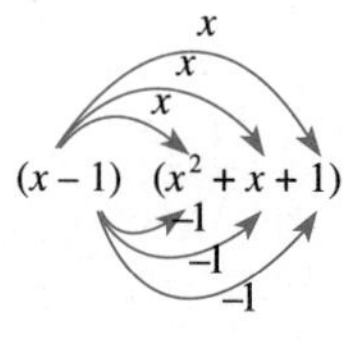

Each term of the first polynomial is multiplied by each term of the second polynomial.

EXAMPLES 7–10

Multiply these polynomials.

7. $(x + 4)(x + 4)$

8. $(x - 4)(x - 4)$

9. $(x - 7)(x + 7)$

10. $(2x - 3y)(2x + 3y)$

Solutions

7. $(x + 4)(x + 4) = x^2 + 4x + 4x + 16$
$= x^2 + 8x + 16$ — **Notice this is $(x + 4)^2$.**

8. $(x - 4)(x - 4) = x^2 - 4x - 4x + 16$
$= x^2 - 8x + 16$ — **$(-4) \cdot (-4) = 16$. Compare this with Example 7. Notice this is $(x - 4)^2$.**

9. $(x - 7)(x + 7) = x^2 - 7x + 7x - 49 = x^2 - 49$

10. $(2x + 3y)(2x - 3y) = 4x^2 - 6xy + 6xy - 9y^2$
$= 4x^2 - 9y^2$

In Examples 9 and 10, we multiply the difference of two numbers by the sum of those numbers.

Examples 7 and 8 are special cases in which a binomial is squared. Examples 9 and 10 are also special cases. The product is the difference of two squares.

Special Polynomial Products

Squaring a binomial:

$$(a + b)^2 = a^2 + 2ab + b^2$$

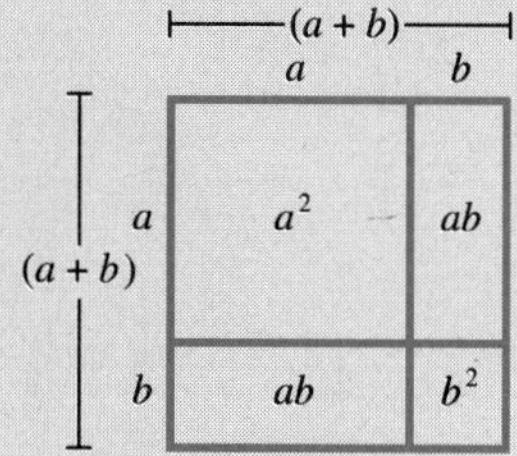

$a^2 + ab + ab + b^2$

Figure 2.20

$$(a - b)^2 = a^2 - 2ab + b^2$$

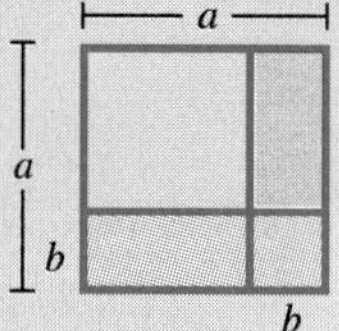

$a^2 - ab - ab + b^2$
Here we must add back the b^2 that we subtracted off twice.

Figure 2.21

Multiplying the sum and difference of two numbers:

$$(a + b)(a - b) = a^2 - b^2$$

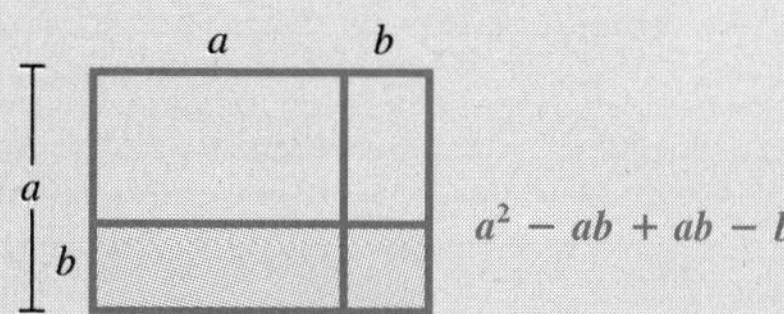

$a^2 - ab + ab - b^2$

Figure 2.22

EXAMPLES 11–15 Multiply these polynomials.

11. $(2x - 3)^2$ **12.** $(5x + 2y)^2$ **13.** $(2x - 7)(2x + 7)$
14. $(ay - c)(ay + c)$ **15.** $(2x - 3)^3$

Solutions

11. $(2x - 3)^2 = (2x)^2 - 2(2x)(3) + 9$ — **Squaring a binomial.**
$= 4x^2 - 12x + 9$ — **Second term is negative.**

12. $(5x + 2y)^2 = (5x)^2 + 2 \cdot 5x \cdot 2y + (2y)^2$
$= 25x^2 + 20xy + 4y^2$

13. $(2x - 7)(2x + 7) = 4x^2 - 14x + 14x - 49$ — **Sum and difference of two numbers.**
$= 4x^2 - 49$

14. $(ay - c)(ay + c) = (ay)^2 - c^2 = a^2y^2 - c^2$ — **Sum and difference of two numbers.**

15. $(2x - 3)^3 = (2x - 3)^2 \cdot (2x - 3)$ — **First do two factors.**
$= (4x^2 - 12x + 9)(2x - 3)$ — **The square of a binomial.**
$= (4x^2 - 12x + 9) \cdot 2x + (4x^2 - 12x + 9) \cdot (-3)$
$= 8x^3 - 24x^2 + 18x - 12x^2 + 36x - 27$
$= 8x^3 - 36x^2 + 54x - 27$ — **Combine like terms.** ■

Polynomial Representations of Quantitative Relationships

Examples 16 through 18 illustrate applications of polynomial expressions.

EXAMPLE 16 Suppose your current salary is x dollars per year. In the next two years, you expect raises of 8% and 5%.

(a) Write an algebraic statement representing your annual salary at the end of the second year.

(b) Would there be any difference in your annual salary at the end of the second year if your 5% raise came before your 8% raise?

Solutions

To analyze a problem, it is often beneficial to *try a numerical example first.* Suppose your salary is now \$10,000. Your salary at the end of the first year will equal

$$10{,}000 + 0.08(10{,}000) = (1 + .08) \cdot 10{,}000 = (1.08)10{,}000 = \$10{,}800$$

Your salary at the end of the second year will equal

$$10{,}800 + 0.05(10{,}800) = (1 + .05) \cdot 10{,}800 = \$11{,}340$$

(a) In algebraic language, we can say that your salary at the end of the second year equals

$$\underbrace{(x + 0.08x)}_{\substack{\text{salary at}\\ \text{end}\\ \text{of first year}}} + \underbrace{0.05(x + 0.08x)}_{\substack{5\%\text{ increase of}\\ \text{first-year salary}}} = x + 0.08x + 0.05x + 0.004x$$
$$= 1.134x$$

where x is your initial salary.

We can shorten the calculation somewhat if we realize that an 8% increase is equivalent to multiplying by 1.08. Thus another way to think of the salary at the end of the second year is

$$\underbrace{1.05(1.08)x}_{5\%\text{ increase of salary after an }8\%\text{ increase the first year}} = 1.134x$$

(b) If we try the example of \$10,000 again, we see that your salary at the end of the first year equals $10{,}000 + 0.05(10{,}000) = \$10{,}500$. Your salary at the end of the second year equals $10{,}500 + 0.08(10{,}500) = \$11{,}340$. Algebraically,

$$(x + 0.05x) + 0.08(x + 0.05x) = x + 0.05x + 0.08x + 0.004x$$
$$= 1.134x$$

Or, in a more compact form,

$$(1.08)(1.05x) = 1.134x$$

Thus the order of the raises makes *no* difference in your salary at the *end of the second year,* but the 8% raise first gives a larger salary at the end of the first year. Thus you will make \$300 more *over* the two-year period if your initial salary is \$10,000. ■

EXAMPLE 17

Find an algebraic statement to describe the pattern. Is your statement always true?

(a) $5 + 7 + 9 = 3 \cdot 7$

$8 + 10 + 12 = 3 \cdot 10$

$4.27 + 6.27 + 8.27 = 3 \cdot (6.27)$

(b) $\left(\frac{1}{4}\right)^2 + \left(\frac{3}{4}\right) = \left(\frac{1}{4}\right) + \left(\frac{3}{4}\right)^2$

$\left(\frac{2}{5}\right)^2 + \left(\frac{3}{5}\right) = \left(\frac{2}{5}\right) + \left(\frac{3}{5}\right)^2$

$(0.23)^2 + 0.77 = 0.23 + (0.77)^2$

(c) $2^2 + 6(1^2) = 5 \cdot 2 \cdot 1$

$6^2 + 6(2^2) = 5 \cdot 6 \cdot 2$

$10^2 + 6(5^2) = 5 \cdot 10 \cdot 5$

Solutions

(a) If we let n be the middle number, then we have $(n - 2) + (n) + (n + 2) = 3n$. By noticing that $(n - 2) + n + (n + 2) = 3n - 2 + 2 = 3n$, we see that this statement is always true.

(b) Let the numbers in the parentheses on the left side be x. In general, $x^2 + y \neq x + y^2$. However, we observe that the second number y is very special; it is $(1 - x)$. That is, if $x = \frac{1}{4}$, then $1 - x = \frac{3}{4}$. If $x = \frac{2}{5}$, then $1 - \left(\frac{2}{5}\right) = \frac{3}{5}$, and so on. Thus $x^2 + (1 - x) = x + (1 - x)^2$. Does the left side equal the right side?

By rewriting the right side,

$$\begin{aligned} x + (1 - x)^2 &= x + (1 - 2x + x^2) \qquad (1 - x)^2 = 1 - 2x + x^2 \\ &= x^2 + 1 - x \end{aligned}$$

we obtain the left side, which means the statement is always true.

(c) $a^2 + 6 \cdot b^2 = 5 \cdot ab$. By letting $a = 1$ and $b = 2$, we see that $1 + 6 \cdot 2^2 \neq 5 \cdot 1 \cdot 2$. Therefore the statement is not always true. ■

EXAMPLE 18

Suppose you receive $1,000 a year from a trust fund for each of the four years you are in college obtaining your B.S. degree. You decide to invest it at 10% annual interest. If no money is withdrawn from the account, how much will you have at the end of four years? Assume you receive the money at the beginning of each year.

Solution

Recall from Section 2.1 that the formula for the total amount of money A with principal P and interest I compounded over n years annually is $A = P(1 + I)^n$. Each year we add another $1,000 to the current amount and then multiply by 1.10 to compute the new total amount.

At the end of	*the amount of money plus interest*	*= Total*
One year	$1{,}000(1.10)$	= $1,100
Two years	$[1{,}000(1.10) + 1{,}000](1.10)$ $= 1{,}000(1.10)^2 + 1{,}000(1.10)$	= $2,310
Three years	$[1{,}000(1.10)^2 + 1{,}000(1.10) + 1{,}000](1.10)$ $= 1{,}000(1.10)^3 + 1{,}000(1.10)^2 + 1{,}000(1.10)$	= $3,641
Four years	$[1{,}000(1.10)^3 + 1{,}000(1.10)^2 + 1{,}000(1.10) + 1{,}000]1.10$ $= 1{,}000(1.10)^4 + 1{,}000(1.10)^3 + 1{,}000(1.10)^2 + 1{,}000(1.10)$	= $5,105.10

Table 2.15

At the end of four years, you saved $5,105.10. ■

In general, if we replace 1.10 with x in Example 18, then the amount of money at the end of four years is equal to

$$1{,}000x^4 + 1{,}000x^3 + 1{,}000x^2 + 1{,}000x$$

We recognize this expression as a polynomial of degree 4, and by evaluating this expression for $x = 1.10$, we obtain the answer to the question in Example 18.

Division of a Polynomial by a Monomial

In the first two sections of this chapter, we divided monomials by monomials. For example,

$$3x^2 \div x = \frac{3x^2}{x} = 3x$$

and

$$6x^5y^2 \div 2x^3y^3 = \frac{6x^5y^2}{2x^3y^3} = \frac{3x^2}{y}$$

We can also divide any polynomial by a monomial. For example,

$$(4x^2 + 6x) \div 2x = \frac{4x^2 + 6x}{2x} = \frac{2x(x + 3)}{2x} = x + 3$$

When we divide a polynomial by a monomial, we use the *fundamental property of fractions*

$$\frac{ax}{bx} = \frac{a}{b}, \text{ where } x \neq 0 \qquad b \neq 0$$

One reason for using equivalent fractions is that $\frac{(1.6x^7y)}{0.32x^7} = 5y$, and $5y$ is easier to use than $\frac{(1.6x^7y)}{0.32x^7}$.

> **To Divide by a Monomial:**
>
> **1.** Combine like terms in the numerator as much as possible.
>
> **2.** Factor out the largest common factor from all terms in the numerator.
>
> **3.** Apply the fundamental property of fractions.

EXAMPLES 19–23

Express each of the following fractions in equivalent simplest form (assume $x \neq 0$, $y \neq 0$).

19. $\frac{6x^2y - 3xy}{3xy}$ **20.** $\frac{3x^3 - x(x - 1)}{x^2}$ **21.** $\frac{x^{10} - x^{60}}{x^{10}}$

22. $\frac{4x^2 + 3}{4x^2}$ **23.** $\frac{[4x(x - 3) - x(x^3 + x^2)]}{x}$

Solutions

19. $\frac{6x^2y - 3xy}{3xy} = \frac{3xy(2x - 1)}{3xy}$ Factor the numerator.

$= 2x - 1$ Fundamental property of fractions.

20. $\frac{3x^3 - x(x - 1)}{x^2} = \frac{3x^3 - x^2 + x}{x^2}$ Multiply and combine like terms in the numerator.

$$= \frac{x(3x^2 - x + 1)}{x \cdot x}$$ **Factor the numerator and denominator.**

$$= \frac{3x^2 - x + 1}{x}$$ **Fundamental property of fractions.**

21. $\dfrac{x^{10} - x^{60}}{x^{10}} = \dfrac{x^{10}(1 - x^{50})}{x^{10}}$

$$= 1 - x^{50}$$

22. $\dfrac{4x^2 + 3}{4x^2}$ **The numerator cannot be factored. $4x^2$ is not a factor of the numerator.**

CAUTION There is a great temptation to remove two $4x^2$ in Example 22. Such a move is incorrect! Consider

$$\frac{15 - 6}{6}.$$

One of a host of incorrect computations that could be made is

$$\frac{15 - \not{6}}{\not{6}} = 15.$$

This is false, because in fact,

$$\frac{15 - 6}{6} = \frac{9}{6} = \frac{3}{2}.$$

Even though the "same number" occurs in the numerator and the denominator of a fraction, this is not a license to remove the number. The number must *be a factor of* the entire numerator and the entire denominator. Just as 6 is not a factor of the numerator in

$$\frac{15 - 6}{6},$$

$4x^2$ is not a factor of the numerator in

$$\frac{(4x^2 + 3)}{4x^2}.$$ ■

23. $\dfrac{[4x(x - 3) - x(x^3 + x^2)]}{x}$

$$= \frac{4x^2 - 12x - x^4 - x^3}{x}$$ **Multiply in the numerator.**

$$= \frac{x(4x - 12 - x^3 - x^2)}{x}$$ ***x* is the largest common factor in the numerator.**

$$= 4x - 12 - x^3 - x^2$$ **Fundamental property of fractions.**

$$= -x^3 - x^2 + 4x - 12$$ **Terms are rearranged by descending degree.** ■

2.4 EXERCISES

I. Procedures

1-30. Multiply the following polynomials. Sketch a rectangle model for the multiplication of those binomials that are indicated with an (R).

1. $(2x^2y)(-3xy^2)$
2. $(-2x^3y)(-3xy)^3$
3. $2xy(-3x^2 - y + 1)$
4. $-2xy(3x^2 + y - 1)$
5. $(x + 1)(x - 1)$ (R)
6. $(3 + m)(3 - m)$ (R)
7. $(5x - 2)(5x + 2)$ (R)
8. $(1.2n - 1)(1.2n + 1)$
9. $(x + 1)^2$ (R)
10. $(3 - m)^2$ (R)
11. $(5x + 2)^2$
12. $(1.2n - 1)^2$
13. $(x + 2)(x + 3)$ (R)
14. $(x + 2)(x - 3)$ (R)
15. $(3x + 1)(x + 2)$ (R)
16. $(3x - 1)(x - 2)$ (R)
17. $(2x + 3)(3x + 2)$ (R)
18. $(2x - 3)(3x + 2)$ (R)
19. $(10x - 3)(4x - 2)$
20. $(x + 2)^3$
21. $(x + 2)(2 - x)$ (R)
22. $(3x - 4)(3 - x)$
23. $(6 - 5x)(1 - 2x)$
24. $(ax + b)^2$
25. $(ax - b)(ax + b)$
26. $(ax + b)(cx + d)$
27. $(x - 1)(x^2 + x - 1)$
28. $(x - 1)(x + 1)(x + 2)$
29. $(x^2 + 1)(x^3 - x + 1)$
30. $(2x^2 - 3x + 1)(x^2 - x + 3)$

31–46. Perform the indicated operations. Write an equivalent expression in its simplest form.

31. $\dfrac{6x^3y^5}{x^4y}$
32. $\dfrac{(-2x^{-1}y)^3}{-3(xy^2)^2}$
33. $\dfrac{3a^2 + 6ab}{3a}$
34. $\dfrac{2x^2 - x(x - 1)}{-x}$
35. $\dfrac{3x^2y - 2xy^2}{2xy}$
36. $\dfrac{16x^7y^{12} - 10x^5y^8}{2x^2y^9}$
37. $\dfrac{9x^4 - x^2(x - 1)}{x^2}$
38. $\dfrac{a^3b^2c - ab(ac^3 - bc)}{a^3b^2c}$
39. $\dfrac{2ab(-1 + a^2b)}{-ab}$
40. $\dfrac{3a^2 - 2}{a^2}$
41. $\dfrac{4x + y}{4xy}$
42. $\dfrac{x^5(x^6 - x^{18})}{x^{11}}$
43. $\dfrac{-x + x^2(3x - 1) - (x^3 - x)}{x^4}$
44. $\dfrac{y(x + 1)^2 - y(x - 1)(x + 1)}{2xy}$
45. $\dfrac{(n - 1)^2 - 2(n - 1) + 3(n - 1)}{-n}$
46. $\dfrac{x^n(x^2 - x)}{x^{n+1}}$

II. Concepts

47-50. Illustrate the following multiplications using areas of rectangles. Example: $(x + 1)(x + 3)$

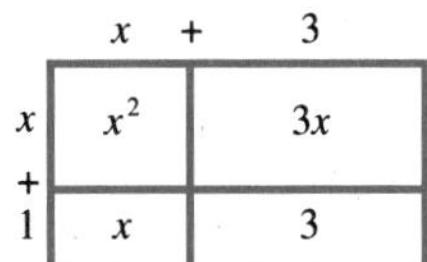

Area = $(x + 1)(x + 3)$
$= x^2 + x + 3x + 3$
$= x^2 + 4x + 3$

47. $(x + 2)(x + 3)$
48. $(2x + 2)(x + 1)$
49. $(x + 3)^2$
50. $(x + 3)(x - 3)$

51. The rectangle $ABCD$ below is divided into two smaller rectangles. The area of each rectangle is given. The length AT of one of the rectangles is also given.

(a) Find the dimensions of the rectangle $ABCD$.

(b) What is the perimeter of the rectangle $ABCD$?

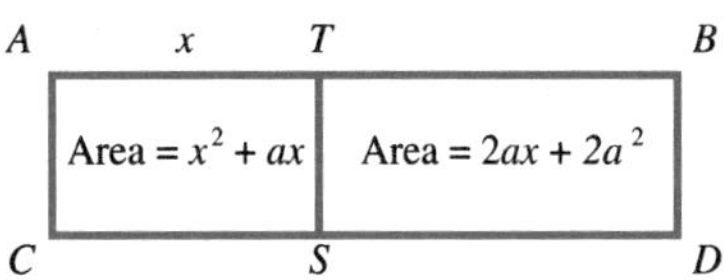

52. Expand each of the following expressions.

(a) $(x + 1)^1$ **(b)** $(x + 1)^2$ **(c)** $(x + 1)^3$
(d) $(x + 1)^4$ **(e)** $(x + 1)^5$ **(f)** $(x + 1)^6$
(g) Can you guess the solution for $(x + 1)^{10}$? For $(x + 1)^n$?

53. Write a description of how to multiply two polynomials. Include some examples. Include sketches of several rectangle models for multiplying two binomials of the first degree.

54–59. What two polynomials could have been multiplied together to get each of the following answers?

54. $5x^2$

55. $x^2 + 5x + 6$

56. $x^2 - 8x + 16$

57. $x^2 - 25$

58. $4x^2 + 20x + 25$

59. $x^3 + 1$

60. How many terms are there in the product $(a + b)(c + d + e)(f + g + h + i)(j + k + m + n + p)$? Explain your reasoning.

III. Applications

61–73. Solve each of the following exercises.

61. The formulas for finding the surface area and volume of several solids are given below.

	Solid	*Surface area*	*Volume*
(a)	Sphere	$4\pi r^2$	$\left(\frac{4}{3}\right)\pi r^3$
(b)	Rectangular prism	$2wh + 2lh + 2lw$	lwh
(c)	Cube	$6e^2$	e^3
(d)	Cylinder	$2\pi r^2 + 2\pi rh$	$\pi r^2 h$
(e)	Cone	$\pi rs + \pi r^2$	$\left(\frac{1}{3}\right)\pi r^2 h$

For each solid, find the ratio of surface area to volume.

62. **(a)** Given the two circles with the same center, find an algebraic expression for the area of the shaded part.

(b) Use the expression to find the area of the shaded part if $x = 10.125$ cm.

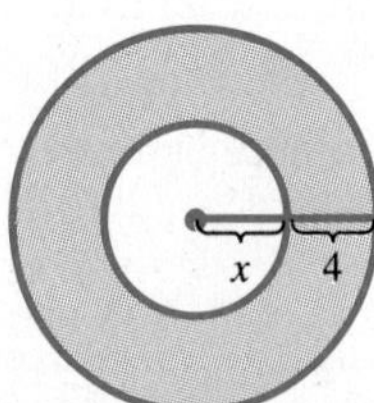

63. **(a)** Show that the area of the shaded part is equal to $\pi(A - B) \cdot (A + B)$.

(b) Find the area of the shaded part when $A = 15$ cm and $B = 12.5$ cm. Use $\pi \approx 3.14$.

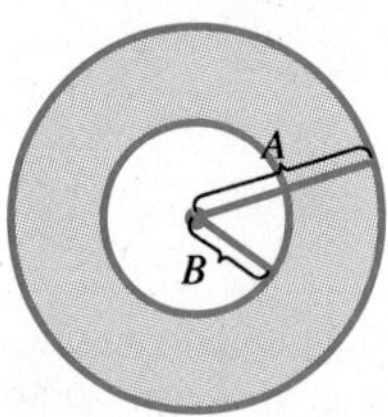

64. The amount of cement needed to make a concrete pipe that is L feet long, has an inside radius B, and has an outside radius A is $\pi LA^2 - \pi LB^2$.

(a) Factor the expression. *Hint:* See Exercise 63.

(b) Find the amount of cement needed to make 1,000 meters of pipe (1,000 meters long) with an inside diameter of 65 cm and an outside diameter of 69.5 cm. Use $\pi \approx 3.14$.

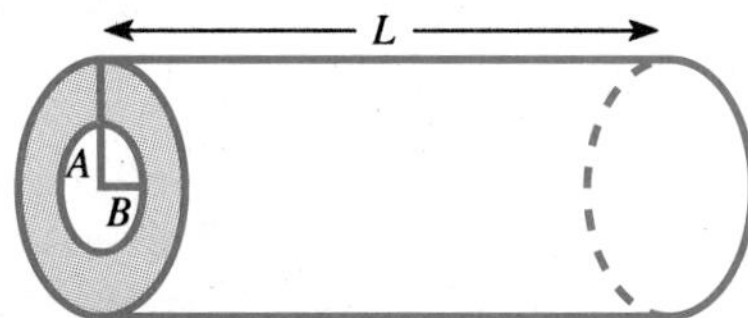

65. **(a)** Find the shaded area.

(b) Solve part (a) another way to check your answer.

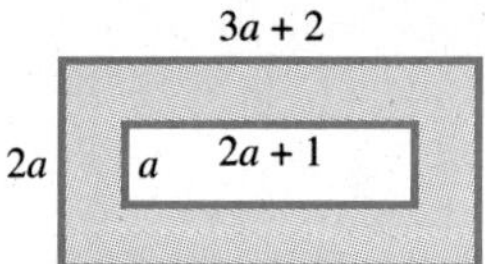

66. One of the formulas approved by the Internal Revenue Service for depreciating business property is

$$v = \frac{C(N - n)}{N}$$

where v is the value at the end of n years, C is the original cost, and N is the total number of years for which it is being depreciated. If an office machine, originally costing $55,000 ($C$) is to be depreciated over a period of 20 years (N), what will the value (v) be at the end of

(a) 5 years?

(b) 10 years?

67. In Exercise 66, what is the value (v) at the end of 20 years? Does this make sense? Why?

68. In constructing a highway, engineers use the following formula to figure out how much expansion due to temperature change to allow for:

$$I = kl(T - t)$$

where I is the expansion at the temperature T in degrees Fahrenheit (°F), t is the temperature at which the highway was built (°F), and l is the length of highway in miles. The constant $k = 0.00012$ per °F is used for most two-lane highways.

(a) How much expansion would there be for a one-mile stretch of highway at 95°F if it was originally built at 65°F?

(b) For the same highway, what is the expansion at 115°F?

69. For Exercise 68, calculate the expansion at 0°F. Does this make sense? Why?

70. Suppose on your twenty-first birthday you inherit \$2,000 a year for life. You invest this money at $11\frac{1}{2}\%$ interest compounded annually. How much money will you have after one year? Two years? Three years? Four years? Five years? x years?

71. In Example 18, how much money would be in the account at the end of four years if the annual interest rate were $14\frac{1}{2}\%$?

72. In Example 18, how much money would be in the account if your estate paid \$2,000 at the start of each year? Assume the annual interest rate is 10%.

73. While you were going to college, suppose you worked each of the preceding four summers. At the start of each year, you put some savings into an account paying 13% annual interest: At the start of the first year, you put in \$500; the second year, \$750; the third year, \$1,100; and the fourth year, \$1,700. If no money were withdrawn, how much money would be in the account at the end of the fourth year?

IV. Extensions

74. Square the following numbers ending in 5.

$5^2 \quad 15^2 \quad 25^2 \quad 35^2$

Give a general rule for squaring a number ending in 5. *Hint:* $25^2 = (20 + 5)^2$. If $n5$ is a two-digit number, then $(n5)^2 = (10n + 5)^2$.

75. Calculate the following products of two-digit integers.

$3 \cdot 7 \quad 24 \cdot 26 \quad 12 \cdot 18 \quad 61 \cdot 69$

Give a general rule for calculating such products.

76. In Chapter 1 we developed the formula $2 + 3 + \cdots + n - 2 = \frac{[n(n-3)]}{2}$ for finding the number of diagonals of a polygon. Use this to find the formula for the sum of the following positive integers:

(a) $1 + 2 + 3 + \cdots + n$

(b) $2 + 4 + 6 + \cdots + 2n$

77. Use the results in Exercise 76 to compute the following sums.

(a) $1 + 2 + \cdots + 100$

(b) $2 + 4 + 6 + \cdots + 1{,}000$

(c) $5 + 10 + \cdots + 1{,}000$

(d) $4 + 7 + 10 + \cdots + 2{,}002$

78–81. **(a)** Write an algebraic statement to describe the pattern. **(b)** Is your statement true?

78. $1^2 + 2^2 = 3^2 - 2^2$

$2^2 + 3^2 = 7^2 - 6^2$

$3^2 + 4^2 = 13^2 - 12^2$

$4^2 + 5^2 = 21^2 - 20^2$

79. $2^2 + 6(1^2) = (5)(2)(1)$

$6^2 + 6(2^2) = (5)(6)(2)$

$15^2 + 6(5^2) = 5(15)(5)$

$4.8^2 + 6(1.6^2) = 5(4.8)(1.6)$

$\left(\frac{2}{3}\right)^2 + 6\left(\frac{1}{3}\right)^2 = 5\left(\frac{2}{3}\right)\left(\frac{1}{3}\right)$

80. $3^2 - 1^2 = 2^3$

$6^2 - 3^2 = 3^3$

$10^2 - 6^2 = 4^3$

81. $3^2 + 4^2 = 5^2$

$10^2 + 11^2 + 12^2 = 13^2 + 14^2$

$36^2 + 37^2 + 38^2 + 39^2 + 40^2 = 41^2 + 42^2 + 43^2$

82. Compute the product $(1 + x)(1 + x^2)(1 + x^4)(1 + x^8) \cdots (1 + x^{1,024})$.

83. Consider any four consecutive integers. Find the product of the smallest and largest integers. Then find the product of the two middle integers. Make a conjecture about the relationship between these two products and justify your conjecture using algebraic expressions.

84. Which is larger, the square of the average of two consecutive integers or the average of the squares of two consecutive integers? Explain your answer using algebraic expressions.

CHAPTER 2 SUMMARY

The introductory example on rumor spread demonstrated the need for exponents as an efficient way to represent repeated multiplication. Scientific notation is an efficient way to express very large or very small numbers and to estimate calculations that are required in many applications. Many of the applications in this chapter, such as interest, light-filtering through a medium, bacteria growth, car rental plans, volume, and area, will also occur in later chapters. Algebraic expressions are general expressions, such as polynomials, which represent real numbers. Therefore, all of the properties of real numbers hold in algebra. Relationships between variables can be described by functions and their graphs.

Important Words and Phrases

integral exponents
base
power
scientific notation
polynomial
term
coefficient
constant
monomial
binomial
trinomial
linear function
like terms
factor
products
sums
algebraic expression
quadratic function
exponential function
algebra as a language
degree of a polynomial
equivalent fractions

Important Properties and Procedures

- Properties for exponents
- Equivalent algebraic expressions
- Factoring polynomials
- Multiplying polynomials
- Division by a monomial
- Adding and subtracting polynomials
- Distributive property
- Fundamental property of fractions
- Interest formula
- Translating problem situations into algebraic expressions

REVIEW EXERCISES

I. Procedures

1–15. Find an equivalent expression for each of the following by using the properties of exponents. Express all answers with positive exponents.

1. $\dfrac{3^3 \cdot 3^2}{3}$

2. $\dfrac{3^3 + 3^2}{3}$

3. $\dfrac{(3^3 \cdot 3)^2}{3}$

4. $\dfrac{3^3 \cdot 3^{-2}}{3}$

5. $\dfrac{3^3 + 3^{-2}}{3}$

6. $\dfrac{3^{-2} - 2^{-2}}{3^{-1} - 2^{-1}}$

7. $\dfrac{3^3 \cdot 3^{-2} \cdot 3^4}{3^{-1} \cdot 3^0 \cdot 3^{-3}}$

8. $\dfrac{-2^4 \cdot 8}{(-2^3 \cdot 4)^2}$

9. $\dfrac{(2^2x^3)(4x)}{24x^5}$

10. $\dfrac{(4x^5y)(-2x^3y^3)}{(8x^4y^2)^2}$

11. $\dfrac{15x^{-1}y^2}{5x^4y^{-3}}$

12. $\dfrac{(-2x^{-1})^3}{6x^2}$

13. $\dfrac{-3(xy)^2}{(-xy)^{-2}}$

14. $\dfrac{x^{20} \cdot x^0}{x^{30}}$

15. $\dfrac{a^n(a^3 - a^2)}{a^{n+2}}$

16–28. **(a)** Perform the indicated operations. Write an equivalent expression in its simplest form. **(b)** Indicate where in the process the distributive property is used.

16. $a(a - 1) - 2a^2$

17. $3 - 2(x - 3) - x(6x - 3)$

18. $x^2y - x(3 - y) + y(x^2 - x)$

19. $m - [3m^2n - (m - 1)] - [-n(2m^2 + n)]$

20. $(x + y)^2 - (x + y)(x - y)$

21. $(2x - 3)(3x + 1) - (6x - 1)(x + 2)$

22. $(a + 3)^2 - (a + 3)(a - 3)$

23. $(x + 5)(x - 2) + (2x - 1)(x + 1)$

24. $\dfrac{3a^2b - 2ab^2}{2ab}$

25. $\dfrac{15x^{10}y^5 - 3x^6y^6}{6(x^2y)^3}$

26. $\dfrac{3x^4 - x^2(x - 2)}{x^2}$

27. $\dfrac{x - x(3x + 2) - (x^2 - x)}{x^2}$

28. $\dfrac{x(x - y) - y(y - x)}{xy}$

29 and 30. **(a)** Estimate each of the following by using scientific notation. **(b)** Compute each by using a calculator. Round off your answers to four decimal places.

29. $\dfrac{(3{,}689{,}000)(89{,}675{,}345)}{(111{,}896)(152{,}223)}$

30. $\dfrac{(1{,}005.756)(35.0001)}{(9{,}000)(0.00045)}$

II. Concepts

31–34. Insert parentheses on the left side of the equality sign so that each statement is true.

31. $7 + 5a - a = 11a$

32. $7 + 5a - a = 7$

33. $7 + 5a - a = 0$

34. $7 + 5a - a = 7 + 4a$

35–42. True or false. If false, give an example that shows it is false.

35. $(a - 1) = -(1 - a)$

36. $\dfrac{x^6}{x^3} = x^2, x \neq 0$

37. $(a + b)^2 = a^2 + b^2$

38. $\dfrac{a^2 + b}{a^2} = b, a \neq 0$

39. If x is any real number, then $-x^2$ is always a negative number.

40. $x^n + (-x)^n = 0$ for any odd integer n and $x \neq 0$.

41. If $x \neq 0$, then x^{-2} is always negative.

42. If $x \neq 0$, then $(x^3)^2 = x^5$.

43. In which of the following expressions is either x or $x + 3$ a factor of the entire expression?

(a) $3xy$ **(b)** $3x + y$ **(c)** x **(d)** $3x(x + y)$

(e) $3 + x(x + 1)$ **(f)** $(x + 1)(x + 3)$

(g) $6 - 2(x + 3)$ **(h)** $2x^2(x + 3)(x + 2)$

(i) $x^2 + 3$

44. Write a note to a friend explaining why the special formula for squaring a binomial works.

45. If a polynomial is the difference of two perfect square numbers (algebraic expressions), what are its factors and why?

46–48. Make a table of values for each pair of relationships, and graph a set of x-y pairs for each of the two on the same coordinate system. Use *two* different colored pens/pencils or a graphing calculator.

46. **(a)** $y = 2x$ **(b)** $y = -2x$

47. **(a)** $y = -x^2 - 1$ **(b)** $y = x^2 + 1$

48. **(a)** $y = 2^x$ **(b)** $y = 2^{-x}$

III. Applications

49–62. Solve the following problems.

49. **(a)** If you invest \$10,000 at 10.8% interest per year for five years, how much money will there be if no money is withdrawn from the account?

(b) At this rate, will your investment double by the end of eight years?

50. **(a)** How much acid is in 100 milliliters (ml) of a 30 percent acid solution? How much water is in the 100 ml solution?

(b) If 50 ml of water is added to the solution in part (a), what percentage of the new solution is acid?

(c) If you mix 10 gallons of 30% acid solution with 20 gallons of 15% acid solution, what is the percentage of acid in the mixture?

(d) How much acid is in x ml of a 30% acid solution?

51. Two trains leave a station at the same time, one traveling 65 mph due west and the other traveling 55 mph due east.

(a) After t hours, how far apart will the trains be?

(b) In how many hours will the trains be 600 miles apart?

52. A certain amoeba reproduces by dividing in half every four hours. If we start with one amoeba, how many amoebas will there be at the end of

(a) 4 hours? **(b)** 8 hours? **(c)** 12 hours?
(d) 24 hours? **(e)** $4n$ hours?

53. The intensity of light filtering through a special sheet of colored plastic 1 mm thick is reduced by one-sixth. By how much is the intensity of light reduced if

(a) two sheets of plastic are put together? (*Hint:* How much light comes through each 1 mm of plastic?)

(b) ten sheets of plastic are put together?

(c) n sheets of plastic are put together?

54. Suppose a copy machine enlarges a sheet of paper about 1.10 times as large as the original. If you made copies of copies, and an original chart was 10 cm by 16 cm, what would be the dimensions of the second, third, and eighth copies?

55. Newton's Law of Gravitation between two bodies is given by

$$A = \frac{GM_1M_2}{d^2}$$

where G is the gravitational constant, 0.000000000033, M_1 the mass in pounds of one body, M_2 the mass in pounds of the other body, and d the distance between the two bodies. What is the attraction between two ships weighing 30,000 tons and 50,000 tons, when the distance between their centers of gravity is 100 feet?

56. (a) Write an algebraic expression representing your salary two years from now if in the next two years you expect raises of 6% and 7%, respectively.

(b) Is there any difference in your salary if you first receive a 7% raise and then a 6% raise?

57. Write an algebraic statement for the amount of money in a bag in which there are x dimes and y quarters.

58. Write an algebraic expression for the amount of interest earned in one year from two savings accounts if there are x dollars in the first account with a 6.5% annual interest rate and y dollars in the second account with a 7% annual interest rate.

59. Rectangle $ABCD$ has been subdivided into four rectangles. The areas of three of the smaller rectangles are given.

D		C
(shaded)	$6x$	
x^2	$3x$	
A		B

(a) Find the area of the fourth rectangle.

(b) Find the dimensions of rectangle $ABCD$.

(c) Find the perimeter of rectangle $ABCD$.

60. The average length of time that a motion-picture image is on the screen is approximately 6.4×10^{-2} seconds. How many images are projected during a two-hour movie?

61. A sheet of paper is folded in half, and then in half again. If this halving procedure continues and the paper is opened up, how many regions would there be after one fold? Two folds? Three folds? Five folds? Ten folds? One hundred folds? Realistically, about how many folds can you make with a sheet of paper? Does the size of the paper affect the number of folds?

62. A telephone cable across the Atlantic Ocean connects Newfoundland with Scotland. To keep the sound loud enough to hear, there are 51 amplifiers spaced evenly along the cable. Each one of the amplifiers increases the signal strength about a million times (10^6) to make up for the fading of the signal along the cable. How much has the signal been strengthened by the

(a) first amplifier? **(b)** sixth amplifier?

(c) fifty-first amplifier?

CUMULATIVE REVIEW: CHAPTERS 1 AND 2

I. Procedures

1–12. Perform the indicated operations and express all fractions in equivalent simplest form.

1. $\frac{3}{5} - 2 - \frac{5}{6}$

2. $8 - 6[-1 - 3(5 - 2) - 5] + 10$

3. $\dfrac{\dfrac{5}{24}}{\dfrac{-6-2}{3}}$

4. $3 - \dfrac{2}{5}\left(\dfrac{10}{3} - \dfrac{15}{4}\right)$

5. Multiply $\frac{6}{5}$ by $\frac{10}{18}$. Subtract $\frac{2}{3}$ from the result.

6. $\dfrac{(2)(8^4)}{4^{10}}$

7. $\dfrac{3^2 - 3^{-1}}{3^{-1}}$

8. $\dfrac{-xy(2x^5y^0)}{x^2}$

9. $(a + 3)^2 - (a + 3)(a - 2)$

10. $x - x(3x^4 - 2) - (x - x^5)$

11. $\dfrac{15m^{20}n^{15} - 10m^2n^5}{5m^2n^5}$

12. Evaluate $2xy^2 - x^2 - y^3$ when $x = 3$ and $y = -3$.

13–15. Write an algebraic statement that describes the following situations.

13.

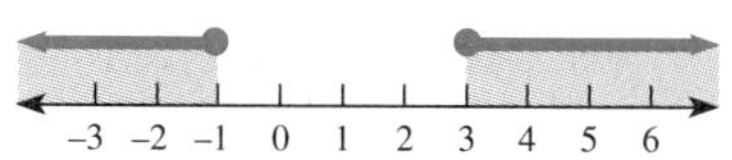

14. x is 3 units from the point 1.

15. x is greater than or equal to 0 and x is less than 10.

16–18. Graph the solution set of each of the following statements on a number line.

16. $x > 4$ or $x < -4$.

17. x is less than 2 and x is greater than 0.

18. x is greater than or equal to 2 units from the point 3.

II. Concepts

19–24. Determine whether each statement is always true, sometimes true, or never true. Explain your answers.

19. $\dfrac{(10.0006 - 7.53924)}{(0.0005)(9.9368)} < 1$

20. The product of two negative integers and one positive integer is always negative.

21. $\dfrac{1}{2}\left(\dfrac{15.9}{7} - \dfrac{8.2}{5}\right) > 0$

22. If $x < 0$, then $x + 5 < 0$

23. If $x \neq 0$, then $\dfrac{2x^4 - x^6}{x^2} = 2x^2 - x^4$.

24. If $x \neq 0$, then $-x^2$ is always negative.

25–28. Multiple choice.

25. $(-3)^{-2} =$

(a) -5 **(b)** 6 **(c)** 9 **(d)** $\frac{1}{9}$ **(e)** $-\frac{1}{9}$

26. $\dfrac{0.00000000105}{1.5 \times 10^{-16}} =$

(a) 7×10^6 **(b)** 7×10^{-6} **(c)** 7×10^{24}
(d) 7×10^7 **(e)** 7×10^{23}

27. If $(2 - a)$ is subtracted from $(a - 2)$, the result is

(a) 0 **(b)** $2a$ **(c)** 4 **(d)** $4 - 2a$
(e) $2a - 4$

28. The time it takes a bicyclist going 7 mph to go d miles is represented by

(a) $7d$ **(b)** $\frac{7}{d}$ **(c)** $\frac{d}{7}$ **(d)** $7 + d$

(e) none of these

29. Write a brief summary explaining why algebra is considered a general language to describe the patterns of arithmetic. Give at least two examples.

30. State the distributive property and give three examples of how it is used in algebra.

III. Applications

31. **(a)** How much salt is in 400 ml of a 30% salt solution?

(b) If 100 ml of water is added to the solution in part (a), what percent of the new solution is salt?

32. A certain strain of amoeba doubles every three hours. If we start with 20 amoebae, how many amoebae will there be at the end of

(a) six hours?

(b) one day?

(c) How long will it take to produce 3,200 amoebae?

(d) Make a graph of this relationship. Describe the pattern.

33. The Sandy Dune Company rents dune buggies for \$35.50 a day plus \$4.25 an hour.

(a) How much will it cost to rent the buggy for 4 hours?

(b) Describe the range of values for the cost if you rent the buggy for two consecutive days and you estimate the number of hours to be between 10 and 15 hours (total for both days).

34. The start-up cost to manufacture figgles is \$750 and it costs \$4.25 to make each figgle. If we plan on selling the figgles for \$8.50, how much profit is made on selling 2,000 figgles?

35. For the years 1982 to 1992 the average January temperature in Deepfrost, Michigan, was −1°C, 0°C, −5°C, −11°C, +10°C, +6°C, −3°C, −20°C, + 15°C, −2°C.

(a) What was the average temperature for January for this period?

(b) What was the median temperature for this period?

(c) What was the range of temperatures for this period?

36. This is your first year at Moo State, where tuition is \$100 per credit. At the end of your first year, tuition goes up by 7%. At the end of the second year, it goes up 11%.

(a) How much does it cost per credit at the start of your third year?

(b) Does it make a difference if the tuition goes up 11% the first year and 7% the second year? Why?

37. The cost and revenue for a certain video company are displayed on the accompanying graph. Using the graph, estimate

(a) the start-up costs for the company.

(b) the break-even point.

(c) the profit on selling 12,000 videos.

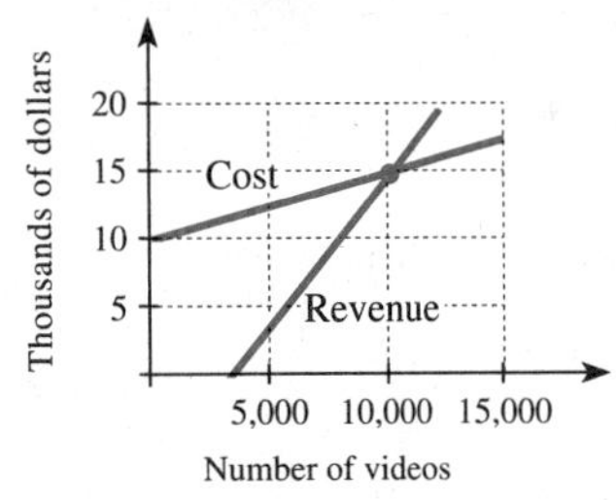

38–41. Solve each problem using three different methods: table, graph, and algebraic expression.

38. If \$5,000 is invested at a simple annual interest rate of $9\frac{1}{2}$%, how much money is in the account at the end of five years? (Assume no money is withdrawn from the account.)

39. Two moped riders, Chris and Pat, leave Lansing an hour apart. Chris leaves at 2:00 P.M. traveling north at a rate of 25 mph. Pat leaves Lansing an hour later traveling north at a rate of 38 mph. How far apart are they after Pat has been riding

(a) five hours **(b)** t hours?

40. Bill receives money as a birthday present from his grandfather. The money is to be paid as follows: \$1 the first day, \$2 the second day, \$4 the third, \$8 the fourth, etc.

(a) How much money did Bill receive on the twelfth day?

(b) If that process were to continue, how much money would Bill receive on the thirtieth day?

(c) If that process were to continue, how much money would Bill receive on the Nth day?

41. Suppose Bill's grandfather offers him a choice of two birthday presents: the one described in Exercise 40 or \$100 a day. How much money does the second plan yield for

(a) 12 days? **(b)** 30 days? **(c)** N days?

(d) Which is the better plan? Give reasons for your choice.

CHAPTER

3 Linear Relationships

The ability to organize and analyze data is essential in solving problems. As we have seen in the last two chapters, such data can be summarized in a table, a graph, or an equation (formula). In this chapter and the next chapter we will use these representations to study linear functions.

Source: Donovan Reese/Tony Stone Images

3.1 EXPLORING LINEAR RELATIONSHIPS: TABLES, GRAPHS, EQUATIONS

We now continue the saga of the car rental problem, which was begun in Example 2 from Chapter 1.

EXAMPLE 1

The daily cost to rent a small car from the We Try Harder agency is given in Table 3.1.

Miles in one day	0	100	200	300	400	500
Cost in dollars	23	43	63	83	103	123

Table 3.1

(a) What is your daily cost if you expect to drive the given number of miles?

(i) 400 **(ii)** 250 **(iii)** 320

(b) If you can spend the given amount of money to rent a small car for one day, how many miles could you drive on that day?

(i) \$50 **(ii)** \$500

Solutions

Using a Table

(a) (i) The cost is easy to read from Table 3.1. We locate 400 miles in the top row and then locate the corresponding cost in the bottom row. The cost is \$103.

(ii) To drive 250 miles in one day, we would estimate that the cost is between \$63 and \$83. Since 250 miles is halfway between 200 and 300 miles, it is reasonable to assume that the cost is halfway between \$63 and \$83, or \$73.

(iii) To drive 320 miles, it would not be as easy to estimate the cost from the chart. We know that it would be between \$83 and \$103 and closer to \$83.

Using a Graph A more convenient way of organizing the data in Table 3.1 is by displaying it in a graph as shown in Figure 3.1. The points (*A, B, C, D, E*) represent five entries in the table. For example, point *A* says that it costs \$23 before any miles are accumulated. Point *B* says that it costs \$43 to drive 100 miles. The locations of the points are approximate. They appear to lie on a line that has been sketched.

(a) (i) The cost to drive 400 miles is \$103, which is point *E* on the graph.

(ii) To find the cost of driving 250 miles, we locate 250 on the horizontal axis and then follow a vertical (perpendicular) line upward until we intersect the line. We

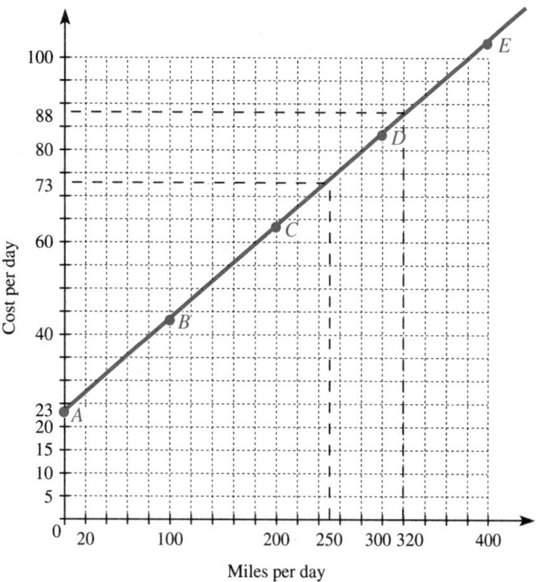

Figure 3.1

then follow a horizontal (perpendicular) line to the left until we intersect the vertical axis. We estimate this point to represent a cost of \$73.

(iii) Similarly, from the graph, we estimate the cost for 320 miles to be \$88.

Use a graphing calculator to graph $Y1 = 23 + 0.2x$. Are you able to see the graph? You probably must change your viewing rectangle to see the graph. If you use the rectangle: $0 \leq x \leq 500$, $0 \leq y \leq 200$, TRACE and perhaps ZOOM, you find that when x is near 400, y is near 103. The numbers on the axes, the labels on the graphs, and the coordinates of points of intersection may not appear on your graphing calculator screen. They are given here to help you visualize the problem and to indicate the size of possible viewing rectangles. We will use $Y1$, $Y2$, etc. to indicate we are using the graphing calculator.

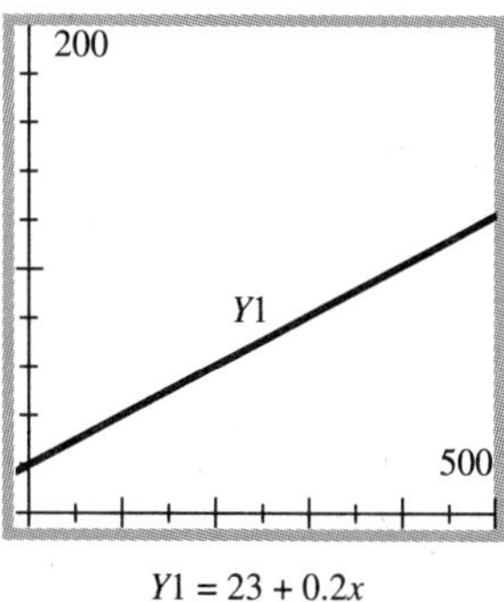

$Y1 = 23 + 0.2x$

Figure 3.2

Using an Equation Another way of analyzing this information is by the following equation:

$$C = 23 + 0.2M$$

where C is the cost of driving M miles in one day. The equation is obtained by observing the pattern in Table 3.1 and noting that it costs \$23 to rent the car, plus \$20 to drive every 100 miles, or 20¢ a mile.

(a) (i) Substituting 400 for M, we have

$$C = 23 + 0.2(400) = 103$$

(ii) Substituting 250 for M, we have

$$C = 23 + 0.2(250) = 73$$

(iii) Substituting 320 for M, we have

$$C = 23 + 0.2(320) = 87$$

Notice that we obtain an accurate cost from the equation, whereas we obtain only an approximate cost from the graph or table. The cost, C, is a *linear function* of the number of miles, M, and could be written as Cost (miles) = Cost (M) = $23 + 0.2M$.

(b) (i) From the table, we estimate for \$50 we can drive between 100 and 200 miles. The graph gives an estimate of 150 miles. Substituting 50 for C in this equation gives

$$50 = 23 + 0.2M$$

We must solve the equation for M:

$$50 - 23 = 23 - 23 + 0.2M$$ **We subtract 23 from both sides of the equation.**

$$27 = 0.2M$$

$$\frac{27}{0.2} = \frac{0.2}{0.2}M$$ **We divide both sides of the equation by 0.2.**

$$135 = M$$

Thus 135 miles can be driven for a cost of \$50.

(ii) Use the equation $C = 23 + 0.2M$ and substitute 500 for C.

$$500 = 23 + 0.2M$$

$$477 = 0.2M$$ **Subtract 23 from both sides.**

$$2{,}385 = M$$ **Divide both sides by 0.2.**

We could drive 2,385 miles, a ridiculous answer for one day! Thus we would never spend \$500 to rent such a car for one day. ■

EXAMPLE 2 Jupiter candy bars cost \$20 per box (of 100 bars) plus \$2,300 in fixed costs to make.

(a) What does it cost to make the given number of boxes of Jupiter bars?

(i) 400 **(ii)** 250 **(iii)** 320

(b) If you can spend the given amount of money to buy boxes of Jupiter bars, how many boxes can you buy (at the wholesale cost)?

(i) \$5,000 **(ii)** \$50,000

Solutions

Using a Table Let us first make a table.

Boxes made	0	100	200	300	400	500
Cost in dollars	2,300	4,300	6,300	8,300	10,300	12,300

Table 3.2

We can read the costs from Table 3.2 as we did in Example 1.

Using a Graph By comparing this table to the one in Example 1, we see that if we change the numbers on the vertical scale of the graph from 5 to 500, 10 to 1,000, 40 to 4,000, and so on, then the graph is the same and the graphical solution is the same.

Using an Equation The solutions to the problems found by using an equation are very similar to Example 1. An equation is $C = 2{,}300 + 20B$, where C is the cost in dollars of making B boxes of Jupiter bars. It could also be written Cost (bars) = Cost (B) = 2,300 + 20B.

(a) (i) Substituting 400 for B, we have

$$C = 2{,}300 + 20(400) = 10{,}300.$$

(ii) Substituting 250 for B, we have

$$C = 2{,}300 + 20(250) = 7{,}300.$$

(iii) Substituting 320 for B, we have

$$C = 2{,}300 + 20(320) = 8{,}700.$$

Notice the similarities between Examples 1 and 2.

(b) (i) Substituting 5,000 for C gives

$$5{,}000 = 2{,}300 + 20B$$

$$2{,}700 = 20B \qquad \textbf{Subtract 2,300 from both sides.}$$

$$135 = B \qquad \textbf{Divide both sides by 20.}$$

You can make 135 boxes of Jupiter bars for \$5,000.

(ii) Use the equation $C = 2{,}300 + 20B$ and substitute 50,000 for C. Then

$$\begin{aligned} 50{,}000 &= 2{,}300 + 20B \\ 47{,}700 &= 20B && \text{Subtract 2,300 from both sides.} \\ 2{,}385 &= B && \text{Divide both sides by 20.} \end{aligned}$$

You can make 2,385 boxes of Jupiter bars for \$50,000. In this case, the solution for $C = 50{,}000$ is plausible. ■

Solving Linear Equations

Statements that involve equality are called equations. The equation $50 = 23 + 0.2M$ from the car rental problem is an example of a linear equation in one variable since the exponent of the variable M is 1.

> **Definition**
>
> The process of finding values of the variable that result in a true statement is called **solving the equation.** These values are called **solutions** or **roots** of the equation.

For example, 135 is a solution of the equation $50 = 23 + 0.2M$, since $50 = 23 + 0.2(135)$ is a true statement.

Two equations are **equivalent** when they have identical solutions. The strategy for solving a linear equation is to transform it into an equivalent equation in which the solution is obvious. That is, in solving a linear equation, we want to isolate the terms with the variable to one side and all else on the other side. For example,

$$x + 3 = 7$$

is transformed into the equivalent equation $x = 4$ by adding -3 to both sides of the equation. The solution of both equations is 4. In another example,

$$2\pi r = 10$$

is transformed into the equivalent equation $r = \frac{5}{\pi}$ by dividing both sides by 2π. The solution of both equations is $\frac{5}{\pi}$.

To transform an equation into a simpler but equivalent equation we can apply the following basic properties.

> **Properties of Equality**
>
> **1.** Adding or subtracting the same quantity to both sides of the equation does not change its solution.
>
> **2.** Multiplying or dividing both sides of an equation by the same nonzero quantity does not change its solution.

EXAMPLES 3–5

Solve each of the following linear equations.

3. $-37x = 111$ **4.** $x - 0.05x = 100$ **5.** $\frac{15x}{2} = -45$

Solutions

3. $-37x = 111$

$\frac{-37x}{-37} = \frac{111}{-37}$ **Divide both sides by -37.**

$x = -3$ **The solution is -3.**

4. $x - 0.05x = 100$

$0.95x = 100$ **Combine like terms.**

$\frac{0.95x}{0.95} = \frac{100}{0.95}$ **Divide both sides by 0.95.**

$x \approx 105.26$ **The solution is 105.26.**

5. $\frac{15x}{2} = -45$

$\left(\frac{2}{15}\right)\frac{15x}{2} = -45\left(\frac{2}{15}\right)$ **Multiply both sides by $\frac{2}{15}$ to make the coefficient of x be 1.**

$x = -6$ **The solution is -6.** ■

In solving equations, we can check to see if we have obtained the correct solution. **If we substitute the solution into the original equation, we must obtain a true statement.** In Example 5, we obtained $x = -6$. To check this solution, we substitute -6 into the original equation:

$\frac{15x}{2} = -45$ **Original equation.**

$\frac{15}{2}(-6) \stackrel{?}{=} -45$ **Substitute -6 for x.**

$-45 \stackrel{\checkmark}{=} -45$ **We obtain a true statement, so -6 is the solution. It checks.**

Check Examples 3 and 4.

EXAMPLES 6–7

Solve for r in each of the following linear equations.

6. $c = 2\pi r$ **7.** $d = rt + s$

Solutions

Although these equations involve symbols instead of numbers, the process of solving them is the same. These equations are called **literal equations.**

6. $c = 2\pi r$

$r = \frac{c}{2\pi}$ **Divide by 2π.**

7. $d = rt + s$

$d - s = rt$ **Subtract s.**

$r = \dfrac{d - s}{t}$ **Divide by $t \neq 0$.** ■

EXAMPLE 8

The price of hamburger rose 10% this month. It now costs \$2.19 per pound.

(a) How much did it cost per pound last month?

(b) If the price rose 8% last month, how much did it cost per pound two months ago?

Solutions

(a) If P was the price last month, then

$$\underset{\text{price last month}}{P} + \underset{\text{10\% increase}}{0.10P} = (1 + 0.10)P = 1.1P = \underset{\text{current price}}{2.19}$$

Since $1.1P = 2.19$,

$$P = \frac{2.19}{1.1} \approx 1.99$$

The cost last month was approximately \$1.99 per pound.

(b) If P was the price two months ago, then with two increases, we have

$$\begin{aligned}(1.1)P + (0.08)(1.1)P &= [1 + (0.08)](1.1)P \\ (1.08)(1.1)P &= 2.19 \\ 1.188P &= 2.19 \\ P &\approx \$1.84 \text{ per pound}\end{aligned}$$ ■

In Examples 3–6, the linear equation has the form $ax = b$, hence the solution is $x = \frac{b}{a}$. Let us turn to a problem similar to the second part of Example 1.

EXAMPLE 9

Suppose Acme Car Rental charges \$20 per day and 22¢ per mile. How far can we travel in one day for \$130?

Solution

We have $\$130 - \$20 = \$110$ for mileage. Since $\frac{\$110}{\$0.22} = 500$, we can travel 500 miles.

Algebraically, if M denotes the number of miles driven, then we can describe the situation by using a linear equation:

$$0.22M + 20 = 130$$

$$0.22M = 110$$ **Subtract 20 from both sides of the equation.**

$$M = \frac{110}{0.22} \quad \text{Divide both sides by 0.22.}$$

$$M = 500 \text{ miles}$$

This example illustrates another form of a linear equation: $ax + b = c$.

EXAMPLES 10–11 Solve for x.

10. $3x + 2 = 119$

11. $0.13x - 0.02 = 27$

Solutions

Using an Equation

10.

$$3x + 2 = 119$$

$$3x = 117 \quad \text{Subtract 2.}$$

$$x = \frac{117}{3} \quad \text{Divide by 3.}$$

$$x = 39$$

Check: $3(39) + 2 \stackrel{?}{=} 119$ Substitute 39 for x.

$$117 + 2 \stackrel{?}{=} 119$$

$$119 \stackrel{\checkmark}{=} 119 \quad \text{Check.}$$

11.

$$0.13x - 0.02 = 27$$

$$0.13x = 27.02$$

$$x = \frac{27.02}{0.13} \approx 207.85$$

Check: $0.13(207.85) - 0.02 \stackrel{?}{=} 27$ Substitute 207.85 for x.

$$27.0205 - 0.02 \stackrel{?}{=} 27$$

$$27.0005 \stackrel{\checkmark}{\approx} 27 \quad \text{Check.}$$

Using a Graph Even though a graphing calculator is not needed to solve these equations, using one to solve Example 10 can help you understand how to solve more complex problems.

We solve $3x + 2 = 119$ in two ways.

10. $3x + 2 = 119$ is equivalent to $3x - 117 = 0$. Graph $Y1 = 3x - 117$. After determining a suitable viewing rectangle, find the x-coordinate of the point where the graph crosses the x-axis; that is, where $y = 0$.

Graph $Y1 = 3x + 2$ and $Y2 = 119$. Find the x-coordinate of the point of intersection of the two lines. The key step is to find an appropriate viewing rectangle. The Y-range must certainly include 119, but the X-range is determined by experimentation. Notice that the answer may not be exact, due to the viewing rectangle chosen.

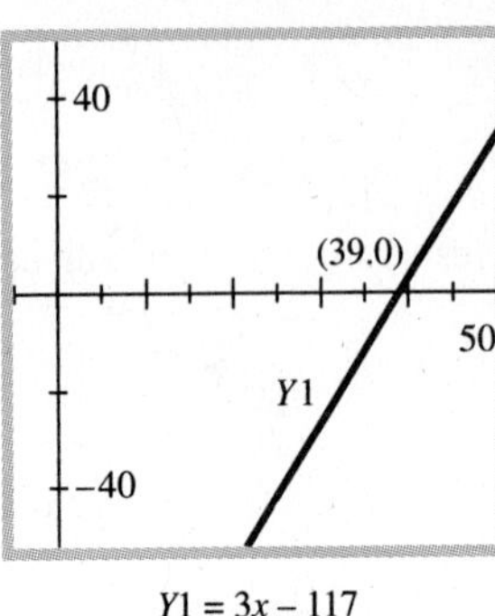

$Y1 = 3x - 117$

Figure 3.3

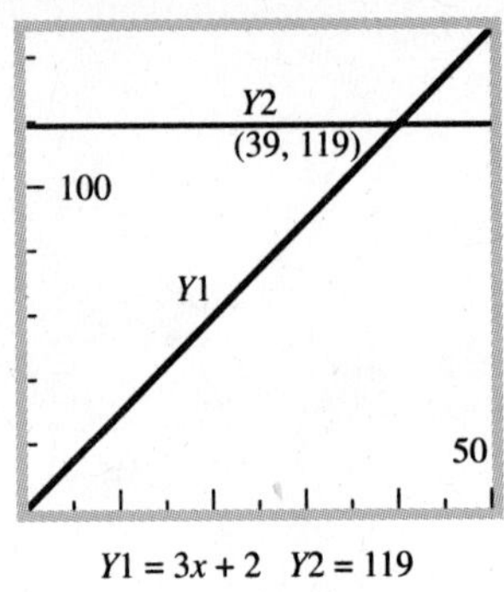

$Y1 = 3x + 2 \quad Y2 = 119$

Figure 3.4

You might also solve Example 11 using a graphing calculator.

EXAMPLE 12

(a) Is 317 in the sequence 2, 5, 8, 11, . . . ?

(b) Is 964 in the sequence 2, 5, 8, 11, . . . ?

Solutions

Integers in this sequence can be represented by the expression $3n + 2$. Thus we must determine whether there is an integer that is a solution to the following equation:

(a) $3n + 2 = 317$ Solve for n.

$$3n = 315$$

$$n = 105$$

Thus $317 = 3(105) + 2$, and 317 is in the sequence.

(b) For 964, we have

$$3n + 2 = 964 \qquad \text{Solve for } n.$$

$$3n = 962$$

$$n \approx 320.6667$$

Since n is not an integer, 964 is not in the sequence.

EXAMPLE 13

The formula $F = \frac{9}{5}C + 32$ gives the relationship between degrees Fahrenheit F and degrees centigrade C (Celsius). Solve for C.

Solution

$$F = \frac{9}{5}C + 32$$

$$F - 32 = \frac{9}{5}C$$

$$\frac{5}{9}(F - 32) = C \quad \text{or} \quad \frac{5}{9}F - \frac{160}{9} = C$$

EXAMPLE 14 Determine whether 315 can be written as the sum of

(a) three, **(b)** four, or **(c)** five

consecutive integers.

Solutions

Let x denote the first integer. Then the next four consecutive integers are $x + 1, x + 2, x + 3, x + 4$.

(a) To see if 315 can be written as the sum of three consecutive integers, we must solve the following equation for x:

$$\begin{aligned} 315 &= x + (x + 1) + (x + 2) \\ 315 &= 3x + 3 \\ 312 &= 3x \\ 104 &= x \end{aligned}$$

Thus $315 = 104 + 105 + 106$

(b) For four consecutive integers, we have

$$\begin{aligned} 315 &= x + (x + 1) + (x + 2) + (x + 3) \\ 315 &= 4x + 6 \\ 309 &= 4x \\ 77.25 &= x \end{aligned}$$

Since x is not an integer, 315 cannot be written as the sum of four consecutive integers.

(c) For five consecutive integers, we have

$$\begin{aligned} 315 &= x + (x + 1) + (x + 2) + (x + 3) + (x + 4) \\ 315 &= 5x + 10 \\ 305 &= 5x \\ 61 &= x \end{aligned}$$

So $315 = 61 + 62 + 63 + 64 + 65$.

EXAMPLE 15 The international airmail rate in 1991 was 45¢ for each half-ounce up to 2 ounces and 42¢ for each additional half-ounce.

(a) Sketch a graph of the cost in terms of the weight.

(b) What is the maximum weight of an item that could be sent for each of the following amounts?

(i) \$1.35 **(ii)** \$3.90 **(iii)** \$4.00

Solutions

Using a Graph

(a)

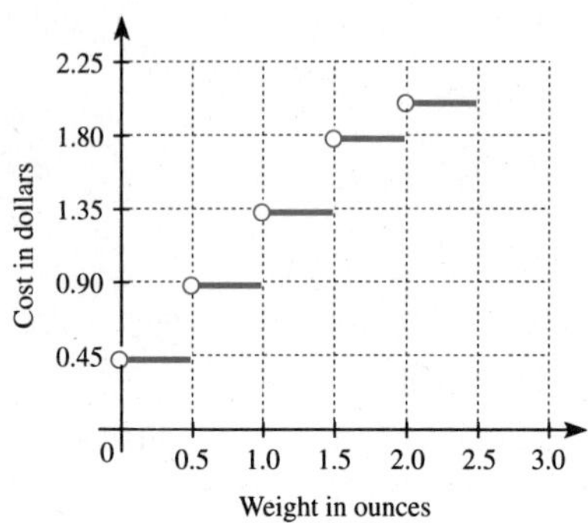

Figure 3.5

Observe that this graph is different from other graphs we have seen so far. It is called a **step** graph. The relationship is called a **step function.**

Using an Equation

(b) (i) An item could weigh 3 half-ounces, or 1.5 ounces, since 3(45¢) = \$1.35.

(ii) Let x = the number of half-ounces the item weighs. Then

$$\underbrace{4(0.45)}_{\text{cost for first 4 half-ounces}} + \underbrace{(x - 4)(0.42)}_{\text{cost for additional half-ounces over 4}} = 3.90$$

$$1.80 + 0.42x - 1.68 = 3.90$$

$$0.12 + 0.42x = 3.90$$

$$0.42x = 3.78$$

$$x = 9$$

Such an item could weigh 9 half-ounces, or 4.5 ounces.

(iii) The additional 10¢ could not buy an additional half-ounce; you would simply pay \$3.90 for 4.5 ounces as in part (ii).

Check the answers to parts (b)(i), (b)(ii), and (b)(iii) on the graph. ■

EXAMPLE 16

The Ajax Manufacturing Co. makes widgets. The start-up costs are \$10,000 (for equipment, etc.). It then costs \$4 to make each widget. If each widget sells for \$5.95, how many widgets must be sold to make a \$30,000 profit?

Solution

Recall that revenue − cost = profit. If x is the number of widgets sold, then revenue = $5.95x$, and cost = $4x + 10,000$. Substitution into the first equation yields

$$5.95x - (4x + 10,000) = 30,000$$

$$5.95x - 4x - 10{,}000 = 30{,}000$$
$$1.95x - 10{,}000 = 30{,}000$$
$$1.95x = 40{,}000$$
$$x \approx 20{,}512.82 \quad \text{or} \quad 20{,}513 \text{ widgets.}$$

Check: If $x = 20{,}513$ widgets, the profit is \$30,000.35. If $x = 20{,}512$ widgets, the profit is \$29,998.40. So, it is impossible to make exactly \$30,000 profit. But, if 20,513 widgets are sold, then we are guaranteed *at least* a \$30,000 profit. ■

3.1 EXERCISES

I. Procedures

1–24. Solve for x in each of the following equations. Check your solution.

1. $47x = 611$
2. $-47x = 611$
3. $-1.23x = 36.9$
4. $-0.123x = 3.69$
5. $2x - 3 = 7$
6. $2x - 3 = -7$
7. $-2x + 3 = 7$
8. $-2x - 3 = 7$
9. $1.03x = 50{,}000$
10. $1.0609x = 50{,}000$
11. $x + 2x + 3x = 36$
12. $x + 2x + 3x = 37$
13. $x + 0.17x = 300$
14. $x - 2x + 4x = 81$
15. $x - 0.23x = 17{,}935$
16. $3x - 1.7346x = 53.257$
17. $12.3x = 0.7263$
18. $1.23x = 72.63$
19. $1.23x + 0.024 = 7.263$
20. $0.123x - 0.024 = 7.263$
21. $0.1x + 0.2x + 0.3x = 36$
22. $0.11x + 0.12x + 0.13x = 36$
23. $x + 2x + 3x + 4x + 5x + 6x + 7x + 8x + 9x + 10x = 110{,}000$
24. $(1.1)\cdot(2.2)\cdot(3.3)\cdot(4.4)\cdot(5.5)x = (2.2)\cdot(3.3)\cdot(4.4)\cdot(5.5)\cdot(6.6)$

25–36. Solve each equation for the indicated variable. Check your solution.

25. $F = ma$ for m
26. $A = lw$ for l
27. $C = \pi d$ for d
28. $I = prt$ for r
29. $I = prt$ for t
30. $I = (1.03)P$ for P
31. $A = bh$ for h
32. $S = 16gt^2$ for g
33. $F = ab + bc$ for b
34. $S = 4Lw + 8w$ for w
35. $L = a + nd$ for a
36. $L = a + nd$ for d

II. Concepts

37 and 38. Solve for x. Check your solution.

37. **(a)** $ax + b = c$ **(b)** $ax + by = c$
38. **(a)** $ax + bx = c$ **(b)** $ay + by = cx$

39 and 40. What is the relationship between the solutions of the two equations?

39. $ax = b$ and $bx = a$
40. $ax + b = c$ and $ax + c = b$
41. Give three examples of linear equations that are equivalent to $2x + 4x - 3 = 15$.
42. Give an example of a problem situation that could be described by a graph similar to the one in Example 15.
43. Determine a suitable viewing rectangle that could be used to solve Example 1 (b)(ii) with a graphing calculator.
44. Solve each of the following equations using a graphing calculator. Explain your method in each case.
(a) $1.0609x = 50{,}000$ **(b)** $x + 0.17x = 300$
(c) $1.23x - 0.024 = 7.263$

III. Applications

45. The car rental charges for Ajax Car Rental are \$30 a day plus 15¢ for each mile over 100. (You get 100 "free" miles.)

(a) Make a table that gives the total cost C of renting a car that you drive for M miles, where $M = 50, 75, 100, 125, 150, 175, 200, 225$, and 250 miles.

(b) Graph these points.

(c) Complete this equation giving C in terms of M:

$$C = \begin{cases} ? & \text{if } M \leq 100 \\ ? & \text{if } M > 100 \end{cases}$$

(d) What is the cost of driving 320 miles?

(e) If the charge were \$75, how many miles would have been driven?

46. It costs \$750 to prepare the type for a pamphlet and 25¢ a copy for production.

(a) Make a table that gives the total cost C of preparing x pamphlets, where $x = 10, 20, 30, 40$, and 50.

(b) Graph these points.

(c) Write an equation that gives C in terms of x.

(d) Graph this equation.

(e) What is the total cost of producing 500 pamphlets?

The pamphlets sell for 75¢ each.

(f) Write an equation that gives the total income I from the sale of x pamphlets.

(g) Graph this equation on the graph in part (d).

(h) How many pamphlets must be sold to break even?

(i) How many pamphlets must be sold to make a \$1,000 profit?

(j) How many pamphlets must be sold to make a 10% (of costs) profit?

47–67. Solve each of the following problems. While you do not really need to set up an algebraic equation to solve all of these problems, write an equation to represent each problem. You might use a graphing calculator on some problems.

47. How far could you travel on \$900 if the mileage charge at Acme Car Rental is

(a) 30¢ per mile (b) 31¢ per mile?

48. Widgets cost \$3.50 each to make, and each sells for \$5.95. How many widgets do you have to sell to make a profit of \$3,000?

49. At an average rate of 55 mph, how long would it take you to travel from Boise, Idaho, to Washington, D.C., a distance of 2,343 miles?

50. An automobile dealer makes a profit of approximately \$312 on each car sold. How many cars must be sold in one month to make a profit of \$10,000 for that month?

51. A woman with an hourly wage of \$7.65 working 40 hours a week with no overtime pays a federal income tax of \$754 for the standard deduction with three exemptions. How many days did she work to pay her federal income tax? Assume she works 8 hours a day.

52. The perimeter of *each* of the following geometric figures is 37.2418 cm. Find the length of s in each case.

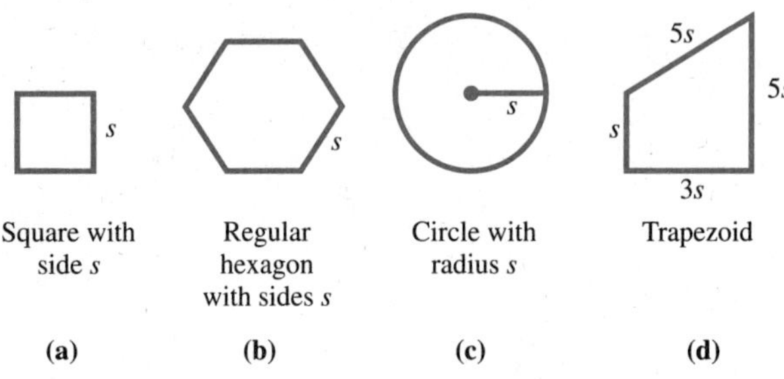

Square with side s	Regular hexagon with sides s	Circle with radius s	Trapezoid
(a)	(b)	(c)	(d)

53. You bought four new tires at a recent sale where the fourth tire was half price. You paid \$168.70 (excluding tax). What was the regular price of one tire?

54. Which of the following integers are in the sequence: 4, 11, 18, 25, . . . ?

(a) 95 (b) 995 (c) 9,995 (d) 99,995

(e) 99,999,995

Hint: Find an expression for the general term of the sequence.

55. Determine whether 1,155 can be written as the sum of

(a) three (b) four (c) five (d) six

(e) seven consecutive integers.

56. What is the heaviest foreign airmail package you could send for \$10? See Example 15.

57. A widget costs \$3.50 to make and sells for \$6. The start-up costs for a widget factory are \$20,000. How many widgets must you sell in order to

(a) break even

(b) make a profit of \$10,000?

58. A real estate salesperson usually works for a small salary plus a percentage of each house she sells. If her salary is \$100 a week plus 3.5% of the price of each house sold, how many dollars' worth of houses must she sell in order to clear \$20,000 in one year? Assume she works 50 weeks per year.

59. Scrooge McTightwad pays his employees \$125 a week and gives a \$5 per week raise every six months. How many years would you have to work for Scrooge to earn \$300 a week?

60. Your first two test scores in Math 101 were 77 and 65. What score do you need on the third test to have an average of

(a) 75? **(b)** 80? **(c)** 85?

Each test is worth 100 points.

61. For a recent tire sale, the advertisement said, "Buy three tires and get the fourth tire for \$1.98." If you paid \$121.92 (excluding tax), what was the regular price of each tire?

62. Currently, the U.S. automakers must achieve a 27-miles-per-gallon (mpg) average for all the cars they sell. Suppose Major Motors sells 350,000 large-size cars that average 15 mpg and 1,500,000 intermediate cars that average 21 mpg. How many small cars averaging 32 mpg must Major Motors sell to achieve the mileage requirement?

63. Encyclopedia sales personnel earn 12% commission for each \$250 encyclopedia set they sell. How many sets must they sell to earn \$300 per week, if they

(a) work by commission only?

(b) receive \$50 per week plus commission?

64. For the past two years, the inflation rate was 6% and 4%. How much would you expect the \$42.95 mathematics textbook you bought for this course to have cost two years ago?

65. The taxi fare in Podunk is \$1.05 for the first $\frac{1}{5}$ mile and 10¢ for each additional $\frac{1}{10}$ mile.

(a) How far can you travel on \$5?

(b) If you give a 10% tip, how far can you travel on \$5?

66. In 1991, Southwestern Bell charged 21¢ for the first minute and 18¢ for each additional minute for calls covering a distance of 23 to 28 miles made between 8:00 A.M. and 5:00 P.M. weekdays.

(a) How long could you talk on a weekday for \$10?

(b) From 5:00 P.M. to 11:00 P.M., there is a 25% discount. How long could you talk for \$10?

(c) On Saturday and part of Sunday, there is a 40% discount. How long could you talk for \$10?

67. For local telephone service, the monthly charge is either \$10.40 for unlimited service or \$5.00 plus 8¢ per call for each call over 25 for measured service. For how many calls per month are the charges the same under the two plans?

IV. Extensions

68–72. Tackle the following more challenging applications.

68. We have seen that the relationship between the Fahrenheit and Celsius temperature scales is given by $F = \left(\frac{9}{5}\right)C + 32$. When the weather forecast is given in Windsor, Ontario, the broadcaster might say the expected high for today is 20° Celsius. In neighboring Detroit, the weather person would predict a high of 68° Fahrenheit for the same day. Is there anywhere in the world where this confusion would not occur, that is, where both forecasters would predict the same temperature in both scales?

69. What (type of) positive integers can be written as the sum of

(a) three **(b)** four **(c)** five

consecutive integers? Why?

70. Find the fraction represented by each of the following repeating decimals.

(a) 0.121212 . . . **(b)** 0.123123123 . . .

(c) 0.12345345345 . . .

(d) 12.12323232323 . . .

(*Hint:* In part [a], let n be the desired fraction. If $n = 0.121212\ldots$, then $100n = 12.12\ldots$. Thus $100n - n = 12.1212\ldots - 0.121212\ldots = 12$. Solve for n.)

71. Suppose you wish to borrow \$2,400 to buy a used car. An (unscrupulous) lender may offer you a one-year "10% loan" computed as follows: You owe $2{,}400 + 0.10(2{,}400) = 2{,}640$. Now $\frac{2{,}640}{12} = 220$, so your monthly payment should be \$220. This process is called add-on interest. However, 10% is not the annual percentage rate (APR). Find the APR. *Hint:* If r is the interest rate, then

$$2{,}640\left(\frac{r}{12}\right) + 2{,}420\left(\frac{r}{12}\right) + \cdots + 220\left(\frac{r}{12}\right) = 240$$

Why?

72. (a) For which whole numbers k is the solution (for x) of $kx + 12 = 3k$ also a whole number?

(b) In what way, if any, could a graphing calculator be helpful in solving this problem?

3.2 PROPORTION AND PERCENT

How often have you experienced, or perhaps heard of, situations such as the ones below?

1. Frozen orange juice is \$2.09 for four cans. How much would ten cans cost?
2. My Taurus gets about 28 miles to a gallon. Can I make it from Los Angeles to San Francisco on one tank of gas (12 gallons)?
3. Nine acres are needed to graze 2 cows. How many acres will I need for the 2,000 head in my herd?

In each of these three situations, *two measures* are being compared, dollars and cans, miles and gallons, acres and cows. We are reminded of the language of ratio from Chapter 1. However, the two measures are being compared *twice* in these statements, and the second time one of the quantities is *missing.* With algebraic notation, these three situations could be translated into

1. \$2.09 for four cans, X dollars for ten cans.
2. 28 miles for 1 gallon, Y miles for 12 gallons.
3. 2 cows on nine acres, 2,000 cows on Z acres.

In each case, it seems to be a reasonable assumption to say that the *ratio* of the two measures in question *stays the same*, even though the quantities change.

1. The ratio $\frac{\$2.09}{4}$ cans is the same as the ratio $\frac{X \text{ dollars}}{10 \text{ cans}}$.
2. The ratio $\frac{28 \text{ miles}}{1 \text{ gallon}}$ is the same as the ratio $\frac{Y \text{ miles}}{12 \text{ gallons}}$.
3. The ratio $\frac{2 \text{ cows}}{9 \text{ acres}}$ is the same as the ratio $\frac{2{,}000 \text{ cows}}{Z \text{ acres}}$.

Indeed, if the pairs of ratios are the same, then we could set up an equation in each case:

1. $\frac{2.09}{4} = \frac{X}{10}$ 2. $\frac{28}{1} = \frac{Y}{12}$ 3. $\frac{2}{9} = \frac{2{,}000}{Z}$

NOTE: Even though we have dropped the measures (dollars, cows, gallons, and so on) when we set up the equations, each answer *does* have a measure associated with it. ■

Equations between two ratios have a wide variety of applications. For this reason, these equations have a special name.

Definition

Two ratios $\frac{a}{b}$ and $\frac{c}{d}$ are **proportional** if

$$\frac{a}{b} = \frac{c}{d}, \text{ where } b \neq 0, d \neq 0.$$

The statement $\frac{a}{b} = \frac{c}{d}$ is called a *proportion.*

We observe that the above proportions can be transformed into linear equations, and we can solve them by using the methods of the previous section.

EXAMPLE 1 Frozen orange juice is $2.09 for four small cans. What can we expect to pay for ten cans?

Solution

X is the amount paid for ten cans.

$$\frac{2.09}{4} = \frac{X}{10} \quad \text{We expect the two ratios to be proportional.}$$

$$\frac{20.9}{4} = X \quad \text{Multiply both sides by 10.}$$

$$\$5.23 \approx X$$

Therefore we would expect to pay $5.23 for ten cans of frozen orange juice if the rate for ten were the same as the rate for four. ■

EXAMPLE 2 My Taurus gets about 28 miles to a gallon. How far can I drive on 12 gallons? Can I make it from Los Angeles to San Francisco (403 miles) without refueling?

Solution

Y is the number of miles for 12 gallons.

$$\frac{28}{1} = \frac{Y}{12} \quad \text{Two equal ratios.}$$

$$28(12) = Y \quad \text{Multiply both sides by 12.}$$

$$Y = 336 \text{ miles}$$

Since it is 403 miles from Los Angeles to San Francisco, the answer is no. ■

EXAMPLE 3 Nine acres are needed to graze 2 cows. How many acres are needed to graze 2,000 cows?

Solution

Z is the number of acres for 2,000 cows.

$$\frac{2}{9} = \frac{2{,}000}{Z} \qquad \text{Two equal ratios. Note that this is not yet a linear equation.}$$

$$2Z = 18{,}000 \qquad \text{Multiply both sides by } 9Z,\ Z \neq 0. \text{ We now have a linear equation.}$$

$$Z = 9{,}000 \qquad \text{Divide both sides by 2.}$$

Therefore it takes 9,000 acres. ■

In setting up a direct proportion, the strategy is to **find two ratios that are equal.** It is possible that more than one choice for selecting equal ratios exists. Consider the following example.

EXAMPLE 4 A certain fertilizer sells for \$5.00 for 3 pounds. To fertilize your lawn, you will need 16 pounds of fertilizer. How much money will it cost you to fertilize your lawn?

Solutions

We look at three equivalent solutions. Let x equal the amount of money for 16 pounds of fertilizer.

(a) We assume the amount of money paid for fertilizer is proportional to the number of pounds purchased. Thus we know the ratio.

$$\frac{\text{amount of money}}{\text{pounds of fertilizer}} = \frac{5.00}{3} \quad \text{is equal to} \quad \frac{\text{amount of money}}{\text{pounds of fertilizer}} = \frac{x}{16}$$

That is,

$$\frac{5.00}{3} = \frac{x}{16} \quad \text{and} \quad x = \$26.67$$

(b) The ratios in part (a) could have been written in inverted form, as

$$\frac{3}{\$5.00} = \frac{16}{x} = \frac{\text{pounds of fertilizer}}{\text{amount of money}}$$

which also yields

$$x = \$26.67$$

(c) As an alternate approach, we may equate the ratio of money to the ratio of pounds:

$$\frac{\text{amount of money for 3 pounds fertilizer}}{\text{amount of money for 16 pounds fertilizer}} = \frac{5.00}{x}$$

That is,

$$\frac{3 \text{ pounds}}{16 \text{ pounds}} = \frac{\$5.00}{\$x}$$

and again,

$$x = \$26.67 \quad ■$$

The Basic Rate Formula (Optional)

The familiar notion that rate × time = distance is an example of the basic rate formula, another method that can be used in many problems involving rates and ratios.

Basic rate formula: Rate × Base = Product

For example,

$$\frac{60 \text{ miles}}{\text{hour}} \times 2 \text{ hours} = 120 \text{ miles}$$

and

$$25\% \times 80 \text{ boys} = 20 \text{ boys}$$

Viewing the basic rate formula indicates the three basic equations:

1. Product = Rate × Base

2. $\text{Rate} = \dfrac{\text{Product}}{\text{Base}}$

3. $\text{Base} = \dfrac{\text{Product}}{\text{Rate}}$

EXAMPLE 5

(Example 1 reconsidered) Small cans of frozen orange juice cost $2.09 for four cans. How much would you expect to pay for ten such cans? We could think of this problem in two ways.

Solution 1

As a rate, four cans for $2.09 is $\dfrac{4 \text{ cans}}{2.09 \text{ dollars}}$, so the base should be expressed in dollars. Let the cost of 10 cans = x dollars. Using the basic rate formula, we can say

$$\text{Rate} \times \text{Base} = \text{Product}$$

$$\frac{4 \text{ cans}}{2.09 \text{ dollars}} \times x \text{ dollars} = 10 \text{ cans}$$

Solving the equation $\frac{4}{2.09}x = 10$ yields $x = \frac{2.09}{4}(10) = 5.23$.
The cost of ten cans is \$5.23.

Solution 2

If we think of the rate as

$$\frac{2.09 \text{ dollars}}{4 \text{ cans}},$$

then the base can be expressed in "cans," and the product is x dollars. Thus, we can say

$$\text{Rate} \times \text{Base} = \text{Product}$$

$$\frac{2.09 \text{ dollars}}{4 \text{ cans}} \times 10 \text{ cans} = x \text{ dollars.}$$

Evaluating the expression $x = \frac{2.09}{4}(10)$, we find $x = 5.23$.
The cost of ten cans is \$5.23. ■

EXAMPLE 6 The ratio of men to women at King High School is 6:5. If there are 1,727 students at the school, how many are men? (*Note:* 6:5 means a ratio of 6 to 5.)

Solution 1

Let the number of women $= x$. We can use the basic rate formula as follows:

$$\text{Rate} \times \text{Base} = \text{Product}$$

$$\frac{6 \text{ men}}{5 \text{ women}} \times x \text{ women} = \frac{6}{5}x \text{ men.}$$

We solve the equation

$$\text{women} + \text{men} = 1{,}727$$

$$x + \frac{6}{5}x = 1{,}727$$

$$\frac{11}{5}x = 1{,}727$$

Thus

$$x = \frac{5}{11}(1{,}727) = 785 \quad \text{and} \quad \frac{6}{5}x = \frac{6}{5}(785) = 942$$

There are 785 women and 942 men at King High School.

Solution 2

If the ratio of men to women is 6:5, then there could be 6 men and 5 women, 18 men and 15 women, etc. The number of men is a multiple of 6 and the number of women is the same multiple of 5. Thus there is a number y such that

$$\begin{aligned}\text{common factor} &= y\\ \text{number of men} &= 6y\\ \text{number of women} &= 5y\end{aligned}$$

Now solve the equation

$$\begin{aligned}6y + 5y &= 1{,}727\\ 11y &= 1{,}727\\ y &= 157 \qquad 6y = 942 \qquad 5y = 785\end{aligned}$$

There are 942 men and 785 women at King High School. ■

NOTE: Another approach is to solve the proportion

$$\frac{6}{5} = \frac{1727 - x}{x}$$ Recall x is the number of women.

We will return to such equations, called rational equations, in Chapter 5. ■

Recall that a percent is a special ratio where the second measure is 100.

EXAMPLE 7 The table gives the approximate popular vote in the presidential elections of 1980 and 1984. Find the entries labeled (a), (b), and (c).

1980	*Vote (in thousands)*	*Percent of vote*	*1984*	*Vote (in thousands)*	*Percent of vote*
Reagan	43,911	(a)	Reagan	54,282	59.17%
Carter	(b)	41.7%	Mondale		
Anderson	5,702	6.7%			
Total	85,099			(c)	

Table 3.3

Solution

Use the basic rate formula as indicated:

$$\text{Rate} \times \text{Base} = \text{Product}$$

The base is the total number of votes.

(a) In this case, the rate (percent), r, is unknown.

$$r \cdot 85{,}099 = 43{,}911 \qquad \text{or} \qquad 85{,}099r = 43{,}911.$$

Thus

$$r = \frac{43{,}911}{85{,}099} \approx 0.515999.$$

Reagan received approximately 51.6% of the popular vote in 1980. We can also think of this as follows: If 43,911 people out of a total 85,099 people voted for Reagan, then $\frac{43{,}911}{85{,}099}$ or 0.516 or 51.6% of the total population voted for Reagan.

(b) The product, P, is unknown. Thus

$$0.417 \cdot 85{,}099 = P, \text{ or } P = 35{,}486.283.$$

Carter received approximately 35,486,000 votes in 1980.

(c) The base, B, is unknown.

$$0.5917B = 54{,}282 \quad \text{or} \quad B = \frac{54{,}282}{0.5917} = 91{,}739.$$

Approximately 91,739,000 votes were cast in the 1984 presidential election. ■

EXAMPLE 8

During a week in 1992, the Nielsen ratings for "Home Improvement" were 14.2|28. The first number represents the percent of 92.1 million homes tuned to this show (see Figure 3.6). The second number represents the percent of homes with TV sets in use at the time "Home Improvement" aired that were tuned to "Home Improvement." In how many homes were the TV sets in use when this episode was aired?

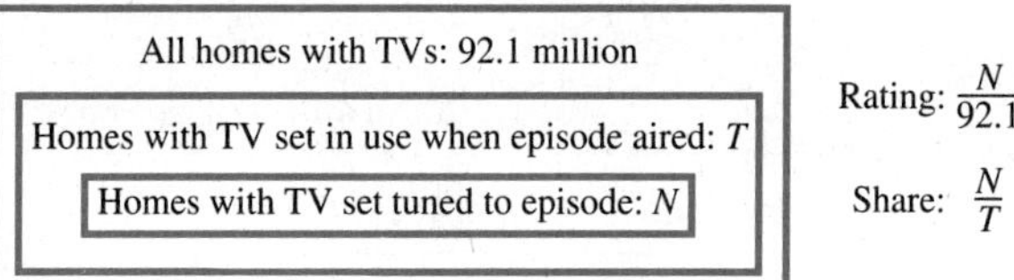

Figure 3.6

Solution

The solution to this problem involves two steps:

Step 1. Find the number, N, of homes tuned to "Home Improvement," and

Step 2. Find the number of homes, T, in which television sets were in use.

Step 1. Using

$$\begin{aligned} \text{Rate} \times \text{Base} &= \text{Product, we can say} \\ 14.2\% \times 92.1 \text{ million homes} &= N \text{ homes.} \\ N = 0.142 \times 92.1 &= 13.0782. \end{aligned}$$

Approximately 13.1 million homes were tuned to the show.

Step 2. Using

$$\text{Rate} \times \text{Base} = \text{Product, we can say}$$
$$28\% \times T \text{ million homes} = 13.1 \text{ million homes.}$$

Solving the equation

$$0.28T = 13.1 \text{ yields } T = \frac{13.1}{0.28} = 46.8$$

In approximately 46.8 million homes, there was at least one TV set in use. ■

EXAMPLE 9

At a game preserve, 200 geese were captured. Red bands were put on the necks of the geese and then the geese were released. Two months later, 350 geese were captured at the preserve. It was determined that 29 of the captured geese had the red bands on their necks. About how many geese are on the game preserve?

Solution

This particular method of estimating wildlife population is called the *capture-recapture* method. The method assumes that the ratio of banded geese to all geese in the preserve is proportional to the ratio in the second capture. If x is the number of geese in the preserve, then

$$\frac{29 \text{ banded}}{350 \text{ total in capture}} = \frac{200 \text{ banded}}{x \text{ total in preserve}}$$

Note that the ratio of units is the same on both sides.

$$29x = 70{,}000$$
$$x \approx 2{,}414 \text{ geese in the preserve}$$

Can you think of some reasons why this estimate for the number of geese may not be accurate? ■

EXAMPLE 10

Sometimes it is difficult to repeat the capture-recapture process. Another method is to allow for a margin of error in the ratio of the form

$$\frac{\text{number of banded geese}}{\text{total number of geese}}$$

and then compute a lower and an upper estimate of the size of the population. Assume there is a 5% margin for error in the ratio $\frac{29}{350}$ expressed as a percent. Estimate the total number of geese in the preserve.

Solution

As a percent, $\frac{29}{350} \approx 8\%$. We assume between $8\% - 5\% = 3\%$ and $8\% + 5\% = 13\%$ of the geese were tagged. Thus

$$0.03 = \frac{200}{N_1} \quad \text{and} \quad 0.13 = \frac{200}{N_2}$$

N_1 is the upper estimate and N_2 the lower estimate for the number of geese.

give two estimates for the number of geese. Now $N_1 = 6{,}667$ and $N_2 = 1{,}538$. The number of geese is between 1,538 and 6,667. ■

In the capture-recapture method, it is assumed that a proportion "models" the real-life situation. Are there any factors not taken into account by this proportional model?

EXAMPLE 11

At 12:00 noon on a bright sunny day, your 5-foot-10-inch body casts a shadow of 3 feet 11 inches. How tall is the incinerator chimney at your school if you measure the chimney's shadow to be about 93 feet?

Solution

We have two similar triangles. Therefore, the ratios of height to shadow length will be proportional. Since we want the chimney's height in feet, we keep the dimensions in feet. Of course, you could convert to inches for each measure if you wish.

If Y is the height of the chimney in feet, then

$$\frac{\text{5 feet 10 inches}}{\text{3 feet 11 inches}} = \frac{Y}{\text{93 feet}}$$

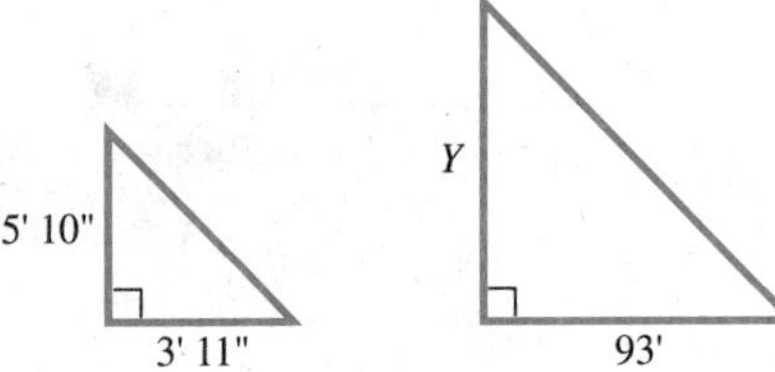

Figure 3.7

Therefore

$$\frac{5\frac{10}{12}}{3\frac{11}{12}} = \frac{Y}{93}$$ Inches converted to fractional feet.

$$\frac{5.83}{3.92} \approx \frac{Y}{93}$$ Decimals are easy to work with. We rounded off to two decimal places.

$$\frac{(93)(5.83)}{3.92} \approx Y$$

$$Y \approx 138.31 \text{ feet}$$ The height of the chimney. ■

3.2 EXERCISES

I. Procedures

1–16. Solve the following proportions.

1. $\frac{X}{18} = \frac{25}{72}$

2. $\frac{20}{5} = \frac{Z}{15}$

3. $\frac{X}{18} = \frac{6}{5}$

4. $\frac{18}{750} = \frac{3{,}500}{X}$

5. $\frac{1.48}{6.2} = \frac{5.4}{Z}$

6. $\frac{X}{0.0036} = \frac{1{,}000}{17}$

7. $\frac{1\frac{1}{2}}{3\frac{1}{4}} = \frac{Y}{4}$

8. $\frac{\frac{7}{8}}{Z} = \frac{\frac{15}{16}}{\frac{3}{2}}$

9. $\frac{1.4 \times 10^3}{6.3 \times 10^{-2}} = \frac{W}{2.8 \times 10^2}$

10. $\frac{3.6 \times 10^{-3}}{Z} = \frac{4.5 \times 10^{-4}}{2 \times 10^3}$

11. $\frac{3}{4} = \frac{z}{68}$

12. $\frac{7}{6} = \frac{63}{x}$

13. $\frac{a}{6} = \frac{19}{4}$

14. $\frac{12}{x} = \frac{108}{9}$

15. $\frac{9}{4} = \frac{z - 1}{6}$

16. $\frac{6}{3 - d} = \frac{1}{4}$

II. Concepts

17–22. Complete the following exercises.

17. Two sweatshirts cost $48.80. How much would X sweatshirts cost?

18. On a map, $\frac{1}{4}$ inch represents ten miles. How many miles would be represented by x inches?

19. An automobile went 84 miles on $6\frac{1}{2}$ gallons of gasoline. How many gallons would be needed to go Y miles?

20. Consider this list of running records for 1992.

Women	*Distance*	*Men*
10.49 seconds	100 meters	9.86 seconds
21.34 seconds	200 meters	19.72 seconds
47.60 seconds	400 meters	43.29 seconds
1 minute, 53.78 seconds	800 meters	1 minute, 41.73 seconds
3 minutes, 52.47 seconds	1,500 meters	3 minutes, 29.46 seconds

(a) Consider the following pairs of races: 100 and 200 meters, 200 and 400 meters, 400 and 800 meters, 800 and 1,500 meters. Are the ratios of distance to time proportional or nearly proportional for any of these pairs of races, for either the men's or women's record?

(b) Suppose the ratio of distance to time for the 100-meter records was proportional to the ratio for 1,500 meters. What would be the "new" men's and women's world records for 1,500 meters?

21. Estimate (or find out) your weight and height at birth. Find the ratios of your present weight to birth weight, and present height to height at birth. Are these ratios proportional?

22. Make up *your own* proportion problem from each of the following excerpts, and then solve it.

(a) There are about 460 words on page 248 of a book that contains 670 pages.

(b) Ten cases of measles were noted in a fifth-grade class of 28 students. There are 320 students in the school.

(c) A punch you are making calls for three parts grape juice to two parts soda. You have 7 gallons of grape juice.

(d) Your younger brother grew two inches in a three-month period at age 3. He is now 6.

(e) Your heart beats about 75 times a *minute* when you are not running. You are 40 years old now.

23–28. In each of the following statements, a proportion has been used. Identify the proportion, and then say whether you think it is appropriate or realistic.

23. If I can eat four pancakes in 15 minutes, then I can eat thirty-two pancakes in two hours.

24. If a turkey weighs 17 pounds standing on one leg, then it weighs 34 pounds standing on both legs.

25. If an 8-inch pizza costs $4, then a 12-inch pizza should cost $6.

26. If I can get 20 mpg driving at 30 mph, then I can get 40 mpg driving at 60 mph.

27. If it takes 15 minutes to cut a log into three pieces, then it takes 20 minutes to cut a log into four pieces.

28. If 100 people die from 6 P.M. Friday to 6 P.M. Saturday on a three-day holiday weekend, then 300 people will probably die from 6 P.M. Friday to 6 P.M. Monday.

29. We can associate a ratio $y{:}x$ with the point (x,y) on a graph. Which of the graphs below could represent ratios equal to 2:3? Give a reason for your answer.

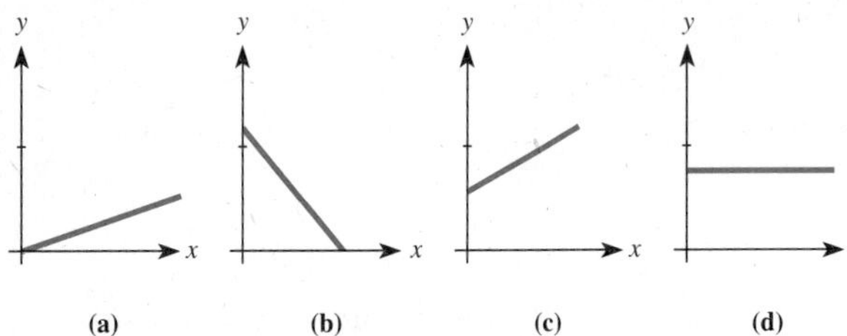

30. Draw a graph showing the indicated ratio for $y{:}x$.

(a) 2:5 **(b)** 5:2 **(c)** 5:5

31. These two proportions have the same solution for x.

$$\frac{x}{6} = \frac{55}{7} \qquad \frac{6}{x} = \frac{7}{55}$$

How many more proportion exercises involving the same four quantities (x, 6, 7, 55) can you construct that have the same solution?

32. Juan was thinking about the problem: Orange juice is advertised at three cans for \$2.69; how much will ten cans cost? He remarked that he didn't need to solve a proportion to find the answer. "I can just think: 'If three cans cost \$2.69, then each can will cost about one-third of \$2.70, or 90¢. So ten cans will cost almost \$9.00.' " Make up another problem that you could solve mentally the way Juan did.

33. Explain how you can be more confident that you are using the basic rate formula or solving a proportion correctly by analyzing the units used to measure each of the quantities involved in the problem.

34. Compare the strategies of (a) using a proportion and (b) using the basic rate formula to solve a problem. In your discussion, give an example of a problem for which each method is preferable.

III. Applications

35 and 36. **(a)** Sketch a graph of the ratio indicated. **(b)** Solve the problem(s).

35. In a bread recipe, the ratio of milk to flour is $\frac{4}{3}$. If eight cups of milk are used, how many cups of flour are used?

36. A certain grade of beef is allowed to have 1.7 grams of fat in every 10 grams of beef.

(a) How many grams of fat could be in a 375-Kg side of beef if it were of this grade?

(b) How heavy would a carcass have to be in order to qualify for this grade of beef if the carcass contained 75 kg of fat?

37–60. Solve each of the following exercises.

37. One season the Cincinnati Reds won 11 of their first 18 baseball games. At this rate, how many games will the Reds win in their 162-game schedule? What would be their final won-lost record for the year? What would be their winning percentage? (Do you think the percentage would be good enough for first place in their division?)

38. Last season, a baseball player made 240 hits in 600 times at bat. This season, his batting average is the same. If he has batted 500 times, how many hits has he had? *Note:* A baseball player's batting average is the ratio of the number of hits to the total times at bat.

39. A man bought a new car. In the first eight months, he drove it 16,000 miles. At this rate, how many miles will he drive it by the end of one year?

40. The coffee beans from 14 trees are required to produce 17 pounds of coffee, which is the average that each person in the United States drinks each year.

(a) How many trees are required to produce 391 pounds of coffee?

(b) How many trees are required to produce enough coffee for the 75 million coffee drinkers in the United States?

41. A quality-control inspector examined 200 light bulbs and found 18 to be defective. At this rate, how many defective bulbs would there be in a lot of 22,000?

42. On a world map, $\frac{1}{8}$ inch represents 100 miles. About how wide is the United States on this map? How wide (or long) is your home state (or county) on this map?

43. The ratio of cement to sand in a concrete mixture is 1 to 3. How much sand is needed in a mixture that has 390 pounds of cement?

44. A river is about 0.089 inches wide (on the average) on an aerial photo that has a scale of 1 inch to 11,500 inches.

(a) How wide is the river?

(b) How wide would this river appear on a map where 1 inch represents 300 feet?

45. At a pace of 8 minutes per mile, a runner burns off 120 calories a mile.

(a) How far would this runner have to run (at 8 minutes a mile) to burn off a 1,350-calorie meal?

(b) How many calories would this runner burn off if she ran a marathon race (26.2 miles)? (Why are marathon runners often very hungry at the end of the race?)

46. A runner loses about 2 pounds of water weight for each 5 miles he runs.

(a) At this rate, how many pounds in water weight would the runner lose in a marathon (26.2 mile) race? In a marathon race, you can drink along the way.

(b) If water loss exceeds 5 pounds per 100 pounds of your body weight, you can become very ill. How much would you have to weigh at the start of a marathon to be within safe limits for the water loss rate given in part (a) of this problem?

47. A company claims that its tablet "absorbs ten times its weight in excess stomach acid."

(a) If the tablet weighs 3 grams, how much acid can it absorb?

(b) Suppose you had a kilogram (1,000 grams) of excess acid in your stomach (unlikely, but just suppose). How many of these tablets would you have to take?

48. Three partners go into the bathtub manufacturing business. Winken puts up \$100,000, Blinken puts up \$80,000, and Nodd puts up \$30,000. At the end of one year, their profit is \$15,000. How much profit should each receive, if they each receive a share in proportion to the money they put into the business?

49. A light-year is 3.1×10^{16} feet.

(a) How many miles are in a light-year?

(b) How many meters are in a light-year?

50. Our nearest neighboring star is 4.3 light-years away from earth.

(a) How many meters away is this star?

(b) How many years would it take to drive there at 88.5 km/hour? At 55 mph?

51. Use your own height to estimate the height of a redwood tree by the shadow method, given that your shadow is $8\frac{3}{4}$ feet long and the tree's shadow is 204 feet long. Will you get the same answer as someone else working this problem? Why or why not?

52. When a tree 18 feet tall casts a shadow 54 feet long, how long a shadow is cast by a person 6 feet tall?

53. Similar triangles can be used to determine the distance across a river, as shown in the diagram.

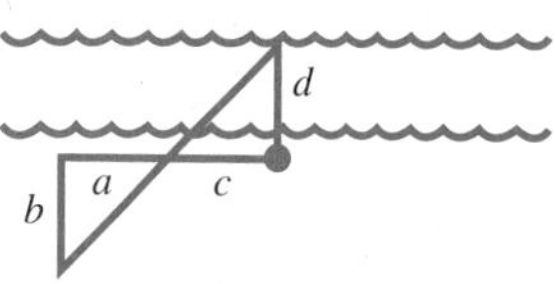

Let d be the distance across the river. Distances a, b, and c can easily be laid out and measured.

(a) If $a = 30$ feet, $b = 70$ feet, and $c = 20$ feet, what is the distance d across?

(b) If $a = 105$ meters, $b = 250$ meters, and $c = 75$ meters, what is the distance d across?

54. In Glacier National Park, 52 mountain goats were caught, tagged, and then set free. One year later, 74 different mountain goats were observed, and 12 of these had tags.

(a) What is an estimate for the mountain goat population in the park?

(b) What assumptions did you make when you set up your "model" in part (a)?

(c) Can you give some reasons why your estimate in part (a) could possibly be inaccurate?

55. Estimate the population of untagged trout in a lake for each of the following situations.

(a) There were 500 tagged trout released; 16 out of 200 trout captured were tagged.

(b) There were 1,000 tagged trout released; the sample yielded 120 tagged and 80 untagged trout.

56. A cattle producer had 350 head of breeding cows in the north pasture. Of these cows, 78 did not produce calves at breeding time last year. The herd has been expanded this year with the purchase of an additional 240 such cows. How many nonproductive cows can be expected this year?

57. The pressure exerted by a gas within a container is proportional to the Kelvin temperature of the gas. A driver puts 20 N/cm^2 pressure in each tire in Death Valley (temperature = 35°C = 308°K) and then drives to the top of Pikes Peak (temperature = 0°C = 273°K). Assuming the walls of the tire are stiff enough to prevent any change in volume, what is the pressure in each tire at the top of Pikes Peak? (N is the abbreviation for Newton. Newton is a unit of force.)

58. The change in volume of gas is proportional to the change in temperature (Boyle's Law). That is, if the temperature increases, the volume increases.

(a) If the volume of a gas at 32°C is 1 liter, what is the volume of the gas at 48°C?

(b) The volume of a gas at 50°C is 2.2 liters. What must be the temperature if 1 liter of the gas is desired?

(c) If the volume of gas is halved, what happens to the temperature?

59. The probability that an event will occur is the ratio of the number of times it occurs to the total number of outcomes. For example, if a die is tossed, the probability of a one coming up is $\frac{1}{6}$.

(a) If a die is tossed 24 times, how many times could we expect a one to occur? A four to occur?

(b) If a pair of dice is tossed, what is the probability the *sum* of 7 will occur?

(c) How many times would you expect the sum of 7 to occur in 120 tosses of the dice?

60. The table below gives the approximate popular vote in the presidential elections of 1860 and 1864. Find the entries labeled (a), (b), (c), (d), and (e).

61–74. A rule of thumb is a guess. It is generally easy to remember and falls somewhere between a precise formula and a shot in the dark. Some of the following exercises use rules of thumb.

61. "One-fifth of the length of a telephone pole should be in the ground." What is the ratio of the length of the part of the pole above the ground to the part of the pole in the ground?

62. "You should expect a speech to take one-third more time than it took when you rehearsed it." What is the ratio of the length of an actual speech to the length of its rehearsed version?

63. "Every two years you can buy a computer that performs twice as well for half the price." If a computer costs $4,000 today and has a performance rating of 60, what will be the price and the performance rating of a computer bought eight years from now?

64. From a small group of people around you at a baseball game, you estimate the ratio of men to women is 3:2. If the announced attendance at the game is 48,000, about how many of the people attending are women?

65. "The closer the proportions of a rectangle are to 5 to 3, the more pleasing it is to the eye." If the ratio of the length to the width of a rectangle is 5:3 and its perimeter is 70 cm, find its length and width.

66. The base of an isosceles triangle is one-fourth its perimeter. Find the ratio of the indicated quantities.

(a) leg to base **(b)** perimeter to leg

67. The width of a rectangle is $\frac{3}{4}$ of its length. If the perimeter is 50 cm, find each side.

68. The sides of a triangle are in the ratio 7:10:11. If the perimeter of the triangle is 336 meters, find the length of each side.

69. (a) The angles of a triangle are in the ratio 2:3:5. Find the measure of each angle.

(b) Can the sides of a triangle be in the ratio 2:3:5?

70. Concrete used for pavement consists of 1 part cement, 2.2 parts sand, and 3.5 parts water. To mix 1,000 pounds of this concrete, how many pounds of sand are needed?

71. There are approximately 2,400 stocks traded on the New York Stock Exchange. Yesterday's stock market report indicated that "advancers led decliners by 5 to 3." If 560 of the stocks were unchanged yesterday, how many were "advancers"?

72. "Seven of every ten exploratory oil wells are dry holes. Of those that hit gas or oil, only one of forty is commercially successful." How many exploratory wells must you drill to end up with ten wells that are commercially successful?

Table for Exercise 60

1860	*Vote (in thousands)*	*Percent of vote*	*1864*	*Vote (in thousands)*	*Percent of vote*
A. Lincoln	1,866	(a)	A. Lincoln	2,216	55.05%
S. Douglas	(b)	29.4%	G. McClellan		
J. Breckenridge	847	(d)			
J. Bell	(e)	12.6%			
Total	4,677			(c)	

73. In a recent week the Nielsen ratings for "The Simpsons" were 14.5|26.

(a) How many homes were tuned to this show?

(b) In how many homes were the TV sets in use when this show was aired? (See Example 8.)

74. The teachers in the Johnstown School District are negotiating to reduce the pupil-teacher ratio from 30:1 to 25:1. There are 8,000 students in the district. If each additional teacher would cost the district $30,000, what will be the added costs to the district if the teachers prevail in the negotiations?

75 and 76. In each case, estimate the size of the population. Then assume a 5% margin for error and find a lower and an upper estimate of the size of the population.

75. A total of 400 penguins is captured and tagged; later 200 are captured, of which 50 are tagged.

76. A high school has 600 students. A random survey of 200 people in the town contains 40 students.

IV. Extensions

77. A photographer takes a 2.4 cm by 3 cm color slide and uses it to make a rectangular color print with the shape similar to the slide. The color print is 28 cm on its longer side.

(a) What is the length of its shorter side?

(b) What is the area of the color slide in part (a)? What is the area of the color print in part (a)?

(c) Find the ratio of the area of the print to the area of the slide. Also find the ratio of the longer side of the print to the longer side of the slide. Are these two ratios proportional?

(d) Can you find *any* relationship between the ratio of the areas of the two rectangles and the ratio of the corresponding sides of the rectangles from part (c)?

(e) In general, suppose that we have two similar rectangles and that the ratio of two corresponding sides equals a number r. What is the ratio of the areas of the two rectangles?

78. We return to the famous question asked in the exercises of Chapter 1: What should the ratio of the width W to the length L of a rectangle be so that if a square whose sides are equal to the width is cut off from the original rectangle, the result will be a rectangle that is the same shape as (that is, similar to) the original one?

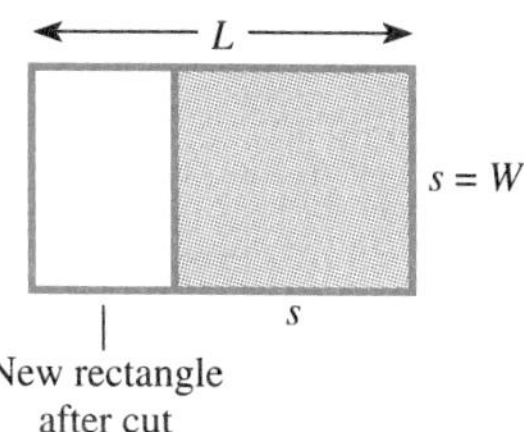

You should be able to get a good start on this problem now. *Hints:* Make the problem easy by putting the longer side, L, equal to 1. In addition, what do you know about the ratios of corresponding sides of similar rectangles?

79. In a standard television screen the ratio length:width = 4.3.

(a) A television set is usually described by the length of the diagonal of its screen. What is the area of the screen of a "19-inch" television set?

(b) *Consumer Reports* said that "the viewing area of a 27-inch television is 80% greater than that of a 20-inch set." Either justify this assertion or show it is false.

80. The gear ratio in a bicycle is

$$\frac{\text{number of teeth in the front sprocket wheel}}{\text{number of teeth in the rear sprocket wheel}}(\text{diameter of the bicycle's tire})$$

In a ten-speed bicycle, there are two front sprocket wheels and five rear sprocket wheels. The diameter of the wheels is usually 27 inches.

(a) What is the gear ratio of a ten-speed bicycle with a front sprocket wheel with 46 teeth and a rear sprocket wheel with 16 teeth?

(b) When you come to a hill, should you switch to a lower or a higher gear ratio? Give a reason for your answer.

(c) If you can turn a gear ratio of 68 at 100 revolutions (of the pedals) per minute or a 72 gear ratio at 84 rpm, which one should you choose to achieve the greater speed?

81. Going into the last day of the major league baseball season, Babe Boggs and Kirby Lockett had identical batting records of 200 hits in 600 at bats. On the last day of the season, Boggs had 7 hits in 8 times at bat and Lockett had 9 hits in 12 times at bat. Who won the batting title? Give a reason for your answer.

3.3 MORE LINEAR EQUATIONS

The solution of the linear equations in the last two sections depended essentially on the ability to apply these basic properties:

1. Adding or subtracting the same quantity to both sides of an equation does not change its solution.
2. Multiplying or dividing both sides of an equation by the same nonzero quantity does not change its solution.

Solving Equations of the Form $ax + b = cx + d$

In this section, we apply these properties to more complicated linear equations that arise in many different applications. We begin by adding another chapter to the saga of the We Try Harder Car Rental Company.

EXAMPLE 1

To rent a compact car, the We Try Harder Car Rental Company (WTH) charges \$19.95 plus 19¢ a mile per day, while the U Drive It Company (UDI) charges \$15.95 plus 23¢ a mile per day. To rent a car for one day which company should you choose?

Solutions

Using an Equation Our choice depends on the number of miles we expect to drive. The first question we might ask is "For what mileage are the two rates the same?" If M denotes the number of miles, then We Try Harder would charge

$$19.95 + 0.19M \quad \text{or} \quad \text{Cost WTH}(M) = 19.95 + 0.19M$$

while U Drive It would charge

$$15.95 + 0.23M \quad \text{or} \quad \text{Cost UDI}(M) = 15.95 + 0.23M$$

The cost would be the same if

$$15.95 + 0.23M = 19.95 + 0.19M$$

Solving this equation, we have

$$\begin{aligned} 15.95 + 0.04M &= 19.95 && \text{Subtract } 0.19M \text{ from both sides.} \\ 0.04M &= 4 && \text{Subtract 15.95 from both sides.} \\ M &= 100 && \text{Divide both sides by 0.04.} \end{aligned}$$

Therefore the cost will be the same if we drive 100 miles.

Using a Table We check several other possibilities using Table 3.4. We would probably choose the U Drive It company's car if we expect to drive less than 100 miles and the We Try Harder company's car if we expect to drive more than 100 miles.

	50 miles	*75 miles*	*100 miles*	*125 miles*	*150 miles*
WTH cost	\$29.45	\$34.20	\$38.95	\$43.70	\$48.45
UDI cost	\$27.45	\$33.20	\$38.95	\$44.70	\$50.45

Table 3.4

Using a Graph We can sketch the graphs suggested by the table of data and see that the answer is reasonable.

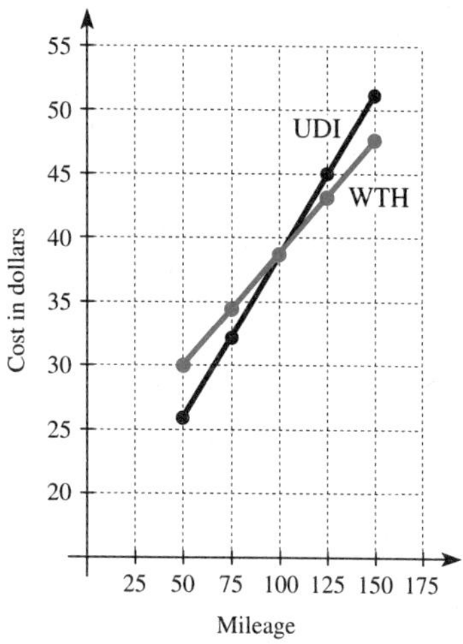

Figure 3.8

In the preceding example, we needed to solve a linear equation of the form

$$ax + b = cx + d$$

More equations of this form are solved in the next several examples.

EXAMPLES 2–4

Solve for x.

2. $3x + 2 = 2x - 4$

3. $32x + 14 = 42x - 57$

4. $m_1x + b_1 = m_2x + b_2, m_1 \neq m_2$

Solutions

2. Using an Equation

$$3x + 2 = 2x - 4$$

$$3x = 2x - 6 \qquad \text{Subtract 2 from both sides.}$$

$$x = -6 \qquad \text{Subtract } 2x \text{ from both sides.}$$

Check the solution!

Using a Graph To solve $3x + 2 = 2x - 4$ on the graphing calculator, we first graph $Y1 = 3x + 2$ and $Y2 = 2x - 4$. Then, after determining an appropriate viewing rectangle, find the x-coordinate of the point of intersection of the two lines.

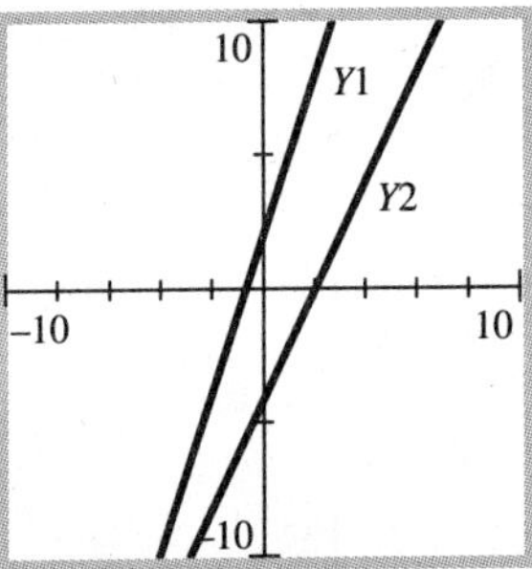

The intersection point is off the screen. We need another window.
$Y1 = 3x + 2 \quad Y2 = 2x - 4$

Figure 3.9(a)

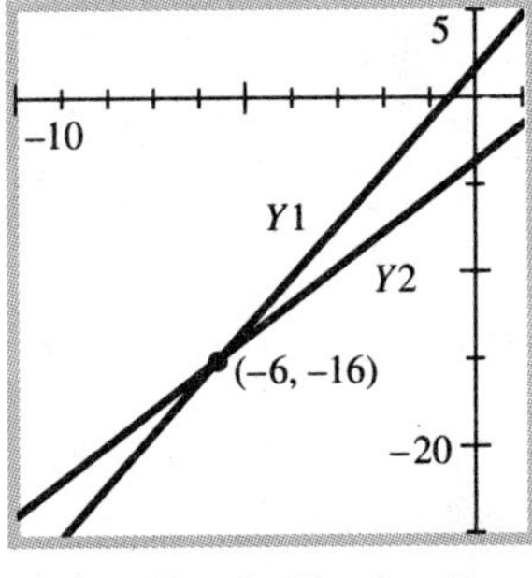

$Y1 = 3x + 2 \quad Y2 = 2x - 4$

Figure 3.9(b)

The x-values could be slightly different, depending on the viewing rectangles. The graphs cross at about $x = -6$.

3. $32x + 14 = 42x - 57$

$$32x = 42x - 71 \qquad \text{Subtract 14 from both sides.}$$

$$-10x = -71 \qquad \text{Subtract } 42x \text{ from both sides.}$$

$$x = \frac{-71}{-10} = \frac{71}{10} = 7.1 \qquad \text{Divide both sides by } -10.$$

To reduce the number of negative signs involved, we may wish to solve the problem so that the coefficient of x is always positive, that is,

$$32x + 14 = 42x - 57$$

$$14 = 10x - 57 \qquad \text{Subtract } 32x \text{ from both sides.}$$

$$71 = 10x \qquad \text{Add 57 to both sides.}$$

$$7.1 = x \qquad \text{Divide both sides by 10.}$$

You could also solve this equation using a graphing calculator.

4.
$$
\begin{aligned}
m_1x + b_1 &= m_2x + b_2 && m_1 \neq m_2\\
m_1x &= m_2x + b_2 - b_1 && \text{Subtract } b_1 \text{ from both sides.}\\
m_1x - m_2x &= b_2 - b_1 && \text{Subtract } m_2x \text{ from both sides.}\\
(m_1 - m_2)x &= b_2 - b_1 && \text{Distributive property.}\\
x &= \frac{b_2 - b_1}{m_1 - m_2} && \text{Divide both sides by } m_1 - m_2\text{, which is a nonzero quantity since } m_1 \neq m_2.
\end{aligned}
$$

EXAMPLE 5

Solve for x.

$$2.2143x - 4.8631 = 7.4198x + 8.2137$$

Solution

The distinguishing feature of this equation is the "messiness" of the numbers. The fundamental question is: How should we do the computations (with our calculator)? It is usually easiest to do the algebra first and then save the calculator for one computation at the end.

Observe:

$$
\begin{aligned}
2.2143x - 4.8631 &= 7.4198x + 8.2137\\
2.2143x &= 7.4198x + (8.2137 + 4.8631)\\
2.2143x - 7.4198x &= 8.2137 + 4.8631\\
x(2.2143 - 7.4198) &= 8.2137 + 4.8631\\
x &= \frac{8.2137 + 4.8631}{2.2143 - 7.4198}
\end{aligned}
$$

We still have not done any computing. The procedure thus far was identical to that used in Example 4. Notice that $m_1 = 2.2143$, $m_2 = 7.4198$, $b_1 = -4.8631$, and $b_2 = 8.2137$. We can now compute the solution. Before we do, however, let us *estimate* it first. Thus

$$x \approx \frac{8 + 5}{2 - 7} \approx \frac{13}{-5} \approx -2.6$$

By calculator, we obtain $x \approx -2.5121$. Check both the estimate and the calculator answer.

NOTE: This equation can also be solved with a graphing calculator. See Exercise 61.

EXAMPLES 6–7

Solve for x.

6. $\frac{2}{3}x - \frac{3}{4} = 2x + \frac{1}{2}$

7. $\frac{x}{a} - \frac{x}{b} = a - b$, where $a, b \neq 0$, $a \neq b$.

Solutions

6. To avoid computations involving fractions, we multiply both sides by a common denominator of $\frac{2}{3}, \frac{3}{4}, \frac{1}{2}$.

$$\frac{2}{3}x - \frac{3}{4} = 2x + \frac{1}{2}$$

$$12\left(\frac{2}{3}x - \frac{3}{4}\right) = 12\left(2x + \frac{1}{2}\right)$$ Multiply both sides by 12, the least common denominator of $\frac{2}{3}, \frac{3}{4}, \frac{1}{2}$.

$$12\left(\frac{2}{3}x\right) - 12\left(\frac{3}{4}\right) = 12(2x) + 12\left(\frac{1}{2}\right)$$ Distributive property.

$$8x - 9 = 24x + 6$$

$$-15 = 16x$$

$$-\frac{15}{16} = x$$

7.
$$\frac{x}{a} - \frac{x}{b} = a - b$$ A common denominator is ab.

$$ab\left(\frac{x}{a} - \frac{x}{b}\right) = ab(a - b)$$ Multiply both sides by ab.

$$ab\left(\frac{x}{a}\right) - ab\left(\frac{x}{b}\right) = ab(a - b)$$ Distributive property.

$$bx - ax = ab(a - b)$$

$$(b - a)x = ab(a - b)$$ Distributive property. Divide both sides by $b - a$.

$$x = \frac{ab(a - b)}{(b - a)} = -ab$$ Remember, $\frac{a - b}{b - a} = -1$.

The next example furnishes an oversimplified illustration of the basic economic law of supply and demand. **Supply** is the amount of a particular item *supplied* by a merchant when the item sells for a certain price. Consequently, when the price of the item is high, the merchant will supply more of it since she will make more profit. When the price is low, she will hold back on the amount of the item she supplies.

Demand is the amount of a particular item *demanded* by consumers when the item sells for a certain price. When the price is high, they will want less of the item; when the price is low, they will want more.

The ideal situation occurs when the supply equals the demand. In this case, there is an **equilibrium price** that is "high enough" for the supplier and "low enough" for the consumer.

EXAMPLE 8

Suppose the demand for hamburger is

$$D = 3{,}600 - 1{,}500p \quad \text{or} \quad \text{Demand }(p) = 3{,}600 - 1{,}500p$$

and the supply is given by

$$S = -1{,}200 + 1{,}800p \quad \text{or} \quad \text{Supply }(p) = -1{,}200 + 1{,}800p$$

where p is the price per pound of hamburger (in dollars). At $1 a pound, the consumers want to buy 2,100 pounds, but the suppliers will supply only 600 pounds. At $2 per pound, the consumers want to buy only 600 pounds, while the suppliers will supply 2,400 pounds.

Notice that the supply will continue to grow as the price rises, but the demand drops to zero when $p = \$2.40$ ($D = 3{,}600 - 1{,}500(2.40) = 0$).

What is the equilibrium price, that is, the price per pound of hamburger when supply equals demand?

Solution

Using an Equation

$$\begin{aligned} \text{Demand} &= \text{Supply} \\ 3{,}600 - 1{,}500p &= -1{,}200 + 1{,}800p \\ 12 - 5p &= -4 + 6p \\ 16 &= 11p \\ \frac{16}{11} &= p \quad \text{or} \quad p \approx \$1.45 \text{ per pound} \end{aligned}$$

Divide by 300. (We could add and subtract first, but we still obtain $p = \frac{4{,}800}{3{,}300} = \frac{16}{11}$.)

At \$1.45 per pound, the demand is about 1,425 pounds.

Using a Graph Notice the similarity between this example and Example 1 on car rentals. Many different types of problem situations can be represented by similar algebraic models.

To view the solution on a graphing calculator, use the viewing rectangle $0 \le x \le 5$; $-1{,}500 \le y \le 4{,}000$ and graph $Y1 = 3{,}600 - 1{,}500x$ and $Y2 = -1{,}200 + 1{,}800x$. Then find the x-coordinate of the point of intersection of the two lines.

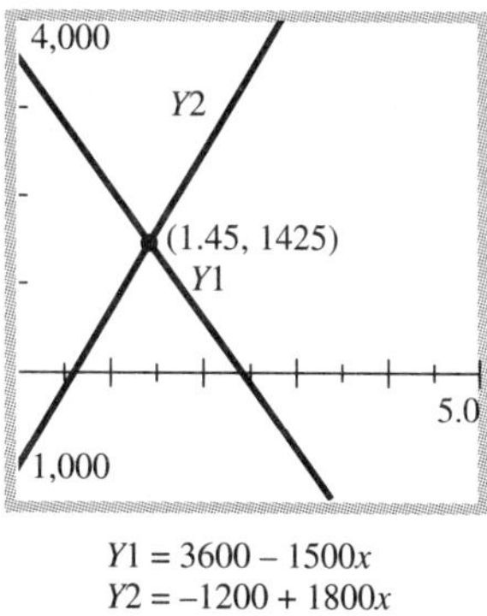

$Y1 = 3600 - 1500x$
$Y2 = -1200 + 1800x$

Figure 3.10

EXAMPLE 9 Fast Freddy can run at the rate of 5 meters per second, while Slower Sam can run only 3 meters per second. They wish to have a race, but to make it fair, Sam gets a 40-meter head start. What is the longest race Sam can win?

Solution

Suppose d denotes the length of the race. Drawing a picture might help to see the relationships (Figure 3.11).

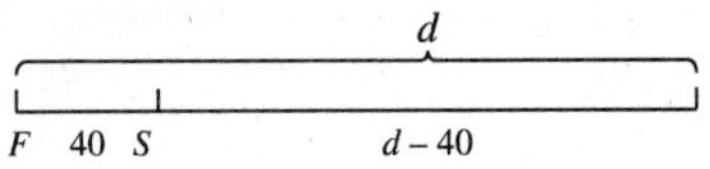

Figure 3.11

Let us find the length of a dead-heat race. Sam can pick a slightly shorter distance. In a dead-heat race, both runners would run for the same amount of time.

$$\text{Fred's time} = \frac{d}{5} \qquad \text{Time} = \frac{\text{distance}}{\text{rate}}$$

$$\text{Sam's time} = \frac{d - 40}{3}$$

$$\frac{d}{5} = \frac{d - 40}{3} \qquad \text{Times are equal. We solve for } d.$$

$$3d = 5d - 200 \qquad \text{Multiply by 15.}$$

$$200 = 2d$$

$$100 = d$$

Therefore if Sam chooses any distance shorter than 100 meters, he will win. ■

Problem-solving Strategy

The *essential step* in solving an *applied problem* is the ability to do either one of two things to set up an equation:

1. Find two equal quantities, or
2. Find two equivalent ways to express the same quantity.

For instance, in Example 1, we found the two costs of renting a car that were equal. In Example 8, the supply and demand were equal. In Example 9, the times of Fred and Sam were equal.

Linear Equations Involving Parentheses

We now try our skills at solving slightly more complicated linear equations using parentheses.

EXAMPLES 10–11

Solve for x in each of the following equations.

10. $4 - 2(x - 3) = x - 5(x + 1)$

11. $3 - 2[x - 2(x - 1)] = 4 + 3x - 2[7 - (1 - x)]$

Solutions

10. $4 - 2(x - 3) = x - 5(x + 1)$

$$
\begin{aligned}
4 - 2x + 6 &= x - 5x - 5 && \text{Distributive property.}\\
10 - 2x &= -4x - 5 && \text{Collect like terms. Add } 4x \text{ and } -10 \text{ to both sides.}\\
2x &= -15\\
x &= -\frac{15}{2} = -7.5 && \text{Divide both sides by 2.}
\end{aligned}
$$

Since the expressions are more complicated, it is advisable to check the answer to be sure we have not made a careless mistake. Indeed, it is always advisable to check answers.

Check:

$$
\begin{aligned}
4 - 2\left(-\frac{15}{2} - 3\right) &\stackrel{?}{=} -\frac{15}{2} - 5\left(-\frac{15}{2} + 1\right) && \text{Substitute } -\frac{15}{2} \text{ for } x.\\
4 - 2\left(-\frac{21}{2}\right) &\stackrel{?}{=} -\frac{15}{2} - 5\left(-\frac{13}{2}\right)\\
4 + 21 &\stackrel{?}{=} -\frac{15}{2} + \frac{65}{2}\\
25 &\stackrel{?}{=} \frac{50}{2}\\
25 &\stackrel{\checkmark}{=} 25 && \text{Check.}
\end{aligned}
$$

11. If you can solve this equation with all its parentheses, you can probably solve any linear equation. *Start by working from the inside out.*

$$
\begin{aligned}
3 - 2[x - 2(x - 1)] &= 4 + 3x - 2[7 - (1 - x)]\\
3 - 2[x - 2x + 2] &= 4 + 3x - 2[7 - 1 + x] && \text{Distributive property.}\\
3 - 2[-x + 2] &= 4 + 3x - 2[6 + x] && \text{Combine terms.}\\
3 + 2x - 4 &= 4 + 3x - 12 - 2x && \text{Distributive property.}\\
2x - 1 &= x - 8 && \text{Combine terms.}\\
x &= -7 && \text{Add } -x \text{ and 1 to both sides.}
\end{aligned}
$$

Check:

$$
\begin{aligned}
3 - 2[-7 - 2(-7 - 1)] &\stackrel{?}{=} 4 + 3(-7) - 2[7 - (1 - (-7))]\\
3 - 2[-7 + 16] &\stackrel{?}{=} -17 - 2(-1)\\
3 - 2[9] &\stackrel{?}{=} -17 + 2\\
-15 &\stackrel{\checkmark}{=} -15
\end{aligned}
$$

Identities, Contradictions, and Conditional Equations

In the next example, three similar-looking equations yield remarkably different solutions.

EXAMPLE 12 Solve for x.

(a) $3(2x - 5) = 2(3x - 1) + x$

(b) $3(2x - 5) = 2(3x - 1) + 7$

(c) $3(2x - 5) = 2(3x - 1) - 13$

Solutions

Using an Equation

(a) $3(2x - 5) = 2(3x - 1) + x$

$6x - 15 = 6x - 2 + x$

$6x - 15 = 7x - 2$

$-13 = x$

This equation has just one solution. It is a *conditional equation.*

(b) $3(2x - 5) = 2(3x - 1) + 7$

$6x - 15 = 6x - 2 + 7$

$6x - 15 = 6x + 5$

$-15 = 5$ **False!**

The statement, $-15 = 5$, is obviously false. The two basic properties of equalities stated that performing the same operation to both sides of an equation does not change its solution. Since $-15 = 5$ is an equation with no solution, the original equation has no solution. There is no value of x that will make the equation true. This equation is called a *contradiction.*

(c) $3(2x - 5) = 2(3x - 1) - 13$

$6x - 15 = 6x - 2 - 13$

$6x - 15 = 6x - 15$ or $0 = 0$ **Always true!**

This time, we arrived at the same quantity on both sides of the equation. No matter what real number we substitute for x, the statement $6x - 15 = 6x - 15$ is true. Thus *every real number is a solution to the original equation.* Such an equation is called an *identity.*

Definition

An equation is called

1. A **conditional equation** if it has only a *finite* number of solutions. (A linear equation has one solution.)
2. An **identity** if it has *all* the numbers of a specified infinite set as solutions.
3. A **contradiction** if it has *no* solution; it implies a contradiction of some known fact.

Using a Graph Notice what happens when we try to solve each of the three equations in the previous example using the graphing calculator with an appropriate viewing rectangle.

(a) There are two lines that intersect in a point whose x-coordinate is -13. (Conditional equation)

(b) There are two parallel lines. (A contradiction: there are *no* solutions)

(c) Only one line appears to be graphed. (Identity: *all* real numbers are solutions.)

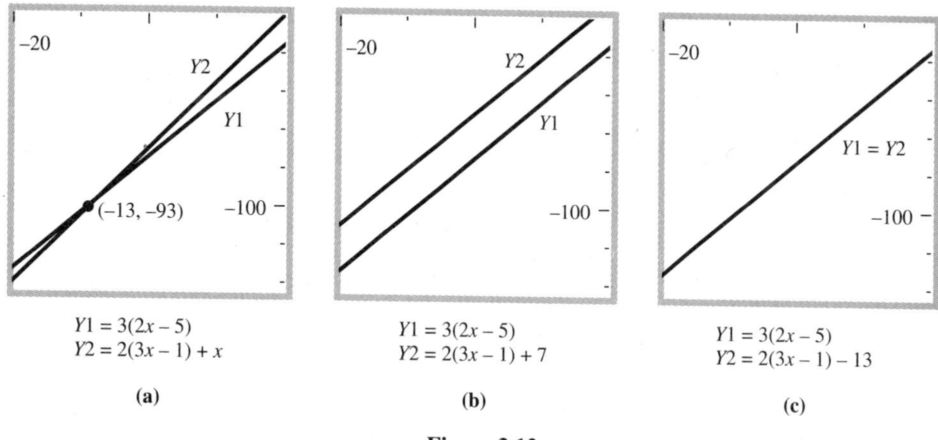

Figure 3.12

This idea will be reconsidered in the next chapter when we study systems of linear equations.

Most of the equations we have solved in this chapter are conditional. Properties such as $x^n x^m = x^{n+m}$ and $a(b + c) = ab + ac$ are identities. Writing equivalent expressions leads to a set of identities. For example, when we rewrite $3x + 5(2 + x)$ as $8x + 10$, we use

$$3x + 5(2 + x) = 3x + 10 + 5x, \text{ an identity; true for all } x.$$
$$3x + 5(2 + x) = 8x + 10, \text{ an identity; true for all } x.$$

Here are a few more applications.

EXAMPLE 13

Return to the Ajax Manufacturing Company, which continues to make widgets. Recall that after the initial cost of \$10,000, it costs \$4 to make a widget that sells for \$5.95. How many widgets must Ajax sell to make a 10% profit (10% of its costs)?

Solution

As before, let x denote the number of widgets. Since profit = revenue − cost, we have

Revenue	$= 5.95x$	or	Revenue (x)	$= 5.95x$
Cost	$= 10{,}000 + 4x$	or	Cost (x)	$= 10{,}000 + 4x$
Profit	$= 0.1(10{,}000 + 4x)$	or	Profit (x)	$= 0.1(10{,}000 + 4x)$

Consequently,

$$\text{Revenue} - \text{Cost} = \text{Profit}$$
$$5.95x - (10{,}000 + 4x) = 0.1(10{,}000 + 4x)$$

$$\begin{aligned} 5.95x - 10{,}000 - 4x &= 1{,}000 + 0.4x \\ 1.95x &= 11{,}000 + 0.4x \\ 1.55x &= 11{,}000 \\ x &= \frac{11{,}000}{1.55} \approx 7{,}097 \text{ widgets} \end{aligned}$$

EXAMPLE 14

You are promoting a concert for the popular rock group "War and Peace" in a 10,000-seat auditorium. Most concerts charge \$10 for reserved seats and \$7 for general admission, but it is up to the promoter to determine the number of each type of seat. War and Peace charges \$40,000 for its appearance. Your expenses are \$20,000. Assuming all seats will be sold, how many seats should be designated reserved seats if you wish to

(a) break even? **(b)** make a \$30,000 profit?

Solutions

Let r equal the number of reserved seats.

(a) In order to break even, you want

$$\text{Revenue} = \text{Costs}$$

Revenue equals $10r$ for reserved seats, but how much for general admission? If there are r reserved seats, there must be $(10{,}000 - r)$ general-admission seats. The revenue from these seats is, then, $7(10{,}000 - r)$. Thus

$$\begin{aligned} \text{Revenue} &= 10r + 7(10{,}000 - r) \\ \text{Costs} &= \$20{,}000 + \$40{,}000 = \$60{,}000 \end{aligned}$$

To break even,

$$\begin{aligned} 10r + 7(10{,}000 - r) &= 60{,}000 \\ 10r + 70{,}000 - 7r &= 60{,}000 \\ 3r &= -10{,}000 \\ r &= \frac{-10{,}000}{3} \end{aligned}$$

A negative solution is not possible. You cannot break even; you must make a profit with those ticket prices. Even if every seat is general admission, you take in \$70,000 and make a profit of \$10,000.

(b) To make a \$30,000 profit, we recall

$$\begin{aligned} \text{Revenue} - \text{Cost} &= \text{Profit} \\ 10r + 7(10{,}000 - r) - 60{,}000 &= 30{,}000 \\ 10r + 70{,}000 - 7r - 60{,}000 &= 30{,}000 \\ 3r + 10{,}000 &= 30{,}000 \\ 3r &= 20{,}000 \\ r &= \frac{20{,}000}{3} \approx 6{,}667 \end{aligned}$$

So two-thirds, or 6,667, of the 10,000 seats must be reserved to make a \$30,000 profit. ■

EXAMPLE 15 Given the circles below,

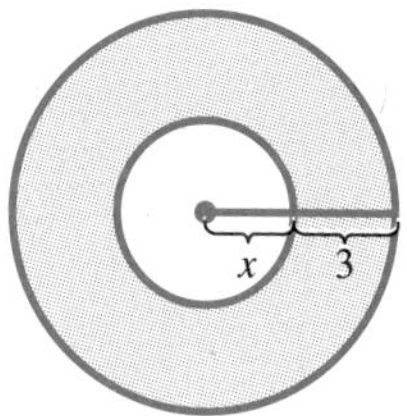

Figure 3.13

find x if the shaded area is 36 cm^2.

Solution

The shaded area A is the difference of two circles. That is,

$$A = \text{area of outer circle} - \text{area of inner circle}$$
$$A = \pi(x+3)^2 - \pi x^2$$

Thus

$$\pi(x+3)^2 - \pi x^2 = 36 \qquad A = 36.$$
$$\pi[(x+3)^2 - x^2] = 36$$
$$(x+3)^2 - x^2 = \frac{36}{\pi}$$
$$x^2 + 6x + 9 - x^2 = \frac{36}{\pi} \qquad \text{Multiplication of binomials: } (x+3)^2 = (x+3)(x+3) = x^2 + 6x + 9.$$
$$6x + 9 = \frac{36}{\pi}$$
$$2x + 3 = \frac{12}{\pi} \qquad \text{Divide by 3.}$$
$$2x = \frac{12}{\pi} - 3$$
$$x = \frac{\left(\frac{12}{\pi} - 3\right)}{2} \approx 0.41 \text{ cm} \qquad \pi \approx 3.14$$ ■

EXAMPLE 16 An auto supply company wants to make 900 gallons of battery solution that is 50% acid. It has a supply of solution that is 70% acid, some that is 20% acid, and some distilled water without acid.

(a) How many gallons of the 70% solution and the 20% solution should be mixed together to make the 900 gallons of 50% solution?

(b) How many gallons of the 70% solution and the distilled water should be mixed together to make the 900 gallons of 50% solution?

Solutions

(a) To find an equation, we note that

$$\text{amount of acid in 70\% solution} + \text{amount of acid in 20\% solution} = \text{amount of acid in 50\% solution}$$

We can express the amount of acid in two different ways. If x denotes the number of gallons of 70% solution, then $.70x$ is the number of gallons of acid. Similarly, $(900 - x)$ gallons of 20% solution are used, and they contain $.20(900 - x)$ gallons of acid. The 900 gallons of 50% solution contain $.5(900) = 450$ gallons of acid. Thus

$$\begin{aligned} 0.70x + 0.20(900 - x) &= 450 \\ 0.70x + 180 - 0.20x &= 450 \\ 0.50x + 180 &= 450 \\ 0.50x &= 270 \\ x &= 540 \\ 900 - x &= 360 \end{aligned}$$

Thus 540 gallons of 70% solution are mixed with 360 gallons of 20% solution.

(b) Again let x denote the number of gallons of 70% solution. Then, since the distilled water has 0% acid,

$$\begin{aligned} 0.70x + 0(900 - x) &= 450 \\ 0.70x &= 450 \\ x &\approx 642.86 \approx 643 \\ 900 - x &= 257 \end{aligned}$$

Thus it takes approximately 257 gallons of distilled water to dilute 643 gallons of 70% solution into 900 gallons of 50% solution. ■

You have solved many application problems in this chapter. You will solve many more in succeeding chapters. Let us summarize some of the tips we have used in solving such problems:

Summary of Problem-solving Strategies

1. Read the problem carefully.
2. Draw a picture (if you can).
3. Make a table or chart to summarize the data (if you can).
4. Observe the patterns. Guess an answer and check it. These procedures will usually help with steps 5 and 6.

5. Choose a variable and state what quantity the variable represents.
6. Find an equation using the variable: Look for two equal quantities or two ways to express the same quantity. Observing patterns in a table may help.
7. Solve the equation.
8. Check your answer with the original problem.

NOTE: In solving an application, the wrong equation might be obtained. Therefore, to check answers go back to the *original statements in the problem* to see if the answer makes sense. ■

3.3 EXERCISES

I. Procedures

1–18. Solve each of the following equations for x and check your answers.

1. $3x + 7 = 5x + 13$

2. $3x - 7 = 5x + 13$

3. $-3x + 7 = 5x + 13$

4. $3x + 7 = 5x - 13$

5. $18x - 42 = 7x - 37$

6. $101x + 102 = 103x + 104$

7. $8x + 9x + 10 = 7x - 4x - 3$

8. $30x - 33 - 36x - 39 = 39 + 36x + 33 - 30x$

9. $0.3x - 0.24 = 0.2x + 0.09$

10. $0.02x + 3.75 = 0.8x - 0.15$

11. $\frac{x}{2} - \frac{1}{4} = \frac{x}{4} - \frac{1}{2}$

12. $\frac{2x}{3} + \frac{1}{3} = \frac{5x}{6} - 1$

13. $\frac{5x}{6} - \frac{7x}{15} = 1 + \frac{3x}{10}$

14. $34 - \frac{5}{6}x = 32 - \frac{1}{2}x$

15. $ax - b = cx + d$

16. $ax + bx + c = dx + ex - f$

17. $\frac{x}{a} + \frac{x}{b} = a + b$

18. $\frac{x}{a} + \frac{x}{b} + \frac{x}{c} = d$

19–24. Solve the proportion for the indicated variable.

19. $\frac{y}{y - 3} = \frac{3}{7}$

20. $\frac{4 - x}{x} = \frac{17}{3}$

21. $\frac{2}{3} = \frac{x + 4}{x + 13}$

22. $\frac{x + 6}{x - 6} = \frac{7}{2}$

23. $\frac{5}{7} = \frac{3x + 6}{5x - 2}$

24. $\frac{3x - 1}{6x + 3} = \frac{4}{9}$

25–39. Solve each of the following equations for x. State whether the equation is a conditional equation, an identity, or a contradiction. See Example 12.

25. $3(2x - 5) = 5(x - 4)$

26. $3(2x - 5) = 5(x - 4) + 6$

27. $3(2x - 5) = 5(x - 4) + 5$

28. $3(2x - 5) = 5(x - 4) + x$

29. $3(2x - 5) = 5(x - 4) + x + 5$

30. $5 - 2(x - 1) = 2(x - 3)$

31. $5 - 2(x - 1) = 2(x - 3) - 7$

32. $5 - 2(x - 1) = 2(3 - x) + 7$

33. $\frac{3}{4} - \frac{2}{3}(x - 1) = \frac{x}{12} + \frac{2}{3}$

34. $x(x + 1) + 3x - 2 = x^2 - 4$

35. $x^3 + x^2 + x + 1 = x(x^2 + x + 2)$

36. $x^3 + x^2 + x + 1 = x(x^2 + x + 1)$

37. $x^3 + x^2 + x + 1 = x(x^2 + x + 1) + 1$

38. $5 - 2[3 - 2(x - 4)] = 3x - 4$

39. $1 - [1 - (1 - x)] = x - [x - (x - 1)]$

40–43. Solve each of the following equations, doing one calculator calculation at the end. Estimate your answer before using the calculator. Also try to solve some of these with a graphing calculator. Which method do you prefer? Why?

40. $2.0421x - 1.9517 = 0.8164x + 0.1239$

41. $2.0421x - 1.9517 = -0.8164x + 0.1239$

42. $9.8820 - 4.1421(x - 3.6274) = 0.0123x + 18.6732$

43. $9.8820 + 4.1421(x - 3.6274) = 0.0123x - 18.6732$

44–47. Solve each of the following equations for the variable indicated.

44. $S = a + (n - 1)d$ for n

45. $A = \frac{1}{2}h(b_1 + b_2)$ for b_1

46. $(y + a)(y + b) - y^2 = c$ for y

47. $(y + a)(y + b) - y^2 = a + b + ab$ for y

II. Concepts

48–54. Find all values of x for which the following statements are true. You might also compare the graphs of both sides using a graphing calculator.

48. $5(x + 2) = 5x + 10$

49. $5(x + 2) = 5x + 2$

50. $(x + 3) - 8 = (x - 8) + 3$

51. $4 - 2(x - 3) = (4 - 2)(x - 3)$

52. $x - \frac{10}{3} = \frac{x - 10}{3}$

53. $\frac{x}{3} - 10 = \frac{x - 10}{3}$

54. $(x + 3)^2 = x^2 + 3^2$

55 and 56. For what value of c is $x = 3$ the solution to the equation?

55. $3x + c = 2x - 2c$

56. $3x + c = cx - 2$

57 and 58. For what value(s) of c is each of the following equations (a) an identity? (b) a contradiction?

57. $5(x + 3) = 5x + c$

58. $5 - 2(x + 3) = c - 2x$

59. Does any equation of the form $ax + b = cx + d$ (a, b, c, and d are real numbers) have a solution? Explain your answer.

60. Solve the problem in Example 1 using a graphing calculator. Be sure you have an appropriate viewing rectangle. Do you think the algebraic solution or the graphical solution is easier for you? Why?

61. Try to solve the equation of Example 5 with a graphing calculator. Describe your method. Did you get the same answer? Why or why not? Do you think the algebraic or the graphical solution is easier? Why?

62. Try to solve the equations of Examples 10 and 11 with a graphing calculator. Describe your method. Did you get the same answer as you did without using the calculator? Why or why not?

63. From Exercises 25–39, choose a contradiction and an identity. Use a graphing calculator to solve them. In each case, explain why you obtained the graph you did.

64. In Example 14, what do you expect to happen when you graph $y = 10x + 7(10{,}000 - x)$ and $y = 60{,}000$ on the same coordinate system? Check your conjecture by graphing, and explain the result.

III. Applications

65–83. Solve each of the following problems. Check your answers. Sketch a graph in Exercises 65 and 66.

65. Suppose Acme Car Rental charges \$17.95 a day plus 21¢ per mile, while Hurts Car Rental charges \$21.95 a day plus 15¢ per mile. If you want to rent a car for one day, which one would you choose? Why?

66. Suppose the demand for hamburger is $D = 4{,}800 - 2{,}400p$, and the supply is $S = -3{,}600 + 1{,}800p$, where p is the price per pound. At what value of p is the equilibrium price (where supply equals demand) found?

67. Often, to harvest timber, it is necessary to construct a logging road. There are two types of roads to consider: Type A has an initial cost of \$100,000, and type B has an initial cost of \$80,000. Since type A is a better road, hauling costs are \$6.50 per million board feet as compared to \$7.50 per million board feet on the type B road. Which type of road should be constructed? Determine the "break-even point" in million board feet hauled on the two roads. Then analyze other cases as in Example 1, including drawing a graph. Cost is the major consideration. When would type A be better? Type B?

68. As a prospective employee in a clothing store, you are offered a choice of salary: \$100 per week plus 30% of everything you sell, or \$150 per week and 15% of everything you sell. How much would you have to sell to get the same weekly salary under both plans? Which one would you choose?

69. John starts at point A and walks due east at 5 mph. Mary simultaneously starts at point B and walks due west at 3 mph.

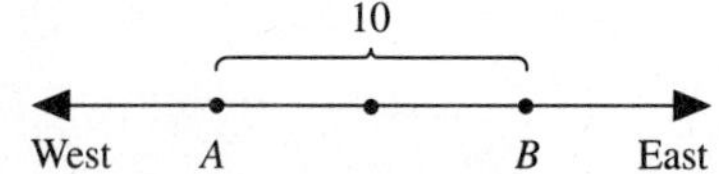

(a) In how much time will they meet if they are 10 miles apart at the beginning?

(b) Assuming they keep walking after they meet, in how much time will they be 30 miles apart?

70. Find the value of x for which the perimeters of the two figures are equal:

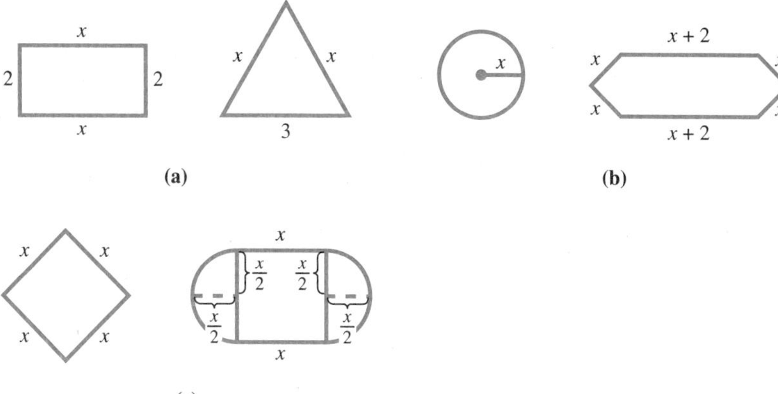

71. In each case, find x if the shaded area is 17.54 cm^2.

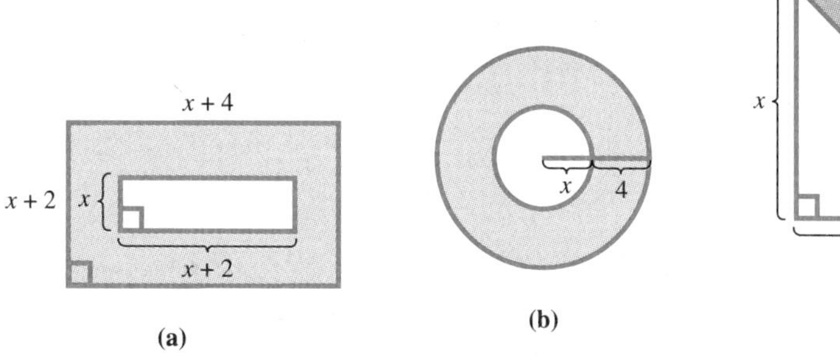

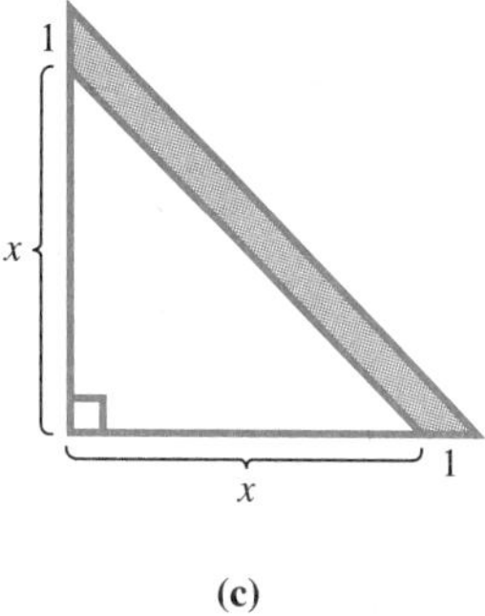

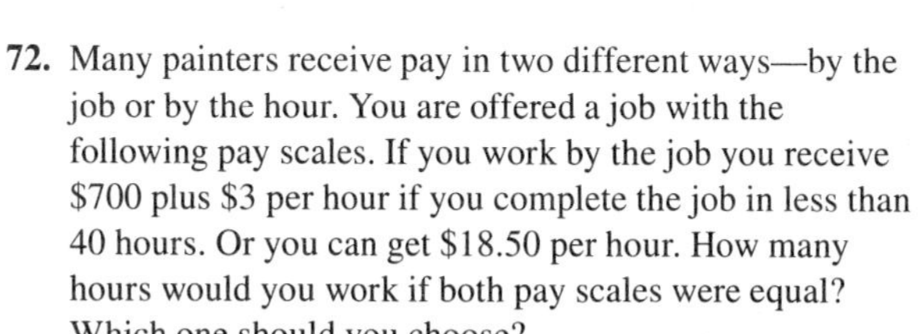

72. Many painters receive pay in two different ways—by the job or by the hour. You are offered a job with the following pay scales. If you work by the job you receive \$700 plus \$3 per hour if you complete the job in less than 40 hours. Or you can get \$18.50 per hour. How many hours would you work if both pay scales were equal? Which one should you choose?

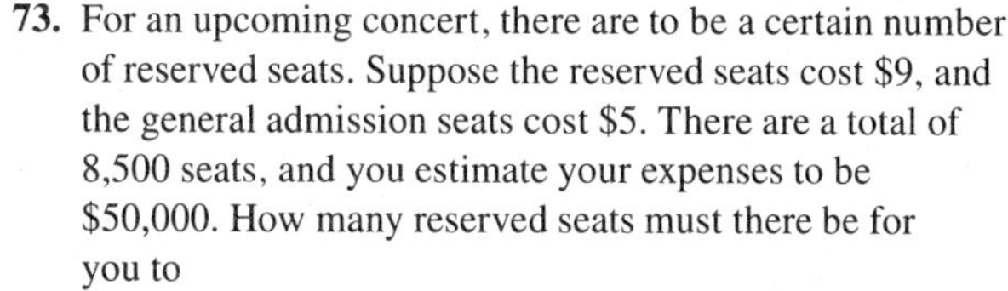

73. For an upcoming concert, there are to be a certain number of reserved seats. Suppose the reserved seats cost \$9, and the general admission seats cost \$5. There are a total of 8,500 seats, and you estimate your expenses to be \$50,000. How many reserved seats must there be for you to

(a) break even? (b) make a \$20,000 profit?

74. To manufacture widgets, suppose the start-up costs were \$29,000, and the cost per widget were \$4.00. How many widgets must you sell at \$6.95 each to

(a) break even? (b) make a 15% (of costs) profit?

75. You walked to town at 4 mph and returned home on the bus, which traveled 40 mph. If the round trip took 1 hour and 15 minutes, how far did you walk?

76. A fertilizer mixture containing 32% nitrogen is desired for a certain crop. The farmer has fertilizer with a 17% nitrogen content and another fertilizer with 43% nitrogen content. How much of each fertilizer is needed to make 100 kg of fertilizer with the desired 32% nitrogen content?

77. Mr. Smith lives 15 miles from work, where he has free parking. Some days, he must drive to work so that he can call on clients, but other days, he can take the bus. It costs him 20¢ per mile to drive the car and 80¢ round trip to take the bus. Assuming there are 22 working days in a month, how many days can he drive and still hold his transportation costs to $30 a month?

78. A typical scoring system on a multiple-choice test where each item has five alternative answers is: (number correct) $-\frac{1}{4}$(number incorrect). Your score on a 50-question test was 35, and you answered every question. How many did you get correct?

79. The owner of the Acme Manufacturing Co. must decide which widget-making machine to buy. The first machine costs $100,000, and each widget will cost $1.95 to make. The second, more expensive machine costs $250,000, but it is more efficient, because each widget costs only $1.35 to make. How many widgets must be manufactured for the cost of the two machines to be the same? Which machine should be bought?

80. Two joggers start simultaneously from opposite ends of the one-mile-long Empire State Bridge. The first jogs 5 mph, and the second jogs 7 mph. They pass each other and continue to the ends of the bridge, turn around, and jog until they meet. How far has the first jogger run? The second?

81. Suppose you drive to work. If you drive at 45 mph, you arrive one minute early. If you drive at 40 mph, you arrive one minute late. How far do you live from work?

82. The daily rental charges for a car at Hartz Car Rental are given by the equation $C = \$25 + \$0.22M$, where M is the number of miles driven and C is the cost.

(a) What will be the charge for driving 280 miles in one day?

(b) How far can you drive for $50?

The daily rental charges at Evis Car Rental are $50 plus 20¢ for each mile over 100 miles.

(c) Write an equation that gives the cost per day in terms of the miles driven. (Assume the daily mileage is more than 100 miles.)

(d) If you planned on driving 280 miles today, from which of the two companies would you choose to rent a car? Why?

(e) For what daily mileage is the charge for both companies the same?

(f) Graph both equations on the same coordinate axes. Compare both the equations and the graphs. How are they similar? Which would you prefer to use? Why?

83. A six-lane running track is in the shape of a rectangle with a semicircle at each end. The length of the rectangle is 1.5 times its width. Each lane is to be 1 meter wide.

(a) What are the length and width of the rectangle if the inside lane is 1,500 meters long?

(b) For a 1,500-meter race, the inside runner would start at the finish line. Where should the runners in the other five lanes start if all are to run 1,500 meters, staying in their lanes? *Hint:* First draw a picture.

IV. Extensions

84. You have three job offers and you must choose one of them.

Sales Job (#1)	*Supervisor's Job (#2)*	*Office Job (#3)*
• $15,000 per year to start • 5% raise every six months • much traveling • three weeks annual vacation	• $16,200 per year to start • 10% raise every year • four 10-hour shifts per week • one weekend shift per month • two weeks annual vacation	• $15,800 per year to start • annual raises of $1,000, $1,200, $1,400, etc. • regular hours, no traveling • two weeks annual vacation

After how long is the salary of the two indicated jobs equal?

(a) #1 and #2 **(b)** #2 and #3 **(c)** #1 and #3

(d) List one advantage and one disadvantage for each job.

(e) Which job would you choose? Why?

(f) Describe your methods for solving parts (a), (b), and (c).

85. A jogger is training for a marathon. She wants to run 20 miles in two hours. She runs the first 10 miles at 8 mph. How fast must she run the last 10 miles to achieve her goal?

86. (a) If the six-digit integer $abcde2$ is three times as large as the six-digit integer $2abcde$, find the other digits in the integer.

(b) Solve the same problem if the known digit is 1 instead of 2.

87. If you "walked" around the earth at the equator, your head would travel farther than your feet. If the distance your head travels is 36 feet more than the distance your feet traveled, how tall are you?

88. Find all three-digit whole numbers satisfying all the following criteria:

(a) the tens digit is 7

(b) the hundreds digit is 4 less than the ones digit

(c) if the digits are reversed, the new number exceeds the old by 396.

89. Find all angles whose supplement exceeds their complement by 90°.

90. A giant watermelon weighed 100 pounds and was 99% water. While standing in the sun, some of the water evaporated, so it was only 98% water. How much did the watermelon weigh then?

91. A test was passed by 76% of all students in the class of juniors and sophomores. Among juniors, 80% passed the test, as did 70% of the sophomores. What is the percent of juniors in the class?

3.4 LINEAR INEQUALITIES AND ABSOLUTE VALUE INEQUALITIES

In the last section we discussed problems that asked the following questions:

1. Given two different rental plans for a car, how many miles can be driven so that the two plans are of equal cost?
2. In the manufacturing of widgets, how many widgets must be produced so that revenue equals cost?
3. What is the length of a dead-heat race between Fast Freddy (running at 5 meters per second) and Slower Sam (running at 3 meters per second) that ends in a tie, if Slower Sam gets a 40-meter head start?

We were able to solve these problems by using linear equations. However, more realistic questions might be:

1. Given two car rental plans, which is the cheaper, or when is plan A *less than* plan B?
2. In manufacturing widgets, how many widgets must be produced to make a profit, or when is revenue *greater than* cost?
3. What is the *longest* race Slow Sam can win?

The models for the latter problems are *inequalities* rather than equalities. Many other words can suggest an inequality: "The average is *over* 80," "produce *more than* 100,000 television sets," "the time is *under* five minutes per mile," "the opinion poll is accurate to *within* a 2% error." We have solved many problems about inequalities by finding a linear equation and then solving for x. We then looked at numbers larger or smaller than x to make our decisions. However, it is more direct to solve an inequality. The procedure for solving an inequality is similar to that for solving an equation.

Solving Inequalities

An **inequality** is a sentence that says two quantities are unequal; that is, one quantity is smaller (or larger) than another quantity.

From Chapter 1, we know that $-3 < -2$ is a true inequality, while $4 < 3$ is a false inequality. The sentence $x + 7 < 9$ is also an inequality, but it is neither true nor false. *Finding numbers that make this inequality true is called solving the inequality. The numbers that make the inequality true are called the solutions of the inequality.* For example,

If $x = 1$, then $1 + 7 < 9$ is true; so 1 is a solution to $x + 7 < 9$.

If $x = -4$, then $-4 + 7 < 9$ is true; so -4 is a solution to $x + 7 < 9$.

If $x = 3$, then $3 + 7 < 9$ is *false;* so 3 is *not* a solution to $x + 7 < 9$.

Many values of x are solutions to $x + 7 < 9$. Notice that a linear equation that is not an identity has only one solution (or perhaps none), whereas a linear inequality generally has an infinite number of solutions.

To solve the equation $x + 7 = 9$, we added -7 to both sides, obtaining $x = 2$ as the solution. The same procedure works for inequalities.

For example, $3 < 4$, so $3 + 2 < 4 + 2$ and $3 + (-2) < 4 + (-2)$. The relative direction of the inequality symbol does not change; that is, another true inequality is obtained. This concept is illustrated on the number line. See Figure 3.14.

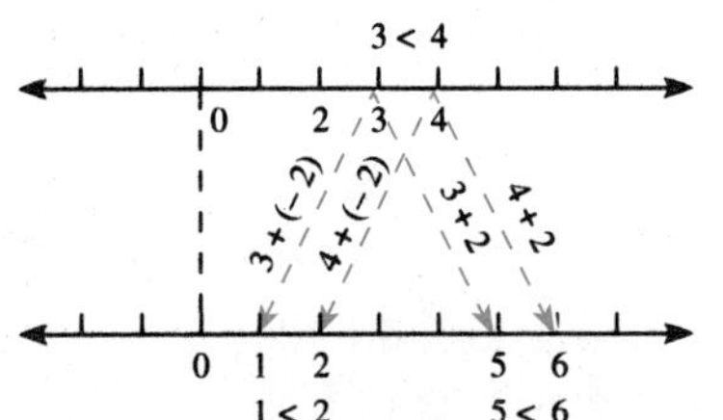

Figure 3.14

The Additive Property of Inequality

Adding (or subtracting) the same number to both sides of an inequality does not alter its *relative* direction.

Algebraically, we say: If $a < b$, then $a + k < b + k$, for any real numbers a, b, and k.

EXAMPLES 1–2

Solve the following inequalities, and graph the solutions on a number line.

1. $x + 7 < 9$

2. $x + 12 \geq 3$

Solutions

1.

$$x + 7 < 9$$

$$x + 7 + (-7) < 9 + (-7) \qquad \text{Add } -7 \text{ (or subtract 7) to both sides of the inequality.}$$

$$x < 2$$

Therefore the solution $\{x \mid x < 2\}$ seems to be correct. Graphing the solution on the number line, we get

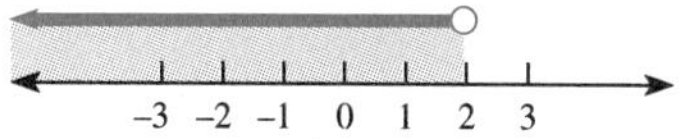

Figure 3.15

It is always wise to check a few values of the solution to see if you have made any careless errors.

Check: Pick several numbers from the solution set and substitute into the original inequality. Also try some numbers that are not in the solution set to see if they make the inequality false.

If $x = 1$, then $1 + 7 < 9$ is true.

If $x = -9$, then $-9 + 7 < 9$ is true.

If $x = 3$, then $3 + 7 < 9$ is false.

Therefore the solution $\{x \mid x < 2\}$ seems to be correct.

2.

$$x + 12 \geq 3$$

$$x + 12 + (-12) \geq 3 + (-12) \qquad \text{Add } -12 \text{ to both sides of the inequality.}$$

$$x \geq -9$$

The solution is $\{x \mid x \geq -9\}$, or graphically,

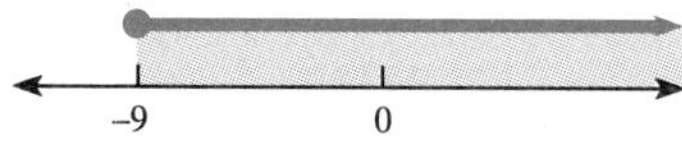

Figure 3.16

Check the solution. ■

We can add or subtract any number to both sides of an inequality without changing the relative direction of the inequality. Is the same true for multiplication and division? Let us try multiplying and dividing by a positive number:

$4 < 6$ Multiply by 2: $2(4) \; ? \; 2(6)$ $8 < 12$ Inequality stays the same	$15 > 9$ Multiply by $\frac{1}{3}$: $\frac{1}{3}(15) \; ? \; \frac{1}{3}(9)$ $5 > 3$ Inequality stays the same
$-12 < -8$ Divide by 2: $\frac{-12}{2} \; ? \; \frac{-8}{2}$ $-6 < -4$ Inequality stays the same	$-9 > -15$ Divide by $\frac{1}{3}$: $\frac{-9}{\frac{1}{3}} \; ? \; \frac{-15}{\frac{1}{3}}$ $-27 > -45$ Inequality stays the same

From these examples, we see that *multiplying or dividing by a positive number does not change the direction of the inequality.*

Let us now try a few examples with negative numbers:

$4 < 6$ Multiply by -2: $-2(4) \; ? \; -2(6)$ $-8 > -12$ Inequality changes	$15 > 9$ Multiply by $-\frac{1}{3}$: $-\frac{1}{3}(15) \; ? \; -\frac{1}{3}(9)$ $-5 < -3$ Inequality changes
$-12 < -8$ Divide by -2: $\frac{-12}{-2} \; ? \; \frac{-8}{-2}$ $6 > 4$ Inequality changes	$-9 > -15$ Divide by $-\frac{1}{3}$: $\frac{-9}{-\frac{1}{3}} \; ? \; \frac{-15}{-\frac{1}{3}}$ $27 < 45$ Inequality changes

From these examples, we see that *multiplying or dividing* both sides of an inequality *by a negative number changes the direction of the inequality.*

Multiplication Property for Inequality

1. Multiplying or dividing both sides of an inequality by a *positive* number leaves the relative direction of the inequality the same.

> **2.** Multiplying or dividing both sides of an inequality by a *negative* number changes the relative direction of the inequality.
>
> Algebraically,
>
> **1.** If $a < b$ and $k > 0$ (k is *positive*), then $ak < bk$ and $\left(\frac{a}{k}\right) < \left(\frac{b}{k}\right)$.
>
> **2.** If $a < b$ and $k < 0$ (k is *negative*), then $ak > bk$ and $\left(\frac{a}{k}\right) > \left(\frac{b}{k}\right)$.

For example, $3 < 5$, so $2(3) < 2(5)$ and $-2(3) > -2(5)$. This concept can be illustrated with number lines (see below).

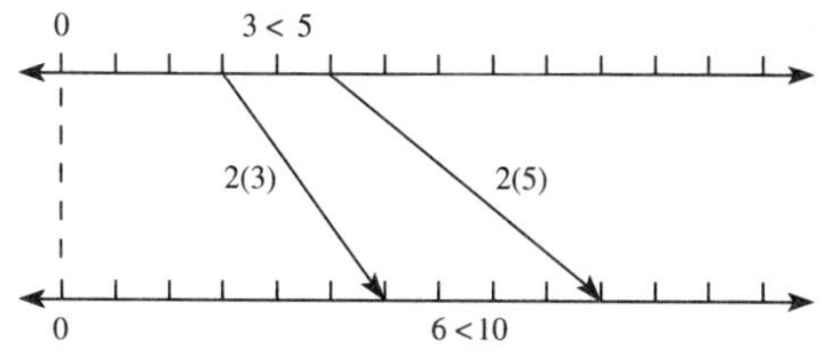

Figure 3.17(a)

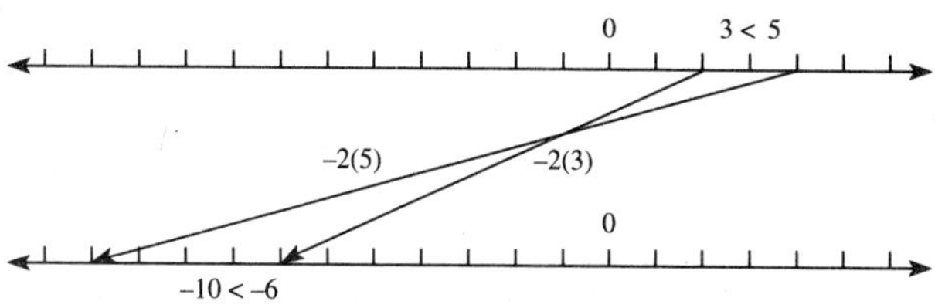

Figure 3.17(b)

EXAMPLES 3–5

Solve each of the following inequalities, and graph the solutions.

3. $-\frac{1}{3}x < -2$ **4.** $2x + 3 \leq 5$ **5.** $-x - 4 < 8$

Solutions

3.

$$-\frac{1}{3}x < -2$$

$$(-3)\left(-\frac{1}{3}x\right) > (-3)(-2) \quad \text{Multiply by } -3.$$

$$x > 6 \quad \text{Notice the change in direction of the inequality.}$$

The solution is $\{x|x > 6\}$, or graphically,

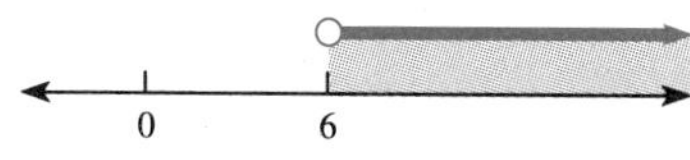

Figure 3.18

Check: If $x = 8$, then $-\frac{8}{3} < -2$ is true.

If $x = 24$, then $-\frac{24}{3} < -2$ is true.

If $x = 3$, then $-\frac{3}{3} < -2$ is false.

For the following problems, we leave the checking to you.

4. $2x + 3 \leq 5$

$2x \leq 2$ **Add -3 to both sides.**

$x \leq 1$ **Divide by 2.**

The solution is $\{x|x \leq 1\}$, and here is the graph:

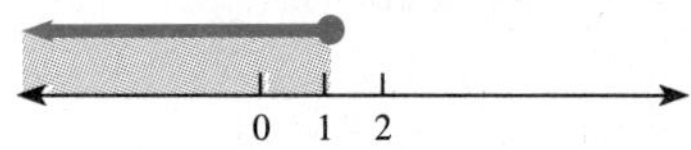

Figure 3.19

Graphing $Y1 = 2x + 3$ and $Y2 = 5$ in a standard viewing rectangle, we see that $2x + 3 \leq 5$ when $x \leq 1$.

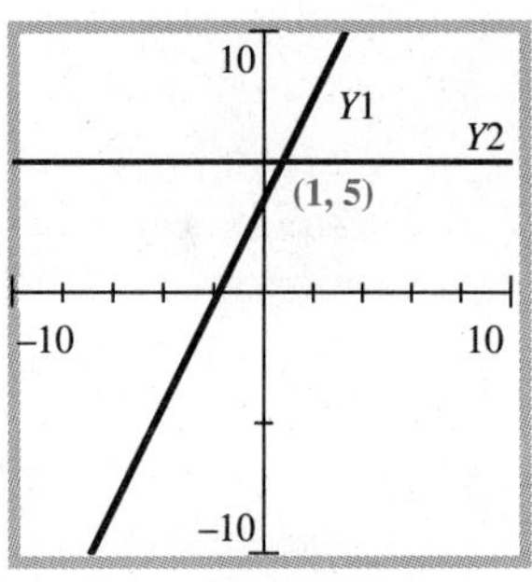

$Y1 = 2x + 3 \quad Y2 = 5$

Figure 3.20

5. $-x - 4 < 8$

$-x < 12$ **Add 4 to both sides.**

$x > -12$ **Divide by -1. Notice the change in the inequality.**

The solution is $\{x|x > -12\}$, and graphically,

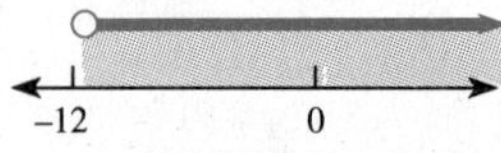

Figure 3.21

You might confirm this solution on a graphing calculator. ■

EXAMPLES 6–8

Solve each of the following inequalities.

6. $-3(x + 4) + 2 \geq 8 - x$

7. $2(x - 1) + 5x < 7(x - 4)$

8. $\frac{1}{3}(5x + 6) \geq x + 2\left(\frac{x}{3} + 1\right)$

Solutions

6. $-3(x + 4) + 2 \geq 8 - x$

$$-3x - 12 + 2 \geq 8 - x \quad \text{Distributive property.}$$
$$-3x - 10 \geq 8 - x$$
$$-3x \geq 18 - x \quad \text{Add 10.}$$
$$-2x \geq 18 \quad \text{Add } x.$$
$$x \leq -9 \quad \text{Divide by } -2. \text{ Direction changes.}$$

So the solution is $\{x | x \leq -9\}$. Check the answer.

7. $2(x - 1) + 5x < 7(x - 4)$

$$2x - 2 + 5x < 7x - 28$$
$$7x - 2 < 7x - 28$$
$$-2 < -28 \quad \text{A contradiction!}$$

Therefore there are no values of x which make the inequality true. There is *no solution*.

8. $\frac{1}{3}(5x + 6) \geq x + 2\left(\frac{x}{3} + 1\right)$

$$\frac{5}{3}x + 2 \geq x + \frac{2}{3}x + 2$$
$$\frac{5}{3}x + 2 \geq \frac{5}{3}x + 2$$
$$2 \geq 2 \quad \text{Always true!}$$

Therefore we have an *identity*. Thus the solution is the set of *all real numbers*. ■

We now consider two applications of inequalities.

EXAMPLE 9

A car rental agency offers two plans: Plan A is a straight 27¢ a mile and plan B is \$27 plus 12¢ a mile. Which plan is cheaper?

Solution

Using an Equation Our answer will depend on the number of miles we drive. Let M be the number of miles we expect to drive.

$$\text{Cost of plan A} = 0.27M$$
$$\text{Cost of plan B} = 27 + 0.12M$$

We want to know when

$$\text{Cost of plan A} \leq \text{Cost of plan B}$$

Thus we need to solve

$$\begin{aligned} 0.27M &\le 27 + 0.12M \\ 0.15M &\le 27 \\ M &\le \frac{27}{0.15} \\ M &\le 180 \text{ miles} \end{aligned}$$

Our results suggest that plan A is cheaper, if we drive less than 180 miles. If we drive more than 180 miles, plan B is cheaper. We might want to check this for a specific number of miles. If we let $M = 100$, then the cost of plan A $= 0.27(100) = \$27.00$, and the cost of plan B $= 27 + 0.12(100) = \$39.00$. Let $M = 300$: The cost of plan A is \$81.00, and the cost of plan B is \$73.00.

One further comment about our solution: We chose to find when plan A was cheaper than plan B. We could also have chosen to find when plan B is cheaper than plan A. We would then solve

$$\begin{aligned} 27 + 0.12M &\le 0.27M \\ 180 &\le M \quad \text{or} \quad M \ge 180 \end{aligned}$$

Plan B is cheaper if we drive more than 180 miles. The answer is, of course, the same as in our first solution.

Using a Graph

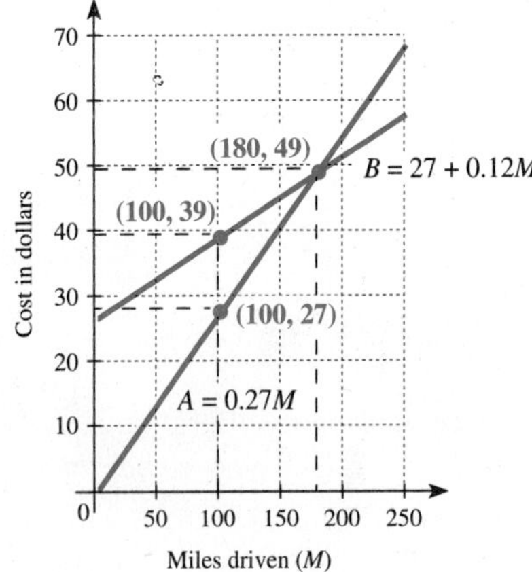

Figure 3.22

$$\begin{aligned} \text{Cost of } A &= 0.27M \\ \text{Cost of } B &= 27 + 0.12M \end{aligned}$$

From the graph, A is a better buy below 180 miles and B is better above 180 miles. ■

EXAMPLE 10

The sum of three consecutive integers is between 30 and 60. What are the possible values for the set of integers?

Solution

Let x be the first integer. Then, $x + 1$ is the second consecutive integer, and $x + 2$ is the third consecutive integer. The sum of the three integers equals $x + x + 1 + x + 2 = 3x + 3$. But the sum is *between* 30 and 60.

$30 < \text{sum} < 60$	Definition of "between" (see Chapter 1).
$30 < 3x + 3 < 60$	Solve the double inequality.
$27 < 3x < 57$	Add -3 to both inequalities.
$9 < x < 19$	Divide by 3.

Therefore, x must be between 9 and 19. Our solution, then, is any set of three consecutive integers where the first integer is between 9 and 19. Some possible sets are {10, 11, 12}, {11, 12, 13}, and {16, 17, 18}. ■

Compound Statements

The above example imposes two conditions on the sum: sum > 30 *and* sum < 60. This is an example of a **compound statement.** We solved some simple compound statements in Chapter 1. The following statements are also compound.

EXAMPLES 11–13

11. Find all x such that $2x - 1 < 3$ and $x - 1 > -3$.

12. Find all x such that $6x - 4 \le 2x$ or $-3x \le -9$.

13. Find all x such that $-4 < 3x - 1 < 5$.

Solutions

11. The word "and" implies we are seeking *all* values of x that are solutions to *both* inequalities, that is, the **intersection** of the solution sets:

$$2x - 1 < 3 \quad \text{and} \quad x - 1 > -3$$
$$x < 2 \quad \text{and} \quad x > -2$$

Graphing the solutions on a number line, we have

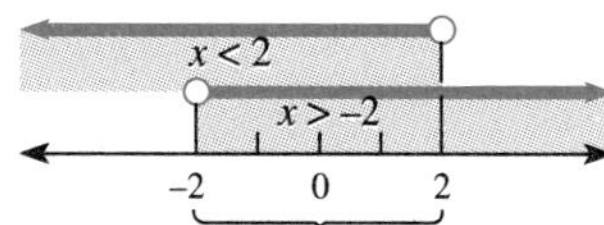

Figure 3.23

Another way to write the solution is

$$-2 < x < 2$$

which is read "x is between -2 and 2" or "-2 is less than x *and* x is less than 2."

12. The word "or" implies we are seeking values of x that are solutions to one inequality *or* the other, that is, the **union** of the solution sets:

$$6x - 4 \leq x \quad \text{or} \quad -3x \leq -9$$
$$x \leq 1 \quad \text{or} \quad x \geq 3$$

Graphing the solutions on the number line, we have

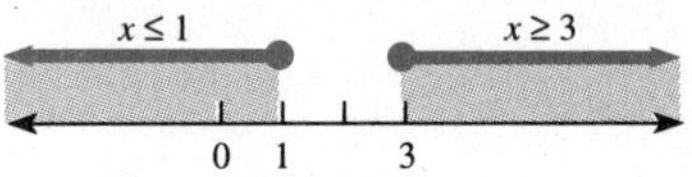

Figure 3.24

The solution is the union of both solution sets. The only numbers left out are those between 1 and 3.

13. $-4 < 3x - 1 < 5$ is the same as the compound statement $-4 < 3x - 1$ and $3x - 1 < 5$. Both must be true at the same time. Thus we can solve the double inequality:

$$-4 < 3x - 1 < 5$$
$$-4 + 1 < 3x - 1 + 1 < 5 + 1 \qquad \textbf{Add 1 to both inequalities.}$$
$$-3 < 3x < 6$$
$$\frac{-3}{3} < \frac{3x}{3} < \frac{6}{3} \qquad \textbf{Divide both inequalities by 3.}$$
$$-1 < x < 2$$

Therefore, x is between -1 and 2. Graphing the inequality, we have

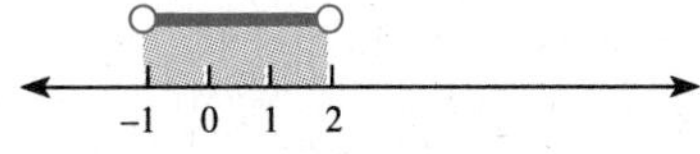

Figure 3.25

EXAMPLE 14

The lengths of the three sides of a triangle are x, $2x - 1$, and 5. What are the possible values of x?

Solution

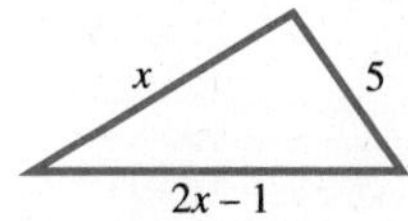

Figure 3.26

We recall that the sum of any two sides of a triangle must be greater than the third.

Therefore, the following three inequalities must *all* be *true* simultaneously for the triangle to exist:

$$x + 5 > 2x - 1$$
$$x + (2x - 1) > 5$$
$$(2x - 1) + 5 > x$$

Solving each inequality, we have, respectively,

$$x < 6$$
$$x > 2$$
$$x > -4$$

The solution is the intersection of all three solutions, which is $2 < x < 6$. A number line is useful to picture the intersection (see Figure 3.27).

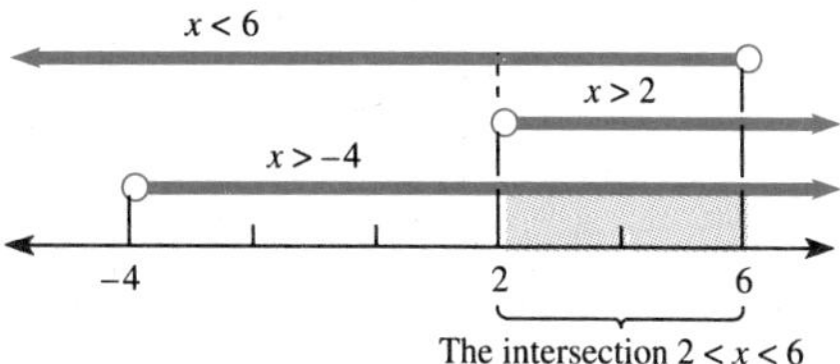

Figure 3.27

Solving Absolute Value Inequalities

In Chapter 1 we defined the distance between a point x and the origin as the absolute value of x. Thus,

$|3|$ is the distance between the origin and the point 3.

$|-3|$ is the distance between the origin and the point -3.

Notice that $|3| = |-3| = 3$. See Figure 3.28.

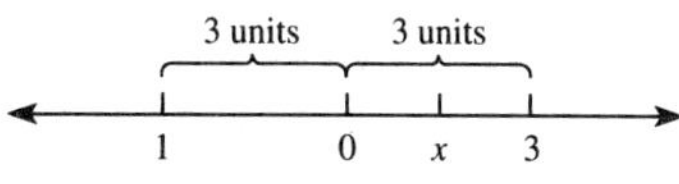

Figure 3.28

If we pick a point x that is between the origin and the point 3, then *the distance between the point x and the origin is less than* 3 *units.* This can be written algebraically as

$$|x| < 3$$

Notice that the above inequality is also valid for any point x between the origin and the point -3. Thus all the points x that are less than 3 units from the origin are *between the points* -3 *and* 3. This can be expressed as $-3 < x < 3$.

Absolute Value Inequality

If the distance between a point x and the origin is less than a units, we write

$|x| < a, a \geq 0$

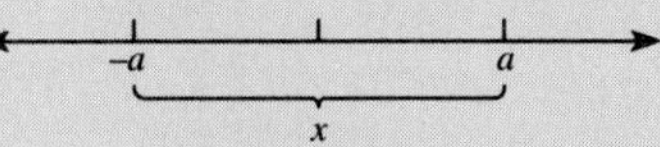

Figure 3.29

If the distance between two points x and a is less than b units, we write

$|x - a| < b, b \geq 0$

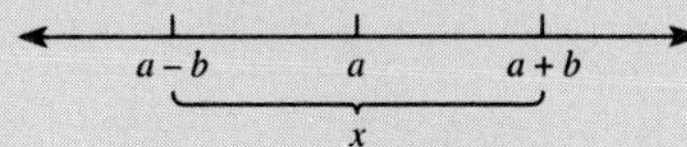

Figure 3.30

We can use the definition of absolute value and the properties of inequalities to solve absolute value inequalities.

EXAMPLES 15–18

Solve the following, and graph the solution set on a number line.

15. $|x| < 5$ **16.** $|x - 3| < 5$ **17.** $|x + 3| \leq 5$ **18.** $|3 - x| < 5$

Solutions

15. $|x| < 5$

$-5 < x < 5$ By definition x is a point that is less than 5 units from the origin.

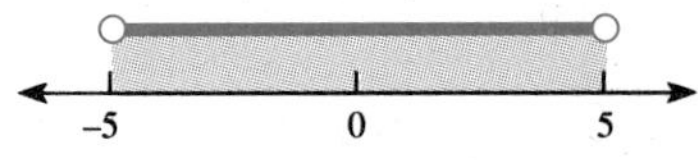

Figure 3.31

In words, x is any point between -5 and 5, or $\{x | -5 < x < 5\}$.

16. $|x - 3| < 5$

Using an Inequality

$-5 < x - 3 < 5$

$-5 + 3 < x - 3 + 3 < 5 + 3$ Add 3 to both inequalities.

$-2 < x < 8$

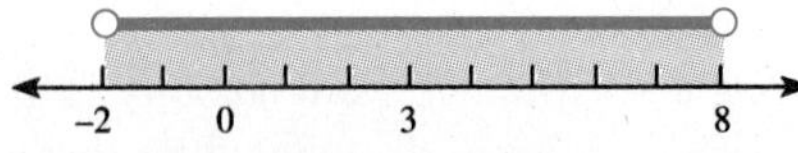

Figure 3.32

Therefore, all the points that are less than 5 units from the point 3 are between -2 and 8, or $\{x \mid -2 < x < 8\}$.

Using a Graph Graphing $Y1 = \text{ABS}(x - 3)$ and $Y2 = 5$ in a standard viewing rectangle, we see that $|x - 3| < 5$ when $-2 < x < 8$.

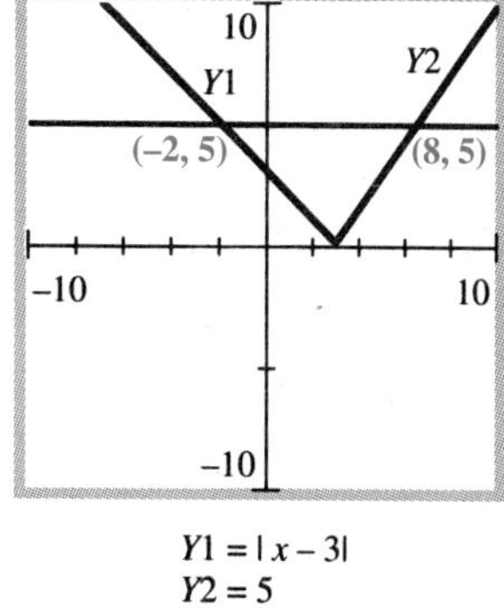

$Y1 = |x - 3|$
$Y2 = 5$

Figure 3.33

17. $|x + 3| \leq 5$

$-5 \leq x + 3 \leq 5$

$-8 \leq x \leq 2$ Add -3 to both inequalities.

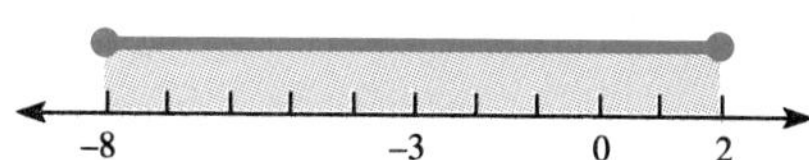

Figure 3.34

Since $x + 3 = x - (-3)$, $|x + 3| = |x - (-3)|$. Therefore, $|x + 3| \leq 5$ is the set of all points x that are less than or equal to 5 units from the point -3, or $\{x \mid -8 \leq x \leq 2\}$.

18. $|3 - x| < 5$

$-5 < 3 - x < 5$

$-8 < -x < 2$ Subtract 3 from both inequalities.

$8 > x > -2$ Multiply through by -1.

or

$-2 < x < 8$

Figure 3.35

If x is not a point between -3 and 3, then the distance from x to the origin is greater than 3 units. We write

$$|x| > 3$$

By the definition of absolute value, this is equivalent to saying

$$x > 3 \quad \text{or} \quad -x > 3$$

If we multiply both sides of $-x > 3$ by -1, we get $x < -3$. Thus the solution set is $\{x|x > 3 \text{ or } x < -3\}$. The solution is shown in Figure 3.36.

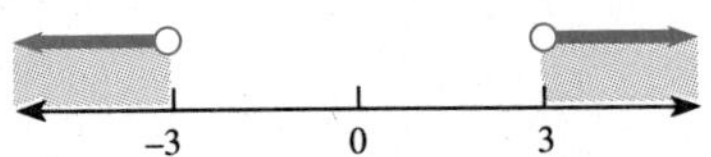

Figure 3.36

EXAMPLES 19–20

Solve each of the following, and graph the solution set.

19. $|x| \geq 5$ **20.** $|x - 3| \geq 5$

Solutions

19. $|x| \geq 5$

This is equivalent to solving the following compound sentence:

$$-x \geq 5 \quad \text{or} \quad x \geq 5$$
$$x \leq -5 \quad \text{or} \quad x \geq 5$$

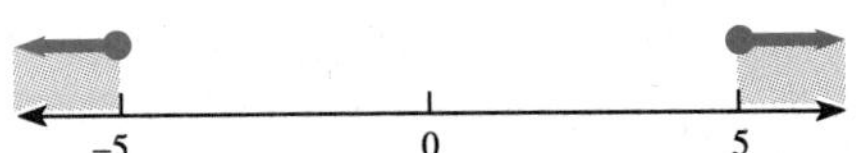

Figure 3.37

20. $|x - 3| \geq 5$

This is equivalent to solving the following compound sentence:

$$-(x - 3) \geq 5 \quad \text{or} \quad (x - 3) \geq 5 \qquad \text{Solve both inequalities.}$$
$$x - 3 \leq -5 \quad \text{or} \quad x - 3 \geq 5$$
$$x \leq -2 \quad \text{or} \quad x \geq 8$$

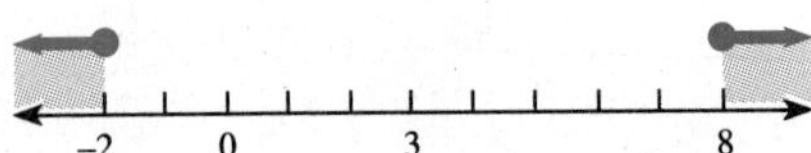

Figure 3.38

We now summarize the results for absolute-value inequalities. For $k \geq 0$,

Absolute Value Inequality	*Equivalent Inequalities*
$\|x\| \leq k$	$-k \leq x \leq k$
$\|x - a\| \leq k$	$-k \leq x - a \leq k$
$\|x\| \geq k$	$x \geq k$ or $x \leq -k$
$\|x - a\| \geq k$	$x - a \geq k$ or $x - a \leq -k$

3.4 EXERCISES

I. Procedures

1–40. **(a)** Solve each of the following inequalities. **(b)** Graph each solution on a number line. **(c)** Check your answers.

1. $x + 2 < 8$ **2.** $x + 2 > 8$ **3.** $-x + 2 > 8$

4. $-x + 2 < 8$ **5.** $|x| < 8$ **6.** $|x| \geq 8$

7. $|x - 2| \leq 8$ **8.** $|x - 2| \geq 8$ **9.** $|x + 2| \leq 8$

10. $|x + 2| \geq 8$ **11.** $|2x| \geq 6$ **12.** $|2x - 1| \geq 7$

13. $|3x + 1| \leq 7$ **14.** $\left|\frac{1}{2}x - 4\right| \leq 1$

15. $6 < x + 1 < 10$ **16.** $-1 < x - 5 < 8$

17. $-\frac{3}{4}x \leq 2$ **18.** $2x > -\frac{3}{4}$ **19.** $\frac{2}{3}x \leq \frac{8}{3}$

20. $\frac{2}{3}x \leq \frac{3}{8}$ **21.** $2x - 1 < 5$ **22.** $3x + 2 \geq 6$

23. $-2x - 1 \leq 4$ **24.** $-3x - 7 > 8$

25. $4x - \frac{2}{5} > \frac{3}{5}$ **26.** $-4x + \frac{2}{5} > -\frac{3}{5}$

27. $3(x + 2) - 5x < x$

28. $2(3x - 5) + 7 > x + 12$

29. $-2(t - 5) - 1 \leq 5t + 7(1 - t)$

30. $3(y - 4) - (6y + 2) \geq 5 - 3y$

31. $-3x + \frac{1}{2} + x < -\frac{1}{3}(4x + 1)$

32. $2x + \frac{2}{3} - x > \frac{5}{6}(2x - 1)$

33. $x + 2 > 8$ and $x < 10$

34. $x - 5 \leq 6$ or $x > 6$

35. $7 - 2x \leq 1$ or $x \leq 0$

36. $3 - x \geq 0$ and $x \geq 0$

37. $|5 - x| < 7$ **38.** $|5 - x| > 7$

39. $|3 - 2x| \leq 4$ **40.** $|3 - 2x| > 1$

II. Concepts

41–46. If $a \leq b$, then mark each of the following statements with AT for always true, ST for sometimes true, or NT for never true. If a statement is sometimes true (ST), give an example when it is true and one when it is false.

41. $2a \leq 2b$ **42.** $a - 3 \leq b - 3$

43. $-2a \leq -2b$ **44.** $\frac{1}{a} \leq \frac{1}{b}$

45. $\frac{a}{b} \leq 1$ **46.** $a^2 \leq b^2$

47. For each of the following graphs, pick *all* the statements A through J that describe the graphs.

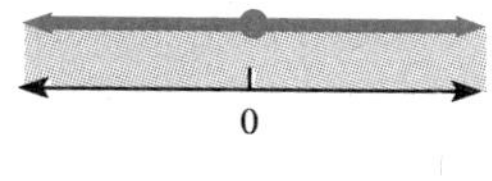

(a)

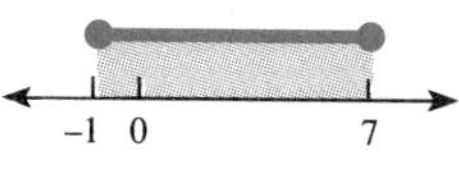

(b)

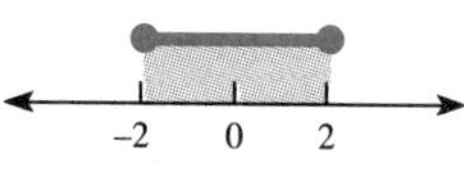

(c)

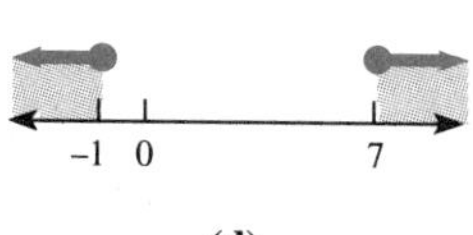

(d)

A. The point x is less than 2 units from the origin.

B. The point x is greater than or equal to 4 units from the point 3.

C. $x \geq -1$ and $x \leq 7$

D. $|x - 3| \geq 4$

E. $-2 < x < 2$

F. $|x| \geq 0$

G. $|x| < 2$

H. $|x - 3| \leq 4$

I. $x < -1$ or $x \geq 7$

J. $x - 3 \geq 0$ or $2x - 1 < 7$

48. Solve the problem in Example 9, Section 3.3, using an inequality.

49. Explain why the inequalities $|x - 3| < 5$ and $|3 - x| < 5$ have the same solution.

50. What type of inequalities have solutions whose graph is of the form

Give several examples. Explain your reasoning.

51. Solve each of the following exercises using a graphing calculator. Describe your method in each case.

(a) Exercise 3

(b) Exercise 7

(c) Exercise 10

(d) Exercise 15

52. Try to solve the inequality in the indicated example in this section using a graphing calculator. Explain your method and your result in each case.

(a) Example 20

(b) Example 7

(c) Example 8

53 and 54. For what values of c are the following inequalities **(a)** an identity? **(b)** a contradiction?

53. $-3(x - 1) \geq -3x + c$

54. $3 - 2(x + 2) \leq c - 2x$

55. If $x < y$, what are the possible values for p such that

(a) $x + p < y$?

(b) $x < y - p$?

III. Applications

56 and 57. Solve each of the following exercises. Express your answer using an inequality.

56. If I average between 26 and 30 miles per gallon on the highway and my gas tank holds between 12 and 12.5 gallons, what values are possible for the "driving range" of my car?

57. My odometer showed I drove exactly 300 miles, and I needed 10 gallons of gas to fill up the gas tank. However, my odometer may be "off" by 1%, and the "full tank" could be off by $\frac{1}{2}$ gallon. What values are possible for my actual gas mileage? (Assume I started with a full tank of gas.)

58–81. Solve each of the following problems.

58. Mavis Rent-A-Car charges \$95.70 per week plus 16 cents a mile under plan A and \$150.50 per week with unlimited mileage under plan B. How many miles must be driven for plan B to be the best choice?

59. A bank has two checking account plans: Plan A charges \$2.00 a month plus 4¢ a check. Plan B charges \$1.50 a month plus 5¢ a check. When is it cheaper to have plan B?

60. In the city, teachers are paid \$20,000 to start and \$500 for each year of teaching experience. In the suburbs, teachers are paid \$17,500 to start and \$700 for each year of teaching experience. How many years of teaching experience must a teacher have in order to make the salary in the suburbs higher?

61. The taxi fare from Here to There is \$1.20 for the first $\frac{1}{5}$ mile and 10¢ for each additional $\frac{1}{10}$ mile.

(a) What is the longest distance you can travel for \$5?

(b) If you give a 15% tip, what is the longest distance you can travel for \$5?

62. A vacuum salesperson earns 15% commission for each \$250 vacuum he sells. How many vacuums must he sell to earn at least \$300 per week, if he

(a) works by commission only?

(b) receives \$50 per week plus commission?

63. The vacuum salesperson in Exercise 62 is offered a choice of a straight commission of 18% of all sales or \$50 plus 15% of all sales. How many vacuum cleaners must be sold to make the straight commission plan the better choice?

64. Your club wants to hire a band for its spring festival. Band A will play for \$300 plus 40% of the revenue. Band B will play for \$600. If you expect 500 people to attend, what is the most you could charge each person if you want to make Band A the better buy?

65. The selling price of an item is the cost to the store plus a markup amount. The X-mart store has a minimum of a 30% markup on all items. A color television costs the X-mart store \$195. What are the possible values for the amount of markup if the X-mart selling price is to be below the competitor's selling price of

(a) \$275? **(b)** \$250?

66. **(a)** What are the possible values for the first of three consecutive integers, if the sum of any two integers is larger than the remaining consecutive integer?

(b) Can any three positive consecutive integers be the lengths of the sides of a triangle? Why?

67. The sum of three consecutive integers is between 50 and 80. What are the possible values for the three consecutive integers?

68. Find the range of values of x in each figure below.

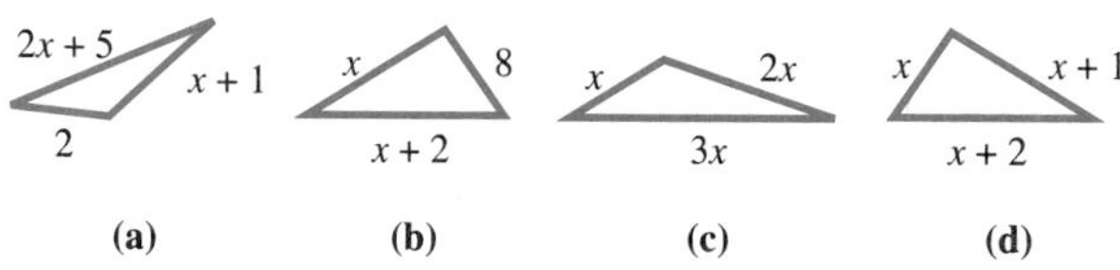

69. Find x such that the perimeter of the trapezoid is

(a) greater than 50.

(b) greater than 10 but less than 50.

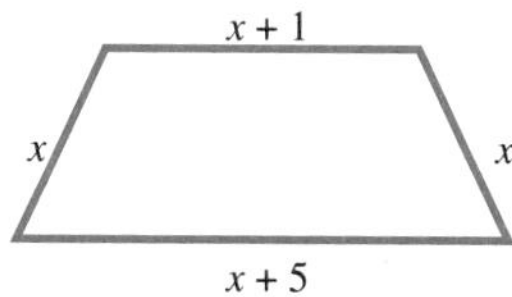

70. Three towns, A, B, and C, do not lie on a straight line. If A is 10 miles from B, and B is 15 miles from C, how far could A be from C?

71. Ted can walk to school in 10 minutes and to his friend's house in 25 minutes. Assuming Ted and his friend walk at the same rate, how long does it take Ted's friend to walk to school?

72. Your first two test scores in Math 101 were 77 and 65 (each out of 100 points). What score do you need on the third test (100 points) to have an average of at least

(a) 75? **(b)** 80? **(c)** 85?

73. Currently the U.S. automakers must achieve a 27 mpg average for all the cars they sell. Suppose Major Motors sells 350,000 large-size cars that average 21 mpg. What is the least number of small cars averaging 32 mpg that Major Motors must sell to achieve the mileage requirement?

74. Often, in a timber sale, it is necessary to construct a logging road. There are two types of roads to consider: Type A has an initial cost of \$100,000, and type B has an initial cost of \$80,000. Since type A is a better road, hauling costs are \$6.50 per million board feet as compared to \$7.50 per million board feet on the type B road. If cost is the major consideration, how many board feet must be hauled to make the type A road the better buy?

75. The minimum amount of food (F) in pounds per day that must be eaten by a person of weight (w) to maintain a 98.6°F body temperature is given by the inequality

$$F \geq \frac{1}{50}w$$

(a) If a person weighs 180 pounds, how many pounds of food must he eat to maintain a temperature of 98.6°F?

(b) A child weighing 68 pounds eats a pound of food a day. Is this enough food to maintain a normal body temperature?

76. The weight of a mouse (w) and the amount of food (F) in ounces it must eat a day to maintain normal body temperature are given by

$$F \geq \frac{1}{2}w$$

(a) If a mouse weighs approximately 8 ounces, how much food should it eat a day?

(b) Which has the higher ratio of F to w, the mouse or a human?

(c) Can you give a reason for this phenomenon?

77. The relationship between Centigrade (C) and Fahrenheit (F) temperature is

$$F = \frac{9}{5}C + 32$$

(a) If the temperature is over 92°F, what is the Centigrade temperature?

(b) If the temperature is less than 25°C, what is the Fahrenheit temperature?

(c) A quick rule of thumb for converting from Centigrade to Fahrenheit is to double the Centigrade temperature and add 30. For what range of Centigrade temperatures will this rule be accurate to 2°F?

78. Parts for an electronic box are mass-produced. Thus the sizes of the parts differ to a small degree. The assembly of the parts will work if the measurement of each part falls within a certain range, as shown below.

Parts	*Allowable error in tenths of a millimeter*
a	$-2 \le a \le 2$
b	$-4 \le b \le 4$
c	$-3 \le c \le 3$
d	$-1 \le d \le 1$
e	$-2 \le e \le 2$

The total error must be between -6 and 6, otherwise the electronic box will not function properly.

(a) If the error of part a is -1, b is 3, c is -2, and d is $\frac{1}{2}$, what can the error of e be for there to still be an acceptable electronic box?

(b) If the error of part a is 2, b is 4, c is -1, and d is 0.7, what is an acceptable error for e?

(c) If the error of part a is 0, b is 4, c is 3, and d is 0, what is an acceptable error for e?

79. The sides of a rectangle are measured accurately to within 0.001 cm. If the length is 3.2 ± 0.001 cm, and the width is 2.4 ± 0.001, write an inequality for the area. *Note:* 3.2 $\pm$ 0.001 means $3.199 \le \text{length} \le 3.201$.

80. The Tenth City Bank gives the following options for monthly charges for checking accounts—option A: $2 plus 25 cents per check; option B: $3 plus 20 cents per check; and option C: $8 with unlimited checks. You estimate your monthly check usage to be M checks. For what values of M would you choose

(a) option A?

(b) option B?

(c) option C?

Check your results by graphing.

81. The Rent-A-Jalopy Car Rental Agency gives the following options—option A: $15 per day plus 20¢ per mile; option B: $25 per day plus 15¢ per mile; and option C: $60 per day with unlimited mileage. You estimate your mileage for today to be M miles. For what values of M would you choose

(a) option A?

(b) option B?

(c) option C?

Check your results by graphing.

IV. Extensions

82. You want to buy a better car. The dealer suggested your present car is worth $1,600. You figure you can afford $150 per month. The dealer would allow you 48 months to pay at an interest rate of 8 percent per year. What is the price range of the car you can afford?

83. In 1993 a chocolate company raised the price of a $1.00 chocolate bar to $1.25. The president's economic council said price increases were limited to 9.5% increases. To comply, the company increased the weight of the chocolate bar so the new price could remain at $1.25. If the candy bar weighed 1.05 ounces, how much should it increase in weight?

84. Suppose we used the simplified formula $F = 2C + 30$ instead of $F = \frac{9}{5}C + 32$ to change from Celsius to Fahrenheit temperatures. For what values of C is this simplified formula reasonably accurate? (Indicate your interpretation of "reasonably accurate" and try to use an inequality in your solution.)

CHAPTER 3 SUMMARY

In this chapter we studied many linear relationships, some of which are linear functions. For example, the cost to drive a car is a function of the number of miles driven. By discussing the cost of driving a car a given number of miles, we are led to three methods for displaying a linear function: a table, a graph, and an equation. There are advantages to each method. In the function, $\text{cost}(M) = 23 + 0.2M$, we were given either the cost or the number of miles. For example, if the cost is

\$200, then we need to find a solution for M in the equation $200 = 23 + 0.2M$. This is called solving a linear equation in one unknown. Linear equations in one variable of the form $ax = b$, $ax + b = c$, and $ax + b = cx + d$ were solved in this chapter both algebraically and graphically. Graphing linear functions will be studied in more detail in Chapter 4.

Proportions were defined and solved as an example of a linear equation of the form $ax = b$. This completes the idea of ratio introduced in Chapter 1.

Comparing the costs of *two* car rental plans created the need for inequalities. Many of the applications of linear equations may be stated as linear inequalities. For example, when are the costs of two car rental plans *equal*, or when is the cost of one plan *less than* that of the other? When does cost *equal* revenue, or when is revenue *greater than* cost (creating a profit)? Equations and inequalities can be explored with a graphing calculator.

Important Words and Phrases

linear equation	identity
linear function	contradiction
solutions of an equation	an inequality
equivalent equations	absolute value inequality
literal equation	intersection
proportions	union
conditional equation	

Important Properties and Procedures

- Solving an equation in one variable
- Properties of equality
- Properties of inequalities
- Solving an inequality
- Graphing the solutions of an inequality in one variable
- Solving an absolute-value inequality
- Checking a solution
- The basic rate formula

REVIEW EXERCISES

I. Procedures

1–25. Solve each of the following equations for x, and check your answers.

1. $5 = \frac{x}{15}$

2. $-5 = \frac{15}{x}$

3. $5(x - 2) = 15$

4. $-5 = 15(x - 2)$

5. $5x - 2 = 15$

6. $5(x - 2) = 15(2x - 5)$

7. $5 - 3(x - 2) = 15$

8. $5x - 3(x - 2) = 15$

9. $5x - 3(x - 2) = 15(x + 1)$

10. $(5x - 3)(3x - 2) = 15x^2$

11. $\frac{3}{x - 2} = \frac{15}{2}$

12. $\frac{2}{3} = \frac{x - 2}{15}$

13. $14x + 1 = 6 - 3(2x - 5)$

14. $6 - 4(2 - 3x) = 4(x + 2)$

15. $\frac{x - 5}{2} = \frac{x}{3} - 2$

16. $7 - 2x = -(3x - 2) + 2(x - 5)$

17. $3(2x - 3) + 5 = 1 - [8 - 2(1 + 2x)]$

18. $5x + 2(3x - 1) = -1 - 5(x - 3)$

19. $\frac{3x}{2} - \frac{2(x + 1)}{3} = 3 - \frac{x}{2}$

20. $3 = \frac{-x}{5} + \frac{2(x + 1)}{3}$

21. $2(4x - 1) = -(10 + x)$

22. $\frac{6x + 20}{4} = 3 + \frac{5x}{6}$

23. $\frac{x}{3} - \frac{2x - 1}{4} + 1 = \frac{5x}{6}$

24. $-4(2 - 3x) = 18 - x$

25. $4\{2 + 3[x - 2 - (-2x)]\} = 36x - 17$

26–29. Solve each of the following equations by either performing one calculator calculation at the end or by graphing. Estimate your answer before using a calculator.

26. $1.0412x - 0.1234 = 0.8123x + 1.9526$

27. $3(2.3614x - 2.1823) = 2x - 6.0321$

28. $1.6251(2.3621x - 4.1283) = 0.9725x - 5.6213$

29. $7.4291 + 2.1235(x - 3.716) = 0.0231x - 20.7025$

30–33. Solve each equation for the indicated variable.

30. $F = ab + ac$ for a

31. $S = \frac{1}{2}gt^2$ for g

32. $S = 2ab + 2ac + 2bc$ for b

33. $A = 2\pi r^2 + 2\pi rh$ for h

34–49. **(a)** Solve the following inequalities. **(b)** Graph your solutions on a number line. **(c)** Check your answers.

34. $3x + 5 \geq 8$

35. $4x - \frac{2}{3} > \frac{1}{3}$

36. $|x| - 1 \leq 5$

37. $3(x - 4) - (6x + 2) \geq 5 - 3x$

38. $|x| \leq 8$

39. $-3(2 - 2x) < 6x - 30$

40. $-5(3x - 2) > 4 - 15x$

41. $\frac{3}{5}(2x - 1) - \frac{1}{3}(4x + 3) \geq 3$

42. $\frac{3}{4}(3x + 2) - \frac{1}{2}(2x - 3) < \frac{1}{8}$

43. $|x - 3| \leq 6$

44. $|x + 5| \geq 8$

45. $|x - 1| \leq 5$

46. $7 - 2x \leq 3$ or $3x \leq 9$

47. $5 - 3x > 2$ or $-x < 5$

48. $x - 3 > 5$ and $3 - x < 1$

49. $5 - x > 2$ and $4 + x > 3$

II. Concepts

50–57. True or false. Explain why.

50. The equation $\frac{x}{2} + \frac{x}{3} = \frac{x}{5}$ is a contradiction.

51. The equation $\frac{5}{x + 5} = 0$ has one solution.

52. The equation $\frac{x}{3} - 10 = \frac{x - 30}{3}$ is an identity.

53. The number 5 is a solution for the equation

$$3 = -\frac{x}{5} + \frac{2(x + 1)}{3}.$$

54. Both $\frac{x}{2} - \frac{x}{3} = 1$ and $3x - 2x = 6$ have the same solution.

55. If $a < b$, then $\frac{1}{a} < \frac{1}{b}$.

56. $|x - 3| = |x| - |-3|$

57. If r is the solution of $ax = b$ and s is the solution of $cx = d$, then rs is the solution of $acx = bd$.

58. Describe, in your own words, the process of finding the solution(s) to a linear equation or inequality or showing that no solutions exist. In what ways can a graphing calculator be useful?

59. How does the strategy "use a proportion" relate to solving a linear equation?

III. Applications

60–82. Solve the following problems. Be sure to check your answers.

60. Find three consecutive odd integers whose sum is 105.

61. Two skateboarders start toward each other from opposite ends of a six-mile road. Linda rides at a steady 9 mph, and Mark rides $\frac{2}{3}$ as fast. How far does each person ride before they meet?

62. Find x if the perimeter of the hexagon shown here is 36 centimeters.

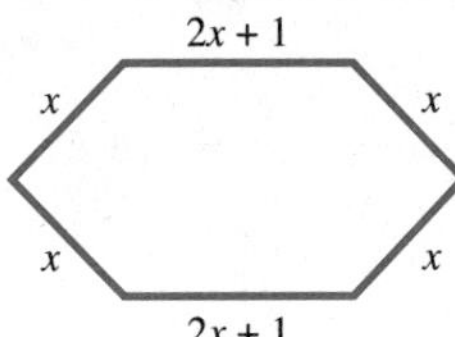

63. From a small group of people around you at a parade, you estimate the ratio of men to women is 2:5. If the announced attendance at the parade is 35,000, about how many of the people attending are men?

64. If the ratio of the length to the width of a rectangle is 5:3 and its perimeter is 84.48 cm, find its length and width.

65. The angles of a triangle are in the ratio 1:2:3. Find the measure of each angle.

66. There are approximately 2,400 stocks traded on the New York Stock Exchange. Yesterday's stock market report indicated that "advancers led decliners by 5 to 3." If 720 of the stocks were unchanged yesterday, how many were "decliners"?

67. The ratio of the number of hits to the number of times at bat is called a batting average. If Ted had 180 hits in his first 413 "at bats," how many hits must he obtain in his next 137 at bats in order to average .450 overall?

68. The ticket-taker at a movie theater took a total of 22 tickets after the movie had started. The total cost of the tickets was \$32.25. If adults' tickets cost \$2.50 and children's tickets cost 75¢, how many children entered the theater?

69. Two companies, A and B, offer you a sales position. Both jobs are essentially the same, but A pays a straight 11% commission and B pays \$75 a week plus 7% commission. Commission is based on the total amount of sales.

(a) Make a table.

(b) Draw a graph.

(c) For what amount of sales will company A pay the same as company B?

(d) For what amount of sales will company A pay more money?

(e) If you knew in any one week the best salesperson rarely had sales over \$1,700, how would this affect your decision? Explain.

70. How many quarts of a 30% salt solution should be added to 24 quarts of a 50% salt solution in order to produce a 42% salt solution?

71. The approximate volume of the payload of this truck is given by

$$V = \frac{d(a + b)}{2}w.$$

(a) What is the depth, d, of the material being hauled if $V = 1{,}000$ cubic feet, $w = 8$ feet, $a = 20$ feet, and $b = 25$ feet?

(b) Solve for w (in terms of a, b, and V).

(c) Solve for b (in terms of a, w, and V).

72. The car rental charges for Ajax Rental are \$25 a day plus 15¢ for each mile greater than 125 (you get 125 "free" miles).

(a) Make a table that gives the total cost C of renting a car that you drive for m miles where $m = 75$, 100, 125, 150, 175, 200, 225, 250, 275, and 300 miles.

(b) Graph these points.

(c) Complete this equation giving C in terms of m:

$$C = \begin{cases} ? & \text{if } m \leq 125 \\ ? & \text{if } m > 125 \end{cases}$$

(d) What is the cost of driving 320 miles?

(e) If the charge were \$75, how many miles would have been driven?

73. Find x if the shaded area is 72 cm^2.

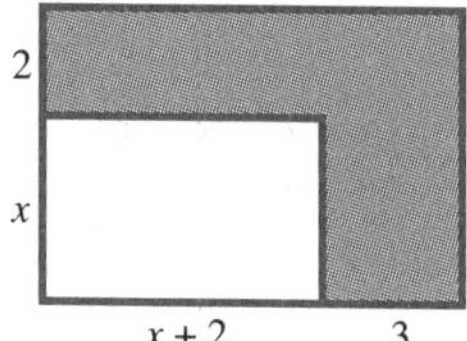

74. An automobile cooling system presently contains a mixture of water and antifreeze that freezes at 12°F. The owner wants protection against temperatures as low as −20°F. The antifreeze manufacturer publishes the following table:

Percentage of antifreeze in mixture	*Minimum safe temperature*
0	32°F
.	.
.	.
.	.
20	12°F
.	.
.	.
.	.
90	−20°F

How much of the mixture should the owner drain from his cooling system and replace with pure antifreeze if the cooling system holds 8 liters of coolant?

75. Suppose a television network can rent a two-hour movie for two showings during prime time for \$600,000 and can produce two hours of a regular series on tape for \$400,000. Television revenue is based on advertising; one network charges \$2.00 per thousand sets tuned in for each 30-second commercial. (The NAB [National Association of Broadcasters] Code allows six 30-second commercials per half-hour.) If 9 million sets would be tuned to the regular series the first time and 6 million would be tuned to reruns, how many would need to be tuned to the movie altogether to give as much profit as the series?

76. It is customary to represent the speed of supersonic aircraft by the ratio of the speed of the aircraft A to the speed of sound s, called the Mach number M (named after Ernst Mach, an Austrian physicist). That is, $M = \frac{A}{s}$. If the speed of sound is about 740 mph and the aircraft is designed to operate between Mach 1.7 and Mach 2.4, what is the designed speed range of the aircraft in miles per hour?

77. Linear depreciation is one of several methods approved by the Internal Revenue Service for depreciating business property. If the original cost of the property is C dollars and it is depreciated linearly over N years, its value V at the end of n years is given by

$$V = C\left(1 - \frac{n}{N}\right)$$

A machine having an original cost of \$10,000 is depreciated linearly over 20 years. When will its value be \$6,500?

78. As dry air moves upward, it expands. In so doing, it cools at a rate of about 1°C for each 100 m of rise, up to about 12 km.

(a) If the ground temperature is T°C, write a formula for the temperature T at height h. (A small plane pilot would need such a formula to judge how high to fly before icing would be a problem.)

(b) If the ground temperature is 20°C, what is the temperature at a height of 800 m?

79. Last year, truck drivers struck for 30 days. Assuming these drivers made \$10.50 an hour and worked 260 eight-hour days a year before the strike, what percent increase is needed in yearly income to make up for the lost time within one year?

80. You can estimate the height of a pole by placing a mirror on level ground so you can just see the top of the pole in the mirror. In the following illustration, the girl is 172 cm tall and her eyes are 12 cm from the top of her head. By her measurement, AM = 120 cm and BM = 4.5 meters. How tall is the pole? (Assume the two angles labeled α are equal.)

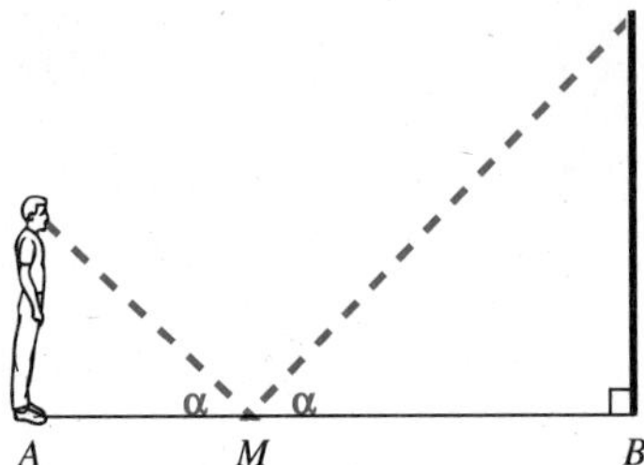

81. Many dental insurance plans pay 80% of the charges after a \$25 deductible. After your first visit of the year, you receive a reimbursement check from the insurance company for \$48.60. What was the amount of your original bill?

82. Part of the sale of certain "luxury" items is subject to a 10% tax. The bill for your new Avantus, including a 6% sales tax on the entire sale price and an additional 10% tax on the amount of the sale price exceeding \$30,000, was \$39,400. What was the sale price of this car?

CHAPTER

4 Lines and Linear Systems

In this chapter we turn to a more detailed study of graphs of linear functions and finding linear equations (functions) as models in various applications.

Source: Kathleen Campbell / Allstock

4.1 GRAPHS OF LINES

We begin by looking at the latest developments in the saga of the We Try Harder Car Rental Company.

EXAMPLE 1

In a price war, the U Drive It Company (UDI) drops the initial charge for renting a car, but it raises the mileage charge to 30¢ a mile. To be competitive, the We Try Harder Company (WTH) lowers its initial charge to $15 and raises the mileage charge to 20¢ a mile. Now which company should we choose?

Solution

We know our choice depends on (is a function of) the numbers of miles we expect to drive.

Using a Table To analyze our choices, we can make a table:

Miles	0	10	20	50	75	100	125	150	175	200
WTH cost	15	17	19	25	30	35	40	45	50	55
UDI cost	0	3	6	15	22.5	30	37.5	45	52.5	60

Table 4.1

From the table, we begin to get an overall picture of the relative costs. For example, we can see that UDI is less expensive up to 150 miles. We can now draw a graph.

Using a Graph Let the vertical axis represent the cost, and let the horizontal line represent the number of miles. (See Figure 4.1.)

The points representing the entries in the table for WTH lie on a line l_1, and the points representing the entries in the table for UDI lie on line l_2 (see the graph). We also observe that the intersection of l_1 and l_2 represents the same cost and mileage for both plans, that is, point (150, 45). UDI is cheaper below 150 miles, but the line representing its data has a steeper incline, or it is rising faster than the line for WTH. The graph can be used to *estimate* various costs. For example, the cost for 230 miles is $61 for WTH and $69 for UDI. If we imagine the line continuing, we can see that for long distances, WTH is substantially cheaper.

Using an Equation We can write an equation to represent the relationship between the cost of renting a car and the number of miles for each company: Let M be the number of miles we expect to drive.

$$\text{cost}_{\text{UDI}} = 0.30M \quad \text{or} \quad \text{cost}_{\text{UDI}}(M) = 0.30M$$
$$\text{cost}_{\text{WTH}} = 15 + 0.20M \quad \text{or} \quad \text{cost}_{\text{WTH}}(M) = 15 + 0.20M$$

If we know the miles when the costs are equal, we can pick a smaller (or larger) number of miles and calculate the cost of each to determine the less expensive plan.

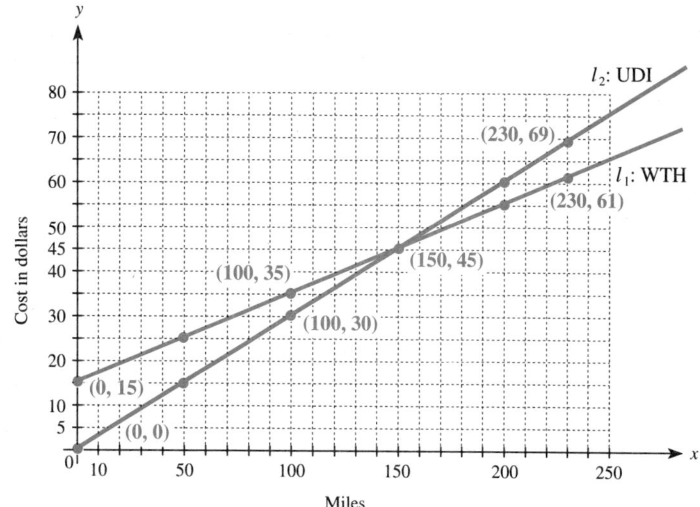

Figure 4.1

$$\begin{aligned} \text{cost}_{\text{UDI}} &= \text{cost}_{\text{WTH}} \\ 0.30M &= 15 + 0.20M \\ 0.10M &= 15 \\ M &= \frac{15}{0.10} = 150 \text{ miles} \end{aligned}$$

At 150 miles, the two plans are equal in cost.

Trying a smaller number of miles, say 100, the cost of UDI is \$30, and the cost of WTH is \$35. Therefore we would choose the U Drive It Company if we plan on driving less than 150 miles, but we would choose the We Try Harder Company if we plan to drive over 150 miles.

In Example 1, we observed that the cost of renting a car from each company could be represented by linear functions (The cost is a function of the number of miles driven.):

$$\text{cost}_{\text{UDI}} = 0.30M$$
$$\text{cost}_{\text{WTH}} = 15 + 0.20M$$

If the coordinates of a point on line l_1 are substituted into the equation $\text{cost}_{\text{WTH}} = 15 + 0.20M$, the result is a *true* statement. Thus the *coordinates of the points* on line l_1 are *solutions* to the equation $\text{cost}_{\text{WTH}} = 15 + 0.20M$. Conversely, the *solutions* to $\text{cost}_{\text{WTH}} = 15 + 0.20M$ are the *coordinates* of the points on line l_1. A similar relationship exists for line l_2 and the equation, $\text{cost}_{\text{UDI}} = 0.30M$. A graph can be thought of as a geometric representation (a picture) of the solutions of an equation. An equation can be thought of as an algebraic representation of a graph. This observation leads to two fundamental problems:

1. Given an equation, find its graph.

2. Given a graph, find its equation.

In this section, we will look at linear equations and determine their graphs. In the next section, we will determine the equation of a given line.

Graphing Linear Equations and Linear Functions

EXAMPLES 2–4

Sketch a graph representing the solutions of each of the following linear relationships.

2. $y = 2x$ **3.** $y = 2x + 3$ **4.** $\frac{x}{2} + \frac{2y}{3} = 4$

Solutions

The fundamental strategy we have been using is to find a number of points on the graph and then sketch a line through them. To find these points, we select at random some solutions to the equation and arrange them in a table.

2. $y = 2x$

x	0	1	2	4	−1	−2	−3
y	0	2	4	8	−2	−4	−6

Table 4.2

To determine the solutions, pick a value for x and substitute it into the equation to find a corresponding value of y. For example, if $x = 0$, then $y = 0$. If $x = 5$, then $y = 2(5) = 10$. We could also have selected a value for y and then determined the corresponding value of x.

Graphing the above points and sketching a line through them, we have

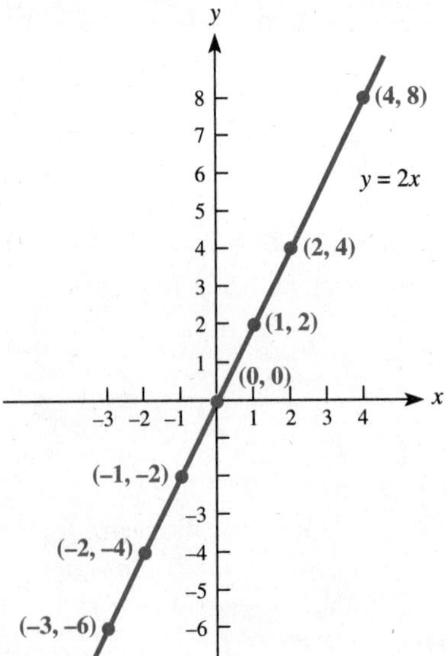

Figure 4.2

We use the same strategy for determining the graphs of the remaining equations.

3. $y = 2x + 3$

x	0	1	2	-1	-2	$\frac{-3}{2}$
y	3	5	7	1	-1	0

Table 4.3

4. $\frac{x}{2} + \frac{2y}{3} = 4$

x	0	8	4	2	-1
y	6	0	3	$\frac{9}{2}$	$\frac{27}{4}$

Table 4.4

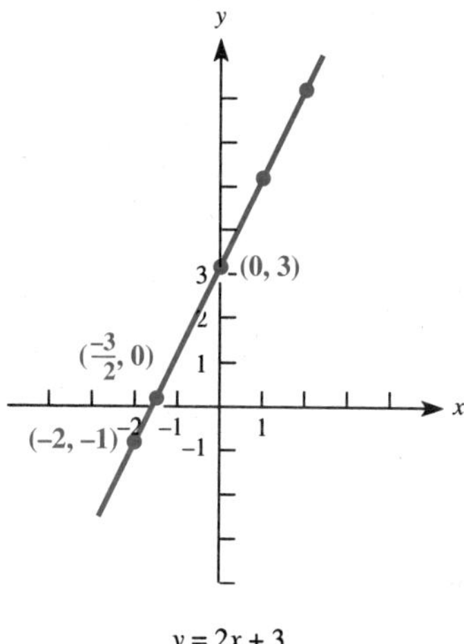

Figure 4.3

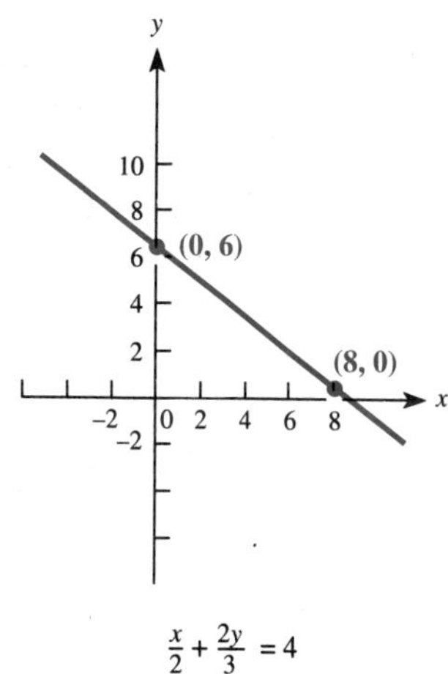

Figure 4.4

NOTE: We can also use a graphing calculator, but it is very useful to be able to sketch the graph of a straight line by hand. ■

We notice that to find a solution to a linear equation, we can let x (or y) be any number and then determine the corresponding value of y (or x). Thus, there are an infinite number of such pairs *and* an infinite number of solutions. The solutions to a linear equation in two variables lie on a line.

Graphs of Linear Equations

The graph of the solutions of any equation of the form

$$y = mx + b$$

or more generally,

$$Ax + By = C$$

where A and B are not both equal to zero, is a straight line.

NOTE: The equation $y = mx + b$ represents a linear function; y is dependent on x (y is a function of x). The equation $Ax + By = C$ represents a linear relationship. ■

The x- and y- Intercepts

Since two points determine a straight line, we need to locate only two points to sketch its graph. The easiest points to find are the **intercepts,** the points where the line crosses (intercepts) an axis. The ***x*-intercept** $(x, 0)$ is found by setting $y = 0$ and solving for x. The ***y*-intercept** $(0, y)$ is found by setting $x = 0$ and solving for y.

EXAMPLES 5–6

Find the intercepts and sketch the graph for each equation.

5. $y - 2x = 4$ **6.** $3y + \frac{2}{3}x = 6$

Solutions

5. $y - 2x = 4$

To find the x-intercept: Let $y = 0$; then $0 - 2x = 4$ and $x = -2$.

To find the y-intercept: Let $x = 0$; then $y - 0 = 4$ and $y = 4$.

Thus the x-intercept is $(-2, 0)$, and the y-intercept is $(0, 4)$.

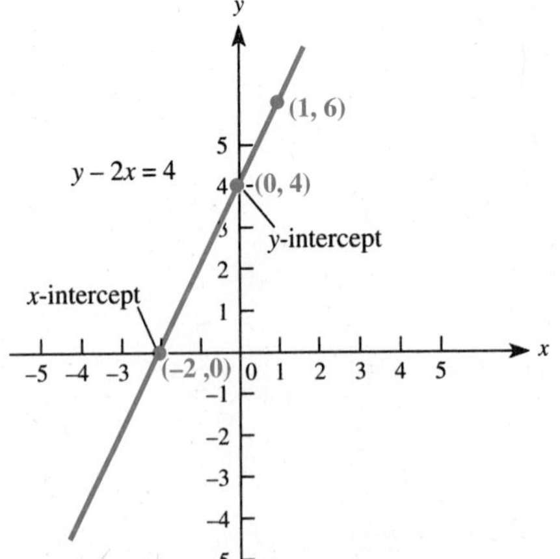

Figure 4.5

We should pick a third point to make sure we have not made a careless mistake. We select one at random. For example, if $x = 1$, then $y = 6$, and the point $(1, 6)$ lies on the line.

6. $3y + \frac{2}{3}x = 6$

To find the x-intercept: Let $y = 0$; then $0 + \frac{2}{3}x = 6$ and $x = 9$.

To find the y-intercept: Let $x = 0$; then $3y + 0 = 6$ and $y = 2$.

Thus the x-intercept is (9, 0), and the y-intercept is (0, 2).

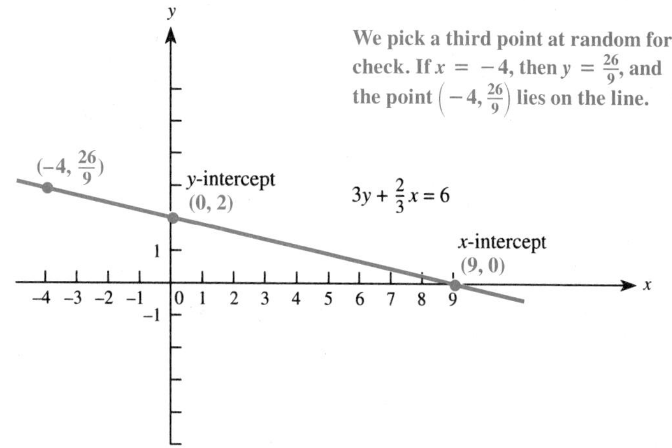

Figure 4.6

Notice that to find the intercepts, we let one variable equal zero and solve for the other. For example, in the equation $3y + \frac{2}{3}x = 6$, to find the x-intercept

let $y = 0$

solve $3(0) + \frac{2}{3}x = 6$ for x

We are solving a linear equation in one variable—the main task in Chapter 3. Similarly, to find the y-intercept,

let $x = 0$

solve $3y + \frac{2}{3}(0) = 6$ for y

Again we are solving a linear equation in one variable. The linear equations in one variable from Chapter 3 are special cases of linear equations in two variables, namely, where one of the coefficients of one of the variables is equal to zero. ■

We now consider some special cases of $Ax + By = C$, where A, B or C equals 0.

EXAMPLES 7–9

Sketch the graph of the following equations. Label the intercepts, if they exist.

7. $x = y$ **8.** $x = 2$ **9.** $y = -2$

Solutions

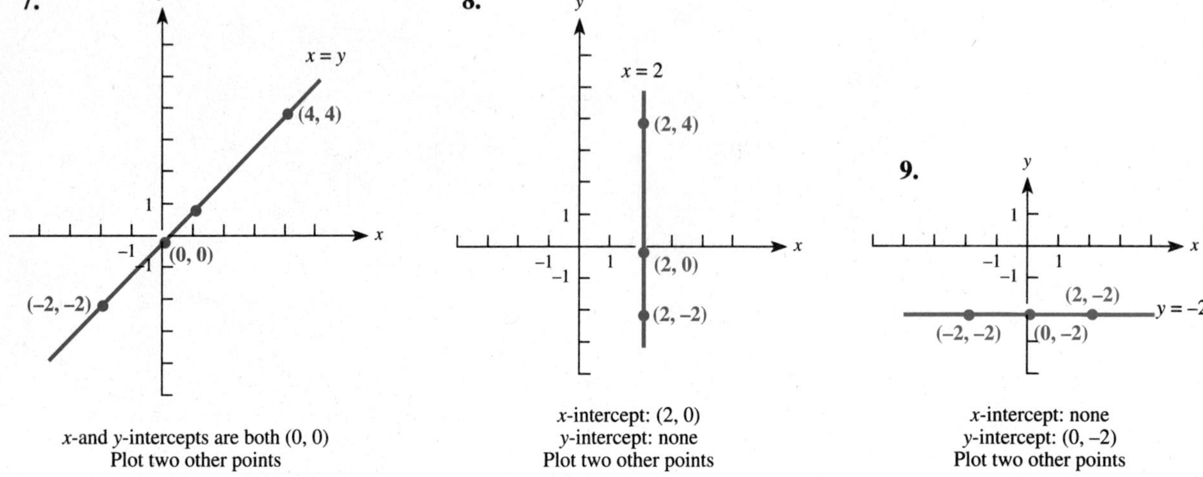

Figure 4.7 **Figure 4.8** **Figure 4.9**

The lines represented by $y = k$ and $x = k$, where $k \neq 0$, have only one intercept:

- The line of $x = k$ is a vertical line intercepting the x-axis at $(k, 0)$.
- The line $y = k$ is a horizontal line intercepting the y-axis at $(0, k)$.
- The line $y = ax$, where $a \neq 0$, is a line through the origin.
- The line $y = mx + b$; where $m \neq 0$ and $b \neq 0$, is a slanting line with two distinct intercepts. ■

In summary, we have the following:

Strategy for Graphing the Solutions of a Linear Function by Hand.

1. Determine and locate the x- and y-intercepts (if both exist).
2. Draw the line determined by the two intercepts.
3. Pick a third point, and check to see that it lies on the line.

Recall that to "sketch" the graph of a line on many graphing calculators, we must express the equation in the form $y =$ expression in x. Check the graphs obtained in Examples 5 and 6 by entering

$$Y1 = 2x + 4 \quad \text{and} \quad Y2 = -\frac{2}{9}x + 2$$

on your calculator.

Many situations can be described by graphs made up of parts of two or more lines.

EXAMPLE 10

(a) Sketch the graph of $y = \begin{cases} 30 & \text{for } 0 \le x \le 100 \\ 30 + 0.15x & \text{for } x > 100 \end{cases}$

(b) Give an example of a situation that could be described by such a graph.

Solutions

(a)

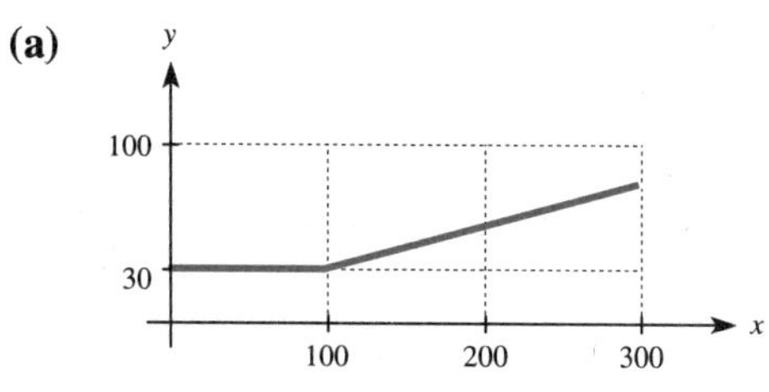

Figure 4.10

(b) For example, the daily cost to rent a car at Ajax Rental is \$30 plus 15¢ for each mile over 100. ■

Many of the applications that we examined in previous sections have actually involved two variables. In some cases a specific value of one variable was given. Otherwise we could set two quantities equal, thereby eliminating one variable. Many of the examples and exercises on applications that follow should be familiar to you.

EXAMPLE 11

The following table gives the population of Suburbia at the end of some of the past years.

Year	*Population*	*Year*	*Population*
1980	50,000	1987	85,000
1981	55,000	1988	90,000
1982	60,000	1990	100,000
1985	75,000	1992	110,000

Table 4.5

(a) Make a graph to illustrate the data in the table.

(b) Find a mathematical model (equation or function) to describe the data. Assume the pattern in the table continues.

(c) When will the population reach 140,000?

(d) What will be the population in 2000?

Solutions

(a)

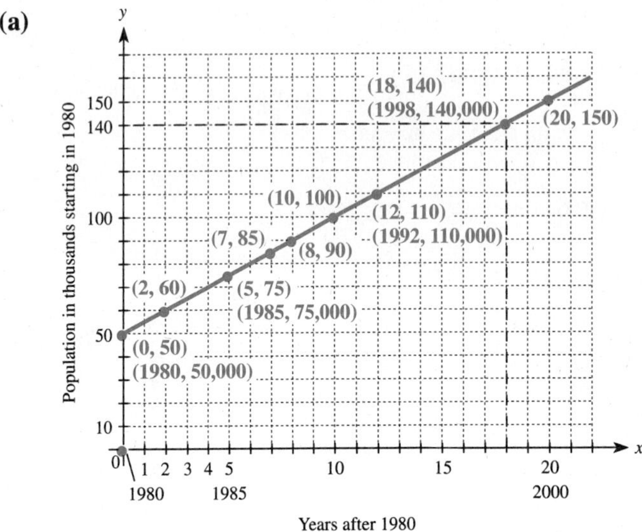

Figure 4.11

It is sometimes necessary to have the unit lengths on the two axes of a graph represent different scales. In the graph above, we let the origin correspond to 1980, since that is where we *began* counting the population. For this reason, *all* graphs require that the scale be identified.

(b) To find an equation representing the data, we observe that our graph appears to be a straight line, so our equation is a linear equation of the form $P = mx + b$ where x = number of years after 1980 and P is the population for a given number of years after 1980. From the table and the graph, we observed that the population was 50,000 in 1980 (0) and increased 5,000 a year. Thus, we write $P = 50{,}000 + 5{,}000x$.

(c) From the graph, we locate the point on the line that has the coordinates $(x, 140)$. It is $(18, 140)$. The population will reach 140,000 in 1998 $(1980 + 18)$.

We could have obtained this information from our equation by letting $P =$ 140,000 and solving for x:

$$140{,}000 = 50{,}000 + 5{,}000x$$
$$x = 18$$

Thus, 18 years after 1980, or 1998, the population will reach 140,000.

(d) To determine the population in 2000, we look for the point on the line with coordinates $(20, P)$. From the graph, we observe that $P = 150$, or 150,000. This information could also be obtained from the equation by letting $x = 20$ and solving for P:

$$P = 50{,}000 + 5{,}000(20)$$
$$P = 150{,}000$$ ■

EXAMPLE 12

The BJAX company is planning on producing small, square pizzas. It costs \$2 to make each pizza, which sells for \$5.

(a) Write an equation to describe the profit from selling the pizzas.

(b) Graph the equation.

(c) How much profit can be made on 50 pizzas?

(d) How many pizzas should be sold a day to make a profit of \$300?

Solutions

(a) Let x be the number of pizzas sold. The profit is a function of the number of pizzas sold. We write Profit (number of pizzas) or more briefly $P(x)$.

$$\text{Profit} = \text{Revenue} - \text{Cost}$$
$$P(x) = 5x - 2x$$
$$P(x) = 3x$$

(b)

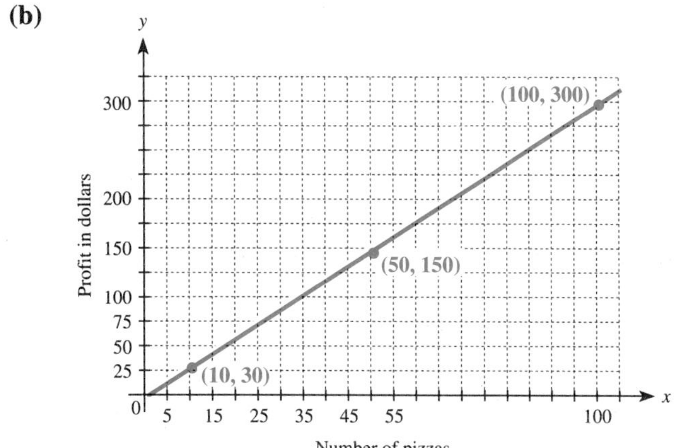

Figure 4.12

(c) From the graph, we observe that 50 pizzas will yield a profit of \$150. The equation gives $P(x) = 3(50) = \$150$.

(d) From the graph, we observe that selling 100 pizzas will create a profit of \$300, and from the equation, we have

$$300 = 3x \quad \text{or} \quad x = 100$$

EXAMPLE 13

A competitor of the BJAX Pizza Company estimates its revenue and cost of making pizzas by the graph below.

(a) What are the start-up costs for this company?

(b) What is the cost of producing 20 pizzas?

(c) What is the revenue from selling 20 pizzas?

(d) How many pizzas must be made each day to make a profit?

(e) What are the cost of and profit from selling 80 pizzas?

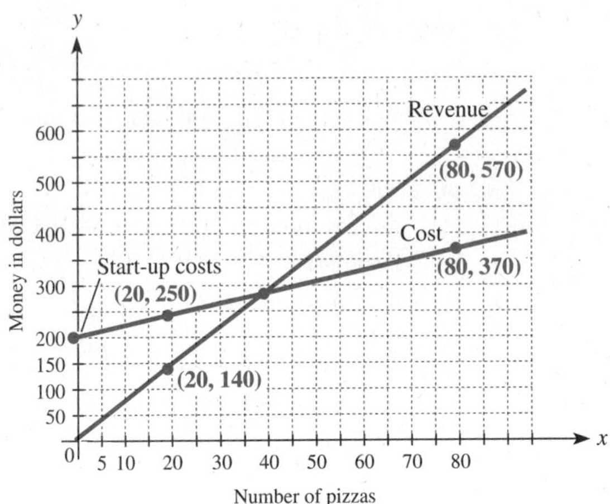

Figure 4.13

Solutions

(a) The start-up costs are what it costs to set up before any pizzas are produced. This corresponds to the vertical (y) intercept, which is \$200.

(b) Twenty pizzas cost \$250.

(c) Twenty pizzas sell for about \$140. Notice that no profit is made on 20 pizzas.

(d) A profit will be made when the revenue exceeds the costs. The intersection of the two lines is the *break-even point,* which occurs at 40 pizzas. More than 40 pizzas must be sold to realize a profit.

(e) Eighty pizzas sell for about \$570 and cost about \$370. The profit is approximately \$200. ■

4.1 EXERCISES

I. Procedures

1–4. For each equation, determine which pairs of numbers (x, y) are solutions to the equation by using first graphing and then substitution.

1. $3y = 3x - 1$ **(a)** $\left(0, \frac{1}{2}\right)$ **(b)** $\left(\frac{3}{2}, \frac{2}{3}\right)$ **(c)** $(0, 1)$ **(d)** $(0.5, 0)$

2. $x + y = 0$ **(a)** $(0, 0)$ **(b)** $(1, -1)$ **(c)** $(1, 1)$ **(d)** $\left(\frac{1}{2}, -\frac{17}{34}\right)$

3. $2y - 3x = 6$ **(a)** $(-2, 0)$ **(b)** $\left(1, \frac{9}{2}\right)$ **(c)** $(2.1, 6.15)$ **(d)** $(2, 3)$

4. $x = 5$ **(a)** $(5, 0)$ **(b)** $(5, 6)$ **(c)** $(3, 5)$ **(d)** $(0, 5)$

5–24. For each equation below, **(a)** sketch the graph. **(b)** Label the intercepts and your "checkpoints."

5. $3x - 4y = 12$

6. $-3x + 4y = 12$

7. $-3x - 4y = 12$

8. $y = 2x$

9. $y = -x$

10. $y = -2x + 3$

11. $y = -2x - 3$

12. $x + y = 0$

13. $x - 2y = 0$

14. $3x - 5y = 15$

15. $-5x + 6y = 15$

16. $x = 3$

17. $x = -3$

18. $x = 0$

19. $y = -2$

20. $y = 0$

21. $\frac{x}{2} - \frac{y}{3} = 1$

22. $\frac{x}{2} + \frac{2y}{3} = 6$

23. $\frac{x}{4} + y = 8$

24. $\frac{2x}{5} - \frac{2y}{3} = 1$

II. Concepts

25–29. **(a)** Sketch each of the following graphs on the same set of coordinate axes. **(b)** Record any observations you make about each set of graphs.

25. $y = x, y = 2x, y = -3x, y = 10x$

26. $y = x, y = x + 1, y = x - 1, y = x + 5$

27. $y = 3x + 1, y = 4x + 1, y = -2x + 1$

28. $x + 2y = 3, 2x + 4y = 7, 3x + 6y = 10$

29. $y = 2x, y = -\frac{x}{2}$

30–34. Mark the following statements as AT for always true, ST for sometimes true, or NT for never true. If sometimes true (ST), give an example of when it is true and when it is false.

30. If $(-a, b)$ is a point in quadrant II, then $(a, -b)$ is a point in quadrant IV.

31. The line $y = ax + b$ has an x-intercept.

32. The line $y = ax + b$ has a y-intercept.

33. The line $y = a$ is perpendicular to the y-axis at the point $(0, a)$.

34. If (a, b) is a solution to a linear equation, then (b, a) is also a solution.

35. Do you think a linear equation (function) is a reasonable model for the population growth of a city (see Example 11)? Give reasons for your answer.

36. Explain the relationship between the equations $2x + 3 = 0$ and $y = 2x + 3$.

37. In Example 12, see if you can find a viewing rectangle on your graphing calculator so you can trace to the point (50, 150).

38. In Example 13, see if you can find a viewing rectangle on your graphing calculator so you can find the break-even point.

39–40. Sketch the graph and give an example of a situation that could be described by such an equation and graph.

39. $y = \begin{cases} 0 & \text{if } 0 \le x \le 100 \\ 0.15x & \text{if } x > 100 \end{cases}$

40. $y = \begin{cases} 4 & \text{if } 0 \le x < 1 \\ 4 + 2x & \text{if } 1 \le x < 8 \\ 20 & \text{if } x \ge 8 \end{cases}$

III. Applications

41–43. For each problem, **(a)** Make a graph to correspond to the values in the table. **(b)** Find a mathematical model (equation) to describe the table. **(c)** Answer the question(s) posed in each problem. Indicate (locate) the answer on the graph.

41. The table represents the cost of renting a car for one day from the We Try Harder Company.

Miles driven in one day	*Cost*
0	\$ 23
100	\$ 43
200	\$ 63
300	\$ 83
400	\$103
500	\$123

(a) What are the intercepts? Do they both make sense? Explain.

(b) What is the cost of driving 350 miles?

(c) What is the number of miles that can be driven for \$150?

42. The table on the next page represents the distance traveled in 5.5 hours at various average speeds.

Distance (miles)	*Average speed (mph)*
110	20
165	30
220	40
275	50
330	60

(a) What are the intercepts? Do they both make sense? Explain.

(b) How many miles can be driven at 55 mph?

(c) What is the average speed required to make a journey of 300 miles?

43. The table represents the federal income tax (in dollars) paid for various salaries (in dollars) by a married couple filing jointly in 1991.

Salary	*Tax*
$15,200	$2,284
$15,700	$2,359
$16,200	$2,434
$16,700	$2,509
$17,200	$2,584
$17,700	$2,659
$18,200	$2,734
$18,700	$2,809
$19,200	$2,884

(a) What are the intercepts? Do they both make sense? Explain.

(b) What is the tax for a married couple with a joint income of $18,000? A joint income of $14,000?

(c) What is the income of a couple who paid $2,600 in taxes?

(d) Is a linear equation an accurate model? Why?

44–52. For each problem below, **(a)** Write a linear equation in two variables. **(b)** Graph the equation. **(c)** Answer the question(s) posed in the problems.

44. The Ajax Manufacturing Company makes widgets. It costs the company $5 to make each widget. The company sells the widgets for $7.50.

(a) What is the profit (P) on a total of (w) widgets?

(b) How many widgets must be sold to make a $10,000 profit?

(c) What is the profit on 1,000 widgets?

45. Acme Car Rental charges $22 per day and 20¢ a mile. Assume the car is rented for one day for parts (a), (b), and (c) below.

(a) What is the cost (C) for driving a total of (M) miles?

(b) How many miles can be driven for $30?

(c) How much will it cost to drive 200 miles?

(d) How much will it cost to drive 1,000 miles in two days?

46. The maximum weight allowed in an elevator is 1,500 pounds.

(a) How many adults (A) and children (C) can it hold if the average weight per adult is 150 pounds and the average weight per child is 40 pounds?

(b) If ten children get on, how many adults can ride?

(c) If no adults get on, how many children can get on?

47. A train averages 80 mph and a bus averages 50 mph. You ride a train for x hours and a bus for y hours.

(a) Write an equation that says you took an 800-mile trip by train and bus.

(b) If you were on a bus for two hours, how long were you on the train?

48. The Top Hat Company's management decides to manufacture widgets. The start-up cost is $15,000 (for equipment, etc.). They estimate that the cost of each widget (w) is $5.25. They plan on selling the widgets for $9.00 each.

(a) How many widgets must they sell to make a profit P?

(b) How many widgets must they sell to make a profit of $10,000?

(c) What is the profit on 300 widgets?

(d) How many widgets must they sell to make a profit?

49. For the past two years, suppose the inflation rate was 12% and 10%.

(a) What was the price (P) of an item two years ago, if it costs (C) dollars today?

(b) If the item is priced at $100 today, what did it cost two years ago?

(c) What is the cost of an item today if it cost $15.75 two years ago?

50. A real estate salesperson works for a salary of $100 a week plus 3% of the price of each house he or she sells and works 50 weeks a year.

(a) How many dollars worth of houses (h) must the salesperson sell to realize a salary of S dollars a year?

(b) What is his or her income per year if he or she sells \$500,000 worth of houses?

(c) If the salesperson's salary is \$35,000 a year, how many dollars worth of houses must he or she sell?

51. The long-distance telephone rate for some areas of Massachusetts is 55¢ for the first minute and 23¢ for each additional minute. Assume you make one call.

(a) How many minutes (M) could you talk for a sum of \$$S$?

(b) How much will it cost for a 28-minute call?

(c) How long could you talk for \$7?

52. In Exercise 51, there is a 35% discount for calls made between 5:00 P.M. and 11:00 P.M. on weekdays. Assume you made one call on Monday evening beginning around 7:00 P.M.

(a) How many minutes (m) could you talk for a sum of \$$S$?

(b) How much will it cost for a 28-minute call?

(c) How long could you talk for \$7?

53. The formula $F = \frac{9}{5}C + 32$ gives the relationship between Fahrenheit and Centigrade temperatures. A quick rule of thumb for converting from Centigrade to Fahrenheit temperature is "double the Centigrade temperature and add 30."

(a) Write an equation for the rule-of-thumb conversion.

(b) Graph both equations for converting from Centigrade to Fahrenheit temperature.

(c) At what point do the lines intersect?

(d) Interpret the point of intersection.

(e) At 0°C, what is the difference in the two calculated Fahrenheit temperatures?

(f) For what values of the Centigrade temperature will the rule-of-thumb method be too low?

54. It has been found that the number of chirps that a cricket makes in a minute is a function of the temperature. The formula is

$$t = \frac{n}{4} + 40$$

where t is the temperature in degrees Fahrenheit and n is the number of cricket chirps in one minute.

(a) Graph the equation.

(b) At 60°F, how many chirps a minute does a cricket make?

(c) What is the temperature if a cricket makes 150 chirps per minute?

(d) At what temperature does the cricket stop chirping?

55. The speed of sound is approximately 1,100 feet per second. The distance from you that lightning strikes is a function of the time between when you see the flash and when you hear the thunder. For example, if you hear the thunder after 3 seconds, then the lightning is 3(1,100) = 3,300 feet away.

(a) Write a formula for the distance in terms of time.

(b) How many seconds have elapsed if the lightning is one mile away?

(c) How far away is the lightning if you hear the thunder after $6\frac{1}{2}$ seconds?

56. The accompanying graph represents the cost of renting a car for a day from company A and company B.

(a) Which company has no initial charge for renting a car?

(b) How many miles can be driven in one day for \$50 for each company?

(c) How many miles must be driven for the costs to be the same? What is the cost?

(d) Which company is the least expensive for 100 miles? What is the difference in the costs?

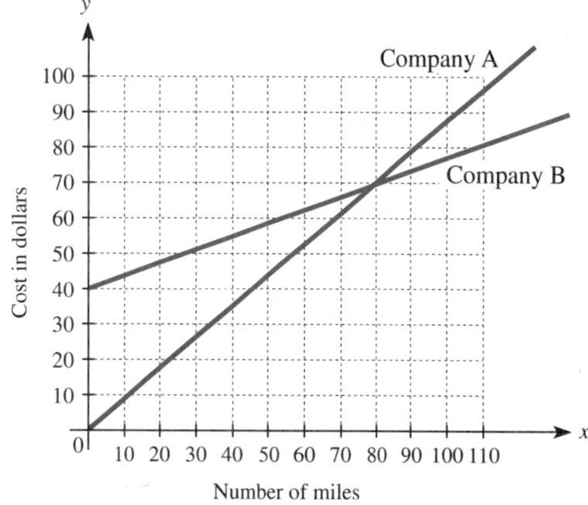

57. The accompanying graph represents the distance that Ann and Betty run in a race after a given time. Since Ann is slower, she gets a head start.

(a) How much of a head start does Ann have?

(b) How far has each woman run after 5 units of time?

(c) Who wins the race if the race is 60 meters long? One hundred meters long?

(d) At what time does Betty overtake Ann?

(e) What is the longest race Ann could win?

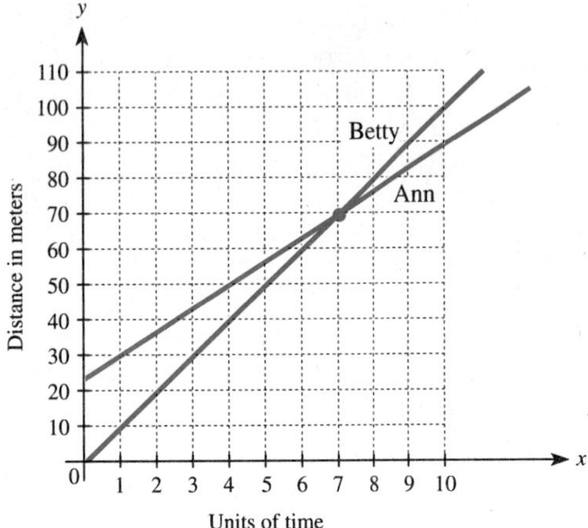

58. The Acme Widget Company has displayed its cost and profit of making widgets on the accompanying graph.

(a) What is the start-up cost?

(b) What is the cost of producing 20 widgets?

(c) What is the profit on producing 20 widgets? Is this reasonable?

(d) How many widgets must be made before a profit is realized?

(e) What are the cost and profit on 150 widgets?

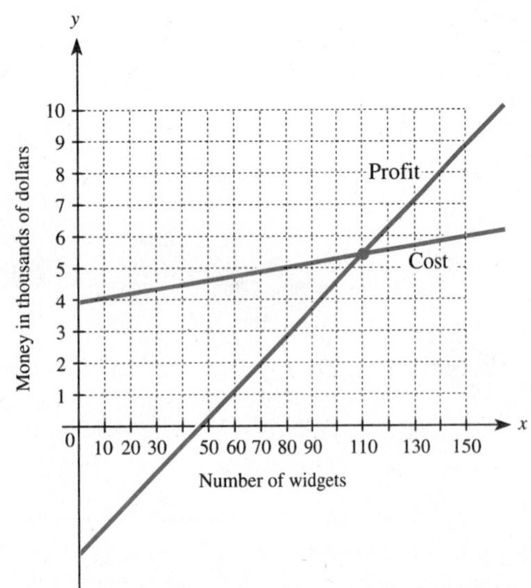

59. The Top Hat Company decides to display its cost and revenue of producing widgets on the accompanying graph.

(a) What are the start-up costs for making widgets?

(b) How many widgets must be sold to break even?

(c) What is the cost of producing 70 widgets?

(d) What is the revenue from selling 70 widgets?

(e) What profit is made on selling 200 widgets?

(f) How many widgets must be sold to realize a profit of \$5,000?

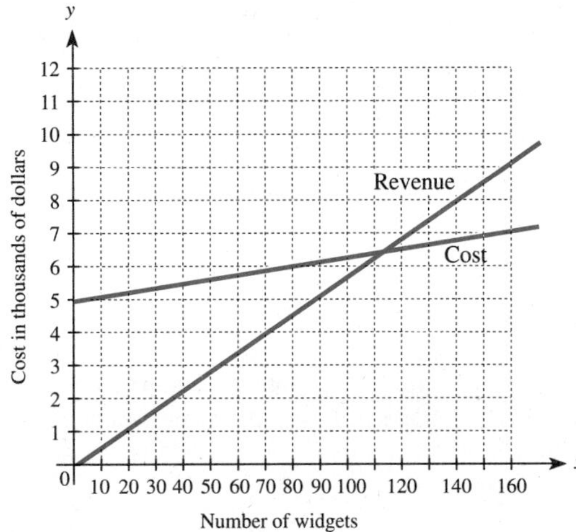

IV. Extensions

60. A *lattice* point is one in which both coordinates are integers. Find all the lattice points in quadrant I that lie on each of the following lines.

(a) $4x + 5y = 77$ **(b)** $3x + 6y = 77$

(c) $x - y = 0$

61. Suppose the scoring in a football game is simplified to 7 points for a touchdown, 3 points for a field goal, and no other scoring.

(a) Write an equation for representing the total score.

(b) What combinations of touchdowns and field goals would give a score of 46? A score of 42?

(c) What scores are impossible to achieve?

62. Following is a table representing the average worker's salary in the Acme Widget Company for the past four years. Four graphs illustrating this data are shown.

Year	1	2	3	4
Salary	\$20,000	\$20,600	\$21,300	\$22,100

Which of the graphs below do you think

(a) was prepared by the union negotiator?

(b) was prepared by the management negotiator?

(c) is the most accurate representation of the data?

(d) What is the moral illustrated by these four graphs?

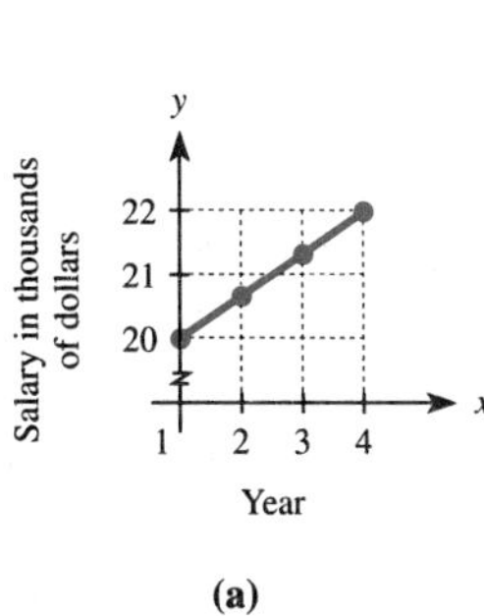

(a)

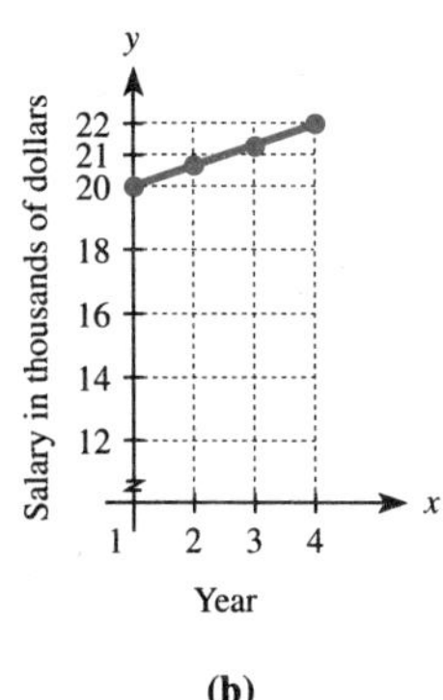

(b)

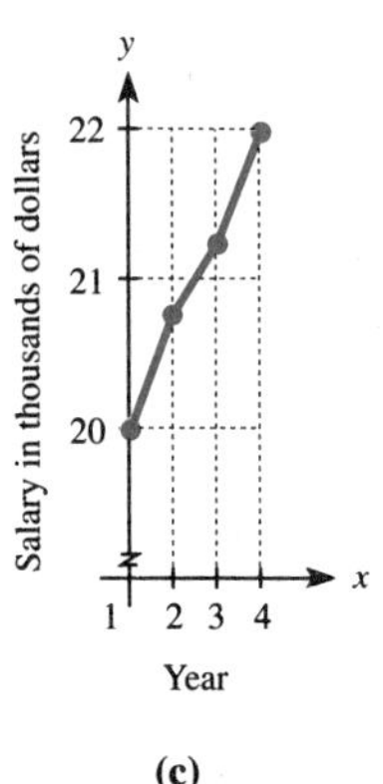

(c)

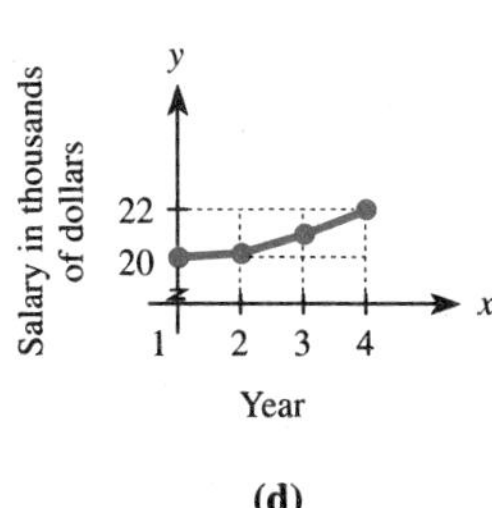

(d)

63. Sketch a graph for Exercise 59 using the axes indicated in the accompanying figure.

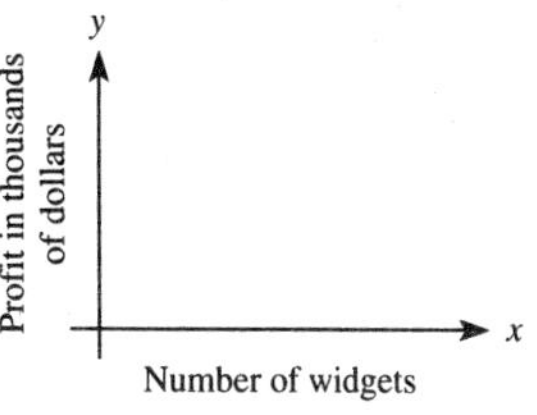

4.2 EQUATIONS OF LINES AND LINEAR MODELS

In this section we consider a problem posed in the last section: Given the graph of a line or some of its characteristics, find its equation.

Slope

Let us look at the graph of the population of Suburbia from Example 11 of the last section.

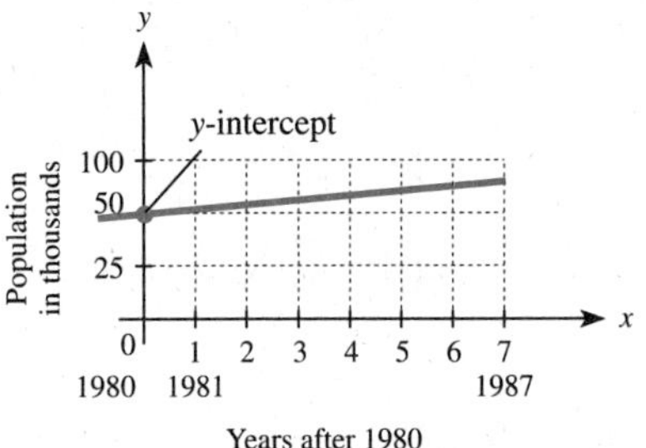

Figure 4.14

The graph was constructed from the data in the following table:

Years x	1980	1981	1982	1983	1984	1987
Population y	50,000	55,000	60,000	65,000	70,000	85,000

Table 4.6

The graph of the population for the years from 1980 to 1987 is a line. Since we know the graph is a line, we know that every point on the line is a solution. Using the graph, we not only know that the points from the table lie on the line, but we also know many other points.

From the values, we deduced the equation of the line to be

$$y = 50{,}000 + 5{,}000x \qquad \textit{y is the population, and x is the number of years after 1980.}$$

Since this equation has the form $y = mx + b$ ($m = 5{,}000$ and $b = 50{,}000$), we know it is the linear equation of the straight line we seek.

The two key numbers in this equation are 50,000, *the constant term,* and 5,000, *the coefficient of x.* The 50,000 is the y-coordinate of the y-intercept. The 5,000 is the rate of increase of population each year or, in more mathematical terms, the rate of increase of y with respect to x. If x increased by 1, y increased by 5,000. If x increased by 2, y increased by 10,000. If x increased by 3, y increased by 15,000, and so on. This constant rate of increase is given by the following ratio:

$$\frac{\text{increase in } y}{\text{increase in } x} = \frac{5{,}000}{1} = \frac{10{,}000}{2} = \frac{15{,}000}{3}$$

This rate of change is constant between any two points on the line. We call this constant rate (or ratio) of change between the y-coordinates and x-coordinates the *slope* of the line:

$$\text{slope} = m = \frac{\text{change in } y}{\text{change in } x}$$

More generally, if we take a line l (Figure 4.15) with any two points $P(x_1, y_1)$ and $Q(x_2, y_2)$, we can express the ratio of change in y-coordinates to change in x-coordinates as

$$\frac{y_2 - y_1}{x_2 - x_1}$$

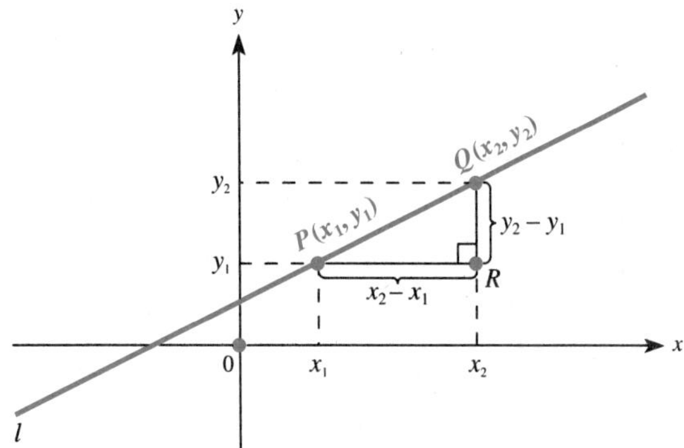

Figure 4.15

where we go from P to Q. We start at P and go to the right on line PR and then up line RQ to point Q. The length PR is sometimes called the *run,* or *horizontal change,* and the length RQ is called the *rise,* or *vertical change.* The constant ratio of vertical change to horizontal change, or rise to run, is called the slope of the line. The slope defines the steepness of the line.

Definition

For any nonvertical line, the **slope** of the line is

$$m = \frac{\text{change in } y\text{-coordinates}}{\text{change in } x\text{-coordinates}} = \frac{y_2 - y_1}{x_2 - x_1} = \frac{\text{vertical change}}{\text{horizontal change}} = \frac{\text{rise}}{\text{run}}$$

for any two points (x_1, y_1) and (x_2, y_2) on the line.

EXAMPLES 1–4 Find the slope of the lines indicated in each case.

1.

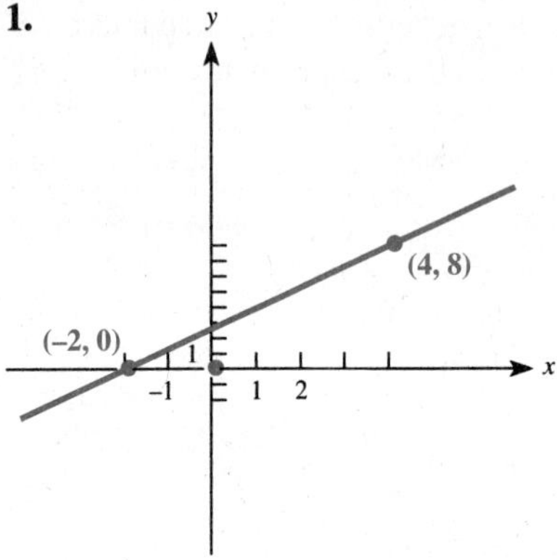

Figure 4.16

2.

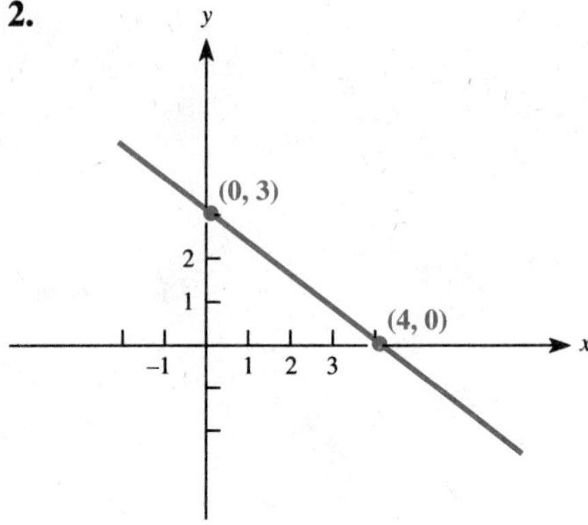

Figure 4.17

3. $y = 2x + 1$

4. $3y + 6x = 12$

Solutions

1. Let

$$(-2, 0) = (x_1, y_1)$$
$$(4, 8) = (x_2, y_2)$$

then

$$\text{slope} = \frac{y_2 - y_1}{x_2 - x_1}$$

$$\text{slope} = \frac{8 - 0}{4 - (-2)}$$

$$\text{slope} = \frac{8}{6} = \frac{4}{3}$$

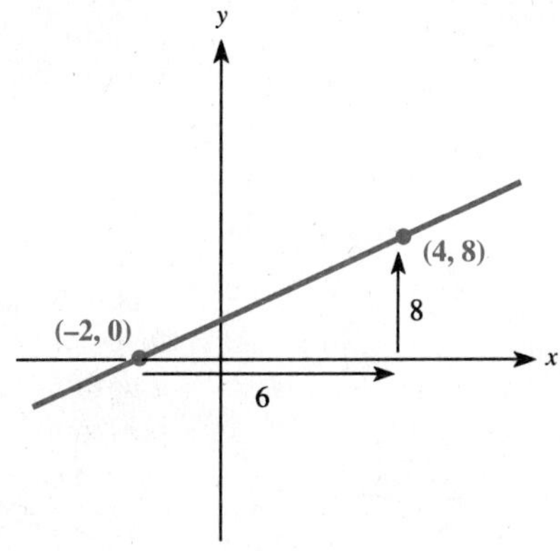

Figure 4.18

NOTE: It makes no difference which point we call (x_1, y_1). That is, if $(x_1, y_1) = (4, 8)$ and if $(x_2, y_2) = (-2, 0)$, then

$$\text{slope} = \frac{0 - 8}{-2 - 4} = \frac{-8}{-6} = \frac{4}{3} \quad ■$$

2. Let

$$(x_1, y_1) = (4, 0)$$
$$(x_2, y_2) = (0, 3)$$

then

$$\text{slope} = \frac{y_2 - y_1}{x_2 - x_1}$$

$$\text{slope} = \frac{3 - 0}{0 - 4} = \frac{3}{-4} = -\frac{3}{4}$$

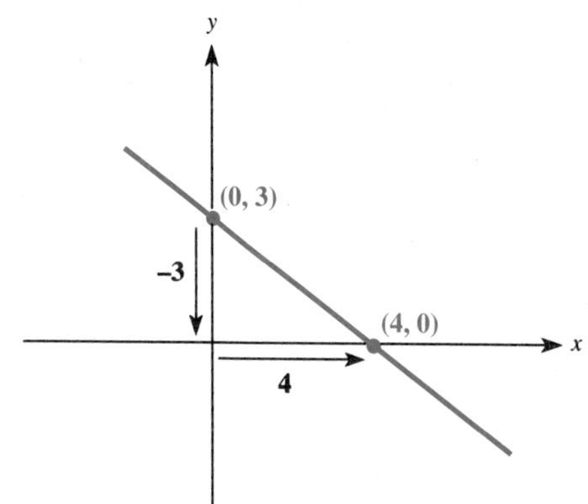

Figure 4.19

3. We can pick any two points on the line $y = 2x + 1$. The intercepts will do nicely, (0, 1) and $\left(-\frac{1}{2}, 0\right)$.

Let

$$(x_1, y_1) = (0, 1)$$
$$(x_2, y_2) = \left(\frac{-1}{2}, 0\right)$$

then

$$\text{slope} = \frac{0 - 1}{\frac{-1}{2} - 0} = \frac{-1}{\frac{-1}{2}}$$

$$\text{slope} = 2$$

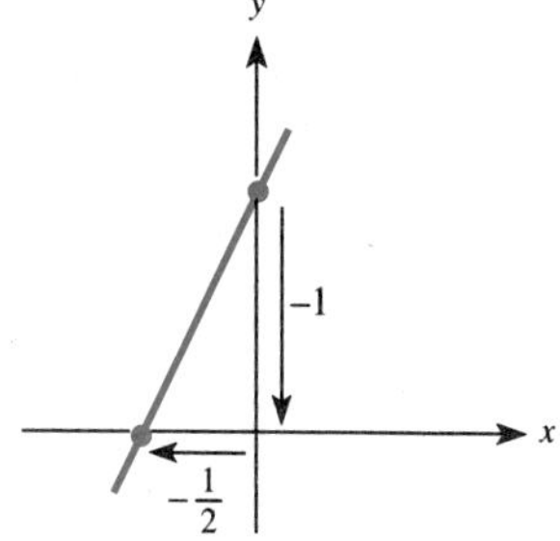

Figure 4.20

Note the slope, m, is the coefficient of x in the term $2x$, and the y-coordinate of the y-intercept is 1.

4. We pick any two points of the line $3y + 6x = 12$, say, (0, 4) and (1, 2).
Let

$$(x_1, y_1) = (0, 4)$$
$$(x_2, y_2) = (1, 2)$$

then

$$\text{slope} = \frac{2 - 4}{1 - 0} = \frac{-2}{1} = -2$$

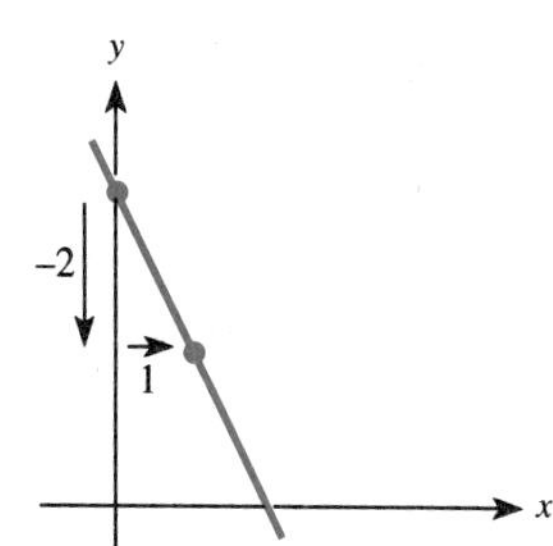

Figure 4.21

Let us put $3y + 6x = 12$ into the form $y = mx + b$ by solving for y:

$$3y + 6x = 12$$
$$3y = -6x + 12$$
$$y = -2x + 4$$

Note again that the slope is the coefficient of x. Slope $= m = -2$. The y-intercept is $(0, 4)$, and $b = 4$. ■

From the above examples, we can infer the following:

Slope and y-Intercept of a Line

1. If the equation of a line is in the form $y = mx + b$, then *m is the slope* and *$(0, b)$ is the y-intercept.*
2. If m is the slope and $(0, b)$ is the y-intercept for a given line, then *the equation of the line* is $y = mx + b$.

For this reason, the linear equation $y = mx + b$ is called the **slope-intercept form** of the line.

Finding an Equation of a Line

EXAMPLES 5–8

For the line in each graph, find

(a) the y-intercept. **(b)** the slope. **(c)** the equation of the line.

5.

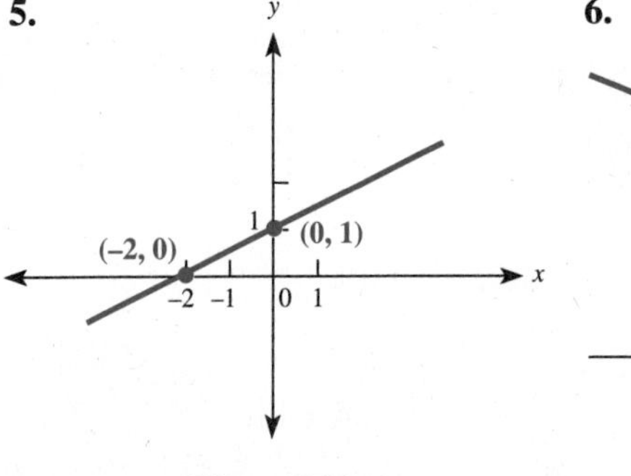

Figure 4.22

6.

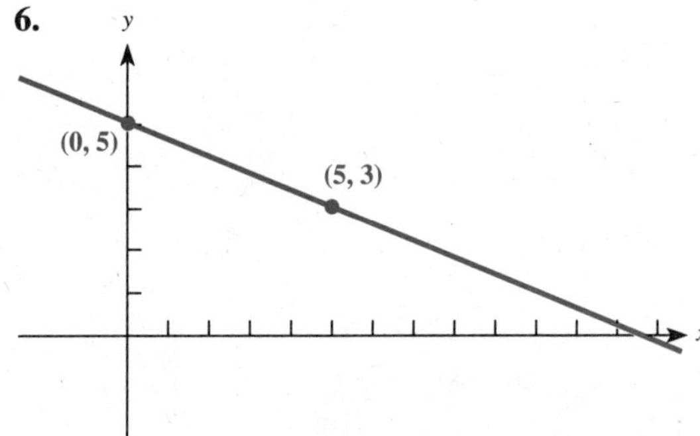

Figure 4.23

7.

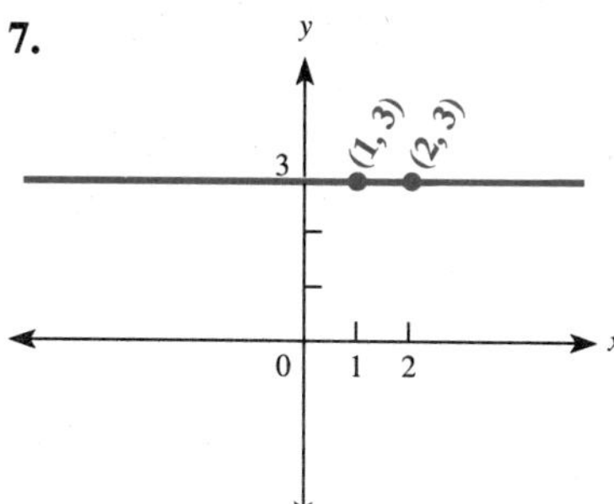

Figure 4.24

8.

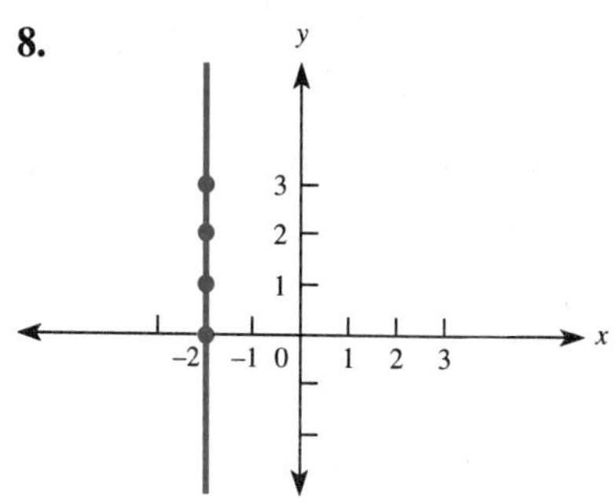

Figure 4.25

Solutions

5. **(a)** By inspection, the y-intercept is $(0, 1)$. Thus $b = 1$.

(b) By considering the two points $(0, 1)$ and $(-2, 0)$, the slope equals

$$m = \frac{1 - 0}{0 - (-2)} = \frac{1}{2}$$

(c) Hence the equation is $y = \frac{1}{2}x + 1$. **We substituted $m = \frac{1}{2}$ and $b = 1$ into the equation $y = mx + b$.**

6. **(a)** $(0, b) = (0, 5)$

(b) $m = \dfrac{5 - 3}{0 - 5} = \dfrac{2}{-5}$

(c) The equation is $y = -\frac{2}{5}x + 5$.

7. **(a)** $(0, b) = (0, 3)$

(b) $m = \dfrac{3 - 3}{2 - 1} = 0$

(c) The equation is $y = 0(x) + 3$, or $y = 3$. **Recall that the equation of a horizontal line is $y = k$.**

8. **(a)** There is no y-intercept.

(b) $m = \dfrac{3 - 1}{-2 - (-2)} = \dfrac{2}{0}$ **Division by zero is undefined.**

The slope is undefined.

(c) The equation is $x = -2$. **Recall that the equation of a vertical line is $x = k$.** ■

Looking at Examples 5–8, we observe the following: If a line is rising from left to right, it has a *positive slope,* $m > 0$ (see Example 5). If a line is falling from left to right, it has a *negative slope,* $m < 0$ (see Example 6). If a line is horizontal, it has *zero slope,* $m = 0$ (see Example 7). If a line is vertical, it has an *undefined slope* (see Example 8).

EXAMPLES 9–12 In each case, find the slope-intercept equation of the line, given

9. Slope $= -3$, and y-intercept $= (0, 0)$

10. Points $\left(0, -\frac{1}{2}\right)$ and $(1, 2)$

11. Slope $= 2$, and point $(3, 4)$

12. Points $(-2, -3)$ and $(-1, 4)$

Solutions

In each case, we will find the values of m and b to substitute into the equation $y = mx + b$.

9. Since $m = -3$ and $b = 0$, we can substitute directly into the equation:

$$y = -3x + 0 \quad \text{or} \quad y = -3x$$

10. Since $(0, b) = \left(0, \frac{-1}{2}\right)$, $b = \frac{-1}{2}$.

$$\text{slope} = \frac{\frac{-1}{2} - 2}{0 - 1} = \frac{\frac{-5}{2}}{-1} = \frac{5}{2}$$

The equation is

$$y = \frac{5}{2}x + \frac{-1}{2} = 2.5x - 0.5$$

11. Here $m = 2$, so $y = 2x + b$. Since the point $(3, 4)$ is on this line, we can find b by substitution:

$$4 = 2(3) + b$$

and solve for b

$$b = 4 - 6 = -2$$

The equation is $y = 2x - 2$.

12. We use the two given points to find the slope:

$$m = \frac{-3 - 4}{-2 - (-1)} = \frac{-7}{-1} = 7$$

Thus $y = 7x + b$. To determine b we will proceed as in Example 11 above. Substitute one of the points, say, $(-2, -3)$, into our equation and solve for b:

$$-3 = 7(-2) + b$$
$$b = 11$$

The equation is $y = 7x + 11$. ■

The strategy for *finding* the *slope-intercept form of a line* $y = mx + b$:

1. Find the slope m. Substitute it into the equation for m.
2. Find the y-intercept, $(0, b)$. Substitute b into the equation.
   (a) If it is given, we substitute b directly into the equation.
   (b) If it is not given, substitute one of the given points into the equation $y = mx + b$ (where m is known), and solve for b.

EXAMPLE 13

(Example 11 revisited) Find an equation of the line with slope 2 that contains the point (3, 4).

Solution

We apply the strategy of finding two equivalent ways of expressing the slope of the line. It is given that $m = 2$. If (x, y) is any point on the line, then

$$m = \frac{y - 4}{x - 3}$$

Thus

$$\frac{y - 4}{x - 3} = 2 \quad \text{or} \quad y - 4 = 2(x - 3).$$

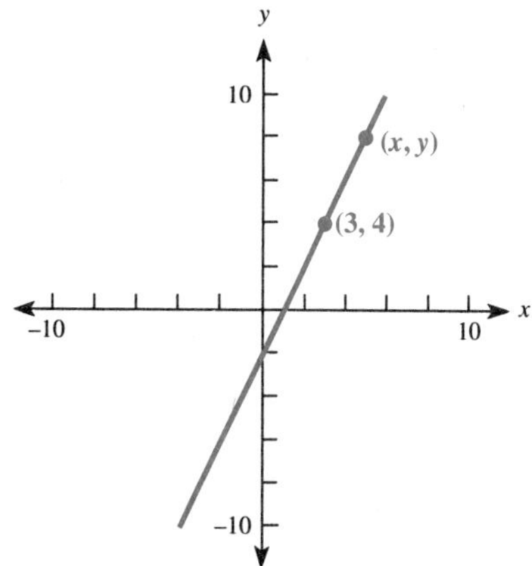

Figure 4.26

A line with slope m and containing the point (x_1, y_1) can be written in **point-slope form** as $y - y_1 = m(x - x_1)$.

EXAMPLE 14

(Example 12 revisited) Find an equation of the line containing the points $(-2, -3)$ and $(-1, 4)$.

Solution

As in Example 12, $m = 7$. Then $y - (-3) = 7[x - (-2)]$, or $y + 3 = 7(x + 2)$ or $y = 7x + 11$. ■

Relating Graphs and Equations of Lines

EXAMPLE 15

The lines in the graph below (see Exercise 57 in Section 4.1) represent the distance that Ann and Betty run in a race. Since Ann is slower, she gets a head start.

(a) For each woman, find a linear equation or linear function representing the distance in terms of time.

(b) Explain the significance of the slope of each line.

(c) Explain the significance of the y-intercept of each line.

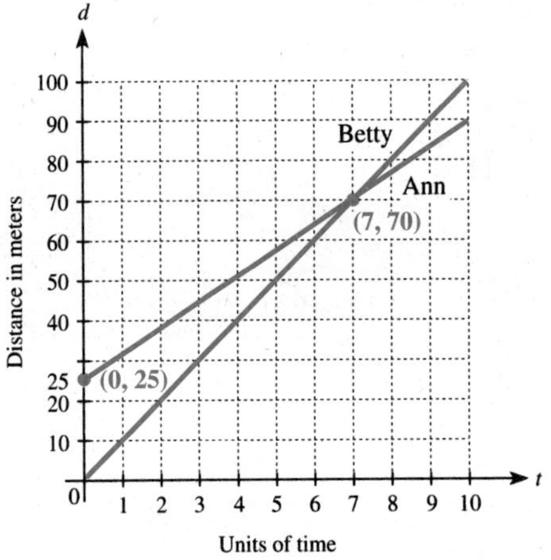

Figure 4.27

Solutions

Let d be the distance and t be the time.

(a) Equation (function) for Betty: We observe that the points (0, 0) and (7, 70) lie on the line.

$$\text{slope} = \frac{0 - 70}{0 - 7} = \frac{-70}{-7} = 10$$

The y-intercept is given as (0, 0). Thus

$$d_B = 10t \qquad \textbf{d_B is Betty's distance.}$$

Equation (function) for Ann: The points (0, 25) and (7, 70) lie on the line.

$$\text{slope} = \frac{70 - 25}{7 - 0} = \frac{45}{7}$$ It makes no difference which point we let be (x_1, y_1) or (x_2, y_2).

The y-intercept is (0, 25). Thus

$$d_A = \frac{45}{7}t + 25$$ d_A is Ann's distance.

(b) The slope is the rate at which each woman runs. Betty runs 10 meters for every unit of time, and Ann runs $\frac{45}{7}$ meters for every unit of time. Note also that the slope is the coefficient of t (the equation is in slope-intercept form).

(c) The y-intercept for Betty represents the fact that she starts at the starting line. The y-intercept for Ann means she gets a head start of 25 meters. ■

EXAMPLE 16

The population of the United States in 1980 was 225 million, and in 1990, it was 249 million.

(a) Graph the data.

(b) Find a linear equation (function) that relates the year x to the population.

(c) Estimate the population in 1987, 1930, and 2025 by using this equation (or function).

(d) Is this model realistic for all values of x?

Solutions

(a) To plot the graph, we first need to determine a unit for the x-axis and the y-axis. One possible choice is to let x equal the number of years after 1980 and let y equal the population in millions.

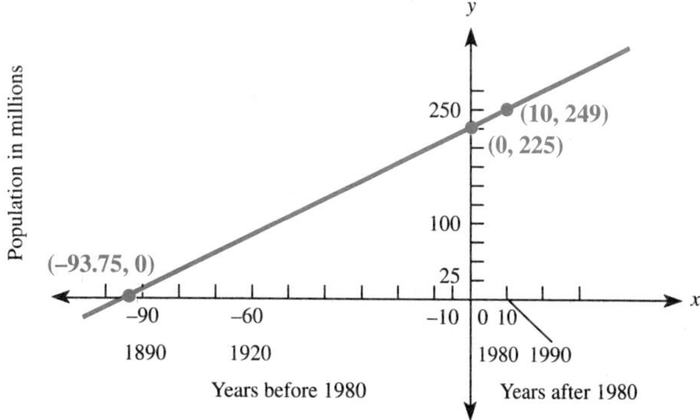

Figure 4.28

(b) With this interpretation, the points (0, 225) and (10, 249) lie on the graph. Then (0, b) = (0, 225), and

$$m = \frac{249 - 225}{10 - 0} = \frac{24}{10} = 2.4$$

Note the slope measures the *rate of growth* of the population—2.4 million people per year. The equation of the line is

$$y = 2.4x + 225$$

(c) The year 1987 corresponds to $x = 7$, so $y = 2.4(7) + 225 = 241.8$ million people. Of course, an *estimate* could have been read from the graph.

The year 1930 corresponds to $x = -50$, so $y = 2.4(-50) + 225 = 105$ million people.

The year 2025 corresponds to $x = 45$, so $y = 2.4(45) + 225 = 333$ million people.

(d) One way to see that the model is unrealistic for all values of x is to observe that the x-intercept is $(-93.75, 0)$ (if $y = 0$, then $2.4x = -225$, so $x = -93.75$). This means that about 94 years before 1980, that is, in 1886, the U.S. population was zero. ■

In many applications we can assume that the data given lie on a line, and then we seek a linear equation (function) $y = mx + b$, as in Example 16. Data do not always lie on a straight line. We will investigate nonlinear models in later chapters. For the present, we will assume a linear relationship for our applications.

4.2 EXERCISES

I. Procedures

1–6. For each equation, **(a)** find the slope. **(b)** find the y-intercept. **(c)** sketch the graph.

1. $y = 4x - 1$

2. $4y + x = 8$

3. $\frac{2}{3}x - \frac{1}{5}y - 2 = 0$

4. $1.1x = 2.2y + 1.21$

5. $x = 6$

6. $y = 5$

7–12. For each line, find **(a)** the slope using the points A and B. **(b)** the y-intercept. **(c)** the equation of the line. If necessary, use approximate values.

7.

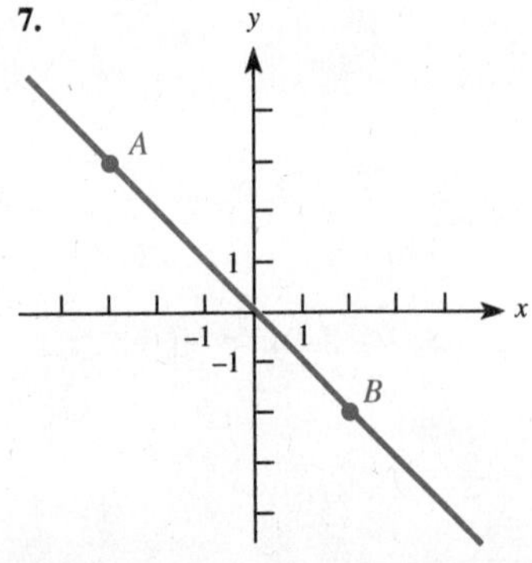

8.

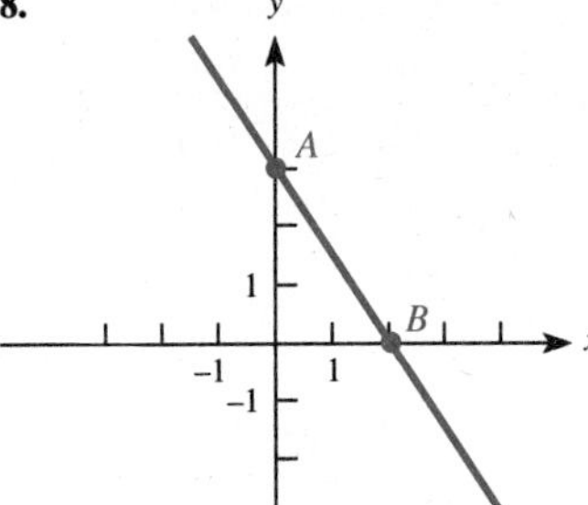

9.

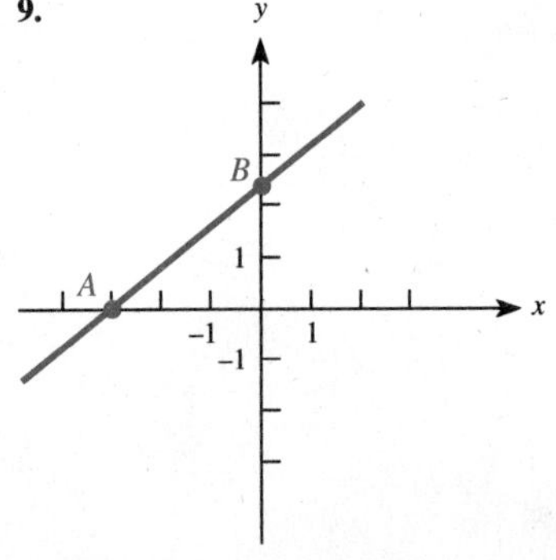

10.

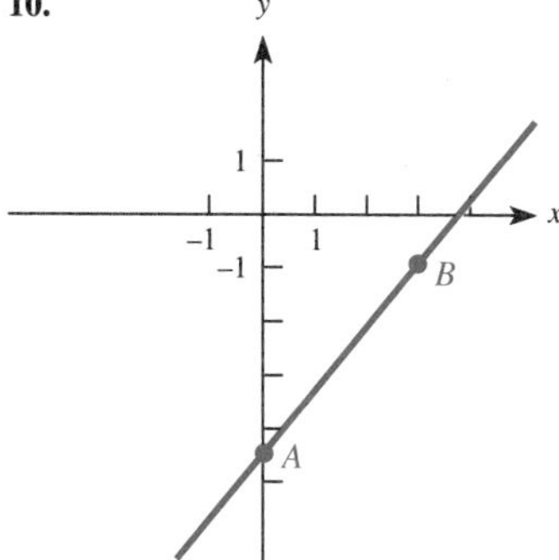

11.

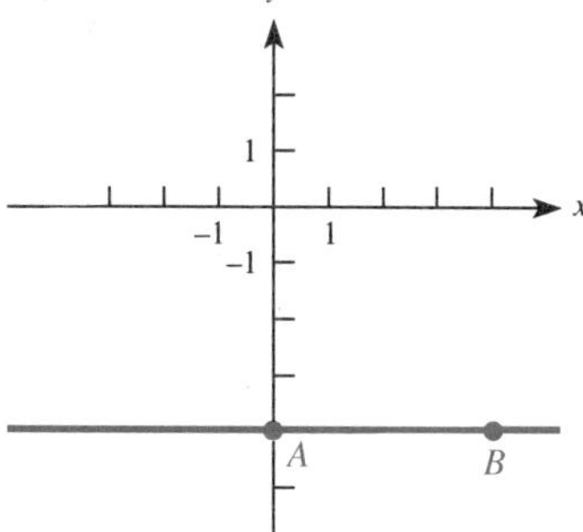

12.

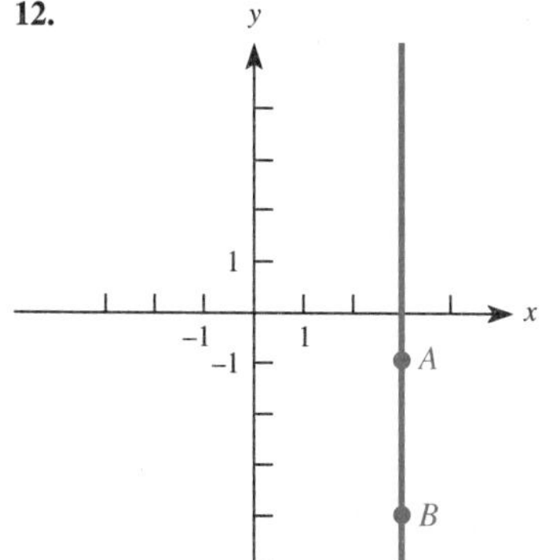

13–32. **(a)** Sketch the graph of the line. **(b)** Find an equation for the line that satisfies the given conditions.

13. Slope $= 2$, y-intercept $= (0, -3)$.

14. Slope $= \frac{1}{3}$, y-intercept $= (0, -1)$.

15. Slope $= -\frac{8}{5}$, y-intercept $= \left(0, \frac{7}{10}\right)$.

16. Slope $= 0$, y-intercept $= (0, 2)$.

17. Slope is undefined, x-intercept $= (2, 0)$.

18. Slope $= 2$, and line contains $(-3, 4)$.

19. Slope $= -\frac{1}{2}$, and line contains $\left(\frac{1}{4}, -\frac{3}{4}\right)$.

20. Line contains the points $(0, 3)$ and $(1, 2)$.

21. Line contains the points $(0, -1)$ and $(2, -4)$.

22. Line contains the points $(2, 0)$ and $(0, 4)$.

23. Line contains the points $(-1, 0)$ and $(0, -1)$.

24. Line contains the points $(0, 0)$ and $(0, 5)$.

25. Line contains the points $(0, 0)$ and $(5, 0)$.

26. Line contains the points $(2, 1)$ and $(8, 6)$.

27. Line contains the points $(-3, 2)$ and $(4, 1)$.

28. Line contains the points $\left(-\frac{3}{5}, \frac{2}{10}\right)$ and $\left(\frac{2}{5}, -\frac{7}{10}\right)$.

29. Slope $= 0.234$, line contains $(7.231, -2.894)$.

30. Slope $= -4.236$, line contains $(0.12, 4.674)$.

31. x-intercept $= 5$; slope $= 4$.

32. x-intercept $= 2$; y-intercept $= -3$.

II. Concepts

33–43. Give the equations of three lines that satisfy the given condition.

33. Contains the point $(0, 0)$.

34. Contains the point $(1, 2)$.

35. Intersects the x-axis at $(-1, 0)$.

36. Intersects the y-axis at $(0, -1)$.

37. Slope $= 2$.

38. Slope $= 0$.

39. Line has undefined slope.

40. y-intercept $= (0, 3)$.

41. x-intercept $= (-1, 0)$.

42. Line with positive slope intersects the circle in two points:

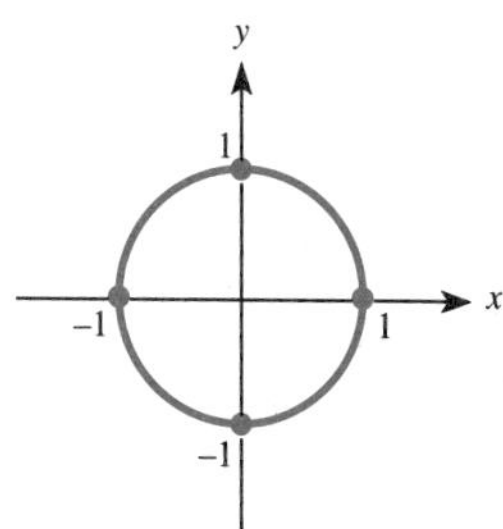

43. Line with negative slope intersects the circle in two points:

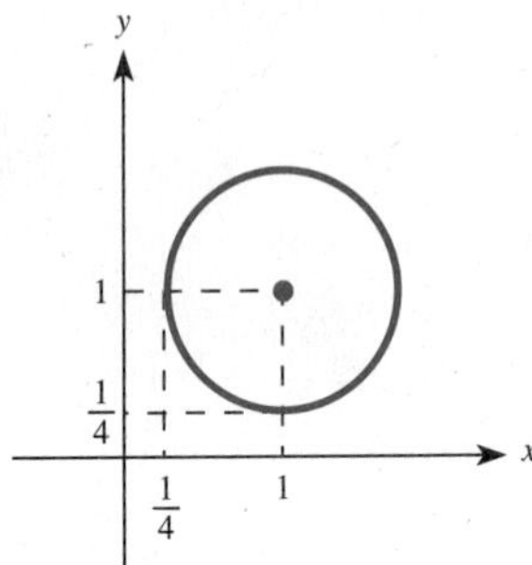

44. In Example 14, is $y - 4 = 7[x - (-1)]$ the same line as $y + 3 = 7(x + 2)$? Give reasons for your answer.

45. Do any two lines, one with positive slope and the other with negative slope, always intersect? Explain.

46. Investigate the relationship between the graphs of $y = ax$ and $y = ax + b$ for various values of a and b. Write a summary of your results. (You might use a graphing calculator.)

47. Investigate the relationship between the graphs of

$$y = ax \quad \text{and} \quad y = -\frac{1}{a}x$$

for various values of a. Write a summary of your results. (You might use a graphing calculator.)

III. Applications

48–62. In each of the following problems, **(a)** Find a linear equation or function relating the data. **(b)** Sketch its graph (a graphing calculator may help). **(c)** State the significance of the slope and y-intercept of the line. **(d)** Assuming the linear relationship is valid, answer the other questions by using the linear equation (function). **(e)** If the line is written in the form $y = mx + b$, for what values of x do you think the linear model is reasonable? (No exact answer is possible; just try to give a reasonable one.)

48. The population of Corvallis, Oregon, was about 41,000 in 1980 and 45,000 in 1990. Estimate the population during the following years.

(a) 1975 **(b)** 1987 **(c)** 2005

49. A car-rental agency charges 25¢ for each mile for the first 100 miles and 17¢ for each mile after 100 miles.

(a) Find the cost to drive 676 miles.

(b) How many miles can be driven for $125?

50. It costs $72 to drive a car 200 miles per month and $256 to drive 1,000 miles per month. Find the cost to drive the following number of miles per month.

(a) 500 miles per month **(b)** 1,500 miles per month

51. In 1985, the average number of cable channels received by a television subscriber was 18.8. In 1990, this average was 33.2. How many channels can the average subscriber expect to receive in 2001?

52. According to the *Wilderness Trail Book,* at 50°F crickets make 76 chirps per minute, and at 65°F they make 100 chirps per minute.

(a) What is the chirping rate at 90°F?

(b) What is the temperature if you count 125 chirps per minute?

53. The average weight of a person who is 70 inches tall is 165 pounds. A person 67 inches tall weighs 145 pounds.

(a) What is the weight of an average person who is 75 inches tall?

(b) What is the height of an average person whose weight is 180 pounds?

54. A 400-gram weight attached to a spring will stretch the spring 25 mm, while a 300-gram weight will stretch the spring 18.75 mm.

(a) How much will a 150-gram weight stretch the spring?

(b) If the spring stretches 30 mm, what weight is attached to it?

55. A long-distance call after 5 P.M. costs $1.25 for 5 minutes and $2.30 for 10 minutes.

(a) How much will a 25-minute call cost?

(b) How long could you talk for $5.00?

56. A family of four with an income of $8,000 paid $300 in state taxes while a family of four with an income of $13,000 paid $800 in state taxes. How much state tax will a family of four pay if its income is

(a) $10,000? **(b)** $15,000? **(c)** $5,000?

57. At $2.20 a case for chocolate sardines, a wholesaler will supply 40 cases. At $1.60 a case, he supplies 25 cases. How much will he supply at

(a) $1.20 a case?

(b) $3.00 a case?

58. At \$2.20 a case for chocolate sardines, the quantity demanded is only 10 cases, while at \$1.20 a case, the demand is 35 cases.

(a) At what price will there be no demand for chocolate sardines?

(b) Using the data in Exercise 57, at what price will the supply equal the demand? *Hint:* Graph both equations on the same set of axes.

59. In 1895, Fred Bacon of Scotland ran the mile in 4 minutes, 17 seconds. In 1954, John Landry of Australia ran the mile in 3 minutes, 58 seconds. Assume we started keeping records of the mile run in 1895.

(a) Find an equation representing year and time.

(b) Check your equation to see how close it comes to the following records: In 1942, both Sydney Wooderson of Britain and Arne Andersson of Sweden ran the mile in 4 minutes, 6.2 seconds. In 1966, Jim Ryun of the United States ran the mile in 3 minutes, 51.3 seconds.

(c) At this rate, what will the record for the mile be in the year 2000?

(d) Check your equation for the current mile record.

60. A teacher wants to rescale the scores on a difficult test so that the maximum possible is still 100 but the mean (average) is 80 instead of 66. If 60 on the new scale is the lowest passing mark, what was the lowest passing mark on the original scale?

61. Gases expand when heated and contract when cooled. A particular gas has a volume of 500 cc at 27°C and a volume of 600 cc at 87°C.

(a) What temperature (called the *absolute temperature*) will give the gas a volume of 0 cc?

(b) What is the volume of the gas at 50°C?

62. The population of the United States was (approximately) 205 million in 1970, 225 million in 1980, and 249 million in 1990.

(a) Using the data for 1970 and 1980, find a linear model that expresses the population in terms of the number of years after 1970.

(b) Was the population for 1990 accurately predicted by the linear model in part (a)? If not, what was the percent error?

(c) According to the linear model in part (a), when will the U.S. population have reached 249 million?

(d) How does the model in part (a) compare to the one in Example 16?

63. Suppose you own a car that is presently 24 months old. From the automobile "Blue Book" you find that its present trade-in value is \$4,900. From an old "Blue Book," you find it had a trade-in value of \$6,250 twelve months ago. Assume the trade-in value decreases linearly with time (straight-line depreciation).

(a) Write a linear equation giving the trade-in value of your car in terms of its age in months.

(b) Sketch its graph.

(c) You plan to get rid of the car before its trade-in value drops below \$2,000. How much longer can you keep the car?

(d) What is the interpretation of the slope of this line?

(e) When do you predict the car will be worthless?

(f) According to your model, what was the car's trade-in value when it was new?

(g) Suppose you actually paid \$12,960 for the car when it was new. How would you explain the difference between \$12,960 and your answer in part (f)?

64. Use the data in the following table to answer the questions.

Year	1965	1979	1983	1987
Percent of males who were smokers	51.6	37.2	34.7	31.0

(a) Using the data for 1965 and 1979, find a linear model that expresses the percent of male smokers x years after 1965.

(b) Was the percent for 1983 accurately predicted by the linear model in part (a)? If not, what was the percent error?

(c) Was the percent for 1987 accurately predicted by the linear model in part (a)? If not, what was the percent error?

(d) According to the linear model in part (a), when will the percent of male smokers reach 15%?

(e) Using the data for 1983 and 1987, find a linear model that expresses the percent of male smokers x years after 1983.

(f) According to the linear model in part (e), when will the percent of male smokers reach 15%?

65. The following table gives the life expectancy of people born since 1920.

Year	*All*	*Males*	*Females*
1920	54.1	53.6	54.6
1930	59.9	58.1	61.6
1940	62.9	60.8	65.2
1950	68.2	65.6	71.1
1960	69.7	66.6	73.1
1970	70.8	67.1	74.7
1975	72.6	68.8	76.6
1980	73.7	70.0	77.4
1985	74.7	71.2	78.2
1990	75.6	72.1	79.0

(a) Using the data for 1940 and 1980, find a linear model for the life expectancy of males born since 1920.

(b) What is the interpretation of the slope and y-intercept of your model?

(c) According to the model, what should be the life expectancy for a male born in 1985?

(d) According to the model, when will the life expectancy of a male reach age 75?

66. A portion of the 1991 Tax Tables for a single taxpayer contains the data:

Taxable income	*Tax*
\$0	\$0
\$20,350	\$3,052.50
\$49,300	\$11,158.50
over \$49,300	\$11,158 + 31% of the amount over \$49,300

(a) Plot the three points representing the first three entries in the table on a graph. Choose appropriate scales for your axes, with income on the horizontal axis and tax on the vertical axis.

(b) Find the equation for each of the two segments joining successive points of the graph.

(c) Sketch the graph.

(d) What is the interpretation of the slope of each of these two segments?

(e) Suppose we stop the table at an income of \$100,000. Plot the point on the graph corresponding to this income, and find the equation of the segment joining it and the previous point representing an income of \$49,300.

67. The three lines on the graph represent bids from three companies to process bills to patients for a large medical group practice. They are:

- Service Bureau (SB): Its bid was a flat fee plus a certain charge per bill.
- Computer Data Management (CDM): Its bid was also a flat fee plus a certain charge per bill, but its flat fee was less than SB's and its price per bill was higher.
- In-House Dept. (ID): Its bid was a certain charge per bill with no flat fee.

(a) Match each of the lines l, m, n with the initials of the firm whose bid the line represents.

(b) What was the approximate amount of the flat fee of the Service Bureau bid?

(c) What was the approximate amount of the charge per bill of the In-House Dept. bid?

(d) To process 80,000 bills per year, which firm submitted the lowest bid?

(e) You expect to process x bills next year. For approximately what values of x would the CDM bid be the lowest?

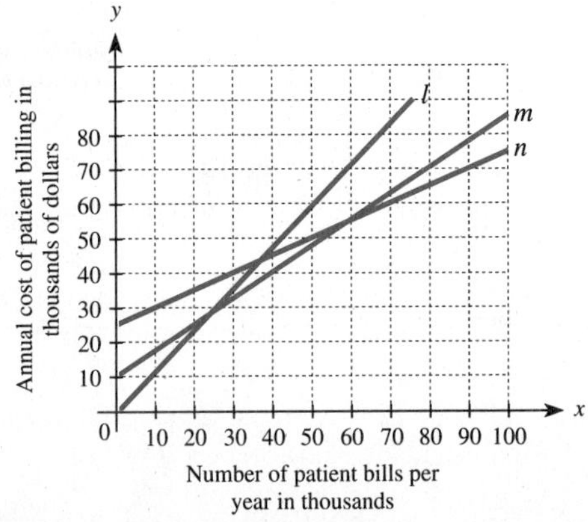

IV. Extensions

68. Graph Codes: Write a series of linear equations for $0 \le y \le 2$ whose graphs will spell out your first name or some other short word. For example, TIM can be spelled by graphing the following:

First letter T	Second letter I	Third letter M
$y = 2$	$x = 0$	$x = 1$
$-3 \le x \le -1$		$y = -2x + 4$
$x = -2$		$y = 2x - 4$
		$x = 3$

69. Show that the line determined by the points (x_1, y_1) and (x_2, y_2) is

$$y - y_2 = \left(\frac{y_2 - y_1}{x_2 - x_1}\right)(x - x_2)$$

This formula is called the *two-point* form of a linear equation.

70. The value of y in the linear equation $y = 24x + 1$ is a square when $x = 0, 1,$ or 2. Find a linear equation of the form $y = mx + b$ so that y is a square when $x = 3, 4,$ and 5.

71. Find an equation for each of the segments AB, BC, CD, and DE of the bridge truss shown. Be sure to specify where you are placing the origin.

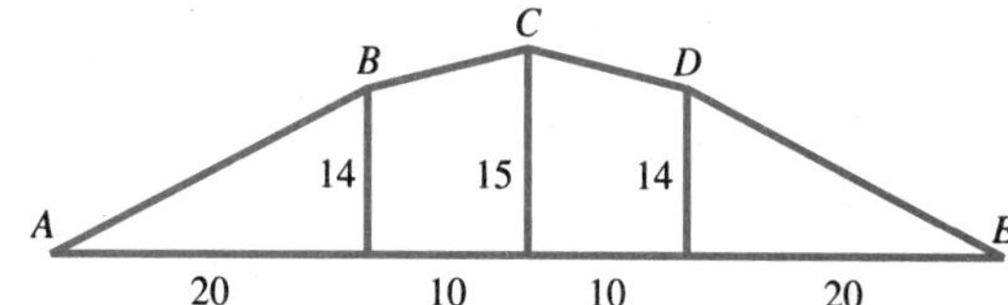

4.3 PROPERTIES OF LINEAR GRAPHS AND FITTED LINES

Parallel and Perpendicular Lines

The study of graphs establishes the close relationship between algebra and geometry. Graphs are a geometric representation of algebraic statements. We will look at two examples of this relationship: *Parallel* and *perpendicular* lines.

The graphs of linear equations of the form

$$y = ax$$

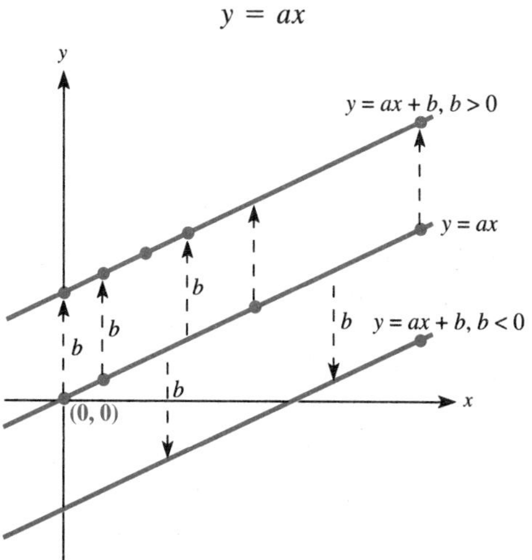

Figure 4.29

are lines that pass through the origin with a y-intercept equal to (0, 0) and slope equal to a. If we consider the equation $y = ax + b$, where $b \neq 0$, we have a line with slope a and y-intercept $(0, b)$. The constant b does not contribute to the slope of the straight line. To go from $y = ax$ to $y = ax + b$ (in Figure 4.29), we slide the line of $y = ax$ up b units (if $b > 0$) or down b units (if $b < 0$) to form the line $y = ax + b$. From the graph, it appears that the *lines with the same slope are parallel.*

Conversely, if two lines are parallel, then they have the same slope.

EXAMPLES 1–2

(a) Graph each set of equations.

(b) Decide whether each set of lines is parallel.

1. $3x = y$
$6y - 18x = 24$

2. $x + y = 1$
$2x + y = 2$

Solutions

1. $3x = y$
$6y - 18x = 24$

Writing each equation in slope-intercept form, we have

(a) $y = 3x + 0$
$y = 3x + 4$

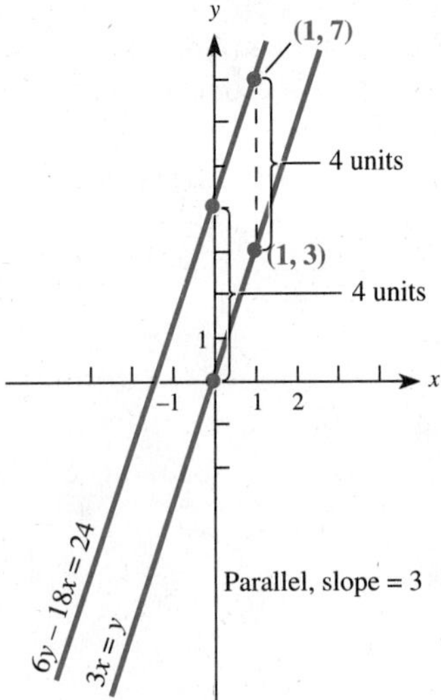

Figure 4.30

(b) The lines are parallel, since the slope of both equals 3. Note that to transform the line of $y = 3x$ to the line of $y = 3x + 4$, we add four units to the y-coordinate of each point on line $y = 3x$.

2. $x + y = 1$

$2x + y = 2$

Writing each equation in slope-intercept form, we have

(a) $y = -x + 1$

$y = -2x + 2$

(b) The lines are not parallel, since the slopes are not equal.

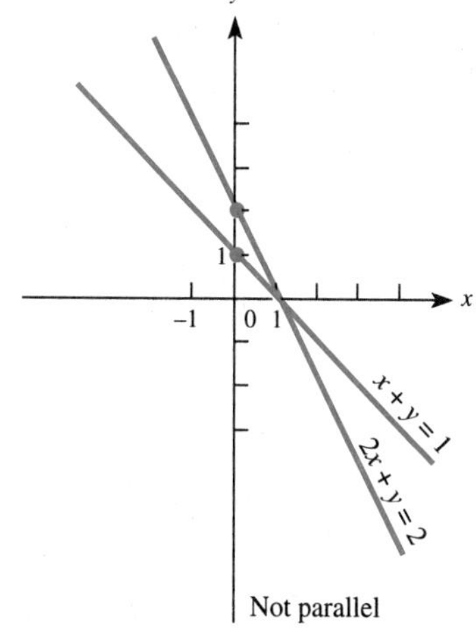

Figure 4.31

To determine a criterion for lines to be **perpendicular,** we examine the lines $y = 2x$ and $y = -\frac{1}{2}x$ (Figure 4.32). We observe that each line passes through the origin, and the lines appear to be perpendicular to each other. The slope of $y = 2x$ is 2, and the slope of $y = -\frac{1}{2}x$ is $-\frac{1}{2}$. The slope $-\frac{1}{2}$ is the negative reciprocal of the slope 2. (Recall

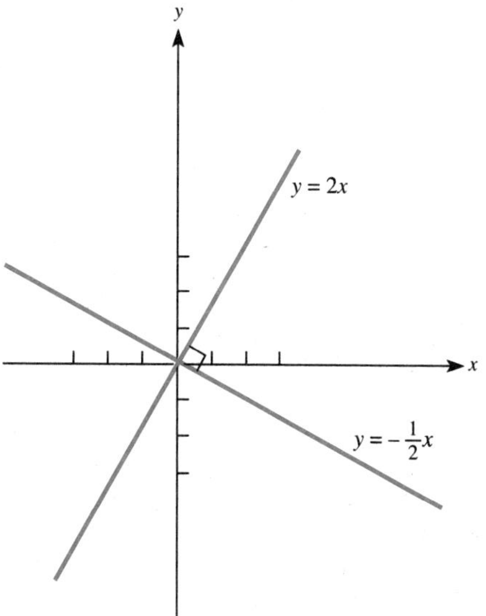

Figure 4.32

that the reciprocal of a is $\frac{1}{a}$, and in general, the reciprocal of $\frac{a}{b}$ is $\frac{b}{a}$). Two lines are *perpendicular* if the *slope* of one line is the *negative reciprocal* of the slope of the other line.

EXAMPLES 3–4

(a) Graph each set of equations.

(b) Decide whether the lines of each set are perpendicular.

3. $y = x$

$y + x = 3$

4. $x + 3y = 1$

$3x + y + 4 = 0$

Solutions

3. (a)

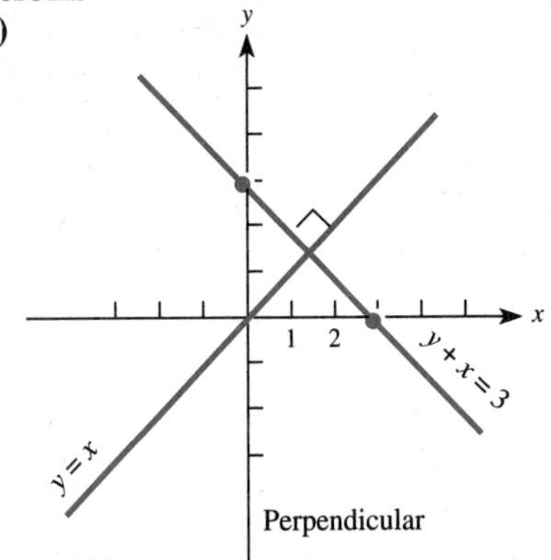

Figure 4.33

(b) We write each set of lines in the slope-intercept form and determine the slope.

$y = x$, slope $= 1$

$y = -x + 3$, slope $= -1$

The negative reciprocal of 1 is -1, so the two lines are perpendicular.

4. (a)

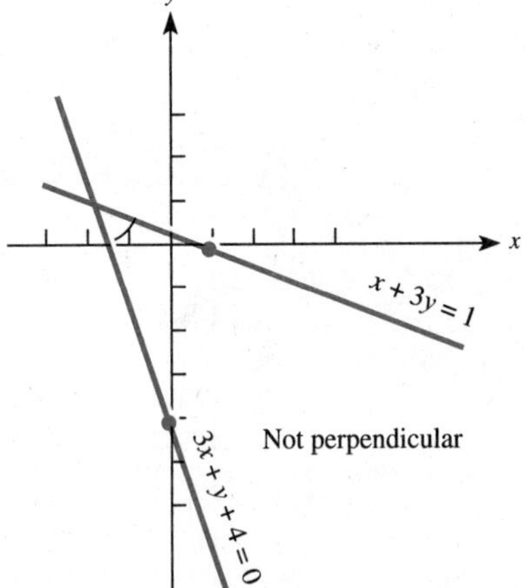

Figure 4.34

(b) $y = \frac{-1}{3}x + \frac{1}{3}$, slope $= -\frac{1}{3}$

$y = -3x - 4$, slope $= -3$

The *negative* reciprocal of $\frac{-1}{3}$ is $\frac{-1}{\frac{-1}{3}} = +3$. The lines are *not* perpendicular. ■

We can summarize the results of Examples 1–4.

Parallel Lines and Perpendicular Lines

Let $y = m_1x + b_1$ and $y = m_2x + b_2$ be any two linear equations.

1. If $m_1 = m_2$, then the lines are *parallel.* Conversely, if two lines are *parallel,* then $m_1 = m_2$.

2. If $m_1 = \frac{-1}{m_2}$, then the lines are *perpendicular.* Conversely, if two lines are *perpendicular*, then $m_1 = \frac{-1}{m_2}$, $m_2 \neq 0$.

EXAMPLES 5–6

Find the equation of the line that both contains the point (1, 2) and satisfies the given condition. Sketch the graph of both lines.

5. Parallel to the line $3x - 4y = 7$.

6. Perpendicular to the line $3x - 4y = 7$.

Solutions

5. We first rewrite $3x - 4y = 7$ in the form $y = mx + b$ to determine its slope:

$$y = \frac{3}{4}x - \frac{7}{4}$$

The slope of the line is $\frac{3}{4}$, so the slope of any line parallel to it is also equal to $\frac{3}{4}$, and it has an equation of the form

$$y = \frac{3}{4}x + b$$

Since (1, 2) lies on the line, we substitute it into the equation and solve for b:

$$2 = \frac{3}{4} \cdot 1 + b$$

$$\frac{5}{4} = b$$

So $y = \frac{3}{4}x + \frac{5}{4}$ (or $3x - 4y = -5$) is the line parallel to $3x - 4y = 7$ and passing through the point (1, 2). (See Figure 4.35.)

6. The perpendicular line has slope $-\frac{4}{3}\left(-\frac{4}{3}\right.$ is the negative reciprocal of $\left.\frac{3}{4}\right)$ and an equation of the form $y = -\frac{4}{3}x + b$. Again (1, 2) lies on the line, so we find b:

$$2 = -\frac{4}{3} \cdot 1 + b$$

$$\frac{10}{3} = b$$

So $y = -\frac{4}{3}x + \frac{10}{3}$, or $4x + 3y = 10$, is the line perpendicular to $3x - 4y = 7$ that passes through the point (1, 2).

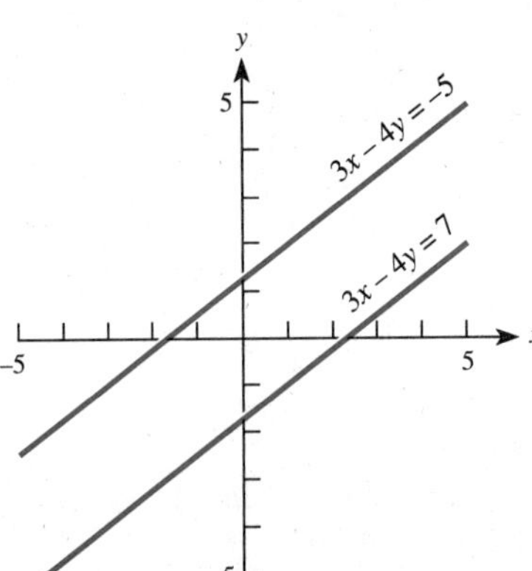

Figure 4.35

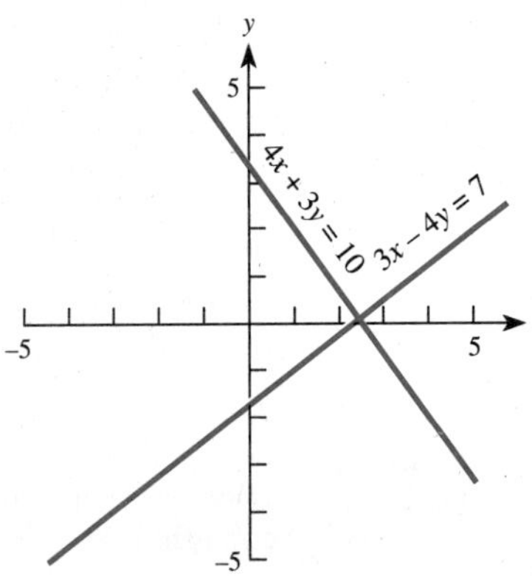

Figure 4.36

Fitted Lines

In Example 16 of Section 4.2, we constructed a model representing the population growth of the United States based upon knowing the population for two years and *assuming* the situation could be described using a linear equation. As we remarked at the time, this linear model was not valid for a long time span, since it led to the conclusion that there was zero population in 1886. One of the basic reasons for obtaining this incorrect result was that we arrived at our conclusions by using only two pieces of data (2 points on the graph). The population growth might better be described by looking at 10 points, as in Figure 4.37. If we now draw the line we drew before, we see that it is reasonably close to the curve for $0 \leq x \leq 10$, but far away for other values of x. For those values of $x > 10$, it is not even close enough for a "ball-park estimate." It is possible, however, to encounter a situation in which many data points appear to lie close to some straight line. Figure 4.38 graphically represents the profit on widgets for prices less than \$10.

The obvious difficulty is that all of the points do not lie on one line. How are we, then, to determine the equation of one line that is close to most of these points? First we will try to plot the points carefully and then select two of them so that the line through these two points is close to all other points. Clearly, such a procedure involves guesswork, but we hope that the line is close enough so that it can be used to predict future behavior. If the line is carefully drawn for the values of x that we do know ($0 \leq x \leq 10$ in Figure 4.38),

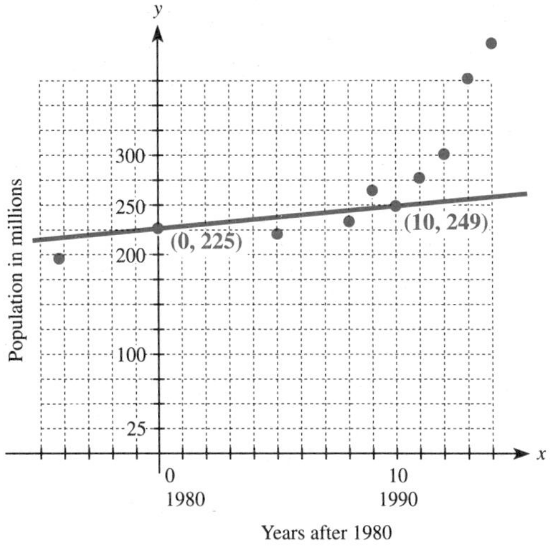

Figure 4.37

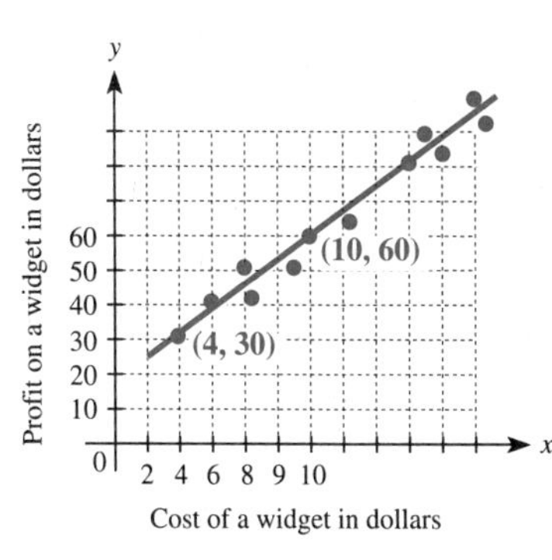

Figure 4.38

then it is hoped that it is also accurate for the values of x we do not know ($x > 10$ in Figure 4.38) and can be used to predict what *will* happen there (for $x > 10$ in Figure 4.38). To make this discussion more concrete, consider the next example.

EXAMPLE 7

The marketing research department of Continental Rubber and Tire Company compiled the data shown in this table:

x	43	43.2	44.8	46	47.5	50	53.8	55	56.1	58	61	64	66
y	1.5	1.8	2.5	2.8	2.4	3.3	3.2	3.6	4.0	4.2	4.3	4.8	5.4

Table 4.7

This information represents the experience of 13 years, where x is the yearly number of passenger cars in the United States in millions, and y is the corresponding tire sales of the company in millions.

(a) Plot the points carefully.

(b) Find an equation of a line that is close to these points.

(c) Use the equation to predict the number of tires sold when there are 72 million passenger cars in the United States.

Solutions

(a) It would be easier to let x equal the number of passenger cars over 40 million and to plot the following points:

x	3	3.2	4.8	6	7.5	10	13.8	15	16.1	18	21	24	26
y	1.5	1.8	2.5	2.8	2.4	3.3	3.2	3.6	4.0	4.2	4.3	4.8	5.4

Table 4.8

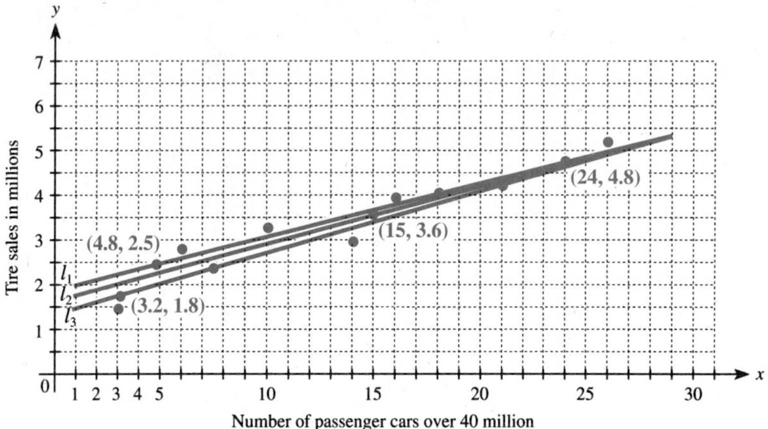

Figure 4.39

(b) After trying three possibilities, all of which appear close, the line l_2 containing (15, 3.6) and (24, 4.8) appears to be a reasonable choice. The slope of the line is

$$m = \frac{4.8 - 3.6}{24 - 15} = \frac{1.2}{9} \approx 0.13$$

The equation of the line is

$$y = 0.13x + b$$

Since the point (15, 3.6) lies on the line, we have

$$3.6 = 0.13(15) + b$$
$$1.65 = b$$

Therefore

$$y = 0.13x + 1.65$$

(c) When the passenger car population reaches 72, $x = 32$ (over 40), so

$$y = 0.13(32) + 1.65$$

or

$$y = 5.81 \text{ million tires}$$

These answers are, of course, approximate. Slightly different answers would be obtained if another line were used. ■

Recall the definition of *median* from Chapter 1. Another fitted line that is close to this data is the **median-median line** or MM–line. It is found as follows:

1. Separate the data into three groups as close to equal in size as possible, according to the values of the horizontal coordinate (the "x-values").
2. Find the **summary point** for the set of points in each of the three groups. The coordinates of the summary point are:

   x-coordinate: the median of the x-coordinates of the points in the set

   y-coordinate: the median of the y-coordinates of the points in the set
3. Find the equation of the line, l, through the two outer summary points.
4. Slide l one-third of the way to the middle summary point by
   - **(a)** finding the y-coordinate on l corresponding to the x-coordinate of the middle summary point.
   - **(b)** finding the distance between the y-coordinates of the point on l found in (a) and the middle summary point.
   - **(c)** adding one-third of the distance in (b)—which may be negative—to the y-intercept of l to obtain a y-intercept b'.
   - **(d)** writing the equation whose slope is that of l and whose y-intercept is b'.

Let us apply this procedure to Example 7.

Step 1: Separate the data into three groups
There are 13 points. We divide them into groups of 4, 5, and 4. The points are already given in the order of their horizontal coordinates in the table.

Step 2: Find the summary points

Table 4.9

x	3	3.2	4.8	6	7.5	10	13.8	15	16.1	18	21	24	26
y	1.5	1.8	2.5	2.8	2.4	3.3	3.2	3.6	4.0	4.2	4.3	4.8	5.4

median x = 4.0 median y = 2.15

median x = 13.8 median y = 3.3

median x = 22.5 median y = 4.55

Summary points:

(4.0, 2.15) (13.8, 3.3) (22.5, 4.55).

Step 3: Find the equation of the line, l, through (4.0, 2.15) and (22.5, 4.55)

$$m = \text{slope of } l = \frac{4.55 - 2.15}{22.5 - 4.0} = 0.13$$

$$y = 0.13x + b$$

$$2.15 = 0.13(4.0) + b \qquad \text{(4.0, 2.15) is on } l$$
$$b = 1.63$$
$$\text{Equation of } l\text{: } y = 0.13x + 1.63$$

Step 4: Find the equation of the median-median (MM) line

(a) $y = 0.13(13.8) + 1.63$ or $y = 3.42$ — y-coordinate of point on l whose x-coordinate is 13.8

(b) $3.42 - 3.30 = 0.12$ — distance between the y-coordinates on l and MM–line

(c) $1.63 + \dfrac{.12}{3} = 1.67$ — y-intercept of MM–line

(d) $y = 0.13x + 1.67$ — equation of MM–line—very close to our "eyeball" estimate

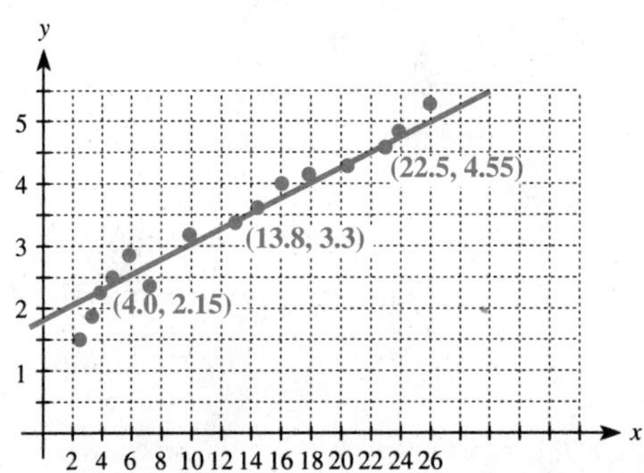

Figure 4.40

Many graphing calculators allow you to find the equation of the MM–line or another fitted line, the least square line, by entering the data and pressing an appropriate key. In this example, the least square line is $y = 0.14x + 1.51$.

4.3 EXERCISES

I. Procedures

1–8. To solve each problem, perform the following steps: **(a)** Graph each set of the equations on the same coordinate axes. **(b)** Decide whether the set of lines is parallel, perpendicular, or neither. **(c)** Give the slope of each line.

1. $x + 1 = y, x + y = 2$

2. $3x + 2y = 10, 8 + 2y = 6x$

3. $y = 3, x = 3$

4. $y = 2, y = -1$

5. $3x - y = 2, 3y - x = 5$

6. $2x = y, 4y + 2x = 6$

7. $5x + 1 = x, 10x + 2 = 2x$

8. $3y = \dfrac{x}{2} + 3, y - 2x = 8$

9–20. Find an equation of the line that satisfies the given conditions. Sketch its graph.

9. Parallel to $y = 3x - 7$ and contains the point (0, 5).

10. Parallel to $y = -\frac{x}{2} + 3$ and contains $(6, 0)$.

11. Parallel to $2x - 3y = 4$ and contains $(-1, 2)$.

12. Parallel to $3x + 4y = 8$ and contains $(-7, 3)$.

13. Parallel to $x = 2$ and contains $(-4, 8)$.

14. Parallel to $y = 4$ and contains $(-4, 8)$.

15. Perpendicular to $y = 3x$ and contains $(0, -4)$.

16. Perpendicular to $y = \frac{x}{2} + 1$ and contains $(4, 0)$.

17. Perpendicular to $2x - 3y = 4$ and contains $(1, 2)$.

18. Perpendicular to $3x + 4y = 8$ and contains $(-7, 3)$.

19. Perpendicular to $x = \frac{7}{2}$ and contains $(3, 4)$.

20. Perpendicular to $y = -3$ and contains $(-2, -1)$.

II. Concepts

21–24. Give equations of three lines that satisfy the given condition. Sketch the graphs.

21. Parallel to $y = 2x$.

22. Parallel to $y = ax$.

23. Perpendicular to $y = 2x$.

24. Perpendicular to $y = ax$.

25. Three vertices of a rectangle are $(0, 0)$, $(2, 1)$, and $(2, 0)$. Find the fourth vertex.

26. Three vertices of a rectangle are $\left(2, \frac{3}{2}\right)$, $(2, -6)$, and $\left(\frac{1}{2}, -6\right)$. Find the fourth vertex.

27. Two vertices of a rectangle are $(0, 0)$ and $(3, 4)$. What are possible values for the coordinates of the other two vertices of the rectangle?

28. Three vertices of a parallelogram are $(0, 0)$, $(2, 1)$, and $(3, 1)$. Find the fourth vertex.

29. Write a set of four equations of lines such that the points of intersection are the corners of a square.

30. Write a set of four equations of lines such that the points of intersection are the corners of a square that has an area of 4 square units.

31. Write a set of three equations of lines such that the three points of intersection are the vertices of a triangle.

32. Write a set of three equations of lines such that the three points of intersection are the vertices of a triangle with area of 6 square units.

33. Find the equation of the line satisfying the given condition(s).

(a) $m = \frac{1}{2}, b = -\frac{1}{2}$.

(b) contains the points $(5, 2)$ and $(7, 3)$.

(c) parallel to $y = \frac{1}{2}x$ and containing the point $(11, 5)$.

(d) perpendicular to $y = -2x$ and containing the point $(5, 2)$.

34. Describe, in your own words, how to construct the median-median line.

III. Applications

35–37. For each set of data, perform the following steps: **(a)** Plot the points carefully. **(b)** Find an equation of a line that is close to most of the points. **(c)** Use both the equation and the graph to answer the questions posed in each problem.

35. The management of Toy Products, Inc., is planning to market a new mechanical toy for the forthcoming Christmas season. One important problem in marketing a new product is pricing. Even though the ultimate consumers of the product are considered by the company not to be very price conscious, the retailers who sell the toy are very influenced by price in their wholesale purchasing. The company, therefore, surveyed the retailers and produced the following data, where p is a suggested price and q is the number of units the retailers would be willing to order at that price (that is, the demand):

p	2.90	3.90	4.50	4.90	5.50	5.90	6.75	7.25	8.90
q	6,850	6,500	5,800	5,250	4,500	4,400	3,700	2,600	600

(a) How many units would they expect to sell at $7.00?

(b) How many units would they expect to sell at $4.25?

(c) At what price would there be no demand?

36. The Ajax Rubber Company has a set of data available through its research department:

x	50	50.5	50.8	51.4	51.9	52.5	53.2	53.8	54.5	55.1
y	117	138	142	154	170	182	200	215	235	264

Here x represents the U.S. annual passenger-car population (in millions), and y indicates corresponding tire sales (in thousands) of the company.

(a) How many tires will be sold if there are 65 million passenger cars in the United States?

(b) If 300,000 tires are sold, how many passenger cars are there in the United States?

37. A recreational specialist decided to estimate the number of people visiting a ski area by establishing a relationship between the number of vehicles in the parking lots and the number of people using the resort at any given time. For several days, a crew actually counted the vehicles and their occupants as they entered and departed from the parking lots. The data were then represented on the following chart:

x	10	15	18	25	55	70	100	130	170	205
y	15	25	20	40	205	210	350	360	650	610

where x is the number of vehicles in the parking lot and y is the number of occupants.

(a) If an aerial photo shows an estimate of 190 cars in the parking lot, how many people are in the resort area?

(b) If there is an estimate of 250 cars, how many people are in the resort area?

38–40. Consider the data in the columns of the following table that give the life expectancy for people born since 1920. **(a)** Find a fitted line for the data on females using the indicated method. Then answer the two questions using this line. **(b)** According to the model, what should be the life expectancy for females born in 1985? **(c)** According to the model, when will the life expectancy of a female reach age 85?

Year	*All*	*Males*	*Females*
1920	54.1	53.6	54.6
1930	57.9	58.1	61.6
1940	62.9	60.8	65.2
1950	68.2	65.6	71.1
1960	69.7	66.6	73.1
1970	70.8	67.1	74.7
1975	72.6	68.8	76.6
1980	73.7	70.0	77.4
1985	74.7	71.2	78.2
1990	75.6	72.1	79.0

38. "eyeball" line

39. median-median line

40. least square line

41–43. Find a fitted line for this data using the indicated method. Then answer the questions: According to the model, **(a)** what will be the record in June 2000? **(b)** when do you expect the mile record to reach 3.67?

Date	*Person*	*Time (in minutes)*
1954	Roger Bannister	3.99
1954	John Landy	3.967
1957	Derek Ibbotson	3.953
1958	Herb Elliot	3.9083
1962	Peter Snell	3.9067
1964	Peter Snell	3.90167
1965	Michael Jazy	3.8933
1966	Jim Ryun	3.855
1967	Jim Ryun	3.85167
1975	John Walker	3.8233
1979	Sebastian Coe	3.8167
1980	Steve Ovett	3.8133
1981	Sebastian Coe	3.7888
1985	Steve Cram	3.772

World Records for One-Mile Run

41. "eyeball" line

42. median-median line

43. least square line

44–46. Find a fitted line for this data using the indicated method. Then answer the questions: According to the model, **(a)** what will be the cost of a postage stamp in 2000? **(b)** when do you expect the cost of a postage stamp to reach 50¢?

Year	*Cost*	*Year*	*Cost*
1960	.04	1978	.15
1963	.05	1981	.18
1966	.06	1982	.20

(Table continued on next page.)

Year	*Cost*	*Year*	*Cost*
1971	.08	1985	.22
1974	.10	1988	.25
1975	.13	1991	.29

Postage Stamp Rates: 1960–1991

44. "eyeball" line

45. median-median line

46. least square line

IV. Extensions

47–48. For each problem, find the equation of a line that satisfies the given condition.

47. A line that is halfway between $y = 3x - 4$ and $y = 3x + 2$.

48. A line that is halfway between $y = \frac{x}{2} + \frac{1}{2}$ and $y = \frac{x}{2} - \frac{1}{2}$.

4.4 LINEAR SYSTEMS

We return briefly to an old problem and examine it in a new light.

EXAMPLE 1

The We Try Harder (WTH) car rental company charges \$19.95 plus 19¢ a mile per day. The U Drive It Company (UDI) charges \$15.95 and 23¢ a mile. How many miles must we drive for the costs to be equal?

Solution

Using an Equation The cost that each company charges can be represented by linear equations in terms of miles (M):

$$\text{For WTH,} \quad C = \$19.95 + 0.19M \qquad \textbf{(Equation 1)}$$

$$\text{For UDI,} \quad C = \$15.95 + 0.23M \qquad \textbf{(Equation 2)}$$

To determine when the costs are equal, we can *substitute* the expression for cost from one equation into the other equation

$$19.95 + 0.19M = 15.95 + 0.23M$$

and solve for M:

$$4 = 0.04M$$

$$M = 100 \text{ miles}$$

Check: When $M = 100$, $C = 19.95 + 0.19(100) = \$38.95$ for WTH; and $C = 15.95 + 0.23(100) = \$38.95$ for UDI. Thus, at 100 miles, the costs are the same, \$38.95.

Using a Graph We also know that each equation can be represented by a line. See Figure 4.41.

We can rephrase our problem in terms of the graph. The costs are equal at the point of intersection of the two lines. On the graph (Figure 4.41), this is point A with coordinates

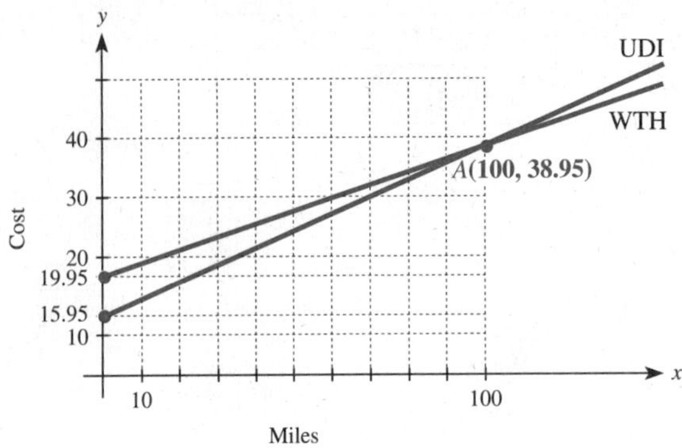

Figure 4.41

(100, 38.95). Thus we need to find values of M and C that simultaneously satisfy two linear equations. ■

Another way to find this solution algebraically is to use the properties of equality stated in Chapter 3 and the following more general property of equality:

Property of Equality

If equals are added or subtracted from equals, then the results are equal: If $a = b$ and $c = d$, then $a \pm c = b \pm d$.

Applying the above property to the system from Example 1, we have

$$\begin{array}{rll} C &= 19.95 + 0.19M & \textbf{(Equation 1)} \\ C &= 15.95 + 0.23M & \textbf{(Equation 2)} \\ \hline 0 &= 4.00 - 0.04M & \textbf{Equals subtracted from equals.} \\ M &= 100 & \textbf{Solve for } M. \\ C &= 38.95 & \textbf{Substitute } M = 100 \textbf{ into one of the } \textit{original} \textbf{ equations and solve for } C. \end{array}$$

2 × 2 Linear Systems

Many of the problems we have studied have actually involved *two* linear equations. To solve each of these problems, we need to find values of x and y that simultaneously satisfy two linear equations.

Definition

Consider the following set of two linear equations:

$$ax + by = e$$
$$cx + dy = f$$

The graphs of each of these equations are straight lines. A system of two equations in two variables is called a **2 × 2 linear system** (read "two by two"). A pair of values (x, y) that *simultaneously* satisfies the two equations is called a **solution** of the system. These values of x and y are the coordinates of the point of intersection of the two lines.

We now discuss several algorithms (step-by-step procedures) for determining the solution of a linear system or showing that no solution exists.

As we did in the car-rental example, we could draw a careful graph of each line and estimate the coordinates of the point of intersection. Two algebraic methods can yield more precise answers.

EXAMPLE 2

Solve the following linear system

$$3x - y = 7$$
$$2x + 3y = 12$$

(a) by graphing and estimating the solutions.

(b) by algebraic methods—substitution and addition.

Solutions

Using a Graph

(a) Graph both lines. The point of intersection is estimated to be (3, 2).

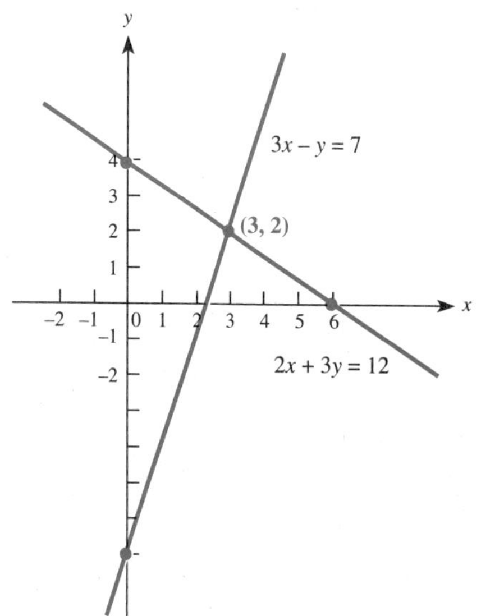

Figure 4.42

Using an Equation

(b) *Substitution method:*

$$\begin{cases} 3x - y = 7 \\ 2x + 3y = 12 \end{cases}$$

$y = 3x - 7$ — **Solve for y in the first equation.**

$2x + 3(3x - 7) = 12$ — **Substitute this expression for y into the second equation. We could also have solved both equations for y and set the two expressions equal.**

$2x + 9x - 21 = 12$

$11x = 33$

$x = 3$ — **Solve for x.**

$3(3) - y = 7$ — **Substitute $x = 3$ into one of the original equations.**

$y = 2$ — **Solve for y.**

The solution is $x = 3$ and $y = 2$, or $(3, 2)$.

Addition method:

$$\begin{cases} 3x - y = 7 \\ 2x + 3y = 12 \end{cases}$$ — **We choose to eliminate y first. We could just as well eliminate x first.**

$$\begin{cases} 9x - 3y = 21 \\ 2x + 3y = 12 \end{cases}$$ — **Multiply the first equation by 3. Write the second equation below it.**

$11x + 0 = 33$ — **Add the two equations.**

$x = 3$ — **Solve for x.**

$3(3) - y = 7$ — **Substitute $x = 3$ into one of the *original* equations to obtain an equation with y as the only variable.**

$y = 2$ — **Solve for y.**

The solution is $x = 3$ and $y = 2$, or $(3, 2)$.

Check: To guard against a careless mistake, we check the solution using both *original* equations.

$$2(3) + 3(2) \stackrel{?}{=} 12 \qquad 3(3) - 2 \stackrel{?}{=} 7$$
$$12 \stackrel{\checkmark}{=} 12 \qquad 7 \stackrel{\checkmark}{=} 7$$

■

Comments on Example 2

The graphing method is a practical way to solve a 2×2 linear system. When you graph

$$Y1 = 3x - 7 \quad \text{and} \quad Y2 = -\frac{2}{3}x + 4,$$

you get a figure similar to the one shown. You might find the coordinates of the point of intersection are $x = 3.05$ and $y = 2.15$. After using the ZOOM key several times, you might obtain $x = 3.01$ and $y = 2.02$. The moral is that the way your calculator processes numbers and the viewing window you use may affect your ability to find an exact solution such as $x = 3$ and $y = 2$.

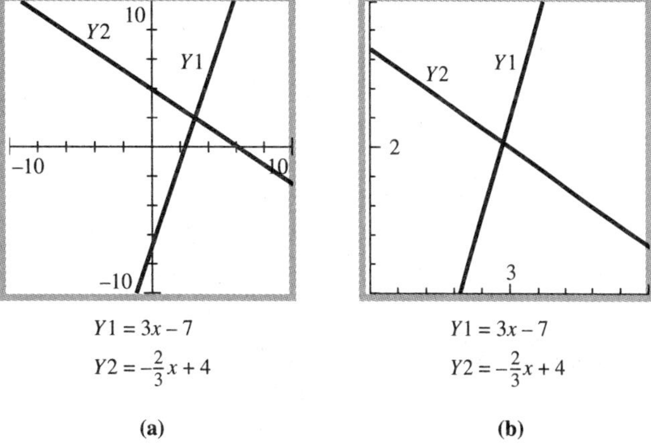

$Y1 = 3x - 7$
$Y2 = -\frac{2}{3}x + 4$

(a)

$Y1 = 3x - 7$
$Y2 = -\frac{2}{3}x + 4$

(b)

Figure 4.43

In using the addition method, we replace one linear system with an equivalent one (one with the same solution) until we reach a system for which we can simply "read off" the solution. In this example, we essentially considered five equivalent systems. In the last one, the solution is easy to "read off."

(a)	(b)	(c)	(d)	(e)
$3x - y = 7$ $2x + 3y = 12$	$9x - 3y = 21$ $2x + 3y = 12$	$9x - 3y = 21$ $11x = 33$	$3x - y = 7$ $x = 3$	$y = 2$ $x = 3$

The graphs of these five systems are:

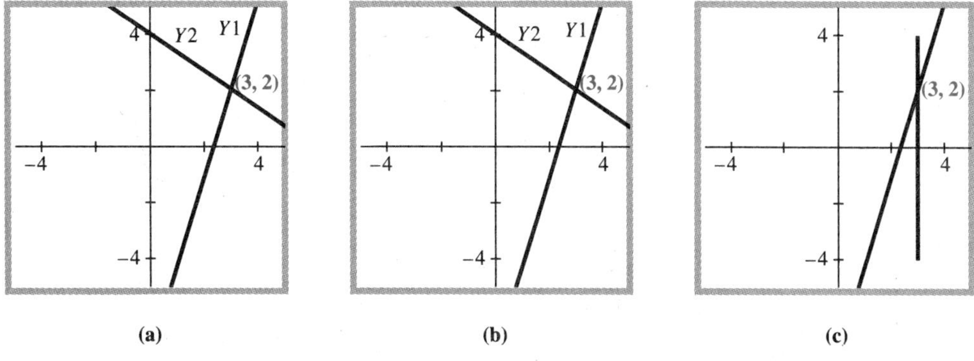

(a) (b) (c)

Figure 4.44 (continued on next page)

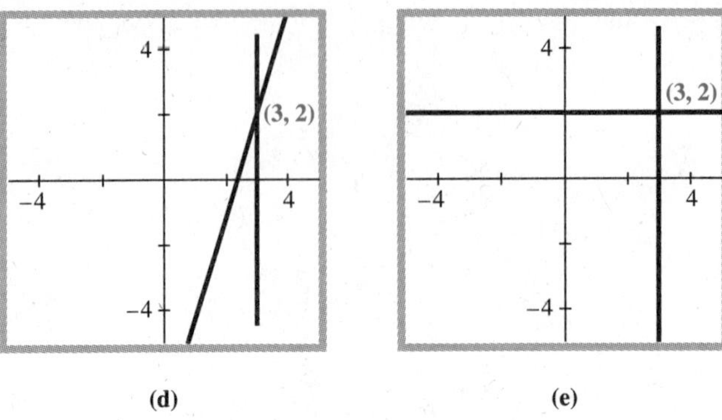

Figure 4.44

In summary, we can solve a 2×2 linear system using one of the following methods.

Strategies for Solving a 2×2 Linear System

A. *Graphing:*

1. Graph each linear equation. **May yield only approximate solutions. This may be done on a graphing calculator.**
2. Estimate the coordinates of the point of intersection.

B. *Substitution:*

1. Solve one of the equations for one variable. **May involve fractions.**
2. Substitute this expression for this variable into the other equation.

or

1. Solve each equation for the same variable. **Easy to use if one coefficient equals 1.**
2. Set the two quantities equal.

Then

3. Solve the resulting linear equation.
4. Substitute the solution to step 3 into one of the original equations, and solve for the second variable.
5. Check the values of x and y in both of the original equations.
6. State the solutions.

C. *Addition (Elimination):*

1. Put both equations in the form $Ax + By = C$.
2. Multiply one or both equations by a number such that the coefficients of one variable will be the same or opposites of each other. **Most general method.**
3. Add or subtract the two equations to obtain a new equation in one variable.
4. Solve the new equation for the one variable.
5. Substitute this value into one of the original equations and solve for the second variable.
6. Check the values for x and y in both of the original equations.
7. State the solutions.

Note: a version of C using matrices and the concept of matrix inverse is possible on a graphing calculator.

EXAMPLES 3–4

Solve each of the linear systems by the substitution or the addition method.

3. $2x - 3y = 7$
$-2x + 4y = -8$

4. $3x - y = 2$
$x - 3y = -10$

Solutions

3. $2x - 3y = 7$
$-2x + 4y = -8$

In this problem, solving for either x or y in one of the equations to substitute into the second equation involves calculating with fractions. For example, in the first equation, $x = \frac{(3y + 7)}{2}$ and $y = \frac{(2x - 7)}{3}$. We will try the addition method:

$$\begin{cases} 2x - 3y = 7 \\ -2x + 4y = -8 \end{cases}$$ **The coefficients of the x terms are opposites.**

$0 + y = -1$ **Add the two equations.**

$y = -1$ **Solve for y.**

$2x - 3(-1) = 7$ **Substitute $y = -1$ into the first equation and solve for x.**

$2x = 4$

$x = 2$

Check: If $x = 2$ and $y = -1$, then

$$2(2) - 3(-1) \stackrel{?}{=} 7 \quad \text{and} \quad -2(2) + 4(-1) \stackrel{?}{=} -8$$
$$4 + 3 \stackrel{?}{=} 7 \qquad\qquad -4 - 4 \stackrel{?}{=} -8$$
$$7 \stackrel{\checkmark}{=} 7 \qquad\qquad -8 \stackrel{\checkmark}{=} -8$$

The solution is $x = 2$ and $y = -1$, or $(2, -1)$.

4. $3x - y = 2$
$x - 3y = -10$

We observe that the coefficient of x in the second equation is 1. Therefore it should be easy to try the substitution method:

$$\begin{cases} 3x - y = 2 \\ x - 3y = -10 \end{cases}$$

$x = 3y - 10$ **Solve for x in the second equation.**

$3(3y - 10) - y = 2$ **Substitute for x in the first equation and solve for y.**

$9y - 30 - y = 2$

$8y = 32$

$y = 4$

$3x - (4) = 2$ **Substitute $y = 4$ into the first equation and solve for x.**

$3x = 6$

$x = 2$

Check: If $x = 2$ and $y = 4$, then

$$\begin{aligned} 3(2) - 4 &\overset{?}{=} 2 \\ 6 - 4 &\overset{?}{=} 2 \\ 2 &\overset{\checkmark}{=} 2 \end{aligned} \quad \text{and} \quad \begin{aligned} 2 - 3(4) &\overset{?}{=} -10 \\ 2 - 12 &\overset{?}{=} -10 \\ -10 &\overset{\checkmark}{=} -10 \end{aligned}$$

The solution is $x = 2$ and $y = 4$, or $(2, 4)$. ■

The Geometry of Two Lines

Given two lines, one of three possibilities must occur:

1. The two lines are different and intersect in one point.
2. The two lines are different and parallel, and therefore they do not intersect.
3. The two lines are the same, therefore they have an infinite number of points in common.

The algebraic counterparts of these phenomena occur in the following example.

EXAMPLE 5 Solve each of the following systems by graphing and by an algebraic method. Use a graphing calculator if possible.

(a) $3x + 2y = 4$ | **(b)** $3x + 2y = 4$ | **(c)** $3x + y = 3$
$6x + 4y = 21$ | $6x + 4y = 8$ | $5x + 2y = 5$

Solutions

(a) $\begin{cases} 3x + 2y = 4 \\ 6x + 4y = 21 \end{cases}$

$\begin{cases} 6x + 4y = 8 \\ 6x + 4y = 21 \end{cases}$ **Multiply the first equation by 2.** **Subtract.**

$0 = -13$ **A contradiction.**

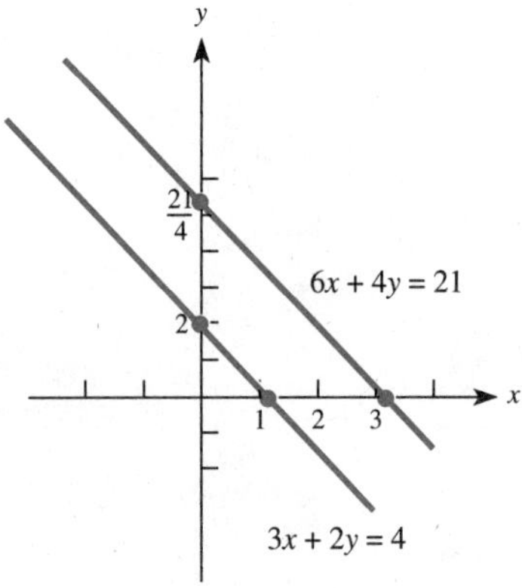

Figure 4.45

Since a contradiction was obtained, there is no solution, as is seen in the graph. The lines are parallel $\left(\text{the slopes are } -\frac{3}{2}\right)$, so there is *no point of intersection*, or *no solution*. If a system of equations has no solution, the system is called **inconsistent.**

(b) $\begin{cases} 3x + 2y = 4 \\ 6x + 4y = 8 \end{cases}$

$\begin{cases} 6x + 4y = 8 \\ 6x + 4y = 8 \end{cases}$ Multiply the first equation by 2. Subtract.

$$0 = 0$$

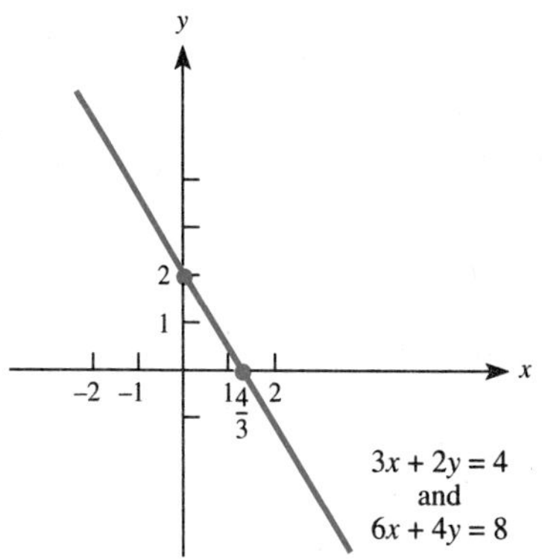

Figure 4.46

However, $0 = 0$ is always true. Hence any pair of numbers that is a solution to one equation is also a solution to the other equation. This can also be seen from the graph, which shows the lines coincide. Hence they have an *infinite number of points* in common, or an *infinite number of solutions.*

(c) $\begin{cases} 3x + y = 3 & \textbf{(Equation 1)} \\ 5x + 2y = 5 & \textbf{(Equation 2)} \end{cases}$

$y = 3 - 3x$ Solve for y in Equation 1.

$5x + 2(3 - 3x) = 5$ Substitute into Equation 2.

$5x + 6 - 6x = 5$

$-x = -1$ Solve for x.

$x = 1$

$5(1) + 2y = 5$ Substitute $x = 1$ into Equation 2.

$y = 0$ Solve for y.

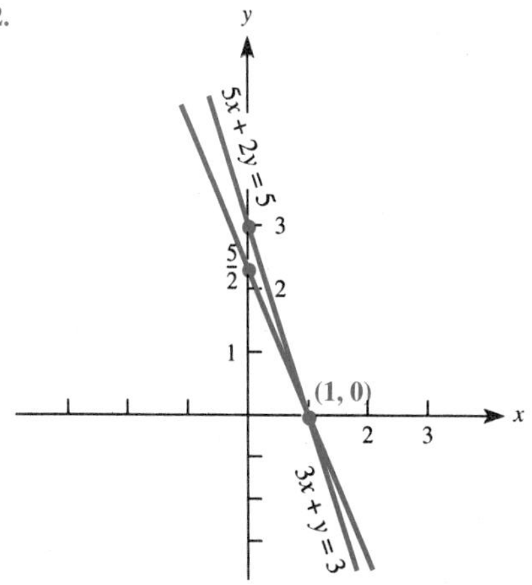

Figure 4.47

The solution is $x = 1$ and $y = 0$, or $(1, 0)$. The lines are close together, but they intersect in one point. ■

Two *different* lines can intersect in at most one point, so such a 2×2 linear system has at most one solution. A 2×2 linear system may have no solution if the two lines are parallel, or an infinite number of solutions if the two equations are really two names for the same line.

EXAMPLE 6

For what values of c does the linear system

$$\begin{aligned} cx - 2y &= 7 \\ 3x + 4y &= 8 \end{aligned}$$

have exactly one solution?

Solution

Method 1 There is a unique solution if the two lines are different and not parallel. Therefore the slopes must be unequal. Calculating these slopes, we find

$$\begin{aligned} 2y &= cx - 7 \\ y &= \frac{c}{2}x - \frac{7}{2} \end{aligned}$$

So $m = \frac{c}{2}$ for the first line. For the second line, we have

$$\begin{aligned} 4y &= -3x + 8 \\ y &= -\frac{3}{4}x + 2 \end{aligned}$$

So $m = \frac{-3}{4}$. Thus there is a unique solution if $\frac{c}{2} \neq \frac{-3}{4}$, or $c \neq \frac{-3}{2}$.

Method 2 If we try to solve for x using the addition method, we have

$$\begin{cases} 2cx - 4y = 14 & \text{Multiply by 2.} \\ 3x + 4y = 8 \end{cases}$$

$$\begin{aligned} 2cx + 3x &= 22 && \text{Add the two equations.} \\ (2c + 3)x &= 22 && \text{Distributive property.} \\ x &= \frac{22}{2c + 3} && \text{Divide both sides by } (2c + 3). \end{aligned}$$

Now $\frac{22}{2c+3}$ is defined, and consequently, a unique solution exists, as long as $2c + 3 \neq 0$, or $c \neq \frac{-3}{2}$. (Recall that we cannot divide by zero.) ■

EXAMPLE 7

Solve the system

$$\begin{aligned} 3.241x - 1.802y &= 6.314 \\ 1.014x + 2.763y &= 12.825 \end{aligned}$$

Solution

Using an Equation We proceed as in the previous examples. Thus using the addition method, we would have the following steps:

$$(3.241)(2.763)x - (1.802)(2.763)y = (6.314)(2.763)$$
$$(1.802)(1.014)x + (1.802)(2.763)y = (1.802)(12.825)$$

NOTE: It is somewhat easier to save the calculations until the last step.

$$[(3.241)(2.763) + (1.802)(1.014)]x = (6.314)(2.763) + (1.802)(12.825)$$

Add the two equations.

$$x = \frac{(6.314)(2.763) + (1.802)(12.825)}{(3.241)(2.763) + (1.802)(1.014)} = \frac{40.556}{10.782} \approx 3.761$$

Solve for x. Use a calculator.

$$(3.241)(3.761) - 1.802y = 6.314$$
$$-1.802y = 6.314 - (3.241)(3.761)$$

Substitute for x and solve for y.

$$y = \frac{6.314 - (3.241)(3.761)}{-1.802}$$
$$y \approx 3.261$$

Use a calculator again.

Checking these values in the second equation, we find $(1.014)(3.761) + (2.763)(3.260) \approx 12.821$. This is close; the discrepancy is due to error in rounding off.

Using a Graph To use a graphing calculator, we first must put both equations into the form $y = mx + b$.

$$Y1 = (3.241x - 6.314) \div 1.802$$

or

$$Y1 = \frac{3.241x}{1.802} - \frac{6.314}{1.802}$$

$$Y2 = (-1.014x + 12.825) \div 2.763$$

or

$$Y2 = \frac{-1.014x}{2.763} + \frac{12.825}{2.763}$$

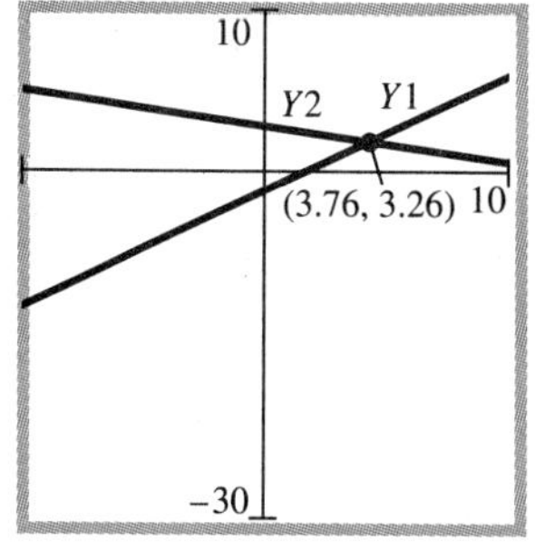

$Y1 = \dfrac{3.241x - 6.314}{1.802}$

$Y2 = \dfrac{-1.014x + 12.825}{2.763}$

Figure 4.48

Now we can enter $Y1$ and $Y2$ into a graphing calculator.

3×3 Linear Systems

The graph of the equation $ax + by + cz = m$ is not a line. It is a plane in three dimensions. We still, however, refer to the following system of the equations in three variables

$$\begin{aligned} ax + by + cz &= m \\ dx + ey + fz &= n \\ gx + hy + iz &= p \end{aligned}$$

as a 3×3 *linear system*. Each variable is of the *first* degree.

Such a system can be solved by using the addition method, as we did for a 2×2 linear system.

EXAMPLE 8

Solve the 3×3 linear system.

8. $$\begin{aligned} 2x - y + z &= -1 \\ x + y - 2z &= 5 \\ x - 4y - 3z &= -4 \end{aligned}$$

Solution

The strategy for a 3×3 linear system is similar to the strategy for a 2×2 linear system.

Strategy for Solving a 3×3 Linear System

1. Use *any* two equations, and eliminate one of the three variables by using the addition method. Call this Equation 4.
2. Repeat step 1 using a different pair of equations to eliminate the *same* variable. Call this Equation 5.
3. Solve the 2×2 system formed by Equations 4 and 5.
4. Substitute the solutions from step 3 into an original equation to obtain the value of the third variable.
5. Check the solutions in all original equations.

$$\begin{aligned} 2x - y + z &= -1 && \textbf{(Equation 1)} \\ x + y - 2z &= 5 && \textbf{(Equation 2)} \\ x - 4y - 3z &= -4 && \textbf{(Equation 3)} \end{aligned}$$

Step 1:

$$\begin{cases} 2x - y + z = -1 \\ x + y - 2z = 5 \end{cases}$$

We eliminate y to get a 2×2 system in x and z.

$$3x - z = 4 \qquad \textbf{(Equation 4)}$$

Add Equation 1 and Equation 2.

Step 2:

$$\begin{cases} 4x + 4y - 8z = 20 \\ \underline{\ x - 4y - 3z = -4} \end{cases} \quad \textbf{(Equation 3)}$$

$$5x - \quad 11z = 16 \quad \textbf{(Equation 5)}$$

Multiply Equation 2 by 4.

Add these two equations.

Step 3:

Solve for x and z.

$$\begin{cases} 3x - \quad z = 4 & \textbf{(Equation 4)} \\ 5x - 11z = 16 & \textbf{(Equation 5)} \end{cases}$$

$$\begin{cases} -33x + 11z = -44 \\ \underline{\quad 5x - 11z = 16} \end{cases} \quad \textbf{(Equation 5)}$$

Multiply Equation 4 by −11.

$$-28x \qquad = -28$$

Add these two equations.

$$x \qquad = 1$$

Solve for x

$$z = -1$$

Solve for z by substituting x into Equation 4.

Step 4:

$$2(1) - y + (-1) = -1$$
$$y = \quad 2$$

Substitute x and z into Equation 1 to solve for y.

Step 5: *Check:* If $x = 1, y = 2$, and $z = -1$, then

$$2(1) - 2 + (-1) \stackrel{?}{=} -1 \quad \text{and} \quad 1 + 2 - 2(-1) \stackrel{?}{=} 5$$
$$-1 \stackrel{\checkmark}{=} -1 \qquad\qquad\qquad 5 \stackrel{\checkmark}{=} 5$$

$$\text{and} \quad 1 - 4(2) - 3(-1) = -4$$
$$-4 \stackrel{\checkmark}{=} -4$$

The solution is $x = 1, y = 2$, and $z = -1$, or $(1, 2, -1)$. ■

Solving 3×3 linear systems is more practically done on a graphing calculator using matrix methods. In this book we give only this short introduction to 3×3 systems.

We close this section with some examples of applications using linear systems. Some of these we have seen before.

EXAMPLE 9

Last week, when I went to Sandy's (a hamburger place), I paid \$8.80 for 8 hamburgers and 4 orders of fries. The next day, I treated my Little League baseball team to dinner and paid \$27 for 20 hamburgers and 20 orders of fries. How much does 1 hamburger and 1 order of fries cost? (There is no sales tax.)

Solution

If we let x equal the price of one hamburger and y equal the price of one order of fries, then

$$8x + 4y = 8.80 \quad \textbf{(Equation 1)} \quad \text{My bill.}$$
$$20x + 20y = 27.00 \quad \textbf{(Equation 2)} \quad \text{The team's bill.}$$

We solve the above 2 × 2 linear system:

$$\begin{cases} 2x + y = 2.20 \\ x + y = 1.35 \end{cases} \quad \text{Divide Equation 1 by 4; divide Equation 2 by 20; then subtract.}$$
$$x = 0.85$$
$$0.85 + y = 1.35 \quad \text{Substitute } x \text{ into the second equation to solve for } y.$$
$$y = 0.50$$

Thus a hamburger costs \$0.85, and an order of fries costs \$0.50. Check the answer. ■

Some Thoughts on Applications

We have examined many applications requiring the solution of a linear equation or a linear system. Applications in which the solution of other types of equations is necessary will be discussed in future chapters. Solving an application requires choosing a reasonable model for the situation.

For now, the question is more specific: Should I try to use a linear equation or a linear system to solve this problem? For many problems, the answer is, "It does not matter." We will now offer more examples to support this premise.

The following examples were discussed earlier. We will solve them in two ways: with a linear equation and with a linear system.

EXAMPLE 10

You are promoting a concert for the popular rock group "War and Peace" in a 10,000-seat auditorium. Most concerts charge \$10 for reserved seats and \$7 for general admission, but it is up to the promoter to determine the number of each type of seat. War and Peace charges \$40,000 for its appearance. Your expenses are \$20,000. How many seats should be designated reserved seats if you wish to make a \$30,000 profit?

Solutions

There are two possibilities:

1. A linear equation: The expenses are \$40,000 + \$20,000 = \$60,000. If r equals the number of reserved seats, then $10{,}000 - r$ is the number of general admission tickets. Therefore,

$$\underbrace{10r + 7(10{,}000 - r)}_{\text{Revenue}} - \underbrace{60{,}000}_{\text{Expenses}} = \underbrace{30{,}000}_{\text{Profit}}$$
$$10r - 7r = 30{,}000 - 10{,}000$$
$$3r = 20{,}000$$
$$r \approx 6{,}667$$

2. A linear system: There are really two variables here:

r = number of reserved seats

g = number of general admission seats

The linear system is

$$
\begin{array}{rcll}
r + g & = & 10{,}000 & \textbf{(Equation 1 for seats)} \\
10r + 7g - 60{,}000 & = & 30{,}000 & \textbf{(Equation 2 for profit)} \\
7r + 7g & = & 70{,}000 & \textbf{(Equation 1)} \quad \text{Multiply Equation 1 by 7.} \\
10r + 7g & = & 90{,}000 & \textbf{(Equation 2)} \quad \text{Restate Equation 2.} \\
\hline
-3r & = & -20{,}000 & \text{Subtract.} \\
r & \approx & 6{,}667 & \\
g & \approx & 3{,}333 & \text{Substitute } r \text{ into the first equation to obtain } g.
\end{array}
$$

Hence $g = 3{,}333$ general admission seats, and $r = 6{,}667$ reserved seats. Check the solution. ■

EXAMPLE 11

An auto supply company wants to make 900 gallons of battery solution that is 50% acid. It has a supply of solution that is 70% acid, some that is 20% acid, and some distilled water without acid. How many gallons of the 70% solution and the 20% solution should be used to make the 900 gallons of 50% solution?

Solutions

1. Using x as the number of gallons of 70% acid solution, we solved in Chapter 3 the linear equation

$$0.70x + 0.20(900 - x) = 450$$

2. Again, there are two variables here: x = number of gallons of 70% acid solution and y = number of gallons of 20% acid solution. Then we have the system

$$
\begin{array}{rcll}
x + y & = & 900 & \text{Gallons of solution} \\
0.70x + 0.20y & = & 0.50(900) & \text{Amount of acid}
\end{array}
$$

Note: Substituting $y = 900 - x$ in the second equation yields the linear equation above. By the addition method, we obtain

$$
\begin{array}{rcll}
0.20x + 0.20y & = & 180 & \text{Multiply the first equation by 0.20.} \\
0.70x + 0.20y & = & 450 & \\
\hline
-0.50x & = & -270 & \text{Subtract.} \\
x & = & 540 & \text{Gallons.} \\
y & = & 360 & \text{Gallons.}
\end{array}
$$

Check the solution: 540 gallons of 70% acid solution and 360 gallons of 20% acid solution are needed to make 900 gallons of 50% acid solution. ■

For the applications we have studied thus far, we could solve the problems using either a linear system in two variables or one linear equation in one variable. However, there are problems where two variables are mandatory. The Leontief economic model, which follows, is an example. Other examples occur in the exercises.

EXAMPLE 12

(Optional) **Leontief input-output economic model:** To give an extremely oversimplified model of the economy, let us suppose there are three sectors: agriculture, industry, and consumers. Agriculture and industry produce the goods; the consumers buy them. Agriculture and industry, however, also use each other's goods, thus they must produce enough goods to satisfy both their needs and the consumers' needs.

Suppose that for each \$1 of agricultural goods produced, agriculture used \$0.20 worth of its own goods and \$0.10 worth of industrial products. For each \$1 of industrial products made, industry uses \$0.35 of its own products and \$0.15 worth of agricultural goods. What is the dollar value of the agricultural goods and industrial products that the two sectors must produce to satisfy a consumer demand of \$20 billion for agricultural goods and \$110 billion for industrial products?

Solution

Here,

x = dollar value (in billions) of *all* goods produced by agriculture

y = dollar value (in billions) of *all* products made by industry

Then, for each of the two sectors,

Value of amount produced − Value of amount used by the two sectors
= Value of amount needed by consumers

For agriculture,

Value produced $= x$

Value used by agriculture $= 0.20x$

Value used by industry $= 0.15y$

Value needed by consumers $= 20$

Thus

$$x - 0.20x - 0.15y = 20$$

For industry,

Value produced $= y$

Value needed by agriculture $= 0.10x$

Value used by industry $= 0.35y$

Value needed by consumers $= 110$

Thus

$$y - 0.10x - 0.35y = 110$$

We must find the solution to the system:

$$\begin{cases} x - 0.20x - 0.15y = 20 \\ y - 0.10x - 0.35y = 110 \end{cases}$$

$$\begin{cases} 0.8x - 0.15y = 20 \\ -0.1x + 0.65y = 110 \end{cases}$$

$$\begin{cases} 0.8x - 0.15y = 20 \\ -0.8x + 5.20y = 880 \end{cases}$$ **Multiply the second equation by 8.**

$$5.05y = 900$$ **Add the two equations.**

$$y \approx 178.22$$ **Solve for y.**

$$x \approx 58.43$$ **Substitute into the first equation, and solve for x.**

Thus agriculture must produce \$58.43 billion worth of goods, while industry must make \$178.22 billion worth of products. Check these values. ■

NOTE: In actual practice, there are more than two variables to represent different kinds of agriculture and industry. More difficult systems of equations need to be solved, but the process is the same. ■

4.4 EXERCISES

I. Procedures

1–14. Solve each of the following linear systems by graphing and by an algebraic method. Be sure to check your answers. Use whichever method you find easier first.

1. $x + y = 10$
$x - y = 4$

2. $x - y = 10$
$x + y = 4$

3. $2x - y = 7$
$3x + 4y = 16$

4. $3x + y = 7$
$4x - 5y = 3$

5. $-2y = -6$
$5x + 7y = 106$

6. $2x = 10$
$7x - 5y = -30$

7. $3x - 2y = 1$
$5x + 4y = 9$

8. $4x - 3y = 11$
$11x + 6y = 16$

9. $3x + 2y = 9$
$2x - 3y = 19$

10. $4x - 3y = -1$
$-3x + 4y = 20$

11. $5x + 3y = 13$
$7x - 5y = 18$

12. $3x - 7y = 6$
$5x + 11y = -7$

13. $2x - 5y = 7$
$-8x + 20y = 13$

14. $x + 2y = 8$
$5x + 10y = 8$

15–24. Solve each of the following linear systems by an algebraic method. Be sure to check your answers.

15. $x + y = a$
$x - y = b$

16. $x - y = 2c$
$x + y = 2d$

17. $ax - by = m$
$cx + dy = n$

18. $ax + by = m$
$cx - dy = n$

19. $3.7x + 1.5y = 8.41$
$1.6x + 2.2y = 7.36$

20. $3.5x - 4.4y = 8.58$
$7.2x - 5.9y = 17.02$

21. $3.742x + 1.573y = 8.621$
$1.602x + 2.237y = 7.431$

22. $0.980x + 0.43y = 0.91$
$-0.61x + 0.23y = 0.48$

23. $3x - 4y = x + 1$
$x + 2y = 3y + 1$

24. $x = 3x - 2y + 4$
$y = x - 3y + 14$

25–32. Solve each of the following systems for x, y, and z. Be sure to check your answers.

25. $x + y = 5$
$y + z = 7$
$x + z = 6$

26. $x - y = 4$
$y - z = 1$
$x - z = 5$

27. $\begin{aligned} x - 4y \quad &= 11 \\ 2x + y \quad &= 4 \\ -x - 3y + z &= 3 \end{aligned}$

28. $\begin{aligned} 2y - z &= 6 \\ y + 3z &= -4 \\ 2x - y + z &= -6 \end{aligned}$

29. $\begin{aligned} x + y - z &= 6 \\ x + y + z &= 12 \\ x - 3y - z &= 10 \end{aligned}$

30. $\begin{aligned} x + y + z &= 6 \\ x - y - z &= 0 \\ x + 3y + 2z &= 11 \end{aligned}$

31. $\begin{aligned} 2x - 3y + 5z &= 11 \\ 5x + 4y - 6z &= -5 \\ -4x + 7y - 8z &= -14 \end{aligned}$

32. $\begin{aligned} 3x - y + 2z &= 5 \\ 2x + 3y + z &= 1 \\ 5x + y + 4z &= 8 \end{aligned}$

33 and 34. For what value(s) of c do the following linear systems have exactly one solution?

33. $\begin{aligned} cx - 3y &= 4 \\ 4x + 6y &= 7 \end{aligned}$

34. $\begin{aligned} cx - 3y &= 4 \\ 4x + 6y &= 8 \end{aligned}$

II. Concepts

35–38. For each of the following, give an example of a 2×2 linear system that has the indicated property.

35. Solution is $(0, 0)$.

36. Solution is $(2, -3)$.

37. Linear system has no solution.

38. Linear system has infinitely many solutions.

39–42. For each of the following, given an example of a 3×3 linear system that has the following property.

39. Solution is $(0, 0, 0)$.

40. Solution is $(1, 2, 3)$.

41. Linear system has no solution.

42. Linear system has infinitely many solutions.

43. Using your graphing calculator, sketch the graph of these four lines on the same screen: $x + 2y = 3$, $2x + 3y = 4$, $4x + 5y = 6$, and $-3x - 2y = -1$.

(a) What do you observe?

(b) Make a generalization.

44. Using your graphing calculator, sketch the graph of these four lines on the same screen: $x + 3y = 5$, $2x + 5y = 8$, $4x - y = -2$, and $-3x + 2y = 7$.

(a) What do you observe?

(b) Make a generalization.

45. When you use the addition method of solving a 2×2 linear system and obtain the given equation, describe the nature of the solution(s) (if any) of the system.

(a) $0 = 0$ **(b)** $x = 2$ **(c)** $0 = 2$

46. Which of the three algebraic methods of solving a 2×2 linear system is the easiest to use? Give a reason for your answer. Would your answer change if you used a graphing calculator? Why or why not?

III. Applications

47–52. Solve each of the following problems using a linear system. Sketch a graph of each system.

47. Find the dimensions of a rectangle of perimeter 92 cm whose length is 4 cm longer than its width.

48. Last week, when I went to Sandy's, it cost \$8.25 for 5 hamburgers and 5 orders of french fries. A few days later, I bought 20 hamburgers and 10 orders of french fries for \$27. Find the cost of one hamburger and one order of french fries.

49. One phogg and two moggs cost \$11, while three phoggs and five moggs cost \$29. What is the cost of one phogg and of one mogg?

50. A bag of 800 coins worth \$90 contains quarters and nickels. Find the number of each type of coin in the bag.

51. A bag of 2,900 coins worth \$211.25 contains nickels and dimes. How many of each type of coin are in the bag?

52. The Acme Car Rental Company charges \$17.95 per day plus 21¢ per mile, while the U Drive It Company charges \$19.95 plus 15¢ per mile. Which company should you choose if you wish to rent a car for one day? Cost is the major consideration.

53–64. Solve, using any method.

53. The cruising speed of an airplane is 200 mph. You wish to hire the plane for a four-hour sightseeing trip. You tell the pilot to fly as far north as possible and then return.

(a) How far is the round trip if there is no wind?

(b) How far is the round trip if there is a 30-mph wind blowing from the north?

54. Airplanes usually take longer to fly from east to west than from west to east, since the wind blows from west to east. On a recent trip from Cleveland to Phoenix, a distance of 1,950 miles, the flying time was 3.5 hours. That same day, a plane flying from Phoenix to Cleveland made the trip in 3 hours. Assuming both planes had the same airspeed, find the airspeed and wind speed that day.

55. In preparing a diet for some experimental animals, a biologist determines that the animals need 20 ounces of protein and 6 ounces of fat. She is able to purchase two types of food, one with 20% protein and 2% fat, the other with 10% protein and 6% fat. How many ounces of each food should go into the diet mix?

56. You want to invest $10,000, part in high-risk stock that pays 12% interest and the rest in a savings account paying 5.5% interest. How much money should you invest in stock and in a savings account to realize a 9% return on your investment?

57. Given two feeds with one containing 45% crude protein and the other containing 17% crude protein, how much of each should be used to yield a 1,000-pound mixture containing 27% crude protein?

58. It costs a book publisher $20,000 to purchase a book for publication. Printing costs are $10 per book, and the book will sell for $16 wholesale and $21.95 retail.

(a) How many books must the publisher sell to make a $30,000 profit?

(b) How many books must the publisher sell to make a 12% (of costs) profit?

(c) How many books must the retailer sell to make a $20,000 profit if its expenses are $1.25 per book sold?

59. The State College football stadium holds 80,000 people. A regular seat costs $10, while a student ticket is $2.50. If the expenses for a game are $20,000, what is the maximum number of student seats that could be allotted to make a $300,000 profit with a "full house"?

60. Joe Investor owns ST Oil and Zerox stock. The closing prices for each day of a recent week are given in the table. If the value of his portfolio was $19,500 on Monday and $20,300 on Friday, how many shares of each stock did Joe own?

	Zerox	*ST Oil*
Monday	45	60
Tuesday	$45\frac{1}{2}$	$62\frac{1}{8}$
Wednesday	46	63
Thursday	$47\frac{1}{4}$	$61\frac{1}{2}$
Friday	47	62

61. The total number of registered Democrats, Republicans, and Independents in Hooterville is 100,000. In a recent election, 50% of the registered Democrats, 60% of the registered Republicans, and 70% of the registered Independents voted. A total of 55,200 votes was cast. The ratio of registered Democrats to registered Independents is 9 to 1. How many registered Democrats, Republicans, and Independents voted in the election?

62. See Example 12. Suppose that for each dollar of agricultural goods produced, agriculture uses $0.25 of its own products and $0.15 worth of industrial products. For each dollar of industrial products produced, suppose industry uses $0.05 worth of agricultural goods and $0.40 worth of industrial products. If the consumers require $15 billion worth of agricultural products and $100 billion worth of industrial products, what is the value of goods and products each sector must produce to meet the consumer demand?

63. Suppose it takes 0.05 units of coal and 0.3 units of electricity to make 1 unit of coal, while it takes 0.4 units of coal and 0.1 units of electricity to make 1 unit of electricity. If the consumers demand 10 million units of coal and 30 million units of electricity, how many units of coal and electricity must each industry produce to meet this demand?

64. Two engineers form a consulting firm. For each dollar of business the electrical engineer does, she "buys" $0.25 of the mechanical engineer's services. For each dollar of business, the mechanical engineer buys $0.30 of the electrical engineer's time. If for the given week the electrical engineer has $800 worth of business, while the mechanical engineer has $650 worth, what is the value of the work each engineer must do for that week?

IV. Extensions

65 and 66. Solve each of the following linear systems for x and y.

65. $3{,}679x + 1{,}321y = 13{,}679$

$1{,}321x + 3{,}679y = 11{,}321$

66. $ax + by = c$

$dx + ey = f$, when $a, b, c, d, e,$ and f are six consecutive integers.

67. A magic square is a square array of integers, all different, such that the sum (or product) of the elements in each row, in each column, and on both diagonals is the same.

For example, the box below is a 3×3 additive magic square, since the sum in each

$$\begin{aligned} \text{Row} &= 5 + 3 + 10 = 11 + 6 + 1 = 2 + 9 + 7 = 18 \\ \text{Column} &= 5 + 11 + 2 = 3 + 6 + 9 = 10 + 1 + 7 = 18 \\ \text{Diagonal} &= 5 + 6 + 7 = 10 + 6 + 2 = 18 \end{aligned}$$

5	3	10
11	6	1
2	9	7

(a) Given a set of numbers and the common sum, make a 3×3 additive magic square. You will need to supply more numbers.

1. Sum = 51

Numbers: 13, 14, 15, 16, 18, 20

2. Sum = 42

Numbers: 11, 15, 16, 17

3. Sum = 21

Numbers: 2, 6, 13

(b) In a 3×3 additive magic square, what values are possible for the common sum? What values are possible for the sum of the four corner entries? Explain your reasoning in each case.

Hint: How does the common sum S depend on the middle entry? Suppose the following is a magic square:

a	b	c
d	e	f
g	h	i

Then we have the system:

$$\begin{aligned} S &= a + e + i \\ S &= d + e + f \\ S &= g + e + c \end{aligned}$$

CHAPTER 4 SUMMARY

Many of the applications in earlier chapters are special cases of linear equations in two variables. For example, the equation for determining the cost C to drive M miles is $C = 15 + 0.20M$. This is a linear equation in two variables, C and M. In Chapter 3, we examined special cases of linear equations in two variables by knowing a specific value of one of the variables and then solving for the unknown variable.

The solutions to a linear equation in two variables can be represented by a graph; the solutions lie on a line. In this chapter we have discussed two fundamental problems: Given a linear equation, find its graph. Given a graph of a straight line, find its equation. We explored these concepts using a graphing calculator.

Finally, we showed that some applications can be modeled by either one linear equation (function) or a system of linear equations.

Important Words and Phrases

x-intercept
y-intercept
slope
vertical lines
horizontal lines
parallel lines
perpendicular lines
slope-intercept form of a linear equation
point-slope form of a linear equation
fitted lines
median-median line
2×2 linear system
3×3 linear system
simultaneous solution
inconsistent linear system
Leontief input-output economic model

Important Properties and Procedures

- Locating (plotting) points on a graph
- Determining the coordinates of a given point
- Graphing the solutions of a linear equation in two variables
- Finding an equation of a given line
- Determining whether lines are parallel, perpendicular, or neither
- Finding the simultaneous solutions to a set of linear equations

REVIEW EXERCISES

I. Procedures

1–8. **(a)** Sketch the graph for each equation below. **(b)** Label the intercepts and any other points you may plot. **(c)** Find the slope.

1. $y = 3x - 5$

2. $y = -3x + 4$

3. $3x - 4y = 12$

4. $-5x + 6y = 30$

5. $2x + 4y = 7.2$

6. $0.8x - 0.4y = 2$

7. $\frac{x}{3} - \frac{y}{2} = 1$

8. $x - \frac{1}{3}y = \frac{1}{2}$

9–20. **(a)** Find an equation of the line which satisfies the given condition. **(b)** Sketch the graph of the line.

9. Slope $= -2$, y-intercept $= 3$.

10. Slope $= \frac{3}{4}$, y-intercept $= 0$.

11. Contains the points $(0, 2)$ and $(2, 3)$.

12. Contains the points $(0, -1)$ and $\left(\frac{3}{4}, \frac{1}{2}\right)$.

13. Slope $= -1$, contains $(-2, -3)$.

14. Slope $= \frac{1}{2}$, contains $(4, -3)$.

15. Contains the points $(0.123, 1.657)$, $(-2.136, 7.240)$.

16. Contains the points $(-1.426, 4.189)$, $(-2.814, 6.340)$.

17. Parallel to $y = 2x - 5$, contains the point $(0, 6)$.

18. Parallel to $y = -\frac{1}{3}x + 1$, contains the point $(6, 0)$.

19. Perpendicular to $y = 5x - 2$, contains the point $(0, 1)$.

20. Perpendicular to $2x + 3y = 6$, contains the point $(0, 5)$.

21–24. Find the simultaneous solution to the linear systems by graphing. Use a graphing calculator, if you wish.

21. $y = 4 + 2x$
$y = x + 3$

22. $x = 4 - y$
$x + 2y = 0$

23. $y - x = -3$
$4 - y = -x$

24. $2x - y = 6$
$x + y = -3$

25–30. Solve each of the following systems by an algebraic method. Be sure to check your answers.

25. $x + 1 = 3x - 4y$
$3y + 1 = x + 2y$

26. $x + 2y = 3x + 4$
$y - x = 14 - 3y$

27. $3x - 5y = 12$
$5x - 3y = 4$

28. $2x + 3y = 12$
$3x - y = 1$

29. $\frac{x}{3} + \frac{y}{4} = 10$
$\frac{x}{3} - \frac{y}{2} = 4$

30. $\frac{2x}{3} + \frac{3y}{4} = 2$
$\frac{x}{6} + \frac{y}{2} = -2$

31 and 32. Solve each of the following systems for x, y, and z. Be sure to check your answers.

31. $x + y + z = 1$
$2x - 3y = 1$
$x + 4y - z = -8$

32. $x - 2y + 3z = -4$
$4x + 3y - 2z = 20$
$x + 3y - 5z = 14$

33. Find the coordinates of the five corner points of this region.

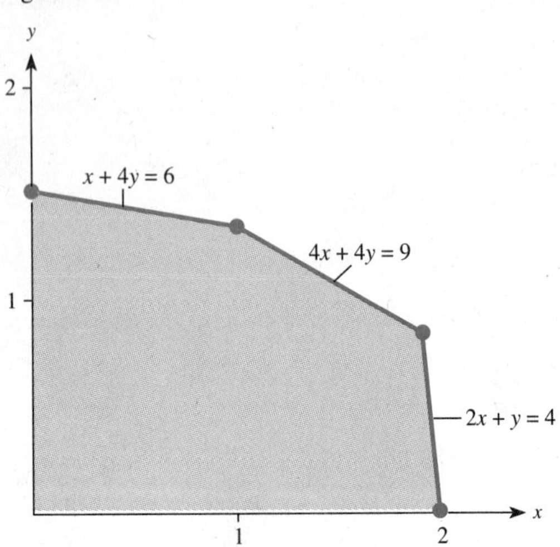

II. Concepts

34–43. Given the line $cx + 4y = 12$, find a value of c such that the line satisfies the given condition.

34. Contains the point $(3, -3)$.

35. x-intercept $= -4$.

36. Has slope $-\frac{1}{2}$.

37. Is horizontal.

38. Is parallel to $y = 12x - 17$.

39. Is perpendicular to $y = 12x - 17$.

40. Forms an inconsistent linear system with $6x + 12y = 37$.

41. Passes through the square with vertices (2, 3), (4, 3), (2, 5), (4, 5).

42. Is parallel to $Y1$.

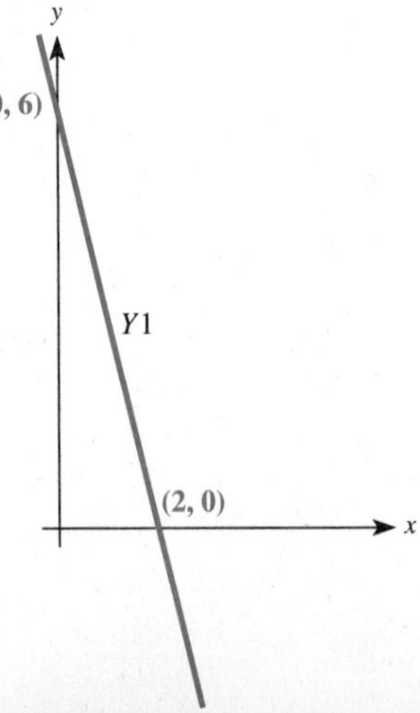

43. Is perpendicular to $Y1$.

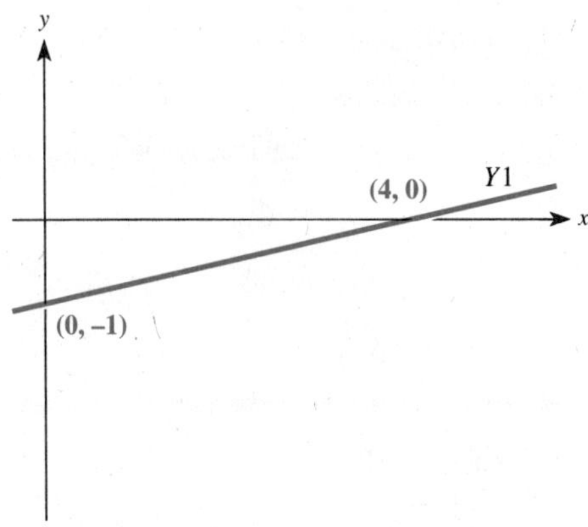

III. Applications

44–49. Solve each of the following problems.

44. The UDI car rental agency charges \$26 per day and 21¢ a mile.

(a) What is the cost C for driving a total of M miles in one day?

(b) Graph the equation to part (a).

(c) How many miles can be driven for \$45 in one day?

(d) How much will it cost to drive 150 miles in one day?

45. The WTH car rental agency charges \$19 and 27¢ a mile.

(a) What is the cost C of driving M miles?

(b) How many miles must be driven for the cost of renting a car from the WTH agency to be equal to that of the UDI agency in Exercise 44?

(c) Graph the equations of part (a) of Exercise 44 and Exercise 45.

(d) Use the graph to determine which agency is the least expensive if 100 miles are driven. Which is the cheapest for 500 miles?

46. For the past two years, the inflation rate was 5% and 4%.

(a) What was the price p of an item two years ago, if it costs c dollars today?

(b) If the item is priced at \$10 today, what did it cost two years ago?

(c) What is the cost of an item today if it cost \$8.75 two years ago?

(d) Graph the equation you found in part (a). Check your answers to parts (b) and (c) with the graph.

(e) Use the graph to estimate the cost of an item today if it cost $10 two years ago.

47. A 300-gram weight attached to a spring will stretch the spring 22.25 mm, while a 200-gram weight will stretch the spring 16 mm.

(a) How much will a 150-gram weight stretch the spring?

(b) If the spring stretches 28 mm, what weight is attached to it?

48. A book of stamps contains both 25¢ and 35¢ stamps.

(a) Write an equation for the cost of a book of 100 stamps.

(b) Graph the equation.

(c) Find six possible combinations of 25- and 35-cent stamps that could go into a book.

(d) Use the graph to estimate the cost of the book of stamps for each combination in part (c). Check the estimate with the equation in part (a).

49. The sum of three numbers is 48. If two of the numbers are consecutive positive even integers,

(a) find an equation in two variables that could be used to find the three integers.

(b) graph the equation.

(c) list three solutions.

50–56. Solve each of the following. Describe how you found the solution—by graphing or by an equation.

50. Ms. Smith receives $100,000 from her husband's estate. She invests part at 8% yearly and the remaining amount in a safe 5% bond. Her total annual income from interest is $7,250. How much did she invest in bonds?

51. A flower garden is enclosed by 54 feet of fencing. If the flower garden is rectangular and 3 feet longer than it is wide, what are the dimensions of the plot?

52. A rock concert sold 5,300 tickets. Tickets could be purchased in advance for $3.50 or at the door for $6.00. If the concert collected a total of $20,550, how many advanced-sale tickets were sold?

53. A freight train travels $\frac{3}{5}$ as fast as a passenger train. Both leave the station together. After traveling for five hours, the passenger train has traveled 150 miles farther than the freight train. Find the average speed of each train.

54. The equation relating Fahrenheit (F) and Centigrade (C) temperature is $F = \frac{9}{5}C + 32$, and the equation relating Centigrade to Kelvin (K) temperature is $K = C + 273$.

(a) An experiment calls for 412°K, but you only have a Fahrenheit thermometer. At what Fahrenheit temperature should you conduct the experiment?

(b) To what Kelvin temperature does 212°F correspond?

55. A hockey team figures it needs 60 points for the season to make the playoffs. A win is worth 2 points, and a tie is worth 1 point.

(a) Write an equation for the number of wins w and ties t it will take to get the team into the playoffs.

(b) If there are eight games left and the team has 50 points, for what combinations of wins and ties can the team get into the playoffs?

56. The following table gives body measurements (in inches) for misses' pattern sizes as specified by the Measurements Standard Committee of the Pattern Fashion Industry.

	Misses sizes							
	6	8	10	12	14	16	18	20
Bust b	$30\frac{1}{2}$	$31\frac{1}{2}$	$32\frac{1}{2}$	34	36	38	40	42
Waist w	23	24	25	$26\frac{1}{2}$	28	30	32	34
Hips h	$32\frac{1}{2}$	$33\frac{1}{2}$	$34\frac{1}{2}$	36	38	40	42	44

(a) Find a formula giving h in terms of b.

(b) Graph the ordered pairs (w, b), and draw a line that seems to fit these eight points as well as possible. Find an equation for your line.

(c) Repeat part (b) for the ordered pairs (w, s).

(d) Suppose you used the equation in part (c). Could you reliably find misses' sizes by applying this equation to the given waist measurements and rounding to the nearest even integer?

(e) Would you save room by writing the formulas instead of the table?

57. In the following graph, line l_1 represents the cost of producing n figgles, and line l_2 represents the revenue on selling n figgles.

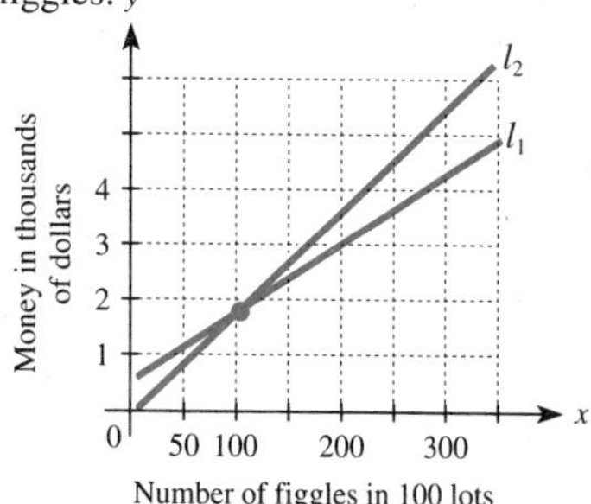

(a) What are the start-up costs (costs before any figgle is produced)?

(b) What are the cost and revenue on producing and selling 50, 100, 300 figgles?

(c) What is the break-even point? Give both the number and cost.

(d) What is the profit on producing and selling 50, 100, and 300 figgles?

(e) How many figgles must be produced to have $3,000 in revenue?

58. The smallest adult shoe is size 5 and fits a 9-inch foot. An 11-inch foot requires size 11.

(a) If your foot is one foot long, what size shoe do you need?

(b) Shaquille O'Neal, the famous basketball player, wears a size 27 shoe. How long is his foot?

59. Suppose you want to sell T-shirts on campus. You are trying to decide on a price per shirt, so you conduct market research and find that 800 students will buy a shirt for $10, but only 100 will buy it for $17. You find a supplier who can supply 200 shirts if they are sold for $10 and 600 if they are sold for $17. At what price per shirt will the supply equal the demand, assuming both can be modeled by linear equations?

(a) What equations must you find to solve this problem?

(b) What is the interpretation of the y-intercept of each equation? Is it valid?

(c) What is the interpretation of the x-intercept of each equation? Is it valid?

(d) What is the interpretation of the slope of each equation used? Is it valid?

(e) What would be a reasonable domain for the demand equation?

60. There are 10,000 seats in Rosebud Stadium. Box seats are $25 each, reserved seats are $15 each, and general admission seats are $10 each. Designating how many seats are to be considered each type is left up to the manager of the event using the stadium. Suppose the manager wants twice as many reserved seats as general admission seats. For the War and Peace concert, the manager estimates expenses at $125,000 and wishes to make a $20,000 profit. How many seats of each type should there be?

61–63. The table gives the cost of a 30-second commercial during the Super Bowl games from 1977 to 1992. Find a fitted line for this data using the indicated method. Then answer the questions: According to the model, **(a)** what will be the cost of a 30-second commercial in 2000? **(b)** when do you expect the cost of a 30-second commercial to reach $1.5 million?

Year	*Cost (in thousands of dollars)*	*Year*	*Cost (in thousands of dollars)*
1977	125	1985	525
1978	185	1986	550
1979	180	1987	590
1980	234	1988	650
1981	275	1989	680
1982	345	1990	700
1983	400	1991	800
1984	375	1992	850

Super Bowl Advertising Costs

61. "eyeball" line

62. median-median line

63. least square line

CUMULATIVE REVIEW: CHAPTERS 1–4

I. Procedures

1–8. Compute the following. Express all answers that are fractions in simplest equivalent form.

1. **(a)** $\dfrac{3-5}{8} - 1 + \dfrac{5}{3}$ **(b)** $\dfrac{3}{5} - 2 - \dfrac{5}{6}$

2. $8 - 6[-1 - 3(5 - 2) - 5] + 10$

3. $\dfrac{(-12)(-4)}{(-6)(-4)}$

4. $\dfrac{(-2^3)(3)^2}{12^2}$

5. $3^{-2} - 3^0$

6. $\dfrac{2^{-2} + 1}{2}$

7. $\dfrac{(3^2 \cdot 2)^3}{120}$

8. Subtract -3 from 12. Multiply your answer by $\frac{3}{5}$.

9–14. Perform the following operations. Express all answers that are fractions in simplest equivalent form.

9. $5 - 2(4 - 2x) + x(6 - x)$

10. $\dfrac{3a^{10}b^6c^{-2}}{6a^8b^7c}$

11. $(2a^2b^3c^{-2})(-8ab^{-2}c)$

12. $(-2x^{-2}y)^3(x^5y^{-3})$

13. $(x - 1)^2 - (x + 1)(x + 3)$

14. $(3 - x)(3 + x) - (x + 3)^2$

15–18. Solve each of the following equations for x.

15. $3 + 2(x - 1) = 8 - x$

16. $3x - (2 - x) = 3(2x - 4)$

17. $\dfrac{x + 2}{5} - \dfrac{x}{3} = 2$

18. $3 - \dfrac{x + 1}{2} = x$

19–21. Find the solution to the following system of equations.

19. $y = 3x + 3$
$2y = 2x - 1$

20. $x = 8 - 2y$
$2x = 6 + y$

21. $4y - 2x = 9$
$2y = x + 6$

22. If $x = -2$ is a solution to the following equation, find c.

$$3x + 2(5 - x) = 4x + c$$

23–26. Find the solution to the following inequalities and graph the solution on a number line.

23. $\dfrac{x - 2}{3} \le 2x - 1$

24. $x > 4$ or $x < -4$

25. $|x - 4| = 9$

26. $|3 - x| < 6$

27–29. Write an algebraic statement that describes the following situations.

27. x is less than 2 and x is greater than 0.

28.

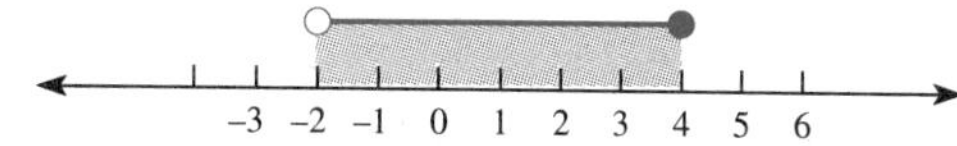

29. x is 3 units from the point 1.

II. Concepts

30–35. Matching. Select the letter of the graph that *best* matches the description of each problem.

30. The equation of the line $y = 2$.

31. A line parallel to the line $y = x + 1$.

32. A line with the equation $2x - y = 4$.

33. A line with slope equal to -3.

34. A line with negative slope and an x-intercept of (2, 0).

35. A line perpendicular to the line $2y = x + 1$.

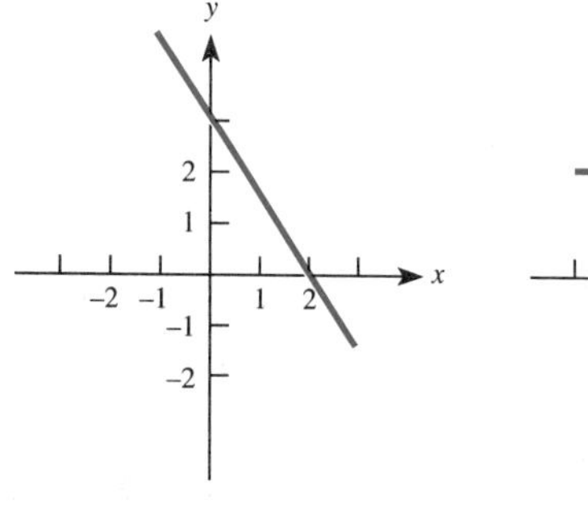

(a)

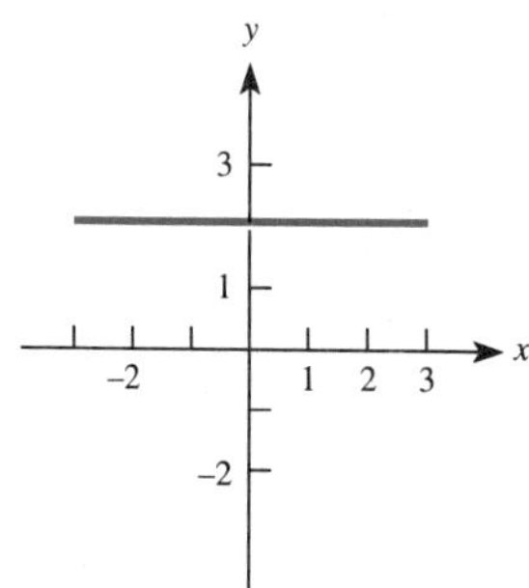

(b)

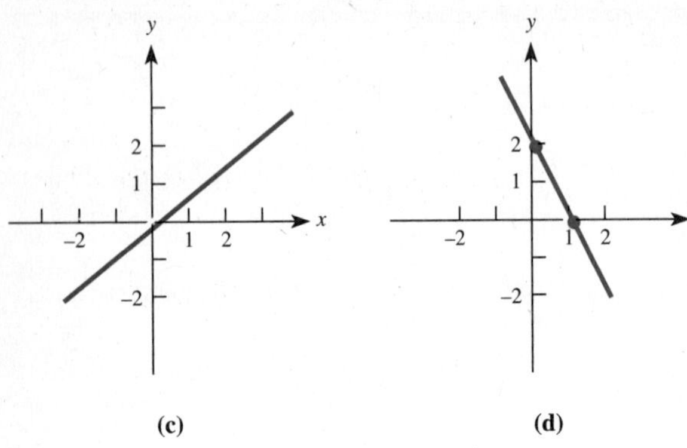

(c) (d)

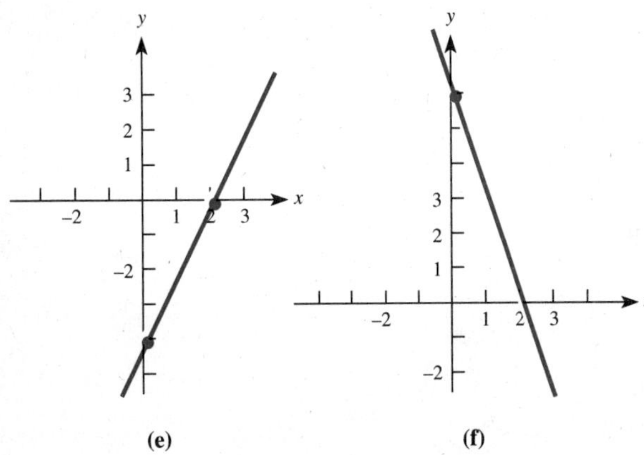

(e) (f)

36–42. True or false: Determine whether each statement is true or false. Give reasons for your answers.

36. $\dfrac{(10.0006 - 7.53924)}{(0.0005)(9.9368)} < 1$

37. $\dfrac{1}{2}\left(\dfrac{15.9}{7} - \dfrac{8.2}{4}\right) > 0$

38. The product of three negative integers is always positive.

39. If $3 + x > 0$, then $x > 0$

40. For all x, $-x$ is negative.

41. If $x \neq 0$, then x^{-2} is positive.

42. If $x < 0$, then $|x| < 0$.

43–53. Multiple Choice. Select the correct letter of the response for each problem.

43. Which of the following statements best describes the graph below?

−4 −3 −2 −1 0 1 2 3 4

(a) $x < 3$ or $x > -2$ **(b)** $x > 4$ or $x < -2$
(c) $x < -2$ or $x > 3$ **(d)** $x > 3$
(e) $x > -2$ and $x < 3$

44. Which of the following points lies on the line $x + 3y = 6$?

(a) $(0, 0)$ **(b)** $(0, 2)$ **(c)** $(1, 0)$ **(d)** $(1, 1)$
(e) $(2, 2)$

45. The equation $3x + 8 = 4(x + 1) - (x - 2)$

(a) has exactly one solution.
(b) has no solution.
(c) has infinitely many solutions.
(d) There is not enough information to decide.

46. A runner who runs at the rate of 8 minutes per mile will burn off 120 calories for each mile run. How far would this runner have to run to burn off a chocolate sundae containing 1,320 calories?

(a) 165 miles **(b)** 15 miles **(c)** 11 miles
(d) 88 miles **(e)** none of these

47. If $x \neq 0$, then $\dfrac{4x^6 + 2}{x^3} =$

(a) $4x^6 + 1$ **(b)** $4x^3 + 1$ **(c)** $4x^2 + 1$
(d) $4x^6$ **(e)** none of these

48. $\dfrac{2.01 \times 10^{-12}}{4.12 \times 10^{-13}}$ is

(a) greater than one **(b)** less than one
(c) equal to one **(d)** none of these

49. For what values of c will the following equation be a contradiction?

$$x - 4(2 - x) = 5(x - 1) + c$$

(a) $c = 3$ **(b)** $c = -3$ **(c)** all $c \neq -3$
(d) all $c \neq 3$ **(e)** none of these

50. Which of the following equations represents a line that passes through the origin and is perpendicular to the line $y - 2x = 6$?

(a) $y = x$ **(b)** $-y = 2x$ **(c)** $y = -2x$
(d) $2y = -x$ **(e)** none of these

51. An algebraic expression that represents the average rate of speed of a bicyclist who traveled x miles in 6 hours is

(a) $6x$ **(b)** $\dfrac{6}{x}$ **(c)** $\dfrac{x}{6}$ **(d)** $6 + x$
(e) none of these

52. The equation $3x + 2 = 4(x + 1) - (x - 2)$

(a) has exactly one solution.

(b) has no solution.

(c) has infinitely many solutions.

(d) There is not enough information to decide.

53. A man walks 14 miles in three hours. At this rate, how far can he walk in ten hours?

(a) $\frac{3}{140}$ mile **(b)** $\frac{30}{14}$ miles **(c)** $\frac{52}{10}$ miles

(d) $\frac{140}{3}$ miles **(e)** none of these

54–57. Find the next two terms in each of the following sequences. Find an expression to describe the general term.

54. $1, 5, 9, 13, 17, \ldots$

55. $34, 24, 14, 4, -6, \ldots$

56. $\frac{1}{2}, \frac{2}{3}, \frac{3}{4}, \frac{4}{5}, \frac{5}{6}, \ldots$

57. $\frac{1}{3}, \frac{1}{9}, \frac{1}{27}, \frac{1}{81}, \ldots$

58. Explain the relationship between the equations $y = x$ and $y = ax + b$. Explain the effects of a and b in the equation. Give examples to illustrate your discussion.

59. Describe the set of all lines that are

(a) parallel to $y = 2x - 1$.

(b) perpendicular to $y = 2x - 1$.

60. Describe the relationship between the following equations: $6 = 3x + 1$, $-8.5 = 3x + 1$, $0 = 3x + 1$, $y = 3x + 1$.

III. Applications

61. Find the value(s) of x such that the perimeter of the figure below is

(a) equal to 100.

(b) greater than 20 but less than 60.

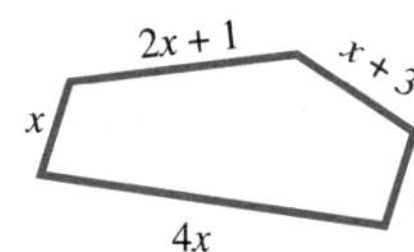

62. For what values of x will the triangle and rectangle have the same perimeter?

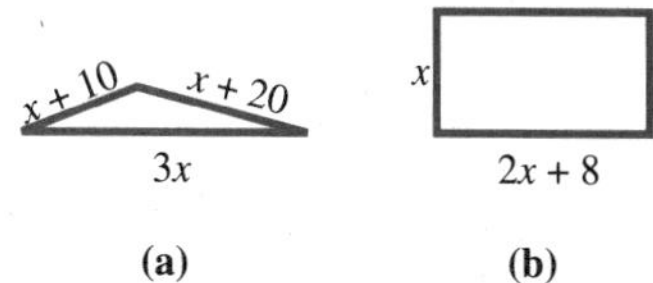

63. A city in Florida has a population that is tripling every year. If there are 1,000 people today, what will the population be

(a) three years from now?

(b) N years from now?

64. You need to rent a truck. Agency A has an initial fee of \$25 and charges \$18 per hour. Agency B has an initial charge of \$40 and charges \$5 less per hour than A.

(a) What is the charge from each agency if you rent the truck for six hours?

(b) For what number of hours will the two agencies charge the same?

65. Two buses leave Chicago. One bus leaves at 6:00 P.M. traveling east at 55 mph. An hour later the second bus leaves Chicago traveling west at 60 mph. At what time will the buses be 630 miles apart?

66. A canoeist paddles at the rate of 15 miles per hour in still water. The rate of the current is 8 miles per hour.

(a) How long will it take her to go 10 miles against the current?

(b) How long will it take her to paddle back to where she started?

67. For the years 1976 to 1985 the average January temperature in Deepfrost, Michigan, was -1°C, 0°C, -5°C, -11°C, $+10$°C, $+6$°C, -3°C, -20°C, $+15$°C, -2°C. What was the average temperature for January for this period?

68. At the beginning of a trip the odometer on a car (with a full tank of gas) read 118,750. At the end of the trip, which took $4\frac{1}{2}$ hours, the odometer read 119,020. The driver needed 9.5 gallons of gas to fill the gas tank. (The odometer measures the number of miles driven.)

(a) How many miles to the gallon did the car get?

(b) What was the average rate of speed on the trip?

69. At the game reserve 120 elk were captured, tagged, and released. A year later 255 elk were captured. Of these 45 had tags. About how many elk are on the game reserve? (Assume the number of elk remains constant.)

70–73. Try to solve each problem three different ways: with a table, a graph, and an equation. Discuss the advantages and disadvantages of each method.

70. The Sandy Dune Company rents dune buggies for \$35.50 a day plus \$4.25 an hour.

(a) How much will it cost to rent the buggy for 4 hours?

(b) Describe the range of values for the cost if you rent the buggy for two consecutive days and you estimate the number of hours to be between 10 and 15 hours (total for both days).

71. The start-up cost to manufacture figgles is \$750, and it costs \$4.25 to make each figgle. If we plan on selling the figgles for \$8.50, how much profit is made on selling 2,000 figgles?

72. Your club wants to advertise its fund-raising event with a brochure. To print brochures, Company A charges \$82 plus $6\frac{1}{2}$¢ a copy. Company B charges \$6 plus 8¢ a copy. If you order 7,500 brochures, what is the charge from each company? Which is the least expensive?

73. The amount of light filtering through a special sheet of plastic 1-mm thick is 80% of the original light. How much light comes through if

(a) four sheets of plastic are put together?

(b) n sheets of plastic are put together?

74. Suppose a bookstore buys a textbook from the publisher for \$35. It marks it up 35% and sells it new to the students. At the end of the year, the bookstore buys back the book from the students at half the price the students paid for it. The bookstore marks up the book 75% and sells the book again. It then buys the book back at half the new price and then marks it up again 75%. If this process continues, how much will the textbook cost at the start of the fifth year? How much profit has the store made on the book?

75. Linear depreciation is one of several methods approved by the Internal Revenue Service for depreciating business property. If the original cost of the property is C dollars and it is depreciated linearly over N years, its value V at the end of n years is given by

$$V = C\left(1 - \frac{n}{N}\right).$$

If computer equipment for a small business costs \$50,000 today and it is depreciated linearly over 20 years,

(a) When will its value reach \$10,000?

(b) What is its value after 6 years?

76. Forensic scientists use the lengths of certain bones to calculate the height of the living person. The bones that are used and the relationship of the bone length to the height of the person are given below:

Male	Female
$h = 69.089 + 2.238\text{ F}$	$h = 61.412 + 2.317\text{ F}$
$h = 81.688 + 2.392\text{ T}$	$h = 72.572 + 2.533\text{ T}$
$h = 73.570 + 2.970\text{ H}$	$h = 64.977 + 3.144\text{ H}$
$h = 80.405 + 3.650\text{ R}$	$h = 73.502 + 3.876\text{ R}$

F is the femur, T is the tibia, H is the humerus, and R is the radius.

(a) If the femur of a 26-year-old female measures 46.2 cm, what was the height of the person?

(b) If the tibia of a 30-year-old male is 50.1, what was the height of the person?

(c) If a male and female are the same age, what is the length of the radius if they both have the same height?

(d) Graph the set of equations for the male on the same set of axes. What is the significance of the intercepts?

CHAPTER

Rational Expressions, Equations, and Functions

In the past four chapters, we have examined exponential and quadratic functions briefly, and linear functions and linear systems in depth. A linear relationship between two quantities can be modeled by a straight line. What happens if we examine the reciprocal of a linear function, for example the **rational equation** $y = \frac{1}{x}$? In this chapter we will study rational equations and rational expressions in more detail.

Source: S. Green–Armytage/The Stock Market

5.1 EXPLORING RATIONAL RELATIONSHIPS: TABLES, GRAPHS, EQUATIONS

Recall that we studied the relationship $y = \frac{1}{x}$ in Example 17 of Chapter 2 as $y = x^{-1}$. We begin by reviewing the function $y = \frac{1}{x}$ and then examining two situations that make use of this function.

EXAMPLE 1 Examine the relationship between x and y if $y = \frac{1}{x}$.

Solution

Using a Table

x	y
-3	$-\frac{1}{3}$
-2	$-\frac{1}{2}$
-1	-1
$-\frac{1}{2}$	-2
$-\frac{1}{3}$	-3
$-\frac{1}{9}$	-9
$-\frac{1}{10}$	-10
$\frac{1}{10}$	10
$\frac{1}{9}$	9
$\frac{1}{4}$	4
$\frac{1}{3}$	3
$\frac{1}{2}$	2
1	1
2	$\frac{1}{2}$
3	$\frac{1}{3}$
10	$\frac{1}{10}$

Table 5.1

Using a Graph

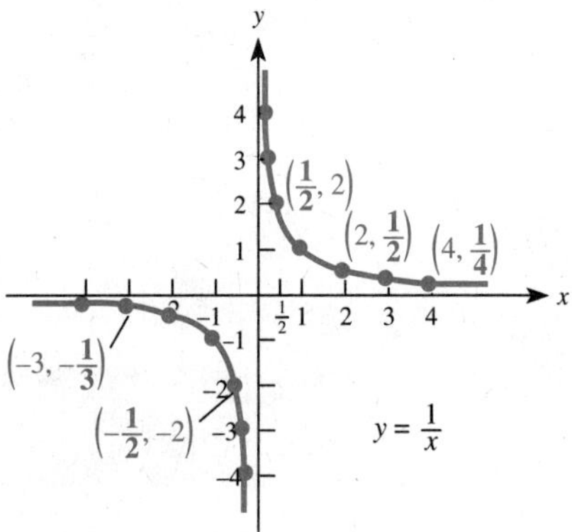

Figure 5.1

We observe that as $|x|$ increases, y decreases and approaches zero. We also notice that as x approaches zero from the right side, y increases without bound. We say y is approaching "infinity." Similarly, as x approaches zero from the left side, y decreases and is approaching "negative infinity." The graph is a way to visualize the fact that division by zero is not defined. The relationship $y = \frac{1}{x}$ is an example of a rational equation. ■

Another way to express the pattern in Table 5.1 is $xy = 1$ or to say that x is the reciprocal of y. We have looked at this relationship in three ways: with a table, with a graph, and with an equation. This relationship is an example of a **rational function.** We could also write $f(x) = \frac{1}{x}$.

Many applications can be modeled using rational functions. Consider the following two examples.

EXAMPLE 2

If the rate is constant, say $r = 50$ miles per hour, then the relationship between distance and time is given by $D = 50t$. The graph is a line through the origin whose slope is equal to the rate 50. Now suppose the distance is constant, say $D = 1{,}000$ miles. What is the relationship between rate and time?

Investigate the relationship between the rate and time when the distance is 1,000 miles.

Solution

Let's try a few values for rate and time and examine the graph:

Using a Table

$D = rt$

Distance = 1,000 miles

r (mph)	*t (in hours)*	*Distance*
10	100	1,000
20	50	1,000
40	25	1,000
50	20	1,000
100	10	1,000
200	5	1,000
1,000	1	1,000

Table 5.2

Using a Graph

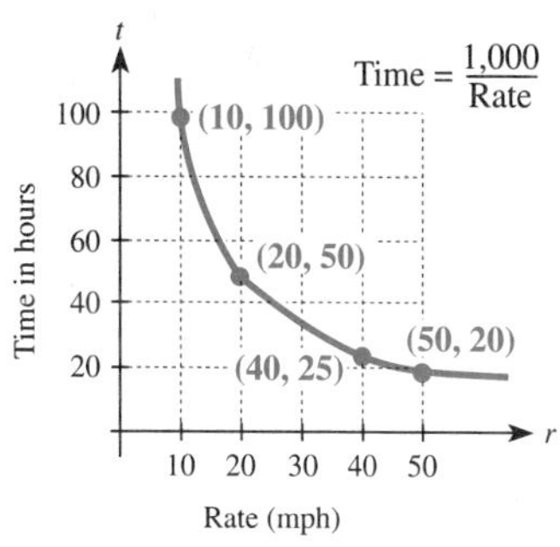

Figure 5.2

Notice the similarity between $y = \frac{1}{x}$ and $t = \frac{1{,}000}{r}$. The graph in Figure 5.2 is very similar to the graph in Figure 5.1. As the rate increases, the time decreases and approaches zero. Also, as the rate decreases or approaches zero, the time gets larger and larger and approaches infinity. Notice that negative values for time and rate are meaningless in this case, so our graph is only in the first quadrant. ■

EXAMPLE 3

A glider is launched with an initial velocity of 18 mph. The glider can travel 63 miles downwind in the same amount of time that it takes to travel 45 miles against the wind. What is the wind speed?

Solution

To solve this problem we need to find an equation which expresses the relationships in the problem.

Using a Table We recall the model

$$D = rt$$
$$\text{Distance} = \text{rate} \times \text{time}$$

for objects moving at a constant rate. We are looking for the speed of the wind. Let x equal the speed of the wind. Then $18 + x$ is the speed of the glider *with* the wind, and $18 - x$ is the speed of the glider *against* the wind. We can summarize our information in a table.

	Distance in miles	*Rate miles per hour*	*Time in hours*
Downwind (with wind)	63	$18 + x$	?
Upwind (against wind)	45	$18 - x$	?

Table 5.3

What about the "time" slots in the box above? Since

$$D = rt$$

then

$$\frac{D}{r} = t$$

Thus the *downwind* time is

$$\frac{D}{r} = \frac{63}{18 + x}$$

Filling in the box by using $\frac{D}{r} = t$, we obtain

	Distance in miles	*Rate miles per hour*	*Time in hours*
Downwind	63	$18 + x$	$\frac{63}{18 + x}$
Upwind	45	$18 - x$	$\frac{45}{18 - x}$

Table 5.4

Since the glider travels 63 miles downwind *in the same time* as 45 miles upwind, we can set the two times equal and we have the following equation:

$$\frac{63}{18 + x} = \frac{45}{18 - x}$$
$$\text{time downwind} = \text{time upwind}$$

Using a Graph We can use a graph to find a reasonable solution. One way to graph this equation is to think of it as two separate equations that describe the time downwind (T_{downwind}) and the time upwind (T_{upwind}). The two equations are:

$$T_{\text{downwind}} = \frac{63}{18 + x}$$

and

$$T_{\text{upwind}} = \frac{45}{18 - x}$$

We can graph each of these equations and find the intersection point—the point where the times are equal—and then find the speed of the wind (x). (See Figure 5.3.)

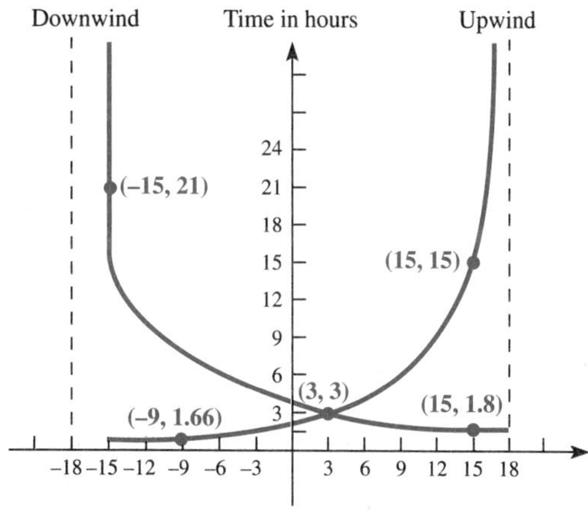

Figure 5.3

Since only positive values of x are meaningful in this situation, we graph $Y1 = \frac{63}{18 + x}$ and $Y2 = \frac{45}{18 - x}$ in a window containing reasonable values for x and Y, say $[0 \le x \le 6, 0 \le Y \le 6]$ and find their intersection point.

From the graph we find that $x = 3$ mph, which is the speed of the wind. The corresponding value for Y is 3 mph. This is the speed of the glider.

By looking at a larger window we can graphically investigate such questions as (1) what happens when the wind speed is 18 mph? 20 mph? and (2) what are reasonable values for x?

Using an Equation The above equation is a **rational equation** in the variable x, the wind velocity, and we will learn how to solve for x in such an equation. ■

The equations in the preceding example are rational equations in one variable. Rational equations frequently occur in such fields as business, economics, and forestry, as well as in the natural and biological sciences and engineering. While we can use graphs to solve these problems, we sometimes want to transform rational expressions to a simpler form before we enter the equation into a calculator. It is also useful to be able to solve simple rational equations algebraically. For these reasons we will investigate operations on rational expressions and equations.

Factoring Polynomials: A General Case

In this section we will continue the investigation of factoring that was introduced in Chapter 2. The distributive property can be used to express a polynomial as a product of factors. The process of rewriting a polynomial expression as a product of factors is a useful tool for solving equations and for transforming algebraic expressions into equivalent expressions. Consider this example.

EXAMPLE 4

The height h (in feet) above the ground of a golf ball depends on (is a function of) the time t (in seconds) it has been in flight. One golfer hits a tee shot that has a height approximately given by

$$h = 80t - 16t^2$$

How long does it take for the ball to hit the ground?

Solution

Using a Graph When the ball hits the ground, the height $h(t) = 0$. Thus we obtain the solution to $0 = 80t - 16t^2$ by graphing $h = 80t - 16t^2$ and finding values of t when $h(t) = 0$ (Figure 5.4).

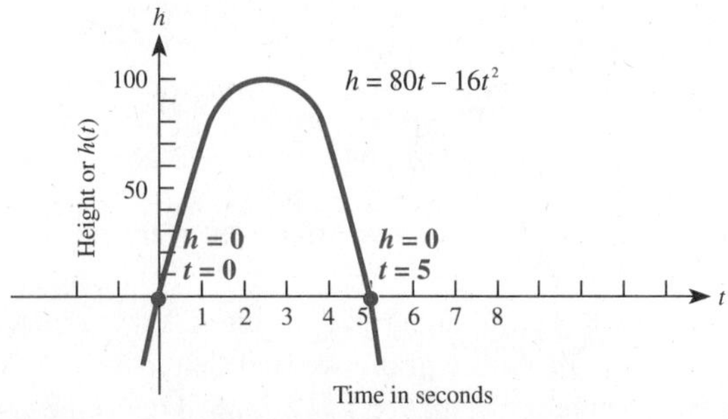

Figure 5.4

Using an Equation We solve $h = 80t - 16t^2$ when $h = 0$.

$$0 = 80t - 16t^2 \quad \text{Substitute } h = 0.$$

To solve this nonlinear equation for t, we can rewrite the right side as a *product of factors:*

$$0 = t(80 - 16t) \quad \text{Distributive property in reverse.}$$

So either

$$t = 0 \quad \text{or} \quad (80 - 16t) = 0 \quad \text{If the product is 0, one of the factors must be 0.}$$

Thus

$$t = 0 \quad \text{or} \quad t = 5$$

The ball is on the ground at 0 (still on the tee) and at 5 seconds. It takes the ball 5 seconds to hit the ground. ■

Example 4 indicates the power of factoring or graphing to obtain a solution to a nonlinear polynomial equation. We will return to the solution of such equations in the next chapter.

Let us now investigate factoring in more detail. In Chapter 2 we were introduced to the process of removing monomial factors from polynomials.

The Distributive Property

By using the distributive property:

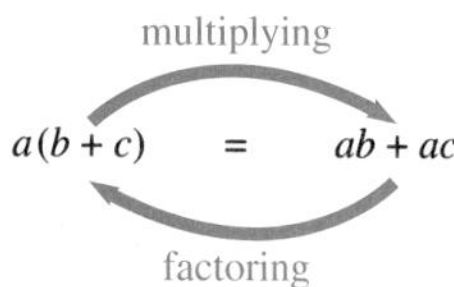

Going from left to right is called *multiplying*, going from right to left is called *factoring*. We include several examples here for the sake of completeness.

EXAMPLES 5–7

Factor as much as possible.

5. $6x^3 - 2x^2$

6. $3x^2y^2 - 6x^3y + 12x^2y^3 + 48x^4y^3$

7. $4xy - 3z + 6$

Solutions

5. $6x^3 - 2x^2 = x^2(6x - 2)$ The distributive property: The monomial x^2 is a common factor in both terms.

$= 2x^2(3x - 1)$ The number 2 is also a common factor.

Check:

$2x^2(3x - 1) = 2x^2(3x - 1)$ Multiply.

$\stackrel{\checkmark}{=} 6x^3 - 2x^2$

6. $3x^2y^2 - 6x^3y + 12x^2y^3 + 48x^4y^3$

$= 3x^2y(y - 2x + 4y^2 + 16x^2y^2)$ **Use the distributive law to check the answer.**

7. $4xy - 3z + 6$ is already factored as much as possible. ■

EXAMPLE 8

Choose any three different digits and form the six three-digit integers containing the chosen digits. (For example, if 2, 5, and 7 are the chosen digits, then the six integers are 257, 275, 527, 572, 725, and 752.) Find the sum of these six integers. What is the largest integer that is a factor of this sum (regardless of which three digits are chosen)?

Solution

If x, y, and z represent the three digits, then the six integers are

$$\begin{array}{ll} 100x + 10y + z & 100y + 10z + x \\ 100x + 10z + y & 100z + 10x + y \\ 100y + 10x + z & 100z + 10y + x \end{array}$$

Their sum is

$$200x + 200y + 200z + 20x + 20y + 20z + 2x + 2y + 2z$$

or

$$222x + 222y + 222z = 222(x + y + z)$$

since 222 is a common factor. Thus no matter what digits x, y, z are chosen, the integer 222 is a factor of the sum of the six integers. ■

Factoring Trinomials

In Chapter 2 we used the distributive property to multiply two binomials. The product was often a trinomial. We review how to use the distributive property to multiply and then we reverse the process to write a trinomial as the product of two binomial factors, if it is possible to do so.

EXAMPLE 9A

Multiply the binomials.

(a) $(y + 7)(y + 1)$ **(b)** $(3x + 8)(2x - 3)$ **(c)** $(x - 6)^2$

Solutions

We use the distributive property to multiply binomials.

(a) $(y + 7)(y + 1) = y(y + 1) + 7(y + 1)$ **Distribute $(y + 1)$ over the sum $y + 7$.**

combine

$= y^2 + y + 7y + 7$ **Add $y + 7y$.**

$= y^2 + 8y + 7$ **Answer.**

(b) $(3x + 8)(2x - 3) = 6x^2 \underbrace{- 9x + 16x}_{\text{combine}} - 24$ Apply the distributive property twice.

$= 6x^2 + 7x - 24$ Answer.

(c) $(x - 6)^2 = (x - 6)(x - 6) = x^2 \underbrace{- 6x - 6x}_{\text{combine}} + 36$

$= x^2 - 12x + 36$ Answer.

EXAMPLE 9B

Factor each trinomial into the product of two binomials.

(a) $y^2 + 8y + 7$ **(b)** $x^2 - 12x + 36$

Solutions

The process is the reverse of that in Example 9A. We use a strategy called **systematic trial and error**.

(a) $y^2 + 8y + 7 = (? \pm ?)(? \pm ?)$

Our job is to fill in the question marks.

1. What are the possible *first* terms of the binomials?

$$y^2 + 8y + 7 = (y \pm ?)(y \pm ?) \quad y \cdot y = y^2$$

2. What are the possible *last* terms?

$$y^2 + 8y + 7 = (y \pm 7)(y \pm 1) \quad \text{7 and 1 are the only integer factors of 7.}$$

You can see that in order to obtain $+8y$ as the middle term of the trinomial, both binomials must have positive *last* terms. Thus,

$$y^2 + 8y + 7 = (y + 7)(y + 1) \quad \text{Answer.}$$

(b) $x^2 - 12x + 36 = (? \pm ?)(? \pm ?)$

1. What are the possible *first* terms?

$$x^2 - 12x + 36 = (x \pm ?)(x \pm ?) \quad \text{The only possibility with integer coefficients.}$$

2. What are the possible *last* terms? Since $36 = 1 \cdot 36$, $36 = 2 \cdot 18$, $36 = 3 \cdot 12$, $36 = 4 \cdot 9$, and $36 = 6 \cdot 6$, there are a number of possibilities for the two last terms. Which of these works?

$$(x \pm 1)(x \pm 36)$$
$$(x \pm 2)(x \pm 18)$$
$$(x \pm 3)(x \pm 12)$$
$$(x \pm 4)(x \pm 9)$$
$$(x \pm 6)(x \pm 6)$$

A little inspection will show that the last pair, with negative signs, yields the correct result:

$$x^2 - 12x + 36 = (x - 6)(x - 6) \quad \text{Answer.}$$

In this example, we encountered a large number of possible factors for $x^2 - 12x + 36$; 20 possibilities in all. We could multiply all these out to see which one works; however, we can narrow down our search *by using some information about the signs* in the trinomial. The last term, 36, is *positive*, and therefore the *last* terms of our two binomials must have the *same* sign, either both positive or both negative. Indeed, we can narrow our search *still* further by noting that the second term of $x^2 - 12x + 36$, $-12x$, is negative. Where does the second term come from? It's always the *sum* of the *outside* and *inside* partial products. (Refer back to Example 9A.) The only way we can get $-12x$ for the middle term in the trinomial is if both binomials have negative last terms.

Our strategy is called *systematic trial and error* because we systematically narrow down the possibilities for the binomial factors of a trinomial until there are only a few cases left. We may have to test these few cases one at a time to see which one is correct. Let's consider some more examples.

EXAMPLES 10–12

Factor as much as possible.

10. $-z^3 - 5z^2 - 6z$ **11.** $2x^2 + 10x - 48$ **12.** $6y^2 - 7y + 2$

Solutions

10. $-z^3 - 5z^2 - 6z = -z(z^2 + 5z + 6)$ **Look for common monomial factors first. Apply the distributive property.**

$= -z(z \pm ?)(z \pm ?)$ **z is the only possibility for both first terms.**

$= -z(z + ?)(z + ?)$ **Both *last* terms must be *positive* to give the trinomial a positive middle term ($+5z$) and a positive last term ($+6$).**

What are the possible values for the *last* terms?

$$6 = 6 \cdot 1$$
$$6 = 2 \cdot 3$$

The factors 2 and 3 will yield the middle term ($5z$), so

$$-z^3 - 5z^2 - 6z = -z(z + 2)(z + 3)$$ **Completely factored.**

NOTE: The trinomial $z^2 + 5z + 6$ has 1 as the coefficient of the squared term, z^2.

■

Check: Suppose we multiply the factors:

$$-z(z + 2)(z + 3) = -z(z^2 + \underbrace{2z + 3z}_{\text{combine}} + 2 \cdot 3) = -z(z^2 + 5z + 6) = -z^3 - 5z^2 - 6z$$

For such trinomials, we can look at the coefficients of the middle and last terms, 5 and 6, respectively, and ask, "What integers have product 6 and sum 5?"

11. $2x^2 + 10x - 48$

$= 2(x^2 + 5x - 24)$ Monomial factors first.

$= 2(x + ?)(x - ?)$ The last terms must have opposite signs. What integers have the product -24 and the sum $+5$?

$= 2(x + 8)(x - 3)$ Answer.

Check:

$$2(x + 8)(x - 3) = 2(x^2 + 5x - 24) = 2x^2 + 10x - 48$$

12. $6y^2 - 7y + 2$

$= (? - 1)(? - 2)$ In this case, the last terms are evident, but the first terms are not.

Here are the possible first terms:

$$6y^2 = 6y \cdot y \quad \text{or} \quad 6y^2 = 2y \cdot 3y$$

Thus the possible factors are

$$(6y - 1)(y - 2) \quad \text{or} \quad (y - 1)(6y - 2)$$

or

$$(3y - 1)(2y - 2) \quad \text{or} \quad (2y - 1)(3y - 2)$$

Trial and error will show that

$$6y^2 - 7y + 2 = (2y - 1)(3y - 2)$$

(outer product: $-4y$; inner product: $-3y$) The correct middle term is $-7y$.

Check:

$$(2y - 1)(3y - 2) = 6y^2 - 7y + 2$$

EXAMPLES 13–15

Factor as much as possible. Check the answer by multiplying.

13. $x^2 + 4x + 2$ **14.** $24x^2 + 11x - 18$ **15.** $x^2 - 16$

Solutions

13. The only possible factors with integer coefficients are

$$x^2 + 4x + 2 = (x + 2)(x + 1)$$

(inner product: $2x$; outer product: x) Does not generate the correct middle term.

These factors will not work. Thus $x^2 + 4x + 2$ *cannot* be factored any further using integers.

14. $24x^2 + 11x - 18$ $\quad 24 = 24 \cdot 1, 12 \cdot 2, 8 \cdot 3, 6 \cdot 4$
$18 = 18 \cdot 1, 9 \cdot 2, 6 \cdot 3$

There are a *lot* of possible factors for this trinomial. Sometimes the middle term can provide some hints for which factors to try first. For example, $11x$ is not a very large middle term. Consider these factors.

$432x$

$(24x - 1)(x + 18)$ Generates $431x$ for a middle term.

$-x$

Therefore we might first try $24 = 6 \cdot 4$ and $18 = 6 \cdot 3$.

$-12x$

$(6x - 3)(4x + 6)$ Generates the wrong middle term.

$36x$

Our initial attempt did not work, but what can we learn from it? Notice that $(4x + 6) = 2(2x + 3)$. We could not remove a factor of 2 from $24x^2 + 11x - 18$. Clearly, our first attempt was doomed. The same thing occurs with a factor of 6 if we switch the position of the two last terms. Let's try another pair of factors:

$$24x^2 + 11x - 18 \stackrel{?}{=} (8x - 3)(3x + 6) \quad \text{No!}$$

Notice that $3x + 6 = 3(x + 2)$, and there was no factor of 3 in $24x^2 + 11x - 18$. Any other placement of 6 and 3 also leads to a monomial factor. Let's try another pair of factors, $2 \cdot 9$, for the last term.

$-16x$

$24x^2 + 11x - 18 = (8x + 9)(3x - 2)$ Answer. Generates $11x$ for the middle term.

$27x$

Example 14 shows how our search for factors can be somewhat systematic if we try to use information gained from each of our previous trials.

$4x$

15. $x^2 - 16 = (x - 4)(x + 4)$ Answer. The "middle" term is 0. ■

$-4x$

Factoring a Difference of Perfect Squares

Example 15 is a special case. It is not a trinomial, but it does factor into the product of two binomials. This special case is called the **difference of two perfect squares.**

The Difference of Two Perfect Squares

$$A^2 - B^2 = (A - B)(A + B)$$

The second terms of the binomial factors are the same size but opposite in sign, so the "middle" term of their product will drop out. This factorization is the "reverse" of the special product $(a + b)(a - b) = a^2 - b^2$ given in Section 2.4.

EXAMPLES 16–19

Factor as much as possible.

16. $4x^2 - 9y^2$ **17.** $25x^2z - 49y^2z$ **18.** $x^2 + y^2$ **19.** $x^4 - y^4$

Solutions

16. $4x^2 - 9y^2 = (2x - 3y)(2x + 3y)$ Difference of two squares.

17. $25x^2z - 49y^2z = z(25x^2 - 49y^2)$ Look for any monomial factors first.

$= z(5x - 7y)(5x + 7y)$ Difference of two squares.

Multiply and check.

18. $x^2 + y^2$ *cannot* be factored further. Notice that the possible factors

$$(x + y)(x + y) \quad \text{or} \quad (x - y)(x - y)$$

$$2xy \qquad\qquad -2xy$$

generate a middle term. The *sum* of two squares cannot be factored with integer coefficients.

19. $x^4 - y^4 = (x^2)^2 - (y^2)^2$ Both x^4 and y^4 are squares.

$= (x^2 - y^2)(x^2 + y^2)$ Difference of squares.

$= (x - y)(x + y)(x^2 + y^2)$ Difference of squares. ■

One application of the difference of two squares is illustrated in the following example.

EXAMPLE 20

Compute *exactly*.

$$(1{,}000{,}003)(999{,}997)$$

Solution

If we use a calculator, we may obtain 10^{12}, which is not an exact answer, since we know the last digit in the product is 1. To find the exact value, we rewrite

$$1{,}000{,}003 = 1{,}000{,}000 + 3 = 10^6 + 3$$
$$999{,}997 = 1{,}000{,}000 - 3 = 10^6 - 3$$

Thus the product is

$$(10^6 + 3)(10^6 - 3) = (10^6)^2 - 3^2$$
$$10^{12} - 9 = 999{,}999{,}999{,}991$$

The round-off error in the calculator was the culprit that gave us the incorrect value, 10^{12}. ■

Factoring Perfect Square Trinomials

In Section 2.4 we developed a special product rule: $(a + b)^2 = a^2 + 2ab + b^2$. Let us consider the squares of some other binomials.

$$(x + 3)^2 = (x + 3)(x + 3) = x^2 + \overbrace{3x + 3x}^{2 \cdot 3x} + 3^2 = x^2 + 6x + 9$$

$$(x - 9)^2 = (x - 9)(x - 9) = x^2 - \overbrace{9x - 9x}^{2 \cdot (-9x)} + 9^2 = x^2 - 18x + 81$$

$$(3x - y)^2 = (3x - y)(3x - y) = (3x)^2 - \overbrace{3xy - 3xy}^{2 \cdot (-3xy)} + (-y)^2$$
$$= 9x^2 - 6xy + y^2$$

In general,

$$(\text{first term} + \text{second term})^2$$
$$= (\text{first term})^2 + \begin{pmatrix} \text{twice the product} \\ \text{of the first and second terms} \end{pmatrix} + (\text{second term})^2$$

The trinomials that result from these products are called **perfect square trinomials** because they are equal to the square of a binomial. We use the preceding equation in reverse in order to factor trinomials that are perfect squares.

EXAMPLES 21–23

Factor as much as possible.

21. $x^2 - 10x + 25$ **22.** $16x^2 + 40xy + 25y^2$ **23.** $9x^3 + 18x^2 + 9x$

Solutions

21. $x^2 - 10x + 25$

Notice that the first and last terms are squares. This suggests the possibility of a perfect square trinomial.

$$x^2 - 10x + 25 = (x - 5)(x - 5) \quad \text{or} \quad (x - 5)^2 \qquad \text{Answer.}$$

22. $16x^2 + 40xy + 25y^2$ First and last terms are squares.

$$= (4x + 5y)(4x + 5y) \quad \text{or} \quad (4x + 5y)^2$$

23. $9x^3 + 18x^2 + 9x = 9x(x^2 + 2x + 1)$ Look for monomial factors.

$$= 9x(x + 1)(x + 1) \quad \text{or} \quad 9x(x + 1)^2 \qquad \text{Answer.}$$ ■

Here is a summary of perfect square trinomials. They play an important role in the solution of quadratic equations, which we will discuss in Chapter 6.

Perfect Square Trinomials

$$A^2 + 2AB + B^2 = (A + B)(A + B) = (A + B)^2$$
$$A^2 - 2AB + B^2 = (A - B)(A - B) = (A - B)^2$$

Although we have devoted most of our discussion about factoring to the factoring of trinomials, there are instances when we can use factoring to rewrite, or simplify, other polynomial expressions.

EXAMPLES 24–26

Factor as much as possible.

24. $3(x - 1) + x(x - 1)$

25. $(x - 1)^2(2x + y) - (x - 1)^2(x - 2y)$

26. $ax + ay + bx + by$

Solutions

24. $3(x - 1) + x(x - 1)$ The binomial $x - 1$ is a common factor.

$= (x - 1)(3 + x)$ Answer. Use of the distributive property.

Multiply and check.

25. $(x - 1)^2(2x + y) - (x - 1)^2(x - 2y)$ $(x - 1)^2$ is a common factor.

$= (x - 1)^2[(2x + y) - (x - 2y)]$

$= (x - 1)^2(2x + y - x + 2y)$ Watch signs.

$= (x - 1)^2(x + 3y)$ Answer.

26. $ax + ay + bx + by$ Try factoring monomials from pairs.

$= a(x + y) + b(x + y)$ Notice $x + y$ is now a common factor.

$= (x + y)(a + b)$ Answer. Multiply and check.

NOTE: We could have paired ax with bx and ay with by to obtain $(a + b)x + (a + b)y = (a + b)(x + y)$. ■

Finding Equivalent Algebraic Fractions

Recall our discussion of the *fundamental property of fractions* from Chapters 1 and 2.

Fundamental Property of Fractions

$$\frac{ax}{bx} = \frac{a}{b} \text{ for } b \neq 0, x \neq 0$$

NOTE: For the remainder of this chapter we will assume that denominators in algebraic fractions are not equal to zero, unless otherwise specified. ■

The process of factoring will also help us to express algebraic fractions in simplest form. We say that *a fraction is in simplest form if there are no common factors in the numerator and denominator.*

EXAMPLES 27–30

Express fractions in simplest form.

27. $\dfrac{63x^4y^3}{42x^3y^6z}$ **28.** $\dfrac{6x^2y - 2y}{4xy}$ **29.** $\dfrac{3x^2 + x - 10}{x^2 + 5x + 6}$ **30.** $\dfrac{25 - x^2}{x^2 - 3x - 10}$

Solutions

In each case, we factor the numerator and denominator, and then we apply the fundamental property of fractions if possible.

27. $\dfrac{63x^4y^3}{42x^3y^6z} = \dfrac{21x^3y^3(3x)}{21x^3y^3(2y^3z)}$ Look for all common monomial factors.

$= \dfrac{\cancel{21x^3y^3}(3x)}{\cancel{21x^3y^3}(2y^3z)} = \dfrac{3x}{2y^3z}$ Answer. Fundamental property.

28. $\dfrac{6x^2y - 2y}{4xy} = \dfrac{\cancel{2y}(3x^2 - 1)}{\cancel{2y}(2x)} = \dfrac{3x^2 - 1}{2x}$ Factor and then use the fundamental property.

29. $\dfrac{3x^2 + x - 10}{x^2 + 5x + 6} = \dfrac{\cancel{(x+2)}(3x - 5)}{\cancel{(x+2)}(x + 3)} = \dfrac{3x - 5}{(x + 3)}$ Answer.

30. $\dfrac{25 - x^2}{x^2 - 3x - 10} = \dfrac{(5 - x)(5 + x)}{(x - 5)(x + 2)}$ Recall that $5 - x = -(x - 5)$ so that $\frac{(5 - x)}{(x - 5)} = -1$.

$= \dfrac{-\cancel{(x-5)}(5 + x)}{\cancel{(x-5)}(x + 2)} = -\dfrac{5 + x}{x + 2}$ Answer. ■

We will now apply these algebraic skills (the fundamental property of fractions and the distributive property) to recognize equivalent expressions that involve algebraic fractions.

Equivalent Rational Expressions

EXAMPLE 31

Which of the following expressions are equivalent?

(a) $\dfrac{4x^3 - 16x}{x^3 + 3x^2 + 2x}$ **(b)** $\dfrac{4(x - 2)}{x + 1}$ **(c)** $\dfrac{4x(x - 2)(x + 2)}{x(x + 2)(x + 1)}$

(d) $\dfrac{4x^2 - 16}{x^2 + 3x + 2}$ **(e)** $\dfrac{4(x + 2)}{x + 1}$

Solution

Start with one of the expressions and use algebraic operations to transform the expression into an equivalent expression. Let's start with the expression in part (a).

(a) $\dfrac{4x^3 - 16x}{x^3 + 3x^2 + 2x} = \dfrac{4x(x - 2)(x + 2)}{x(x + 2)(x + 1)}$

Start with the expression in part (a). Factor the numerator and denominator. Thus the expressions in (a) and (c) are equivalent.

$= \dfrac{4(x - 2)}{x + 1}$

Remove common factors. Fundamental property of fractions. Thus the expression in (b) is equivalent to the expression in (c) and hence (a).

(a) $\dfrac{4x^3 - 16x}{x^3 + 3x^2 + 2x} = \dfrac{x(4x^2 - 16)}{x(x^2 + 3x + 2)}$

$= \dfrac{4x^2 - 16}{x^2 + 3x + 2}$

If we start with the expression in part (a) and factor out only the x (since x is a factor of every term in both the numerator and denominator), then we have the expression in (d). Thus all of the expressions (a)-(d) are equivalent. If we substitute for x in expression (e) and one of the others, say (b), we will not get the same value for y. Try $x = 0$, then (b) yields -8 and (e) yields 8. Expression (e) is not equivalent to the expressions (a)-(d). ■

It is important to remember that when we operate on algebraic expressions or, in this case, find the simplest form of an algebraic fraction, we are in fact finding a set of equivalent expressions. Depending on the situation, one form may be more desirable than another form. But they are all equivalent.

NOTE: If two expressions are equivalent, then each expression will generate the same table of values for corresponding values of x and y whenever the expressions are defined. Also, the graphs of the two expressions are identical for all values for which the expression is defined. ■

Consider the following example:

EXAMPLE 32

When $x \neq 1$, show that the expression

$$\frac{x^2 - 1}{x - 1}$$

is equivalent to the expression $x + 1$ by graphing each expression.

Solution

$$\text{Let } Y1 = \frac{x^2 - 1}{x - 1} \quad \text{and} \quad Y2 = x + 1$$

We note that $Y1$ is not defined for $x = 1$. So we will exclude $x = 1$ from both graphs and see if the resulting graphs are equivalent. By graphing with $Y1 = \frac{x^2 - 1}{x - 1}$ and $Y2 = x + 1$ we see that their graphs appear identical.

In fact, when $x \neq 1$, the graph of $Y1$ is a line with a hole, so the two expressions are equivalent if $x \neq 1$. The hole in the graph of $Y1$ will show up using an appropriate

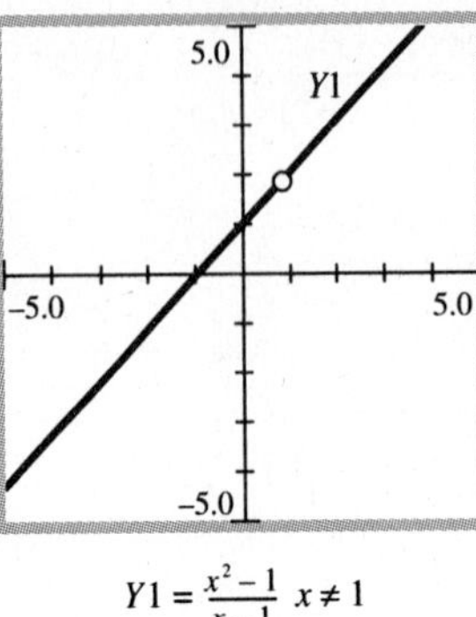

$Y1 = \frac{x^2 - 1}{x - 1} \quad x \neq 1$

Figure 5.5 (a)

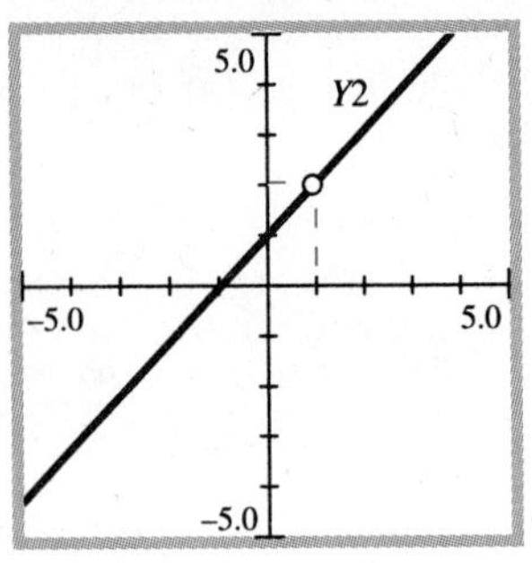

$Y2 = x + 1 \quad x \neq 1$

Figure 5.5 (b)

ZOOM on some graphing calculators around $x = 1$. However, the hole will not show up on the graph of $Y2$. But we agreed to omit $x = 1$ on this graph.

EXAMPLE 33

The cost, C, in thousands of dollars to provide mammograms to x percent of the women in a particular state is given by the formula

$$C = \frac{150x}{100\% - x}$$

(a) What will it cost to provide mammograms to 50% of the women?

(b) What happens to the cost as x increases?

Solution

Using an Equation

(a) If $x = 50\%$, then

$$C = \frac{150(50\%)}{100\% - 50\%} \qquad 50\% = .50$$

$$= \frac{75}{0.50}$$

$$= 150.0, \text{ or } \$150{,}000.$$

(b) As x increases, the denominator gets smaller and smaller, which means that the costs will get larger and larger. There is no theoretical maximum for the cost. It does not make sense for x to be greater than 100.

Using a Graph

(a) It will cost $150,000 to provide mammograms to 50% of the women.

(b) As the number of women increases, the cost increases. But the number of women cannot exceed 100%.

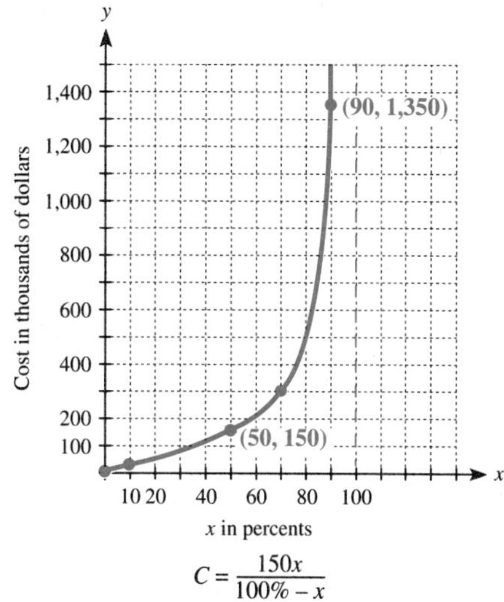

$$C = \frac{150x}{100\% - x}$$

Figure 5.6

5.1 EXERCISES

I. Procedures

1–50. Use the distributive property to factor each of the following expressions as much as possible. (Use the distributive property in reverse to check your answer.)

1. $y^2 + 3y + 2$

2. $x^2 - 6x + 8$

3. $w^2 - 2w - 35$

4. $x^2 + 7x + 6$

5. $x^2 + x + 1$

6. $x^2 + 3x^2 + 4$

7. $x^2 - 25$

8. $0.09y^2 - 100$

9. $81x^2 - 49$

10. $x^2 + 4$

11. $x^2 + 14x + 49$

12. $x^2 - 16x + 64$

13. $0.16q^2 + 0.8q + 1$

14. $0.0009x^2 - 0.03x + 0.25$

15. $2x^2 + 7x - 15$

16. $3y^2 - 8y - 3$

17. $12x^2 - 11x - 15$

18. $12x^2 - 7x - 12$

19. $3y^2 + 10y + 3$

20. $2x^2 + 2x + 3$

21. $4x^2 + x - 21$

22. $27x^2 - 6x - 8$

23. $3x^2y - 2xy^2$

24. $16x^2 - 8x$

25. $y^4 + 3y^3 + y^2$

26. $2x^2s^2 + x^2s - 10x^2$

27. $24x^2 + 6x - 45$

28. $3x^2 + 24x + 48$

29. $0.3x^2 - 7.5$

30. $12x^2z^2 + 10x^2yz - 12x^2y^2$

31. $14y^4 - 7y^3 - 21y^2$

32. $4x^2 + 8x + 4$

33. $50x^2y^2 - 50xy + 1$

34. $18x^2 + 9xvw + v^2w^2$

35. $12y^4 + 11y^3 - 15y^2$

36. $15x^2 + 11x^3 - 12x^4$

37. $x(x - 1) + y(x - 1)$

38. $(x - 1)^2y - (x - 1)^2z$

39. $(x + 1)(x - 1) - (x + 1)(x + 2)$

40. $(x + y)x^3 - (2x - y)x^3$

41. $2x - 2y + xz - yz$

42. $ac + bc + ad + bd$

43. $x^2 + 5x - 2xy - 10y$

44. $bc - d + c - bd$

45. $(x - 1)a^2 - (x - 1)2ab + (x - 1)b^2$

46. $x^6 - 36$ (*Hint:* write x^6 as $(x^3)^2$ and apply the formula for the difference of two squares.)

47. $16x^8 - y^4$

48. $8x^4 - 50$

49. $x^4 - 8x^2 + 16$ (*Hint*: write x^4 as $(x^2)^2$ and apply the formula for factoring a perfect square trinomial.)

50. $x^4 + 5x^2 + 6$

51–68. Write each rational expression in simplest form.

51. $\dfrac{15a^3b^2c^4}{27a^2b^3c}$

52. $\dfrac{144y^2z^3}{48xy^7z}$

53. $\dfrac{3x^3 - 2x^2}{x^2}$

54. $\dfrac{x^3 - 2x}{2x^2}$

55. $\dfrac{12x^6y^2z^5 - 6x^5y^3z^2 + 10x^2y^4z^9}{2x^2y^2z^2}$

56. $\dfrac{6x^3y^2z - 12x^2y^2z^4 + 18x^3y^4z^2}{3x^3y^3z^6}$

57. $\dfrac{x^2 - xy}{y - x}$

58. $\dfrac{16 - x^2}{x^2 - 2x + 8}$

59. $\dfrac{x^4 - x^2}{x^4 + 2x^3 + x^2}$

60. $\dfrac{9x^2 - 16}{9x^2 + 24x + 16}$

61. $\dfrac{x^2 + 4x + 3}{x^2 + 6x + 5}$

62. $\dfrac{2y^2 + 3y - 9}{2y^2 + 11y + 15}$

63. $\dfrac{6x^2 - 13x - 5}{6x^2 - 15x}$

64. $\dfrac{8x^2 + 4xy - 2x}{12x^2 - 3x + 6xy}$

65. $\dfrac{(x + 1)^2(3x - y) + (x + 1)(x + y)}{x^2 + x}$

66. $\dfrac{(x - 2)x^3 - (2 - x)x^3}{x^2 - 2x}$

67. $\dfrac{(x + 1)^2(3x - y)}{(x + 1)^3}$

68. $\dfrac{(3x + 5)^2(2x - 7)^3}{(3x + 5)(4x^2 - 28x + 49)}$

II. Concepts

69 and 70. In each case, find an expression for the length of the side of the square.

69. Area = $y^2 + 6y + 9$

70. Area = $36x^2 - 60x + 25$

71 and 72. Given expressions for the area of a rectangle and the length of one of its sides, find an expression for the other side.

71. $x + 3$; Area = $x^2 + 7x + 12$

72. $y - 2$; Area = $y^2 + 3y - 10$

73. **(a)** In the figure, the *unshaded* region is a geometrical model of the algebraic expression $x^2 - y^2$ (take away the shaded region). Divide the unshaded region into pieces and rearrange the pieces to show *geometrically* that $x^2 - y^2 = (x + y)(x - y)$.

(b) What is an expression for the area of the shaded region in the figure given the areas of the subrectangles? What factoring problem, or multiplication problem, is suggested?

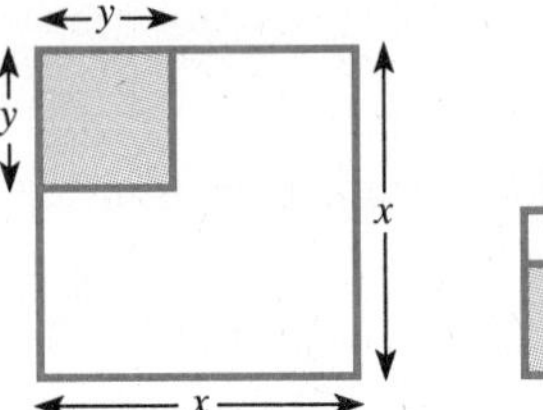

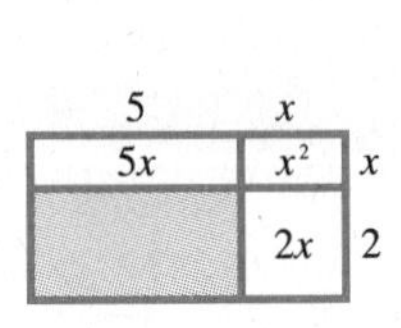

74. Draw a square of side $x + y$, and divide it into rectangular pieces which show geometrically that $(x + y)^2 = x^2 + 2xy + y^2$.

75–80. Make a perfect square trinomial by filling in the box. In each case, state the binomial that is squared to yield the trinomial.

75. $x^2 + 2x + \square = (\quad)^2$

76. $x^2 - 4x + \square = (\quad)^2$

77. $x^2 - \square + 9 = (\quad)^2$

78. $x^2 + \square + 100 = (\quad)^2$

79. $\square - 20x + 25 = (\quad)^2$

80. $\square + 12xy + 4x^2y^2 = (\quad)^2$

81–84. In each of the exercises below, the student has made an error in the solution (*all* given answers are wrong). Describe the error and find the correct solution. Describe how you would help this student.

81. $\dfrac{x^2 + 2x}{x^2 + 3x + 2} = \dfrac{x\cancel{(x + 2)}}{\cancel{(x + 2)}(x + 1)} = \dfrac{x}{x + 1} = 1$ **Incorrect answer.**

82. $\dfrac{x^2 + 4}{x^2 + 5x + 6} = \dfrac{\cancel{(x + 2)}(x + 2)}{\cancel{(x + 2)}(x + 3)} = \dfrac{x + 2}{x + 3}$ **Incorrect answer.**

83. $\dfrac{x^2 - 1}{1 - x}$ is already in simplest form. **Incorrect answer.**

84. $\dfrac{x^2 + y^2}{x + y} = \dfrac{\cancel{x^2}^{\,x} + \cancel{y^2}^{\,y}}{\cancel{x} + \cancel{y}} = x + y$ **Incorrect answer.**

85–88. Determine the answers to the following problems both algebraically and graphically. Substitute values for x to see if the equivalent expressions yield the same value.

85. Which of the following are equivalent to $3x$?

(a) $\frac{3x + 2}{2}$ (b) $\frac{3x^3}{x^2}$ (c) $\frac{6x^2 - 3x}{3x}$

(d) $\frac{6x^2 - 3x}{2x - 1}$ (e) $\frac{3x^2 - 6x}{x - 2}$

86. Which of the following are equivalent to -1?

(a) $\frac{x^2 - 4}{4 - x^2}$ (b) $\frac{-3x^3}{(3x)^3}$ (c) $\frac{-x + 3}{x + 3}$

(d) $\frac{x - 1}{x}$ (e) $\frac{-x^2 - 1}{x^2 + 1}$

87. Show that $x + 1$ is equivalent to

$$\frac{x^4 - x^2}{x^3 - x^2}$$

if $x \neq 1$ or 0.

88. Which of the following expressions are equivalent?

(a) $-2x(x + 2)$ (b) $\frac{-2x(x^2 - 4)}{x - 2}$

(c) $\frac{2x(x^2 - 4)}{2 - x}$ (d) $-2x(x - 2)$

(e) $\frac{2x^3 - 8x}{2 - x}$ (f) $\frac{x^3 - 4x}{-x}$

89 and 90. Find three expressions that are equivalent to each expression. Graph all four expressions on a graphing calculator to demonstrate equivalence.

89. $\frac{3x}{2}$ **90.** $\frac{4x^2 - 16}{x^3 - 2x^2}$

91. Tamara and Rose were trying to find the simplest form of a given algebraic fraction. Tamara had $2x - 6$ as an answer and Rose had $2(x - 3)$.

(a) Make a case for each of them being correct.

(b) Write two different possible algebraic expressions for the original fraction.

(c) Write an explanation of the difference between factors and terms.

III. Applications

92. In Example 3,

(a) what happens to time when the wind speed is 18 mph? When it is 20 mph?

(b) what are feasible (realistic) values for the wind speed?

93. Graph the equation in Example 3. How much time will it take the glider in Example 3 to go 50 miles

(a) against a 3-mph wind?

(b) with a 3-mph wind?

(c) against a wind of x mph?

(d) with a wind of x mph?

(e) Make a table and graph for the expressions in (c) and (d). Use the table or graph to find the time it takes to go against a 4-mph wind and a 10-mph wind.

94. **(a)** If a bicyclist can cover 75 miles in 5.5 hours in still air, what is her rate of speed in miles per hour? If there is a 5-mph wind, how much time will it take her to go the 75 miles

(b) against the wind?

(c) with the wind?

95. The time t in hours it takes to distribute voter registration forms to x percent of the students at Mu State is given by the formula

$$t = \frac{250x}{150 - 150x} \quad (x \text{ is expressed as a decimal in the formula})$$

(a) Graph this relationship.

(b) How much time will it take to distribute the forms to 40% of the students?

(c) What happens to the time as x approaches 100%?

(d) Use the graph to find out how much time it takes to distribute the forms to 20%, 50%, and 80% of the population.

(e) Use the graph to find out how much of the population can be reached in one hour and in ten hours.

96. The cost C in thousands of dollars to distribute voter forms to x percent of the students at Mu State is given by the formula

$$C = \frac{300x}{125 - 125x} \quad (x \text{ is expressed as a decimal in the formula})$$

(a) Graph this relationship.

(b) What is the cost to distribute forms to 40% of the students? 90% of the students?

(c) Use the graph to find how much it costs to distribute the forms to 20%, 50%, and 80% of the population.

(d) Use the graph to find out how much of the population can be reached for \$10,000 and for \$15,000.

97. Compute *exactly*. (See Example 20.)
(a) (10,000,004)(9,999,996)
(b) (1.00005)(0.99995)

98. Choose any four digits and form all possible four-digit integers using these four digits. (There are 24 possibilities in all!) Find the sum of these 24 numbers. What is the largest integer that is a factor of the sum? (See Example 8.)

99. What is the largest integer that is a factor of the sum of any four-digit integer and its reversal (for example, the reversal of 1,234 is 4,321)?

100. Use the equation below and the figure to find $B_{\frac{1}{2}}$. This is the cross-sectional area of a log at the midpoint of a log, given that $V = 165\text{ ft}^3$, $L = 16$ feet, the radius of the large end is 2 feet, and the radius of the small end is 1 foot. (You need the formula for area of a circle to calculate B and b.) Here,

B = area of base at large end of the log
b = area of base at small end of the log
L = length of the log
V = volume of the log

$$B_{\frac{1}{2}} = \frac{6V - (B + b)L}{4L}$$

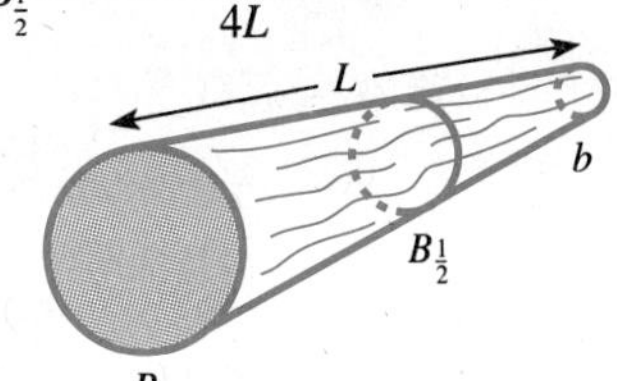

IV. Extensions

101. Sum and difference of two perfect cubes: The difference of two *perfect cubes* can be factored as follows:

$$x^3 - y^3 = (x - y)(x^2 + xy + y^2) \qquad \textbf{Equation (1)}$$

Similarly, the sum of two perfect cubes can be factored:

$$x^3 + y^3 = (x + y)(x^2 - xy + y^2) \qquad \textbf{Equation (2)}$$

(a) Multiply the right sides of Equations 1 and 2 above to show that they are the correct factors.

(b)–(g) Use Equations 1 and 2 to help factor the following expressions. In each case, multiply to check.

(b) $x^3 - 27$ **(c)** $8x^3 + 125$
(d) $64x^3 + y^3$ **(e)** $0.008x^6 - y^6$
(f) $54x^3y^3 - 16z^6$ **(g)** $0.000003x^3 + 0.081y^3$

102. (a) Do there exist positive integers x and y satisfying the following equations?

1. $x^2 - y^2 = 48$ **2.** $x^2 - y^2 = 23$
3. $x^2 - y^2 = 45$ **4.** $x^2 - y^2 = 90$

Hint: $x^2 - y^2 = 48$. Now $x^2 - y^2 = (x + y)(x - y)$. How many ways can 48 be expressed as a product of two positive integers?

$48 = 1 \cdot 48$ $48 = 2 \cdot 24$ $48 = 3 \cdot 16$
$48 = 4 \cdot 12$ $48 = 6 \cdot 8$

For which of these factorizations can $x + y$ represent the larger factor and $x - y$ represent the smaller factor? That is, which of the following linear systems

$x + y = 48$, $x - y = 1$ $x + y = 24$, $x - y = 2$ $x + y = 16$, $x - y = 3$
$x + y = 12$, $x - y = 4$ $x + y = 8$, $x - y = 6$

have a solution? What is the solution?

(b) In general, for what type of positive integers does $x^2 - y^2 = n$ have at least one solution?

5.2 MULTIPLICATION AND DIVISION OF RATIONAL EXPRESSIONS

This is the first of several sections in which we deal with arithmetic operations on algebraic fractions (rational expressions). As we have noted previously, computation with algebraic fractions can facilitate our work in writing equivalent algebraic expressions and our efforts to solve equations. We look first at multiplication and division.

Multiplication

EXAMPLE 1

Recall Example 1 in Section 1.3 of Chapter 1. The Gold Diggers paint half of the football field of the opposing team gold. The next night they paint half of the remaining unpainted field gold. Each night they continue to paint half of what is left. After seven nights, how much of the field will have been painted? The answer is $\frac{255}{256}$ of the field will have been painted. This time we want to know how much of the field is painted each night.

Solution

Let's look at the field after each of the first three nights.

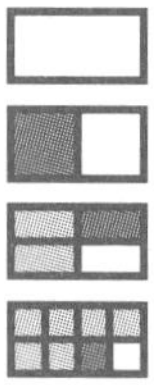

Night	Amount Painted Each Night		Total Amount Painted
start		0	0
1		$\frac{1}{2}$	$\frac{1}{2}$
2		$\frac{1}{4}$	$\frac{3}{4}$
3		$\frac{1}{8}$	$\frac{7}{8}$

Figure 5.7

On the first night, one-half of the total field was painted. That is,

$$\frac{1}{2} \text{ of } 1 = \frac{1}{2} \times 1 = \frac{1}{2}.$$

On the second night, one-half of one-half of the field was painted. That is,

$$\frac{1}{2} \text{ of } \frac{1}{2} = \frac{1}{2} \times \frac{1}{2} = \frac{1}{4}.$$

On the third night, one-half of one-fourth of the field was painted. That is,

$$\frac{1}{2} \text{ of } \frac{1}{4} = \frac{1}{2} \times \frac{1}{4} = \frac{1}{8}.$$

This example illustrates the process of multiplying two fractions. (See Section 1.3 of Chapter 1 for more examples.) ■

To multiply two fractions, we multiply the numerators and multiply the denominators. Arithmetic with numerical fractions provides us with a model for arithmetic with algebraic fractions. We multiply algebraic fractions in the same way.

Definition of Product of Fractions

Given two algebraic fractions $\frac{a}{b}$ and $\frac{c}{d}$, we define their **product**

$$\frac{a}{b} \cdot \frac{c}{d}$$

$$\text{to be } \frac{ac}{bd}, \text{ where } b \neq 0, d \neq 0$$

Let's see how our definition works.

EXAMPLES 2–4

Multiply and express the answer in simplest form.

2. $\dfrac{3xy^2}{4wz} \cdot \dfrac{6x^3}{5w^2y}$ **3.** $\dfrac{x}{x+1} \cdot \dfrac{3x+5}{x-1}$ **4.** $\dfrac{x^3}{x+3} \cdot \dfrac{x^2-9}{x}$

Solutions

2. $\dfrac{3xy^2}{4wz} \cdot \dfrac{6x^3}{5w^2y} = \dfrac{3xy^2 \cdot 6x^3}{4wz \cdot 5w^2y}$ Definition of multiplication.

$= \dfrac{18x^4y^2}{20w^3yz}$ Properties of exponents.

Now the multiplication is done, but perhaps the resulting fraction has a simpler equivalent form. Indeed, this one does, since

$\dfrac{18x^4y^2}{20w^3yz} = \dfrac{\cancel{2} \cdot 9x^4 \cdot \cancel{y} \cdot y}{\cancel{2} \cdot 10w^3\cancel{y}z}$ Fundamental property of fractions.

$= \dfrac{9x^4y}{10w^3z}$ Answer in simplest form.

3. $\dfrac{x}{x+1} \cdot \dfrac{3x+5}{x-1} = \dfrac{x(3x+5)}{(x+1)(x-1)}$ Definition of multiplication.

$= \dfrac{3x^2+5x}{x^2-1}$ There are no common factors in the numerator and denominator. Answer in simplest form.

It is also perfectly correct to leave the answer to this problem in factored form:

$$\frac{x(3x+5)}{(x+1)(x-1)}$$

4. $\dfrac{x^3}{x+3} \cdot \dfrac{x^2-9}{x} = \dfrac{x^3 \cdot (x^2-9)}{(x+3) \cdot x}$ Definition of multiplication.

Now, *look before you leap!* We want to put the answer in simplest form. We ask, "Are there any common factors in the numerator or denominator?"

$$\frac{x^3(x^2-9)}{(x+3)x} = \frac{x^3(x+3)(x-3)}{(x+3)x}$$ Factor as much as possible.

$$= \frac{\overset{x^2}{\cancel{x^3}}\cancel{(x+3)}(x-3)}{\cancel{(x+3)}\cancel{x}}$$ Fundamental property of fractions (or "remove common factors").

$$= x^2(x-3) \quad \text{or} \quad x^3 - 3x^2$$ Answer in simplest form. ■

NOTE: Example 4 suggests that we should factor as much as possible *before* multiplying out the numerators and denominators. In this way, we may discover a simpler form of the answer. ■

EXAMPLES 5–7

Multiply. Express answers in simplest form.

5. $\dfrac{x^2-5x+6}{4x^2-1} \cdot \dfrac{2x^2-5x-3}{x^2-6x+9}$

6. $\dfrac{3x^3-6x^2}{x^2y} \cdot \dfrac{4xy}{4-x^2}$

7. $\dfrac{4y+2}{y+2} \cdot \dfrac{y^2+2y}{2y^3+3y^2+y}$

Solutions

5. $\dfrac{x^2-5x+6}{4x^2-1} \cdot \dfrac{2x^2-5x-3}{x^2-6x+9}$

$$= \frac{(x^2-5x+6)(2x^2-5x-3)}{(4x^2-1)(x^2-6x+9)}$$ Definition of multiplication.

$$= \frac{(x-2)\cancel{(x-3)}\cancel{(2x+1)}\cancel{(x-3)}}{\cancel{(2x+1)}(2x-1)\cancel{(x-3)}\cancel{(x-3)}}$$ Factor completely and remove common factors.

$$= \frac{x-2}{2x-1}$$ Answer in simplest form.

6. $\dfrac{3x^3-6x^2}{x^2y} \cdot \dfrac{4xy}{4-x^2}$

$$= \frac{(3x^3-6x^2)\cdot 4xy}{x^2y(4-x^2)}$$ Definition of multiplication.

$$= \frac{3\cancel{x^2}\overset{(-1)}{\cancel{(x-2)}}\cdot 4x\cancel{y}}{\cancel{x^2}\cancel{y}\cdot\cancel{(2-x)}(2+x)}$$ Factor numerator and denominator to look for simple form.

$$= \frac{-12x}{2+x}$$ Answer. Remember that $(x-2) = (-1)(2-x)$.

7. $\dfrac{4y+2}{y+2} \cdot \dfrac{y^2+2y}{2y^3+3y^2+y}$

$$= \frac{(4y + 2)(y^2 + 2y)}{(y + 2)(2y^3 + 3y^2 + y)}$$ **Definition of multiplication.**

$$= \frac{2(2y+1)(y)(y+2)}{(y+2)(y)(2y+1)(y + 1)}$$

$$= \frac{2}{y + 1}$$ **Answer in simplest form.** ■

Division

To divide $\frac{3}{5}$ by $\frac{8}{7}$, we write

$$\frac{\frac{3}{5}}{\frac{8}{7}}$$

To eliminate the denominator, we multiply both the numerator and denominator by $\frac{7}{8}$, because $\frac{7}{8}$ is the multiplicative inverse of $\frac{8}{7}$ $\left(\frac{8}{7}\cdot\frac{7}{8} = 1\right)$. Thus,

$$\frac{\frac{3}{5}}{\frac{8}{7}} = \frac{\frac{3}{5}\cdot\frac{7}{8}}{\frac{8}{7}\cdot\frac{7}{8}} = \frac{\frac{3}{5}\cdot\frac{7}{8}}{1} = \frac{3}{5}\cdot\frac{7}{8}$$

So

$$\frac{3}{5} \div \frac{8}{7} = \frac{3}{5}\cdot\frac{7}{8}$$ **An equivalent problem.**

To divide one numerical fraction by another, say, $\frac{3}{5} \div \frac{8}{7}$, we change the division problem to a multiplication problem (invert the divisor and multiply). Thus $\frac{3}{5} \div \frac{8}{7}$ becomes $\frac{3}{5}\cdot\frac{7}{8} = \frac{21}{40}$.

Division of algebraic fractions proceeds in exactly the same way.

Definition of Quotient of Fractions

Given two algebraic fractions,

$$\frac{a}{b} \text{ and } \frac{c}{d}, \quad \text{where } b, d, c \neq 0$$

we define their quotient as follows:

$$\frac{a}{b} \div \frac{c}{d} = \frac{a}{b}\cdot\frac{d}{c}$$

EXAMPLES 8–10

Perform the indicated operations. Express answers in simplest form.

8. $\dfrac{6x^2y^3}{4wz^2} \div \dfrac{2w^3z^3}{10x^4y^4}$

9. $\dfrac{25x^2 - 16y^2}{x^2 + 7x + 10} \div \dfrac{x^2 - 4}{x^2 + 3x - 10}$

10. $\dfrac{2xy - 6y^2}{3x^3 + 6x^2y} \div \dfrac{4xy - 12y^2}{x + 2y}$

Solutions

8. $\dfrac{6x^2y^3}{4wz^2} \div \dfrac{2w^3z^3}{10x^4y^4}$

$= \dfrac{6x^2y^3}{4wz^2} \cdot \dfrac{10x^4y^4}{2w^3z^3}$ We change the division problem to a multiplication problem, according to our definition above.

$= \dfrac{(6x^2y^3)(10x^4y^4)}{(4wz^2)(2w^3z^3)} = \dfrac{60x^6y^7}{8w^4z^5}$ Properties of exponents and definition of multiplication.

$= \dfrac{\cancel{4} \cdot 15x^6y^7}{\cancel{4} \cdot 2w^4z^5} = \dfrac{15x^6y^7}{2w^4z^5}$ Answer in simplest form.

9. $\dfrac{25x^2 - 16y^2}{x^2 + 7x + 10} \div \dfrac{x^2 - 4}{x^2 + 3x - 10}$

$= \dfrac{25x^2 - 16y^2}{x^2 + 7x + 10} \cdot \dfrac{x^2 + 3x - 10}{x^2 - 4}$ Change to a multiplication problem.

$= \dfrac{(25x^2 - 16y^2)(x^2 + 3x - 10)}{(x^2 + 7x + 10)(x^2 - 4)}$

$= \dfrac{(5x - 4y)(5x + 4y)\cancel{(x + 5)}\cancel{(x - 2)}}{\cancel{(x + 5)}(x + 2)\cancel{(x - 2)}(x + 2)}$ Factor completely and remove common factors.

$= \dfrac{(5x - 4y)(5x + 4y)}{(x + 2)(x + 2)}$ Answer in simplest form. No need to multiply out.

10. $\dfrac{2xy - 6y^2}{3x^3 + 6x^2y} \div \dfrac{4xy - 12y^2}{x + 2y}$

$= \dfrac{2xy - 6y^2}{3x^3 + 6x^2y} \cdot \dfrac{x + 2y}{4xy - 12y^2}$ The equivalent multiplication problem.

$= \dfrac{(2xy - 6y^2)(x + 2y)}{(3x^3 + 6x^2y)(4xy - 12y^2)}$

$= \dfrac{\cancel{2y}\,\cancel{(x - 3y)}\cancel{(x + 2y)}}{3x^2\cancel{(x + 2y)}\underset{2}{\cancel{(4y)}}\cancel{(x - 3y)}}$ Look for common factors.

$= \dfrac{1}{6x^2}$ Answer in simplest possible form. ■

Now let's take a look at Example 10 again. How can division of algebraic fractions help us? Suppose that these fractions came from an application involving a real-world problem, and suppose that $x = 1.5$ and $y = 2$. We would have a choice of *either* putting the values of x and y in the first line of Example 10 and obtaining

$$\frac{2(1.5)(2) - 6(2)^2}{3(1.5)^3 + 6(1.5)^2(2)} \div \frac{4(1.5)(2) - 12(2)^2}{1.5 + 2(2)}$$

(which involves much computation), *or* we could do the algebra first (as in Example 10) and then substitute in the values of x and y:

$$\frac{1}{6x^2} = \frac{1}{6(1.5)^2} = \frac{1}{6(2.25)} = \frac{1}{13.5} = 0.074$$ To three-decimal accuracy.

The arithmetic is much easier in the second case, and this is another reason for finding equivalent algebraic expressions—to make computation easier.

Occasionally, operations on fractions appear in tandem, that is, several operations appear in the same problem. Consider this example.

EXAMPLE 11

Perform the indicated operations. Express the answer in simplest form.

$$\left(\frac{y^2}{2x + 4} \div \frac{y}{x + 2}\right) \cdot \frac{4x^2}{12y^3}$$ *Two* operations in same problem.

According to our order of operations, we work first inside parentheses. So let us do the division problem first, and then multiply that quotient by the fraction outside the parentheses.

Solution

$$\left(\frac{y^2}{2x + 4} \div \frac{y}{x + 2}\right) \cdot \frac{4x^2}{12y^3} = \left(\frac{y^2}{2x + 4} \cdot \frac{x + 2}{y}\right) \cdot \frac{4x^2}{12y^3}$$ Definition of division of fractions.

$$= \left(\frac{y^2(x + 2)}{(2x + 4)y}\right) \cdot \frac{4x^2}{12y^3} = \left(\frac{\overset{y}{\cancel{y^2}}\cancel{(x + 2)}}{2\cancel{(x + 2)}\cancel{y}}\right) \cdot \frac{4x^2}{12y^3}$$ Fundamental property of fractions (remove common factors).

$$= \frac{y}{2} \cdot \frac{4x^2}{12y^3}$$ Now do the multiplication.

$$= \frac{\cancel{y} \cdot \cancel{4}x^2}{2 \cdot \underset{3y^2}{\cancel{12y^3}}} = \frac{x^2}{6y^2}$$ Answer in simplest form. ■

Complex Fractions

Division of fractions is sometimes written in the form of a **complex fraction.** In a complex fraction, both the numerator and the denominator can themselves be fractions. Here are some examples.

EXAMPLES 12–13

Perform the indicated operations. Express the answer in simplest form.

12. $\dfrac{\frac{1}{x}}{\frac{1}{x^2}}$

13. $\dfrac{\frac{5x}{x^2-1}}{\frac{x}{1-x}}$

Solutions

12. $\dfrac{\frac{1}{x}}{\frac{1}{x^2}} = \dfrac{1}{x} \div \dfrac{1}{x^2}$

The complex fraction indicates division. We have merely rewritten the problem, since $\dfrac{A}{B}$ also means $A \div B$.

$= \dfrac{1}{x} \cdot \dfrac{x^2}{1}$

The equivalent multiplication problem.

$= \dfrac{x^2}{x} = \dfrac{x}{1} = x$

Answer in simplest form.

NOTE: We could have multiplied the numerator and denominator by x^2:

$$\dfrac{\frac{1}{x}}{\frac{1}{x^2}} \cdot \dfrac{x^2}{x^2} = x.$$

13. $\dfrac{\frac{5x}{x^2-1}}{\frac{x}{1-x}} = \dfrac{5x}{x^2-1} \div \dfrac{x}{1-x}$

Rewritten.

$= \dfrac{5x}{x^2-1} \cdot \dfrac{1-x}{x}$

Changed to multiplication.

$= \dfrac{5x(1-x)}{(x-1)(x+1) \cdot x}$

Factor.

$= \dfrac{5\cancel{x}\overset{(-1)}{\cancel{(1-x)}}}{\cancel{(x-1)}(x+1)\cancel{x}} = \dfrac{-5}{x+1}$

Recall $(1-x) = (-1)(x-1)$.
Answer in simplest form.

EXAMPLE 14

Which of the following expressions are equivalent?

(a) $\dfrac{x^2-y^2}{8a^3} \div \dfrac{(x-y)^2}{16a^4}$

(b) $\dfrac{2a(x+y)}{x-y}$

(c) $\dfrac{(x-y)(x+y)}{8a^3} \div \dfrac{(x-y)^2}{16a^4}$

(d) $\dfrac{16a^4(x-y)(x+y)}{8a^3(x-y)^2}$

(e) $\dfrac{(x^2-y^2)(x-y)^2}{8a^3 16a^4}$

Solution

Start with one of the expressions and try to transform it into one of the other expressions. Let's start with the expression in (a).

(a) $$\frac{x^2 - y^2}{8a^3} \div \frac{(x-y)^2}{16a^4} = \frac{x^2 - y^2}{8a^3} \cdot \frac{16a^4}{(x-y)^2}$$ Definition of division.

$$= \frac{16a^4(x-y)(x+y)}{8a^3(x-y)^2}$$ Factor. This is part (d).

$$= \frac{2a(x+y)}{x-y}$$ Equivalent fraction. This is part (b).

Note we could first factor the expression in part (a):

$$\frac{x^2 - y^2}{8a^3} \div \frac{(x-y)^2}{16a^4} = \frac{(x-y)(x+y)}{8a^3} \div \frac{(x-y)^2}{16a^4}$$ This is part (c).

The expression in part (e) is not equivalent to the expression in part (a) because

$$\frac{(x^2 - y^2)(x-y)^2}{8a^3 16a^4} = \frac{x^2 - y^2}{8a^3} \cdot \frac{(x-y)^2}{16a^4} \neq \frac{x^2 - y^2}{8a^3} \div \frac{(x-y)^2}{16a^4}$$

Therefore, the expressions in (a), (b), (c), and (d) are all equivalent to one another.

NOTE: We can substitute the same values for a, x, and y in each expression. The equivalent expressions yield the same values. For example, if $a = 1$ and $x = 2$ and $y = 0$, then expression (a) yields

$$\frac{4-0}{8} \div \frac{4}{16} = 2$$

Expressions (b), (c), and (d) also yield 2, while expression (e) yields $\frac{1}{4}$.

5.2 EXERCISES

I. Procedures

1–40. Perform the indicated operations. Express the answer in simplest form.

1. $\frac{5}{7} \cdot \frac{8}{15}$
2. $\frac{5}{8} \div \frac{25}{24}$
3. $\frac{x}{y} \cdot \frac{3}{x}$
4. $\frac{xy}{3} \div \frac{x^2y^2}{12}$
5. $\frac{15x^3y^2}{4x^9y^2} \cdot \frac{8x^2y^8}{10xy^6}$
6. $\frac{x^3y}{18z^2} \div \frac{7xy^3}{42yz^3}$
7. $\frac{x+1}{3x} \cdot \frac{x-3}{5x}$
8. $\frac{x}{x-1} \div \frac{x+1}{x}$
9. $\frac{3-x}{x} \cdot \frac{x^3}{x^2-9}$
10. $\frac{4x^2 - y^2}{x^2 - 2x + 1} \div \frac{2y - x}{1 - x}$
11. $\frac{x^2 + xy}{3xy} \cdot \frac{6y}{x^2 - y^2}$
12. $\frac{x^2 - 7x + 10}{x - 5} \div \frac{1}{x - 2}$
13. $\frac{x^2 - 5x + 6}{x - 3} \div (x - 2)$
14. $\frac{2x^2 - 13x - 28}{2x^2 - 7} \div (16 - 9x^2)$

15. $\dfrac{y^2 + y}{y^2 + 2y + 1} \div \dfrac{y^2 - 1}{y^2 - y - 2}$

16. **(a)** $\dfrac{\frac{1}{x}}{\frac{5a}{x}}$ **(b)** $\dfrac{\frac{ab}{b}}{\frac{a^2b}{ba}}$ **(c)** $\dfrac{\frac{x^2 - y^2}{x + 2}}{\frac{y - x}{x^2 + 4x + 4}}$

17. $\dfrac{x^2 - y^2}{a^2 - b^2} \cdot \dfrac{bx - ax}{y - x}$

18. $\dfrac{a^2 - 16}{a^2 + a - 12} \cdot \dfrac{a^2 + 5a + 6}{8 + 2a - a^2}$

19. $\dfrac{6x^3 + x^2 - x}{6x^2 + 5x + 1} \div \dfrac{3x^2 + 2x - 1}{3x^2 + 4x + 1}$

20. $\dfrac{a}{x^2 - 5x + 6} \div \dfrac{ax - 2a}{x^2 - 9}$

21. $\dfrac{x^2 - 4}{x^2 - 4x + 4} \cdot \dfrac{x^2 - 5x + 6}{2x - 6}$

22. $\dfrac{a^2 - 7a + 6}{a - 1} \div \dfrac{a^2 - 4a - 12}{a^2 - 4}$

23. $\dfrac{x^2 - 5x - 6}{3x} \cdot \dfrac{-9x^2}{6 - x}$

24. $\dfrac{x^2y^3}{3x^2y^6} \cdot \dfrac{6xy^5}{5x^3y^2} \cdot \dfrac{xy}{3x^2y}$

25. $\dfrac{4m^2}{m - 1} \cdot \dfrac{m^2 - 2m + 1}{m + 1} \cdot \dfrac{m^2 - 1}{m^3 - m^2}$

26. $\left(\dfrac{x^2 - 5x - 6}{4x} \cdot \dfrac{-8x^2}{2x - 4}\right) \div \dfrac{2x^3 - 18x}{2x^2}$

27. $\left(\dfrac{m^2 - m - 12}{18m^2} \div \dfrac{m - 4}{m^2 + 6m + 9}\right) \cdot \dfrac{9m^3}{m + 3}$

28. $\left(\dfrac{x - 3x^2}{x^3 - x} \cdot \dfrac{x^2 + 2x + 1}{6x + 6}\right) \div \dfrac{x^2 - 9}{3x^3}$

29. $\left(\dfrac{x + 1}{x} \div \dfrac{3x^2}{x - 1}\right) \cdot \dfrac{6x^2 - 6}{2x + 4}$

30. $\dfrac{x - y}{x + y} \cdot \dfrac{3x + 7y}{x - 4y}$

31. $\dfrac{x^2 - 2xy + y^2}{25x^2 - y^2} \div \dfrac{3x^2 - 6xy}{2xy}$

32. $\dfrac{(x + 1)^2}{3(x + 1)} \cdot \dfrac{6x^2}{(x - 1)(x + 1)}$

33. $\dfrac{25(x + 3)(x + 2)^2}{15(x - 7)(x - 3)} \div \dfrac{5(x + 2)(x + 3)^2}{3(x - 7)^3(x - 3)}$

34. $\dfrac{(x + 1)^3(x - 1)}{x(x - 3)} \cdot \dfrac{(x + 3)(x - 3)^2}{(x - 1)^2(x + 1)^2}$

35. $\dfrac{2x^2(x - 1)(2x + 7)^2}{4x(x + 1)} \div \dfrac{6x(5x - 3)(2x + 7)}{(x + 1)(x - 1)}$

36. $\dfrac{(x - y)(x + y)^3}{(3x - 2y)^2} \cdot \dfrac{(3x - 2y)(3x + 2y)}{(x + y)^2(x + y)}$

37. $\dfrac{\frac{x}{2x + 1}}{\frac{x}{4x^2 - 1}}$

38. $\dfrac{\frac{1}{x + y}}{\frac{1}{x + y}}$

39. $\dfrac{\frac{x^2 - y^2}{3x^2}}{\frac{(x + y)^2}{6x}}$

40. $\dfrac{\frac{3y^2 + 2y - 1}{3y^2 + 4y + 1}}{\frac{6y^3 + y^2 - y}{6y^2 + 5y + 1}}$

II. Concepts

41 and 42. In each problem below, a student has made an error in the solution. (The answers are wrong.) Find the error and give the correct solution. How would you help the student understand the error?

41. $\dfrac{x^2 - 4}{x^2 + 4} \cdot \dfrac{x^2 + 5x + 6}{2x - 4}$

$= \dfrac{\cancel{(x - 2)}\cancel{(x + 2)}\cancel{(x + 2)}(x + 3)}{\cancel{(x + 2)}\cancel{(x + 2)}\cancel{(x - 2)}(2)}$

$= \dfrac{x + 3}{2}$ Incorrect answer.

42. $\dfrac{9x^2 - 1}{x + 3} \div \dfrac{x^2 - 9}{9x - 3}$

$= \dfrac{\cancel{(3x - 1)}(3x + 1)(x - 3)\cancel{(x + 3)}}{\cancel{(x + 3)}(3)\cancel{(3x - 1)}}$

$= \dfrac{(3x + 1)(x - 3)}{3}$ Incorrect answer.

43. Explain why the equation

$$\frac{5}{(x - 2)(x + 3)} = 0$$

has no solution.

44–47. What rational expression must be multiplied by each expression to get the indicated product? Explain your solutions.

44. $\dfrac{3}{4} \cdot ? = 1$

45. $\dfrac{3}{4} \cdot ? = \dfrac{2}{3}$

46. $\dfrac{x^2 - 1}{x} \cdot ? = x + 1$

47. $4x^2 - 25 \cdot ? = 14(2x - 5)$

48–51. What rational expression must each expression be divided by to get the indicated quotient? Explain your solutions.

48. $\dfrac{2}{3} \div ? = 1$

49. $\dfrac{2}{3} \div ? = \dfrac{3}{5}$

50. $\dfrac{x^2 - 5x + 6}{x} \div ? = 1$

51. $\dfrac{x^2 - 5x + 6}{x} \div ? = x - 3$

52–54. Express each fraction as the product of two fractions. Will your answers be the same as another student's? Why or why not?

52. $\dfrac{1}{2}$

53. $\dfrac{3x}{2}$

54. $x - 1$

55. Which of the following expressions are equivalent to $x + 1$? Explain why they are equivalent.

(a) $(x^2 + x) \div x$

(b) $\dfrac{x^2 + 7x + 6}{x} \cdot \dfrac{x}{x + 6}$

(c) $\dfrac{x^2 - 1}{x} \div \dfrac{x}{x - 1}$

(d) $\dfrac{x^2 + 7x + 6}{x} \div \dfrac{x + 6}{x}$

III. Applications

56. Find an expression for

(a) the product of the reciprocals of two consecutive integers.

(b) the quotient of the reciprocals of two consecutive integers.

57–59. In Example 1 of this section, graph the amount of the field that is painted each successive night. On the same set of axes graph the total amount of the field painted after each night. Use the graphs to answer the following questions:

57. How much of the field was painted on the tenth night?

58. On what night did the total amount painted exceed $\frac{99}{100}$ths of the field?

59. As the number of nights gets very large, describe what happens to each of the two graphs.

IV. Extensions

60. Give the final product.

(a) $\left(1 - \frac{1}{2}\right)\left(1 - \frac{1}{3}\right)\left(1 - \frac{1}{4}\right)$

(b) $\left(1 - \frac{1}{2}\right)\left(1 - \frac{1}{3}\right)\left(1 - \frac{1}{4}\right)\left(1 - \frac{1}{5}\right)\left(1 - \frac{1}{6}\right)\left(1 - \frac{1}{7}\right)\cdots\left(1 - \frac{1}{10}\right)$

(c) $\left(1 - \frac{1}{2}\right)\left(1 - \frac{1}{3}\right)\left(1 - \frac{1}{4}\right)\left(1 - \frac{1}{5}\right)\cdots\left(1 - \frac{1}{n}\right)$

(d) $\left[1 - \left(\frac{1}{2}\right)^2\right]\left[1 - \left(\frac{1}{3}\right)^2\right]\left[1 - \left(\frac{1}{4}\right)^2\right]\left[1 - \left(\frac{1}{5}\right)^2\right]\cdots\left[1 - \left(\frac{1}{n}\right)^2\right]$

(e) Graph the patterns described in parts (c) and (d). Describe what happens to the graph as n gets very large.

61. What is wrong with this argument? Suppose $x = y$ $(x \neq 0, y \neq 0)$. Then

$$xy = y^2$$

So

$$-xy = -y^2$$

So

$$x^2 - xy = x^2 - y^2$$

So

$$x(x - y) = (x + y)(x - y)$$

So

$$x = x + y$$

but since $y = x$

$$x = x + x = 2x$$

So

$$x = 2x$$

So

$$1 = 2$$

We have just proved that $1 = 2$. Which step above is false? Why?

5.3 ADDITION AND SUBTRACTION OF RATIONAL EXPRESSIONS

In this section we continue our discussion of operations on fractions. These computational techniques can help us to write equivalent algebraic expressions and to solve equations.

Addition and Subtraction

EXAMPLE 1

Reexamine Example 1 in Section 1.3 of Chapter 1. The Gold Diggers paint half of the football field of the opposing team gold. The next night they paint half of the remaining unpainted field gold. Each night they continue to paint half of what's left. After seven nights, how much of the field has been painted?

Solution

Let's look at the field after each of the first three nights.

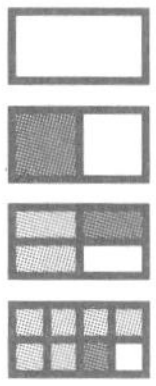

Night	Amount Painted Each Night		Total Amount Painted
start		0	0
1		$\frac{1}{2}$	$\frac{1}{2}$
2		$\frac{1}{4}$	$\frac{3}{4} = \frac{1}{2} + \frac{1}{4}$
3		$\frac{1}{8}$	$\frac{7}{8} = \frac{1}{2} + \frac{1}{4} + \frac{1}{8}$

Figure 5.8

To find the total amount of the field painted after each night, we subdivide the field each night into equivalent units. Thus, on the second night, we have

$$\frac{3}{4} = \frac{1}{2} + \frac{1}{4} = \frac{2}{4} + \frac{1}{4} = \frac{3}{4}$$

and on the third night, we have

$$\frac{7}{8} = \frac{1}{2} + \frac{1}{4} + \frac{1}{8} = \frac{4}{8} + \frac{2}{8} + \frac{1}{8} = \frac{7}{8}$$

This example illustrates the process of adding fractions. You will recall that in order to add or subtract numerical fractions, the denominators must be identical. This process involves finding a common denominator. The **least common denominator,** abbreviated **LCD,** will usually make the computations easier. Let us review.

EXAMPLES 2–3

Perform the indicated operations.

2. $\frac{3}{8} + \frac{7}{8}$

3. $\frac{7}{12} - \frac{9}{20}$

Solutions

2. Consider $\frac{3}{8} + \frac{7}{8}$. The denominators *are* the same. Therefore we can simply add the numerators:

$$\frac{3}{8} + \frac{7}{8} = \frac{3+7}{8} = \frac{10}{8} = \frac{5}{4} \qquad \text{Answer in simplest form.}$$

3. Consider $\frac{7}{12} - \frac{9}{20}$. In this case, the denominators *are not* the same. We want to rename each fraction so that the denominators are the same. How do we find the LCD (least common denominator) for 12 and 20? Both 12 and 20 must divide this LCD evenly. Thus *every factor* of 12 and 20 must also divide the LCD. Let us factor 12 and 20 to see what is needed for the LCD:

$$12 = 4 \cdot 3 = 2 \cdot 2 \cdot 3$$
$$20 = 4 \cdot 5 = 2 \cdot 2 \cdot 5 \qquad \text{Factored into prime factors.}$$

The LCD will need two factors of 2, a factor of 3, and a factor of 5. Thus the LCD is $2 \cdot 2 \cdot 3 \cdot 5 = 60$. So,

$$\frac{7}{12} - \frac{9}{20} = \frac{35}{60} - \frac{27}{60} = \frac{8}{60} = \frac{2}{15} \qquad \text{Answer in simplest form.}$$

NOTE: The denominator 12 needed a factor of 5 to make the LCD. The denominator 20 needed a factor of 3 to make the LCD. We could also use any common denominator; for example, $12 \times 20 = 240$ could be used as a common denominator. ■

When we create a common denominator using the LCD, it may help to think of the renaming process as *multiplying by* 1 *in a convenient way.* For instance,

$$\frac{7}{12} - \frac{9}{20} = \frac{7}{12} \cdot (1) - \frac{9}{20} \cdot (1) \qquad \text{Multiplying by 1 does not change anything.}$$
$$= \frac{7}{12} \cdot \left(\frac{5}{5}\right) - \frac{9}{20} \cdot \left(\frac{3}{3}\right)$$
$$= \frac{35}{60} - \frac{27}{60} = \frac{8}{60} = \frac{2}{15}$$

We have rewritten 1 in a convenient way by using what is missing from the LCD. This involves the fundamental property of fractions:

$$\frac{7}{12} = \frac{7}{12} \cdot \frac{5}{5} = \frac{7 \cdot 5}{12 \cdot 5} = \frac{35}{60}$$
$$\frac{9}{20} = \frac{9}{20} \cdot \frac{3}{3} = \frac{9 \cdot 3}{20 \cdot 3} = \frac{27}{60}$$

That is, we multiply each fraction by 1 in the form of

$$1 = \frac{\text{factors in LCD missing from denominator}}{\text{factors in LCD missing from denominator}}$$

NOTE: Writing fractions with a common denominator or LCD involves the reverse of the fundamental property of fractions:

Equivalent Fractions

$$\frac{a}{b} = \frac{a}{b} \cdot 1 = \frac{a}{b} \cdot \frac{c}{c} = \frac{ac}{bc} \quad \text{if } b, c \neq 0.$$

We add (or subtract) algebraic fractions as we did in the arithmetical model reviewed in Example 1.

Let's try some examples.

EXAMPLES 4–5

Perform the indicated operations.

4. $\frac{5y}{3x} + \frac{7z}{3x}$

5. $\frac{3}{x^2} - \frac{x}{2y}$

Solutions

4. $\frac{5y}{3x} + \frac{7z}{3x} = \frac{5y + 7z}{3x}$ The denominators are the same, so add the numerators. Answer in simplest form.

5. $\frac{3}{x^2} - \frac{x}{2y}$ First find a common denominator (or LCD):

$$x^2 = x \cdot x$$
$$2y = 2 \cdot y$$
$$\text{LCD} = 2x^2y$$

$$\frac{3}{x^2} - \frac{x}{2y} = \frac{3}{x^2} \cdot \left(\frac{2y}{2y}\right) - \frac{x}{2y} \cdot \left(\frac{x^2}{x^2}\right)$$ Multiply by 1 in a "convenient" way to rename the fractions.

$$= \frac{6y}{2x^2y} - \frac{x^3}{2x^2y}$$ Now the fractions have the same denominator.

$$= \frac{6y - x^3}{2x^2y}$$ Subtract the numerators to get the answer.

NOTE: The steps we followed are exactly the same steps we used when adding numerical fractions. Let us summarize them.

Strategies to Add (Subtract) Fractions

Case 1: If the fractions already have the same denominator, add (subtract) the numerators and put the sum (difference) over the denominator. If needed, express the answer in simplest form:

$$\frac{a}{b} + \frac{c}{b} = \frac{a + c}{b}, \, b \neq 0$$

Case 2: If the fractions do not have the same denominator, then:

a. Find a common denominator, preferably the LCD, by

  i. factoring each denominator completely and

  ii. forming a common denominator or the LCD.

b. Rename each fraction with a common denominator by multiplying by 1 "in a convenient way."

c. The denominators are now the same, so proceed as in Case 1.

$$\frac{a}{b} + \frac{c}{d} = \frac{a}{b} \cdot \left(\frac{d}{d}\right) + \frac{c}{d} \cdot \left(\frac{b}{b}\right) = \frac{ad}{bd} + \frac{cb}{bd} = \frac{ad + cb}{bd}, \, b \neq 0, d \neq 0$$

Since bd is a common multiple of b and d, it can be used as a common denominator. ■

This algorithm (above) will enable us to add (subtract) any algebraic fractions.

Comment: It might be helpful to review the sections on factors and terms in Chapter 2 and on adding and subtracting fractions in Chapter 1. While any common denominator will work when adding or subtracting fractions, the LCD is usually easier to work with for algebraic fractions; it generally results in less symbol manipulation.

EXAMPLES 6–8

Perform the indicated operations. Express the answer in simplest form.

6. $\dfrac{5x}{y^2z} + \dfrac{2y}{xz^2}$ These denominators are just monomials.

7. $\dfrac{3}{x + 7} - \dfrac{5}{x}$ Some of these denominators have more than one term.

8. $\dfrac{a}{a - b} + \dfrac{b}{a + b}$

Solutions

6. $\dfrac{5x}{y^2z} + \dfrac{2y}{xz^2}$

Find a common denominator: $y^2z = y \cdot y \cdot z$

$xz^2 = x \cdot z \cdot z$

$\text{LCD} = xy^2z^2$

$$= \frac{5x}{y^2z} \cdot \left(\frac{xz}{xz}\right) + \frac{2y}{xz^2} \cdot \left(\frac{y^2}{y^2}\right)$$ **Multiply by 1 to rename the fractions.**

$$= \frac{5x^2z}{xy^2z^2} + \frac{2y^3}{xy^2z^2}$$ **Same denominators.**

$$= \frac{5x^2z + 2y^3}{xy^2z^2}$$ **Answer. (Add numerators to get the answer.)**

7. $\dfrac{3}{x+7} - \dfrac{5}{x}$ **Find a common denominator: The only factor of $x + 7$ is $x + 7$. The only factor of x is x. Therefore the LCD is $x(x + 7)$. (Careful, x is not a factor of $x + 7$.)**

$$= \frac{3}{x+7} \cdot \left(\frac{x}{x}\right) - \frac{5}{x} \cdot \left(\frac{x+7}{x+7}\right)$$ **We multiply by 1 in a convenient way.**

$$= \frac{3x}{x(x+7)} - \frac{5(x+7)}{x(x+7)}$$ **The denominators are the same.**

$$= \frac{3x - 5(x+7)}{x(x+7)}$$ **Subtract the numerators.**

$$= \frac{3x - 5x - 35}{x(x+7)}$$ **Combine like terms.**

$$= \frac{-2x - 35}{x(x+7)}$$ **Answer in simplest form.**

8. $\dfrac{a}{a-b} + \dfrac{b}{a+b}$ LCD $= (a - b) \times (a + b)$

$$= \frac{a}{a-b} \cdot \left(\frac{a+b}{a+b}\right) + \frac{b}{a+b} \cdot \left(\frac{a-b}{a-b}\right)$$ **Multiply by 1 to rename.**

$$= \frac{a(a+b)}{(a-b)(a+b)} + \frac{b(a-b)}{(a+b)(a-b)}$$ **The fractions have the same denominators.**

$$= \frac{a(a+b) + b(a-b)}{(a-b)(a+b)}$$ **Add the numerators.**

$$= \frac{a^2 + ab + ab - b^2}{(a-b)(a+b)}$$ **Combine like terms in the numerator.**

$$= \frac{a^2 + 2ab - b^2}{(a-b)(a+b)} \quad \text{or} \quad \frac{a^2 + 2ab - b^2}{a^2 - b^2}$$ **Answer in simplest form.** ■

EXAMPLES 9–10

Perform the operations. Express the answer in simplest form.

9. $\dfrac{a^2 + ab}{a^2 - b^2} + \dfrac{ab + b^2}{a^2 - b^2}$

10. $\dfrac{x}{x^2 + 5x + 6} - \dfrac{2}{x^2 - 4}$ **The denominators are polynomials that can be factored into products of binomials.**

Solutions

9. $\dfrac{a^2 + ab}{a^2 - b^2} + \dfrac{ab + b^2}{a^2 - b^2}$ Denominators are the same.

$= \dfrac{a^2 + 2ab + b^2}{a^2 - b^2}$ Add the numerators.

$= \dfrac{(a + b)(a + b)}{(a + b)(a - b)}$ This fraction can be reduced to simpler terms.

$= \dfrac{a + b}{a - b}$ Answer in simplest form.

10. $\dfrac{x}{x^2 + 5x + 6} - \dfrac{2}{x^2 - 4}$

Find the LCD. Factor the denominators:
$x^2 + 5x + 6 = (x + 2)(x + 3)$
$x^2 - 4 = (x - 2)(x + 2)$
$\text{LCD} = (x + 2)(x + 3)(x - 2)$

$= \dfrac{x}{(x + 2)(x + 3)} \cdot \left(\dfrac{x - 2}{x - 2}\right) - \dfrac{2}{(x - 2)(x + 2)} \cdot \left(\dfrac{x + 3}{x + 3}\right)$ We multiply each fraction by 1 in the form: [(missing factor)/(missing factor)]

$= \dfrac{x(x - 2)}{(x + 2)(x + 3)(x - 2)} - \dfrac{2(x + 3)}{(x - 2)(x + 2)(x + 3)}$ Same denominators.

$= \dfrac{x(x - 2) - 2(x + 3)}{(x + 2)(x + 3)(x - 2)}$ Subtract numerators.

$= \dfrac{x^2 - 2x - 2x - 6}{(x + 2)(x + 3)(x - 2)}$ Perform operations in the numerator.

$= \dfrac{x^2 - 4x - 6}{(x + 2)(x + 3)(x - 2)}$ Answer. Note that the numerator cannot be factored any further. ■

Note: We will leave denominators in factored form, such as $(x + 2)(x + 3)(x - 2)$, rather than multiplying them (too messy). Also, if the denominators are left factored, you may see a way to express the answer in simplest form, as in Example 9.

EXAMPLES 11–12 Perform the indicated operations. Express the resulting fraction in simplest form.

11. $x + 1 + \dfrac{2x^2 - x}{x - 3}$

12. $\dfrac{3x - 2}{(x - 2)(x + 1)} - \dfrac{x - 1}{x^2 - 1}$

Solutions

11. $x + 1 + \dfrac{2x^2 - x}{x - 3}$ The LCD is $x - 3$.

$= \dfrac{x + 1}{1} + \dfrac{2x^2 - x}{x - 3}$ Rewrite $x + 1$ in fractional form: $x + 1 = \dfrac{x + 1}{1}$

$$= \frac{x+1}{1} \cdot \left(\frac{x-3}{x-3}\right) + \frac{2x^2 - x}{x-3}$$ Multiply by 1.

$$= \frac{(x+1)(x-3)}{x-3} + \frac{2x^2 - x}{x-3}$$ The denominators are the same.

$$= \frac{(x+1)(x-3) + 2x^2 - x}{x-3}$$ Add the numerators.

$$= \frac{x^2 - 2x - 3 + 2x^2 - x}{x-3}$$ Combine like terms in the numerator.

$$= \frac{3x^2 - 3x - 3}{x-3} = \frac{3(x^2 - x - 1)}{x-3}$$ Answer in simplest form.

12. $$\frac{3x-2}{(x-2)(x+1)} - \frac{x-1}{(x-1)(x+1)}$$ Factor the denominators to find the LCD. The LCD is $(x-2)(x+1)(x-1)$.

$$= \frac{3x-2}{(x-2)(x+1)} \cdot \left(\frac{x-1}{x-1}\right) - \frac{x-1}{(x-1)(x+1)} \cdot \left(\frac{x-2}{x-2}\right)$$

Multiply by 1 in the form of (missing factor)/(missing factor).

$$= \frac{(3x-2)(x-1)}{(x-2)(x+1)(x-1)} - \frac{(x-1)(x-2)}{(x-1)(x+1)(x-2)}$$ The denominators are now the same.

$$= \frac{(3x-2)(x-1) - (x-1)(x-2)}{(x-2)(x+1)(x-1)}$$ Subtract the numerators.

Careful!

$$= \frac{(3x^2 - 5x + 2) - (x^2 - 3x + 2)}{(x-2)(x+1)(x-1)}$$ Combine like terms in the numerator.

Change *all* signs (distributive property).

$$= \frac{3x^2 - 5x + 2 - x^2 + 3x - 2}{(x-2)(x+1)(x-1)}$$ Now remove parentheses and subtract.

$$= \frac{2x^2 - 2x}{(x-2)(x+1)(x-1)}$$ Combine like terms.

$$= \frac{2x\cancel{(x-1)}}{(x-2)(x+1)\cancel{(x-1)}}$$ $x - 1$ is a common *factor* in numerator and denominator.

$$= \frac{2x}{(x-2)(x+1)}$$ Answer in simplest form. ■

Combining Operations on Algebraic Functions

The operations of addition, subtraction, multiplication, and division can occur in combinations. Consider these examples.

EXAMPLES 13–15

13. $\left(\dfrac{3}{x-1} - \dfrac{x}{x+1}\right) \div \dfrac{x}{x-1}$

In Example 13, subtraction is done first (in parentheses), then division.

14. $\dfrac{1 - \dfrac{1}{a}}{1 - a^2}$

Examples 14 and 15 involve complex fractions. There are operations on fractions in the numerator or denominator or both.

15. $\dfrac{\dfrac{1}{x} - \dfrac{1}{y}}{\dfrac{2}{y^2} - \dfrac{2}{x^2}}$

Solutions

13. $\left(\dfrac{3}{x-1} - \dfrac{x}{x+1}\right) \div \dfrac{x}{x-1}$ Do the part inside the parentheses first.

$= \left[\dfrac{3}{x-1} \cdot \left(\dfrac{x+1}{x+1}\right) - \dfrac{x}{x+1}\left(\dfrac{x-1}{x-1}\right)\right] \div \dfrac{x}{x-1}$ The LCD for subtraction problem in parentheses is $(x+1)(x-1)$.

$= \left[\dfrac{3(x+1) - x(x-1)}{(x-1)(x+1)}\right] \div \dfrac{x}{x-1}$ Denominator is the same, so subtract numerators in the brackets.

$= \left[\dfrac{3x + 3 - x^2 + x}{(x-1)(x+1)}\right] \div \dfrac{x}{x-1}$ Distributive property is used in the numerator.

$= \dfrac{-x^2 + 4x + 3}{(x-1)(x+1)} \div \dfrac{x}{x-1}$ Subtraction problem is done; now do the division.

$= \dfrac{-x^2 + 4x + 3}{(x-1)(x+1)} \cdot \dfrac{x-1}{x}$ Invert and multiply.

$= \dfrac{(-x^2 + 4x + 3)(x-1)}{(x-1)(x+1)(x)}$

$= \dfrac{-x^2 + 4x + 3}{x(x+1)}$ Answer in simplest form.

14. $\dfrac{1 - \dfrac{1}{a}}{1 - a^2}$ Perform any operations in numerator and denominator first.

$= \dfrac{1 \cdot \left(\dfrac{a}{a}\right) - \dfrac{1}{a}}{1 - a^2}$ LCD for subtraction problem in the numerator is a.

$= \dfrac{\dfrac{a-1}{a}}{1 - a^2} = \dfrac{a-1}{a} \div \dfrac{1 - a^2}{1}$ We can rewrite the complex fraction as division: $1 - a^2 = \dfrac{(1 - a^2)}{1}$.

$$= \frac{a-1}{a} \cdot \frac{1}{1-a^2}$$ **Definition of division.**

$$= \frac{(a-1)}{a(1-a)(1+a)}$$ **Multiply and factor.**

$$= \frac{\overset{(-1)}{(a-1)}}{a(1-a)(1+a)} = \frac{-1}{a(1+a)}$$ **Answer in simplest form.**

15. Here is an *alternate* solution method for a complex fractions problem.

$$\frac{\frac{1}{x} - \frac{1}{y}}{\frac{2}{y^2} - \frac{2}{x^2}}$$

The LCD of the numerator is xy. The LCD of the denominator is x^2y^2. Note that x^2y^2 is the LCD of xy and x^2y^2 or the LCD of the numerator and denominator together.

$$= \frac{\left(\frac{1}{x} - \frac{1}{y}\right)}{\left(\frac{2}{y^2} - \frac{2}{x^2}\right)} \cdot \frac{\frac{x^2y^2}{1}}{\frac{x^2y^2}{1}} = \frac{xy^2 - x^2y}{2x^2 - 2y^2}$$

Multiply numerator and denominator by the common LCD. Use the distributive property.

$$= \frac{xy\overset{(-1)}{(y-x)}}{2(x-y)(x+y)}$$ **Factor numerator and denominator.**

$$= \frac{-xy}{2(x+y)}$$ **Answer in simplest form.** ■

5.3 EXERCISES

I. Procedures

1–56. Perform the indicated operations. Express the answers in simplest form.

1. $\frac{5}{7} + \frac{3}{8}$
2. $\frac{2}{9} + \frac{6}{13}$
3. $\frac{2}{3} - \frac{1}{7}$
4. $\frac{2}{9} - \frac{7}{15}$
5. $\frac{3}{x} + \frac{2}{y}$
6. $\frac{x}{y} - \frac{y}{x}$
7. $\frac{5}{y^2} - \frac{3}{4y}$
8. $\frac{x}{y^2x} - \frac{z}{x^2y}$
9. $\frac{3}{xy} - \frac{y}{yz} + \frac{y}{xz}$
10. $\frac{x}{x^2y} - \frac{x}{y^2z} + \frac{y}{x^2z^2}$
11. $\frac{3a}{a-b} + \frac{1}{b}$
12. $\frac{ab}{c} - \frac{b}{c-a}$
13. $\frac{3x}{x-y} + \frac{2y}{x+y}$
14. $\frac{4x^2}{x-1} - \frac{2}{x-2}$
15. $\frac{5}{x^2-x-12} - \frac{x}{x-4}$
16. $\frac{x}{x^2-5x+6} + \frac{3}{x-2}$
17. $\frac{8x}{x^2-4} + \frac{x}{x+2}$
18. $\frac{2}{x-1} - \frac{3}{x^2-1}$
19. $\frac{x-1}{x+1} - \frac{x+1}{x-1}$
20. $\frac{2x+3}{x} + \frac{5x-7}{2x+3}$
21. $\frac{8a-3b}{2a-b} - \frac{2a-b}{2b-a}$
22. $\frac{x-y}{2x+y} - \frac{2x-y}{x+y}$
23. $\frac{3}{x-1} - \frac{x}{x+1}$
24. $\frac{5}{2x} - \frac{x+1}{3x}$
25. $\frac{2}{x^2-4} + \frac{3}{x+2} - \frac{1}{2x-4}$
26. $\frac{1}{1-x} - \frac{x}{1+x} - \frac{1+x^2}{1-x^2}$
27. $\frac{x}{x^2-9} + \frac{3}{x^2+6x+9} + \frac{3}{6-2x}$

28. $\dfrac{2}{a^2 - 4} + \dfrac{3}{a^2 - 2a}$

29. $\dfrac{a + 2}{a^2 - 4} + \dfrac{3a}{a^2 - 2a}$

30. $\dfrac{x^3 - x}{3x} - \dfrac{3x - 2}{9x^2 - 4}$

31. $1 + \dfrac{1}{x + 1}$

32. $4 - \dfrac{3}{a^2 - 2a}$

33. $\dfrac{8}{a^2 - 2a} + 1$

34. $\dfrac{1}{a^2 - 4} - 1$

35. $\dfrac{x - 2}{x^2 + 10x + 16} + \dfrac{x + 1}{x^2 + 9x + 14}$

36. $\dfrac{x - 1}{x^2 - 5x + 6} - \dfrac{x}{x^2 - 9}$

37. $a + \dfrac{2}{a^2 - 1} + \dfrac{4}{3a + 3}$

38. $\dfrac{2x - 1}{x^2 - 2x - 15} - \dfrac{3}{x - 5} + 2x$

39. $\dfrac{x + 3}{x^2 - x} - \dfrac{12}{x^2 + x - 2}$

40. $\dfrac{4}{x^2 - 5x + 6} - \dfrac{x + 1}{x^2 - 9}$

41. $\dfrac{1}{1 - x} + \dfrac{x}{x^2 - 1}$

42. $\dfrac{a}{a^2 - b^2} - \dfrac{b}{b - a}$

43. $\dfrac{x - 2}{x + 2} - \dfrac{3}{4 - x^2} + 6x$

44. $\dfrac{6x}{x^2 - x^4} - \dfrac{3x^2}{x - 1}$

45. $\left(\dfrac{x + 2}{x} + \dfrac{1}{3}\right) \div \dfrac{2}{3x^2}$

46. $\left(\dfrac{1}{a + b} - \dfrac{1}{a - b}\right) \div \dfrac{4b}{a^2 + 2ab + b^2}$

47. $\left(\dfrac{x^2 + y^2}{y} - 2x\right) \div \left(\dfrac{1}{x} - \dfrac{1}{y}\right)$

48. $\left(\dfrac{x - 1}{x + 1} + \dfrac{2 - x}{x + 2}\right) \div \left(\dfrac{x}{x - 1} - \dfrac{x - 1}{x + 2}\right)$

49. $\dfrac{2a^2 - 2b^2}{4a} \cdot \left(\dfrac{1}{a + b} + \dfrac{1}{a - b}\right)$

50. $\dfrac{9 - \dfrac{1}{x^2}}{3 - \dfrac{1}{x}}$

51. $\dfrac{\dfrac{16}{x} - x}{\dfrac{4}{x} - 1}$

52. $\dfrac{1 + \dfrac{1}{x}}{x - \dfrac{1}{x}}$

53. $\dfrac{\dfrac{4}{y^2} - 1}{y - \dfrac{4}{y}}$

54. $\dfrac{2 + \dfrac{3}{x}}{4x - \dfrac{9}{x}}$

55. $\dfrac{\dfrac{1}{x} - \dfrac{1}{y}}{\dfrac{1}{x - y}}$

56. $\dfrac{\dfrac{1}{a^2} - \dfrac{1}{ab}}{\dfrac{1}{a} - \dfrac{1}{b}}$

II. Concepts

57–60. In each of the following problems, the student has made an error in the solution. All answers are incorrect. Find the error, and give the correct solution. How would you help this student?

57. $$\frac{1}{x} - x^2 = \frac{1}{x} - \frac{x^2}{1}$$
$$= \frac{1 - x^2}{x - 1}$$
$$= \frac{\overset{(-1)}{\cancel{(1 - x)}}(1 + x)}{\cancel{(x - 1)}}$$
$$= -(1 + x)$$ **Incorrect answer.**

58. $$\frac{3x}{x - 1} - \frac{2x}{x + 1} = \frac{3x}{x - 1} \cdot \left(\frac{x + 1}{x + 1}\right) - \frac{2x}{x + 1}\left(\frac{x - 1}{x - 1}\right)$$
$$= \frac{3x(x + 1) - 2x(x - 1)}{(x - 1)(x + 1)}$$
$$= \frac{3x^2 + 3x - 2x^2 - 2x}{(x - 1)(x + 1)}$$
$$= \frac{x^2 + x}{(x - 1)(x + 1)}$$
$$= \frac{x\cancel{(x + 1)}}{(x - 1)\cancel{(x + 1)}}$$
$$= \frac{x}{x - 1}$$ **Incorrect answer.**

59. $$\frac{2x}{x - 2} - \frac{x}{x + 2} = \frac{2x}{x - 2}\left(\frac{x + 2}{x + 2}\right) - \frac{x}{x + 2}\left(\frac{x - 2}{x - 2}\right)$$
$$= \frac{2x(x + 2) - x(x - 2)}{(x - 2)(x + 2)}$$
$$= \frac{2x^2 + 4x - x^2 + 2x}{(x - 2)(x + 2)}$$
$$= \frac{\cancel{x^2} + 6x}{\cancel{x^2} - 4} = \frac{6x}{-4}$$
$$= -\frac{3x}{2}$$ **Incorrect answer.**

60. $$\frac{3}{x + 2} - \frac{x}{x^2 + 9x + 14} = \frac{3}{x + 2} - \frac{x}{(x + 2)(x + 7)}$$

$$= \frac{3}{x+2}\left(\frac{x+7}{x+7}\right) - \frac{x}{(x+2)(x+7)}$$

$$= \frac{3(\cancel{x+7}) - x}{(x+2)(\cancel{x+7})}$$

$$= \frac{3-x}{(x+2)} \quad \textbf{Incorrect answer.}$$

61–64. Express each fraction as the sum of two fractions.

61. $\frac{2}{3}$ **62.** $\frac{x+1}{x}$

63. $\frac{x^2}{x+1}$ **64.** $\frac{1}{x(x+1)}$

65–68. Express each fraction as the difference of two fractions.

65. $\frac{1}{3}$ **66.** $\frac{x^2-1}{x}$

67. $\frac{1}{x}$ **68.** $\frac{1}{x(x+1)}$

69–71. **(a)** Write an algebraic statement to describe each of the following patterns. **(b)** Is your statement true for all numbers? Explain your answers: give an argument for why a pattern is always true or give conditions for when it will be true.

69. $\frac{1}{2} - \frac{1}{3} = \frac{1}{6}$

$\frac{1}{4} - \frac{1}{5} = \frac{1}{20}$

$\frac{1}{9} - \frac{1}{10} = \frac{1}{90}$

70. $\frac{4}{1} - 2 = \frac{\frac{4}{1}}{2}$

$\frac{9}{2} - 3 = \frac{\frac{9}{2}}{3}$

$\frac{16}{3} - 4 = \frac{\frac{16}{3}}{4}$

71. $\left(\frac{1}{3}\right)^2 + \frac{2}{3} = \frac{1}{3} + \left(\frac{2}{3}\right)^2$

$\left(\frac{1}{5}\right)^2 + \frac{4}{5} = \frac{1}{5} + \left(\frac{4}{5}\right)^2$

$\left(\frac{2}{7}\right)^2 + \frac{5}{7} = \frac{2}{7} + \left(\frac{5}{7}\right)^2$

72. Describe the process of adding and subtracting algebraic fractions, including typical errors that can occur and how to avoid these errors. Use two examples to illustrate your discussion.

III. Applications

73. Write an algebraic expression for "a number plus its reciprocal."

74. Write an algebraic expression for "a number divided by the sum of it with its reciprocal."

75. According to Einstein's theory of relativity, we should "add" two velocities, V_1 and V_2, to obtain a velocity V according to the formula

$$V = \frac{V_1 + V_2}{1 + \frac{V_1 V_2}{C^2}}$$

where C is the speed of light. Suppose we use the estimate 6.696×10^8 mph for the speed of light.

(a) Let $V_1 = 60$ mph and $V_2 = 50$ mph (automobile speeds). Add V_1 and V_2 according to Einstein's formula. How close is the result to 110 mph?

(b) Do the same for $V_1 = 600$ mph and $V_2 = 500$ mph (airplane speeds). How close is the result to 1,100 mph?

(c) Do the same for $V_1 = 6{,}000$ mph and $V_2 = 5{,}000$ mph (lunar spacecraft speed). Is the result near 11,000 mph?

(d) Do the same for $V_1 = 60{,}000$ mph and $V_2 = 50{,}000$ mph. For what speeds does Einstein's method of "adding" velocities make a significant difference?

76. In Example 1 of this section, we have the following pattern.

$$\frac{1}{2} = \frac{1}{2}$$

$$\frac{1}{2} + \frac{1}{4} = \frac{3}{4}$$

$$\frac{1}{2} + \frac{1}{4} + \frac{1}{8} = \frac{7}{8}$$

(a) Write the next three rows in the pattern.

(b) What is the sum of the tenth row? Fiftieth row? Nth row?

(c) In what row is the sum equal to $\frac{32{,}767}{32{,}768}$?

(d) What is the first row whose sum exceeds $\frac{99}{100}$?

(e) Graph the sum with the corresponding row number. What happens to the sum as the row number becomes larger and larger?

77–80. Examine the following patterns. For each pattern **(a)** write the next three rows in the pattern. **(b)** What is the sum of the tenth row? Fiftieth row? *N*th row? **(c)** Graph the sum with its corresponding row number. What happens to the sum as the row number gets larger and larger?

77.
$$\frac{1}{3} = \frac{1}{3}$$
$$\frac{1}{3} + \frac{1}{9} = \frac{4}{9}$$
$$\frac{1}{3} + \frac{1}{9} + \frac{1}{27} = \frac{13}{27}$$

78.
$$\frac{1}{4} = \frac{1}{4}$$
$$\frac{1}{4} + \frac{1}{16} = \frac{5}{16}$$
$$\frac{1}{4} + \frac{1}{16} + \frac{1}{64} = \frac{21}{64}$$

79.
$$\frac{1}{5} = \frac{1}{5}$$
$$\frac{1}{5} + \frac{1}{25} = \frac{6}{25}$$
$$\frac{1}{5} + \frac{1}{25} + \frac{1}{125} = \frac{31}{125}$$

80.
$$\frac{1}{10} = \frac{1}{10}$$
$$\frac{1}{10} + \frac{1}{100} = \frac{11}{100}$$
$$\frac{1}{10} + \frac{1}{100} + \frac{1}{1{,}000} = \frac{111}{1{,}000}$$

IV. Extensions

81. $\dfrac{1}{2\cdot 3} + \dfrac{1}{3\cdot 4} + \dfrac{1}{4\cdot 5} + \cdots + \dfrac{1}{n(n+1)} = ?$

82. This problem appeared in an algebra book written by Werl and Hill in 1883. Perform the indicated operations.

$$\frac{\dfrac{1+x}{1-x} + \dfrac{4x}{1+x^2} + \dfrac{8x}{1-x^4} - \dfrac{1-x}{1+x}}{\dfrac{1+x^2}{1-x^2} + \dfrac{4x^2}{1+x^4} - \dfrac{1-x^2}{1+x^2}}$$

The authors wrote in their preface, "It is believed that these exercises will be well adopted to American wants and will provide a rich source of problems." What do *you* think about this?

83. Can you find the *integers* x and y that will satisfy the following equations?

(a) $\dfrac{1}{x} + \dfrac{1}{y} = \dfrac{1}{12}$ **(b)** $\dfrac{1}{x} + \dfrac{1}{y} = \dfrac{1}{48}$

84. You wish to approximate $\sqrt{N}$ for some positive number N. If x_1 is your first guess at the value of $\sqrt{N}$, then x_2 will be a closer approximation to $\sqrt{N}$, where

$$x_2 = \frac{x_1 + \dfrac{N}{x_1}}{2}$$

and

$$x_3 = \frac{x_2 + \dfrac{N}{x_2}}{2}$$

will be even better, and so on.

(a) Suppose $N = 8$, and your first guess for $\sqrt{8}$ is $x_1 = 3$. Calculate x_2, x_3, and x_4. After each successive approximation, square your value to see how close to 8 you come.

(b) Suppose $N = 43$, and your first guess for $\sqrt{43}$ is $x_1 = 7$. Calculate x_2, x_3, and x_4. Check your approximations by squaring them.

(c) Repeat part (b), except this time, use $x_1 = 6$. Which initial value yields a closer approximation to $\sqrt{43}$ after four steps, 6 or 7?

(d) Suppose $N = 80$. Pick your own value of x_1, and calculate x_2, x_3, and x_4. Do the same for $N = 800$ and for $N = 8{,}000$. Check your approximation with a calculator.

5.4 RATIONAL EQUATIONS AND FUNCTIONS

In Example 3 of the first section of this chapter, we discussed an algebraic model that involved a rational equation. Let us return now to that example. As in previous chapters, we solve it by using algebra after we have expressed the equation in one variable. Then we will solve by graphing (using two variables).

EXAMPLE 1 A glider is launched with an initial velocity of 18 mph. The glider can travel 63 miles downwind in the same amount of time that it takes to travel 45 miles against the wind. What is the wind speed?

Solution

Recall that we used the following model: Distance = rate × time.

Using a Table If we let x equal the speed of the wind, then we can obtain the following table:

	Distance in miles	*Rate miles per hour*	*Time in hours*
Downwind (with wind)	63	$18 + x$	?
Upwind (against wind)	45	$18 - x$	?

Table 5.5(a)

Since $t = D/r$, we can fill in the time boxes as follows:

	Distance in miles	*Rate miles per hour*	*Time in hours*
Downwind	63	$18 + x$	$\dfrac{63}{18 + x}$
Upwind	45	$18 - x$	$\dfrac{45}{18 - x}$

Table 5.5(b)

Using an Equation Since the glider travels 63 miles downwind in the *same amount of time* as it travels 45 miles upwind, we obtain this equation:

$$\frac{63}{18 + x} = \frac{45}{18 - x} \qquad \textbf{This is a rational equation.}$$

$$\text{time downwind} = \text{time upwind}$$

Solving the equation using algebra, we can eliminate denominators by multiplying both sides by the LCD. The LCD is $(18 + x)(18 - x)$.

$$(18 + x)(18 - x)\cdot\frac{63}{18 + x} = \frac{45}{18 - x}\cdot(18 + x)(18 - x) \qquad \textbf{Multiply by the LCD.}$$

$$\cancel{(18 + x)}(18 - x)\frac{63}{\cancel{(18 + x)}} = \frac{45}{\cancel{18 - x}}(18 + x)\cancel{(18 - x)} \qquad \textbf{Eliminate denominators.}$$

$$(18 - x)63 = 45(18 + x) \qquad \textbf{The equivalent equation \textit{without} fractions.}$$

$$1{,}134 - 63x = 810 + 45x$$

$$324 = 108x$$

$$3 = x \qquad \textbf{Answer: The wind speed is 3 mph.}$$

Using a Graph We next solve the equation by graphing. Once we have determined the equation for a problem (usually the most difficult part), we can use a graph to find the solution.

Graph

$$Y1 = \frac{63}{18 + x} \quad \text{and} \quad Y2 = \frac{45}{18 - x}$$

$Y1$ is the time downwind

$Y2$ is the time upwind

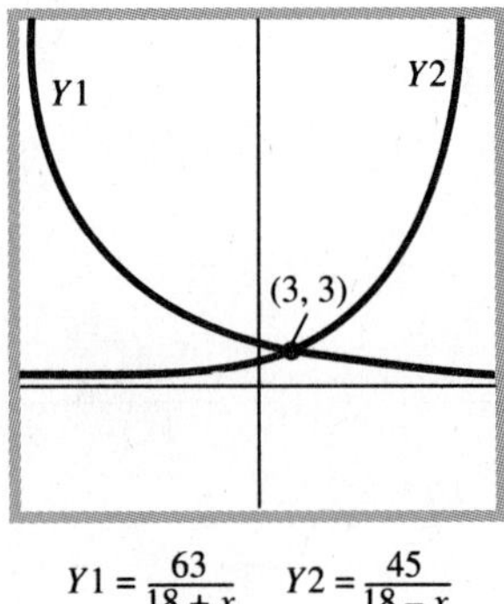

$Y1 = \frac{63}{18 + x} \quad Y2 = \frac{45}{18 - x}$

Figure 5.9

From the graph, we find that $x = 3$, which is the speed of the wind. The corresponding value for y is 3. This is the speed of the glider. ■

Rational Equations in One Variable

Before attempting this more general problem, let us first take a deeper look at the solution of rational equations that involve only one variable.

EXAMPLES 2–3 Solve for x in each equation.

2. $\frac{x}{6} - \frac{1}{2} = \frac{(3 - x)}{4}$ **3.** $\frac{1}{x} = \frac{5}{x} + 1$

Our first goal in solving any equation containing fractions will be to *express it as an equivalent equation without fractions.* To accomplish this, we must eliminate the denominators. Using the properties for solving equations (discussed in Chapter 3), we can multiply both sides of an equation by the same thing and not change the solution. If we multiply both sides of an equation by the least common denominator, then we will succeed in eliminating denominators. We can also use a graphing calculator to solve such equations.

Solutions

2. $\frac{x}{6} - \frac{1}{2} = \frac{(3 - x)}{4}$ The LCD is $3 \cdot 2 \cdot 2 = 12$.

Using an Equation

$$12 \cdot \left(\frac{x}{6} - \frac{1}{2}\right) = 12 \cdot \frac{(3 - x)}{4}$$ Multiply both sides by a common denominator. Property 2 for equations (Chapter 3).

$$12 \cdot \frac{x}{6} - 12 \cdot \frac{1}{2} = 12 \cdot \frac{(3 - x)}{4}$$ Distributive property on the left side.

$$\frac{\overset{2}{\cancel{12}} \cdot x}{\cancel{6}} - \overset{6}{\cancel{12}} \cdot \frac{1}{\cancel{2}} = \frac{\overset{3}{\cancel{12}}(3 - x)}{\cancel{4}}$$

$$2x - 6 = 3(3 - x)$$ An equivalent equation without fractions.

$$2x - 6 = 9 - 3x$$

$$5x = 15$$

$$x = 3$$ Answer.

Check: Substituting $x = 3$,

$$\frac{3}{6} - \frac{1}{2} \overset{?}{=} \frac{(3 - 3)}{4}$$

$$0 \overset{\checkmark}{=} 0$$

Using a Graph To solve this equation by graphing, graph

$$Y1 = \frac{x}{6} - \frac{1}{2} \quad \text{and} \quad Y2 = \frac{3 - x}{4}$$

Find the x-coordinate of the point of intersection.

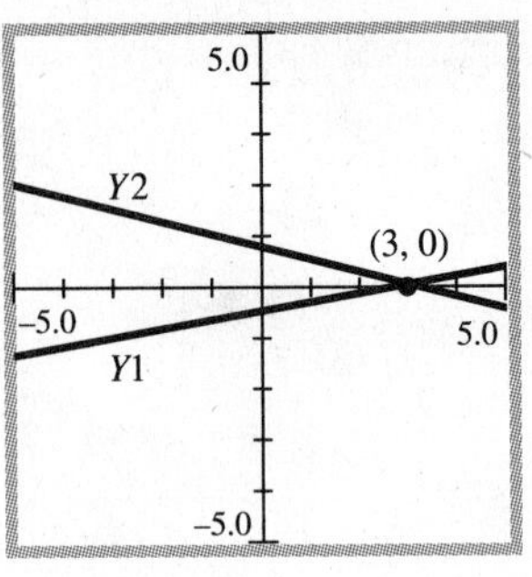

$Y1 = \frac{x}{6} - \frac{1}{2}$ $Y2 = \frac{3-x}{4}$

Figure 5.10

Note both graphs are lines since both $Y1$ and $Y2$ can be expressed in the form $mx + b$.

3. $\frac{1}{x} = \frac{5}{x} + 1$

Using an Equation The LCD is x. We may proceed in exactly the same way as in Example 2, but notice that $x = 0$ is *not* a possible solution for the equation, because it makes the denominator of $\frac{1}{x}$ and of $\frac{5}{x}$ equal to 0.

$$x \cdot \frac{1}{x} = x \cdot \left(\frac{5}{x} + 1\right)$$ Multiply both sides by the LCD.

$$\frac{\overset{1}{\cancel{x}}}{\cancel{x}} = \frac{5\cancel{x}}{\cancel{x}} + x$$

$$1 = 5 + x$$ We obtain an equivalent linear equation.

$$-4 = x$$ Answer. Substitute $x = -4$, and check.

Using a Graph Let

$$Y1 = \frac{1}{x} \quad \text{and} \quad Y2 = \frac{5}{x} + 1$$

and graph $Y1$ and $Y2$.

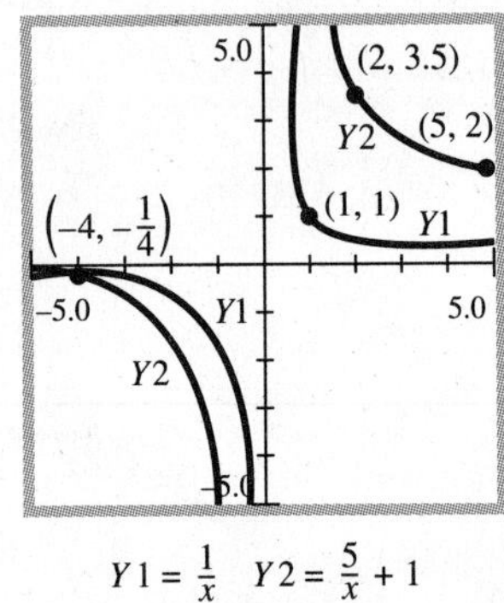

$Y1 = \frac{1}{x}$ $Y2 = \frac{5}{x} + 1$

Figure 5.11

EXAMPLES 4–5 Solve the following rational equations.

4. $\frac{3}{x} - \frac{2}{x-1} = \frac{1}{2x}$

5. $2 - \frac{1}{y-2} = \frac{y-3}{y-2}$

Solutions

4. $\frac{3}{x} - \frac{2}{x-1} = \frac{1}{2x}$ **The LCD is $2x(x-1)$. Here $x \neq 0, x \neq 1$. (Why?)**

Using Equations

$$2x(x-1)\cdot\left(\frac{3}{x} - \frac{2}{x-1}\right) = 2x(x-1)\cdot\frac{1}{2x}$$ **Multiply both sides by the LCD.**

$$\frac{6x(x-1)}{x} - \frac{4x(x-1)}{x-1} = \frac{2x(x-1)}{2x}$$

$$\frac{6\cancel{x}(x-1)}{\cancel{x}} - \frac{4x\cancel{(x-1)}}{\cancel{x-1}} = \frac{\cancel{2x}(x-1)}{\cancel{2x}}$$

$$6(x-1) - 4x = (x-1)$$ **An equivalent linear equation.**

$$6x - 6 - 4x = x - 1$$

$$2x - 6 = x - 1$$

$$x = 5$$ **Answer.**

Check: Substituting $x = 5$ in the equation,

$$\frac{3}{5} - \frac{2}{5-1} \stackrel{?}{=} \frac{1}{2\cdot 5}$$

$$\frac{3}{5} - \frac{2}{4} \stackrel{?}{=} \frac{1}{10}$$

$$\frac{6}{10} - \frac{5}{10} \stackrel{\checkmark}{=} \frac{1}{10}$$

Using a Graph Let

$$Y1 = \frac{3}{x} - \frac{2}{x-1} \quad \text{and} \quad Y2 = \frac{1}{2x}$$

Graph $Y1$ and $Y2$ and find the point of intersection.

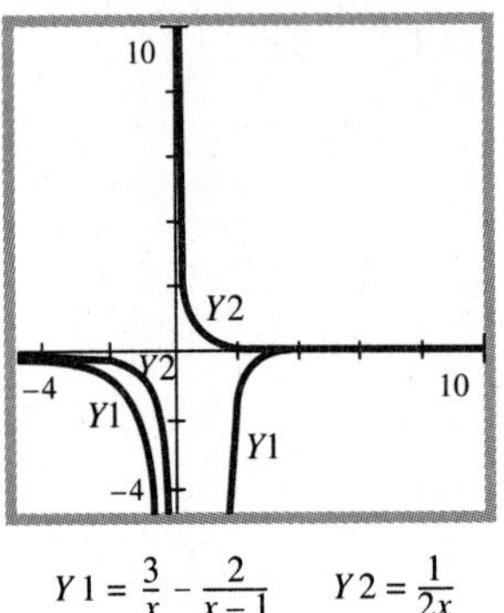

$Y1 = \frac{3}{x} - \frac{2}{x-1} \qquad Y2 = \frac{1}{2x}$

Figure 5.12 (a)

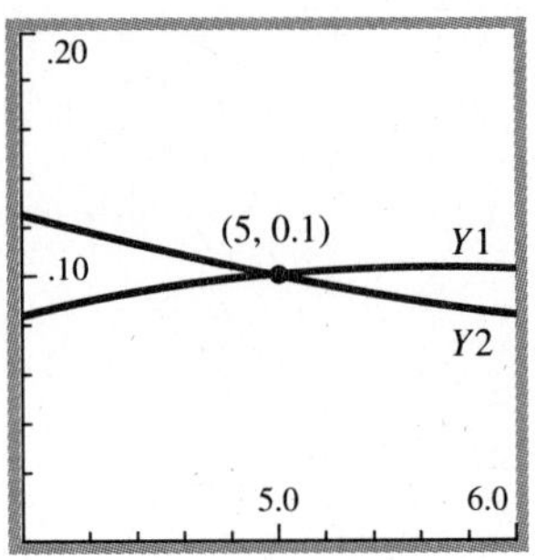

Figure 5.12 (b)

The window in Figure 5.12(b) zooms in on the solution, $x = 5$, $Y = \frac{1}{10}$.

5. $2 - \dfrac{1}{y-2} = \dfrac{y-3}{y-2}$ $\quad$ $y \neq 2$
A common denominator is $y - 2$.

Using an Equation

$$(y-2)\left(2 - \frac{1}{y-2}\right) = (y-2)\left(\frac{y-3}{y-2}\right) \qquad \textbf{Multiply both sides by } y - 2.$$

$$2(y-2) - \frac{\overset{1}{\cancel{(y-2)}}}{\cancel{(y-2)}} = \frac{\cancel{(y-2)}(y-3)}{\cancel{(y-2)}}$$

$$2(y-2) - 1 = y - 3$$
$$2y - 4 - 1 = y - 3$$
$$2y - 5 = y - 3$$
$$y = 2 \qquad \text{BUT...}$$

Here, $y = 2$ is *not* a possible solution to our original equation in Example 5, because the value $y = 2$ makes the denominators equal to 0. (Indeed, not every fractional equation must have a solution.) Therefore, there is no value of y that satisfies Equation 5. This equation is a contradiction. When this occurs, we say there is no solution.

Using a Graph To use a graphing calculator we need to express all equations in terms of Ys and xs; Y is a function of x. We will rewrite our original equations by replacing y with x as follows:

$$Y1 = 2 - \frac{1}{x-2}$$

$$Y2 = \frac{x - 3}{x - 2}$$

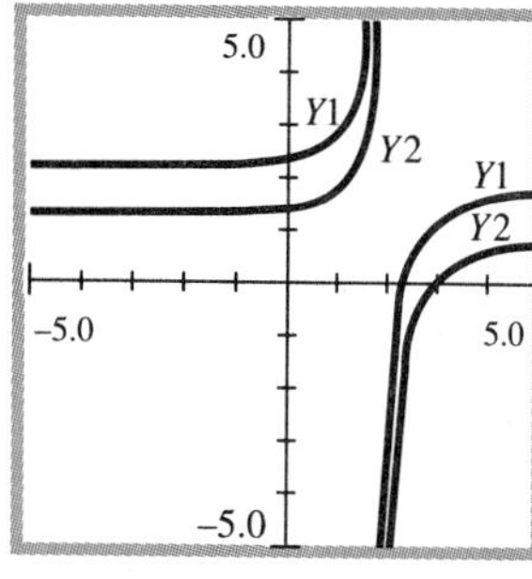

$Y1 = 2 - \frac{1}{x-2}$ $\quad Y2 = \frac{x-3}{x-2}$

Figure 5.13

The graphs appear to intersect when $x = 2$. Use TRACE and ZOOM to see that the two graphs do not intersect. ■

Now let us summarize what we have learned so far about solving rational equations. In addition to the steps listed below, recall the properties for solving equations in Chapter 3, which involve adding (or subtracting) a quantity to both sides of an equation, or multiplying (or dividing) both sides of an equation by a nonzero quantity.

Strategies for Solving a Rational Equation in One Variable

Algebraically

The overall strategy is to find an equivalent equation without fractions:

1. Find all values of the variable that make any denominator equal to 0.
2. Eliminate the denominators by multiplying both sides by a common denominator or the LCD.
3. Solve the resulting equation, which now has no fractions.
4. Reject any value (or values) obtained that make a denominator equal to zero.
5. Substitute the solutions into the original fractional equation, and check your results.

By graphing

1. Graph the functions representing each side of the equation.
2. Find the coordinates of the point of intersection.

Rational (fractional) equations occur in many fields and have many important uses. Here are some examples.

EXAMPLE 6

At a 7-minute-per-mile pace, a long-distance runner burns 150 calories per mile. At this rate, how many calories would be burned off by the runner during an entire marathon race (26.2 miles)?

Solution

This is a proportion, similar to the equations covered in Section 3.2 of Chapter 3. If we let x equal the number of calories burned during the marathon, and assume that the ratio of the number of miles to the number of calories remains constant during the race, then we can obtain

$$\frac{1 \text{ mile}}{150 \text{ calories}} = \frac{26.2 \text{ miles}}{x \text{ calories}}$$

Thus

$$\frac{1}{150} = \frac{26.2}{x} \qquad \text{The LCD is } 150x\text{; exclude } x = 0.$$

$$150x \cdot \frac{1}{150} = \frac{26.2}{x} \cdot 150x \qquad \text{Multiply both sides by the LCD.}$$

$$x = (26.2)(150) = 3{,}930 \text{ calories} \qquad \text{Answer.}$$

EXAMPLE 7

Edith drives from Indianapolis to Syracuse, a distance of 620 miles, in the same time as Bill drives from Indianapolis to Kansas City, 500 miles. Edith is traveling 10 mph faster than Bill. How fast is each person driving?

Solution

Using a Table Once again, the model $D = rt$ will enable us to set up the problem. Let x be the rate of speed in mph of Bill's car. Then $x + 10$ is the rate of speed of Edith's car. We can summarize the data in a table. Recall $t = \frac{D}{r}$.

	Distance	*Rate*	*Time*
Bill	500 miles	x	$\frac{500}{x}$
Edith	620 miles	$x + 10$	$\frac{620}{x + 10}$

Table 5.6

Using an Equation The time traveled is the *same* for both cars. So Bill's time equals Edith's time, or

$$\frac{500}{x} = \frac{620}{x + 10}$$ Resulting fractional equation. LCD $= x(x + 10)$

$$\cancel{x}(x + 10)\frac{500}{\cancel{x}} = \frac{620}{\cancel{(x + 10)}}x\cancel{(x + 10)}$$ Multiply by the LCD, and eliminate denominators.

$$500(x + 10) = 620x$$

$$500x + 5{,}000 = 620x$$

$$5{,}000 = 120x$$

$$x = \frac{5{,}000}{120} \approx 41.67 \text{ mph}$$ Answer. This is Bill's speed.

Edith's speed is then $x + 10 \approx 51.67$ mph.

Using a Graph We can also solve this problem by graphing. We let

$$Y1 = \frac{500}{x} \quad \text{and} \quad Y2 = \frac{620}{x + 10}$$ Graph $Y1$ and $Y2$ and find the point of intersection.

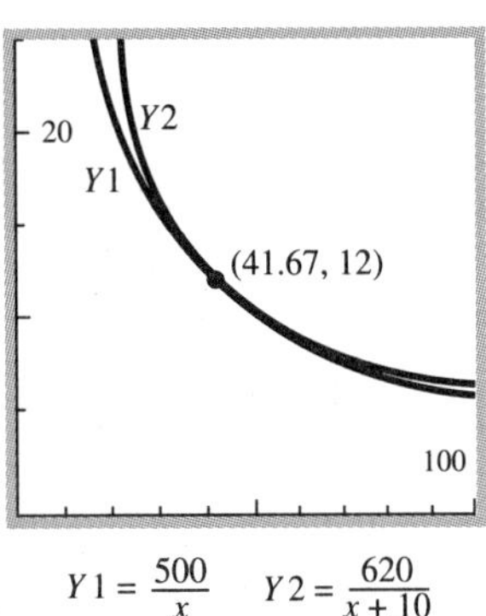

$Y1 = \frac{500}{x}$ $\quad Y2 = \frac{620}{x + 10}$

Figure 5.14

EXAMPLE 8 John and Fred have a lawn mowing service. John estimates he can cut the lawn in front of city hall in 2 hours and 30 minutes. Fred estimates he can do the same job with his mower in 3 hours. How long will it take if they decide to mow the lawn together?

Solution

Let x equal the number of hours it will take both of them working together to do the job. One strategy for solving this type of problem is to think of it as a rate problem. The key is to think of the total job as 1 unit (this is distance in distance $=$ time $\times$ rate). So

$1 = \text{time} \times \text{rate}$. Thus, the combined **rate** to complete this job is 1 divided by the time, or $\frac{1}{x}$. But John's rate is

$$\frac{1 \text{ job}}{2.5 \text{ hours}} \quad \text{or} \quad \frac{1}{2.5}$$

and Fred's rate is

$$\frac{1 \text{ job}}{3 \text{ hours}} \quad \text{or} \quad \frac{1}{3}$$

So the combined rate is

$$\frac{1}{2.5} + \frac{1}{3}$$

Thus

$$\frac{1}{2.5} + \frac{1}{3} = \frac{1}{x}$$

Solving this equation, we have

$$\frac{5.5}{7.5} = \frac{1}{x}$$

$$x = 1.36 \text{ hours, approximately.}$$

Literal Equations

Algebraic models in many variables are sometimes called **literal equations.** We can solve a literal equation for any of its variables. If the literal equation is a rational equation, we can eliminate the denominators, then solve for the desired variable.

EXAMPLE 9

In Exercise 100 of Section 5.1, we considered the equation

$$B_{\frac{1}{2}} = \frac{6V - (B + b)L}{4L} \qquad \textbf{(Equation 1)}$$

where

B = area of cross section at the large end of the log
b = area of cross section at the small end of the log
L = length of the log
V = volume of the log
$B_{\frac{1}{2}}$ = area of cross section at the midpoint of a log

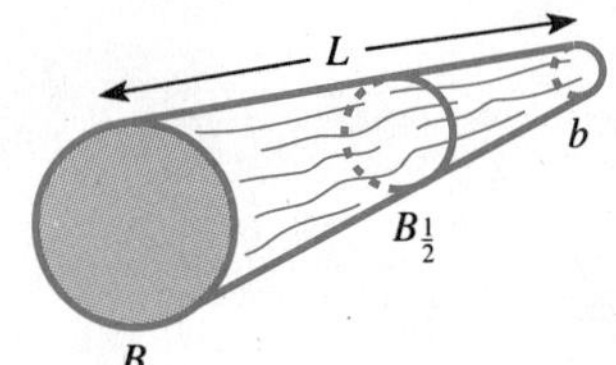

Figure 5.15

Equation 1 is presented in such a way that we can find the area of the cross section in the middle if we know all the other values. However, suppose we wanted to know the volume V of the log. We wish to solve Equation 1 for V.

Solution

$$B_{\frac{1}{2}} = \frac{6V - (B + b)L}{4L} \qquad \text{LCD} = 4L$$

$$4L \cdot B_{\frac{1}{2}} = \frac{6V - (B + b)L}{4L} \cdot 4L \qquad \text{Multiply both sides by the LCD.}$$

$$4LB_{\frac{1}{2}} = \frac{[6V - (B + b)L]\cancel{4L}}{\cancel{4L}} \qquad \text{Denominators are eliminated.}$$

$$4LB_{\frac{1}{2}} = 6V - BL - bL \qquad \text{Eliminate parentheses.}$$

Now, since we wish to solve for the volume V, we will isolate all terms involving V on one side of the equation and get everything else on the other side:

$$4LB_{\frac{1}{2}} + BL + bL = 6V$$

Dividing both sides by 6, we obtain

$$\frac{4LB_{\frac{1}{2}} + BL + bL}{6} = V \qquad \text{Answer.}$$

EXAMPLE 10

The formula for the *average* velocity V (or speed) for a round trip between two destinations is given by

$$V = \frac{2}{\frac{1}{V_1} + \frac{1}{V_2}} \qquad \textbf{(Equation 2)}$$

where V_1 is the average velocity going, and V_2 is the average velocity returning.

Suppose you drive from Eugene to Seattle, a distance of 280 miles. You leave at 7:00 A.M. and wish to be back in Eugene by midnight. You must attend a four-hour meeting in Seattle. Owing to heavy traffic, you expect to average 45 mph for the trip from Eugene to Seattle. What velocity must you average on the return trip to Eugene in order to arrive by midnight?

Solution

If we had values for V and V_1, we could substitute them in and compute our return velocity V_2. $V_1 = 45$ mph, but we do not yet know the value of V, the overall trip velocity. However, we can determine V. For the entire trip, we know that

$$D = V \times t \qquad D = 2 \times 280 = 560 \text{ miles round trip.}$$

so

$$V = \frac{D}{t} \qquad \text{We can drive from 7 A.M. until midnight, except for our four-hour meeting, so } t = 17 - 4 = 13 \text{ hours.}$$

$$V = \frac{560}{13} \approx 43.08 \text{ mph}$$

Now that we have values for $V \approx 43.08$ mph and $V_1 = 45$ mph, we could substitute different values for V_2 into Equation 2 and try to find the value of V_2; however, we could first solve the model algebraically for V_2.

$$V = \frac{2}{\frac{1}{V_1} + \frac{1}{V_2}} \qquad \text{LCD} = \left(\frac{1}{V_1} + \frac{1}{V_2}\right)$$

$$\left(\frac{1}{V_1} + \frac{1}{V_2}\right) \cdot V = \frac{2}{\left(\frac{1}{V_1} + \frac{1}{V_2}\right)} \cdot \left(\frac{1}{V_1} + \frac{1}{V_2}\right) \qquad \text{Multiply by the LCD.}$$

$$\frac{V}{V_1} + \frac{V}{V_2} = 2 \qquad \text{A new rational equation arises. Its LCD is } (V_1V_2).$$

$$(V_1V_2) \cdot \frac{V}{V_1} + (V_1V_2)\frac{V}{V_2} = 2(V_1V_2) \qquad \text{Multiply by this LCD.}$$

$$V_2V + V_1V = 2V_1V_2 \qquad \text{The equivalent equation without fractions.}$$

$$V_1V = 2V_1V_2 - V_2V \qquad \text{Isolate } V_2 \text{ terms on one side, and solve for } V_2.$$

$$V_1V = V_2(2V_1 - V)$$

$$\frac{V_1V}{2V_1 - V} = V_2 \qquad \text{Equation 2 is now solved for } V_2\text{, the desired variable.}$$

Thus

$$V_2 = \frac{V_1V}{2V_1 - V} = \frac{(45)(43.08)}{2(45) - 43.08} \approx 41.32 \text{ mph} \qquad \text{Answer.}$$

You may drive the return trip at an even more leisurely pace than your arrival pace and still arrive by midnight. ■

Now we summarize what we have learned so far about solving literal fractional equations:

Strategies for Solving a Rational Equation in Many Variables for a Particular Desired Variable

1. Eliminate the denominators by multiplying both sides by a common denominator or the LCD.
2. Get all the terms involving the desired variable on one side of the equation.
3. If the desired variable appears in more than one term, *factor* it out.
4. Divide both sides of the equation by any other factors to isolate the desired variable.

NOTE: Our algorithms will work for rational equations in which the exponent of the desired variable is 1. We will solve more general rational equations in Chapter 6. ■

EXAMPLES 11–12

Warning: Consider these two problems:

11. Perform the indicated operations.

$$\frac{3}{x}+\frac{5}{x+1}-\frac{x}{x^2+x}$$

12. Solve for x.

$$\frac{3}{x}=\frac{5}{x+1}-\frac{x}{x^2+x}$$

Solutions

Sometimes there is great confusion over these two types of problems. *Look closely! These are not the same kind of problem.*

11. This problem involves only addition (and subtraction) of fractions or writing an equivalent algebraic expression. The solution to this problem is *an algebraic expression.* To solve this problem, we find the LCD and multiply the *top and bottom* of each fraction by the missing factors to *create common denominators:*

$$\frac{3}{x}\cdot\frac{(x+1)}{(x+1)}+\frac{5}{x+1}\cdot\frac{(x)}{(x)}-\frac{x}{x(x+1)} \qquad \text{The LCD} = x(x+1).$$

Then the numbers can be combined, and everything can be put over $x(x+1)$. The answer is $\frac{7x+3}{x(x+1)}$.

12. *This is an equation!* We use the LCD for a *different purpose* here. The solution here may be *one or more numbers.* To solve, we multiply both sides of the equation by the LCD $x(x+1)$ *to eliminate denominators* in the equation.

$$\cancel{x}(x+1)\frac{3}{\cancel{x}}=\frac{5}{\cancel{x+1}}\cdot x(\cancel{x+1})-\frac{x}{\cancel{x(x+1)}}\cancel{x(x+1)}; \qquad x\neq 0,-1$$

$$(x+1)\cdot 3=5x-x$$

$$x=3 \quad ■$$

5.4 EXERCISES

I. Procedures

1–28. Find the solutions to the equations below algebraically or by graphing, if they exist. Check your answers by substituting.

1. $\frac{x}{2}=5+\frac{x}{3}$

2. $\frac{2x}{3}-5=\frac{x}{4}$

3. $\frac{3}{4}y-\frac{1}{2}y=-2$

4. $\frac{2}{3}x-\frac{2}{5}=\frac{2}{5}x$

5. $x-\frac{1}{2}=3x+\frac{1}{3}$

6. $2-\frac{x}{5}=x-\frac{1}{2}$

7. $x-\frac{x+7}{3}=\frac{1}{2}+3x$

8. $2-\frac{x}{2}=3-\frac{x-1}{3}$

9. $\frac{3}{x} - \frac{5}{2} = -2$

10. $\frac{2}{5b} - \frac{1}{10} = \frac{1}{2b}$

11. $\frac{1}{x} - \frac{2}{x} = \frac{3}{x}$

12. $\frac{1}{x} - \frac{2}{x} = \frac{3}{2}$

13. $\frac{x+1}{x} = \frac{1}{3}$

14. $\frac{3}{5} = \frac{-x+7}{x-3}$

15. $\frac{2x-1}{2x+1} = \frac{x+2}{x-3}$

16. $\frac{4x-2}{2x-7} = \frac{6x}{3x+1}$

17. $2 - \frac{1}{y-1} = \frac{y-2}{y-1}$

18. $\frac{y-7}{y-3} + 1 = \frac{4y}{y-3}$

19. $\frac{1}{z+2} + 1 = \frac{6}{z+2}$

20. $2 - \frac{2}{x-1} = \frac{x-3}{x-1}$

21. $\frac{1}{a-3} + \frac{2}{a+3} = \frac{3}{a^2-9}$

22. $\frac{3}{x-2} + \frac{2}{x+2} = \frac{12}{x^2-4}$

23. $\frac{1}{a} + \frac{1}{3a} = \frac{1}{3}$

24. $\frac{1}{r-1} + \frac{2}{3r-3} = \frac{-5}{12}$

25. $\frac{3}{4a-8} - \frac{2}{3a-6} = \frac{1}{36}$

26. $\frac{2}{y-5} + \frac{1}{y+5} = \frac{11}{y^2-25}$

27. $\frac{2m}{3m+3} - \frac{m+2}{6m+6} = \frac{m-6}{8m+8} + \frac{5}{12}$

28. $\frac{a+3}{a} - \frac{a+4}{a+5} = \frac{15}{a^2+5a}$

29–40. In each equation below, solve for the indicated variable.

29. Solve for l: $V = lw^2$ (volume of a square prism).

30. Solve for l: $A = 2w^2 + 4lw$ (surface area of a square prism).

31. Solve for h: $V = \pi r^2 h$ (volume of a right circular cylinder).

32. Solve for h: $V = \frac{1}{3}\pi r^2 h$ (volume of a right circular cone).

33. Solve for R: $W = I^2R$ (power in watts of an electrical circuit).

34. Solve for K: $H = Kt - \frac{gt^2}{2}$ (height of an object thrown upward after time t).

35. Solve for g: $L = \frac{gp^2}{4\pi^2}$ (length of a pendulum).

36. Solve for B: $A = \frac{(B+b)}{2} \cdot h$ (area of a trapezoid with bases B and b).

37. Solve for x: $m = \frac{2xy}{x+y}$ (the harmonic mean m of two numbers x and y. Also see Exercise 76).

38. Solve for L: $S = \frac{n}{2}(A + L)$

39. Solve for M: $\frac{1}{M} + w = \frac{C}{M}$

40. Solve for t: $\frac{at + bt}{m} = y$

II. Concepts

41–45. In the following exercises, the student who solved the problem *has made an error.* Find the error, and solve the problem correctly. What advice would you give this student?

41. Solve for x: $\frac{5x}{x-3} = \frac{1}{x+3} + \frac{5x^2}{x^2-9}$

Solution: The LCD is $(x - 3)(x + 3)$:

$$\frac{5x}{x-3}\cdot\frac{(x+3)}{(x+3)} - \frac{1}{x+3}\cdot\frac{(x-3)}{(x-3)} + \frac{5x^2}{x^2-9}$$

$$= \frac{5x^2 + 15x - x + 3 + 5x^2}{x^2-9}$$

$$= \frac{10x^2 + 14x + 3}{x^2-9}$$ **Incorrect answer.**

42. *Perform the operations:* $\frac{1}{x+2} + \frac{x+3}{x^2-4}$

Solution: The LCD is $(x + 2)(x - 2)$:

$$(x+2)(x-2)\frac{1}{x+2} = \frac{x+3}{x^2-4}\cdot(x+2)(x-2)$$

$$x - 2 = x + 3$$

$0 = 5$, so there is no solution.

Incorrect answer.

43. Solve for t:

$$\frac{at - bt}{m} = 6$$

Solution: The LCD is m.

$$\not{m}\cdot\frac{at - bt}{\not{m}} = 6m$$
$$at - bt = 6m$$
$$t(a - b) = 6m$$
$$t = 6m - (a - b) \quad \textbf{Incorrect answer.}$$

44. Solve for x:

$$\frac{3}{x-2} - 7 = \frac{x}{x-2}$$

Solution: The LCD is $(x - 2)$.

$$\cancel{(x-2)}\frac{3}{\cancel{x-2}} - 7 = \frac{x}{\cancel{x-2}}\cancel{(x-2)}$$
$$3 - 7 = x$$
$$-4 = x \quad \textbf{Incorrect answer.}$$

45. Solve for x:

$$\frac{x}{3x} - \frac{4}{3} = \frac{2x}{3x}$$

Solution: The LCD is $3x$.

$$3x\left(\frac{x}{3x} - \frac{4}{3}\right) = \frac{2x}{3x}\cdot 3x$$
$$\frac{3x\cdot x}{3x} - \frac{3x\cdot 4}{3} = \frac{2x\cdot 3x}{3x}$$
$$x - 4x = 2x$$
$$-3x = 2x$$
$$0 = 5x$$
$$0 = x \quad \textbf{Incorrect answer.}$$

46. **(a)** Explain why each step works in the strategy box for solving a rational equation in one variable algebraically and by graphing (this box occurs right after Example 5). Illustrate your discussion with an example.

(b) Compare the two solutions: Which method is easier to use? Which is more accurate?

(c) How can you use a graphing calculator in Exercises 44 and 45?

47 and 48. Find a rational equation whose solution is

47. $x = 1$ **48.** $x = \frac{2}{3}$

49 and 50. Find a rational equation that is not defined for

49. $x = 1$ **50.** $x = \frac{2}{3}$

III. Applications

51. In macroeconomic theory, models of the following type are often useful for analyzing the relationships between spending and savings:

$$K = \frac{1}{1 - m - mt}$$

where K is the expenditure multiplier, m is the marginal propensity to consume, and t is a constant tax on income, expressed as a decimal.

(a) Solve this equation for t.

(b) Suppose $m = 0.8$ and $K = -5.4$. What is the value of t? How about for $m = 0.78$ and $K = -5.6$?

(c) Solve this same equation for m.

52. A woman who is in training for a marathon averages 65 miles a week for ten weeks. She trains at an 8-minute-per-mile pace and burns 110 calories per mile (see Example 6).

(a) How many calories must she consume per day just to replenish the calories she burns while running?

(b) How many 500-calorie pieces of carrot cake would she burn up during her training period?

53. Recall the equation for logs:

$$B_{\frac{1}{2}} = \frac{6V - (B + b)L}{4L}$$

(This was discussed in Example 9 of this section.) How long must a log be to yield 450 cubic feet of wood if its cross-sectional areas are $B = 23$ square feet, $B_{\frac{1}{2}} = 14$ square feet, and $b = 5$ square feet?

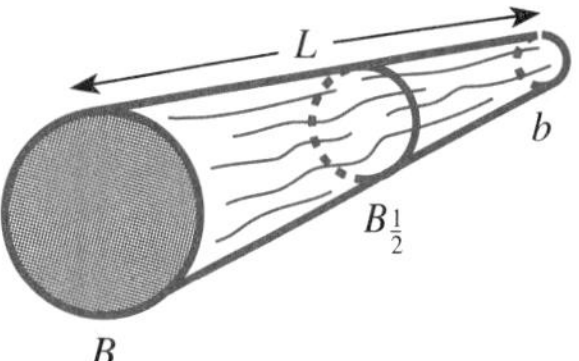

54. What is the volume of the log (in Exercise 53) if $L = 20$ feet, $B = 7$ square feet, $B_{\frac{1}{2}} = 5$ square feet, and $b = 2.3$ square feet?

55. The batting average for a baseball player is the ratio of the total number of hits to the total number of times at bat. (Batting averages are expressed as decimals.) On July 5, Puckett had a record of 105 hits out of 320 times at bat.

(a) What was Puckett's batting average on July 5?

(b) How many consecutive hits must Puckett make to reach Kruk's batting average of .353?

56. Two hoses are used to fill a swimming pool. One hose can fill the pool in seven hours and the other hose can fill the pool in ten hours. How long will it take to fill the pool if both hoses are used?

57. Jack and Jill are stuffing envelopes for the upcoming election. Based on past experience, Jack can stuff the given set of envelopes in four hours and Jill can stuff the envelopes in three hours. How long will it take to do the job if both of them work together?

58. A cyclist rides 10 km at a speed of 20 km/h. She returns at a speed of 10 km/h.

(a) What is the total time for the round trip?

(b) What is her average speed for the round trip?

59. Dick drives at a speed of 65 mph from Lansing to Detroit, a distance of 120 miles. On the return trip he runs into traffic and drives at a speed of 50 mph. What is his average speed for the trip?

60. If Terry drives from Madison to Milwaukee, a distance of 100 miles, at a speed of 50 mph, how fast must she drive on the return trip to have an average speed of 60 mph? Of 65 mph? Is this realistic? Why?

61. Refer to Example 10 of this section in order to answer the following questions.

(a) Suppose you took two hours to shop and sightsee in Seattle. What would be your necessary average velocity to return to Eugene by midnight?

(b) How much time could you spend in Seattle without being forced to speed (average more than 55 mph) on the return trip?

62. The volume of a circular cylinder is given by

$$V = \pi r^2 h$$

where r is the radius of the base, h is the height, and $\pi \approx 3.1416$. What is the height of a cylindrical fuel tank that holds 5,000 cubic meters and has a radius of 14 meters?

63. A conical water storage tank has a radius of 4.2 meters and volume of 50 cubic meters. What is the height of the cone? In the case of a cone, $V = \frac{1}{3}\pi r^2 h$.

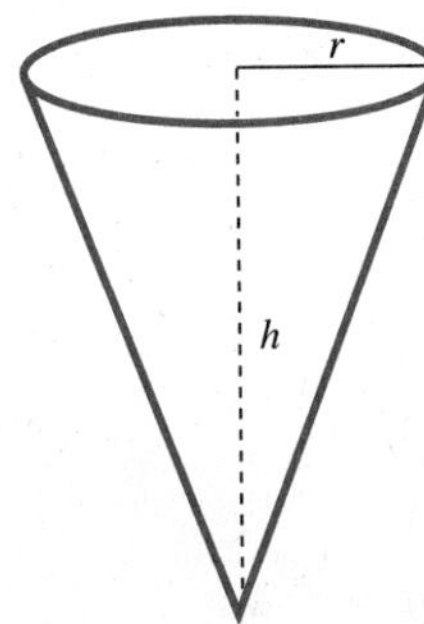

64. Suppose a cone and a cylinder have the same radius and height.

(a) What is the ratio of their volumes?

(b) Suppose a cylinder has a volume of 644 cubic inches. What is the volume of a cone with the same radius and height?

65. One-third of the reciprocal of a number is the same as 23 divided by one less than twice the number. What is the number?

66–70. **(a)** Set up the equation. **(b)** Solve by graphing. **(c)** Solve by using algebra (using equations).

66. Edith and Bill are departing from Indianapolis (see Example 7). They both stop a few times along the way, and this time Edith covers the 620 miles to Syracuse in $\frac{4}{5}$ the time that Bill took to travel the 500 miles to Kansas City. What was the average rate of each person for this trip if Edith drove 20 mph faster than Bill?

67. Suppose it takes Edith *twice as long* to drive the 620 miles from Indianapolis to Syracuse as it takes Bill to drive the 500 miles from Indianapolis to Kansas City.

What is the relationship between Bill's rate of speed and Edith's rate of speed?

68. Beth can fly her private plane 200 km against a 30-km/h wind in the same amount of time that she can fly 300 km with the wind. What is the airspeed of her plane?

69. Jack's motorboat cruises at 20 km/h in still water. It takes him twice as long to go 90 km upstream as it does to return 75 km downstream. Find the speed of the current.

70. It is approximately 2,400 miles from Honolulu to San Francisco. If a transport plane has a cruising speed of 350 mph and a tailwind of 50 mph (wind blows west to east), how many miles from Honolulu is the "point of no return" (the point where the time to fly to San Francisco is the same as the time to return to Honolulu)?

71. In order to obtain an estimate for the required volume V of timber that must be harvested to make expenses, a logging company uses the following formula:

$$P = \frac{SV + FV + Y + L + TV}{V}$$

where

V = required annual logging volume
Y = yarding cost in dollars
L = loading cost in dollars
S = skidding cost in dollars per cubic meter
T = transportation cost in dollars per cubic meter
F = falling cost in dollars per cubic meter
P = selling price in dollars per cubic meter

(a) Solve the equation for V.

(b) How many cubic meters must be harvested if the company estimates that the yarding cost will be \$25,000, the loading cost will be \$55,000, the skidding cost will be \$1.45 per cubic meter, the falling cost will be \$0.40 per cubic meter, transportation will cost \$0.60 per cubic meter, and the selling price will be \$2.50 per cubic meter?

72. In order to arrive at the marketable volume of a log, deductions need to be made for losses from slab s, edgings E, and sawdust S. The following formula is commonly used, where L is the length of the original log and V is the marketable volume of the log after the above losses:

$$V = 0.65(1 - E)(D - S)^2L$$

Solve this equation for E.

73. The total price P that must be paid by a consumer can be expressed as

$$P = N(C + T) + D$$

where N = number of items purchased, C = cost per item, T = tax per item, and D = total delivery charge.

(a) Solve this equation for C.

(b) Solve the equation for N.

(c) If a total price of \$21.20 was paid for a dozen ping-pong balls, the tax was \$0.10 per ball, and the delivery charge was \$3.20, what was the cost of each ball?

IV. Extensions

74. Express $\frac{1}{2}$ as the sum of two different unit fractions. A unit fraction is of the form $\frac{1}{n}$, where n is a positive integer.

75. Find all rectangles with integral sides whose area and perimeter are numerically equal.

76. What pairs of positive integers x and y have a harmonic mean equal to 4? The harmonic mean of x and y is given by

$$\frac{2xy}{x + y}$$

77. Find all pairs of integers whose product is positive and equal to twice their sum.

78. Given a point P, find all integers n such that the plane around P can be covered by nonoverlapping congruent regular n-gons.

79. For which positive integers $n > 2$ is $(n - 2)$ a factor of $2n$?

80. Show that

$$\frac{1}{x + 1} + \frac{1}{x + 2} + \cdots + \frac{1}{2x} > \frac{1}{x}$$

for all integer values of $x > 1$.

81. Find all pairs of fractions $\frac{a}{b}$ and $\frac{c}{d}$ for which the following mistake

$$\frac{a}{b} + \frac{c}{d} = \frac{a + c}{b + d}$$

is actually correct.

CHAPTER 5 SUMMARY

In preceding chapters, we solved time-distance-rate problems. These problems give rise to equations that have variables in the denominator. In this chapter, equivalent forms of fractions were reviewed. We discussed the factoring of polynomials and operations on rational expressions, which culminated in a study of rational equations. Rational equations can be solved either algebraically or by graphing.

Important Words and Phrases

complex fractions
division
quotient
rational expressions
perfect-square trinomial
least common denominator (LCD)
rational equations
rational functions
literal equations

Important Properties and Procedures

- Factoring polynomials
- Factoring polynomials by systematic trial and error
- Factoring the difference of two squares: $a^2 - b^2 = (a + b)(a - b)$
- Factoring a perfect square trinomial: $a^2 - 2ab + b^2 = (a - b)^2$ $a^2 + 2ab + b^2 = (a + b)^2$
- Fundamental property of fractions
- Expressing algebraic fractions in simplest form
- Multiplication and division of rational expressions
- Addition and subtraction of rational expressions
- Finding equivalent rational expressions
- Finding equivalent equations
- Solving rational equations algebraically or by graphing

REVIEW EXERCISES

I. Procedures

1–10. Factor completely.

1. $3x^2y - 6xy^3 + 15x^2y^3$
2. $x^2 + 2x + 1$
3. $2x^3y^2 - 2x$
4. $25x^2 - 1$
5. $x^2 + 2x - 15$
6. $16x^2 - 8x + 1$
7. $10x^2 - 3xy + y^2$
8. $12x^3 - 2x^2 - 4x$
9. $ax - ay - 3x + 3y$
10. $10x^3 + 40x - 2x^2 - 8$

11–26. Perform all indicated operations. Express fractions in simplest form.

11. $\dfrac{a^2 - 2ab - 3b^2}{a^2 + 2ab + b^2}$
12. $\dfrac{3x - 3y}{x - y}$
13. $\dfrac{a^2 + 1}{a + 1} - 1$
14. $\dfrac{4x - y}{xy} + \dfrac{x - y}{2xy^2} + x$
15. $\dfrac{z}{2xy} + \dfrac{3}{x^2} - \dfrac{5z}{12x^3}$
16. $\dfrac{x + 3}{x^2 - x} - \dfrac{12}{x^2 + x - 2}$
17. $\dfrac{b}{a + b} + \dfrac{a}{a - b} - \dfrac{2ab}{a^2 - b^2}$
18. $\dfrac{x - 2}{x^2 + 10x + 16} + \dfrac{x + 1}{x^2 + 9x + 14}$
19. $\dfrac{x^2 - 5x - 6}{3x} \cdot \dfrac{a^2 - 4a - 12}{a^2 - 4}$
20. $\dfrac{a^2 - 7a + 6}{1 - a} \div \dfrac{a^2 - 4a - 12}{a^2 - 4}$
21. $\left(\dfrac{4x^3 - 16x}{12x^3} \cdot \dfrac{x^2 - 5x + 6}{x^2 - 4x}\right) \div \dfrac{3 - x}{2x^2}$
22. $\dfrac{4b^2}{b - 1} \cdot \dfrac{b^2 - 2b + 1}{b + 1} \cdot \dfrac{1 - b^2}{b^3 - b^2}$
23. $\left(\dfrac{x + 1}{x} + \dfrac{5}{6}\right) \div \dfrac{1}{2x}$
24. $\dfrac{5}{2}\left(\dfrac{1}{x + y} - \dfrac{1}{x - y}\right) \div \dfrac{5y}{x^2 + 2xy + y^2}$

25. $\dfrac{x - \dfrac{1}{x}}{\dfrac{1}{x^2} - 1}$

26. $\dfrac{\dfrac{4}{z^2} - 1}{z - \dfrac{4}{z}}$

27–36. Solve algebraically or by graphing. Check your answers by substituting.

27. $\dfrac{3}{x} - \dfrac{5}{2} = -2$

28. $\dfrac{3}{4}y - \dfrac{1}{2}y = -2$

29. $\dfrac{2}{5b} - \dfrac{1}{10} = \dfrac{1}{2b}$

30. $\dfrac{3}{x - 1} = \dfrac{3}{2}$

31. $2 - \dfrac{1}{y + 2} = \dfrac{y + 1}{y + 2}$

32. $\dfrac{1}{z + 3} + 1 = \dfrac{6}{z + 3}$

33. $\dfrac{1}{a - 3} + \dfrac{2}{a + 3} = \dfrac{3}{a^2 - 9}$

34. $\dfrac{3}{x - 2} = \dfrac{12}{x^2 - 4} - \dfrac{2}{x + 2}$

35. $\dfrac{2}{x} = 7 - \dfrac{3}{2x}$

36. $\dfrac{3}{x + 5} + \dfrac{2}{x - 2} = \dfrac{x}{x^2 + 3x - 10}$

37–40. Solve for the indicated variable.

37. Solve for b: $\dfrac{a}{b} = \dfrac{c}{d}$

38. Solve for L: $\dfrac{a}{L} + 2b = d$

39. Solve for M: $\dfrac{1}{M} + W = \dfrac{c}{M}$

40. Solve for L: $S = \dfrac{n}{2}(a + L)$

41. Evaluate

$$\frac{n}{2}(a + L)$$

if $n = 10$, $a = 0.01$, and $L = 10^3$.

42. Evaluate

$$\frac{1}{2}bh$$

if $b = \frac{2}{3}$ and $h = 1\frac{3}{5}$.

43. Find d, given that

$$d = \frac{a}{L} + 2b$$

when $a = 1.2$, $L = 0.6$, and $b = 0.001$.

44. Find M, given that

$$M = \frac{F}{a}$$

when $F = \frac{56}{0.01}$ and $a = 0.007$.

II. Concepts

45. Which of the following pairs of fractions are equal?

(a) $\dfrac{9}{-27}, -\dfrac{1}{3}$ **(b)** $\dfrac{2x - 4}{12}, \dfrac{-2 + x}{6}$

(c) $\dfrac{x - a}{x + b}, \dfrac{-a}{b}$ **(d)** $\dfrac{x - 1}{1 - x}, \dfrac{-1}{1}$

(e) $\dfrac{x - 5}{-2}, \dfrac{5 - x}{2}$ **(f)** $\dfrac{-ax}{bx}, \dfrac{a}{-b}$

46. Which of the following fractions are equal to -1? Check your answers by graphing.

(a) $\dfrac{-x}{x}$ **(b)** $\dfrac{x - 1}{x}$

(c) $\dfrac{x - 1}{1 - x}$ **(d)** $\dfrac{-x + 2}{x + 2}$

(e) $\dfrac{(-2x)^3}{6x^3}$

47–50. Find two fractions that are equivalent to each fraction.

47. $\dfrac{14}{25}$ **48.** $\dfrac{x + 1}{2x}$ **49.** $\dfrac{x^2 - 4}{x^2}$ **50.** $\dfrac{x + 1}{x - 1}$

51 and 52. **(a)** Express each fraction as the sum of two fractions. **(b)** Express each fraction as the difference of two fractions.

51. $\dfrac{x}{x^2 + 1}$ **52.** $\dfrac{x - 1}{x}$

53 and 54. **(a)** Express each fraction as the product of two fractions. **(b)** Express each fraction as the quotient of two fractions.

53. $\dfrac{2xy}{x + y}$ **54.** $\dfrac{mn}{4}$

55. Explain how you can tell if a fraction is in simplest form.

56. Show that -3 is a solution for

$$\frac{1}{x} - \frac{3}{x-2} = \frac{4}{x(x-2)}$$

57. **(a)** Find two numbers whose sum is 12 and whose product is 32.

(b) Factor $x^2 + 12x + 32$.

(c) Explain the similarity between parts (a) and (b).

58. The math instructor for Math 100 forgot to put the instructions next to each problem. The instructions are: "Solve for x" or "Perform the indicated operations and express all fractions in simplest form." Put the correct instructions with each problem and then carry them out.

(a) $\dfrac{1}{x} + \dfrac{x}{x+1} - 1$ **(b)** $\dfrac{1}{x} + \dfrac{x}{x+1} = 1$

(c) $\dfrac{2}{x+2} = \dfrac{3}{x^2 + 4x + 4}$

(d) $\dfrac{3x-1}{x^2 + 2x - 15} - \dfrac{2}{x+5}$

III. Applications

59 and 60. Examine the patterns. For each pattern **(a)** write the pattern for the next three rows. **(b)** What is the sum of the fractions in the tenth row? Thirtieth row? *N*th row? **(c)** Graph the sums and their corresponding row numbers. What happens to the sum as the row number gets larger and larger?

59.

$$\frac{1}{6} = \frac{1}{6}$$

$$\frac{1}{6} + \frac{1}{36} = \frac{7}{36}$$

$$\frac{1}{6} + \frac{1}{36} + \frac{1}{216} = \frac{43}{216}$$

60.

$$\frac{1}{8} = \frac{1}{8}$$

$$\frac{1}{8} + \frac{1}{64} = \frac{9}{64}$$

$$\frac{1}{8} + \frac{1}{64} + \frac{1}{512} = \frac{73}{512}$$

61. John can cut his lawn in 2 hours, but he wants to get done before the rain, so he calls his friend Marsha to help. If they finish the job together in 1 hour and 15 minutes, how long would it have taken Marsha to mow the lawn by herself?

62. The mathematics department has two photocopying machines. One can produce 30 pages a minute and the other can produce 45 pages a minute. The department needs 500 copies of a 5-page final exam for Math 100.

(a) How long will it take each machine to produce the test?

(b) How long will it take the two machines to do the job together?

63. A torrential rain can fill the basin behind a flood-control dam in 6 hours. It is known that the basin with its floodgates open will empty in 18 hours. How long will it take to fill the basin in a torrential rain with the floodgates open? Is this realistic? Why?

64. A man walks five miles in 2 hours and returns in 2.5 hours. What is his average walking speed for the round trip?

65. A cyclist makes a trip of 20 km at a speed of 12 km/h. At what speed must a cyclist pedal on the way back to have an average speed of 15 km/h? Of 20 km/h?

66. Suppose a cyclist with a speed of S_1 mph cycles d miles on the way out and returns at a speed of S_2 on the way back.

(a) Find a general equation for the average speed.

(b) What variables appear to affect the average speed of the cyclist?

(c) Compare this equation with the equation in Example 10 of Section 5.4.

67. If you go somewhere at an average rate of 60 mph and come back the same way at an average rate of 40 mph, what is the average rate for the entire trip? (It is not 50 mph.)

68. The product of a number n and $\frac{3}{4}$ is $\frac{5}{6}$. Find n.

69. If $\frac{1}{3}$ is divided by n, the result is -2. Find n.

70. If 4 is subtracted from n, the result is $-\frac{5}{6}$. Find n.

71. The sum of 10 and the reciprocal of n is $\frac{5}{6}$. Find n.

72. The sum of the reciprocals of two consecutive integers equals 7 divided by the product of the two integers. Find the integers.

73. **(a)** If $2x^2 + 13x + 15$ represents the area of a rectangle, find an expression for the length and width of the rectangle.

(b) Illustrate part (a) with a diagram.

(c) Check your results from part (a) by letting $x = 2$.

74–78. Solve algebraically using equations and by graphing.

74. Arnold can fly his private plane 200 miles against the wind in the same time it takes him to fly 300 miles with the wind. If the wind speed is 30 mph, find the airspeed of the plane.

75. An airplane whose cruising speed in still air is 220 mph can travel 520 miles with the wind in the same time it takes to fly 360 miles against the wind. Find the speed of the wind.

76. Julie's motorboat cruises at 20 mph in still water. It takes her twice as long to go 90 miles upstream as it does to return 75 miles downstream. Find the speed of the current.

77. The Big Muddy has a current of 3 mph. A motorboat takes the same amount of time to go 12 miles downstream as it does to go 8 miles upstream. What is the speed of the boat in still water?

78. Suppose the Detroit Tigers win 15 of their first 20 games of the season. After that, they win only half the time. How many games had they played when their winning percentage was 0.60 (60%)?

79. Symmetrically tapered pieces of wood are used in making furniture. A formula used to determine the taper T (in inches per foot) of each side is

$$T = 6\left(\frac{w - x}{L}\right)$$

where w, x, and L (in inches) are given below:

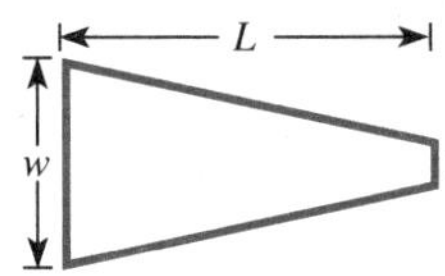

How long is a table leg if $w = 5$ inches, $x = 2$ inches, and $T = \frac{3}{4}$ inch per foot?

80. Solid top decks on houses are generally built with a taper to allow for water runoff. A formula that gives the taper T (in inches per foot) of the top is

$$T = \frac{12(w - x)}{L}$$

where w, x, and L (in inches) appear in the diagram below. Determine w if $x = 20$ inches, $L = 100$ inches, and $T = \frac{1}{4}$ inch per foot.

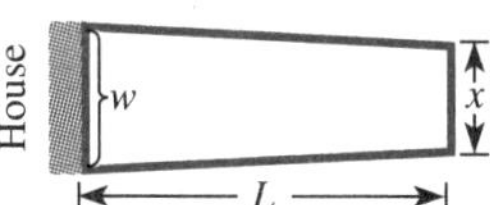

81. The harmonic mean of two numbers x and y is

$$\frac{2xy}{x + y}$$

If the harmonic mean of 4 and another number is 4.8, what is the other number?

82. Many stores offer an installment plan, whereby the customer agrees to pay a certain finance charge. If the loan is paid off early, the customer does not have to pay the entire finance charge. The formula used is

$$C = \frac{k(k + 1)}{n(n + 1)}f$$

where C is the amount of the finance charge you *do not* pay if the loan is paid off early; f is the original finance charge; n is the original number of payments intended; and k is the number of payments remaining when the loan is paid off.

(a) Solve the formula for f.

(b) Find the original finance charge f if there were 36 installments, but the loan was repaid in 12 installments, and the amount C you do not have to pay is $180.18.

CHAPTER

Quadratic Equations and Quadratic Functions

In previous chapters, we investigated several relationships between two variables. Many situations involved "straight-line growth" and could be described by linear functions. Others involving non-linear models were described by exponential functions or rational functions. In this chapter we will study quadratic functions, which were first presented in Chapter 1.

Source: William Waterfall/The Stock Market

6.1 EXPLORING QUADRATIC RELATIONSHIPS: TABLES, GRAPHS, EQUATIONS

In Example 1 of Chapter 1, we searched for the rectangle with the greatest area that had a fixed perimeter. The graph of this pattern was a parabola—the graph of a quadratic function. We now will look at this function in more detail.

EXAMPLE 1

(a) The area A of a circle of radius r is given by

$$A = \pi r^2$$

We can show the dependency of area on radius by Area(r) $= \pi r^2$

A depends on r^2.

(b) The height h of a ball after t seconds thrown upward with an initial velocity V_0 is given by

$$h = \frac{-gt^2}{2} + V_0 t$$

We can show the dependency of height on time by height(t) $= \frac{-gt^2}{2} + V_0 t$.

Here, h depends on t^2, and $g \approx 32.2$ feet/sec^2 is the acceleration due to gravity.

(c) The relationship between the length l of a pendulum chord and the time t that it takes the pendulum to complete a swing through its arc is given by

$$l = \frac{g}{2\pi} t^2$$

We can show the dependency of length on time by length(t) $= \frac{g}{(2\pi)} \cdot t^2$.

Here, l depends on t^2, and g is the acceleration due to gravity.

(d) If D is the (average) diameter of a log, then the number of board feet (a way of estimating usable lumber) in a 16-foot log is given by

$$B = 0.8(D - 1)^2 - \frac{D}{2}$$

We can show the dependency of board feet on diameter by Board feet(D) $= 0.8(D - 1)^2 - \frac{D}{2}$.

Here B depends on D^2.

(e) The surface area A of a sphere depends on the radius r of the sphere according to

$$A = 4\pi r^2$$

We can show the dependency of surface area on the radius by Surface Area(r) $= 4\pi r^2$.

A depends on r^2.

(f) If F is the tension (force) in a string, then the velocity V of the waves in a vibrating string is given by

$$F = \mu V^2$$

We can show the dependency of force on velocity by Force(V) $= \mu V^2$.

F depends on V^2, and μ is the density (mass per unit of length) of the string. ■

Each of the previous relationships differs from our earlier linear models in that one of the variables is of the *second degree.* Relationships between *two* (or more) variables, as in Examples (a) through (f), where at least one of the variables is of second degree, are called **quadratic models.** The graphs of quadratic models can give us a lot of information about the behavior of the variables involved.

This chapter is devoted to the study of quadratic models and their graphs. Let us first spend some time looking at quadratic models and equations in an *informal* way. In the rest of this chapter, we will concentrate on the solution of quadratic equations and their applications.

Quadratic Models: Approximating Solutions from Tables, Graphs, and Successive Approximation

In Chapter 5, we considered the following example:

EXAMPLE 2

The height h (in feet) above the ground of a golf ball depends on the time t (in seconds) it has been in flight. One golfer hits a tee shot that has an approximate height given by

$$h = 80t - 16t^2 \qquad \text{Height }(t) = 80t - 16t^2\text{. Height is a function of time } t.$$

How long does it take for the ball to hit the ground?

Solution

The ball is on the ground when $h = 0$. We set $h = 0$ and obtain a quadratic equation in the variable t:

$$0 = 80t - 16t^2$$

The values of t that yield a height $h = 0$ could be obtained by factoring:

$$0 = t(80 - 16t)$$

so

$$t = 0 \quad \text{or} \quad t = 5$$

Substituting, we find
Height (0) = 0(80 − 16 · 0) = 0
Height (5) = 0(80 − 80) = 0

The ball will hit the ground after 5 seconds.

There is an implicit assumption in this example: The ball lands at the same elevation as that of where it was hit. In other words, the tee and the fairway are the same elevation, as in Figure 6.1.

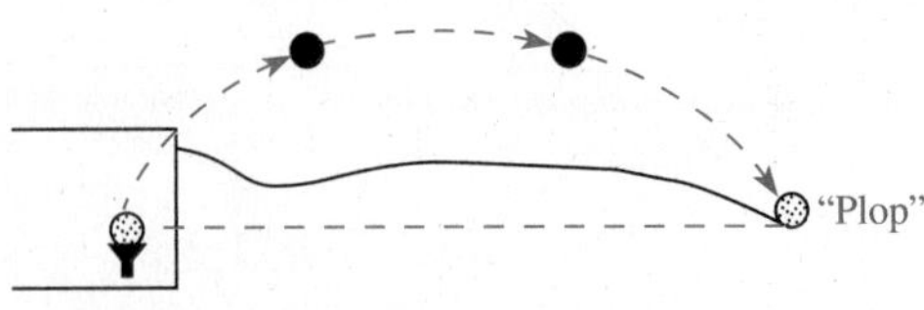

Figure 6.1

However, if the golfer were hitting the ball from a tee that was *elevated 20 feet above the fairway* (as in Figure 6.2), then how long would it take for the ball to hit the ground? How high would the ball go?

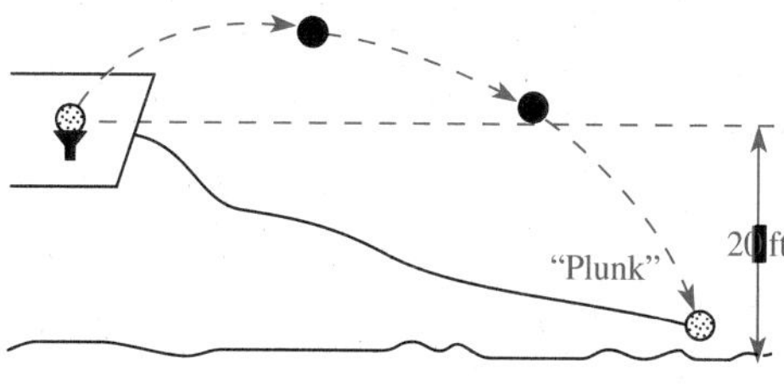

Figure 6.2

EXAMPLE 3

Suppose the height (in feet) of the golf ball above the fairway is given by

$$h = 80t - 16t^2 + 20 \qquad \textbf{Height } (t) = 80t - 16t^2 + 20.$$

after t seconds.

1. How long would it take for the ball to hit the fairway?
2. How high does the ball go?

Solution

Using a Table One possible approach to this problem is to make a table by substituting values of time and computing the corresponding values of height, with a calculator if desired. Substitute t values into $h = 80t - 16t^2 + 20$ to obtain the corresponding h values. For example, at $t = 0$,

$$h = 80(0) - 16(0)^2 + 20 = 20$$

The height of the elevated green is 20 feet at $t = 0$. At $t = 1$,

$$h = 80(1) - 16(1)^2 + 20 = 84$$

and so on.

NOTE: To help generate a table on a graphing calculator, you can use a TABLE command or you can enter $80 * 1 - 16 * 1^2 + 20$. Then use the repeat key and edit the expression—changing 1 to 2, 2 to 3, and so on. ■

At $t = 6$, $h = -76$. The ball does not bury itself 76 feet (unless it hits a deep water hazard), so it has hit the ground before $t = 6$.

t (seconds)	*h (feet)*	
0	20	Ball is going up.
1	84	
2	116	
3	116	Ball is coming down.
4	84	
5	20	
6	−76	

Table 6.1

We can learn several things from this table. The ball rises for at least two seconds, falls for at least two seconds, and hits the ground *before* six seconds. *Exactly when* does it hit the ground? At what time does the ball reach its maximum height? Surely the ball does not level off at 116 feet for the entire time between 2 and 3 seconds. The table only gives us information at a finite number of time points. We would like to be able to tell what the ball is doing at *any time.* A graph may help us get a better picture.

Using a Graph We plot the pairs of values from Table 6.1 on a coordinate axis. The time is used in our graph as the horizontal axis, since height h depends on time t. We calculate some additional points so that our graph is a better approximation (Table 6.2).

(seconds)	*h (feet)*
0.5	56
1.7	109.76
2.6	119.84
3.4	107.04
4.2	73.76
5.5	−24

Table 6.2

We substitute different values of t into $h = 80t - 16t^2 + 20$ to obtain the corresponding values of h. When $t = 0.5$ and $t = 1.7$,

$$h = 80(0.5) - 16(0.5)^2 + 20 = 56$$
$$h = 80(1.7) - 16(1.7)^2 + 20 = 109.76$$

and so forth.

To graph our data, we plot the points (pairs) from Table 6.2. These are pairs of the form (t, h), where t is time and h is height. We then draw a freehand sketch through the plotted points, and we obtain the graph in Figure 6.3. This graph contains the points from Tables 6.1 and 6.2. It indicates that the high point occurs between 2 and 3 seconds, and that the ball hits the ground after 5 seconds but before 5.5 seconds. We want to know *exactly* (if possible) when the ball hits the ground.

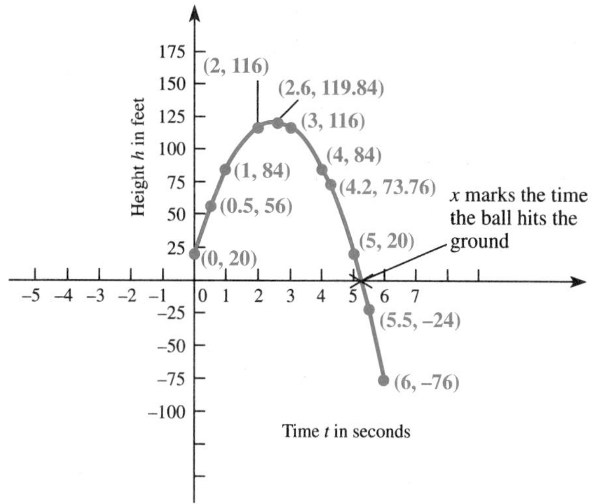

Figure 6.3

One approach would be using a calculator and plugging in values of t in order to approximate the value of t that would make $h = 0$. (The ball hits the gound when height is equal to zero.)

Using Successive Approximation From the graph in Figure 6.3, we see that if $t = 5$, then $h = 20$, and if $t = 5.5$, then $h = -24$. Therefore the ball hit the ground before 5.5 seconds had elapsed. Let us try values of t between 5 and 5.5. If $t = 5.25$, then

$$h = 80(5.25) - 16(5.25)^2 + 20 = -1 \text{ feet}$$

We are very close, but 5.25 seconds is a bit too long; the ball has already hit the ground. If $t = 5.2$,

$$h = 80(5.2) - 16(5.2)^2 + 20 = 3.36 \text{ feet}$$

At $t = 5.2$ seconds, the ball has not quite yet hit the ground. Now we know the ball hits between 5.2 and 5.25 seconds, but probably closer to 5.25 seconds.

If $t = 5.24$, $h = 80(5.24) - 16(5.24)^2 + 20 = -0.12$ feet

If $t = 5.23$, $h = 80(5.23) - 16(5.23)^2 + 20 = 0.75$ feet

If $t = 5.238$, $h = 0.05$ feet

If $t = 5.239$, $h = -0.03$ feet

If $t = 5.2384$, $h = 0.019$ feet

If $t = 5.2385$, $h = 0.0099$ feet

This method is sometimes called *successive approximation*. You see that we *can* get a closer approximation by picking values for t that are between the two previous choices.

Using an Equation What we would really like to do is find the value of t that makes $h = 0$ in $h = 80t - 16t^2 + 20$. That is, we wish to solve

$$0 = 80t - 16t^2 + 20 \qquad \textbf{Equation (1)}$$

Equation 1 is called a quadratic equation in one variable. The expression $80t - 16t^2 + 20$ *is a polynomial of second degree in the variable t.* ■

One of our goals in this chapter is to obtain an algebraic method for determining the time at which the ball hits the ground (when $h = 0$) and a method for determining the high point of the ball's flight (when h achieves its largest value). In this section we will solve problems involving quadratic models using the three methods above: **tables, graphs,** and **successive approximation.** These three methods are useful for solving problems involving *any* type of equation and so they deserve special attention. A detailed study of the algebra of quadratic equations occurs in Sections 6.2 to 6.5.

EXAMPLE 4

What is the height of the tallest tree you can brace with a 250-foot wire? The wire must be anchored to the ground at a distance from the base of the tree that is half the height of the tree. The wire must be tied onto the tree 10 feet down from its top.

Solution

Using an Equation We wish to find the height of the tree. So let x equal the height of the tree.

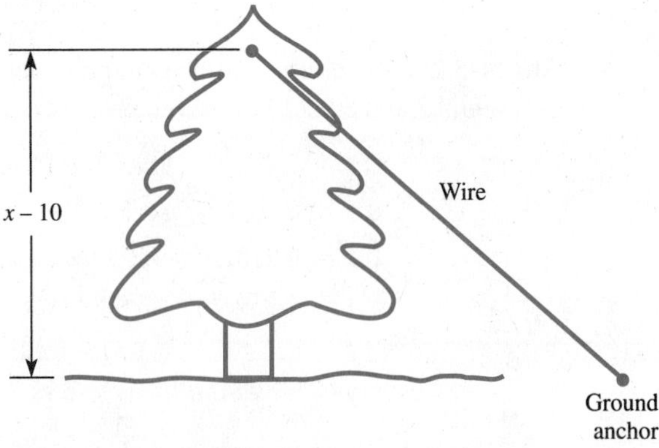

Figure 6.4

We can tie the wire onto the tree 10 feet from the top, or $x - 10$ feet from the ground. We must anchor the wire in the ground at a distance that is half the height of the tree, or $\frac{x}{2}$.

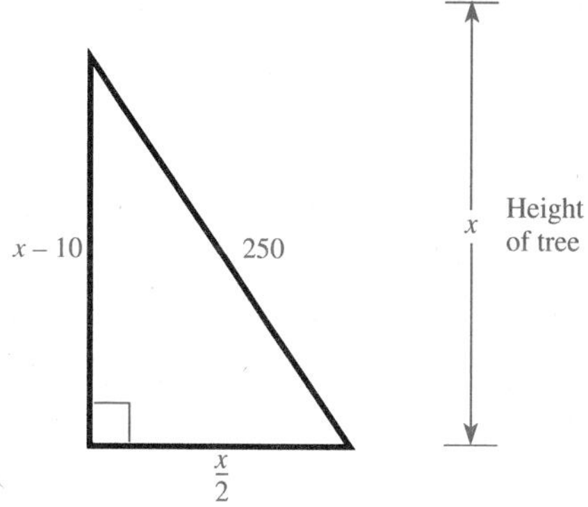

Figure 6.5

We can picture the situation using the figure above. The wire is the hypotenuse of a right triangle. If we use the Pythagorean theorem for the sides of a right triangle, then we obtain

Recall the Pythagorean theorem states that the square of the hypotenuse is equal to the sum of the squares of the two legs: $c^2 = a^2 + b^2$.

$$(x - 10)^2 + \left(\frac{x}{2}\right)^2 = (250)^2$$

$$x^2 - 20x + 100 + \frac{x^2}{4} = 62{,}500 \qquad \textbf{Expanded.}$$

$$\frac{5x^2}{4} - 20x + 100 = 62{,}500 \qquad \textbf{Like terms grouped.}$$

$$5x^2 - 80x = 249{,}600. \qquad \textbf{Subtract 100. Multiply each side by 4.}$$

We may also divide all terms by 5 and get the further equivalent equation

$$x^2 - 16x = 49{,}920$$

The value of x that solves this last equation will give us the height of the tallest tree, because this equation is equivalent to the original equation.

Using a Table

x (feet)	$y = x^2 - 16x$	
		The tree must be at least 16 feet high. Why?
50	3,300	
100	8,400	**$100^2 - (16 \times 100) = 8{,}400$ The value of x is still way too small. The tree is higher than 100 feet.**
200	36,800	**This is getting closer to 49,920.**
250	58,500	**Now x is too big.**

Table 6.3

From the table, we know the tree is between 200 and 250 feet high.

Using a Graph The numbers in this example are quite large. Thus, the graph of this relationship is best seen on a graphing calculator, where you can easily adjust the scale of the graph. The graph below shows a few points in addition to the ones we already obtained in the table. Notice that the graph has negative values until the tree is at least 16 feet high.

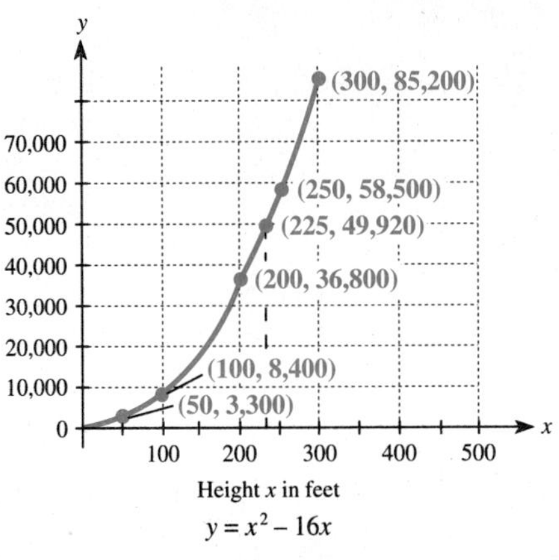

Figure 6.6

Graph $Y1 = x^2 - 16x$ on a graphing calculator. Note that from the table above, we know that the x-axis should be scaled in hundreds.

Using Successive Approximation We use a calculator to compute more accurately the height x of the tallest tree that can be braced with a 250-foot wire. This value of x will make $x^2 - 16x = 49{,}920$. From the table and graph above, x appears close to 225 feet.

x (feet)	$y = x^2 - 16x$	
225	47,025	$225^2 - (225 \times 16) = 47{,}025$
230	49,220	
232	50,112	Here we are getting quite close. The height is between 230 and 232 feet.
231	49,665	
231.5	49,888.25	
231.6	49,933	
231.575	49,921	This is a close approximation. The tallest tree is about 231.575 feet tall. ■

Table 6.4

EXAMPLE 5 The revenue R (in dollars) that a business makes from selling x videotapes in a day is given by $R = 50x - x^2$. How many videotapes should be sold to make the most revenue?

Solution

We will make a table and a graph of the relationship between the number x of videotapes sold and the money R collected, in order to estimate the number of videotapes that will bring in the maximum revenue.

Using a Table

x (videotapes)	$R = 50x - x^2$ *(dollars)*	
0	0	Nothing ventured, nothing gained.
1	49	
2	96	
3	141	$50(3) - 3^2 = 141$
10	400	
20	600	Still increasing.
30	600	What happens between 20 and 30?
25	625	
27	621	Revenue is lower at both 27 and 24 than at 25 videotapes sold.
24	624	

Table 6.5

Using a Graph From the accompanying graph, it appears that the maximum revenue will occur when 25 videotapes are sold.

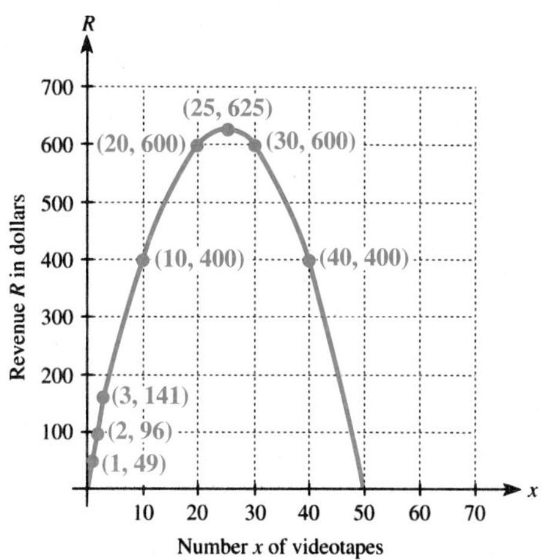

Figure 6.7

Graph of $R = 50x - x^2$. Can you think of some reasons why profit might go down if too many videotapes are sold each day? What happens to R if 50 videotapes are sold? If 60 videotapes are sold?

6.1 EXERCISES

I. Procedures

1–7. In each quadratic model below, compute the value of the indicated variable (use $\pi \approx 3.14$).

1. $A = \pi r^2$. If $r = 3.7$ meters, find A (area of circle).

2. $l = \frac{gt^2}{2\pi}$. If $g = 32.2$ and $t = 1.8$, find l (length of a pendulum chord).

3. $B = \left(\frac{D-4}{4}\right)^2$. If $D = 21.83$, find B (board feet estimate for log diameter D).

4. $B = 0.9(D - 4)^2 - \left(\frac{D}{2}\right)$. If $D = 30$, find B (board feet in a log of diameter D).

5. $A = 4\pi r^2$. If $r = 8$ inches, find A (surface area of a sphere of radius r).

6. $b = 0.22h^2 - 0.71h$. If $h = 7.5$, find b (international rule for board feet in a log of height h).

7. $R = 16.32t - 1.8t^2$. If $t = 5$, find R. If $t = 10$, find R (revenue from a manufacturing process).

8–16. Make up a table and sketch a graph for each of the following relationships. You may wish to use a graphing calculator.

8. $y = 3x^2$

9. $y = -16x^2$

10. $y = x^2 - 36x$

11. $y = 40x - 10x^2$

12. $y = x^2 + 5x + 6$

13. $A = 4\pi r^2$ (surface area of a sphere of radius r).

14. $l = \frac{gt^2}{2\pi}$, $g \approx 32.2$ feet/sec^2 (l is the length of a pendulum chord).

15. $b = 0.22h^2 - 0.71h$ (international rule for board feet in a log of length h).

16. $R = 16.32x - 1.8x^2$ (revenue from a manufacturing process of x items).

II. Concepts

17–20. Find an approximate solution to each of the following equations. Graph each equation and then use the method of successive approximation. Carry out the computations to at least two decimal places. A graphing calculator will help.

17. $x^2 = 32$, given there is a solution between 0 and 8.

18. $x^2 + 5x = 0$, given there is a solution between $x = -3$ and $x = -6$.

19. $x^2 + 5x + 5 = 0$, given there is a solution between $x = -3$ and $x = -4$.

20. $0.7y^2 - 0.35y = 0$, given there is a solution between $y = 0$ and $y = 1$.

21. The following graph describes the flight of a missile fired from the top of a hill. The graph indicates height after t seconds.
From the graph, estimate the following quantities:

(a) At what time does the missile reach its highest point?

(b) What is the maximum height attained by the missile?

(c) How long does it take for the missile to hit the ground?

(d) At what time is the missile 2,000 feet high?

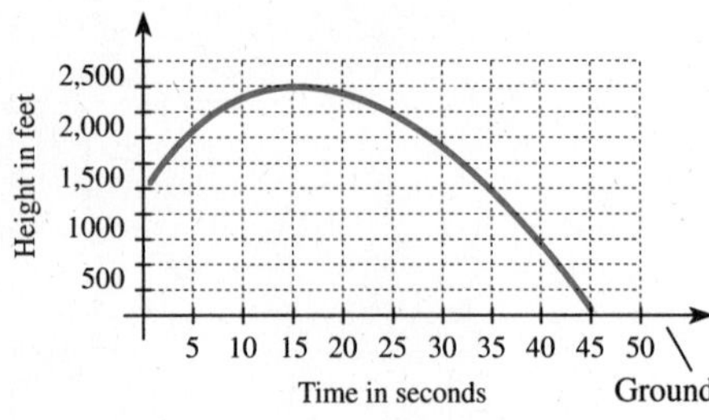

22. The height of a frog t seconds after a jump is described in the graph below.

(a) At what time does the frog reach the highest point?

(b) What is the maximum height attained by the frog?

(c) How long does it take for the frog to land?

(d) At what time is the frog 6 inches high?

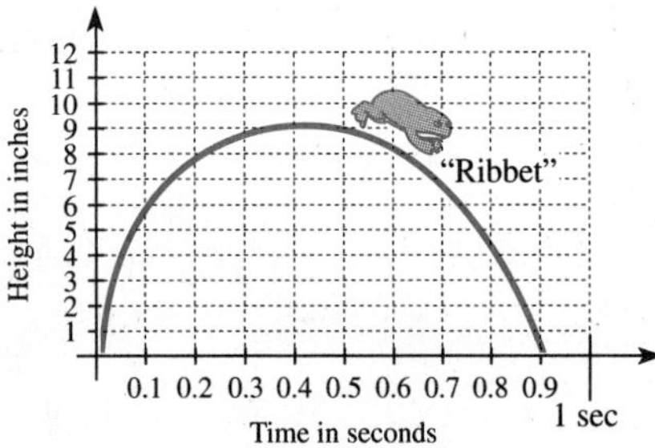

23. Suppose you have the graph of $y = x^2 - 9x$ at your disposal. Write a paragraph explaining how you would use the graph to find the value(s) of x when $y = 1{,}750$.

24. Write a paragraph explaining how you would use a calculator to successively approximate the value(s) of x where $x^2 - 9x = 1{,}750$. Include how you would decide what values of x you would choose first.

25. What are some advantages and disadvantages of approximating a solution to a quadratic equation using a graph?

26. What are some advantages and disadvantages of approximating a solution to a quadratic equation using successive approximation?

III. Applications

27–33. For each of these problems, first write an equation for the quadratic model that represents the problem. Then construct tables and graphs and calculate approximations for a solution. Use a graphing calculator, if possible, to relate the graph to the approximate solution more accurately.

27. A number plus its reciprocal is equal to 9.

28. The dimensions of the rectangle below.

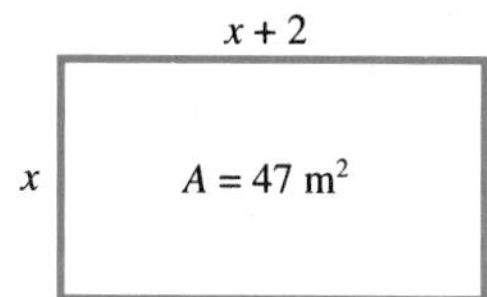

29. The sides of the triangle below.

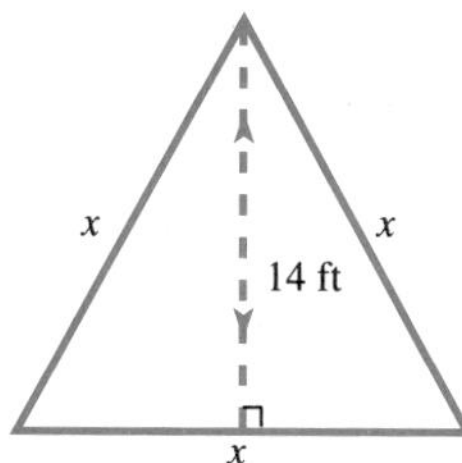

30. The radius of the sphere below (see Exercise 13).

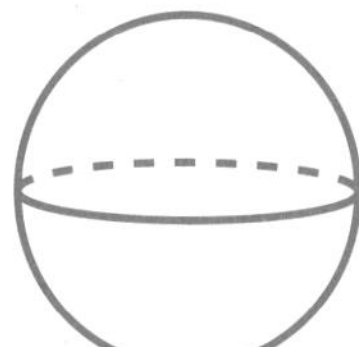

Surface area = 121 m²

31. The cost of manufacturing a number of tubes of suntan lotion is the square of the number of tubes minus 6 times the number of tubes. Determine the number of tubes that can be produced for $628,000.

32. The time it takes for a pendulum with a chord 10 meters long to complete a swing through its arc. See part (c) of Example 1.

33. The width w of an annulus (washer), when its area is 15 cm^2 and the inner circle has a radius of 1 cm.

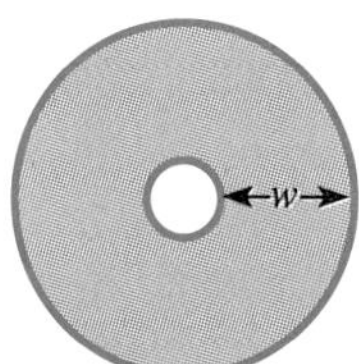

34. A girl throws a stone in the air. The height h in feet of the stone after t seconds is given by

$$h = 50t - 6t^2$$

(a) Graph the relationship between the height and the time.

(b) According to your *graph,* find

1. about what time the stone lands on the ground,
2. about what time the stone reaches its highest point, and
3. the maximum height of the stone.

(c) Starting with the estimates taken from your graph in part (b), use successive approximations accurate to two decimal places to find

1. the time the stone lands on the ground,
2. the time the stone reaches its maximum height, and
3. the maximum height of the stone.

35. The revenue R (in millions of dollars) that a big business makes from selling x franchises is given by

$$R = -10 + 10x - x^2$$

(a) Graph the relationship between the revenue and the number of franchises.

(b) How much revenue is made if the business sells no franchises? If it sells 8 franchises? If it sells 20 franchises?

(c) How many franchises should be sold to make the most revenue?

(d) How many franchises should be sold if the company wants to keep its revenue at about $6 million?

(e) Can you think of a reason why revenue goes down when the company sells too many franchises?

36. A farmer has 2,400 meters of fence. He wants to enclose a rectangular plot with the fence so that the *largest possible* area is enclosed by the fence.

(a) Graph the relationship between the length l of the rectangle (horizontal axis) and the resulting *area* of the rectangle (vertical axis). *Hint:* Recall that the area of a rectangle is $A = lw$. Pick a value for l, say, $l = 300$ meters. Then you can find the corresponding width w, because you know that the perimeter of this rectangle is given by $2{,}400 = 2l + 2w$; so $1{,}200 = l + w$. Therefore $w = 900$ if $l = 300$, and area $= 270{,}000$ m^2.

(b) What type of graph did you obtain?

(c) What are the dimensions of the rectangle that has the greatest area? (The graph and a calculator may help.)

l (meters)	A (square meters)
0	0
100	⋮
200	⋮
300	270,000
⋮	⋮
1,200	0

37. Suppose the farmer of Exercise 36 wishes to use the 2,400 meters of fence to enclose a rectangular plot on three sides, using a creek as the boundary on the fourth side, as in the figure.

(a) Graph the relationship between the length l and the resulting area of the rectangle.

(b) What type of graph did you obtain?

(c) What are the dimensions of the rectangle that has the greatest area in this case? (The graph and a calculator will help.)

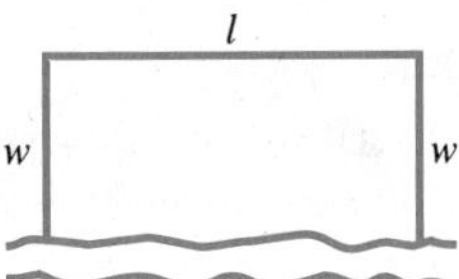

38. Refer to Examples 2 and 3 in this section. In each instance, calculate to two decimal places the maximum height reached by the golf ball. Use a calculator and successive approximation.

39. A frog leaps to a height h (in inches) after time t (in tenths of seconds) given by

$$h = -2t^2 + 10t$$

(a) Graph this relationship, with time on the horizontal axis. Use the graph and calculator approximations to answer the following questions:

(b) At what time does the frog reach its highest point?

(c) What is the maximum height attained by this frog?

(d) At what time is the frog 6 inches high?

IV. Extensions

40. Famous Question: What should the ratio of the length l to the width w of a rectangle be, so that if a square (shaded in figure) is cut off from the original rectangle, then there will be a rectangle (AEFD) that is the same (similar) "shape" as the original rectangle? *Note:* This problem was mentioned previously in Chapters 2 and 3. Suppose, for the sake of simplicity, we say that the length $l = 1$.

(a) Set up the equation that says the rectangle CEFB is similar to rectangle EFDA.

(b) The equation of part (a) will be a proportion. Put this equation in the form $ax^2 + bx + c = 0$. (The proportion turns out to be equivalent to a quadratic equation.)

(c) Use the method of successive approximation to determine the value of x that will yield a similar rectangle. This value of x is called the "golden ratio."

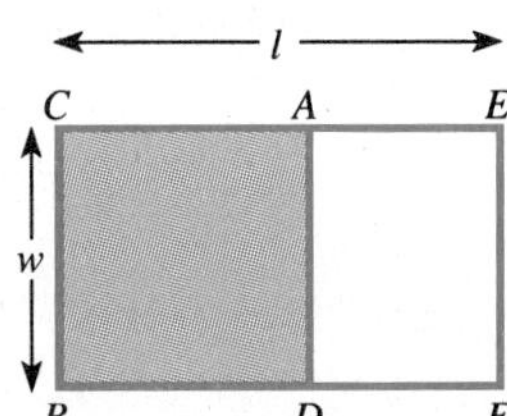

41. The stopping distance d in feet for a car moving at a velocity of V mph is given by the equation

$$d = 0.06V^2 + 1.1V$$

(a) According to this model, what is the stopping distance for a velocity of 20 mph? Forty mph? Fifty mph? Seventy-five mph?

(b) Graph this relationship. Does the model seem reasonable? That is, would you expect the stopping distances that this model predicts for various speeds? How about for very slow speeds? For very fast speeds?

(c) According to this model, what speed is a car going if it takes 100 feet to stop? 250 feet to stop? 350 feet to stop?

6.2 THE SOLUTION OF QUADRATIC EQUATIONS BY FACTORING AND GRAPHING

In the last section, we discussed the need to solve quadratic equations by some algebraic method. In particular, the solution to the equation

$$0 = 80t - 16t^2 + 20$$

would tell us how long it takes for a golf ball to hit the ground (see Example 3 in Section 6.1). We used a table, a graph, and the method of successive approximation to get a good estimate of the solution to this equation. We also want to obtain an algebraic method for the solution of a quadratic equation, whenever solutions exist.

Sometimes a quadratic equation can be factored, and the solution is found very quickly. For example, consider these equations.

EXAMPLES 1–3

Solve the following quadratic equations.

1. $x^2 - 3x + 2 = 0$ **2.** $x^2 = 4x$ **3.** $3x^2 = 2x + 1$

Solutions

1. The left side of this equation can be factored:

$$x^2 - 3x + 2 = 0$$

$$(x - 2)(x - 1) = 0 \quad \text{The product is zero.}$$

Therefore

$$(x - 2) = 0 \quad \text{or} \quad (x - 1) = 0 \quad \text{Therefore one of the factors must be zero.}$$

$$x = 2 \quad \text{or} \quad x = 1 \quad \text{Answer.}$$

The solution of Example 1 recalls an important property of real numbers that is useful in solving quadratic equations:

Property of Zero Products

If a and b are real numbers, and if $a \cdot b = 0$, then either $a = 0$ or $b = 0$.

Notice that there are *two* solutions for this quadratic equation. We will check the solutions by substituting them into the original equation:

Check: $x = 2$ in $x^2 - 3x + 2 = 0$:

$$2^2 - 3(2) + 2 \stackrel{?}{=} 0$$

$$4 - 6 + 2 \stackrel{\checkmark}{=} 0$$

Check: $x = 1$ in $x^2 - 3x + 2 = 0$:

$$1^2 - 3(1) + 2 \stackrel{?}{=} 0$$
$$1 - 3 + 2 \stackrel{\checkmark}{=} 0$$

Both solutions check.

2. When we try to solve a quadratic equation by factoring, we must get all the terms on one side of the equation and zero on the other side.

$$x^2 = 4x$$
$$x^2 - 4x = 0 \quad \text{Get all terms on one side.}$$
$$x(x - 4) = 0 \quad \text{The product is zero.}$$

Therefore

$$x = 0 \quad \text{or} \quad x - 4 = 0 \quad \text{Property of zero products.}$$
$$x = 0 \quad \text{or} \quad x = 4 \quad \text{Answer. Check the solutions.}$$

CAUTION: A common ERROR is often made when solving Example 2.

$$x^2 = 4x \quad \text{Divide both sides by } x. \text{ (This is incorrect.)}$$
$$x = 4$$

When we divide by a variable or an algebraic expression, we must be certain that we are not dividing by zero. In fact, the solution $x = 0$ is lost if you proceed by dividing both sides by x. To minimize the chance of making an error, we suggest that a quadratic equation be put in standard form.

Standard Form of a Quadratic Equation

$$ax^2 + bx + c = 0, a \neq 0$$

3.

$$3x^2 = 2x + 1$$
$$3x^2 - 2x - 1 = 0 \quad \text{Put in standard form.}$$
$$(3x + 1)(x - 1) = 0 \quad \text{Factor the left side.}$$

Therefore

$$3x + 1 = 0 \quad \text{or} \quad x - 1 = 0 \quad \text{Property of zero products.}$$
$$x = -\frac{1}{3} \quad \text{or} \quad x = 1 \quad \text{Answer. Check the solutions.}$$

When a quadratic equation is in the standard form $ax^2 + bx + c = 0$, solutions to the equation are related to the associated graph of the equation $y = ax^2 + bx + c$. Next we reconsider several previous examples, taking a look at their associated graphs.

EXAMPLES 4–7

Put each of these quadratic equations in the standard form, $ax^2 + bx + c = 0$. In each case relate the solutions of the equation to the graph of $y = ax^2 + bx + c$.

4. $x^2 - 3x + 2 = 0$ **5.** $x^2 = 4x$ **6.** $x^2 - 4x = -4$ **7.** $3x^2 + 1 = 0$

Solutions

4. $x^2 - 3x + 2 = 0$ Factor the quadratic equation as in Example 1.

$(x - 2)(x - 1) = 0$

So $x = 2$ or $x = 1$ Property of zero products.

Using a Table

Table of values for $y = x^2 - 3x + 2$.

x	y
0	2
1	0
2	0
1.5	−.25
3	2
−1	6
4	6

Table 6.6

Using a Graph

Graph of $y = x^2 - 3x + 2$.

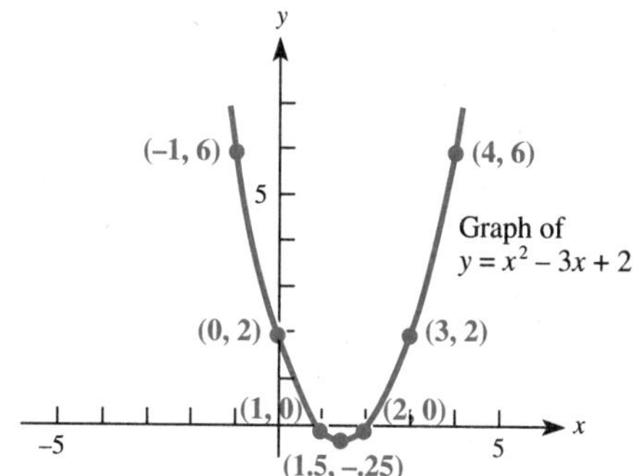

Figure 6.8

Notice that the graph of $y = x^2 - 3x + 2$ crosses the x-axis when $x = 1$ and $x = 2$. These are precisely the solutions to our quadratic equation. The solutions occur when $y = 0$ on the graph; that is, when $0 = x^2 - 3x + 2$.

5. $x^2 = 4x$

$x^2 - 4x = 0$

$x(x - 4) = 0$ Factor the quadratic equation as in Example 2.

So $x = 0$ or $x = 4$ Property of zero products.

Using a Table
Table of values for $y = x^2 - 4x$.

x	y
-2	12
-1	5
0	0
1	-3
2	-4
3	-3
4	0
5	5

Table 6.7

Using a Graph
Graph of $y = x^2 - 4x$.

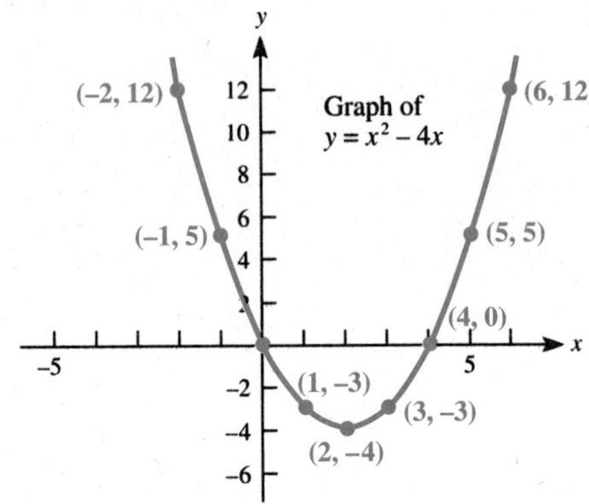

Figure 6.9

Notice that the graph of $y = x^2 - 4x$ crosses the x-axis at the solutions of $0 = x^2 - 4x$, namely, when $x = 0$ and when $x = 4$.

6. $$x^2 - 4x = -4$$

$$x^2 - 4x + 4 = 0 \quad \text{Put in standard form.}$$

$$(x - 2)(x - 2) = 0 \quad \text{Factor.}$$

So either $x = 2$ or $x = 2$. Property of zero products.

Using a Table
Table of values for $y = x^2 - 4x + 4$.

x	y
-3	25
-2	16
-1	9
0	4
1	1
2	0
3	1
4	4
5	9

Table 6.8

Using a Graph
Graph of $y = x^2 - 4x + 4$.

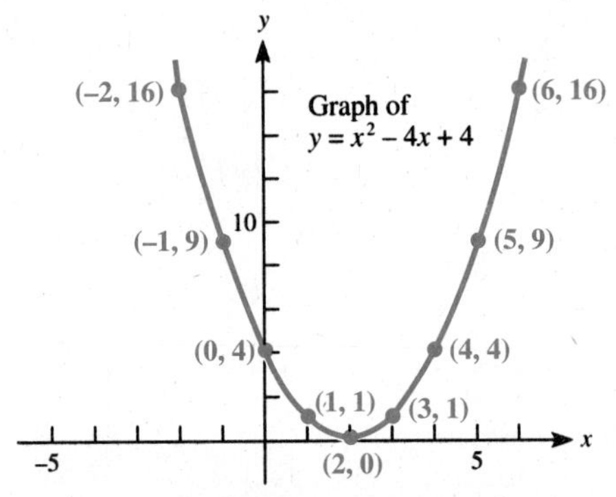

Figure 6.10

Notice that the graph of $y = x^2 - 4x + 4$ meets the x-axis only at $x = 2$, which is precisely the solution of the quadratic equation $x^2 - 4x + 4 = 0$. Because $(x - 2)$ is a repeated factor of this quadratic equation, $x = 2$ is called a "repeated root."

7. $3x^2 + 1 = 0$ This won't factor. Furthermore, the right side can never be zero; it is always positive. There are no real-number solutions.

Using a Table
Table of values for $y = 3x^2 + 1$.

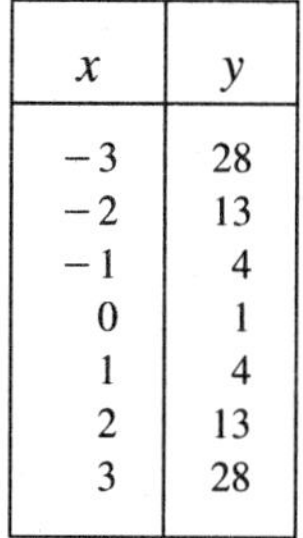

x	y
−3	28
−2	13
−1	4
0	1
1	4
2	13
3	28

Table 6.9

Using a Graph
Graph of $y = 3x^2 + 1$.

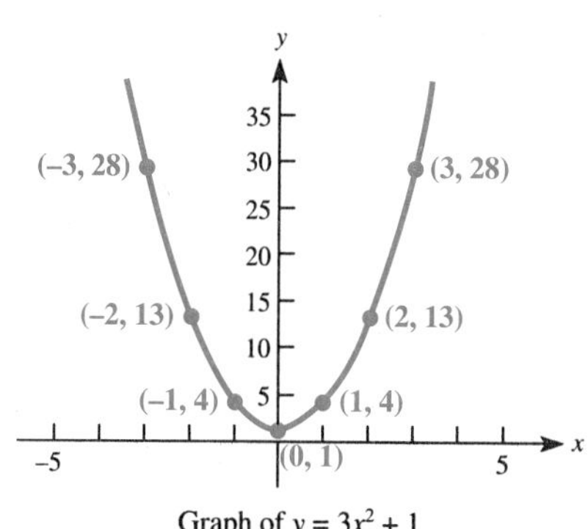

Graph of $y = 3x^2 + 1$

Figure 6.11

Notice that this graph does not cross the x-axis at all *and* that there are *no real-number-valued solutions* to the equation $3x^2 + 1 = 0$. ■

Examples 4–7 demonstrate that *if* there are real numbers r and s so that $(x - r)(x - s) = 0$, then these statements are both true:

1. $x = r$ and $x = s$ are solutions to the quadratic equation.
2. The graph of $y = (x - r)(x - s)$ crosses the x-axis at $x = r$ and $x = s$.

This is a very important relationship between equations of the form $ax^2 + bx + c = 0$ and the graph of the equation $y = ax^2 + bx + c$. The numbers r and s are called *solutions* or *roots* or *zeros* of the original equation. They are called "zeros" because they are the values of x that make $y = 0$ in the equation $y = (x - r)(x - s)$. We will return to a more detailed study of the graphs of quadratic equations over the next several sections.

EXAMPLES 8–9

Solve the following equations.

8. $42w^2 = 2 + 5w$

9. $16x^3 + 60x^2 + 54x = 0$

Solutions

8.

$$42w^2 = 2 + 5w$$

$$42w^2 - 5w - 2 = 0 \quad \text{Put in standard form.}$$

$$(7w - 2)(6w + 1) = 0 \quad \text{Factor.}$$

Therefore

$$7w - 2 = 0 \quad \text{or} \quad 6w + 1 = 0$$ **Property of zero products.**

$$w = \frac{2}{7} \quad \text{or} \quad w = -\frac{1}{6}$$ **Answer.**

9. $16x^3 + 60x^2 + 54x = 0$ **A third-degree equation, a cubic.**

$2x(8x^2 + 30x + 27) = 0$ **A monomial can be factored first.**

$2x(4x + 9)(2x + 3) = 0$ **Factor the binomial.**

Therefore

$$2x = 0 \quad \text{or} \quad 4x + 9 = 0 \quad \text{or} \quad 2x + 3 = 0$$ **Property of zero products.**

$$x = 0 \quad \text{or} \quad x = -\frac{9}{4} \quad \text{or} \quad x = -\frac{3}{2}$$ **Answer.** ■

The method of solution by factoring generalizes to polynomial equations of higher degree, as in Example 9.

Sometimes equations that involve algebraic fractions lead to quadratic equations.

EXAMPLES 10–12

Find all solutions to the equations below. Check your answers.

10. $\frac{6 - x}{6x} = \frac{1}{x + 1}$ **11.** $\frac{4}{y} - 3 = \frac{5}{2y + 3}$ **12.** $\frac{1 - x^2}{x} = \frac{1}{x} + 4x$

Solutions

10. $$\frac{6 - x}{6x} = \frac{1}{x + 1}$$ **LCD = $6x(x + 1)$. Note that $x \neq 0$ and $x \neq -1$, as these values of x would make denominators equal to 0.**

$$\cancel{(6x)}(x + 1)\frac{(6 - x)}{\cancel{6x}} = \frac{1}{\cancel{(x + 1)}}(6x)\cancel{(x + 1)}$$ **Multiply both sides by the LCD.**

$$(x + 1)(6 - x) = 1\,(6x)$$ **Remove common factors in numerators and denominators.**

$$6x + 6 - x^2 - x = 6x$$ **Multiply out. It's a quadratic equation.**

$$x^2 + x - 6 = 0$$ **Put in standard form.**

$$(x + 3)(x - 2) = 0$$ **Factor.**

So $\quad x = -3 \quad \text{or} \quad x = 2.$ **Answer.**

Graph $y = \frac{6 - x}{6x} - \frac{1}{(x + 1)}$ on a graphing calculator. See Exercise 37 at the end of this section.

11. $$\frac{4}{y} - 3 = \frac{5}{2y + 3}$$ **Note that $y \neq 0$ and $y \neq -\frac{3}{2}$, as these values of y would make denominators equal to 0. LCD = $y(2y + 3)$.**

$$y(2y + 3) \cdot \left(\frac{4}{y} - 3\right) = \frac{5}{(2y + 3)} \cdot y(2y + 3)$$ **Multiply both sides by the LCD. Distribute $y(2y + 3)$ on left side.**

$$\not{y}\frac{(2y+3)4}{\not{y}} - 3y(2y+3) = \frac{5y(\cancel{2y+3})}{(\cancel{2y+3})}$$ Remove common factors in numerators and denominators.

$$8y + 12 - 6y^2 - 9y = 5y$$ Multiply. It's a quadratic equation.

$$6y^2 + 6y - 12 = 0$$ Put in standard form.

$$6(y^2 + y - 2) = 0$$

$$6(y-2)(y+1) = 0$$ Factor.

So $y = 2$ or $y = -1$. Answer.

Graph $Y1 = \frac{4}{x-3}$, $Y2 = \frac{5}{2x+3}$.

See Exercise 37.

12. $$\frac{1-x^2}{x} = \frac{1}{x} + 4x$$

$$x \cdot \frac{1-x^2}{x} = \left(\frac{1}{x} + 4x\right) \cdot x$$ LCD $= x$.
Multiply both sides by the LCD.
Note: $x = 0$ makes denominators 0.

$$1 - x^2 = 1 + 4x^2$$

$$0 = 5x^2$$

$$0 = x$$

There is no solution for this equation, since the only possible solution, $x = 0$, makes the denominators zero. Use a graphing calculator to see that these graphs do not meet.

Graph $Y1 = \frac{1-x^2}{x}$ and $Y2 = \frac{1}{x} + 4x$. ■

EXAMPLE 13

You and a friend plan to drive in separate vehicles from Detroit to St. Louis, about 500 miles. Your friend is driving a rental truck. He averages 10 mph less than you do while driving your car. You arrive in St. Louis $2\frac{1}{2}$ hours before your friend. How fast were the vehicles traveling?

Solution

Suppose we let x be your speed in miles per hour. Your friend, then, travels at $x - 10$ mph. The model, Distance = rate × time, will help us to construct a table. Recall that $t = \frac{D}{r}$.

	Distance	*Rate*	*Time*
You	500	x	$\frac{500}{x}$
Your friend	500	$x - 10$	$\frac{500}{x-10}$

Table 6.10

Since you waited $2\frac{1}{2}$ hours for your friend, we know that
(your driving time) + ($2\frac{1}{2}$ hours) = your friend's driving time. Thus

$$\frac{500}{x} + \frac{5}{2} = \frac{500}{x - 10}$$ The LCD is $2x(x - 10)$.

$$2x(x - 10) \cdot \frac{500}{x} + 2x(x - 10) \cdot \frac{5}{2} = \frac{500}{x - 10} \cdot 2x(x - 10)$$ Multiply both sides by the LCD.

$$2\cancel{x}(x - 10) \cdot \frac{500}{\cancel{x}} + \cancel{2}x(x - 10) \cdot \frac{5}{\cancel{2}} = \frac{500}{\cancel{x - 10}} \cdot 2x(\cancel{x - 10})$$ Fundamental property of fractions.

$$1{,}000(x - 10) + 5x(x - 10) = 1{,}000x$$ An equivalent equation without fractions.

$$1{,}000x - 10{,}000 + 5x^2 - 50x = 1{,}000x$$ Expand.

$$5x^2 - 50x - 10{,}000 = 0$$ A quadratic in standard form.

$$5(x^2 - 10x - 2{,}000) = 0$$ Factor.

$$5(x - 50)(x + 40) = 0$$

Therefore

$$x - 50 = 0 \quad \text{or} \quad x + 40 = 0$$ Property of zero products.

$$x = 50 \text{ mph} \quad \text{or} \quad x = -40 \text{ mph}$$

Only $x = 50$ mph makes sense for this application. You averaged 50 mph. Your friend averaged $x - 10 = 40$ mph.

$$\text{Graph } Y1 = \frac{500}{x} + \frac{5}{2} \qquad Y2 = \frac{500}{x - 10}$$

on a graphing calculator and check where these graphs cross. Compare the intersection points to the algebraic solution above. ■

6.2 EXERCISES

I. Procedures

1–15. Put these quadratic equations in standard form. Find the solution to each equation. Try to solve by factoring. Relate the solutions of the equations $0 = ax^2 + bx + c$ to the graph of $y = ax^2 + bx + c$ in each case.

1. $x^2 = 9$

2. $3x^2 = 48$

3. $3x^2 = 7x$

4. $-4x^2 = 9x$

5. $x^2 - 5x = 6$

6. $x^2 = 7x - 6$

7. $2x^2 + x - 15 = 0$

8. $4x^2 + 8x - 21 = 0$

9. $16x^2 = 144$

10. $100x^2 - 49 = 0$

11. $4x^2 + 4x + 1 = 0$

12. $x^2 - 10x = -25$

13. $3y^2 = 10 + 13y$

14. $2w^2 = 9 + 3w$

15. $5z - 1 - 6z^2 = 0$

16–26. Find the solutions to these equations. (Try factoring after the equations have been put in standard form.)

16. $13 - 2x - 15x^2 = 0$

17. $y(y - 1) = 2(y^2 - 6)$

18. $2x(x - 1) = 3(1 - x^2)$

19. $(2x + 1)(x - 1) = (x - 1)(x + 5)$

20. $(x + 3)(3x - 8) = (x + 3)(7 - x)$

21. $x(x + 7) = x(3 - x) - 2$

22. $5 - x(3 - x) = x(x - 3) + 2$

23. $1 + \dfrac{2}{b - 1} = \dfrac{2}{b^2 - b}$

24. $\dfrac{3}{y} - \dfrac{5}{y - 1} = \dfrac{3 - 2y}{y^2 - 1}$

25. $\dfrac{a + 12}{a^2 - 16} - \dfrac{3}{a - 4} = \dfrac{1}{a + 4}$

26. $\dfrac{2x + 11}{x + 4} + \dfrac{x - 2}{x - 4} = \dfrac{12}{x^2 - 16} + \dfrac{7}{2}$

27–32. Find all solutions (if any) to each of these equations.

27. $\dfrac{x^2 + 3}{x^2 + 4} = 0$

28. $\dfrac{x^2 + 3}{x^2 + 3} = 1$

29. $\dfrac{x^2 + 3}{x^2 + 4} = 1$

30. $\dfrac{x^2 - 4}{x^2 + 4} = 0$

31. $\dfrac{x^2 - 4}{x^2 + 4} = 1$

32. $\dfrac{x^2 - 4}{x^2 - 4} = 1$

II. Concepts

33–35. There is an error in each of the following problems. Find the error, and give the correct solution. In each case, how would you help the student?

33. $x - x^2 = 0$
Student's solution:

$$x - x^2 = 0$$
$$x = x^2$$
$$\frac{x}{x} = \frac{x^2}{x}$$
$$1 = x$$

Incorrect answer.

34. $24x^2 + 5x = 36$
Student's solution:

$$24x^2 + 5x - 36 = 0$$
$$(4x + 9)(6x - 4) = 0$$
$$x = -\frac{9}{4} \quad \text{or} \quad x = \frac{2}{3}$$

Incorrect answer.

35. $\dfrac{1}{x} - \dfrac{x}{x - 1} = \dfrac{-3 + 2x}{x^2 - x}$
Student's solution: The LCD is $x(x - 1)$.

$$\frac{1}{x} - \frac{x}{x - 1} = \frac{2x - 3}{x^2 - x}$$
$$x(x - 1)\left(\frac{1}{x} - \frac{x}{x - 1}\right) = \frac{2x - 3}{x(x - 1)} \cdot x(x - 1)$$
$$x - 1 - x^2 = 2x - 3$$
$$0 = x^2 + x - 2$$
$$0 = (x - 1)(x + 2)$$
$$x = 1 \quad \text{or} \quad x = -2$$

Incorrect answer.

36. Write a paragraph explaining the relationship between solutions of the quadratic $0 = ax^2 + bx + c$ and the graph of $y = ax^2 + bx + c$. Give examples and draw sketches.

37. Write a note explaining how you could solve the equations in Examples 10–12 by using a graphing calculator. Then solve each equation by graphing.

38. Write a quadratic equation that has solutions $x = 0$ and $x = 1$. *Hint:* Work backwards.

39. Write a quadratic equation that has solutions $x = 3$ and $x = -7$.

40. Write a quadratic equation that has solutions $x = -2$ and $x = -24$.

41. Write a quadratic equation that has solutions $x = \frac{1}{2}$ and $x = -\frac{2}{3}$.

42. The quadratic equation $ax^2 + 5x + 6 = 0$ has $x = 2$ for one solution. Find the other solution. *Hint:* Substitute $x = 2$ and solve for a.

43. The quadratic equation $4x^2 + bx + 28 = 0$ has $x = 4$ for one solution. Find the other solution.

44. The quadratic equation $6x^2 - 13x + c = 0$ has $x = \frac{2}{3}$ for one solution. Find the other solution.

III. Applications

45–52. Find an appropriate equation and then solve by factoring or graphing.

45. What are the dimensions of the rectangle shown? Its area is 2,100 square feet.

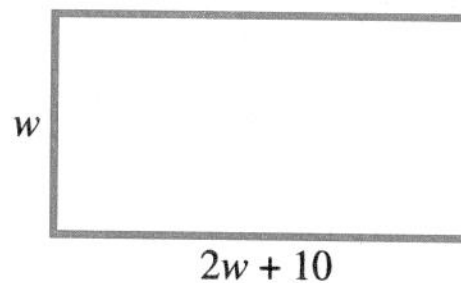

46. Find the unknown lengths of the sides of the triangle.

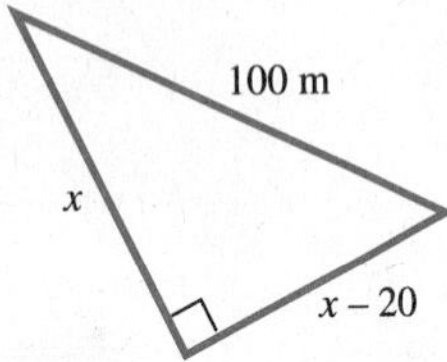

47. The sum of a number and its reciprocal is $\frac{41}{20}$. Find two such numbers.

48. The sum of the reciprocals of two consecutive integers equals 7 divided by the product of the integers. Find the integers.

49. A man drove at a constant speed from his home to a city 180 miles away. On the return trip, he increased his constant speed by 5 mph and made the return trip in 24 minutes less time than had been required to drive to the city. What was his speed from home to the city?

50. A trucker, maintaining a constant speed with his empty truck, traveled 480 miles. The return trip was made with a loaded truck that reduced his constant speed by 20 mph. The return trip took four hours longer than the first trip. What was his speed with the loaded truck?

51. The members of a certain organization decided to donate $2,000 to charity and to share the cost of the contribution equally. If the club had possessed 20 more members, each would have contributed $5 less. How many members now belong to the organization?

52. Each year a certain company divides $7,200 equally among its employees. This year the company employed six more people than last year. Therefore each employee received $60 less. How much did each employee receive last year?

53 and 54. The model that approximates the height h of an object after t seconds if the object is thrown upwards with an initial velocity of v feet per second is

$$h = -16t^2 + vt$$

53. Suppose a baseball is thrown upwards with an initial velocity of 64 feet per second. When will the ball be 48 feet off the ground? Graph the relationship. When will the ball hit the ground?

54. Suppose a ball is tossed upwards at 92 feet per second. When will it be 60 feet high? Graph the relationship. When will the ball hit the ground?

IV. Extensions

55. For what integer value(s) of c could you solve the equation $x^2 + 2x + c = 0$ by factoring?

56. For what whole-number value(s) of b could you solve the equation $3x^2 + bx - 2 = 0$ by factoring?

57. For what integer value(s) of a could you solve the equation $ax^2 - 4x + 6 = 0$ by factoring?

58. Find all solutions to

$$x(x-1)(x-2)(x-3)(x-4)(x-5)(x-6)(x-7)\cdots(x-100) = 0$$

59. Find all solutions to

$$x\left(x-\frac{1}{2}\right)\left(x+\frac{1}{3}\right)\left(x-\frac{1}{4}\right)\left(x+\frac{1}{5}\right)\left(x-\frac{1}{6}\right)\left(x+\frac{1}{7}\right)\cdots\left(x-\frac{1}{100}\right) = 0$$

60. A math teacher wrote a quadratic equation on the board of the form $x^2 + bx + c = 0$ and asked the students to find the two real roots. Tony miscopied one of the coefficients and found that 4 and 1 were roots. Cathy miscopied a different coefficient and found that 3 and -2 were roots. Determine the roots to the equation the teacher wrote.

6.3 THE QUADRATIC FORMULA

In the last section, we discussed a method that enables us to solve *some* quadratic equations by factoring. For example, to solve $x^2 - 5x + 6 = 0$, we factored the left side into $(x-3)(x-2) = 0$. Thus the solutions are $x = 3$ or $x = 2$. However, most quadratic equations with integer coefficients do not have rational factors. For example, consider these quadratic equations:

$$x^2 - 7 = 0$$
$$x^2 + 4x + 2 = 0$$
$$x^2 + 3 = 0$$

Each of these quadratic equations will introduce us to a new concept. The first equation involves calculating a **square root.** The second equation can be solved by a method called **completing the square.** The solutions to the third equation require us to extend our number system to the set of **complex numbers.** We will see that the method of completing the square can be generalized to any quadratic equation to obtain the **quadratic formula,** an algorithm that can be used to solve any quadratic equation. In this section, we will explore equations like the first and second equations above and obtain the quadratic formula. Equations of the third type whose solutions are complex numbers will be studied in Section 6.4.

Quadratic Equations and Square Roots

EXAMPLE 1 How can we solve the first equation above?

$$x^2 - 7 = 0$$

Solution

Using Successive Approximation If $x^2 - 7 = 0$, then $x^2 = 7$ is an equivalent equation. Thus the question becomes, "What number x when multiplied by itself yields 7?" Perhaps we can get an approximation. Since $2^2 = 4$, and $3^2 = 9$, and we want to solve $x^2 = 7$, then x ought to be between 2 and 3. Let's try 2.5:

$(2.5)^2 = 6.25$ Not big enough.

$(2.8)^2 = 7.84$ Too big.

$(2.7)^2 = 7.29$ Still too big.

$(2.6)^2 = 6.76$ Now we know the solution lies between $x = 2.6$ and $x = 2.7$.

Let's try $x = 2.65$:

$(2.65)^2 = 7.0225$ Now we are getting rather close.

We could continue our method of successive approximation. However, calculators can do this tedious job for us.

Recall from Chapter 1 that the positive number which when multiplied by itself yields 7 is called the **positive square root** of 7, and it is denoted by $\sqrt{7}$. (The $\sqrt{\ }$ symbol is called a **radical** symbol.) Thus, $(\sqrt{7})^2 = 7$. In the solution of the first equation, we find that $\sqrt{7} \approx 2.646$.

It is also true that $x \approx -2.646$ is a solution to $x^2 = 7$. This number is called the **negative square root** of 7.

Using a Graph We can also solve the equation by graphing $y = x^2 - 7$ and noting where the graph crosses the x-intercept.

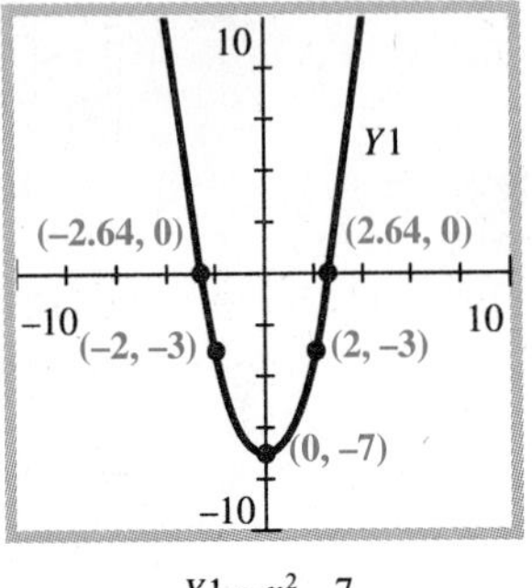

$Y1 = x^2 - 7$

Figure 6.12

We can obtain this graph by plotting points or using a graphing calculator. The graph crosses the x-axis at $x = \pm 2.646$.

The two solutions for the first equation, $x^2 - 7 = 0$, are

$$x \approx 2.646 \quad \text{and} \quad x \approx -2.646$$

Definition of Square Root

Suppose p is a nonnegative number; then the positive number which when multiplied by itself yields p is denoted by $\sqrt{p}$ and is called the **positive square root of p.**

The negative number which when multiplied by itself yields p is denoted by $-\sqrt{p}$ and is called the **negative square root of p.**

Thus $(\sqrt{p})^2 = p$ and $(-\sqrt{p})^2 = p$. The symbol $\sqrt{p}$ always indicates the positive square root.

EXAMPLE 2

Find the positive square root of each of the following numbers.

(a) 9 **(b)** 18.62 **(c)** 64 **(d)** 0 **(e)** 600

Solutions

(a) $\sqrt{9} = 3$ — $3 \cdot 3 = 9$.

(b) $\sqrt{18.62} \approx 4.32$ — Use a calculator.

(c) $\sqrt{64} = 8$ — $8^2 = 64$.

(d) $\sqrt{0} = 0$ — $0 \cdot 0 = 0$.

(e) $\sqrt{600} \approx 24.49$ — Use a calculator.

We can now solve equations like $x^2 - 7 = 0$.

EXAMPLES 3–5 Solve these equations.

3. $x^2 = 49$ **4.** $x^2 - 27 = 0$ **5.** $(x - 3)^2 = 16$

Solutions

3. $x^2 = 49$ What number squared is 49?

Therefore

$$x = \sqrt{49} \quad \text{or} \quad x = -\sqrt{49}$$
$$x = 7 \quad \text{or} \quad x = -7$$

The positive and negative square root are both solutions.

NOTE: Example 3 can also be solved by factoring: $x^2 = 49$, so $x^2 - 49 = 0$, and $(x - 7)(x + 7) = 0$. ■

4. $x^2 - 27 = 0$ Cannot be factored in the set of rational numbers.

$$x^2 = 27$$

Both the positive and the negative square roots are solutions.

$$x = \sqrt{27} \quad \text{or} \quad x = -\sqrt{27}$$
$$x \approx 5.196 \quad \text{or} \quad x \approx -5.196$$

Estimate the solution from the graph of $y = x^2 - 27$. Compare your estimate to the algebraic solution.

5. $(x - 3)^2 = 16$ An expression squared is 16.

Thus

$$(x - 3) = \sqrt{16} \quad \text{or} \quad (x - 3) = -\sqrt{16}$$

So the whole expression equals $\sqrt{16}$ or $-\sqrt{16}$.

$$x - 3 = 4 \quad \text{or} \quad x - 3 = -4$$

$\sqrt{16} = 4$.

$$x = 7 \quad \text{or} \quad x = -1$$

Check: If $x = 7$,

$$(7 - 3)^2 \stackrel{?}{=} 16$$
$$(4)^2 \stackrel{\checkmark}{=} 16$$

Graph $y = (x - 3)^2 - 16$ on your calculator and estimate the solution from the graph. Compare your estimate to the algebraic solution.

If $x = -1$,

$$(-1 - 3)^2 \stackrel{?}{=} 16$$
$$(-4)^2 \stackrel{\checkmark}{=} 16$$ ■

A Property of Square Roots

The subject of square roots, cube roots, and so on will be treated in greater detail in Chapter 7. However, there is a very useful property of square roots that will help to solve quadratic equations.

Notice that $\sqrt{36} = 6$ and that $\sqrt{36} = \sqrt{(4)(9)}$. Also, $\sqrt{4} = 2$, $\sqrt{9} = 3$, and $\sqrt{4} \cdot \sqrt{9} = 2 \cdot 3 = 6$. Thus

$$\sqrt{(4)(9)} = \sqrt{36} = 6 = \sqrt{4} \cdot \sqrt{9}$$

$$\sqrt{(4)(9)} = \sqrt{4} \cdot \sqrt{9}$$

It is always true that the square root of the product of two nonnegative numbers is equal to the product of their positive square roots.

Property of Square Roots

Suppose $a, b \geq 0$; then $\sqrt{ab} = \sqrt{a}\sqrt{b}$.

This property enables us to rewrite some square roots in a convenient equivalent form.

EXAMPLE 6

Use the property of square roots to write the following square roots in an equivalent form with no square factors under the radical.

(a) $\sqrt{72}$ **(b)** $-\sqrt{1{,}000}$ **(c)** $\sqrt{450}$ **(d)** $\sqrt{12{,}480}$ **(e)** $\sqrt{89}$

Solutions

Try to find factors that are perfect squares.

(a) $\sqrt{72} = \sqrt{9 \cdot 8} = \sqrt{9 \cdot 4 \cdot 2}$

$= \sqrt{9} \cdot \sqrt{4} \cdot \sqrt{2} = (3)(2)(\sqrt{2}) = 6\sqrt{2}$ **Answer.**

This says that $\sqrt{72}$ is the same as "six times $\sqrt{2}$." Calculate both on your calculator to check that

$$\sqrt{72} = 6\sqrt{2}$$

NOTE: Our property also works for more than two factors. For example,

$$\sqrt{abc} = \sqrt{a}\sqrt{b}\sqrt{c} \quad \text{for} \quad a, b, c, \geq 0 \quad ■$$

(b) $-\sqrt{1{,}000} = -\sqrt{(100)(10)} = -\sqrt{100} \cdot \sqrt{10} = -\ 10\sqrt{10}$

(c) $\sqrt{450} = \sqrt{(2)(225)} = \sqrt{2} \cdot \sqrt{225} = \sqrt{2}(15) = 15\sqrt{2}$

NOTE: The square root factor is usually written last—thus we write $15\sqrt{2}$ rather than $\sqrt{2}(15)$. ■

(d) $\sqrt{12{,}480} = \sqrt{4 \cdot 3{,}120} = \sqrt{4 \cdot 4 \cdot 780} = \sqrt{4 \cdot 4 \cdot 4 \cdot 195} = 8\sqrt{195}$

(e) The number 89 has no perfect square factors. ■

Completing the Square

EXAMPLE 7

How can we solve an equation like $x^2 + 4x + 2 = 0$?

Solution

Using an Equation If we could get this equation into the form

$$(\text{quantity})^2 = \text{number}$$

then we could solve it by finding the square roots of the number as we did in Examples 3–5.

$$x^2 + 4x + 2 = 0 \qquad \textbf{Equation (1)}$$

What can we do to make the left side of Equation 1 a **perfect square trinomial?** Recall that perfect square trinomials are of the form $x^2 \pm 2ax + a^2$. They factor into a binomial squared:

$$x^2 \pm 2ax + a^2 = (x \pm a)^2$$

If we add $+2$ to both sides of Equation 1, then we obtain a perfect square trinomial on the left side.

$$x^2 + 4x + 4 = 2 \qquad \textbf{Equation (2)}$$

Therefore

$$(x + 2)^2 = 2 \qquad \text{The desired form.}$$

Thus

$$(x + 2) = \sqrt{2} \quad \text{or} \quad (x + 2) = -\sqrt{2}$$
$$x = -2 + \sqrt{2} \quad \text{or} \quad x = -2 - \sqrt{2} \qquad \text{Answer.}$$

We can obtain numerical approximations for these solutions if we use $\sqrt{2} \approx 1.41$:

$$x \approx -2 + 1.41 \quad \text{or} \quad x \approx -2 - 1.41$$
$$x \approx -0.59 \quad \text{or} \quad x \approx -3.41 \qquad \text{Answer.}$$

This solution process is called **completing the square.**

Using a Graph Graph $y = x^2 + 4x + 2$ and observe where the graph crosses the x-axis.

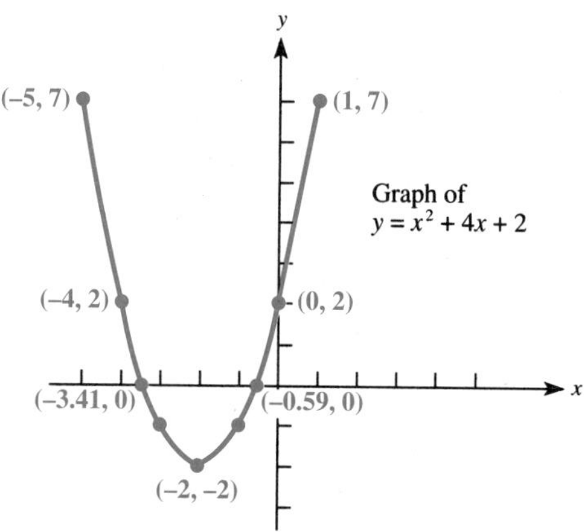

Figure 6.13

The graph crosses the x-axis at $x = -0.59$ and $x = -3.41$.

Notice the relationship between the solutions of $x^2 + 4x + 2 = 0$ and the points where the graph of the equation $y = x^2 + 4x + 2$ crosses the x-axis. ■

EXAMPLE 8

Complete the square by adding the appropriate constant to form a perfect square trinomial.

(a) $x^2 + 6x + \square$ **(b)** $x^2 - 8x + \square$ **(c)** $x^2 + 9x + \square$

Solutions

(a) $x^2 + 6x + \square$

We wish to obtain the form of $x^2 + 2ax + \boxed{a^2}$.

$$x^2 + 6x + \square = x^2 + 2 \cdot 3x + \square \quad \text{Thus } 2a = 6, a = 3, a^2 = 9.$$

$$x^2 + 6x + \boxed{9} = (x + 3)^2 \quad \text{Answer.}$$

Notice that the coefficient of the middle term (6) was *twice* what we needed to square (3). Thus, to complete the square, we need only take $\frac{1}{2}$ of the coefficient of the x term, square it, and add. For example,

$$x^2 + 6x + \square \quad \text{The coefficient of the } x \text{ term is 6, } \left(\frac{1}{2}\right)6 = 3, 3^2 = 9, \text{ add 9.}$$

$$x^2 + 6x + \boxed{9} = (x + 3)^2 \quad \text{Answer.}$$

(b) $x^2 - 8x + \square$ $\quad \frac{1}{2}(-8) = -4. (-4)^2 = 16.$ Add 16.

$$x^2 - 8x + \boxed{(-4)^2}$$

$$x^2 - 8x + 16 = (x - 4)^2 \quad \text{Answer.}$$

(c) $x^2 + 9x + \square$ $\quad \frac{1}{2}(9) = \frac{9}{2}$, add $\left(\frac{9}{2}\right)^2 = \frac{81}{4}$.

$$x^2 + 9x + \frac{81}{4} = \left(x + \frac{9}{2}\right)^2 \quad \text{Answer.}$$ ■

We summarize the process of completing the square below.

Strategy for Completing the Square

To make $x^2 + bx + \square$ a perfect square, add $\left(\frac{b}{2}\right)^2$. Thus

$$x^2 + bx + \left(\frac{b}{2}\right)^2 = \left(x + \frac{b}{2}\right)^2.$$

NOTE: The coefficient of the x^2 term must be 1 to complete the square in this way. ■

EXAMPLES 9–11

Use the method of completing the square to solve the following quadratic equations and then observe where the solutions occur on a graph of the equation.

9. $x^2 + 7x + 4 = 0$ **10.** $x^2 - 6x + 8 = 0$ **11.** $4x^2 + 8x - 1 = 0$

Solutions

9. Using an Equation

$$x^2 + 7x + 4 = 0$$

$$x^2 + 7x = -4$$

$$x^2 + 7x + \square = -4 + \square$$

To facilitate our method we first move the constant term to the right side.

The coefficient of x is 7.

$$\left(\frac{7}{2}\right)^2 = \frac{49}{4}$$

Add $\frac{49}{4}$ to both sides to complete the square.

$$x^2 + 7x + \boxed{\frac{49}{4}} = -4 + \boxed{\frac{49}{4}}$$

$$\left(x + \frac{7}{2}\right)^2 = \frac{33}{4}$$

$-4 = -\frac{16}{4}$

$$x + \frac{7}{2} = \pm\sqrt{\frac{33}{4}}$$

Both the positive and negative square roots will work.

$$x = -\frac{7}{2} \pm \sqrt{\frac{33}{4}}$$

Answer.

Therefore

$$x \approx -3.5 - 2.87 = -6.37 \quad \text{or} \quad x = -3.5 + 2.87 = -0.63$$

Answers.

Check: We substitute $x \approx -6.37$ into Example 9.

$$(-6.37)^2 + 7(-6.37) + 4 \overset{?}{\approx} 0$$

$$40.58 - 44.59 + 4 \overset{?}{\approx} 0$$

$$-0.01 \overset{\checkmark}{\approx} 0$$

We did not get exactly 0, because we have approximated our answer to two decimal places.

Using a Graph Graph $y = x^2 + 7x + 4$.

Notice the relationship between the solutions of $x^2 + 7x + 4 = 0$ and the points where the graph of the equation $y = x^2 + 7x + 4$ crosses the x-axis.

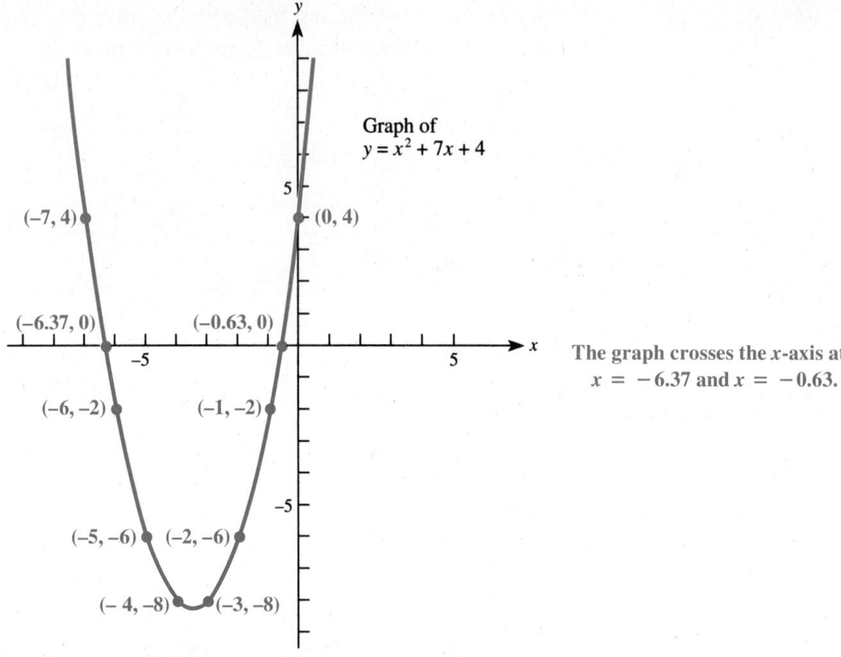

Figure 6.14

10. Using an Equation

$x^2 - 6x + 8 = 0$	
$x^2 - 6x = -8$	Move constant to right side.
$x^2 - 6x + \square = -8 + \square$	Complete the square.
$x^2 - 6x + \boxed{9} = -8 + \boxed{9}$	$\frac{1}{2}(-6) = -3$, add $(-3)^2 = 9$ to both sides.
$(x - 3)^2 = 1$	Write left side as a perfect square.
$x - 3 = \pm 1$	Find the square roots of the right side.
$x = 3 + 1$ or $x = 3 - 1$	
$x = 4$ or $x = 2$	Answer: Check by substitution.

NOTE: Example 10 can also be solved by factoring:

$$x^2 - 6x + 8 = 0$$
$$(x - 2)(x - 4) = 0$$
$$x = 2 \quad \text{or} \quad x = 4$$

Both methods produce the same answers. ■

Using a Graph Graph $y = x^2 - 6x + 8$.

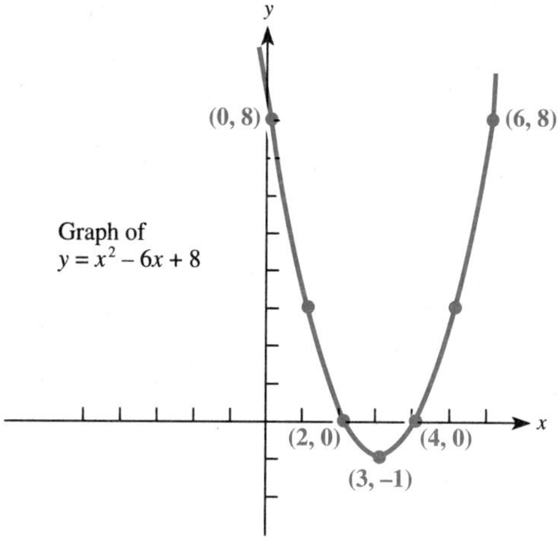

Graph of
$y = x^2 - 6x + 8$

(0, 8)
(6, 8)
(2, 0)
(4, 0)
(3, −1)

The graph crosses the x-axis at $x = 2$ and $x = 4$.

Figure 6.15

Notice the relationship between the solutions of $x^2 - 6x + 8 = 0$ and the points where the graph of the equation $y = x^2 - 6x + 8$ crosses the x-axis.

11. Using an Equation

$$4x^2 + 8x - 1 = 0$$
$$4x^2 + 8x = 1$$

Since the coefficient of the x^2 term is not 1, we must find an equivalent equation with coefficient 1. Divide both sides by 4.

$$\frac{4x^2}{4} + \frac{8x}{4} = \frac{1}{4}$$
$$x^2 + 2x = \frac{1}{4}$$

Now complete the square.

$$x^2 + 2x + \square = \frac{1}{4} + \square$$
$$x^2 + 2x + \boxed{1} = \frac{1}{4} + \boxed{1}$$
$$(x + 1)^2 = \frac{5}{4}$$
$$x + 1 = \pm\sqrt{\frac{5}{4}}$$
$$x = -1 \pm \sqrt{\frac{5}{4}} \approx -1 \pm 1.12$$
$$x \approx 0.12 \quad \text{or} \quad x \approx -2.12$$

Answer. Check by substitution.

Using a Graph Graph $y = 4x^2 + 8x - 1$.

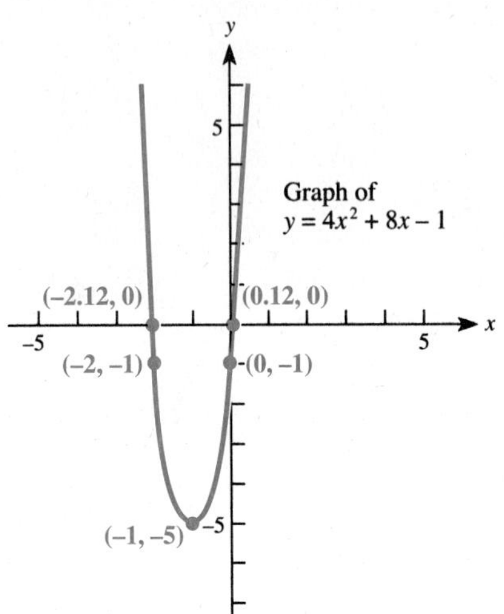

The graph crosses the x-axis at $x \approx -2.12$ and $x \approx 0.12$.

Figure 6.16

Notice the relationship between the solutions of $4x^2 + 8x - 1 = 0$ and the points where the graph of the equation $y = 4x^2 + 8x - 1$ crosses the x-axis.

The Quadratic Formula

The standard form of the quadratic equations we have been studying is

$$ax^2 + bx + c = 0, a \neq 0, \text{ with the corresponding graph } y = ax^2 + bx + c$$

where a, b, and c are the coefficients of the quadratic equation. In the equation $3x^2 + 4x - 2 = 0$, $a = 3$, $b = 4$, $c = -2$. In order to obtain a formula that will quickly produce the solutions to any quadratic equation, *we will complete the square using the coefficients in the general form* of the equation above.

EXAMPLE 12 Find the solutions to the equation

$$ax^2 + bx + c = 0$$ a, b, and c are real numbers. $a \neq 0$.

Solution

We will *assume a is positive.* If it were not positive, we could multiply both sides by -1 and make it positive.

$$ax^2 + bx + c = 0$$

$$ax^2 + bx = -c$$ Isolate the constant term on the right side.

$$\frac{\cancel{a}x^2}{\cancel{a}} + \frac{bx}{a} = \frac{-c}{a}$$ Divide by a. Make the coefficient of x^2 equal to 1.

$$x^2 + \frac{bx}{a} + \square = \frac{-c}{a} + \square$$ Complete the square. $\frac{1}{2}\left(\frac{b}{a}\right) = \frac{b}{2a}, \left(\frac{b}{2a}\right)^2 = \frac{b^2}{4a^2}$.

$$x^2 + \frac{bx}{a} + \boxed{\frac{b^2}{4a^2}} = \frac{-c}{a} + \boxed{\frac{b^2}{4a^2}}$$ Add $\frac{b^2}{4a^2}$ to both sides.

$$\left(x + \frac{b}{2a}\right)^2 = \frac{b^2}{4a^2} - \frac{c}{a}$$ Left side is a perfect square.

$$\left(x + \frac{b}{2a}\right)^2 = \frac{b^2}{4a^2} - \frac{c}{a} \cdot \left(\frac{4a}{4a}\right)$$ Do the subtraction on the right side. The LCD is $4a^2$.

$$\left(x + \frac{b}{2a}\right)^2 = \frac{b^2 - 4ac}{4a^2}$$ *For now,* let us suppose $b^2 - 4ac \geq 0$.

$$\left(x + \frac{b}{2a}\right) = \pm\sqrt{\frac{b^2 - 4ac}{4a^2}}$$ Take square roots.

$$\left(x + \frac{b}{2a}\right) = \pm\sqrt{(b^2 - 4ac) \cdot \frac{1}{4a^2}}$$ Rewrite expression under the radical sign.

$$\left(x + \frac{b}{2a}\right) = \pm\sqrt{b^2 - 4ac} \cdot \sqrt{\frac{1}{4a^2}}$$ Property of square roots.

$$\left(x + \frac{b}{2a}\right) = \pm\sqrt{b^2 - 4ac} \cdot \frac{1}{2a}$$ $\sqrt{\frac{1}{4a^2}} = \frac{1}{2a}$, where a is positive.

$$x + \frac{b}{2a} = \frac{\pm\sqrt{b^2 - 4ac}}{2a}$$ Rewrite the right side.

$$x = -\frac{b}{2a} \pm \frac{\sqrt{b^2 - 4ac}}{2a}$$ Add $-\frac{b}{2a}$ to both sides.

$$x = \frac{-b \pm \sqrt{b^2 - 4ac}}{2a}$$ Answer (put over the common denominator). ■

Thus the solutions to any quadratic equation of the form

$$ax^2 + bx + c = 0, a \neq 0$$

can be obtained by putting the values of the coefficients into the formula

$$x = \frac{-b \pm \sqrt{b^2 - 4ac}}{2a}$$

This formula is called the quadratic formula.

The Quadratic Formula

The solutions to an equation of the form $ax^2 + bx + c = 0$ $(a \neq 0)$ are

$$x = \frac{-b + \sqrt{b^2 - 4ac}}{2a} \quad \text{and} \quad x = \frac{-b - \sqrt{b^2 - 4ac}}{2a}$$

The Discriminant

The corresponding graph of $y = ax^2 + bx + c$ is closely related to the solutions of the quadratic equation $0 = ax^2 + bx + c$. In each example we have done, we have seen that the values of x that are solutions to $0 = ax^2 + bx + c$ are the values of x where the graph of $y = ax^2 + bx + c$ crosses the x-axis, the **x-axis intercepts.** The quantity $b^2 - 4ac$ in the quadratic formula is called the **discriminant.** The discriminant is useful in describing the solutions of a quadratic equation. If the discriminant $b^2 - 4ac > 0$, then we have two *different* solutions to the quadratic equation $0 = ax^2 + bx + c$, and the graph of $y = ax^2 + bx + c$ crosses the x-axis at two different places.

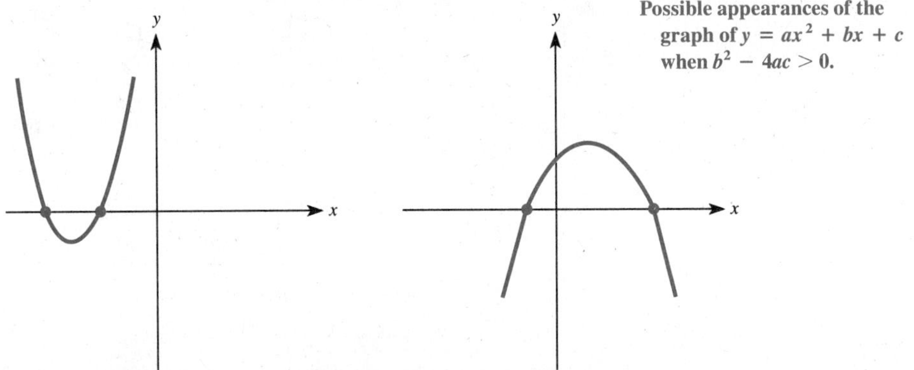

Possible appearances of the graph of $y = ax^2 + bx + c$ when $b^2 - 4ac > 0$.

Figure 6.17

If the discriminant $b^2 - 4ac = 0$, then

$$x = \frac{-b \pm \sqrt{0}}{2a} = -\frac{b}{2a}$$

is the only solution to the quadratic equation $0 = ax^2 + bx + c$. For example, in the equation $0 = x^2 + 4x + 4$, $x = -2$ is the only solution. One way to see this is to rewrite $x^2 + 4x + 4$ as $(x + 2)^2$. Or, you can calculate the discriminant for $x^2 + 4x + 4$ to see that $b^2 - 4ac = 4^2 - (4 \times 1 \times 4) = 16 - 16 = 0$ in this case. When the discriminant is 0, the graph of $y = ax^2 + bx + c$ meets the x-axis at only one point. (See Figure 6.18 for example.)

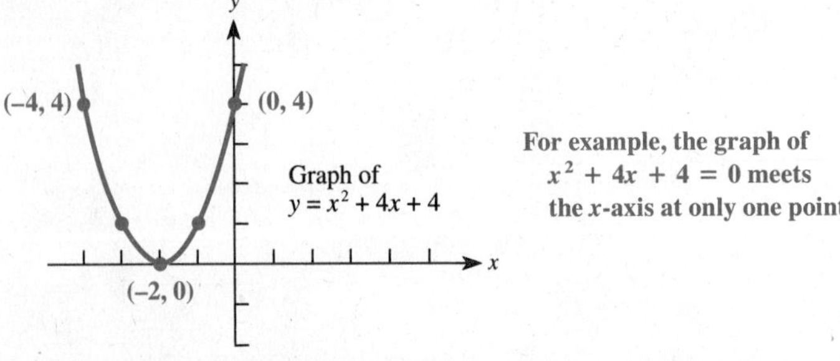

For example, the graph of $x^2 + 4x + 4 = 0$ meets the x-axis at only one point.

Figure 6.18

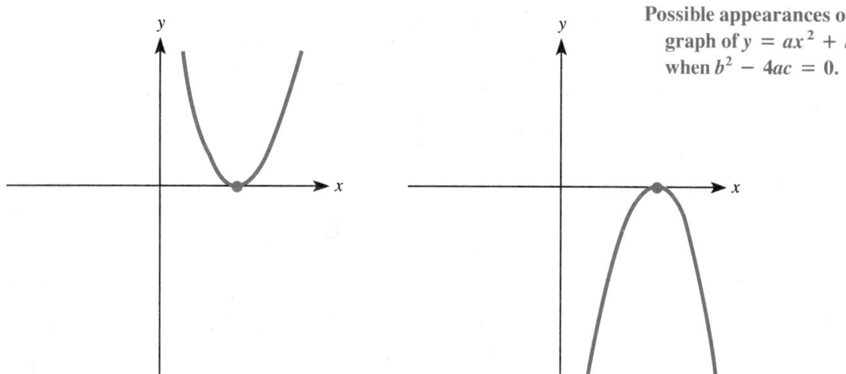

Figure 6.19

Finally, if the discriminant $b^2 - 4ac < 0$, we obtain a negative number under the radical symbol in the quadratic formula. However, we know that when we square any real number, we obtain a *positive* number. Thus the square root of a negative number cannot be a real number. When $b^2 - 4ac < 0$, there are no real numbers that are solutions to $0 = ax^2 + bx + c$, and consequently the graph of $y = ax^2 + bx + c$ does not cross the x-axis at all.

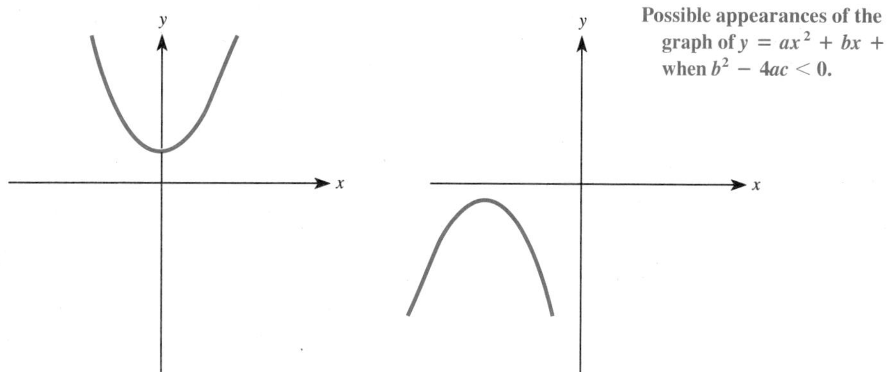

Figure 6.20

We will consider the case $b^2 - 4ac < 0$ in Section 6.4. In this case, the solutions are *complex* numbers.

EXAMPLES 13–16

Use the quadratic formula to solve the following quadratic equations.

13. $3x^2 - 4x = 17$ **14.** $2x^2 = 13x + 7$

15. $3x^2 + 3 = (2x + 1)^2 + 4$ **16.** $4x^2 + 4x = -1$

Solutions

Before we can use the quadratic formula, *each equation must be put in the standard form* $ax^2 + bx + c = 0$.

13. Using an Equation $3x^2 - 4x = 17$, so $3x^2 - 4x - 17 = 0$ is the standard form. Here, $a = 3$, $b = -4$, and $c = -17$. Thus

$$x = \frac{-b \pm \sqrt{b^2 - 4ac}}{2a} = \frac{-(-4) \pm \sqrt{(-4)^2 - 4(3)(-17)}}{2 \cdot 3}$$

$$= \frac{4 \pm \sqrt{16 + 204}}{6} = \frac{4 \pm \sqrt{220}}{6}$$

$\sqrt{220} \approx 14.83$; so

$$x \approx \frac{4 \pm 14.83}{6} = \frac{18.83}{6} \text{ or } \frac{-10.83}{6}$$

Therefore

$x \approx 3.14$ or $x \approx -1.81$ Approximate answer. Check by substitution.

Using a Graph Graph $y = 3x^2 - 4x - 17$.

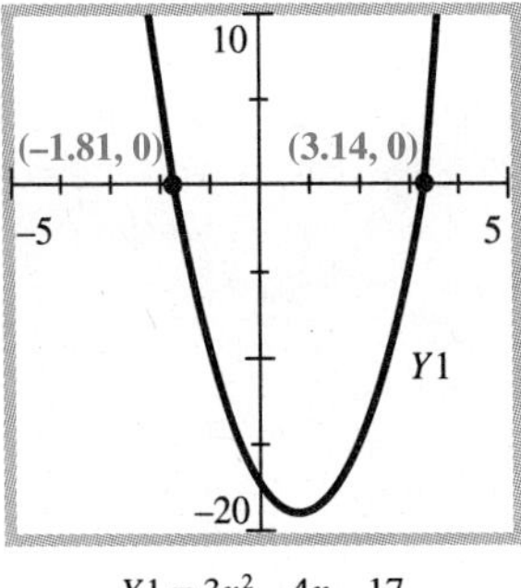

$Y1 = 3x^2 - 4x - 17$

Figure 6.21

We can use a graphing calculator. The graph crosses the x-axis at $x = -1.81$ and $x = 3.14$.

Notice the relationship between the solutions of $3x^2 - 4x - 17 = 0$ and the two points where the graph of the equation $y = 3x^2 - 4x - 17$ crosses the x-axis.

14. Using an Equation $2x^2 = 13x + 7$, so $2x^2 - 13x - 7 = 0$, where $a = 2$, $b = -13$, and $c = -7$.

$$x = \frac{-(-13) \pm \sqrt{(-13)^2 - 4(2)(-7)}}{2(2)}$$ Substitute into the quadratic formula.

$$x = \frac{13 \pm \sqrt{169 + 56}}{4}$$

$$x = \frac{13 \pm \sqrt{225}}{4} = \frac{13 \pm 15}{4}$$

Therefore

$$x = \frac{13 + 15}{4} \quad \text{or} \quad x = \frac{13 - 15}{4}$$

$$x = \frac{28}{4} = 7 \quad \text{or} \quad x = \frac{-2}{4} = -\frac{1}{2}$$

If the solutions are rational numbers (fractions or integers), as in this case, then the quadratic equation will actually factor nicely.

$$(2x^2 - 13x - 7) = 0 \qquad \textbf{Try factoring.}$$

$$(2x + 1)(x - 7) = 0 \qquad \textbf{It works.}$$

Therefore

$$2x + 1 = 0 \quad \text{or} \quad x - 7 = 0$$

$$x = -\frac{1}{2} \quad \text{or} \quad x = 7 \qquad \textbf{Answer (same as we obtained with quadratic formula).}$$

Using a Graph Graph $y = 2x^2 - 13x - 7$.

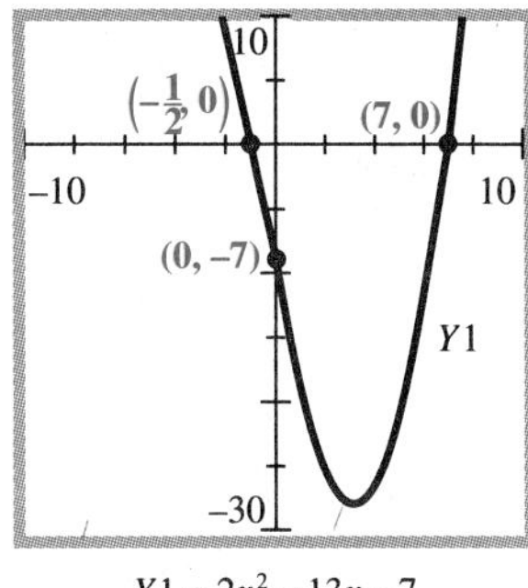

$Y1 = 2x^2 - 13x - 7$

Figure 6.22

We can use a graphing calculator. The graph crosses the x-axis at $x = -\frac{1}{2}$ and $x = 7$.

Notice the relationship between the solutions of $2x^2 - 13x - 7 = 0$ and the two points where the graph of the equation $y = 2x^2 - 13x - 7$ crosses the x-axis.

15. $3x^2 + 3 = (2x + 1)^2 + 4$

$$3x^2 + 3 = 4x^2 + 4x + 1 + 4 \qquad \textbf{Expand and put in the general form.}$$

$$0 = x^2 + 4x + 2 \qquad a = 1, b = 4, c = 2$$

$$x = \frac{-4 \pm \sqrt{(4)^2 - 4(1)(2)}}{2(1)} \qquad \textbf{Substitute into the quadratic formula.}$$

$$x = \frac{-4 \pm \sqrt{16 - 8}}{2} = \frac{-4 \pm \sqrt{8}}{2} \qquad \sqrt{8} = \sqrt{(4)(2)} = \sqrt{4} \cdot \sqrt{2} = 2\sqrt{2}$$

$$x = \frac{-4 \pm 2\sqrt{2}}{2} = \frac{2(-2 \pm \sqrt{2})}{2}$$

$$x = -2 + \sqrt{2} \quad \text{or} \quad x = -2 - \sqrt{2}$$

$$x \approx -0.586 \quad \text{or} \quad x \approx -3.414$$

Answer.

Recall that in Example 7, we solved the equation $x^2 + 4x + 2 = 0$ by completing the square. Check back, and see that we obtained the same answer as in Example 15, which is an equation equivalent to $x^2 + 4x + 2 = 0$.

16. Using an Equation

$$4x^2 + 4x = -1$$

$$4x^2 + 4x + 1 = 0$$

$$x = \frac{-4 \pm \sqrt{4^2 - (4)(4)(1)}}{(2)(4)} = \frac{-4 + 0}{8} = -\frac{1}{2}$$

Using the quadratic formula, we see the discriminant is 0 in this case.

$$x = -\frac{1}{2}$$

Answer.

NOTE: We could also have solved this quadratic equation by factoring: $4x^2 + 4x + 1 = 0$, so $(2x + 1)^2 = 0$, so $x = -\frac{1}{2}$ is the only possible solution. ■

Using a Graph

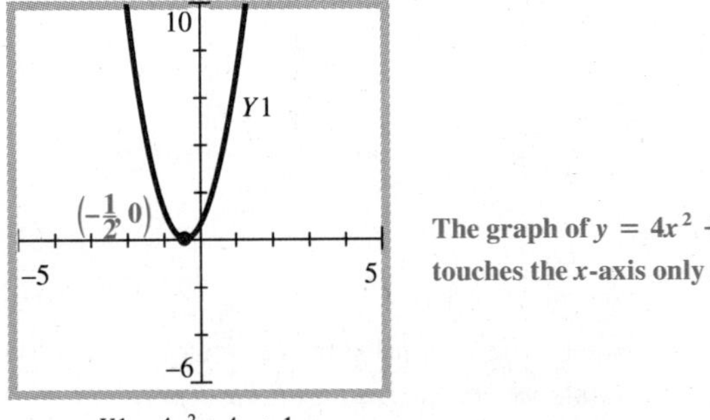

$Y1 = 4x^2 + 4x + 1$

The graph of $y = 4x^2 + 4x + 1$ touches the x-axis only at $x = -\frac{1}{2}$.

Figure 6.23

Let us turn to some applications of quadratic models. First we take another look at the golfing problem from Section 6.1.

EXAMPLE 17

Suppose the height h above ground (in feet) of a golf ball hit off an elevated tee is given by

$$h = 80t - 16t^2 + 20 \quad \text{or} \quad \text{Height}(t) = 80t - 16t^2 + 20$$

after t seconds. How long will it take for the ball to hit the ground?

Solution

Using an Equation The ball is on the ground when the height $h = 0$. Thus we must solve the quadratic equation:

$$0 = 80t - 16t^2 + 20$$

or

$$0 = 16t^2 - 80t - 20$$ Multiply both sides by -1.

$$t = \frac{-(-80) \pm \sqrt{(-80)^2 - 4(16)(-20)}}{2(16)}$$ Use the quadratic formula with $a = 16$, $b = -80$, and $c = -20$.

$$t = \frac{80 \pm \sqrt{6{,}400 + 1{,}280}}{32}$$

$$t = \frac{80 \pm \sqrt{7{,}680}}{32}$$ Using a calculator, $\sqrt{7{,}680} \approx 87.64$.

$$t \approx \frac{80 + 87.64}{32} \quad \text{or} \quad t \approx \frac{80 - 87.64}{32}$$

$$t \approx 5.239 \quad \text{or} \quad t \approx -0.239$$ Answer.

The negative value is a solution to the equation but not a solution for the physical situation. It takes about 5.24 seconds for the ball to hit the ground.

Using a Graph Graph $h = 80t - 16t^2 + 20$.

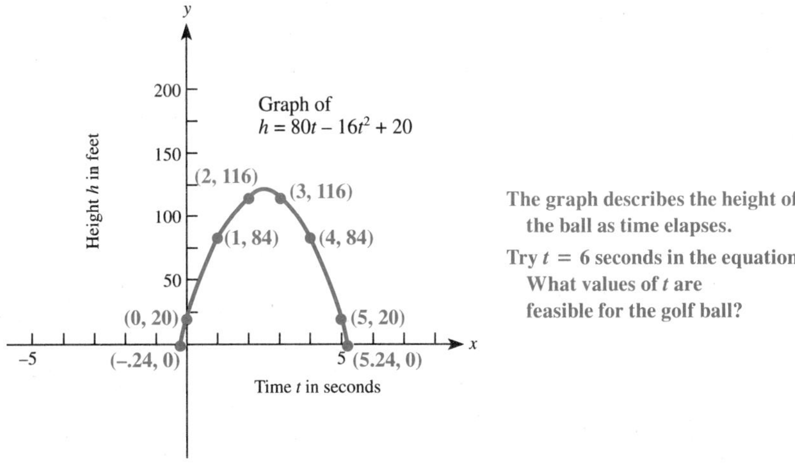

The graph describes the height of the ball as time elapses.

Try $t = 6$ seconds in the equation. What values of t are feasible for the golf ball?

Figure 6.24

In Section 6.1, before we had the power of the quadratic formula, we solved the golf ball problem in Example 3 by successive approximation with a calculator. Look back and see how close we came.

EXAMPLE 18

What is the height of the tallest tree you can brace with a 250-foot wire? The wire must be anchored to the ground at a distance from the base of the tree that is half the height of the tree. The wire must be tied in to the tree 10 feet from its tip. (This problem was first introduced as Example 4 in Section 6.1.)

Solution

We are looking for the height of the tree. So let x equal the height of the tree.

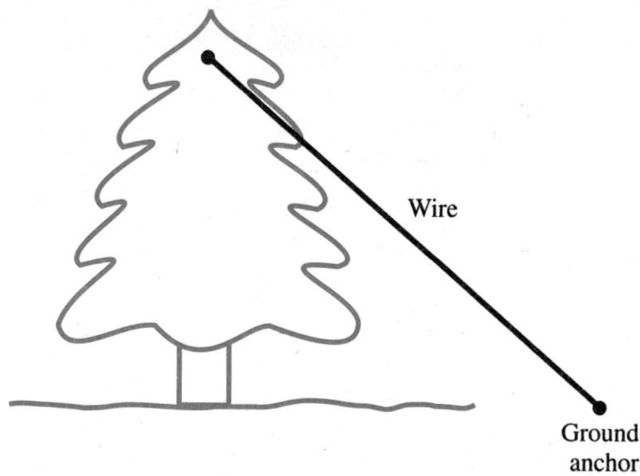

Figure 6.25

We can tie the wire in 10 feet from the top, at $x - 10$ (feet). We must anchor in the ground half this distance from the base of the tree, at $\frac{x}{2}$.

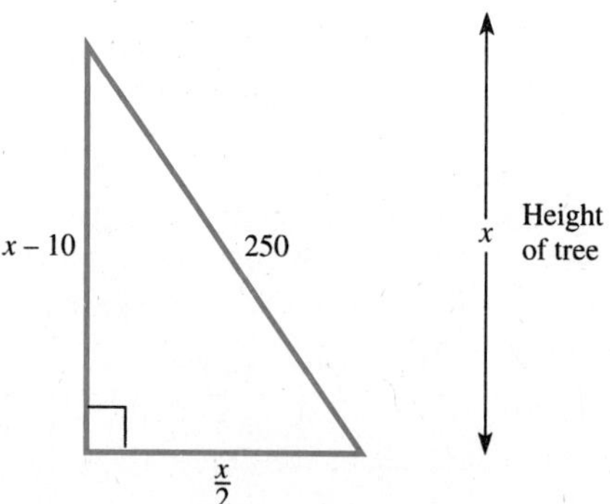

Figure 6.26

Thus we can picture the situation in the figure above. The wire is the hypotenuse of a right triangle. If we use the Pythagorean Theorem, we obtain

$$(x - 10)^2 + \left(\frac{x}{2}\right)^2 = (250)^2$$

$$x^2 - 20x + 100 + \frac{x^2}{4} = 62{,}500 \qquad \textbf{Expanded.}$$

$$\frac{5x^2}{4} - 20x + 100 = 62{,}500$$ Like terms are grouped.

$$1.25x^2 - 20x - 62{,}400 = 0$$ Equation in standard form.

$$x = \frac{-(-20) \pm \sqrt{(-20)^2 - 4(1.25)(-62{,}400)}}{2(1.25)}$$ Use the quadratic formula with $a = 1.25$, $b = -20$, and $c = -62.400$.

$$x = \frac{20 \pm \sqrt{400 + 312{,}000}}{2.5} = \frac{20 \pm \sqrt{312{,}400}}{2.5}$$ Use a calculator.

$$x \approx \frac{20 + 558.93}{2.5} \quad \text{or} \quad x \approx \frac{20 - 558.93}{2.5}$$

$$x \approx 231.57 \text{ feet}$$ Answer. Discard the negative solution since trees have positive height.

Compare our answer for Example 18 to our calculator approximation in Example 4. How close was our approximation?

EXAMPLE 19

Because of heavy traffic, suppose it took you 20 minutes longer to drive the 25 miles from work to home than it did for you to drive from home to work that morning. You estimate that you averaged 10 mph less on the way home. How fast did you drive each way?

Solution

Using a Table A key relationship we are given in the problem is

time for return home = time traveling to work + 20 minutes

If we let V represent the average velocity *to* work; then $V - 10$ represents the average velocity to return home. We can use the model:

Distance = velocity × time

$$D \quad = \quad V \quad \times \quad t$$ Recall $t = \frac{D}{V}$.

	Distance	*Velocity*	*Time* $= \frac{Distance}{Velocity}$
To work	25	V	$\frac{25}{V}$
Return home	25	$V - 10$	$\frac{25}{V - 10}$

Table 6.11

Using an Equation Using the values for time in Table 6.11 and the relationship

time for return home = time for traveling + 20 minutes

we obtain

$$\frac{25}{V-10} = \frac{25}{V} + \frac{1}{3}$$

time for return = time to work + one-third hour (= 20 minutes)

$$3V(V-10)\frac{25}{V-10} = 3V(V-10)\left(\frac{25}{V} + \frac{1}{3}\right)$$ **The LCD is $3V(V-10)$.**

$$75V = 75(V-10) + V(V-10)$$ **Multiply and put in the standard form of a quadratic equation.**

$$75V = 75V - 750 + V^2 - 10V$$

$$0 = V^2 - 10V - 750$$ $a = 1, b = -10, c = -750$

$$V = \frac{-(-10) \pm \sqrt{(-10)^2 - 4(1)(-750)}}{2(1)}$$ **Using the quadratic formula.**

$$V = \frac{10 \pm \sqrt{3{,}100}}{2}$$

$$V \approx \frac{10 \pm 55.68}{2}$$

$V \approx -22.84$ or $V \approx 32.84$ **Answer for velocity to work.**

$V - 10 \approx 22.84$ **Answer for velocity returning home. Some traffic jam!**

Again we can ignore the negative solution, since velocity is a positive quantity. ■

EXAMPLE 20 Suppose you made an investment of \$1,000 in some mutual funds. The funds pay interest compounded annually. After two years, you take your money out of the funds, a total of \$1,150.75. What interest rate did the funds pay?

Solution

Let r be the interest rate. After one year, you have

Investment + interest = one-year total

$$\$1{,}000 + \$1{,}000 \cdot r = (1{,}000 + 1{,}000r)$$

After two years, you have

One year total + interest on one year total = two-year total

$$(1{,}000 + 1{,}000r) + r(1{,}000 + 1{,}000r) = 1{,}150.75$$ **Equation (1)**

$$(1{,}000 + 1{,}000r)(1 + r) = 1{,}150.75$$ **Factor.**

$$1{,}000(1 + r)(1 + r) = 1{,}150.75$$ **Factor.**

Therefore

$$1{,}000(1 + r)^2 = 1{,}150.75$$ **Equation (2)**

We want to solve Equation 2 for r:

$$(1 + r)^2 = \frac{1{,}150.75}{1{,}000} = 1.15075$$

$$1 + r = \pm\sqrt{1.15075} \qquad \text{Take square roots.}$$

$$r \approx -1 \pm 1.073 \qquad \text{A negative rate makes no sense.}$$

$$r \approx 0.073 \quad \text{or} \quad 7.3\% \text{ interest} \qquad \text{Answer.}$$

Interest is often compounded quarterly, or even daily, or continuously. We will return to a further discussion of compound interest in Chapter 7 when we discuss fractional exponents. Also note that we could solve this problem by putting Equation 1 of Example 20 in standard form and using the quadratic formula.

EXAMPLE 21

Sometimes an equation is given in a generalized form without specific numerical coefficients. For example, the surface area A of a square prism is given by

$$A = 2w^2 + 4lw$$

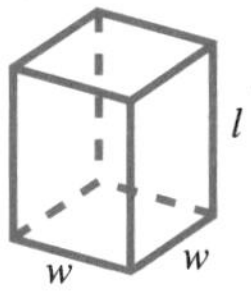

Figure 6.27

where l = length and w = width. If we had numerical values of A and l, then the equation would be an ordinary quadratic equation with variable w. However, sometimes it is useful to solve an equation like $A = 2w^2 + 4lw$ for w and get w equal to something that involves A, l, and some numbers. Solve for w.

Solution

If we rewrite this equation as $0 = 2w^2 + 4lw - A$, then we can use the quadratic formula to solve for w by using the *letters* as **generalized coefficients.** For example, in this equation, we would set $a = 2$, $b = 4l$, and $c = -A$ to solve for w and would get

$$w = \frac{-4l \pm \sqrt{(4l)^2 - 4(2)(-A)}}{2 \cdot 2}$$

$$w = \frac{-4l \pm \sqrt{16l^2 + 8A}}{4} \qquad \text{Answer.}$$

Here, w is equal to an expression involving l and A.

6.3 EXERCISES

I. Procedures

1–8. For each quadratic equation below, put the equation in the general form $ax^2 + bx + c = 0$, and identify the values of a, b, and c.

1. $16x^2 = 3x + 9$

2. $24x = 5 - 6x^2$

3. $(x + 1)^2 = 7$

4. $(2x - 5)^2 = 13x$

5. $18 = 0.9(x - 4)^2 - \frac{x}{4}$

6. $0.65x - 0.5(3 - x)^2 = \frac{1}{2}$

7. $2x^2 = 28$ **8.** $3x^2 = 27x$

9–18. Write an equivalent form for each of the following square roots by searching for factors that are perfect squares and using the property $\sqrt{A \cdot B} = \sqrt{A}\sqrt{B}$.

9. $\sqrt{50}$ **10.** $\sqrt{27}$ **11.** $\sqrt{8}$
12. $\sqrt{12}$ **13.** $\sqrt{250}$ **14.** $\sqrt{600}$
15. $\sqrt{98}$ **16.** $\sqrt{192}$ **17.** $\sqrt{1{,}200}$
18. $\sqrt{1{,}440}$

19–24. Each of these quadratic equations can be solved quickly by using the form **(quantity)² = number.** Use this method to find the solutions to these equations.

19. $x^2 = 9; x = ?$
20. $9x^2 - 18 = 0; x = ?$
21. $16y^2 - 48 = 0; y = ?$
22. $(4y + 3)^2 = 4; y = ?$
23. $(z - 1)^2 - 49 = 0; z = ?$
24. $(2z - 1)^2 - 3 = 0; z = ?$

25–33. Solve the following quadratic equations by the method of completing the square. Check your answers by substituting into the equations.

25. $x^2 + 14x = -10$ **26.** $x^2 - 6x + 2 = 0$
27. $x^2 + 8x = -16$ **28.** $x^2 - 5x + 2 = 0$
29. $2x^2 - 5x + 2 = 0$ **30.** $2x^2 - 6x + 2 = 0$
31. $4x^2 - 8x = 17$ **32.** $-3x^2 + 9x + 20 = 0$
33. $-0.25x^2 + x = 1.8$

34–39. Solve these quadratic equations by using the quadratic formula. Be sure to first put the equation in the form $ax^2 + bx + c = 0$. Graph the corresponding equation $y = ax^2 + bx + c$ in each case and note where the graph crosses the x-axis.

34. $2x^2 + 3x - 1 = 0$ **35.** $x^2 + x = 28$
36. $2x^2 - x = 1$ **37.** $3x^2 = 6x + 10$
38. $3 + x - x^2 = 0$ **39.** $4x - 4 - x^2 = 0$

40–65. Use any method you wish (factoring, completing the square, or the quadratic formula) to solve the following equations. For each problem, graph the corresponding equation $y = ax^2 + bx + c$ on a graphing calculator, and show that the x-axis intercepts are the solutions to $0 = ax^2 + bx + c$.

40. $x^2 = 7x - 12$ **41.** $6 + x = x^2$
42. $6x^2 - 17x = 14$ **43.** $8x^2 + 14x = 15$
44. $x^2 - 49 = 0$ **45.** $3y^2 = 75$
46. $4 - 8x = -4x^2$ **47.** $1 = -10x - 25x^2$
48. $0.8z^2 = 0.16z$ **49.** $49x^2 = 28x$
50. $0.08x^2 - 0.3x = 0.05$ **51.** $0.7y^2 = 0.2y + 0.3$
52. $\frac{x^2}{2} = -3x + 8$ **53.** $6 - \frac{x}{2} - x^2 = 0$
54. $(x + 7)^2 - 29 = 0$ **55.** $(2z - 7)^2 = 25$
56. $(3x - 7)^2 = (4x + 1)^2$ **57.** $(2x - 1)^2 = (x + 5)^2$
58. $3 + \frac{2}{x} - \frac{1}{x^2} = 0\ (x \neq 0)$
59. $1 + \frac{3}{x} = \frac{17}{x^2}\ (x \neq 0)$
60. $3.2x - 0.7 - 0.6x^2 = 0$
61. $0.001 = -0.1x - 0.01x^2$
62. $2x^2 + 3x + 7 = 2x^2 + 3x - 5$
63. $(x - 7)^2 = x^2 - 14x + 50$
64. $3 - \frac{(x + 2)^2}{4} = \frac{x^2}{4} + 7x$
65. $2x^2 + 7x = (2x - 3)(x + 5)$

II. Concepts

66. Write a description explaining how to complete the square on a trinomial.

67. Write a description of the connection between the quadratic equation $ax^2 + bx + c = 0$ and the graph of $y = ax^2 + bx + c$. Give examples and draw pictures for your friend to help him/her understand the connection.

68. Explain how the equations in Examples 3–5 could be solved using a graphing calculator.

69–72. Each of the following solutions to a quadratic equation contains an error. Find the error, and correct the solution in each case. How would you help a student who made this error?

69. $6x^2 - x = 1$
Solution:

$$6x^2 - x = 1$$
$$6x^2 - x - 1 = 0$$
$$(3x - 1)(2x + 1) = 0$$
$$3x - 1 = 0 \quad \text{or} \quad 2x + 1 = 0$$
$$x = \frac{1}{3} \quad \text{or} \quad x = -\frac{1}{2}$$

Incorrect answer.

70. $24z^2 - 16z = 0$
Solution:

$$24z^2 - 16z = 0$$
$$24z^2 = 16z$$
$$24z = 16$$
$$z = \frac{16}{24} = \frac{2}{3}$$ **Incorrect answer.**

71. $3x^2 - 6x - 7 = 0$
Solution:

$$3x^2 - 6x - 7 = 0$$
$$a = 3, b = -6, c = -7$$
$$x = \frac{6 \pm \sqrt{36 - 84}}{6}$$
$$x = \frac{6 \pm \sqrt{-48}}{6}$$ **Incorrect answer.**

72. $2x^2 - 3x + 1 = 0$
Solution:

$$a = 2, b = -3, c = -1$$
$$x = \frac{3 \pm \sqrt{9 + 8}}{2} = \frac{3 \pm \sqrt{17}}{2}$$ **Incorrect answer.**

73–75. For Exercises 73–75, write a quadratic equation that has the given solutions, put the equation in standard form, and sketch the graph of the corresponding equation $y = ax^2 + bx + c$.

73. $x = 3$ and $x = -7$. **74.** $x = 0$ and $x = 7$.

75. $x = \sqrt{2}$ and $x = -\sqrt{2}$.

76. Write a quadratic equation that has only the solution $x = -3$.

III. Applications

77–80. Each of the figures below has an area equal to 24 square meters. Find the indicated dimensions.

77.

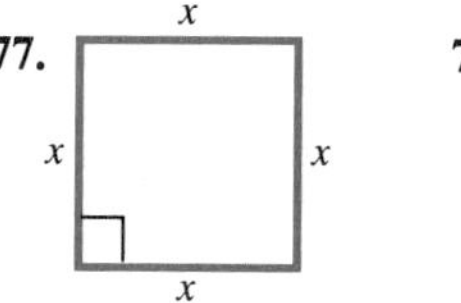

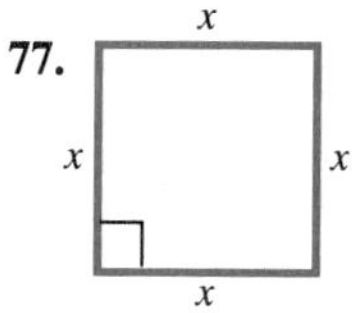

78.

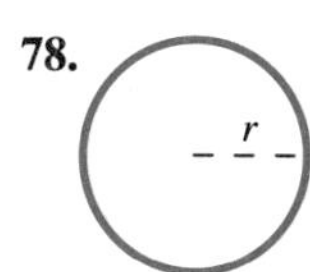

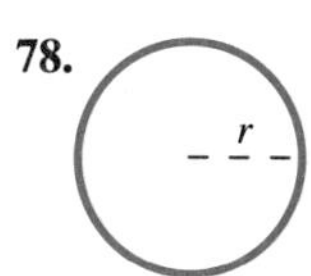

79.

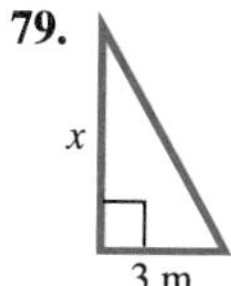

80.

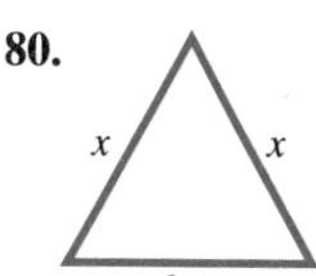

81. For what values of k does $x^2 - kx + k = 0$ have only one solution?

82. The height h above the ground of a golf ball depends on the time t it has been in flight. One golfer hits a tee shot that has a height approximately given by

$$h = 80t - 17t^2$$

where h is in feet and t is in seconds.

(a) Graph this equation, then use the graph to answer parts (b), (c), and (d).

(b) How many seconds have elapsed when the golf ball is 50 feet in the air? In other words, when $h = 50$, what does t equal?

(c) How many seconds have elapsed when the golf ball is 90 feet in the air?

(d) How many seconds have elapsed when the golf ball hits the ground?

83. A missile is shot from the ground with an initial velocity of 44 feet per second. The distance d above the ground after t seconds is given by

$$d = 44t - 16t^2$$

(Note: Missiles have much more mass than golf balls, thus the initial launching velocity is much smaller than the initial velocity of a golf ball when it is hit.)

(a) Graph this equation, then use the graph to answer parts (b) and (c).

(b) How many seconds after the start is the missile at a height of 24 feet?

(c) When will the missile hit the ground?

84. Grace's pool is 40 feet by 30 feet. She wishes to spread a border of barkdust around the pool. How wide can the border be if she has 296 square feet to spread? *Hint:* If the pool was removed from the outermost rectangle in the figure, then only the 296-square-foot border would remain. Write an algebraic expression for the area of the outermost rectangle.

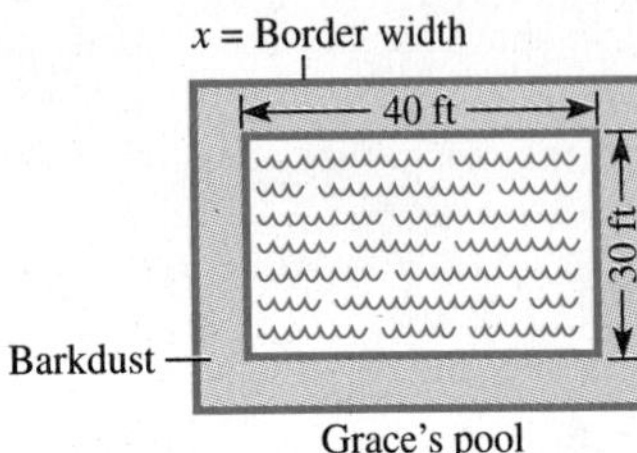

85. An eccentric oil magnate ships oil in spherical tanks that are 3 meters in diameter. The oil magnate now wishes to ship oil in cylindrical tanks that are 4 meters high and have the *same* volume as the spheres that were previously used. What radius must the cylindrical tanks have to meet the demands?

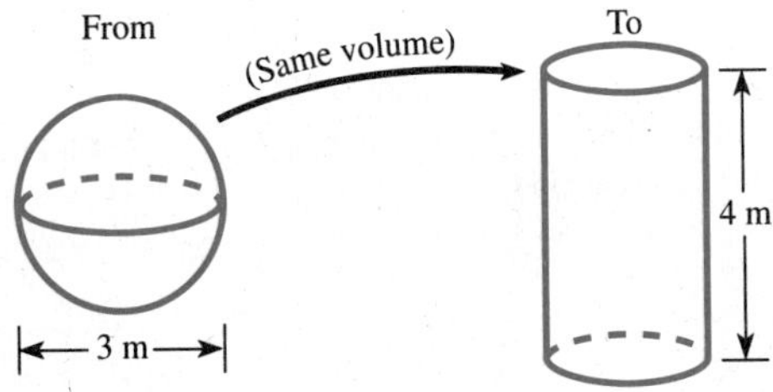

86. Find the missing dimensions of this rectangular box.

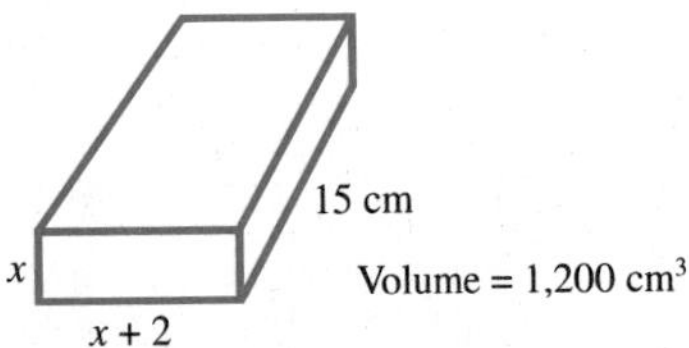

87. Find the dimensions of this rectangular field.

Area = 22,400 m²

Perimeter = 1,200 m

88. What is the height of the tallest telephone pole you can brace with a 50-foot wire cable? You wish to anchor the ground cable at a distance from the base of the pole equal to one-third the height of the pole. You also want to attach the cable to the pole equal to 5 feet from the top of the pole (see Example 18).

89. Suppose it took you 30 minutes less to drive the 40 miles back home after dinner at your parents' house than it did for you to arrive there in rush-hour traffic. You drove about 15 mph faster on the way home. How fast did you drive each way (see Example 19)?

90. Suppose you made an investment of \$500 in some mutual funds that pay interest compounded annually. After two years, you take your money out of the funds and have \$590.25.

(a) What interest rate did the funds pay?

(b) Answer the same question for an investment of \$2,000 that accumulates to \$2,250.85 after two years (see Example 20).

91. A ball is hurled upward with an initial velocity of 96 feet/second. The height h of the object after t seconds is given by

$$h = 96t - 16t^2$$

(a) Graph this equation. Use the graph to answer part (b).

(b) How many seconds have elapsed when the object is 100 feet in the air?

92. The stopping distance d in feet of a car moving at a velocity V mph is given by the equation

$$d = 0.06V^2 + 1.1V$$

(a) Graph this equation. Use the graph to answer parts (b) and (c).

(b) What is the stopping distance for a velocity of 20 mph? Forty mph? Fifty mph? Seventy-five mph?

(c) What speed is a car going if it takes 100 feet to stop? If it takes 250 feet to stop?

93. The radius r of the earth is approximately 4,000 miles.

(a) A television aerial in the figure is 208 feet tall. How far can a person see (distance to horizon) if she is on top of the aerial? *Hints:* A radius is *perpendicular* to the point P that the line of sight hits, because it is a tangent line. The Pythagorean theorem for right triangles will help. Do not forget to make the units the same; that is, 4,000 miles = ? feet (1 mile = 5,280 feet).

(b) Suppose that a person is on top of Mt. Hood (highest mountain in Oregon) at 11,235 feet. How far can he see? Of course, we are assuming clear weather.

(c) Suppose a person is in an airplane at 32,000 feet. How far can she see?

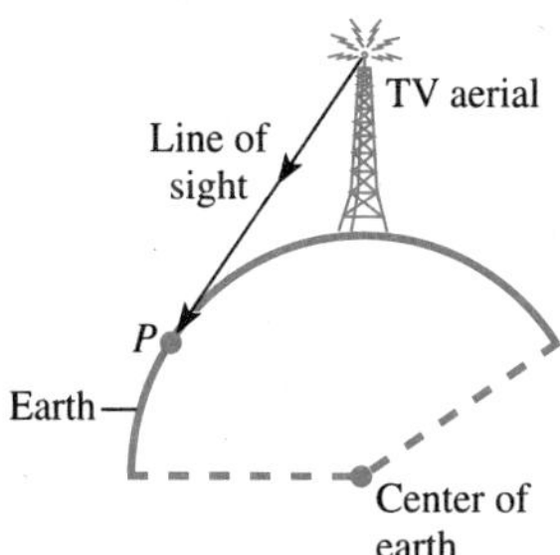

94. Recall that the relationship between the length l of a pendulum chord and the time t that it takes the pendulum to complete a swing through its arc is given by

$$l = \frac{g}{2\pi}t^2$$

where l depends on t^2 and g is the acceleration due to gravity ($g \approx 32.2$ feet/sec). How long will it take a pendulum with a 10 foot pendulum chord to complete a swing through its arc? A 100-foot pendulum chord? A 1-inch pendulum chord?

95. The height h of a body thrown upward with initial velocity V_0 after time t is given by

$$h = V_0 t - \frac{gt^2}{2}$$

where g is the gravitational constant. Solve this equation for t.

IV. Extensions

96. *Famous Question:*

What should the ratio of the length l to the width w of a rectangle be so that if a square (shaded in figure) is cut off from the original rectangle, the result will be a rectangle (AEFD in figure) that is the same (similar) "shape" as the original rectangle?

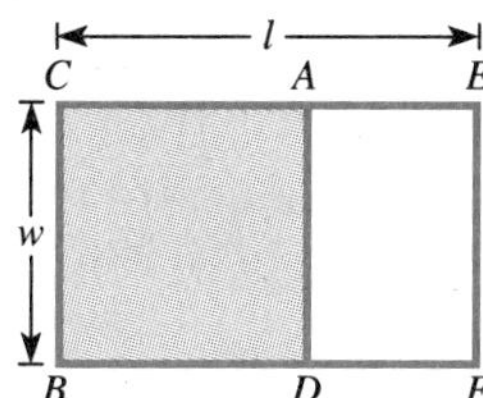

Note: Now that we have the quadratic formula, we can obtain the solution to this problem, which the Greeks sought. Again, for the sake of simplicity, we suggest that you let $l = 1$ (one unit).

(a) Set up the equation that says that rectangle CEFB is similar to rectangle AEFD.

(b) The equation from part (a) is a proportion. Put this equation in the form $ax^2 + bx + c = 0$ (the proportion turns out to be equivalent to a quadratic equation).

(c) Use the quadratic formula to determine the value of x that will yield a similar rectangle when the square is cut off. This value of x is called the Golden Ratio. Find a numerical approximation for your answer with a calculator.

97. The Babylonians wished to solve this problem: Given a number n, find another number x such that x added to its reciprocal yields n.

(a) We have solved this problem for particular values of n. (See Exercise 47 in Section 6.2.) Solve it in *general,* that is, get x in terms of n.

(b) Use your formula from part (a) to find the value for x if $n = 8, -15, 1, 7$, and 100.

(c) What happened when you tried $n = 1$ in part (b)?

98. For what values of a is $a + \frac{1}{a} \geq 2$?

99. Suppose that the interest that was compounded in Example 20 was compounded for three years instead of two. For example, suppose \$1,000 yields \$1,225 when compounded for three years.

(a) Set up the equation to find the interest rate. What kind of equation is it?

(b) Find the interest rate by using a calculator and the method of successive approximation.

(c) Answer the same question for four years, accumulating to \$1,405.60.

100. Compute $7 - 4[7 - 4(7 - 4)]$. Several students rewrote this expression as follows:

$$\overset{3}{\wedge}\quad\overset{3}{\wedge}\quad\overset{3}{\wedge}$$
$$7 - 4[7 - 4(7 - 4)] = [(3)(3)(3)] = 27,$$

which is the correct answer, obtained by an incorrect method. For what pairs of numbers does this incorrect procedure yield the correct answer?

6.4 COMPLEX NUMBERS AS SOLUTIONS TO QUADRATIC EQUATIONS (OPTIONAL)

We have previously mentioned the following problem, posed by the ancient Babylonians. (See Exercise 97 in Section 6.3.)

EXAMPLE 1 The Babylonian Problem: Choose any number n. Is there another number x so that when x is added to its reciprocal we obtain n?

Solution

Using an Equation If we solve the problem posed by the Babylonians for $n = 1$, then we must find a number x such that

$$x + \frac{1}{x} = 1 \qquad \text{The LCD is } x.$$

$$\text{(number)} + \text{(its reciprocal)} = 1$$

$$x\left(x + \frac{1}{x}\right) = 1(x) \qquad \text{Multiply both sides by the LCD.}$$

$$x^2 + 1 = x$$

$$x^2 - x + 1 = 0$$

$$x = \frac{-(-1) \pm \sqrt{(-1)^2 - 4(1)(1)}}{2(1)} \qquad \text{A quadratic equation, } a = 1, b = -1, c = 1.$$

$$x = \frac{1 \pm \sqrt{1 - 4}}{2}$$

$$x = \frac{1 \pm \sqrt{-3}}{2} \qquad \text{Look under the radical sign.}$$

Using a Graph Graph $y = x^2 - x + 1$. See Figure 6.28.

We faithfully carried out the calculations of the quadratic formula. However, you (hopefully) should be somewhat curious about the answer we obtained. What does $\sqrt{-3}$ mean? What number when multiplied by itself yields a *negative* number? In our present system of numbers, the square of a number is always positive.

This phenomenon of a negative number under a square root is not an isolated occurrence. Many quadratic equations yield such a result. Consider an equation as simple as $x^2 + 1 = 0$. What are its solutions? If we proceed as in Section 6.3,

$$x^2 + 1 = 0$$

$$x^2 = -1$$

$$x = \pm\sqrt{-1}$$

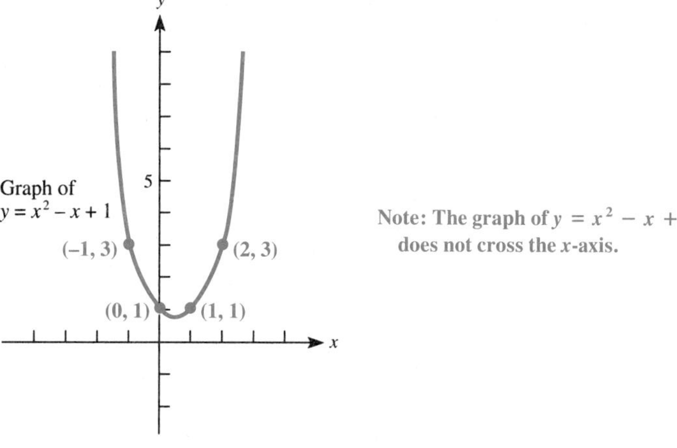

Figure 6.28

The solutions to certain quadratic equations are not in our system of real numbers, but they *are* the solutions. We will need to enlarge our notion of a number in order to incorporate these "new" numbers that can appear as solutions to quadratic equations.

The Complex Numbers

Suppose we *invent* a new number whose square is -1. Then we can put it together with our set of real numbers and the usual operations of addition, multiplication, and so on. The new number is called the **imaginary unit,** and it is denoted by i.

Definition of the Imaginary Unit

If we let i be the number such that $i^2 = -1$, then $i = \sqrt{-1}$.

The equation $x^2 + 1 = 0$, then, has the solutions

$$x^2 = -1 \qquad \textbf{Equation (1)}$$
$$x = \pm\sqrt{-1} = \pm i$$

The imaginary unit, and its negative, are solutions for Equation 1. What, then, should we mean by $\sqrt{-4}$, or by $\dfrac{(1 \pm \sqrt{-3})}{2}$, which are the "solutions" to the Babylonian problem with which we began this section? Recall that we have a property of square roots, which says $\sqrt{a \cdot b} = \sqrt{a}\sqrt{b}$, where $a, b \geq 0$ (see Section 6.3). Let us extend this property for the case of a negative number under the radical.

Definition of the Square Root of a Negative Real Number

Suppose that $a > 0$. Then $\sqrt{-a} = \sqrt{a} \cdot \sqrt{-1} = \sqrt{a}i$

Thus $\sqrt{-4} = \sqrt{4} \cdot \sqrt{-1} = 2i$. In the same way,

$$\frac{1 \pm \sqrt{-3}}{2} = \frac{1 \pm \sqrt{3}i}{2}$$

Thus we can obtain a whole new set of numbers that consists of all the possible combinations of this "new" number i with the *real* numbers we previously possessed. This extended set of numbers is called the set of **complex numbers.**

Definition of Complex Numbers

The set of numbers of the form $a + bi$, where a and b are real numbers and $i = \sqrt{-1}$, is called the set of **complex numbers.**

Some examples of complex numbers are

(a) $7 + 3i$ **(b)** $-4i$ **(c)** $\sqrt{3} + \frac{i}{2}$ **(d)** 17 **(e)** $-\frac{1}{2} - \sqrt{7}i$

All of these numbers are of the form $a + bi$.

(a) $a = 7, b = 3$ **(b)** $a = 0, b = -4$ **(c)** $a = \sqrt{3}, b = \frac{1}{2}$

(d) $a = 17, b = 0$ **(e)** $a = -\frac{1}{2}, b = -\sqrt{7}$

NOTE: Examples (b) and (d) are special cases, because they are *of the form* $a + bi$ with $a = 0$ (in [b]) or $b = 0$ (in [d]). A complex number with $a = 0$, like $-4i$, is called a **pure imaginary number.** If $b = 0$, as in example (d), then the imaginary part of the complex number disappears, and we obtain a real number, 17 in this case. Thus *every real number is also a complex number.* In a complex number $a + bi$, a is called the **real part** and b is called the **imaginary part** of the complex number. Thus the complex number $7 + 3i$ has real part 7 and imaginary part 3. ■

Operations on Complex Numbers

It is possible to perform arithmetic operations on complex numbers, with the provision that $i^2 = -1$.

ADDITION (OR SUBTRACTION) OF COMPLEX NUMBERS

$$(7 - 3i) + (6 - 2i) = 7 + 6 - 3i - 2i = 13 - 5i$$ **Add (subtract) like parts.**

MULTIPLICATION OF COMPLEX NUMBERS

$$\begin{aligned}(6 - 2i)(5 - 7i) &= (6)(5) - (6)(7i) - (2i)(5) - (2i)(-7i) && \text{Multiply just like binomials.}\\ &= 30 - 42i - 10i + 14i^2 \\ &= 30 - 52i + 14(-1) && i^2 = -1\\ &= 16 - 52i && \text{Answer.}\end{aligned}$$

We can also divide complex numbers, but it is not our objective, at present, to perform extended computations with complex numbers. We wish to consider them when they occur as solutions to quadratic equations.

The expansion of our number system to include complex numbers will allow us to solve *any* quadratic equation.

EXAMPLES 2–4

Solve the following equations and observe any patterns in the graphs of the equations.

2. $x^2 + 9 = 0$ **3.** $2 + 3x^2 = 0$ **4.** $x^2 = -28$

Solutions

2. Using an Equation

$$x^2 + 9 = 0$$
$$x^2 = -9$$
$$x = \pm\sqrt{-9} = \pm\sqrt{9}\cdot\sqrt{-1}$$
$$= \pm 3i$$

Check: $(3i)^2 \stackrel{?}{=} 3^2(i)^2 \stackrel{?}{=} 9(-1) \stackrel{\checkmark}{=} -9.$

Using a Graph

Graph $y = x^2 + 9$.

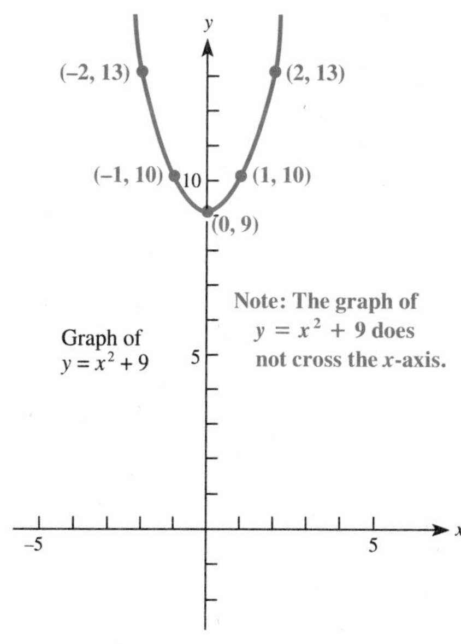

Figure 6.29

3. Using an Equation

$$2 + 3x^2 = 0$$
$$3x^2 = -2$$
$$x^2 = \frac{-2}{3}$$
$$x = \pm\sqrt{\frac{-2}{3}}$$
$$= \pm\sqrt{\frac{2}{3}}\,i \approx \pm 0.817i$$

Using a Graph

Graph $y = 3x^2 + 2$.

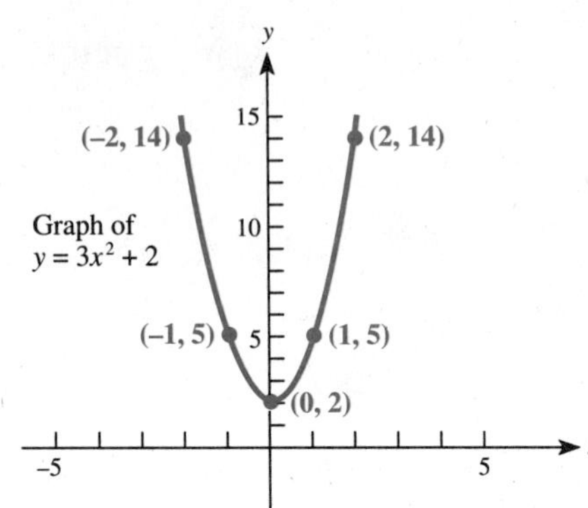

Figure 6.30

Note: The graph of $y = 3x^2 + 2$ *does not* cross the x-axis.

4. $x^2 = -28$, so $x = \pm\sqrt{-28} = \pm\sqrt{28}i \approx \pm 5.29i$

Since $\sqrt{28} = \sqrt{4 \cdot 7} = \sqrt{4} \cdot \sqrt{7} = 2\sqrt{7}$, this solution could also be written as $x = \pm 2\sqrt{7}i$. *All three forms of the solution are correct.* The form you choose for your answer depends upon your preference and whether or not you want to use a decimal approximation for the answer.

The solutions $\pm 2\sqrt{7}i$ can also be written $\pm 2i\sqrt{7}$. This form prevents us from making the mistake of writing the i under the radical symbol. ■

EXAMPLES 5–7

Solve these equations.

5. $3x^2 - 7x = -6$ **6.** $(x + 3)^2 + 16 = (2x + 1)^2$ **7.** $4x^2 + 12x = -9$

Solutions

5.
$$3x^2 - 7x = -6$$
$$3x^2 - 7x + 6 = 0$$

Put in standard form: $a = 3$, $b = -7$, $c = 6$.

$$x = \frac{-(-7) \pm \sqrt{(-7)^2 - 4(3)(6)}}{(2)(3)}$$

Apply the quadratic formula.

$$x = \frac{7 \pm \sqrt{49 - 72}}{6}$$
$$x = \frac{7 \pm \sqrt{(-23)}}{6} = \frac{7 \pm i\sqrt{23}}{6}$$

Two complex-valued solutions.

6.
$$(x + 3)^2 + 16 = (2x + 1)^2$$
$$x^2 + 6x + 9 + 16 = 4x^2 + 4x + 1$$
$$0 = 3x^2 - 2x - 24$$

$a = 3, b = -2, c = -24$

$$x = \frac{-(-2) \pm \sqrt{(-2)^2 - 4(3)(-24)}}{2(3)}$$

Apply the quadratic formula.

$$x = \frac{2 \pm \sqrt{292}}{6}$$ $\sqrt{292} \approx 17.09$

$$x \approx \frac{2 + 17.09}{6} = 3.18 \quad \text{or} \quad x \approx \frac{2 - 17.09}{6} = -2.52$$ Two real-valued solutions.

7. $$4x^2 + 12x = -9$$

$$4x^2 + 12x + 9 = 0$$ This quadratic will factor.

$$(2x + 3)(2x + 3) = 0$$

$$x = -\frac{3}{2}$$ Only one real-valued solution, which occurs twice. ■

The Discriminant

Notice that Example 5 has complex numbers as solutions, while Examples 6 and 7 have real numbers for solutions. Complex solutions will occur if the entry under the radical sign in the quadratic formula is negative. The quadratic formula says that the solutions to $ax^2 + bx + c = 0$ $(a \neq 0)$ are

$$x = \frac{-b \pm \sqrt{b^2 - 4ac}}{2a}$$

The types of solutions to a quadratic equation are determined by the quantity $b^2 - 4ac$. This quantity is called the **discriminant,** which we have already discussed in Section 6.3. We summarize the implications of the discriminant below once again.

Implications of the Discriminant

The quantity $b^2 - 4ac$ is called the *discriminant* for the quadratic equation $ax^2 + bx + c = 0, a \neq 0$.

1. If $b^2 - 4ac > 0$, the equation has two *real*-valued solutions. The corresponding graph of $y = ax^2 + bx + c$ crosses the x-axis twice, at the values of x that are solutions to $0 = ax^2 + bx + c$ (Figure 6.31).

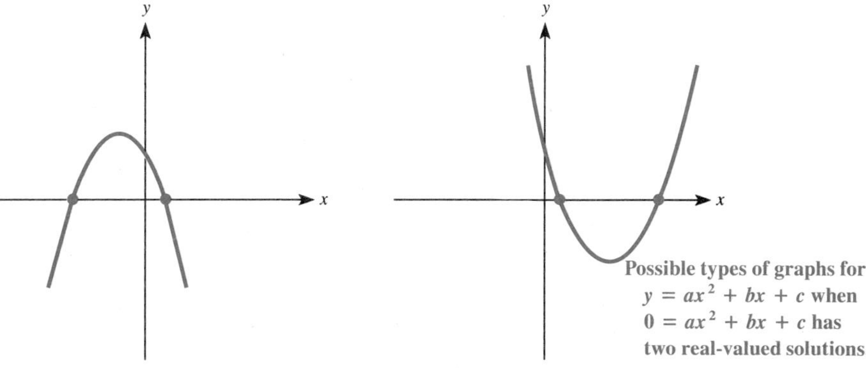

Possible types of graphs for $y = ax^2 + bx + c$ when $0 = ax^2 + bx + c$ has two real-valued solutions.

Figure 6.31

2. If $b^2 - 4ac = 0$, the equation has one (double) real-valued solution. The corresponding graph of $y = ax^2 + bx + c$ crosses the x-axis once, at the value of x that is the solution to $0 = ax^2 + bx + c$ (Figure 6.32).

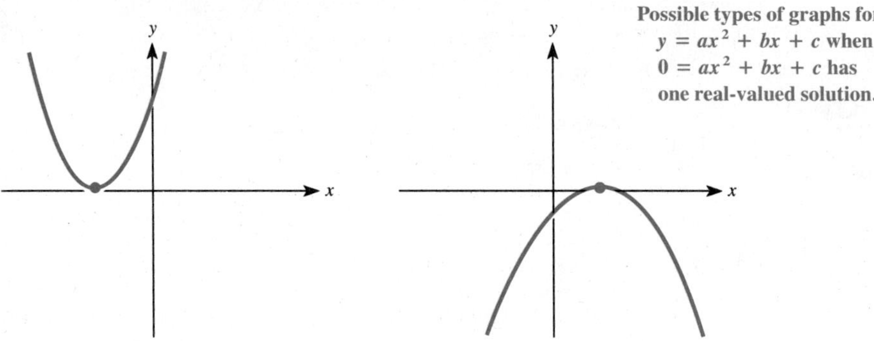
y
x
y
x
Possible types of graphs for $y = ax^2 + bx + c$ when $0 = ax^2 + bx + c$ has one real-valued solution.

Figure 6.32

3. If $b^2 - 4ac < 0$, the equation has two *complex*-valued, nonreal solutions. The corresponding graph of $y = ax^2 + bx + c$ does not cross the x-axis, as no real numbers are solutions to $0 = ax^2 + bx + c$ (Figure 6.33).

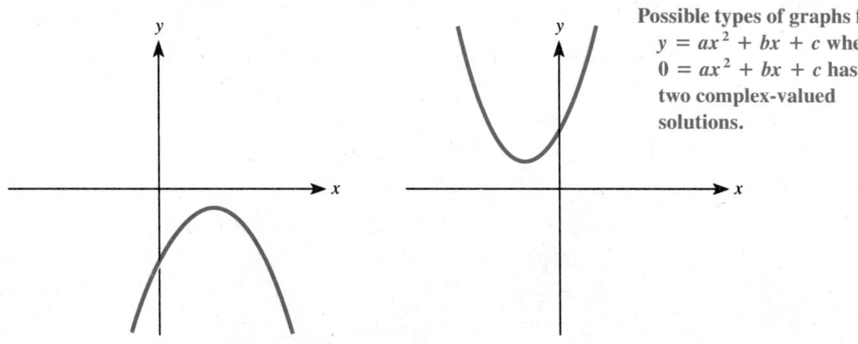
y
x
y
x
Possible types of graphs for $y = ax^2 + bx + c$ when $0 = ax^2 + bx + c$ has two complex-valued solutions.

Figure 6.33

EXAMPLES 8–11

Determine whether these quadratic equations have real- or complex-valued solutions. Indicate the number of times that the corresponding graph of $y = ax^2 + bx + c$ will cross the x-axis.

8. $3x^2 - 7x + 5 = 0$

9. $0.07x^2 - 0.6x = -0.2$

10. $7 = \frac{1}{x^2} - \frac{3}{x}, x \neq 0$

11. $x^4 - 4x^2 - 5 = 0$

Solutions

8. $3x^2 - 7x + 5 = 0$ — $a = 3, b = -7, c = 5$. Check the discriminant.

$$b^2 - 4ac = (-7)^2 - 4(3)(5)$$
$$= 49 - 60 < 0$$

This quadratic has two complex-valued, nonreal solutions. The corresponding graph of $y = 3x^2 - 7x + 5$ does not cross the x-axis. Graph $y = 3x^2 - 7x + 5$ on a graphing calculator.

9. $0.07x^2 - 0.6x = -0.2$ — Put in general form.

$0.07x^2 - 0.6x + 0.2 = 0$ — $a = 0.07, b = -0.6, c = 0.2$

$b^2 - 4ac = (-0.6)^2 - 4(0.07)(0.2)$ — Check the discriminant.

$$= 0.36 - 0.056 > 0$$

This quadratic has two real-valued solutions. The corresponding graph of $y = 0.07x^2 - 0.6x + 0.2$ crosses the x-axis twice. Graph $y = 0.07x^2 - 0.6x + 0.2$ on a graphing calculator.

10. $7 = \dfrac{1}{x^2} - \dfrac{3}{x}$ — LCD $= x^2$

$7x^2 = 1 - 3x$ — Multiply by the LCD.

$7x^2 + 3x - 1 = 0$ — $a = 7, b = 3, c = -1$

$b^2 - 4ac = (3)^2 - 4(7)(-1) > 0$ — Check the discriminant.

This equation is equivalent to a quadratic with two real-valued solutions. The corresponding graph of $y = 7x^2 + 3x - 1$ crosses the x-axis twice. Graph $y = 7x^2 + 3x - 1$ on a graphing calculator.

11. $x^4 - 4x^2 - 5 = 0$ **Equation (1)**

This equation is a *fourth-degree* polynomial equation, but it has no x^3 or x term.

$0 = x^4 - 4x^2 - 5 = (x^2 - 5)(x^2 + 1)$ — Factor.

Thus either $x^2 - 5 = 0$ or $x^2 + 1 = 0$ — Property of zero products.

$x^2 = 5$ or $x^2 = -1$ — First solve for x^2.

$x = \pm\sqrt{5}$ or $x = \pm i$ — Then solve for x.

Thus there are four solutions to Equation 1, $x = \sqrt{5}, -\sqrt{5}, i, -i$, two real-valued solutions and two complex, nonreal-valued solutions. The corresponding graph of $y = x^4 - 4x^2 - 5$ crosses the x-axis twice, at the two real-valued solutions to $0 = x^4 - 4x^2 - 5$. Graph $y = x^4 - 4x^2 - 5$ on a graphing calculator. ■

6.4 EXERCISES

I. Procedures

1–8. In each case, determine whether the quadratic equation has two real solutions, two complex solutions, or one real solution by examining the value of the discriminant. Indicate the number of times that the graph of the corresponding equation $y = ax^2 + bx + c$ will cross the x-axis.

1. $3x^2 - 7x + 5 = 0$
2. $3x^2 - 7x + 2 = 0$
3. $7 - 15x - 6x^2 = 0$
4. $7 - 13x - 6x^2 = 0$
5. $-9 + 6x - x^2 = 0$
6. $-9 - 6x^2 + x^2 = 0$
7. $0.25x^2 = 0.07x - 0.3$
8. $0.25x^2 = 0.07x + 0.3$

9–20. Find the solution to each of the following quadratic equations. Sketch the graph of the corresponding equation $y = ax^2 + bx + c$.

9. $9x^2 + 16 = 0$
10. $24 + 8x^2 = 0$
11. $x^2 = 17x - 8$
12. $2x^2 = 14x + 3$
13. $0.2x^2 - 0.6x + 1 = 0$
14. $0.25x - 0.75x^2 = 0.8$
15. $(x - 4)^2 + 49 = 0$
16. $(3x - 2)^2 - 64 = 0$
17. $(3x + 2)^2 - x^2 + 7 = (x + 5)^2 - 2x$
18. $(3 - 2x)^2 + (x - 7)^2 = 20$
19. $\frac{7}{x^2} - \frac{3}{x} = 5, x \neq 0$
20. $\frac{1}{x} - 25 = \frac{10}{x^2}, x \neq 0$

21–28. Find all solutions to the following equations. In each case, sketch the graph of the corresponding equation $y = ax^4 + bx^2 + c$.

21. $x^4 + 4x^2 + 4 = 0$
22. $x^4 - 1 = 0$
23. $16x^4 + 4x^2 = 0$
24. $16x^4 + 4x^2 - 8 = 0$
25. $x^4 + 5x^2 + 6 = 0$
26. $2x^4 - x^2 = 15$
27. $2x^4 + 3x^2 - 3 = 0$*
28. $x^4 - 3x^2 - 5 = 0$

*Note: If such a fourth-degree equation will not factor, we can use the quadratic formula to solve for x^2.

29–34. Perform the indicated operations.

29. $(3i + 7) + (6 - 4i)$
30. $(\pi + 2i) + (6 - \sqrt{2}i)$
31. $(6 - 4i) - \left(-\frac{1}{2} + \frac{i}{3}\right)$
32. $(0.05 - 0.19i) - (-0.3 - 0.02i)$
33. $(2 - 3i)(7 + 8i)$
34. $(6 - 5i)(6 + 5i)$

II. Concepts

35. For what values of c are the solutions to $x^2 + 4x + c = 0$ *real* numbers?

36. For what values of c are the solutions to $2x^2 - 5x + c = 0$ *complex*, nonreal numbers?

37. For what values of b are the solutions to $x^2 + bx + 10 = 0$ *real* numbers?

38. For what values of b are the solutions to $3x^2 - bx + 5 = 0$ *complex*, nonreal numbers?

39. For what values of a are the solutions to $ax^2 + 12x + 20 = 0$ *real* numbers?

40. For what values of a are the solutions to $ax^2 - 7x + 2 = 0$ *complex*, nonreal numbers?

41. The Babylonian problem asks, "Given n, find a number x such that the sum of x and its reciprocal is n." For what values of n does the Babylonian problem have a *real* number for its solution?

42. Write a note explaining the relationship between the solutions of an equation of the type $0 = ax^2 + bx + c$ and the graph of the corresponding equation $y = ax^2 + bx + c$.

43. Write a note explaining the relationship between the solutions of an equation of the type $0 = ax^4 + bx^2 + c$ and the graph of the corresponding equation $y = ax^4 + bx^2 + c$.

III. Applications

44. Suppose the height h (in feet) of a golf ball after t seconds is given by

$$h = 80t - 16t^2 + 20$$

(a) When will the golf ball be 100 feet high?
(b) When will the golf ball be 125 feet high?
(c) Explain the physical significance of your answers in parts (a) and (b).

45. The height h in feet of a stone you threw into the air is given after t seconds by

$$h = 50t - 6t^2$$

(a) When will the stone be 50 feet high?
(b) When will the stone be 150 feet high?

(c) Explain the physical significance of parts (a) and (b).

(d) Does the stone reach 100 feet?

46. In one theory of inbreeding (formulated over 20 years ago), we find a discussion of the model $8w^2 - 8w + 1 = 0$ in discussing the breeding of animals, where w represents a certain frequency of inbreeding stabilized from one generation to the next. Find the solution to this equation.

47. The equation $2Nw^2 - 2(N - 1)w - 1 = 0$ occurs in the study of population genetics, where N and w represent the population size and a certain frequency of mating, respectively. For what values of the constant N will this quadratic in w have *real* numbers for solutions?

48. The study of chromosome recombination leads to the equation $N = p(1 - p)$, where p represents the probability of combination and is under the constraint $0 \leq p \leq \frac{1}{2}$. For what values of N does this quadratic in the variable p have *real-valued* solutions?

49. In the year 1202 A.D., the Italian mathematician Fibonacci studied the population growth of rabbit colonies and concluded that if he started with one pair of rabbits, the total rabbit population each successive month grew according to the numbers

$$2, 3, 5, 8, 13, 21, 34, 55, 89, \ldots$$

(a) Can you determine a pattern in these Fibonacci numbers?

(b) Fibonacci also investigated successive *ratios* of adjacent month totals to see if he could determine a growth-rate factor. Thus he considered the sequence of numbers

$$\frac{3}{2}, \frac{5}{3}, \frac{8}{5}, \frac{13}{8}, \frac{21}{13}, \frac{34}{21}, \frac{55}{34}, \frac{89}{55}, \ldots$$

Use your calculator to find a decimal equivalent for each ratio. Do you notice anything? You may want to carry the procedure out for several more numbers.

(c) The ratios in part (b) lead to the equation $q = \frac{1}{q} + 1$, where q is the growth-rate factor for the rabbits. Solve the equation for q. Have you seen this number (the solution for the equation) before?

IV Extensions

50. What is the relationship between the roots of $ax^2 + bx + c = 0$ and $cx^2 + bx + a = 0$? Sketch examples of pairs of such equations in order to support your argument for the relationship between the roots.

51. Suppose two players play the following game concerning the equation $ax^2 + bx + c = 0$. Player A chooses a value for one of the coefficients a, b, or c. Player B chooses a value for one of the two remaining coefficients. Player A chooses a value for the third coefficient. Player A wins if both roots of the equation are real, while player B wins if the roots are complex. Is there a winning strategy for either player?

52. In a quadratic equation with *integral* coefficients, the value of the discriminant is D. For how many values of D from 78 to 83 (inclusive) will the roots of this equation be real, irrational, and unequal? Don't be too hasty.

6.5 GRAPHING PARABOLAS: TOWARD A QUICKER APPROACH

The following question was first introduced in Chapter 1.

EXAMPLE 1

Of all rectangles that have a perimeter of 18 inches, which has the largest area?

Solution

Recall that in Chapter 1, Example 1, we began the discussion of this problem with a table and then constructed a graph.

Using a Table Here is a table of values having integer dimensions for the length, width, and corresponding area of such rectangles.

Length in inches	*Width in inches*	*Perimeter in inches*	*Area in square inches*
1	8	18	8
2	7	18	14
3	6	18	18
4	5	18	20
5	4	18	20
6	3	18	18
7	2	18	14
8	1	18	8

Table 6.12

Here are several examples of rectangles with a perimeter of 18 inches. Let P = Perimeter and A = Area.

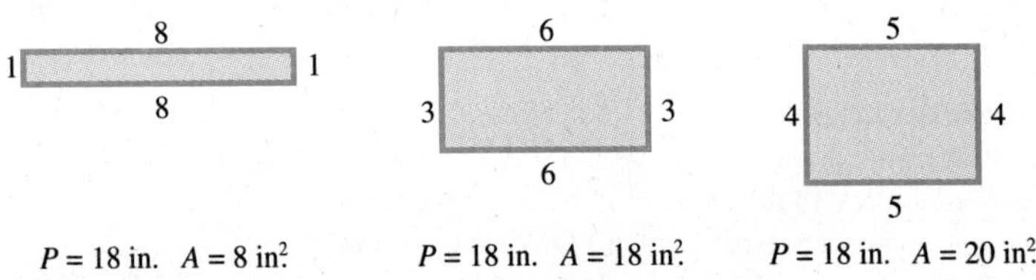

Figure 6.34

Using a Graph Below is the graph of the values in the table.

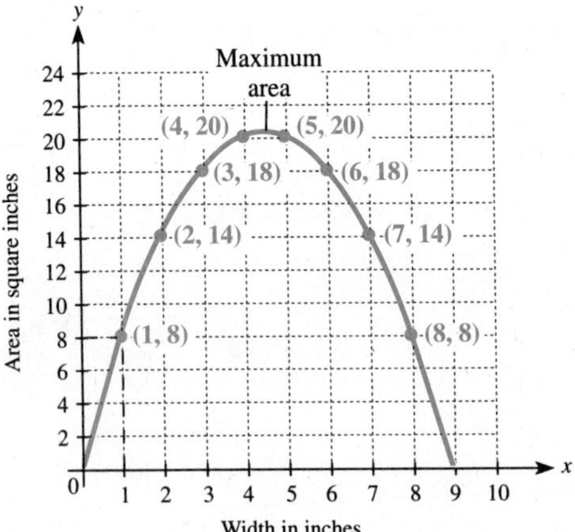

Figure 6.35

Using an Equation When is the area the greatest? What is the value of A at the highest point on the graph? The high point occurs somewhere between $w = 4$ and $w = 5$. Let us find an algebraic model for this problem.

$$A = lw \quad \text{and} \quad P = l + w + l + w = 2(l + w)$$

Figure 6.36

Since perimeter is 18 in this case, we substitute 18 in the perimeter formula.

$$18 = 2(l + w), \text{ so } 9 = l + w$$

Thus

$l = 9 - w$ **We get l in terms of w, then substitute in the area formula.**

$A = lw = (9 - w)w = 9w - w^2$ **Substituting $(9 - w)$ for l.**

Thus, the question is: For what value of w does A reach the high point on the graph?

We will investigate a method for finding the high (or low) point of a parabola. This point is called the **vertex** of the parabola. Its y-coordinate represents the maximum (or minimum) value that a quantity can attain.

Using Symmetry in Parabolas

A graphing calculator is very useful in finding the solutions to quadratic equations. It is also useful to be able to quickly sketch graphs of simpler quadratics. The sketching of these graphs will help us explore some important features of quadratic equations such as symmetry and the vertex.

EXAMPLE 2

Graph the following equation.

$$y = 3x^2$$

Solution

We form a table, then plot some points to estimate the graph.

x	$y = 3x^2$
0	0
1	3
−1	3
2	12
−2	12
0.78	1.83
−0.78	1.83

Table 6.13

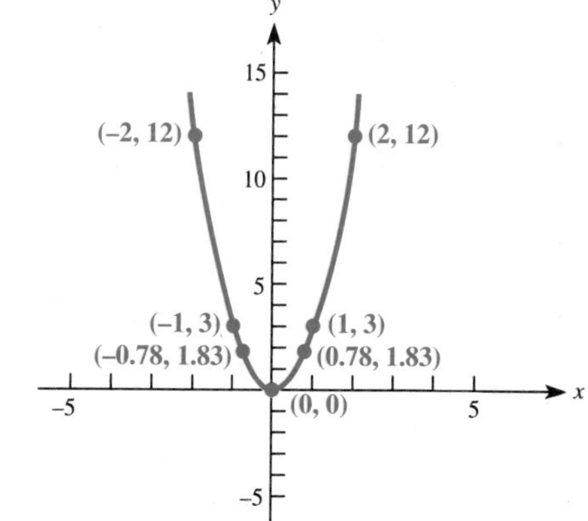

Figure 6.37

The minimum value (low point) of this graph occurs when $x = 0$. The coordinates of the low point are (0, 0). This graph is an example of a **parabola**. The low point is called the **vertex** of the parabola. We say that this graph is **symmetric about the *y*-axis,** because points on the graph with the same y-values, like (1, 3) and $(-1, 3)$, are mirror images of each other when the graph is folded on the y-axis. We will call the points (1, 3) and $(-1, 3)$ **symmetric points.** The x-coordinate of the vertex, in this case $x = 0$, is *exactly halfway* between the x-coordinates of pairs of symmetric points. For example, $x = 0$ is halfway between $x = 1$ and $x = -1$ in the graph. The y-axis (the line $x = 0$) is called the **axis of symmetry** for this parabola. In the parabolas we will study, the *vertical line through the vertex of the parabola is the axis of symmetry* for the parabola. ■

EXAMPLE 3

Graph the following equation.

$$y = 3x^2 - 12$$

Solution

We form a table of values of pairs (x, y). Then we plot the points corresponding to the pairs.

x	$y = 3x^2 - 12$
0	-12
1	-9
-1	-9
2	0
-2	0
3	15
-3	15

Table 6.14

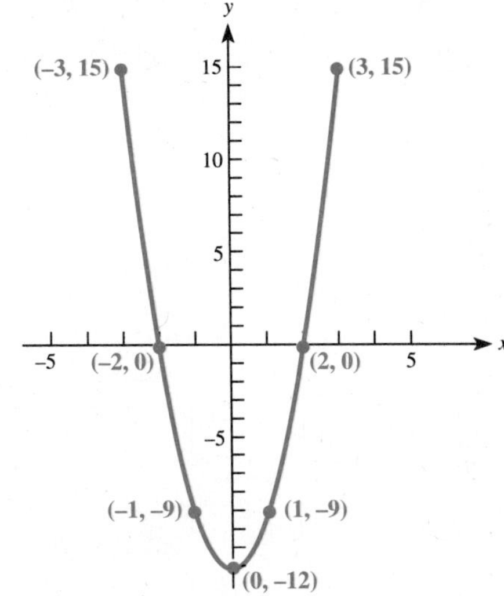

Figure 6.38

This graph is the graph in Example 2, moved 12 units down. (Notice the -12 in this equation.) Once again, the vertex occurs when $x = 0$. The y-axis is again the axis of symmetry, and it occurs midway between symmetric points, such as $(-1, -9)$ and $(1, -9)$. This parabola crosses the x-axis at two points $(-2, 0)$ and $(2, 0)$. The **x-axis intercepts** often aid in graphing a parabola, because they are symmetric points. The x-coordinates of the x-axis intercepts are the solutions to the quadratic equation $0 = 3x^2 - 12$. (Remember, $y = 0$ when the parabola crosses the x-axis.) ■

EXAMPLE 4

Sketch the graph of $y = 4x - x^2$.

Solution

Find the coordinates of several points. Plot the points, and sketch the graph. Notice that this parabola has a *maximum* value, a high point. Its vertex occurs at the point (2, 4). The axis of symmetry is the line $x = 2$.

x	y
−1	−5
0	0
1	3
2	4
3	3
4	0

Table 6.15

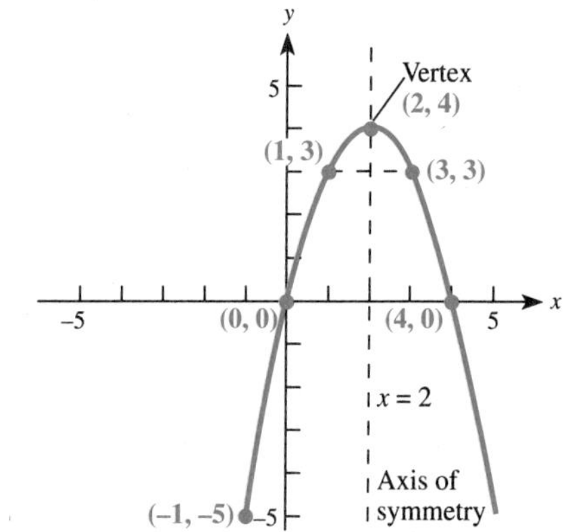

Figure 6.39

EXAMPLE 5

Sketch the graphs of $y = x^2 - 5x + 6$ and $y = x^2 - 5x$ on the same set of axes.

Solution

x	$y = x^2 - 5x$	$y = x^2 - 5x + 6$
0	0	6
1	−4	2
−1	6	12
2	−6	0
3	−6	0
4	−4	2
5	0	6

Table 6.16

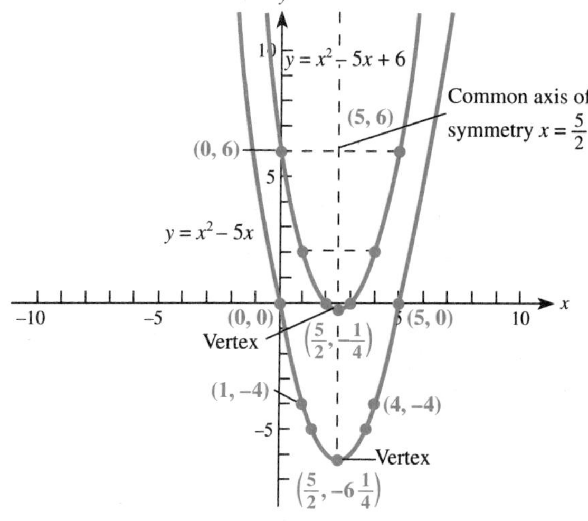

Figure 6.40

The vertex for the graph of $y = x^2 - 5x$ occurs when x is halfway between $x = 0$ and $x = 5$, that is at $x = \frac{5}{2}$. Similarly, the vertex of $y = x^2 - 5x + 6$ occurs halfway between $x = 2$ and $x = 3$, that is, when $x = \frac{3+2}{2} = \frac{5}{2}$. Indeed, the vertices of *both* parabolas in the graph occur when $x = \frac{5}{2}$! In fact, the only difference between the two graphs is that the graph of $y = x^2 - 5x + 6$ is 6 units above the graph of $y = x^2 - 5x$. The graphs share a common axis of symmetry, the line $x = \frac{5}{2}$. The vertex of $y = x^2 - 5x$ occurs at $x = \frac{5}{2}$, while $y = \left(\frac{5}{2}\right)^2 - 5\left(\frac{5}{2}\right) = -6\frac{1}{4}$. The vertex of $y = x^2 - 5x + 6$ occurs at the point $\left(-\frac{5}{2}, -\frac{1}{4}\right)$, 6 units higher. ■

The Vertex of a Parabola

We would like to find the vertex of a parabola in an *easy way*. The parabolas we have been investigating are graphs of equations of the form

$$y = ax^2 + bx + c,\ a \neq 0$$

However, Example 5 suggests that the graph of $y = ax^2 + bx + c$ will be the same as the graph of $y = ax^2 + bx$ "moved" c units vertically (up or down, depending on whether c is positive or negative). So, the x-coordinate of the vertex (high or low point) of $y = ax^2 + bx + c$ will be the same as the x-coordinate of the vertex (high or low point) of $y = ax^2 + bx$.

To find the vertex of $y = ax^2 + bx$, we can find two symmetric points and average them, namely, the two places where $y = ax^2 + bx$ crosses the x-axis, the x-axis intercepts. The y-coordinate is 0 when a graph crosses the x-axis. Therefore, to find the x-intercepts for $y = ax^2 + bx$, we solve the quadratic equation

$$0 = ax^2 + bx$$
$$0 = x(ax + b)$$

Therefore, either $x = 0$ or $ax + b = 0$. So $x = 0$ or $x = -\dfrac{b}{a}$. These x-axis intercepts will yield symmetric points on the graph, so the vertex will occur halfway between the intercepts at

$$x = \frac{0 + \left(\dfrac{-b}{a}\right)}{2} \qquad \text{Average the } x\text{-intercepts.}$$

$$x = \frac{-\dfrac{b}{a}}{2} = -\frac{b}{2a} \qquad \text{The vertex occurs at } x = -\frac{b}{2a}.$$

Here are possible graphs for $y = ax^2 + bx = x(ax + b)$ and for $y = ax^2 + bx + c$.

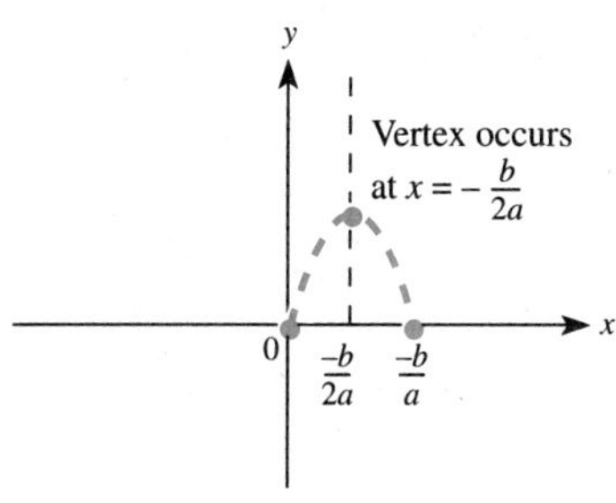

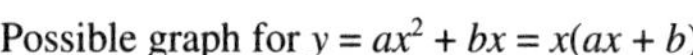

Possible graph for $y = ax^2 + bx = x(ax + b)$

Figure 6.41

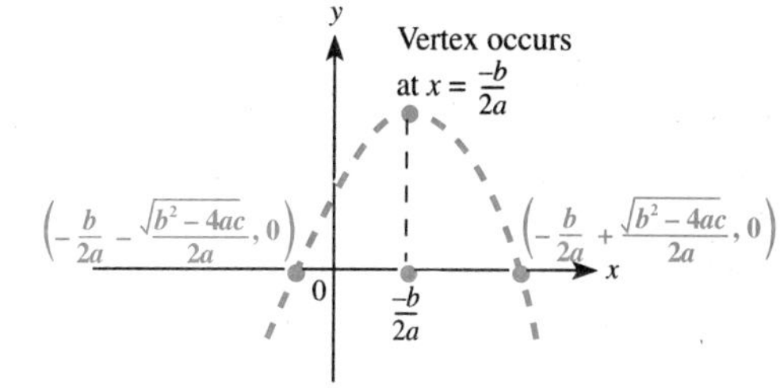

Vertex occurs at $x = \frac{-b}{2a}$

$\left(-\frac{b}{2a} - \frac{\sqrt{b^2 - 4ac}}{2a}, 0\right)$ $\left(-\frac{b}{2a} + \frac{\sqrt{b^2 - 4ac}}{2a}, 0\right)$

Possible graph for $y = ax^2 + bx + c$

Figure 6.42

In general, recall that the roots of $0 = ax^2 + bx + c$ are

$$x = \frac{-b + \sqrt{b^2 - 4ac}}{2a} \quad \text{and} \quad x = \frac{-b - \sqrt{b^2 - 4ac}}{2a}$$

Here we summarize our discussion of graphing quadratic equations in two variables.

Graphs of Quadratic Equations

The graph of an equation of the form

$$y = ax^2 + bx + c, a \neq 0$$

is a parabola. The *vertex* occurs when $x = -\frac{b}{2a}$. The *axis of symmetry* is the vertical line through the vertex, $x = -\frac{b}{2a}$. When $a > 0$, the graph has a *minimum*, a low point (*bends up*). When $a < 0$ the graph has a *maximum*, a high point (*bends down*).

EXAMPLE 6

Let us return to Example 1. Recall that we were trying to find the dimensions of the rectangle with perimeter 18 that had the maximum area. The algebraic model for the problem is $A = 9w - w^2$, where w is the width of the rectangle.

Solution

The graph of the corresponding parabola is given below. The axis of symmetry occurs at the midpoint of the two w-axis intercepts, at $w = 0$ and at $w = 9$.

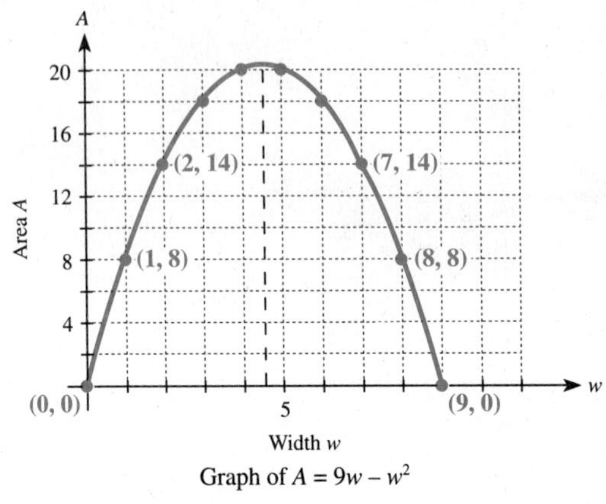

Graph of $A = 9w - w^2$

Figure 6.43

When $w = 9, A = 0$, since length $l = 0$ in this case. Recall that $l = 9 - w$.

Thus, the maximum area occurs at $w = \frac{9+0}{2} = 4.5$. If $w = 4.5$, then $l = 9 - 4.5 = 4.5$. Since $w = 4.5 = l$, the rectangle with the maximum area turns out to be the square with perimeter 18. The maximum area is $A = 4.5^2 = 20.25$ in^2. ■

EXAMPLE 7

(a) Find the vertex of the parabola $y = -3x^2 - 6x + 8$ and sketch the graph of the parabola.

(b) Determine the coordinates of the x-axis intercepts.

Solutions

(a) The following scheme will now help us get a quick sketch of any parabola:

1. Find the coordinates of the vertex and plot it.

2. Find and plot a pair of convenient symmetric pairs.

3. Find and plot a second pair of symmetric points as a check for the graph.

1. For this parabola $a = -3$ and $b = -6$. Thus the vertex occurs when

$$x = \frac{-b}{2a} = \frac{-(-6)}{2(-3)} = \frac{6}{-6} = -1.$$

To find the y-coordinate of the vertex, we substitute $x = -1$ into the equation:

$$y = -3(-1)^2 - 6(-1) + 8 = -3 + 6 + 8 = 11$$

Therefore the vertex of this parabola occurs at the point $(-1, 11)$.

2. To find a pair of symmetric points on the graph, we can move x the same distance left and right of our vertex point and substitute these new values into our equation. Let's move 1 unit left and right of value $x = -1$, to $x = -2$ and $x = 0$.

If $x = 0$, then $y = -3(0)^2 - 6(0) + 8 = 8$.
If $x = -2$, then $y = -3(-2)^2 - 6(-2) + 8 = 8$.

Symmetric points have the same y-value.

3. To determine a second pair of symmetric points for the graph as a check, we move a bit *farther* away from the vertex, say, 3 units either way, to $x = -4$ and $x = 2$.

$$\left.\begin{array}{l}\text{If } x = 2, \text{ then } y = -3(2)^2 - 6(2) + 8 = -16. \\ \text{If } x = -4, \text{ then } y = -16.\end{array}\right\} \quad \textbf{Same } y\textbf{-coordinates.}$$

These five points, the vertex and two pairs of symmetric points, will enable us to sketch the parabola.

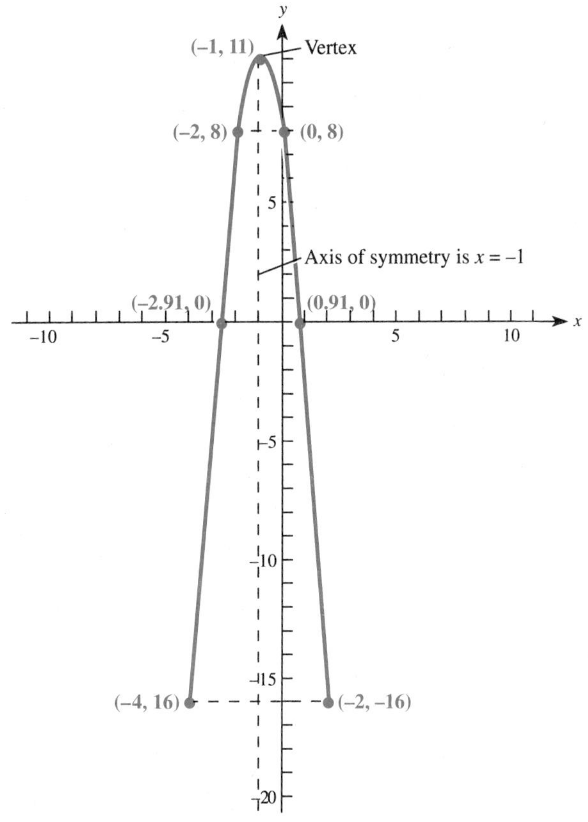

Figure 6.44

Notice that the parabola has a high point (since $a < 0$). This parabola crosses the x-axis in two places.

(b) To find the x-axis intercepts, we set $y = 0$ and solve

$$0 = -3x^2 - 6x + 8 \qquad a = -3, b = -6, c = 8$$

$$x = \frac{-(-6) \pm \sqrt{(-6)^2 - 4(-3)(8)}}{2(-3)} \qquad \textbf{The quadratic formula.}$$

$$x = \frac{6 \pm \sqrt{132}}{-6}$$

$$x \approx -2.91 \quad \text{and} \quad x \approx 0.91$$

The x-axis intercepts occur at $(0.91, 0)$ and $(-2.91, 0)$. Answer.

Here we summarize our algebraic method for sketching the graph of a parabola in a three-step process.

Strategy for Sketching a Parabola

To sketch the graph of $y = ax^2 + bx + c$,

1. Find and plot the vertex, which occurs when $x = -\frac{b}{2a}$; the y-coordinate can be found by substitution.
2. Find and plot a pair of convenient symmetric points.
3. Find and plot another "check" pair of symmetric points.

EXAMPLE 8 Sketch the graph of the parabola

$$y = x^2 + x + 2$$

Solution

1. The vertex occurs when $x = -\frac{b}{2a} = -\frac{1}{2}$.

   If $x = -\frac{1}{2}$, then $y = \left(-\frac{1}{2}\right)^2 - \frac{1}{2} + 2 = \frac{7}{4}$. The vertex is $\left(-\frac{1}{2}, \frac{7}{4}\right)$.

2. We find a pair of symmetric points by moving $\frac{1}{2}$ unit each way from the vertex at $x = -\frac{1}{2}$ (because it will give us easy numbers to work with).

   If $x = 0$, then $y = (0)^2 + 0 + 2 = 2$. $\quad 0 = -\frac{1}{2} + \frac{1}{2}$

   If $x = -1$, then $y = 2$ also. $\quad -1 = -\frac{1}{2} - \frac{1}{2}$

3. For checkpoints, suppose we move $\frac{5}{2}$ units (convenient).

   If $x = -\frac{1}{2} + \frac{5}{2} = 2$, then $y = (2)^2 + 2 + 2 = 8$.

   Therefore,

   if $x = -\frac{1}{2} - \frac{5}{2} = -3$, then $y = 8$ also.

This parabola has a minimum value at the vertex. Notice that $a = 1 > 0$. Also note that the graph *does not* cross the x-axis, that is, there are no x-axis intercepts for this parabola. The parabola does cross the y-axis. The y-axis intercept occurs at (0, 2).

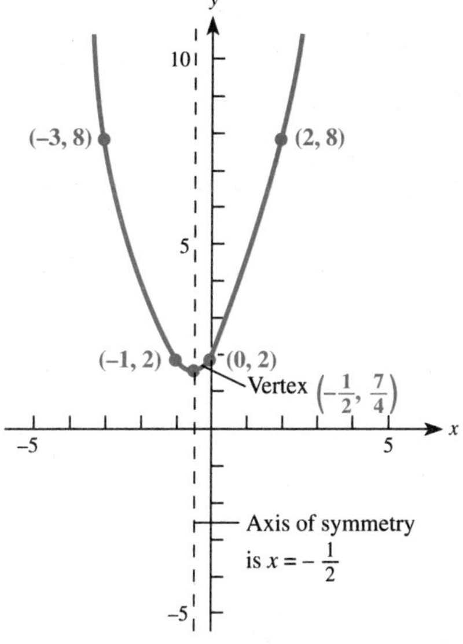

Figure 6.45

EXAMPLE 9

The graph below describes the flight of a missile fired from its silo 200 feet below the earth's surface. The graph indicates height in feet after t seconds.

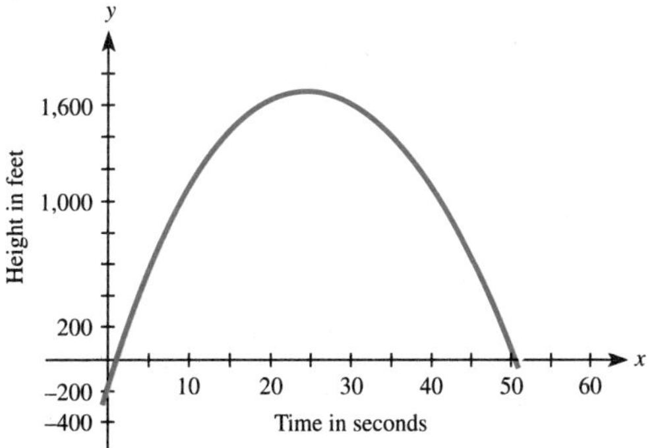

Figure 6.46

(a) Use the graph to estimate

1. the vertex,

2. the x-axis intercepts, and

3. the y-axis intercepts.

(b) What is the physical significance of each of the quantities in part (a)?

Solutions

(a) 1. The vertex (high point) appears to have coordinates at about $x = 27$, $y = 1{,}600$.

2. The x-axis intercepts are at about (2, 0) and (52, 0).

3. The y-axis intercept is $(0, -200)$.

(b) 1. The vertex (27, 1,600) indicates the highest point attained by the rocket. After 27 seconds, the rocket is 1,600 feet high.

2. The x-intercepts occur at ground level when the height y is 0. The intercept (2, 0) means it takes 2 seconds for the rocket to come out of the silo. The intercept (52, 0) indicates that the rocket returns to the ground after 52 seconds.

3. The y-intercept $(0, -200)$ occurs when time $x = 0$. It means that before the rocket is fired, it is 200 feet below the ground. ■

EXAMPLE 10

Recall our problem about the golf ball in Section 6.1. The height h above the ground of a golf ball depends on the time t it has been in flight. A shot hit from an elevated tee has height approximately given by

$$h = 80t - 16t^2 + 20$$

where h is in feet and t is in seconds.

(a) Graph the relationship between h and t, and find the maximum height of the golf ball.

(b) How long will the ball be in flight?

Solutions

t	h
0	20
1	84
2	116
2.6	119.84
3	116
5	20
6	−76

Table 6.17

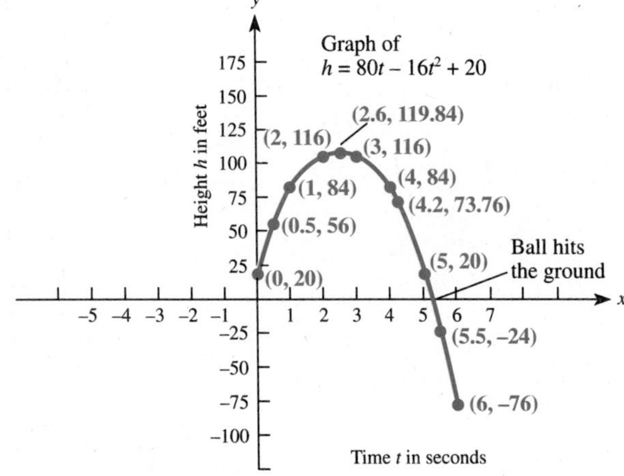

Figure 6.47

The graph suggests that the maximum height of the golf ball occurs between 2 and 3 seconds.

(a) We use our three-step process.

1. The vertex occurs at

$$t = \frac{-80}{2(-16)} = \frac{5}{2} \qquad a = -16, b = 80$$

When $t = \frac{5}{2}$,

$$h = 80\left(\frac{5}{2}\right) - 16\left(\frac{5}{2}\right)^2 + 20 = 120$$

The vertex is the point $\left(\frac{5}{2}, 120\right)$.

2. Here is one pair of symmetric points:

If $t = 2$, then $h = 116$. **Substitute $t = 2$ in the equation.**

If $t = 3$, then $h = 116$.

3. A check pair of symmetric points:

When $t = 0$ and $t = 5$, $h = 20$. **Substitute in the equation.**

(b) To find out when the ball will land, we set height $h = 0$ and solve:

$$0 = 80t - 16t^2 + 20$$

$$t \approx 5.24 \quad \text{or} \quad t \approx -0.24 \text{ seconds} \qquad \textbf{Use the quadratic formula.}$$

The t-intercepts of the graph are then $(5.24, 0)$ and $(-0.24, 0)$.

The vertex indicates the maximum height attained by the ball, 120 feet after $\frac{5}{2}$ seconds. The t-axis intercept $(5.24, 0)$ shows that the ball will land after 5.24 seconds. The h-intercept $(0, 20)$ means that the ball is on an elevated tee 20 feet above the fairway before the golfer hits the shot. We do not sketch the graph for negative height, since it makes no sense for this application. ■

NOTE: It is very useful to be able to quickly sketch a parabola using the vertex and a pair of symmetric points. The concepts of symmetry and transformations will help us study functions in a more general way in Chapter 8. ■

EXAMPLE 11

Suppose the manufacturing cost C in dollars of making x backpacks a day is given by

$$C = x^2 - 12x + 50$$

(a) Graph this cost function.

(b) What is the minimum cost, and how many backpacks are produced at this cost?

(c) Does it cost more to make 4 backpacks or 10 backpacks?

(d) How many packs can be made for $40?

Solutions

(a) The graph will be a parabola that bends up, since $a = 1 > 0$. We perform our three-step process.

1. The vertex occurs at

$$x = \frac{-(-12)}{2} = 6 \qquad a = 1, b = -12$$

Substituting into the equation, we find that when $x = 6$, $C = \$14$. The vertex is (6, 14).

2. We have one pair of symmetric points:

If $x = 3$, then $C = \$23$. **Move 3 units each side of the vertex and substitute.**
If $x = 9$, then $C = \$23$.

3. We have a check pair of symmetric points:

If $x = 0$, then $C = \$50$. **Move 6 units each side of the vertex.**
If $x = 12$, then $C = \$50$.

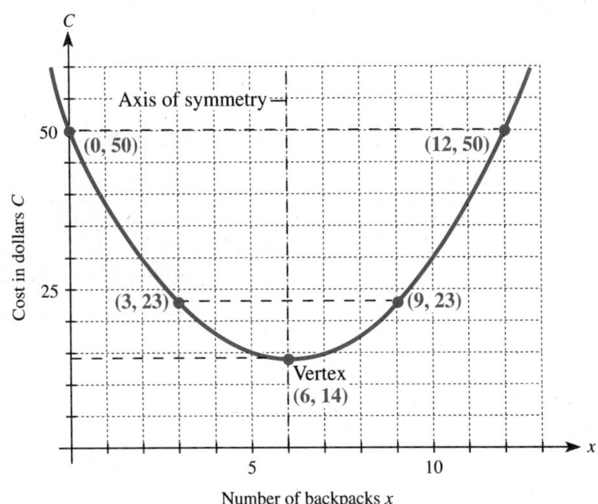

Figure 6.48

The graph indicates that the manufacturing cost can be minimized. If the manufacturer makes too few or too many packs a day, it will cost more, perhaps because of sales loss in the first case and excess labor and inventory costs in the latter.

(b) The minimum cost occurs at the vertex, because that is where the lowest-cost value occurs on the graph. Therefore, the minimum cost is \$14, and the number of packs made for \$14 is 6.

(c) From the graph, it appears to cost more for 10 packs than for 4 packs. Let us substitute in $x = 10$ and $x = 4$, just to make certain:

$$\text{If } x = 4, \text{ then } C = (4)^2 - (12 \cdot 4) + 50 = 16 - 48 + 50 = \$18.$$
$$\text{If } x = 10, \text{ then } C = (10)^2 - (12 \cdot 10) + 50$$
$$= 100 - 120 + 50 = \$30.$$

Therefore it does cost more for 10 packs.

(d) This question gives us a C value, \$40, and asks us to solve for x, the number of packs.

$$\text{If } C = 40, \text{ then } 40 = x^2 - 12x + 50.$$

We obtain a quadratic equation. Using the quadratic formula,

$$0 = x^2 - 12x + 10 \qquad a = 1, b = -12, c = 10$$

Therefore

$$x = \frac{-b \pm \sqrt{b^2 - 4ac}}{2a} = \frac{-(-12) \pm \sqrt{144 - 4(10)}}{2}$$

$$= \frac{12 \pm \sqrt{144 - 40}}{2} = \frac{12 \pm \sqrt{104}}{2}$$

Therefore

$$x \approx \frac{12 + 10}{2} \quad \text{or} \quad x \approx \frac{12 - 10}{2}$$

$$x \approx 11 \quad \text{or} \quad x \approx 1 \quad \textbf{Answer.}$$

Either 11 backpacks or 1 backpack can be made for \$40. *Note:* We can also solve part (d) by graphing. ■

EXAMPLE 12 An international rule for determining the number of board feet in a 16-foot log is given by

$$y = 0.22x^2 - 0.71x$$

where x is the diameter of the log in inches and y is the number of board feet (usable finished lumber).

(a) Graph the equation.

(b) How many board feet can be obtained from a 10-inch-diameter log?

(c) What is the minimum number of board feet, and for what diameter does it occur?

(d) What is the maximum number of board feet, and for what diameter does it occur?

Solutions

(a) and **(b)** **1.** The vertex occurs at $x = -\frac{b}{2a} = \frac{0.71}{0.44} = 1.61$ inches. If $x = 1.61$, $y = -0.57$ board feet. This point is on the graph but it means nothing for the lumber industry, since negative board feet cannot be made into lumber.

2. Two symmetric points, the x-axis intercepts, can be obtained by setting

$$0 = 0.22x^2 - 0.71x = x(0.22x - 0.71)$$

So either

$$x = 0 \quad \text{or} \quad 0.22x - 0.71 = 0$$

$$x = \frac{0.71}{0.22} \approx 3.23 \text{ inches}$$

3. We are really interested in only one side of the parabola, since negative diameters do not count either. If we try $x = 10$ for our checkpoint, then $y \approx 14.9$. This also answers part (b), as a 10-inch-diameter log will yield about 14.9 board feet.

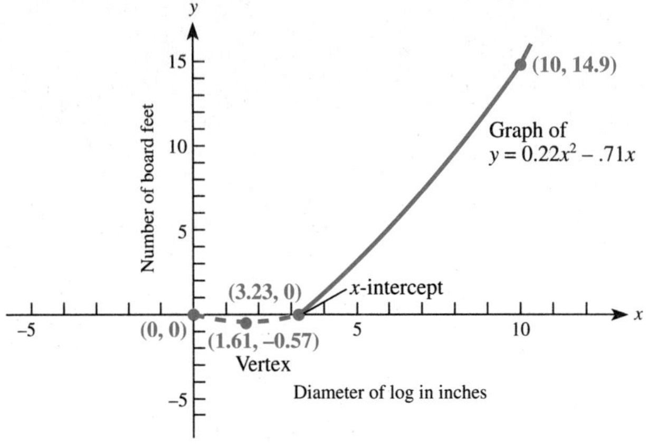

Figure 6.49

The graph is dashed for negative board feet, since it makes no sense for this application.

(c) Look at the graph. The minimum number of board feet is -0.57 on the graph, but in fact, for this application, negative board feet are meaningless. So we get absolutely no board feet out of any log with a diameter between zero inches (no tree) and 3.23 inches (a darn skinny tree).

(d) There is no maximum value, because according to our graph, the larger the diameter of the tree, the more board feet of lumber we obtain (surely this is sensible). The bigger they are, the harder they fall! ■

EXAMPLE 13

Graph $Y1 = 2x^2$ and $Y2 = -2x^2$ on the same set of axes. What effect did multiplying the equation $y = 2x^2$ by -1 have on its graph?

Solution

The graphs are given in Figure 6.50 on the following page. Multiplying by -1 reflects the graph through the x-axis. ■

EXAMPLE 14

Give three examples of equations of parabolas that have $x = 0$ and $x = 4$ for x-intercepts. Sketch these three graphs.

Solution

Since $x = 0$ and $x = 4$ are the intercepts (roots), both x and $(x - 4)$ will be factors of each of the parabola's equations. Thus $y = x(x - 4) = x^2 - 4x$ is one possibility (see Figure 6.51). Other possibilities can be created by moving the vertex of the parabola while

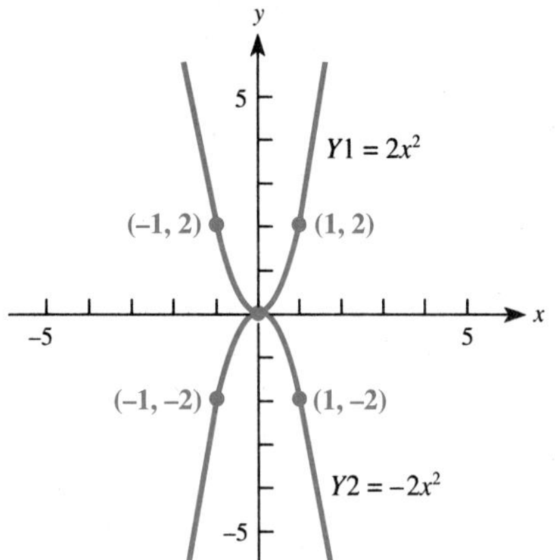

Figure 6.50

keeping the intercepts fixed. Example 13 provides a hint for us. We could reflect the graph through the x-axis by multiplying the equation by -1: $y = -1x(x - 4) = -x^2 + 4x$ will be another possibility. Or, we could "double" the vertex by multiplying by 2: $y = 2x(x - 4) = 2x^2 - 8x$ will also work. (Why doesn't doubling move the zeros—x-intercepts—of the graph, too?)

These two graphs, $Y1 = -x^2 + 4x$ and $Y2 = 2x^2 - 8x$, are presented in Figure 6.52.

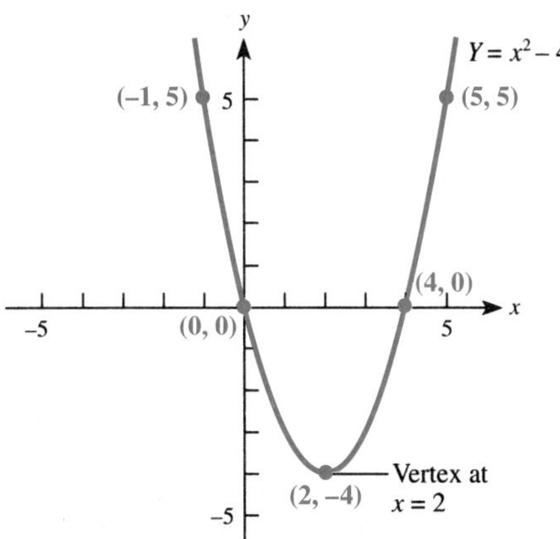

Figure 6.51

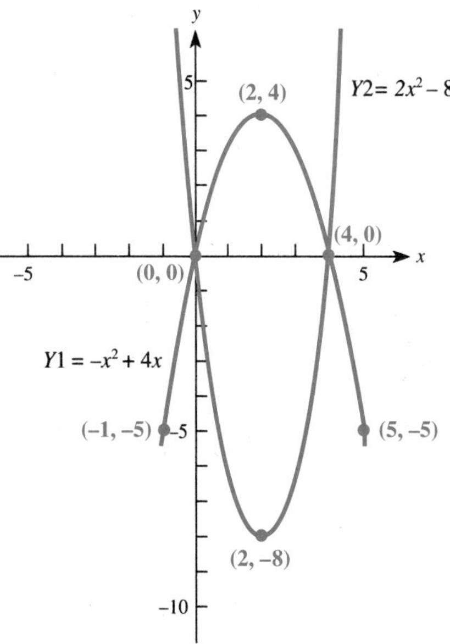

Figure 6.52

6.5 EXERCISES

I. Procedures

1–10. Find the coordinates of the vertex of each of the given parabolas.

1. $y = 3x^2$

2. $y = -5x^2$

3. $y = 6x^2 - 5$

4. $y = -8x^2 + 6$

5. $y = 6x^2 - 5x + 8$

6. $y = 15 - 3x^2 + 7x$

7. $y = \frac{1}{2} + \frac{3}{4}x - \frac{x^2}{2}$

8. $y = 0.07x^2 - 0.03x + 18$

9. $y = 0.85x - 1.23x^2$

10. $y = \frac{1}{3} - \frac{1}{9}x + \frac{5}{12}x^2$

11–30. Sketch the graphs of these equations in the plane. (Recall our three-step method for sketching a parabola.) In each case, find the vertex, the y-axis intercept, and the x-axis intercepts, if they exist. (Check your sketches on a graphing calculator.)

11. $y = x^2 - 9$

12. $y = 5x - 2x^2$

13. $y = x^2 + 6x + 5$

14. $y = x^2 - 10x + 25$

15. $y = -x^2 + 5x - 8$

16. $y = 10 - 7x + x^2$

17. $y = 5x + 6$

18. $y = x^2 - 6x$

19. $y = 0.8x^2 - 0.16x + 4$

20. $y + 8 = 3x^2 + 10x$

21. $3y + 6x = 6$

22. $3y + 6x^2 = 6$

23. $y = 0.85x - 0.25x^2 + 1$

24. $y = 25 - x^2$

25. $y = x^2 + x + 1$

26. $y = 9x^2 - 9x$

27. $y = x^2 + 5x + 10$

28. $2y = 4x + 8$

29. $2y = 4x^2 + 8x + 4$

30. $y = 29$

II. Concepts

31. Graph each of these parabolas on the same set of axes.

(a) $y = x^2$ **(b)** $y = 2x^2$

(c) $y = 4x^2$ **(d)** $y = \frac{1}{2}x^2$

(e) $y = \frac{1}{4}x^2$ **(f)** $y = -x^2$

(g) $y = -2x^2$ **(h)** $y = -4x^2$

(i) $y = -\frac{1}{2}x^2$ **(j)** $y = -\frac{1}{4}x^2$

What effect does the value of a have on the graph of the equation $y = ax^2$?

32. On the same set of axes, graph the following equations.

(a) $y = x^2 - 1$ **(b)** $y = x^2 + 1$

(c) $y = x^2 + 4$ **(d)** $y = 2x^2 + 10$

(e) $y = -x^2 - 1$ **(f)** $y = -x^2 + 1$

(g) $y = -x^2 + 4$ **(h)** $y = -2x^2 + 10$

What effect does c have on the graph of the equation $y = ax^2 + c$?

33. On the same set of axes, graph the following equations.

(a) $y = x^2 + x$ **(b)** $y = x^2 - x$

(c) $y = x^2 + 4x$ **(d)** $y = x^2 + 10x$

(e) $y = -x^2 + x$ **(f)** $y = -x^2 - x$

(g) $y = -x^2 + 4x$ **(h)** $y = -x^2 + 10x$

What effect does b have on the graph of the equation $y = ax^2 + bx$?

34. The graph below describes the flight of a missile fired vertically from the side of a mountain. The graph indicates height in feet after t seconds.

(a) Use the graph to estimate each of the following:

1. The vertex.

2. The x-axis intercepts.

3. The y-axis intercepts.

(b) What is the physical significance of each of the quantities in part (a)? (See Example 9.)

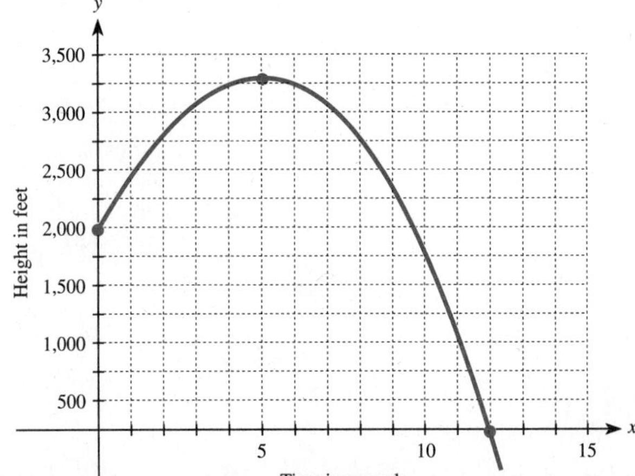

35. Write a description explaining how to find the maximum or minimum value of the graph of the form $y = ax^2 + bx + c$. Explain how you know that such a graph will, indeed, have a maximum or a minimum. How will your response change if you are using a graphing calculator?

36. Write a description of a quick way, using symmetry, to sketch a graph of the form $y = ax^2 + bx + c$.

37. **(a)** Give three examples of the equation of a parabola whose vertex is (0, 0).

(b) Sketch your three parabolas on the same set of axes.

38. **(a)** Give three examples of the equation of a parabola whose vertex is $\left(-\frac{1}{2}, 0\right)$.

(b) Sketch your three parabolas on the same set of axes.

39. **(a)** Give three examples of the equation of a parabola whose vertex is (2, 3).

(b) Sketch your three parabolas on the same set of axes.

40. **(a)** Give three examples of the equations of parabolas that have $(-2, 0)$ and $(6, 0)$ as x-axis intercepts.

(b) Sketch the three parabolas on the same set of axes.

41. **(a)** Give three examples of the equations of parabolas that have (0, 5) as the y-axis intercept.

(b) Sketch the three parabolas on the same set of axes.

42. **(a)** Give three examples of the equations of parabolas that have $(0, -10)$ as the y-axis intercept.

(b) Sketch the three parabolas on the same set of axes.

43–46. Use a graphing calculator to explore these problems.

43. For what values of C will $y = x^2 + 2x + C$ cross the x-axis?

44. For what values of C will $y = 2x^2 - 3x + C$ cross the x-axis?

45. For what values of a will $y = ax^2 - 3x - 5$ cross the x-axis?

46. For what values of a will $y = ax^2 - 2x + 3$ cross the x-axis?

III. Applications

47–58. Solve the problems.

47. In Example 12, we looked at a formula for determining the number of board feet y in a 16-foot log of diameter x (in inches):

$$y = 0.22x^2 - 0.71x$$

(a) In 1770, a European chestnut tree was reported to have a girth (circumference) of 167 feet. The tree is known as the "Tree of the 100 Horses." How many board feet would be in a 16-foot log from this tree? (Remember to calculate the diameter.)

(b) The world's tallest tree is the Howard Libby redwood tree in Redwood Creek Grove of Humboldt County, California. The Libby Tree is 366.2 feet tall and has a girth of 44 feet at the base. About how many board feet are in this tree? *Hint:* If you assume the tree tapers down to essentially zero diameter at the top, what would be a good estimate for the *average* girth of the Libby Tree?

48. Every week, a trash collecting company obtains revenue for each truck loaded with trash that it carts to the dump. The revenue R in dollars for x trucks is given by

$$R = x^2 + 37x - 120$$

(a) Sketch the graph of the revenue in terms of the number of trucks.

(b) If the company has ten trucks on the road, will it make money or lose money? How much does it make or lose? What if the company has two trucks on the road? Locate each of these points on the graph in part (a).

(c) How many trucks must the company have on the road so that it doesn't lose any money? Locate this point on the graph in part (a).

(d) How much does the company lose per week if its workers are on strike and put no trucks on the road? Locate this point on the graph in part (a).

49. C3PO has installed an ejection seat in his land rover for quick escapes. R2D2 is sky high after accidentally bumping the ejection switch. The seat throws R2 upward as described by

$$y = 4 + 100x - 16x^2$$

where x is the number of seconds after the ejection button is hit and y is the number of feet up after x seconds.

(a) Graph the relationship.

(b) How high will R2D2 fly?

(c) How long will it take R2D2 to land?

(d) How high was R2D2 before the button was bumped? What does this mean?

50. A flare is fired vertically upward from ground level. Its height h in feet above the ground at a time t seconds is given by

$$h = 500t - 16t^2$$

(a) How high will the flare go, and when will it reach its maximum height?

(b) When will it hit the ground again?

51. One way to *estimate* the number of board feet of usable lumber in a 16-foot log is

$$B = 0.8\,(D - 1)^2 - \frac{D}{2}$$

where B is the number of board feet and D is the diameter of the log in inches.

(a) Graph the relationship between B and D.

(b) The equation is supposed to give us an *estimate*. How big does the diameter of the log have to be in order to obtain some usable lumber? What happens if we put $D = 0$ into the equation?

Sometimes when a formula provides an *estimate* it may estimate well for some values but not for others. Does our formula appear to estimate better for large diameters or for small diameters?

(c) Suppose a tree has a diameter of 6 feet. How many board feet of usable lumber could be obtained from a 16-foot log?

(d) If a 16-foot log yields 200 board feet, what was the diameter of the tree?

(e) Use this relationship to estimate the number of board feet in a 16-foot log from the "Tree of 100 Horses" (see Exercise 47[a]).

(f) Use this relationship to estimate the total board feet obtainable from the Libby Tree (see Exercise 47[b]).

(g) Compare your answers for Exercise 47(a) with part (e) and for Exercise 47(b) with part (f). Do these two methods of estimating board feet (Exercises 47 and 51) yield similar results? Can you give a reason why this may be so?

52. The quantity q of an agricultural product that will be demanded at a price p dollars is given by

$$q = 20 + 5p - p^2$$

(a) Sketch the relationship, with price as the horizontal axis.

(b) At what price will the most products be demanded, and how many will be demanded at that price?

(c) At what price will there no longer be a demand for the product?

(d) At what price will 30 products be demanded?

(e) What quantity will be demanded if the price is \$2.75?

53. A pole vaulter wishes to clear a height of 17.5 feet. Suppose she takes off with a height h determined by

$$h = 33t - 12t^2$$

where t is the number of seconds since takeoff. Suppose also that her maximum height is attained at the bar.

(a) Will she clear her desired height?

(b) How long is she airborne?

(c) Sketch the graph of the relationship between t and h.

54. A hang glider dives off a cliff and flies with a height given by

$$h = 100 + 10x - 3x^2$$

where x is the number of minutes the glider is airborne and h is in feet.

(a) What is the maximum height that the glider attains?

(b) How long is the glider airborne?

(c) What is the height of the cliff? When will the glider descend back to the level of the cliff?

55. An equation that relates the number of fingerlings y in a lake to the number of catchable trout g is given by

$$y = 0.5g + 0.008g^2$$

(a) Sketch the graph of this relationship.

(b) What part of this graph makes sense for this application?

(c) How many fingerlings are expected if there are 10,000 catchable trout in the lake?

(d) How many catchable trout would have to be planted (even mix of males and females) to ensure a yield of 2 million fingerlings?

56. A 10-meter champion diver leaves the platform, with height h determined by

$$h = 10 - 2t^2$$

where t is the elapsed time in seconds, and h is measured from the water level.

(a) Sketch the graph of this relationship.

(b) What is the maximum height attained, and when will that height be attained?

(c) How long will it be before the diver hits the water?

(d) How long will it take for the diver to *fall* to 5 meters above the water? How about 3 meters?

57. A farmer has 2,400 meters of fence. The farmer wishes to enclose a rectangular plot with a fence so that the largest possible area is enclosed.

(a) Write an equation for the fenced area A in terms of the length l of the rectangular plot.

(b) Graph your relationship between A and l.

(c) What are the dimensions of the rectangle that give the farmer the most area?

58. Suppose the farmer (see Exercise 57) wishes to use the 2,400 meters of fence to enclose a rectangular plot on three sides, using a creek as the boundary on the fourth side, as in the figure below.

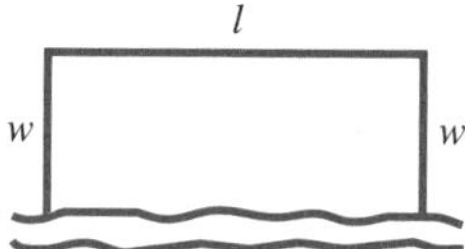

(a)–(c) Answer the same questions as for Exercise 57.

(d) In which case, Exercise 57 or Exercise 58, was the farmer able to enclose more land?

IV. Extensions

59. Using the information you have obtained from Exercises 31–33, first estimate the position of the graphs of each of the equations below. Then graph on a graphing calculator.

(a) $y = x^2 + 3x + 2$ **(b)** $y = 2x^2 + 4x + 1$

(c) $y = 10x^2$ **(d)** $y = x^2 - 10$

(e) $y = \dfrac{-x^2}{10}$ **(f)** $y = 6x - 3x^2$

(g) $y = x^2 + 3x + 3$ **(h)** $y = -x^2 + 5x - 6$

60–62. Use a graphing calculator to explore these problems:

60. For what values of b will $y = x^2 + bx + 3$ cross the x-axis?

61. For what values of b will $y = x^2 + bx - 10$ cross the x-axis?

62. For what values of c will $y = x^2 + cx + c$ cross the x-axis?

63. Here is an alternate method of showing that the vertex of the parabola $y = ax^2 + bx + c$ occurs at $x = -\frac{b}{2a}$. If the parabola crosses the x-axis twice, then there are two solutions to $0 = ax^2 + bx + c$ (because $y = 0$ when the parabola meets the x-axis).

(a) What are the general solutions to $0 = ax^2 + bx + c$? (Think back to Section 6.3.)

(b) These two solutions are *symmetric* points on the graph of $y = ax^2 + bx + c$. The vertex should occur halfway between them. What happens if you average your solutions in part (a)?

64. In this section, we have considered graphs of the form

$$y = ax^2 + bx + c$$

What happens if we *switch x and y*? That is, what do graphs of $x = ay^2 + by + c$ look like? Plot some points for each of the following equations, and sketch the graphs.

(a) $x = y^2$ **(b)** $x = y^2 - 4$

(c) $x = 9 - y^2$ **(d)** $x = 10 - 7y + y^2$

(e) $x = 9y^2 - 9y$ **(f)** $x = y^2 + y + 1$

65. After you have done Exercise 64, answer these questions about horizontal parabolas.

(a) Where does the vertex of a parabola of the form $x = ay^2 + by + c$ occur?

(b) How do you find the places where the graph of $x = ay^2 + by + c$ crosses the y-axis, if it crosses the y-axis?

(c) How can you tell whether the graph of $x = ay^2 + by + c$ opens to the right or to the left?

66. In Chapter 4, when we studied the equations of lines, we found that it was possible to recover the equation of a straight line if we knew *two* points on the line (see the point-slope form).

(a) Sketch the parabolas $y = x^2 - 7x + 12$, $y = -x^2 + 7x - 12$, and $y = 3x^2 - 21x + 36$ on the same set of axes. Notice that all three of these parabolas pass through the points $(4, 0)$ and $(3, 0)$. Thus two points are not enough to determine one unique parabola.

(b) Suppose you are told $(0, 3)$, $(2, 0)$, and $(1, 2)$ are three points on a (vertical) parabola. We know that the equation is of the form $y = ax^2 + bx + c$. If only we knew the values of a, b, and c, we would have the equation. Substitute each of these points into $y = ax^2 + bx + c$. You will obtain three equations in the unknowns a, b, and c. Can you solve this system of three equations in three unknowns?

67. After you have completed Exercise 66, find the equation of the vertical parabola through each of these groups of points.

(a) $(3, 1), (0, 0), (-2, 1)$ **(b)** $(-1, 3), (2, -4), (0, 6)$

(c) $(2, 5), (-1, -1), (6, 3)$

68. For what values of k does the equation

$$x^2 + 2(k - 1)x + (k + 5) = 0$$

have at least one positive root?

6.6 SYSTEMS OF QUADRATIC EQUATIONS

We have already encountered systems of *linear* equations in Chapter 4. In this section, we will investigate systems of equations in which one or more of the equations is a quadratic equation. Consider the following example.

EXAMPLE 1

A television video-game company has the following total cost and total revenue, where x is the number of video units:

$$C = 11x + 200$$
$$R = 120 + x^2$$

The cost in dollars of producing x units is C. The revenue in dollars obtained from the sale of x units is R.

(a) Sketch the graph of both equations on the same set of axes.

(b) Find the "break-even" values of x, that is, the number of units that must be sold for cost to equal revenue.

(c) How many video units must be sold to make a profit?

Solutions

(a) Using a Graph The cost graph is a straight line, with the form of $y = mx + b$. $C = 11x + 200$ has slope 11 and y-intercept 200 ($m = 11$, $b = 200$). The points (0, 200) and (11, 321) are on the graph of the line. The revenue graph is a parabola with the form of $y = ax^2 + bx + c$. We use our three-step process to obtain points on the parabola (see Section 6.5).

1. The vertex for $R = 120 + x^2$ occurs at $x = \frac{-b}{2a} = \frac{-0}{2} = 0$. When $x = 0$, $R = 120$. The vertex is the point (0, 120).
2. Move 5 units left and right of the vertex to obtain a pair of symmetric points.

$$\text{If } x = 5, \text{ then } R = 120 + 5^2 = 145.$$
$$\text{If } x = -5, \text{ then } R = 120 + (-5)^2 = 145.$$

   Here, (5, 145) and (−5, 145) are symmetric points.
3. Move 10 units each way to obtain checkpoints on the graph at (10, 220) and (−10, 220). The graph is shown in Figure 6.53.

(b) The break-even points occur when revenue equals cost. These are the points where the line and parabola cross. We can attempt an "eyeball" estimate for where the graphs cross by studying Figure 6.53. A graphing calculator will give a better estimate. The graphs appear to cross when $x \approx 15$ and when $x \approx -5$. This graphical estimate may be good enough for some applications, but suppose we want to know *exactly* how many video games must be sold to break even.

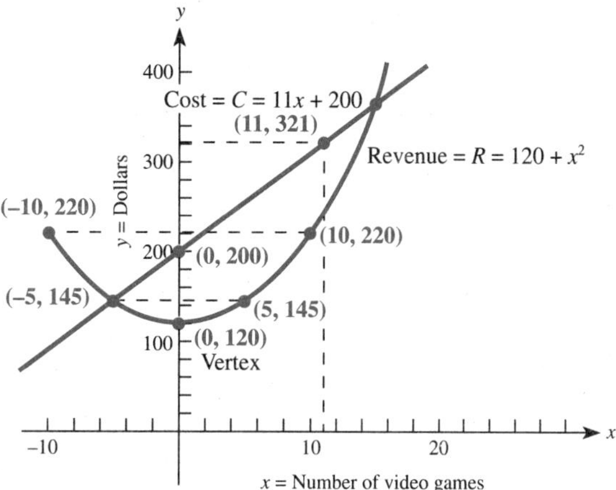

Figure 6.53

Using an Equation We can find the solution *algebraically* by solving the system of equations

$C = 11x + 200$	**Equation (1)**	We break even when $C = R$.
$R = 120 + x^2$	**Equation (2)**	
$11x + 200 = 120 + x^2$		Set Equations 1 and 2 equal.
$0 = x^2 - 11x - 80$		A quadratic equation.
$0 = (x - 16)(x + 5)$		It factors.
$x = 16 \quad \text{or} \quad x = -5$		Answer.

The graphs cross *exactly* when the number of video units is 16 or -5. The negative solution is meaningless in this case. Thus the break-even point occurs when $x = 16$ video units. When $x = 16$, the cost and revenue are both the same:

$$C = 11(16) + 200 = \$376$$
$$R = 120 + (16)^3 = \$376$$

The graphs cross at the point (16, 376). They also cross at (-5, 145).

(c) In order to make a profit, the revenue graph must be above the cost graph so that more money is taken in than is paid out by the video company. In the graph, it is apparent that the company must sell *more* than 16 video units to realize a profit, for if x is bigger than 16, then the parabola *lies above* the cost line (so more money comes in than goes out). ■

The Interplay between Algebraic and Graphical Solution Methods

In Example 1, we solved a system of equations (one equation is a quadratic) by both **graphing** and the **algebraic method of substitution**. The graph of a system of equations gives us information about the solution, and the algebraic solution gives us information about the graph. This interplay between the "algebra" of a system and its analytic "geometry" is attractive and very useful. Let's do several more examples to illustrate our point.

EXAMPLES 2–5

Graph each of the following systems, and find their simultaneous solution.

2. $Y1 = 2x - 3$
   $Y2 = x^2 - 2$

3. $Y1 = 4x^2 - 3x + 2$
   $Y2 = -2x + 7$

4. $Y1 = 2x$
   $Y2 = x^2 + 2$

5. $Y1 = x^2 - 5x + 1$
   $Y2 = 2x^2 + 3x - 7$

Solutions

2. **Using a Graph** $Y1 = 2x - 3$ is a line, with slope $= 2$ and y-intercept at $(0, -3)$. The point $(1, -1)$ is also on the line. The equation $Y2 = x^2 - 2$ is a parabola, with its vertex at

$$x = \frac{-b}{2a} = \frac{-0}{2} = 0$$

So the vertex is $(0, -2)$. The pair $(1, -1)$ and $(-1, -1)$ are symmetric points on the graph. The pair $(3, 7)$ and $(-3, 7)$ are also on the parabola.

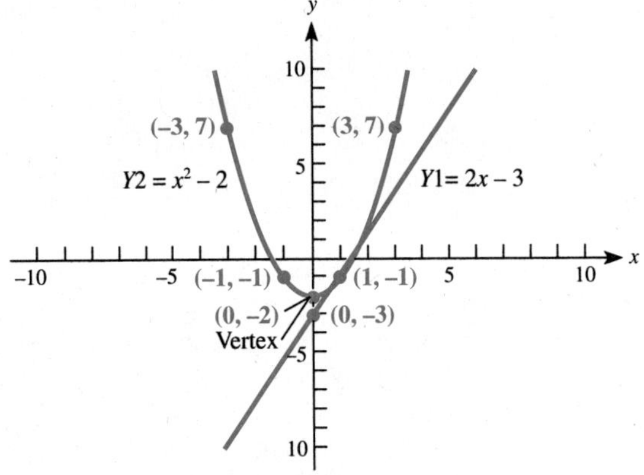

Figure 6.54

Using an Equation In the graph above, the line and parabola appear to intersect at only *one* point. This suggests that the system may only have one algebraic solution. Let's check. Using substitution, we put $Y1 = 2x - 3$ into the second equation and obtain

$$2x - 3 = x^2 - 2 \qquad \text{Solve for } x.$$
$$0 = x^2 - 2x + 1$$
$$0 = (x - 1)(x - 1) \qquad \text{The quadratic factors.}$$

Therefore $x = 1$. Indeed, there is only one solution pair for this system, the point $(1, -1)$.

3. Using a Graph

(a) $Y1 = 4x^2 - 3x + 2$ is a parabola with vertex at

$$x = \frac{-b}{2a} = \frac{-(-3)}{8} = \frac{3}{8} = 0.375$$

When $x = 0.375$, $y = 4(0.375)^2 - 3(0.375) + 2 = 1.4375$. The vertex of this parabola is at the point (0.375, 1.4375).

(b) To find symmetric points on the parabola, it will be *convenient* to move 0.375 units left and right of the vertex.

If $x = 0$ or $x = 0.750$, then $y = 2$.   **We moved 0.375 units, because $x = 0$ is a *convenient* value to substitute.**

If $x = -1$ or $x = 1.750$, then $y = 9$.   **For the check points, we moved 1.375 units each way.**

The other equation of this system, $Y2 = -2x + 7$, is a line, slope -2, y-intercept (0, 7). The point (1, 5) is also on the line.

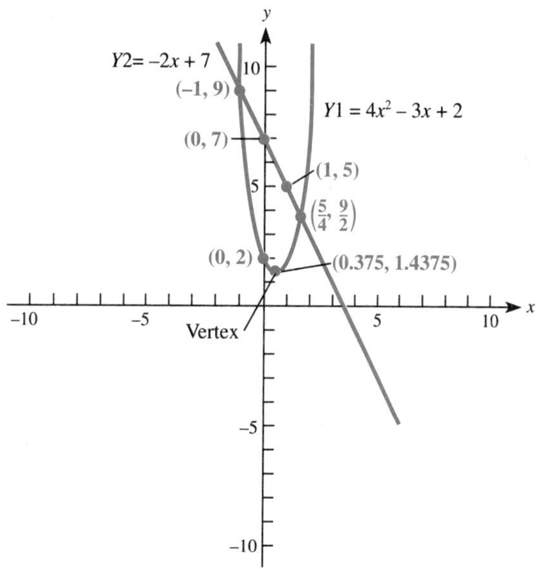

Figure 6.55

Using an Equation In the graph above, the line and the parabola appear to intersect in two points. To confirm that there are two simultaneous solutions, we can use algebra and the method of substitution:

$$4x^2 - 3x + 2 = -2x + 7$$
$$4x^2 - x - 5 = 0$$

We could use the quadratic formula to find the solution. Since $a = 4$, $b = -1$, and $c = -5$, the solutions are

$$x = \frac{-b \pm \sqrt{b^2 - 4ac}}{2a} = \frac{-(-1) \pm \sqrt{(-1)^2 - 4(4)(-5)}}{8}$$

$$= \frac{1 \pm \sqrt{81}}{8} = \frac{1 \pm 9}{8} = -1 \quad \text{or} \quad \frac{10}{8}$$

The solutions are $x = -1$ and $x = \frac{5}{4}$. (We could also have found these solutions by factoring.) When $x = -1$, $y = 9$. When $x = \frac{5}{4}$, $y = \frac{9}{2}$. Thus the simultaneous solutions are $(-1, 9)$ and $\left(\frac{5}{4}, \frac{9}{2}\right)$.

4. **Using a Graph** $Y1 = 2x$ is the equation of a line with slope 2 and y-intercept (0, 0). The point (1, 2) is also on the line. The equation $Y2 = x^2 + 2$ is a parabola with vertex at (0, 2), since $b = 0$. Symmetric points on the parabola are (2, 6) and $(-2, 6)$, as are (3, 11) and $(-3, 11)$. The graphs of these two equations are shown below.

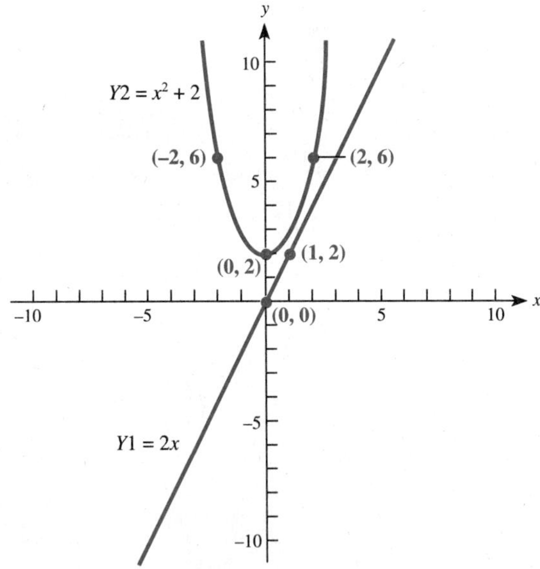

Figure 6.56

Notice that the two graphs *do not* meet. What, then, is the simultaneous solution? It *appears* that *there is no simultaneous solution* for this system.

Using an Equation Let's see what happens if we try to solve such a system algebraically:

$$Y1 = 2x$$
$$Y2 = x^2 + 2$$

Using substitution,

$$2x = x^2 + 2$$
$$0 = x^2 - 2x + 2$$

Using the quadratic formula,

$$x = \frac{-(-2) \pm \sqrt{4 - 4(2)(1)}}{2} = \frac{2 \pm \sqrt{4 - 8}}{2} \quad \text{or} \quad = \frac{2 \pm \sqrt{-4}}{2}$$
$$= \frac{2 \pm 2i}{2} = 1 \pm i$$

The solutions to this system are not real numbers. There are *no pairs of real numbers* that satisfy both equations simultaneously.

5. **Using a Graph** Let's sketch the two parabolas $Y1 = x^2 - 5x + 1$ and $Y2 = 2x^2 + 3x - 7$. First, for $Y1 = x^2 - 5x + 1$, the vertex occurs at

$$x = \frac{-b}{2a} \quad \text{or} \quad x = \frac{-(-5)}{2} = 2.5$$ **When $x = 2.5$, substitute and see that $y = -5.25$.**

Symmetric points occur at

$$x = 2.5 - 1.5 = 1 \text{ and at } x = 2.5 + 1.5 = 4$$ **When $x = 1$ or $x = 4$, $y = -3$.**

Second, let's sketch $Y2 = 2x^2 + 3x - 7$. The vertex occurs at

$$x = \frac{-b}{2a} \quad \text{or} \quad x = \frac{-3}{4} = -0.75$$

Symmetric points occur at

$$x = -0.75 + 0.75 = 0 \quad \text{and} \quad x = -0.75 - 0.75 = -1.5$$ **When $x = 0$ or $x = -1.5$, $y = -7$.**

Other points on this graph, by plotting, include $(0, -7)$, $(1, -2)$, $(-1, -8)$, $(2, 7)$, and $(-3, 2)$.

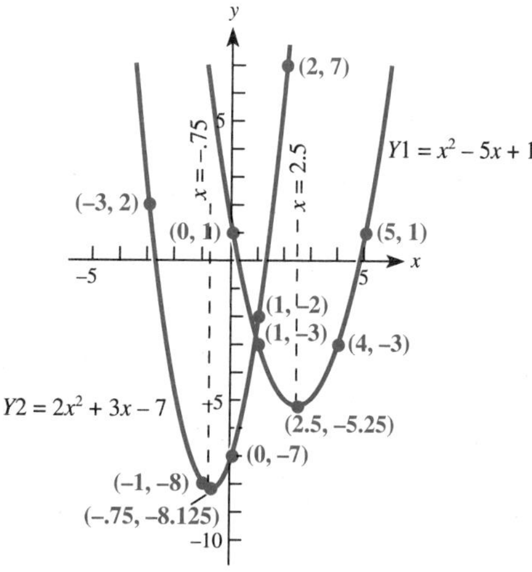

Figure 6.57 (a)

It might be more convenient to use a graphing calculator. (See Figures 6.57(b) and 6.57(c).) From the graph, we see that one solution occurs near $x = 1$, about where the two parabolas cross. It is not clear from our limited graph whether or not the parabolas will meet again, on their left branches.

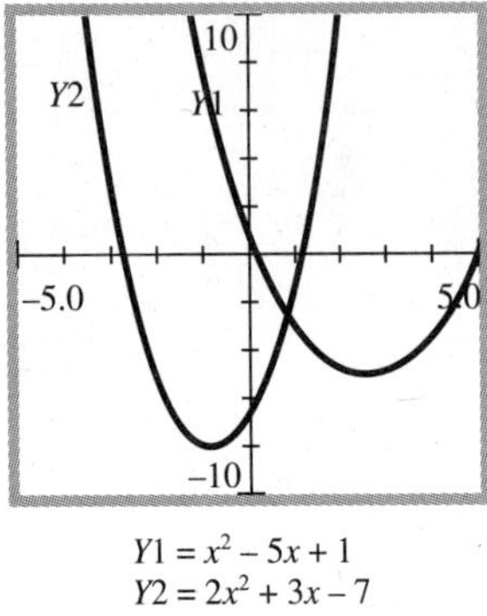

$Y1 = x^2 - 5x + 1$
$Y2 = 2x^2 + 3x - 7$

Figure 6.57 (b)

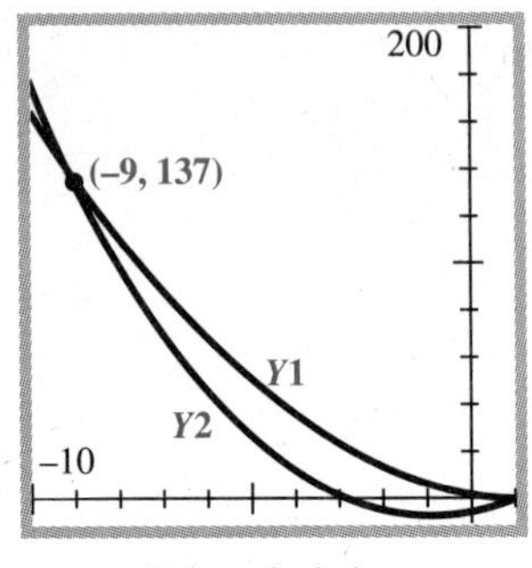

Enlarged window

Figure 6.57 (c)

Using an Equation Using substitution, we look at an algebraic solution to help us answer this question.

$$2x^2 + 3x - 7 = x^2 - 5x + 1$$
$$x^2 + 8x - 8 = 0$$

Using the quadratic formula, $a = 1$, $b = 8$, and $c = -8$. Therefore the solutions are

$$x = \frac{-8 \pm \sqrt{64 - 4(1)(-8)}}{2} = \frac{-8 \pm \sqrt{64 + 32}}{2}$$

$$= \frac{-8 \pm \sqrt{96}}{2} \approx \frac{-8 \pm 9.80}{2} = \frac{1.80}{2} \quad \text{or} \quad \frac{-17.80}{2}$$

Thus $x \approx 0.90$ or $x \approx -8.90$. When $x = 0.90$,

$$y = (0.90)^2 - 5(0.90) + 1 = -2.69$$

When $x = -8.90$,

$$y = (-8.90)^2 - 5(-8.90) + 1 = 124.71$$

Thus the solutions, to two-decimal accuracy, are $(0.90, -2.69)$ and $(-8.90, 124.71)$. The fact that there are two algebraic solutions to this system of equations indicates that the graphs of these two parabolas meet in two places. Therefore, the two graphs in Figure 6.57(a) *must* meet a second time, at $(-8.90, 124.71)$. ■

From Examples 2–5, we can see that a system of equations involving a quadratic can have

1. *two* real-valued solutions (graphs cross twice, as in Examples 3 and 5),
2. *one* real-valued solution (graphs are tangent, as in Example 2), *or*
3. *no* real-valued solutions (the graphs never meet, as in Example 4).

Algebraic Methods in Applications

EXAMPLE 6

You drove 40 miles to your parents' home for dinner. It took you 20 minutes less to return than it did to arrive during rush-hour traffic, because you could drive 10 mph faster on the way to your home. How fast did you drive each way?

Solution

We can use the model

$$\text{Distance} = \text{rate} \times \text{time}$$

Let r be the rate to your parents' house and t the time. Driving *to* your parents' house,

$$40 = rt \qquad \textbf{Equation (1)}$$

Driving *from* your parents' house,

$$40 = (r + 10)\left(t - \frac{1}{3}\right) \qquad \textbf{Equation (2)} \qquad \text{20 minutes} \doteq \tfrac{1}{3}\text{ hour.}$$

$$t = \frac{40}{r} \qquad \text{Solve Equation 1 for } t.$$

$$40 = (r + 10)\left(\frac{40}{r} - \frac{1}{3}\right) \qquad \text{Substitute } t = \frac{40}{r} \text{ into Equation 2.}$$

$$40 = 40 - \frac{r}{3} + \frac{400}{r} - \frac{10}{3} \qquad \text{The LCD is } 3r.$$

$$(3r)40 = (3r)\left(40 - \frac{r}{3} + \frac{400}{r} - \frac{10}{3}\right) \qquad \text{Multiply by the LCD.}$$

$$120r = 120r - r^2 + 1{,}200 - 10r \qquad \text{A quadratic in } r.$$

$$0 = r^2 + 10r - 1{,}200 \qquad \text{Multiply both sides by } (-1).$$

$$0 = (r + 40)(r - 30) \qquad \text{The equation factors.}$$

$$r = -40 \text{ mph} \quad \text{or} \quad r = 30 \text{ mph} \qquad \text{We eliminate the negative solution.}$$

The answer:

$$r = 30 \text{ mph to parents' house}$$
$$r + 10 = 40 \text{ mph upon return}$$

EXAMPLE 7

You and your sister both leave Portland at 9:00 A.M. You drive due south toward California, and your sister drives due east to Idaho. Your sister drives about 15 mph faster than you do (if she can get away with it). After four hours of continuous driving, you are about 325 miles apart. Was your sister speeding?

Solution

Suppose we model the situation in a diagram. You and your sister drive at right angles, as in the figure. The distance 325 miles is the hypotenuse of a right triangle. We use the equation

$$\text{Distance} = \text{rate} \times \text{time}$$

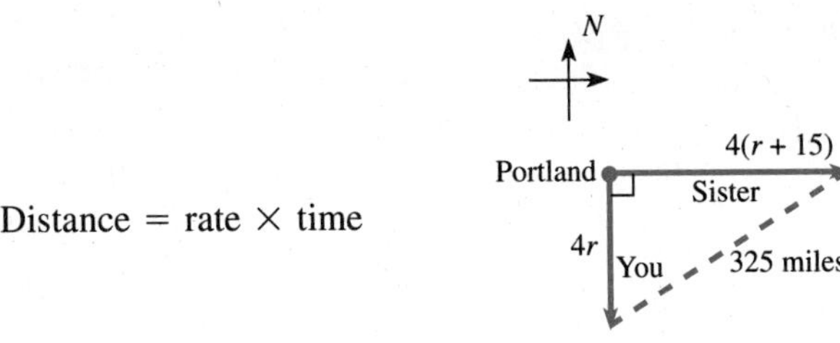

Figure 6.58

If we let r be your average rate of speed, then $4r$ is your distance. Your sister's average rate is then $r + 15$. Since you both traveled for four hours, your sister's distance is $4(r + 15)$. Then,

$$(\text{your distance})^2 + (\text{sister's distance})^2 = (325)^2 \qquad \text{The Pythagorean theorem (Figure 6.58).}$$

$$(4r)^2 + [4(r + 15)]^2 = (325)^2$$

$$16r^2 + 16(r^2 + 30r + 225) = 105{,}625 \qquad \text{A quadratic in } r.$$

$$32r^2 + 480r + 3{,}600 = 105{,}625$$

$$32r^2 + 480r - 102{,}025 = 0 \qquad a = 32, b = 480, c = -102{,}025$$

$$r = \frac{-480 \pm \sqrt{(480)^2 - 4(32)(-102{,}025)}}{2(32)} \qquad \text{Use a calculator.}$$

$$r = 49.46 \quad \text{or} \quad r = -64.46 \qquad \text{Eliminate the negative solution.}$$

$$r + 15 = 64.46 \qquad \text{Your sister's average speed.}$$

Whether your sister was speeding or not speeding depends on where she was driving, on an interstate highway or somewhere else. ■

6.6 EXERCISES

I. Procedures

1–14. Graph each system below. Then find the simultaneous solutions graphically and algebraically.

1. $y = x^2$
$y = 4x$

2. $y = x^2 + 4x + 2$
$y = 3x + 4$

3. $y = x^2 - 5x + 5$
$y = x + 1$

4. $y = 3 - x^2$
$y = x + 3$

5. $y - x^2 = -9$
$y - 3x = 1$

6. $y = x^2 - 3x + 2$
$y = x^2 - 4x + 3$

7. $y = x^2 - 3$
$y = x + 6$

8. $y = x^2 + 3x + 2$
$y = 3x + 5$

9. $3x - 2y + 7 = x^2$
$y - x = x^2$

10. $y = 2x - 7$
$y = x^2 - 2x - 3$

11. $y = x^2 - 1$
$y = x - 4$

12. $y = x^2 - 6x + 8$
$y = x + 2$

13. $y = 4 + x^2$
$y = 4 - x^2$

14. $y = 8 - x^2$
$y = 3x + 5$

15–22. Solve using a graph.

15. $y = 0.5x - 0.8$
$y = 0.04x^2 - 0.4x$

16. $y = 3x^2 + 6x + 3$
$y = -x^2 - 4$

17. $y = 3x^2 + 6x + 3$
$y = -x^2 + 4$

18. $y = x^2 - 4$
$y - 2x + 1 = 0$

19. $y = 0.05x - 0.08$
$y = 0.16x^2 - 3$

20. $0.7x^2 - 0.3x = y - 2$
$0.5x^2 + 0.4x = y$

21. $\dfrac{x^2}{2} - 3x = y$
$y + \dfrac{x}{5} = 6$

22. $y - \dfrac{x^2}{5} = 2x + 7$
$x - y = -3x + \dfrac{x^2}{2} - 3$

II. Concepts

23. Consider the accompanying cost and revenue graphs for producing plywood.

(a) How many board feet should be produced for the cost to equal the revenue?

(b) How many board feet should be produced to make the revenue as large as possible?

(c) Does the company make or lose money when it produces 25,000 board feet?

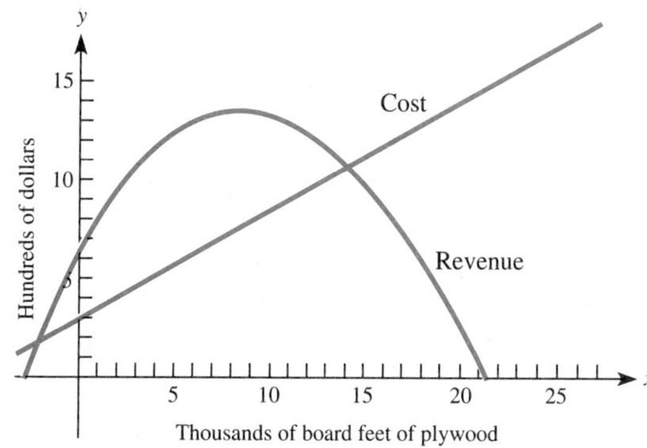

24. Consider the accompanying graphs of cost and revenue.

(a) For what quantity will the cost equal the revenue?

(b) How much must be produced to make a profit?

(c) For what quantity is the cost the lowest? Will the company make a profit when the cost is at its lowest point?

(d) If the company produces nothing on one day, how much money will it lose?

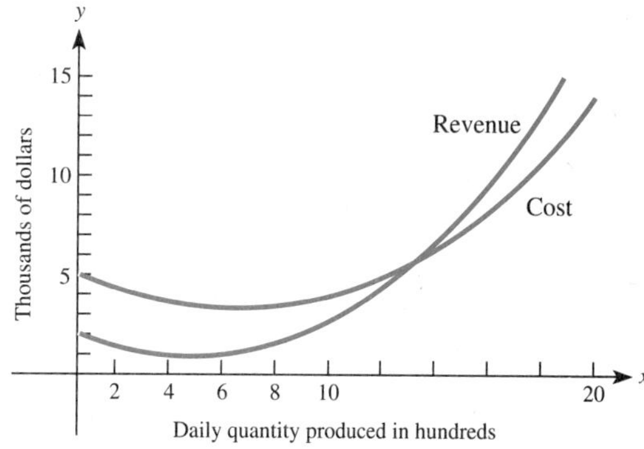

25–32. In each case, create your own system of equations that satisfies the given criteria. Graph each of your systems. Create a system of equations that

25. consists of two lines that meet in one point.

26. consists of a line and a parabola that meet in two points.

27. consists of two parabolas that meet in one point.

28. consists of two parabolas that meet in two points.

29. consists of two parabolas that do *not* intersect.

30. consists of a line and a parabola that do *not* intersect.

31. consists of two lines that do *not* intersect.

32. consists of three equations, a parabola and two lines. The two lines meet in one point, and the parabola intersects *neither* of the two lines.

33. Write a note describing the possibilities for the number of simultaneous solutions to a system of equations of the form

$$y = ax^2 + bx + c$$
$$y = mx + p$$

What are the possibilities for the graph of such a system of equations?

34. Write a note describing the possibilities for the number of simultaneous solutions to a system of equations of the form

$$y = ax^2 + bx + c$$
$$y = dx^2 + ex + f$$

What are the possibilities for the graph of such a system of equations?

35–38. Graph each of these systems. Estimate the solution graphically first, then find the simultaneous solutions algebraically.

35. $y = \dfrac{1}{x}$

$y = -2x + 3$

36. $y - \dfrac{3}{x - 2} = 2x$

$y = x - 3$

37. $y + \dfrac{1}{x + 2} = 7 - x$

$y = x$

38. $2y - \dfrac{1}{x} = 7x$

$y - x = 2$

III. Applications

39. Refer to Example 7 in the text. Suppose you and your sister left as before, but after four hours, you were 225 miles apart. Was your sister speeding? What is the farthest distance you and your sister could be apart after four hours if your sister drives within the speed limit? Use 55 mph as the maximum permissible speed.

40. Refer to Example 6 in the text. Suppose it took you 30 minutes less to return the 40 miles from your parents' house. Did you speed on the return trip?

41. The cost in dollars of canning x tons of pickles a day is given by

$$C = 5x^2 - 15x + 8$$

and the revenue from these pickles is given by

$$R = 2x + 2$$

(a) Graph this system of equations on the same axis.

(b) Find the simultaneous solution to the system algebraically.

(c) How many pickles must be produced for the company to break even? In other words, when will the cost equal the revenue?

(d) Suppose four tons of pickles are produced each day. Will the company realize a profit or a loss? Of how much?

42. The cost in cents of producing x meters of yarn is given by

$$C = 3x^2 - 12x + 14$$

and the revenue function is

$$R = 3x + 2$$

(a) Graph this system of equations on the same axis.

(b) Find the break-even points, where cost equals revenue.

43. The following equations describe a competitive market in agricultural economics:

$$q_s = 6p^2 - 5p - 5$$
$$q_d = 20 + 5p - p^2$$

where p is the price in dollars, q_s is the quantity supplied, and q_d is the quantity demanded.

(a) Graph this system on the same set of axes, with p (price) on the horizontal axis.

(b) Find the equilibrium market price, where $q_d = q_s$.

44. A swine producer bought a number of pasture waterers for $350.00. Later the price of each waterer increased by $20.00. As a result, he purchased two fewer waterers for the same amount of money. How many waterers were bought in the first purchase, and what was the purchase price per unit?

45. You are going to make the frame for a sandbox from a rectangular sheet of steel that is 9 feet by 12 feet. Four equal squares are to be cut out from the corners of the steel sheet.

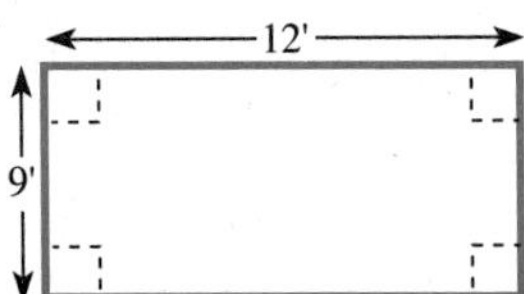

Then the edges are to be bent up. You want the area of the base rectangle of the box to be 60 square feet. Will this box hold the 150 cubic feet of sand that you have ordered?

46. A rectangular field whose dimensions are 30 meters by 50 meters must be plowed. The plow goes around the field and works from the outside to the inside by plowing a strip around. If two-thirds of the area of the field has been plowed when a circuit is finished, how wide is the finished strip?

47. A lawn that is 120 yards long and 50 yards wide is to be mowed by cutting a border around the outside edge. How wide is this border if half the area of the lawn has been mowed?

48. The volumes of two cubes differ by 98 cubic inches, and the lengths of their sides differ by 2 inches. Find the length of the side of the smaller cube.

49. The manager of a real estate firm faces the problem of what monthly rent she should charge for her 60 newly constructed apartments. Her past experience tells her that at \$200 per month, all units will be occupied, but for each \$2 per month increase in rent, one unit is likely to remain vacant. An occupied apartment has \$150-per-month expenses (taxes, maintenance, etc.), while an unoccupied apartment has expenses of \$140 per month. How many apartments must be rented in order to make a \$2,600 profit per month? How many apartments must be rented to obtain the maximum profit?

IV. Extensions

50–53. Graph each of these systems. In each case, you will have to plot points or use a graphing calculator in order to sketch the graph of the second equation. Find the simultaneous solution graphically. Then find the simultaneous solution algebraically, if you can.

50. $y = 7 - x$
$xy = 16$

51. $y = x^2 - x$
$xy = 1$

52. $2y + 3x = 6$
$x^2 + y^2 = 25$

53. $y = -x^2$
$x^2 + y^2 = 25$

54. For what values of m will the line $y = mx - 1$ intersect the parabola $y = x^2$ in

(a) two points? **(b)** one point? **(c)** no points?

55. For what values of m will the line $y = mx + 3$ intersect the parabola $y = -x^2$ in

(a) two points? **(b)** one point? **(c)** no points?

56. For what values of b will

(a) the line $y = 3x + b$ intersect the parabola $y = x^2 + 3x + 7$?

(b) the line $y = -2x + b$ intersect the parabola $y = 2x^2 - 2$?

57. For what values of c will

(a) the line $y = 2x - 3$ intersect the parabola $y = 2x^2 + 3x + c$?

(b) the line $y = -x$ intersect the parabola $y = -x^2 + c$?

58. Give an example of the equations of two parabolas whose graphs intersect in exactly

(a) three points **(b)** four points.

Hint: One of the parabolas could have a horizontal axis of symmetry. See Exercises 64 and 65 in Section 6.5.

CHAPTER 6 SUMMARY

Example 2 of Section 6.1 involves the equation $h = 80t - 16t^2$, which is a quadratic equation. It relates the height h of a golf ball to the time t in flight. If we ask how long it will take for the ball to hit the ground, we let $h = 0$ and solve for t in $0 = 80t - 16t^2$.

One method of solving quadratic equations is to use a calculator to approximate a solution. A more exact method is obtained by observing that $0 = 80t - 16t^2 = 16t(5 - t)$ and applying the property of zero products. Thus, $16t = 0$ or $5 - t = 0$. Our answer is $t = 5$; $t = 0$ when the ball is on the tee.

Another method is to try to factor the quadratic. Since all quadratics are not easily factored, we developed more general procedures for solving quadratics: completing the square, the quadratic formula, and using a graphing calculator.

In addition to solving quadratic equations, this chapter is also concerned with graphs of equations of the form $y = ax^2 + bx + c$. The fundamental questions are:

1. Given a quadratic equation, find its graph (parabola) in an easy way.
2. Given a quadratic equation, *find its vertex* in an easy way.

Many physical situations are modeled with parabolas, such as the path of a projectile, water from a fountain, flight of a high-jumper, and also the cost and revenue of producing a certain number of a product. A variety of applications were discussed, with special attention to maximum and minimum problems. Many concepts can be approached using a graphing calculator.

Important Words and Phrases

successive approximation
quadratic equation in one variable
standard form of a quadratic equation
quadratic formula

Important Words and Phrases (continued)

parabola
maximum point and minimum point
the vertex of a parabola
x- and y-intercepts of a parabola
Pythagorean theorem
discriminant
complex number
real part
imaginary part
imaginary number
imaginary unit
real number
quadratic model
symmetric about the y-axis
axis of symmetry of a parabola
symmetric points

Important Properties and Procedures

- Solving a quadratic equation by using an equation: factoring, completing the square, the quadratic formula
- Solving a quadratic equation by graphing
- Zero product property: If $ab = 0$, then $a = 0$ or $b = 0$
- A property of square roots: $\sqrt{ab} = \sqrt{a}\sqrt{b}$ if $a \geq 0$, $b \geq 0$
- Addition and multiplication of complex numbers
- Classifying the solution (or roots) of a quadratic equation by using the discriminant
- Implications of the discriminant for solutions of quadratics
- Graphing a quadratic equation of the form $y = ax^2 + bx + c, a \neq 0$
- Finding the line of symmetry of a parabola
- Finding the coordinates of the vertex and x- and y-intercepts of a parabola
- Finding the simultaneous solution to a system of equations in which one or more of the equations is a quadratic equation

The table below summarizes the behavior of the graph of $y = ax^2 + bx + c$ under different conditions on a, b, and c. (In the table, we have put $-\dfrac{b}{2a} > 0$, but, of course, it could just as well be a negative number.)

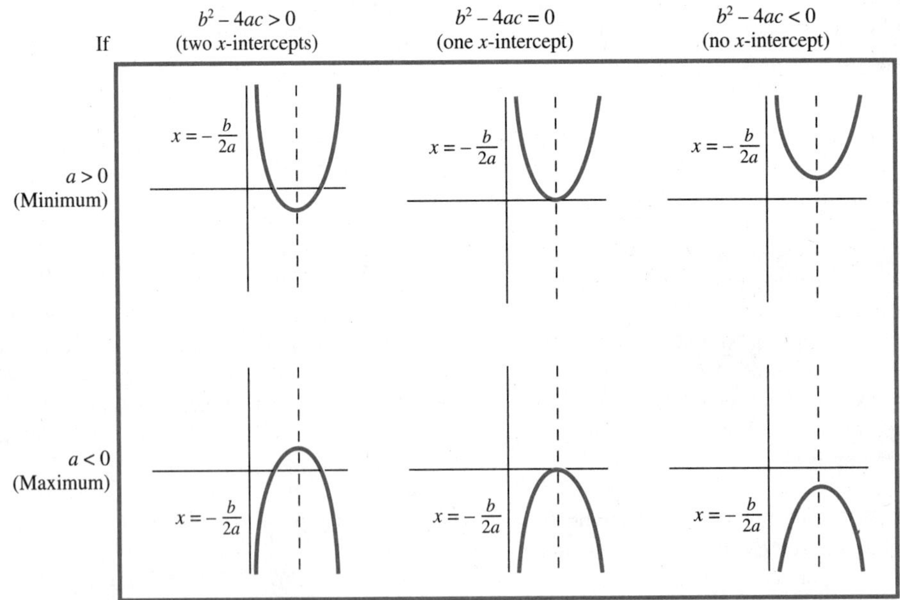

REVIEW EXERCISES

I. Procedures

1–4. Find the solutions to each of the following equations. Try to solve by factoring.

1. $72x^2 - 8x = 0$

2. $72x^2 - 8 = 0$

3. $96x^2 - 8x - 8 = 0$

4. $\dfrac{x^2 + 3}{x^2 + 4} = 1$

5–8. Solve these equations by completing the square.

5. $x^2 + 3x - 3 = 0$

6. $x^2 + 6x + 9 = 0$

7. $2x^2 - 7x + 6 = 0$

8. $2y^2 + 4 = 6y$

9–12. Without solving these equations, determine whether the solutions are real and unequal, real and equal, or complex.

9. $3x^2 - 7x + 12 = 0$

10. $12x(5 - 3x) = 25$

11. $9x^2 - 15x - 13 = 0$

12. $4x^2 - 3 = 6x$

13–28. Find the solutions to the equations below by any method.

13. $x^2 - 4x = -3$

14. $x^2 - 6x = -5$

15. $x^2 = 2(x - 2)$

16. $2x^2 = 3(x + 2)$

17. $\dfrac{x}{2} - \dfrac{3}{4} = \dfrac{5(2 - 3x)}{6}$

18. $\dfrac{2x}{5} - \dfrac{1}{3} = \dfrac{2(4 - x)}{15}$

19. $\dfrac{2}{x} - \dfrac{5}{x + 1} = \dfrac{4}{x^2 + x}$

20. $\dfrac{x}{3x + 4} - \dfrac{5}{2x - 3} = \dfrac{7 - 3x}{6x^2 - x - 12}$

21. $\dfrac{1}{x + 1} - 1 = \dfrac{1}{x - 1}$

22. $5 - \dfrac{1}{x} = \dfrac{x}{x - 3}$

23. $\dfrac{x^2 + 3}{x^2 - 1} = 7$

24. $\dfrac{x}{x^2 - 4} = 0$

25. $0.03x^2 - 0.01x = 0.2$

26. $0.71x - 0.28x^2 = 12$

27. $6x^2 - 14x = -5$

28. $16x^2 - 14x = -5$

29–40. Sketch the graphs of these equations in the plane. Find the coordinates of the vertex, y-intercept, and x-intercepts, if they exist.

29. $y = 3x^2 - 27$

30. $y = 10x - 25x^2$

31. $y = 0.75x - 0.125x^2$

32. $y + 8 = 16x - 24x^2$

33. $3y = -7x + 6$

34. $4xy = 1$

35. $y = -5x^2 - 6$

36. $y = -20x^2 + 15x$

37. $y = -3x^2 - 15x - 18$

38. $y = 7x - 15 + 4x^2$

39. $y = \dfrac{3}{4} - \dfrac{1}{2}x - \dfrac{x^2}{4}$

40. $y = 0.07x^2 + 0.92x - 3$

41–48. Find the simultaneous solution to these systems of equations. Sketch the system.

41. $y = 3x^2 + 6$
$y + x = 1$

42. $y = 10 - 3x^2$
$y = 3x$

43. $2y - x = x^2$
$x^2 - 3x + 6 = y$

44. $y + \dfrac{1}{x} = 7$
$2y + 3x = 4$

45. $4y - 16x^2 = 8$
$2x - 8x^2 = 12y$

46. $0.2x^2 - 0.3x = y + 7$
$0.7x - 0.03x^2 = 2y$

47. $y = x^2 - 5x + 5$
$y = x + 1$

48. $y = x$
$y = \dfrac{200}{x - 1}$

II. Concepts

49. Write a quadratic equation that has solutions $x = 7$ and $x = -2$.

50. Write a quadratic equation that has solutions $x = \frac{1}{2}$ and $x = -5$.

51. The quadratic equation $6x^2 - 5x + c = 0$ has $x = -\frac{2}{3}$ for one solution. Find the other solution.

52. The quadratic equation $4x^2 + bx - 7 = 0$ has $x = \frac{1}{2}$ for one solution. Find the other solution.

53 and 54. For each problem, **(a)** Make a chart. **(b)** Sketch a graph by plotting some points. **(c)** Use a calculator to estimate the solution. **(d)** Solve the problem by using the quadratic formula.

53. A girl throws a javelin. The height h (in feet) of the javelin after t seconds is given by $h = 10t - t^2 + 3$.

(a) How long will it take for the javelin to hit the ground?

(b) Use your graph and a calculator to estimate the maximum height reached by the javelin.

54. The revenue R in millions of dollars that a business makes from selling x franchises is given by

$$R = -8 + 2x + 0.5x^2$$

How many franchises must be sold for the company to gross \$10 million? To gross \$20 million?

55. Give an example of a parabola whose vertex is

(a) $(1, 0)$ **(b)** $(1, 1)$

56. Give an example of a parabola that has

(a) $(-3, 0)$ and $(2, 0)$ as x-axis intercepts.

(b) $\left(\frac{1}{2}, 0\right)$ and $(-5, 0)$ as x-axis intercepts.

57. Give an example of a parabola that has

(a) $(0, 7)$ as the y-axis intercept.

(b) $(0, -5)$ as the y-axis intercept.

58. For what values of c will $y = 2x^2 - 3x + c$ cross the x-axis?

59. For what values of a will $y = ax^2 + 6x - 7$ cross the x-axis?

60. For what values of b will $y = x^2 + bx + 4$ cross the x-axis?

61. Sketch each of these on the same set of axes.

(a) $y = 4x^2$ **(b)** $y = -4x^2$

(c) $y = \frac{1}{4}x^2$ **(d)** $y = -\frac{1}{4}x^2$

What is the effect of the value of a in the equation $y = ax^2$?

62. Sketch each of these on the same set of axes.

(a) $y = 2x^2 + 5$ **(b)** $y = -2x^2 + 5$

(c) $y = 2x^2 - 5$ **(d)** $y = -2x^2 - 5$

What is the effect of the value of c in the equation $y = ax^2 + c$?

III. Applications

63. Suppose you invest \$600 in mutual funds that pay interest compounded annually. After two years, you have accumulated \$712.92. What was the interest rate of the funds?

64. It took you 20 minutes less to drive 35 miles home after work than it did to drive to work. You drove 18 mph faster on the way home. How fast did you drive each way?

65. What is the height of the tallest flagpole you can brace with a 60-foot anchor wire if you wish to tie the anchor wire to the ground at a distance that is half the height of the pole?

66. Safe automobile spacing S in feet is given by

$$S = \frac{V^2}{32} + V + 18$$

where V is the average velocity in *feet per second* on a busy street.

(a) Suppose a car is going 44 feet/second. How far should it be from the car in front of it in order to be safe? ($V = 44$, $S = ?$)

(b) Suppose a car is going 50 mph. How far should it be from the car in front of it in order to be safe? Do not forget to first change miles per hour to feet per second (60 mph = 88 feet/second).

(c) If a car is trailing 100 feet behind its predecessor, what is a safe speed for it to be traveling ($S = 100$, $V = ?$) in feet per second? How fast is this in miles per hour?

Answer the same questions for a car that is 50 feet behind its predecessor and for one that is 200 feet behind.

67. The weight w in *500-pound units* that can be lifted by a rope of diameter d inches is approximated by

$$w = d^2$$

(a) How much weight can be lifted by a 0.5-inch rope (when $d = 0.5$, $w = ?$)?

(b) How thick must a rope be to lift 2,000 pounds? To lift a Cadillac that weighs 4,750 pounds? (Remember—our "unit" weighs 500 pounds.)

68. The area A of a projected picture that is x meters away from the projector is given by

$$A = 0.12x^2$$

How far away must you be from the screen to project a picture that is 25 square meters?

69. The outside measurements of a rectangular picture frame are 14 cm by 16 cm. The frame is the same width x all around, and the picture area is 48 cm^2. How wide is the frame?

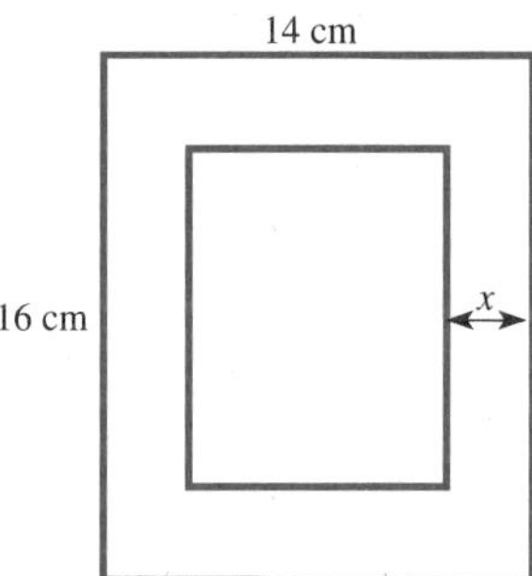

70. Find a number x so that the sum of the number and its reciprocal is 16. (See Exercise 97 in Section 6.3.)

71. A toy missile is fired from a raised silo with an initial velocity of 64 feet/second. The distance d above the ground after t seconds is given by

$$d = 64t - 16t^2 + 8$$

(a) How many seconds after firing will the missile be 50 feet off the ground?

(b) When will the missile hit the ground?

72. The annual yield on a section of forest that is harvested is given by

$$y = 0.4q - 0.3q^2$$

where y represents the annual harvest in thousands of board feet and q represents the volume of stock in thousands of board feet. For what value of q does the maximum value of y occur?

73. The cost function for making ski boots is

$$C = x^2 - 10x + 40$$

where x is the number of pairs of boots and C is the cost in dollars.

(a) What is the minimum cost, and how many pairs must be made to realize the minimum cost?

(b) Does it cost more to make eight pairs or two pairs?

(c) How many pairs can be made for \$50,000?

74. A fireworks rocket travels vertically according to the equation $h = 90t - 16t^2$. The company that manufactures the rockets wants the fireworks to explode at their highest point. For how long should it set the detonator (to set off the fireworks)?

75. One way to obtain an estimate for the number of board feet of usable lumber in a 16-foot log is to use the equation $B = 0.8(D - 1)^2 - \left(\frac{D}{2}\right)$, where B is the number of board feet and D is the diameter of the log in inches. How many board feet of usable lumber could be obtained from a 16-foot log that is $12\frac{1}{2}$ feet in circumference?

76. A diver leaps from a high platform, and her height h in meters is determined by

$$h = 30 + 0.3t - 16t^2$$

where t is time elapsed in seconds.

(a) Sketch this relationship.

(b) How long before the diver is only 10 meters high?

(c) How long before she hits the water?

77. A farmer wishes to use 3,800 meters of fence to enclose a rectangular plot on three sides. The fourth side is bounded by an old blackberry patch.

(a) Write an equation for the area A in terms of the length l of the plot.

(b) What dimension of the plot will yield the most area?

78. The sum of two numbers is 26, and their product is 84. What are the numbers?

79. You drove the 50 miles from work to home. It took you 25 minutes more to return home in rush-hour traffic than it did for you to drive to work in the morning. You estimate you drove 10 mph faster on the way to work. How fast did you drive each way?

80. These equations describe a competitive market for a certain forest product:

$$q_s = 8p^2 + 4p - 7$$
$$q_d = 15 - 6p - 2p^2$$

where p is price in hundreds of dollars, and q_s and q_d are the quantities supplied and demanded, respectively.

(a) Find the equilibrium market price, where $q_s = q_d$.

(b) If the price is \$800, will the quantity supplied or quantity demanded be higher?

CUMULATIVE REVIEW: CHAPTERS 1–6

I. Procedures

1–11. Perform the following operations and express all answers that are fractions in simplest equivalent form.

1. $3 - 6[8 - (2 + 11) - 1]$

2. $\dfrac{8 - 5}{-3 - 5} + \dfrac{15 - 20}{7 + 3}$

3. $x - 3x(4 - x) - x(3x - 1)$

4. $(-2x^6y)(4^{-3}xy)^2$

5. $\dfrac{2^{-1} + 2^{-2}}{2^{-3}}$

6. $\dfrac{3^{-2} - 2^3}{3^{-1} - 2^2}$

7. $\dfrac{x}{1 - x} + \dfrac{1}{x - 1} + 1$

8. $\dfrac{(-xy^4z^{-1})^2}{x^{-2}y^7z^0}$

9. $\dfrac{2x^2 + 5x - 3}{x^2 - 9} \cdot \dfrac{(x - 3)^2}{4x^2 - 4x + 1}$

10. $\left(1 - \dfrac{1}{x}\right) \div \dfrac{x^2 - 2x + 1}{1 - x}$

11. $\dfrac{2x^3 - 4x^2}{x^2 + 7x + 10} \div \dfrac{x - 2}{x + 2} \cdot \dfrac{(x + 5)^2}{2x}$

12–19. Solve each of the following equations or inequalities. Check your answers.

12. $\dfrac{3(x - 1)}{5} + 1 = \dfrac{3x}{2}$

13. $4 - [2 - 2(3 - x) + 4x] = -10$

14. $3 - 2(x + 1) \geq 5 + x$

15. $x - 1 < 3x - 5$ and $x < 6$

16. $2x^2 - 3x = 5$

17. $(x - 3)(x - 2) = 6$

18. $\dfrac{1}{x + 1} - \dfrac{x + 1}{x} = \dfrac{2x - 1}{x^2 + x}$

19. $2 - \dfrac{2x}{x^2 - 4} = \dfrac{5}{x + 2}$

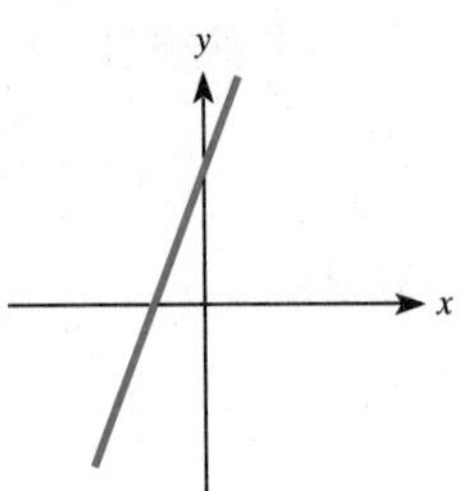

(a)

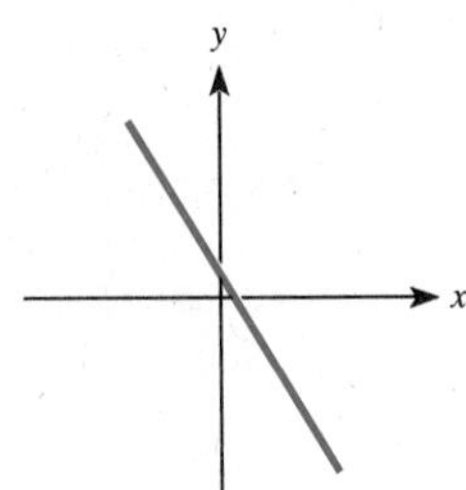

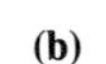

(b)

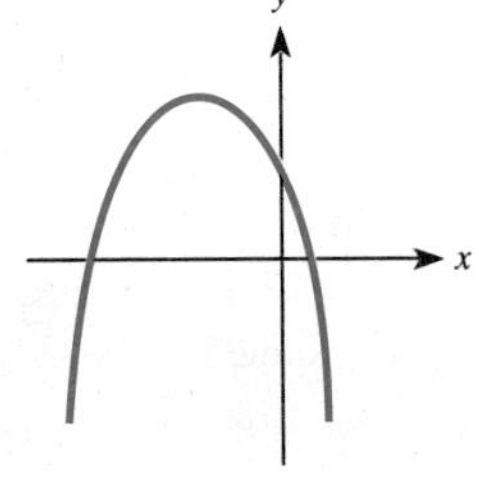

(c)

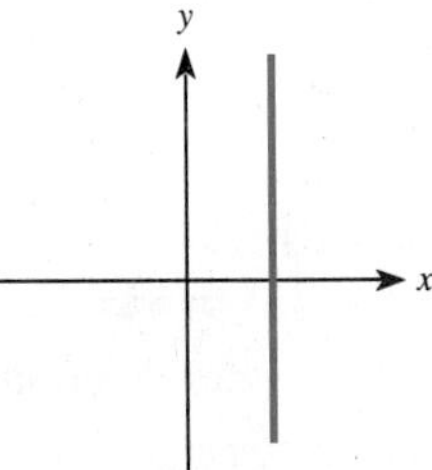

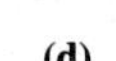

(d)

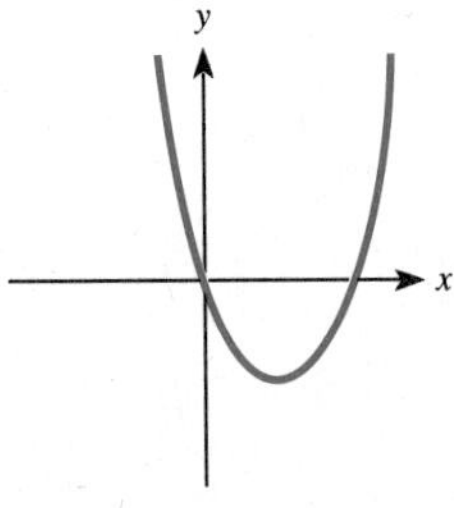

(e)

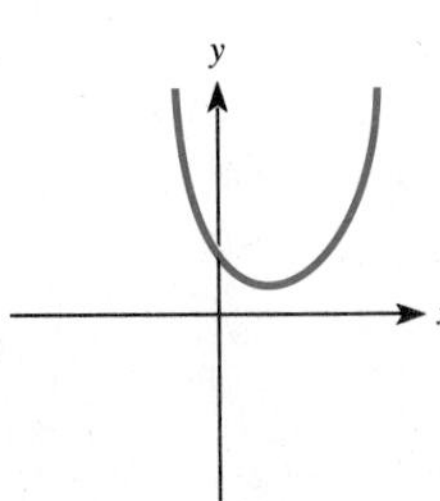

(f)

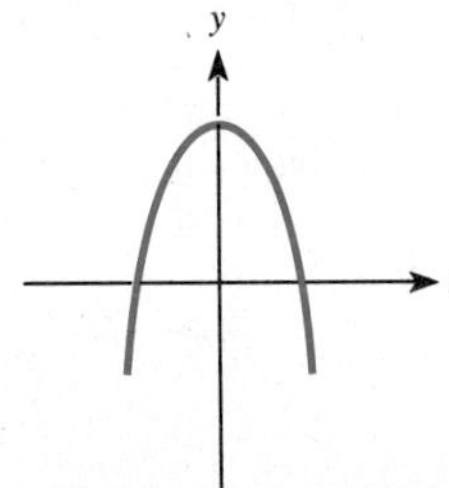

(g)

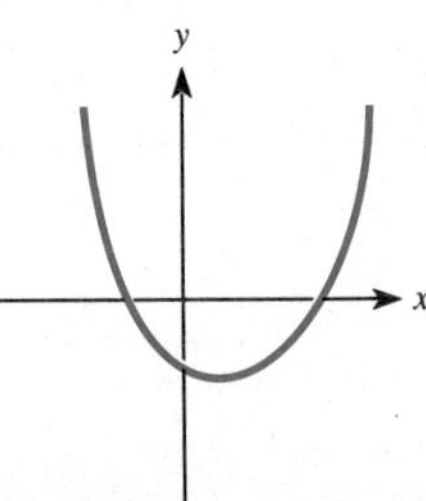

(h)

II. Concepts

20–24. Put the letter of the graph next to the problem that best fits each description.

____**20.** A line with undefined slope.

____**21.** A line with a negative slope and a positive y-intercept.

____**22.** A quadratic equation with a discriminant < 0.

____**23.** A quadratic equation with $a > 0$, $b = 1$, and $c = 0$.

____**24.** The graph of $y = -x^2 + c$ with $c > 0$.

25–27. Determine whether each statement is always true, sometimes true, or never true. Explain why your answers are correct.

25. If $x > 0$, then x^{-4} is always positive.

26. For all real numbers x, $-x$ is always negative.

27. The equation $\dfrac{1}{x+1} = 0$ has a solution.

28–31. Multiple choice: For each problem, select the best answer.

28. If a line has slope $-\frac{2}{3}$ and passes through the origin, then the equation of the line is

(a) $y = x$ **(b)** $y = -3x + 2$ **(c)** $3y - 2x = 0$

(d) $2y + 3x = 0$ **(e)** $3y + 2x = 0$

29. Assuming no denominator is equal to zero, which of the following expressions is equivalent to $\frac{3}{4}$?

(a) $\dfrac{x^2+3}{x^2+4}$ **(b)** $\dfrac{6(x-1)}{9(x-1)}$ **(c)** $\dfrac{9}{16}$

(d) $\dfrac{3+(x+2)}{4(x+2)}$ **(e)** $\dfrac{3-3x}{4-4x}$

30. The slope of a line that passes through the points $(-1, -3)$ and $(2, 0)$ is

(a) 1 **(b)** -1 **(c)** $-\dfrac{1}{3}$ **(d)** -3 **(e)** 2

31. Which of the following statements is *not* true about the graph of $y = 4 - x^2$?

(a) The graph has a maximum point.

(b) The y-intercept is $(0, 4)$.

(c) The x-intercepts are $(2, 0)$ and $(-2, 0)$.

(d) The point $(3, -5)$ lies on the graph.

(e) The graph is symmetric about the line $x = 1$.

32. For what values of c is the following equation an identity?

$$4x + 2(x - 2) + 1 = 6x - c$$

33. For what values of c does the following equation have exactly one solution?

$$x^2 + 8x + c = 0$$

34. For what values of c will the following equation have *real* roots?

$$x^2 - 5x + c = 0$$

35. What is the difference between an algebraic expression and an algebraic sentence (equality or inequality)? How are the two related? Give examples to illustrate your discussion.

36. Describe the various types of equations that have been solved in the past six chapters. Describe their differences and similarities. Describe how to obtain a solution for each of these equations. Illustrate your answers with examples.

III. Applications

37. The We Fly Higher Frisbee Plant decided to display the costs and revenues for producing Frisbees on a graph, as shown here.

(a) Estimate the break-even point (where cost = revenue).

(b) What are the cost and revenue for producing 4,000 Frisbees?

(c) Is there a profit or loss on producing 4,000 Frisbees? Of how much?

(d) Make up another question about the graph and answer it.

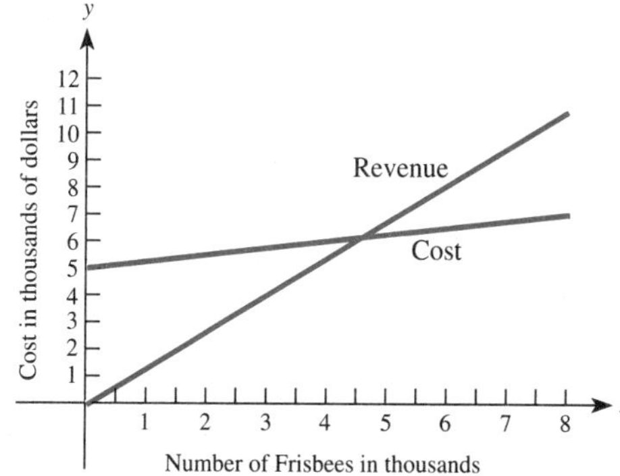

38. The cost and profit for a certain video company are displayed on the accompanying graph. Use the graph to answer the following questions.

(a) What are the start-up costs for the company?

(b) For what number of videos is there no profit? Give a reason why.

(c) What are the costs and profit on selling 400 videos?

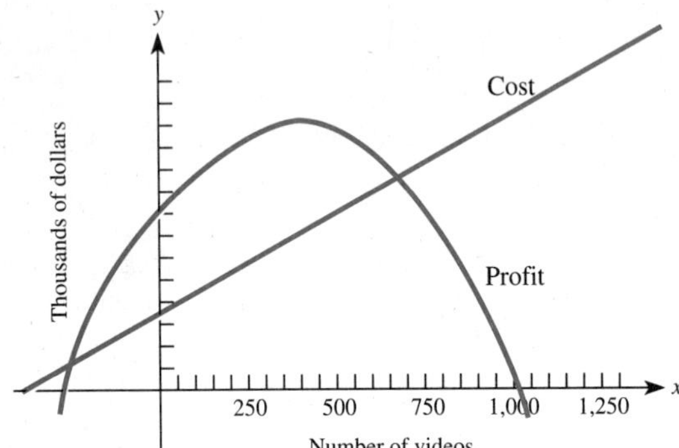

39. Rectangle *ABCD* is subdivided into four smaller rectangles, and the areas of four of them are indicated in the figure.

A		B
	$2x^2$	$8x$
	$6x$	24
D		C

(a) Find expressions for the *dimensions* (length and width) of rectangle *ABCD*.

(b) If the area of rectangle *ABCD* is 100 square centimeters, what are its dimensions?

40. A town with a population of 10,000 is expected to double every 3 years.

(a) What is the population after 24 years?

(b) How many years will it take the population to reach 640,000?

41. Two companies, A and B, offer you a sales position. Both jobs are essentially the same, but A pays a straight 11% commission and B pays $75 a week plus 7% commission. (Commission is a percent of the total amount of sales.) For what amount of sales will company A pay the same as company B?

42. You decide to invest the $10,000 you inherited from your grandfather. You decide to put some in a safe mutual fund that pays $8\frac{1}{2}$% and part in high-risk bonds that pay 11%. If you want $1,000 in interest a year, how much should you invest in each?

43. An airplane whose cruising speed in still air is 240 mph can travel 700 miles with the wind in the same time it takes to fly 500 miles against the wind. Find the speed of the wind.

44. A missile is launched from the ground. The height h, in feet, of the missile from the ground after t seconds is given by the equation

$$h = 168t - 24t^2$$

(a) How high is the missile after 2 seconds?

(b) How many seconds after launching will the missile hit the ground?

(c) At what time does the missile hit its maximum height? What is the maximum height?

45. (a) Find the difference for each row and then find a general statement that describes the resulting pattern.

(b) Prove that your statement is true.

(c) Graph the pattern using the row number as the horizontal axis and the difference as the vertical axis. Describe the graph.

$$\frac{1}{2} - \frac{1}{3} =$$

$$\frac{1}{4} - \frac{1}{5} =$$

$$\frac{1}{5} - \frac{1}{6} =$$

$$\frac{1}{6} - \frac{1}{7} =$$

46. Find a general expression for even integers and for odd integers. Graph both expressions on the same coordinate axes. Describe the two graphs.

47. For each statement below, fill in the blank with the correct answer. Prove your answers.

(a) The sum of two even integers is always an ____________.

(b) The sum of two odd integers is always an ____________.

(c) The sum of an even integer and an odd integer is always ____________.

(d) The product of two even integers is always ____________.

(e) The product of two odd integers is always ______________.

(f) The product of an even integer and an odd integer is always ______________.

48. Bill, who plays for the Indiana Hoops, has made 40 out of his last 72 free throws. How many consecutive free throws must he make to have a free-throw shooting average of 60%?

49. If a swimming pool can be filled in seven hours with one hose and ten hours with a neighbor's hose, how long will it take to fill the pool using both hoses simultaneously?

50. If one machine can stamp 500 cards in six minutes and a second machine can stamp these cards in three minutes, how long would it take the machines to stamp the cards working simultaneously?

51. In a 15-km race, Mary's speed for the first half of the race is 4 km/h. What must her speed be in the last half of the race for her to have an average speed of 6 km/h? Is this realistic? Why?

52. If you were offered a choice between having your salary raised $500 every six months or $1,000 a year, which would you choose if the plan were for five years? *Hint:* Make a table for the salaries under each plan for the first five years.

53. Find all real values of x for which $(x^2 - 5x + 5)^{x^2-9x+20} = 1$.

CHAPTER

Rational Exponents and Exponential Equations

We have already encountered examples of mathematical applications involving integral exponents, such as compound interest, growth of amoebas or germs, inflation, reduction of light filtering through a medium such as glass or water, and decay of radioactive substances. In this chapter we reexamine some of these applications and raise some new questions concerning exponents.

Source: Georg Gerster / Comstock

7.1 EXPLORING RATIONAL EXPONENTS: TABLES, GRAPHS, EQUATIONS

In Example 25 of Section 2.1, we looked at the amount of light that filters through n meters of water when n is a positive integer. In this section we reexamine this problem and ask, "How much light filters through n meters when n is not necessarily a positive integer?"

EXAMPLE 1

As light filters through water, its intensity is reduced. Suppose that in a certain lake the intensity of light is reduced by $\frac{3}{5}$ for each meter of water. How much light filters through at n meters?

Solution

Using a Table The following table illustrates the pattern.

Depth in meters	*Fraction of initial light intensity*	
0	1	**100% of light at surface**
1	$\frac{2}{5}\cdot 1 = 0.40$	**Only $\left(1-\frac{3}{5}\right) = 40\%$ of light at 1-m depth**
2	$\frac{2}{5}\cdot\frac{2}{5} = 0.16$	
3	$\frac{2}{5}\cdot\left(\frac{2}{5}\right)^2 = 0.064$	
4	$\frac{2}{5}\cdot\left(\frac{2}{5}\right)^3 = 0.0256$	
5	$\frac{2}{5}\cdot\left(\frac{2}{5}\right)^4 = 0.0102$	

Table 7.1

At 5 meters, we see that the intensity is about 0.01, or 1%, of the initial light intensity at the surface.

Using an Equation The table suggests the following relationship: The amount of sunlight L is a function of (depends on) the depth, n meters. We write Light(n meters) or Light(n) or, briefer yet, $L(n)$, to express this function.

$$L(n) = \left(\frac{2}{5}\right)^n$$

For example, at 1 meter, the intensity of light is $\left(\frac{2}{5}\right)^1 = \frac{2}{5}$ of what it is at the surface. At 3 meters, the light intensity, $L(n)$, equals $\left(\frac{2}{5}\right)^3 = \frac{8}{125}$ of what it is at the surface.

Using a Graph A visual representation of the function is given by a graph. The graph in Figure 7.1 shows the rapid decline of light intensity as depth of the water increases.

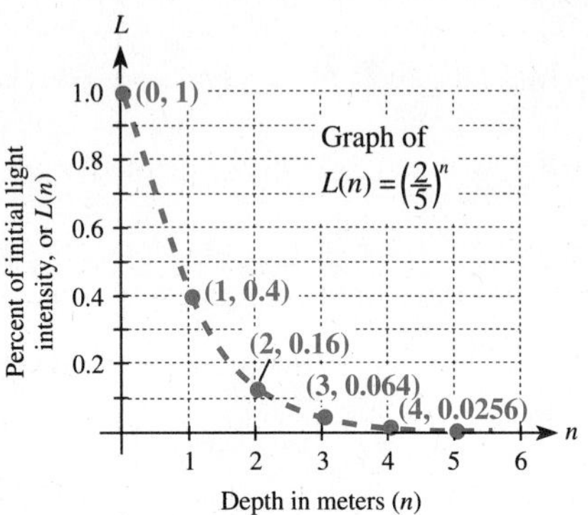

Figure 7.1

In Table 7.1, we have values for light intensity corresponding to integral values of the depth. How would we calculate the light intensity at $\frac{1}{4}$ meter or at $\frac{2}{3}$ meter? By analogy, we would need to calculate $\left(\frac{2}{5}\right)^{\frac{1}{4}}$ and $\left(\frac{2}{5}\right)^{\frac{2}{3}}$. What do $\left(\frac{2}{5}\right)^{\frac{1}{4}}$ and $\left(\frac{2}{5}\right)^{\frac{2}{3}}$ mean? In general, what does $\left(\frac{2}{5}\right)^{\frac{m}{n}}$ mean? We will investigate these questions in the next few pages.

EXAMPLE 2

In 1975, the Earth's population was about four billion and thought to be doubling every 35 years. According to this model, what will be the population in x years? (Since 1975 the rate of population growth has slowed. In 1990, the Earth's population was about 5.25 billion and thought to be doubling every 40 years.)

Solution

Using a Table In 35 years, or by the year 2010, the population will be $4 \cdot 2 = 8$ billion. It will be $4 \cdot 2 \cdot 2 = 4 \cdot 2^2$ or 16 billion in the year 2045.

Year	*Population estimate*
1975	4 billion
2010	$4 \cdot 2 = 8$ billion
2045	$4 \cdot 2^2 = 16$ billion
2080	$4 \cdot 2^3 = 32$ billion
2115	$4 \cdot 2^4 = 64$ billion

At this rate, in what year will the population reach approximately one trillion people?

Table 7.2

Using an Equation Examining the pattern in the table gives us a clue useful in finding a general equation. We can estimate the population in year x by first finding the number of 35-year intervals that have elapsed since 1975. This number is $\frac{x-1975}{35}$. Since the population doubles during each of these intervals, the population in year x is given by

$$P = 4 \cdot 2^{\frac{x-1975}{35}}$$

Using a Graph The graph in Figure 7.2 shows the rapid growth of the population (in billions).

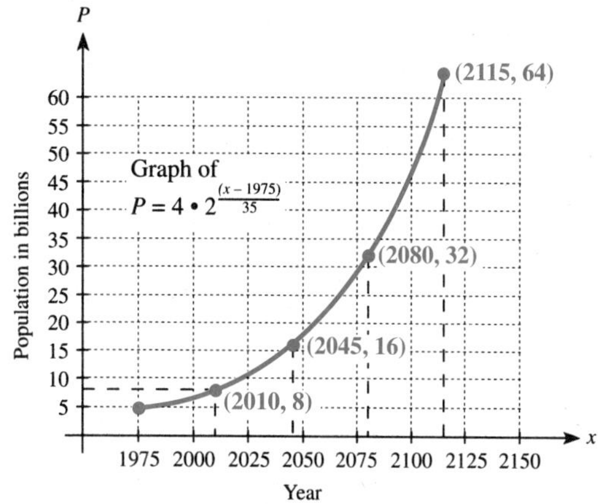

Figure 7.2

According to Example 2, an estimate of the population in 2025 is $P = 4 \cdot 2^{\frac{2025-1975}{35}} = 4 \cdot 2^{\frac{50}{35}} = 4 \cdot 2^{\frac{10}{7}}$. But what does $2^{\frac{10}{7}}$ mean? We need to explore the meaning of an exponent that is a rational number. Before investigating these fractional exponents, let us first review integral exponents. ■

Integral Exponents: A Review

In Chapter 2, we defined

$$x^n = \underbrace{x \cdot x \cdot x \cdots x}_{n\text{-times}} \qquad x^{-n} = \frac{1}{x^n} \qquad x^0 = 1$$

where $x \neq 0$ and n is a positive integer. Recall also that these definitions led to the following properties:

Properties of Integral Exponents

1. $x^m x^n = x^{m+n}$
2. $(x^m)^n = x^{mn}$
3. $(xy)^m = x^m y^m$
4. $\dfrac{x^m}{x^n} = x^{m-n}$

where $x \neq 0$, and m and n are integers.

Using these properties, we can write equivalent algebraic expressions in which there are no common factors in the numerator or denominator. This *simplest form* of an expression is often more convenient to use, especially if we need to evaluate the expression.

EXAMPLE 3

Find an equivalent expression for the following with no common factors in the numerator and denominator, and then evaluate the expression for $x = 3.2$ and $y = -1.7$.

$$\frac{(2xy^2)^4x}{(-2x^2y)^3y^2}$$

Solution

$$\begin{aligned}
\frac{(2xy^2)^4x}{(-2x^2y)^3y^2} &= \frac{16x^4(y^2)^4x}{-8(x^2)^3y^3y^2} && \text{Property 3.}\\
&= \frac{16x^4y^8x}{-8x^6y^3y^2} && \text{Property 2.}\\
&= \frac{16x^5y^8}{-8x^6y^5} && \text{Property 1.}\\
&= \frac{-2y^3}{x} && \text{Property 4.}
\end{aligned}$$

When $x = 3.2$ and $y = -1.7$,

$$\frac{-2y^3}{x} = \frac{-2(-1.7)^3}{3.2} = 3.07$$

It is easier to evaluate $\dfrac{-2y^3}{x}$ than $\dfrac{(2xy^2)^4x}{(-2x^2y)^3y^2}$.

Defining $\sqrt[n]{x}$ and $x^{\frac{1}{n}}$

We now extend the definition of integral exponents to rational exponents such as $\left(\frac{2}{5}\right)^{\frac{1}{4}}$ and $\left(\frac{2}{5}\right)^{\frac{2}{3}}$, which arose in Example 1. Not only do we want a definition for rational exponents that will allow us to calculate the above numbers, but we want one that will also satisfy the same properties that integral exponents satisfy. For example, consider the following examples where we apply the property $x^m \cdot x^n = x^{m+n}$:

$$\left.\begin{aligned} 9^1 \cdot 9^1 &= 9^{1+1} = 9^2 = 81 \\ 9^{\frac{1}{2}} \cdot 9^{\frac{1}{2}} &= 9^{\frac{1}{2}+\frac{1}{2}} = 9^1 = 9 \\ 9^0 \cdot 9^0 &= 9^{0+0} = 9^0 = 1 \end{aligned}\right\} \begin{matrix} \text{Decreases by a factor of } \frac{1}{9}. \\ \text{Decreases by a factor of } \frac{1}{9}. \end{matrix}$$

If this property is to hold true for all rational exponents, then this pattern suggests a way of defining $9^{\frac{1}{2}}$. If $9^{\frac{1}{2}} \cdot 9^{\frac{1}{2}} = 9^1$, then $9^{\frac{1}{2}} = 3$, since $3 \cdot 3 = 9$. But we already know that $\sqrt{9} = 3$. Thus the claim $9^{\frac{1}{2}} = \sqrt{9} = 3$ is consistent with $9^{\frac{1}{2}} \cdot 9^{\frac{1}{2}} = \sqrt{9}\,\sqrt{9} = 3 \cdot 3 = 9$. Therefore we can define $9^{\frac{1}{2}}$ as $\sqrt{9}$; that is, $9^{\frac{1}{2}}$ is the same as "the square root of 9." Similarly, $16^{\frac{1}{2}} = \sqrt{16} = 4$, and $2^{\frac{1}{2}} = \sqrt{2} \approx 1.414$. So far so good. But what does $8^{\frac{1}{3}}$ mean?

To be consistent, $8^{\frac{1}{3}} \cdot 8^{\frac{1}{3}} \cdot 8^{\frac{1}{3}} = 8$, so that $8^{\frac{1}{3}}$ would represent "the cube root of 8," which equals 2, since $2 \times 2 \times 2 = 8$. Similarly, $16^{\frac{1}{4}}$ would represent "the fourth root of 16," which equals 2, since $2 \times 2 \times 2 \times 2 = 16$.

We used radical notation in Chapter 6 to indicate square roots. To emphasize "square root" in the symbol $\sqrt{x}$, we can write $\sqrt[2]{x}$ (convention dictates we leave out the "2" and write $\sqrt{x}$). Thus, to be consistent,

If $\sqrt[2]{x}$ means a number whose second power (square) is x,
then $\sqrt[3]{x}$ means a number whose third power (cube) is x,
$\sqrt[4]{x}$ means a number whose fourth power is x,
and in general, $\sqrt[n]{x}$ means a number whose nth power is x.

For example,

$$\sqrt[3]{8} = 2 \quad \text{since} \quad 2^3 = 8$$
$$\sqrt[4]{81} = 3 \quad \text{since} \quad 3^4 = 81$$
$$\sqrt[5]{32} = 2 \quad \text{since} \quad 2^5 = 32$$

Furthermore, we have a **connection between the *n*th roots and rational exponents,** namely, it makes sense to think about $4^{\frac{1}{5}}$ as $\sqrt[5]{4}$.

In general, we have the following definition.

Definition of *n*th Roots

Given a real number x and a positive integer n greater than 1, the real number b is an ***n*th root** of x if

$$b^n = x.$$

The nth root b is denoted by $\sqrt[n]{x}$, or by $x^{\frac{1}{n}}$.

In $\sqrt[n]{x}$, n is the **index,** x is the **radicand,** and $\sqrt{}$ is the **radical** (symbol).

NOTE: When n is even, there may be more than one real-valued nth root of a number. Since both $2 \times 2 \times 2 \times 2 = 16$ and $(-2) \times (-2) \times (-2) \times (-2) = 16$, both 2 and -2 are fourth roots of 16. Thus, *if x is positive and n is even,* we say that

the **principal *n*th root** of x is b if $\sqrt[n]{x} = b$ and $b > 0$. ■

When n is even, we will use the symbol "$\sqrt{}$" to stand for the principal nth root and the symbol "$-\sqrt{}$" to stand for the other nth root whenever there are two real roots. In the case where x is negative, we have either one real root or no real roots. For example, $\sqrt[3]{-8} = (-8)^{\frac{1}{3}} = -2$, since $(-2) \times (-2) \times (-2) = -8$. However, $\sqrt[4]{-8} = (-8)^{\frac{1}{4}}$ is not a real number, since there is no *real* number that multiplied by itself four times yields -8. (A real number multiplied by itself four times must be positive.)

The symbol $x^{\frac{1}{n}}$ is read as "x (raised) to the $\frac{1}{n}$th power." We should keep in mind, however, that **$\sqrt[n]{x}$ and $x^{\frac{1}{n}}$ are just different names for the same concept; they both mean "find the *n*th root of *x*."**

EXAMPLE 4

Find the indicated real nth roots, if they exist.

(a) $\sqrt[4]{625}$ **(b)** $\sqrt[5]{243}$ **(c)** $\sqrt[6]{64}$ **(d)** $\sqrt{-16}$

(e) $\sqrt[5]{-243}$ **(f)** $-\sqrt{16}$ **(g)** $25^{\frac{1}{2}}$ **(h)** $\left(\frac{243}{32}\right)^{\frac{1}{5}}$

(i) $(-64)^{\frac{1}{6}}$ **(j)** $-64^{\frac{1}{6}}$ **(k)** $(-243)^{\frac{1}{5}}$ **(l)** $(2^4)^{\frac{1}{2}}$

Solutions

(a) $\sqrt[4]{625} = \sqrt[4]{5^4} = 5,$ since $5^4 = 625$.

(b) $\sqrt[5]{243} = \sqrt[5]{3^5} = 3,$ since $3^5 = 243$.

(c) $\sqrt[6]{64} = \sqrt[6]{2^6} = 2,$ since $2^6 = 64$.

(d) $\sqrt{-16}$ does not have a *real* square root.

Of course, from Chapter 6, we can write $\sqrt{-16} = 4i$, which is an imaginary (or complex) number, but we have agreed to consider only *real nth roots* in this chapter.

(e) $\sqrt[5]{-243} = \sqrt[5]{(-3)^5}$

$= -3,$ since $(-3)^5 = -243$.

(f) $-\sqrt{16} = -(\sqrt{16})$

$= -(4)$

$= -4$ Notice the difference between Examples (d) and (f).

(g) $25^{\frac{1}{2}} = \sqrt{25} = 5$ *Check:* $5^2 = 25$

(h) $\left(\dfrac{243}{32}\right)^{\frac{1}{5}} = \sqrt[5]{\dfrac{243}{32}} = \dfrac{3}{2}$ *Check:* $\left(\dfrac{3}{2}\right)^5 = \dfrac{243}{32}$

(i) $(-64)^{\frac{1}{6}} = \sqrt[6]{-64}$ Does not exist. Even roots of negative numbers are not real numbers.

(j) $-64^{\frac{1}{6}} = -(64^{\frac{1}{6}})$
$= -\sqrt[6]{64}$
$= -2$

CAUTION: Recall from Chapter 2 that the exponent is attached only to the number, unless the sign appears in parentheses, as in the next example.

Check: $-2^6 = -64$

(k) $(-243)^{\frac{1}{5}} = \sqrt[5]{-243} = -3$ *Check:* $(-3)^5 = -243$

(l) $(2^4)^{\frac{1}{2}} = \sqrt{2^4} = \sqrt{(2^2)^2} = 2^2$
$= 4$ *Check:* $(2^2)^2 = 2^4$ ■

EXAMPLE 5 Use your calculator to find the indicated real nth roots, if they exist. Express the answer to three decimal places.

(a) $\sqrt{5}$ **(b)** $\sqrt[4]{-10}$ **(c)** $\sqrt[5]{-30}$ **(d)** $-\sqrt[3]{-15}$
(e) $-\sqrt[6]{60}$

Solutions

(a) $\sqrt{5} \approx 2.236$ 2.236 is an approximation to the real number $\sqrt{5}$; $\sqrt{5}$ is a real number whose square is exactly 5.

(b) $\sqrt[4]{-10}$ does not exist. Notice that the index is even and the radicand is negative.

(c) $\sqrt[5]{-30} = (-30)^{\frac{1}{5}} = (-30)^{0.2} \approx -1.974$ On a calculator $(-30)^{0.2}$ may be expressed as (-30) [y^x] .2 or perhaps as (-30) [∧] 0.2.

(d) $-\sqrt[3]{-15} = -(\sqrt[3]{-15})$
$\approx -(-2.466)$
$\approx +2.466$

Be careful of the negative sign in front of the radical. Isolate it with parentheses. We now proceed as in part (c).

(e) $-\sqrt[6]{60} = -(\sqrt[6]{60}) = -(60)^{\frac{1}{6}}$ Isolate the negative sign in front of the radical.
$\approx -(1.979)$
≈ -1.979 ■

We can extend the process of finding nth roots to algebraic expressions. To avoid having to discuss cases for each variable, we will require *all variables to be positive.*

EXAMPLE 6 Find the indicated nth roots. Assume the values of all variables are positive.

(a) $\sqrt{y^4}$ **(b)** $\sqrt[3]{z^9}$ **(c)** $\sqrt[3]{-27x^9y^{12}}$
(d) $\sqrt[6]{64x^6y^{12}}$ **(e)** $\sqrt{x^2 + 6x + 9}$

Solutions

(a) $\sqrt{y^4} = \sqrt{(y^2)^2} = y^2$ since $(y^2)^2 = y^4$ Recall the property for integral exponents, $(x^m)^n = x^{mn}$.

(b) $\sqrt[3]{z^9} = \sqrt[3]{(z^3)^3} = z^3$ since $(z^3)^3 = z^9$

(c) $\sqrt[3]{-27x^9y^{12}} = \sqrt[3]{(-3x^3y^4)^3} = -3x^3y^4$ Since $(-3x^3y^4)^3 = -27x^9y^{12}$.

(d) $\sqrt[6]{64x^6y^{12}} = \sqrt[6]{(2xy^2)^6} = 2xy^2$ Since $(2xy^2)^6 = 64x^6y^{12}$.

(e) $\sqrt{x^2 + 6x + 9} = \sqrt{(x + 3)^2} = (x + 3)$ Since $(x + 3)^2 = x^2 + 6x + 9$. ■

EXAMPLE 7

Write each of the following as an equivalent expression in radical form, and find the indicated root. Assume all variables represent positive quantities.

(a) $(x^6)^{\frac{1}{3}}$ **(b)** $(-32x^{10})^{\frac{1}{5}}$ **(c)** $(125x^3y^6)^{\frac{1}{3}}$

Solutions

(a) $(x^6)^{\frac{1}{3}} = \sqrt[3]{x^6} = \sqrt[3]{(x^2)^3} = x^2$

(b) $(-32x^{10})^{\frac{1}{5}} = \sqrt[5]{-32x^{10}} = \sqrt[5]{(-2x^2)^5} = -2x^2$

(c) $(125x^3y^6)^{\frac{1}{3}} = \sqrt[3]{125x^3y^6} = \sqrt[3]{(5xy^2)^3} = 5xy^2$ ■

We are now in a position to solve the applications raised in Examples 1 and 2 of this chapter.

EXAMPLE 8

Suppose that in a certain lake the intensity of light is reduced by $\frac{3}{5}$ for each meter of water. What is the intensity of light at a depth of $\frac{1}{4}$ meter? At $\frac{1}{10}$ meter? Does the light intensity increase or decrease as we go deeper in the water? Graph the relationship to defend your answer.

Solution

Using an Equation From Example 1, we know that $L(n) = \left(\frac{2}{5}\right)^n$, where $L(n)$ is the intensity of light at n meters. By using a calculator, we find that at $n = \frac{1}{4}$,

$$L(n) = \left(\frac{2}{5}\right)^{\frac{1}{4}} = (0.4)^{\frac{1}{4}} \approx 0.795,$$

or about 80% of the initial light. Similarly, at $n = \frac{1}{10}$,

$$L(n) = \left(\frac{2}{5}\right)^{\frac{1}{10}} = (0.4)^{0.1} = 0.912,$$

or about 91% of the initial light.

Using a Graph The graph below shows that light intensity decreases as water depth increases, as we might expect.

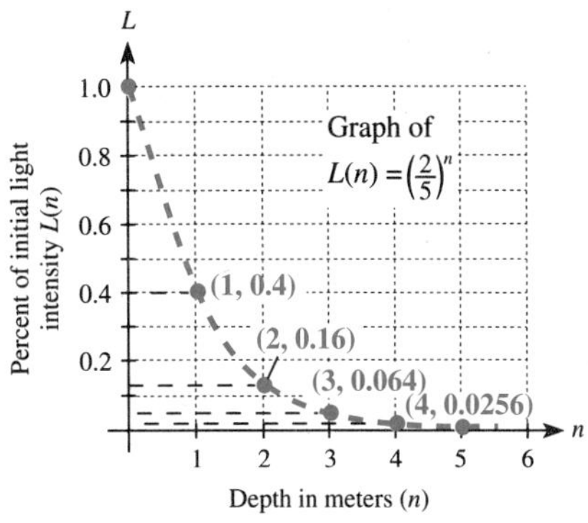

Figure 7.3

So far, the only rational exponents we have investigated are positive "unit" rationals of the form $\frac{1}{n}$ when n is a positive integer. We also raised the question of how much light there is at $\frac{2}{3}$ of a meter; that is, what is $\left(\frac{2}{5}\right)^{\frac{2}{3}}$? This question will be answered in the next section, in which we discuss the meaning of general rational exponents of the form $\frac{m}{n}$ for any integers n and m ($n \neq 0$).

EXAMPLE 9

In 1975, the earth's population was four billion and thought to be doubling every 35 years. Given this model, estimate the population in the year 1982. What will the population be in the year 2050?

Solution

According to the discussion in Example 2, we can estimate the population in year x by checking for the number of 35-year intervals that have elapsed. The expression $\frac{x-1975}{35}$ will give the number of 35-year intervals. Thus, an estimate of population in year x will be

$$P = 4 \cdot 2^{\frac{x-1975}{35}}$$

Thus, the earth's population in the year 1982 was approximately

$$P = 4 \cdot 2^{\frac{1982-1975}{35}} = 4 \cdot 2^{\frac{7}{35}} = 4 \cdot 2^{\frac{1}{5}} \approx 4.59 \text{ billion}$$

The population in 2050 will be approximately

$$P = 4 \cdot 2^{\frac{2050-1975}{35}} = 4 \cdot 2^{\frac{75}{35}} = 4 \cdot 2^{\frac{15}{7}}.$$

We will discuss and calculate $2^{\frac{15}{7}}$ in the next section.

EXAMPLE 10

(a) Which do you think is larger, $\left(\frac{1}{2}\right)^5$ or $\left(\frac{1}{2}\right)^6$? Make a guess.

(b) Which do you think is larger, $\left(\frac{1}{2}\right)^{\frac{1}{5}}$ or $\left(\frac{1}{2}\right)^{\frac{1}{6}}$? Make a guess.

(c) Graph $y = \left(\frac{1}{2}\right)^x$ and use the graph to estimate the answers to parts (a) and (b). Make a generalization.

Solution

(a) $\left(\frac{1}{2}\right)^5 = (0.5)^5 = 0.031$

$\left(\frac{1}{2}\right)^6 = (0.5)^6 = 0.016$ So $\left(\frac{1}{2}\right)^5 > \left(\frac{1}{2}\right)^6$.

(b) $\left(\frac{1}{2}\right)^{\frac{1}{5}} = (0.5)^{0.2} = 0.87$

$\left(\frac{1}{2}\right)^{\frac{1}{6}} = (0.5)^{0.166} = 0.89$ So $\left(\frac{1}{2}\right)^{\frac{1}{5}} < \left(\frac{1}{2}\right)^{\frac{1}{6}}$.

(c)

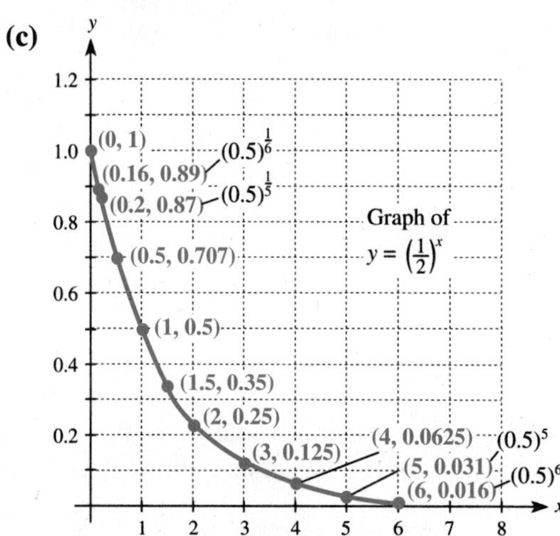

Figure 7.4

As x increases, $\left(\frac{1}{2}\right)^x$ decreases. ■

EXAMPLE 11

(a) Graph $y = x^2$ and $y = \sqrt{x}$, $x \geq 0$, on the same coordinate axes. What do you notice?

(b) Graph $y = x^3$ and $y = \sqrt[3]{x}$ on the same coordinate axes. What do you notice?

Solutions

(a)

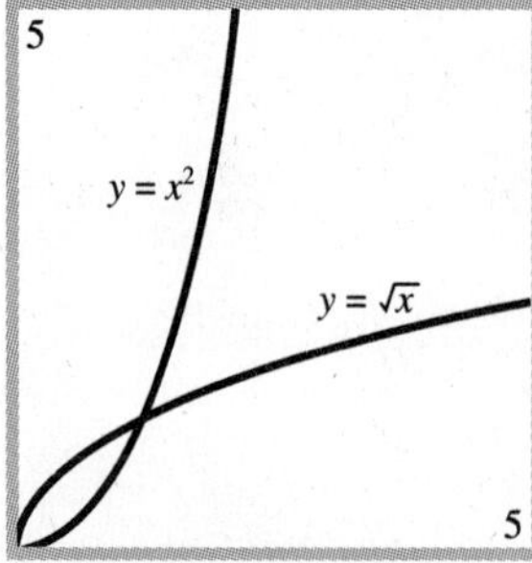

Figure 7.5 (a)

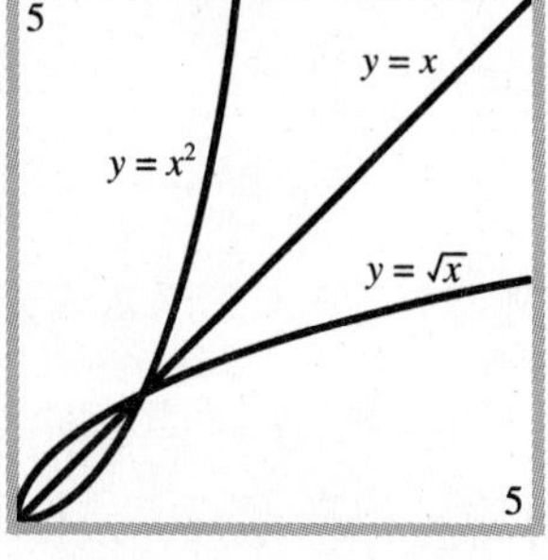

Figure 7.5 (b)

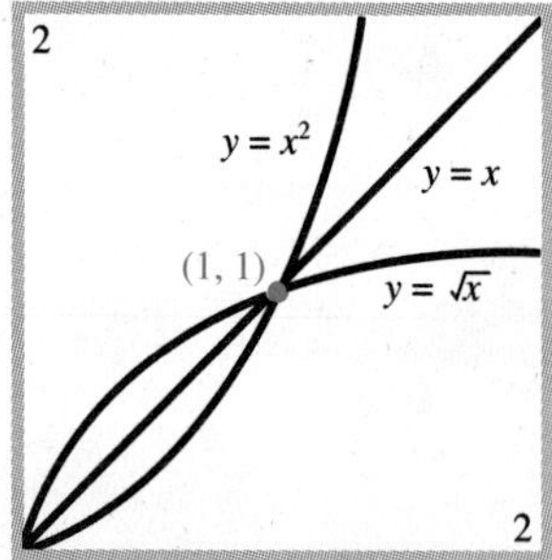

Figure 7.5 (c)

$\sqrt{x}$ is defined only when $x \geq 0$. For $0 \leq x \leq 5$, the graphs appear to be symmetric with respect to the line $y = x$. We draw this line in Figure 7.5(b) and check our conjecture by "zooming in" to the window $[0 \leq x \leq 2, 0 \leq y \leq 2]$ in Figure 7.5(c).

(b)

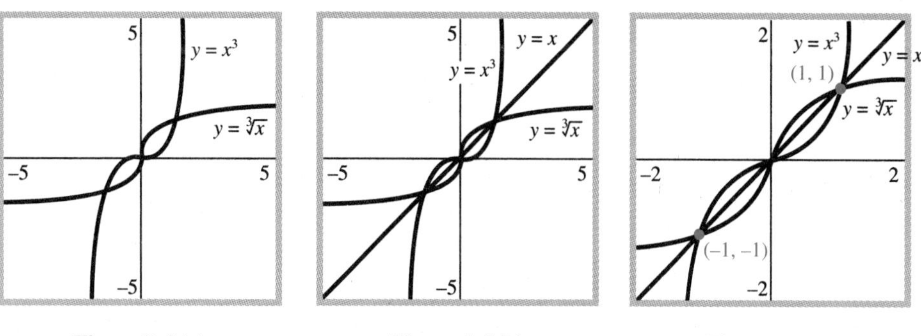

Figure 7.6 (a) **Figure 7.6 (b)** **Figure 7.6 (c)**

Both x^3 and $\sqrt[3]{x}$ are defined for all real x. Again the graphs appear symmetric to the line $y = x$ when $0 \leq x \leq 5$. We draw this line in Figure 7.6(b) and "zoom in" to the window $[-2 \leq x \leq 2, -2 \leq y \leq 2]$ in Figure 7.6(c) to confirm our conjecture.

The Distance Formula

EXAMPLE 12

One important use of square roots occurs in the distance formula. Find the distance between the points $(-1, 2)$ and $(4, 5)$.

Solution

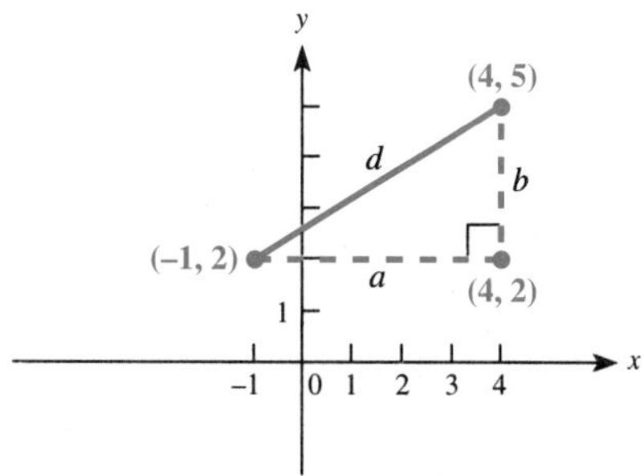

Figure 7.7

The desired distance is d, which is the hypotenuse of the right triangle whose legs are $a = 5 = [4 - (-1)]$ and $b = 3 = (5 - 2)$. By the Pythagorean theorem,

$$d^2 = a^2 + b^2$$

so that

$$d = \sqrt{a^2 + b^2} = \sqrt{5^2 + 3^2} = \sqrt{34} \approx 5.83$$

We can generalize the result of Example 12. Figure 7.8 shows two different points (x_1, y_1) and (x_2, y_2):

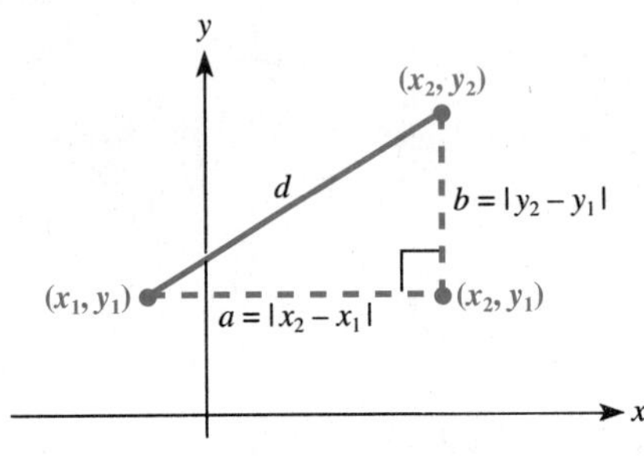

Figure 7.8

Now $a = |x_2 - x_1|$ and $b = |y_2 - y_1|$. Again using the Pythagorean theorem, we have

$$d = \sqrt{(x_2 - x_1)^2 + (y_2 - y_1)^2}$$

since $|x_2 - x_1|^2 = (x_2 - x_1)^2$ and $|y_2 - y_1|^2 = (y_2 - y_1)^2$. This relationship is called the **distance formula.**

EXAMPLE 13

(a) Find the distance between $(-3, -2)$ and $(6, -4)$.

(b) Show that the points $A\ (1, 3)$, $B\ (3, -3)$, and $C\ (14, 4)$ are the vertices of an isosceles triangle.

Solutions

(a) Using the distance formula,

$$d = \sqrt{(-3 - 6)^2 + [-2 - (-4)]^2} = \sqrt{(-9)^2 + 2^2} = \sqrt{85} \approx 9.22$$

(b) The graph on the next page suggests $AC = BC$.

$$AC = \sqrt{(14 - 1)^2 + (4 - 3)^2} = \sqrt{169 + 1} = \sqrt{170}$$

$$BC = \sqrt{(14 - 3)^2 + [4 - (-3)]^2} = \sqrt{121 + 49} = \sqrt{170}$$

Triangle ABC is indeed isosceles.

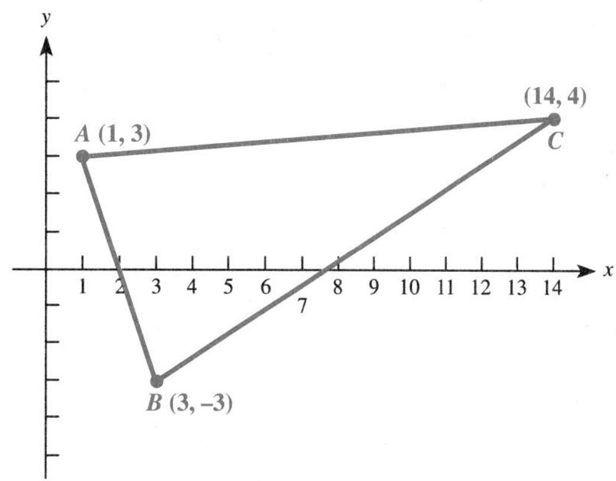

Figure 7.9

7.1 EXERCISES

I. Procedures

1–16. Evaluate each of the following, if possible, by finding the indicated real root. State any properties you used to find the answer. Try to do these mentally or with pencil and paper.

1. $\sqrt{121}$ **2.** $\sqrt[3]{3.375}$

3. $\sqrt[4]{81}$ **4.** $\sqrt[5]{-1{,}024}$

5. $\sqrt[6]{-64}$ **6.** $\sqrt[5]{0.00001}$

7. $\sqrt[5]{-3{,}200{,}000}$ **8.** $\sqrt[4]{\frac{81}{16}}$

9. $16^{\frac{1}{2}}$ **10.** $\left(\frac{-27}{8}\right)^{\frac{1}{3}}$

11. $(2.0736)^{\frac{1}{4}}$ **12.** $(-1)^{\frac{1}{3}}$

13. $(10^{10})^{\frac{1}{5}}$ **14.** $(-1)^{\frac{1}{4}}$

15. $(-1)^{\frac{1}{n}}$ if n is odd **16.** $(-1)^{\frac{1}{n}}$ if n is even

17–22. Use your calculator to approximate each of the following expressions to three decimal places.

17. $\sqrt{39}$ **18.** $1.09^{\frac{1}{2}}$ **19.** $137^{\frac{1}{3}}$

20. $\sqrt[4]{569}$ **21.** $(-753)^{\frac{1}{5}}$ **22.** $-120^{\frac{1}{2}}$

23–34. Write each of the following as an equivalent expression in radical form, and find the indicated root. Assume all variables are positive numbers.

23. $49^{\frac{1}{2}}$ **24.** $(-8)^{\frac{1}{3}}$ **25.** $(25x^2)^{\frac{1}{2}}$

26. $(-125x^6)^{\frac{1}{3}}$ **27.** $(27x^3y^{12})^{\frac{1}{3}}$ **28.** $(-32x^{10})^{\frac{1}{5}}$

29. $-32(x^{10})^{\frac{1}{5}}$ **30.** $(64y^{12})^{\frac{1}{6}}$ **31.** $(64y^{12})^{\frac{1}{3}}$

32. $(64y^{12})^{\frac{1}{2}}$ **33.** $(x^2 + 4x + 4)^{\frac{1}{2}}$ **34.** $[(x + 3)^3]^{\frac{1}{3}}$

35–40. Find the distances between the following pairs of points. Plot the points.

35. (4, 1) and (12, −7)

36. (−9, 6) and (7, −4)

37. (0, 3) and (8, 5)

38. $\left(\frac{1}{2}, \frac{3}{2}\right)$ and $\left(5, -\frac{1}{2}\right)$

39. (3.27, −4.16) and (−10.87, 13.49)

40. (−13.58, π) and (14.42, π)

II. Concepts

41 and 42. Decide whether the three points are vertices of a right triangle.

41. (1, −2), (−1, 2), and (5, 10)

42. (−6, 1), (−2, 9), and (4, −4)

43. **(a)** Find the number whose third root is −2.

(b) Find the number whose principal fourth root is 5.

(c) Find the number whose fifth root is 10.

(d) Find the number whose eleventh root is -1.

44. If $\sqrt[n]{x} = b$, then under what conditions, if any,

(a) can $x = 0$? **(b)** can $b = 0$? **(c)** can $n = 0$?

45. If $(-6)^2 = 36$, then why isn't $\sqrt{(-6)^2} = -6$?

46. Find the coordinates of four points

(a) whose distance from the point (1, 2) is $\sqrt{29}$ units.

(b) that are equidistant from the points (1, 2) and (3, -2).

47. Graph $y = \sqrt{x}$, $y = \sqrt[3]{x}$, $y = \sqrt[4]{x}$, and $y = \sqrt[5]{x}$ on the same set of axes. What do you notice about this family of graphs? (Graphing calculators are helpful for this problem.) Pay special attention to the region, $-1 \le x \le 1$.

48. For what values of n is $\left(\frac{1}{2}\right)^n > \left(\frac{1}{2}\right)^{n+1}$? Explain your reasoning.

49. For what values of n is $\left(\frac{1}{2}\right)^{\frac{1}{n}} > \left(\frac{1}{2}\right)^{\frac{1}{n+1}}$? Explain your reasoning.

III. Applications

50–60. Solve each of the following problems.

50. Estimate the earth's population in 1980 using the formula in Examples 2 and 9.

51. Find the amount of light $\frac{1}{3}$ of a meter below the surface of a lake, if the intensity of light is reduced by $\frac{3}{7}$ for each meter of water. (See Examples 1 and 8.)

52. A certain kind of glass, 1 cm thick, lets in 80% of the light.

(a) How much light will go through a glass that is n cm thick?

(b) How much light will go through a glass that is 2 cm thick? 3 cm thick? 4 cm thick?

(c) How much light will go through a glass $\frac{1}{2}$ cm thick? $\frac{1}{3}$ cm thick? $\frac{1}{4}$ cm thick?

53. A sheet of translucent plastic 2 cm thick allows 60% of the light to come through.

(a) If a sheet of plastic is 6 cm thick, what percent of light passes through?

(b) By how much is the light reduced if the thickness of the plastic is 3 cm, $\frac{1}{2}$ cm, or $\frac{1}{4}$ cm?

54. Suppose there is an initial population of B_0 bacteria and the number of bacteria doubles every hour. The number of bacteria, B, present after t hours, then, is given by $B = B_0 2^t$.

(a) What is the population after three hours if the initial population is 5,000?

(b) What is the population after one-fourth hour if the initial population is 5,000?

(c) Answer (a) and (b) if the initial population is 10,000.

55. Police sometimes use the formula $s = \sqrt{30fd}$ to estimate the speed s (in miles per hour) of a car that skidded d feet upon braking. The variable f is the coefficient of friction determined by the kind of road and the wetness or dryness of the road. The following table gives some values of f:

	Concrete	*Tar*
Wet	0.4	0.5
Dry	0.8	1.0

(a) How fast was a car going on a concrete road if the skid marks on a rainy day were 80 feet long?

(b) How fast was a car going on a dry tar road if the skid marks were 100 feet long?

56. The pulse rate p (in beats per minute) of an adult H meters tall is

$$p = \frac{94}{H^{\frac{1}{2}}}$$

(a) Graph this relationship.

(b) What is the pulse rate of an adult who is 190 centimeters (1.9 m) tall?

(c) Measure yourself in meters to the nearest centimeter. Use the formula to estimate your pulse rate. Now, take your own pulse. How do they compare?

(d) Can you think of any assumptions that are being made about the adult's pulse rate when this formula is used to estimate pulse?

57. The power P (in watts) generated by a windmill is related to the velocity v (in miles per hour) of the wind by the formula:

$$\sqrt[3]{\frac{P}{0.015}} = v$$

(a) Graph this relationship.

(b) Find the wind speed that is necessary to generate 120 watts of power.

(c) Find the wind speed that is necessary to generate 220 watts of power.

58. The initial velocity v of an object projected h meters high is $v = \sqrt{64h}$, assuming there is no air resistance.

(a) Graph this relationship.

(b) What is the velocity needed to shoot an object 625 meters high?

(c) What is the velocity needed to project a missile 1,500 meters high?

59. A quantity of a radioactive substance will decay at a rate that is proportional to the amount present at a given time. The general formula for *decay* is

$$A = A_0 2^{\frac{-t}{H}} \quad \text{or} \quad A = A_0\left(\frac{1}{2}\right)^{\frac{t}{H}}$$

where

A_0 is the amount present initially (at $t = 0$)

A is the amount remaining after time t

H is the half-life, or the time it takes for half of the initial amount to decay

The half-life of radium is 1,600 years. Thus $A = A_0\left(\frac{1}{2}\right)^{\frac{t}{1600}}$. If one starts with 12 grams of radium, how much radium is left after

(a) 40 years? **(b)** 100 years? **(c)** 800 years?

(d) 1,600 years? **(e)** 3,200 years?

60. (Note: This problem is from a Delta Airlines system map.) How far can you see from a Delta jet? Your view depends on the curvature of the earth and also on height above the horizon. If A is your altitude in feet, and V_m is the distance you can see in miles to the horizon (vision in miles) from a plane at A feet high, then Delta says that

$$V_m = 1.22\sqrt{A}.$$

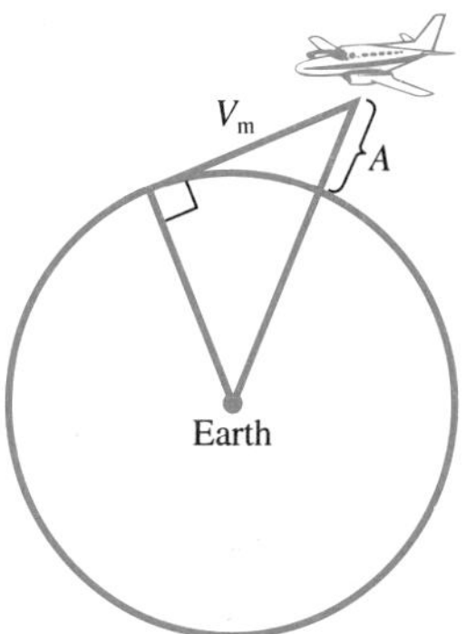

(a) Graph this relationship.

(b) Use this formula to find the distance to the horizon if your altitude is 5,000 feet, 20,000 feet, and 29,000 feet. (Note: This last height is close to the height of Mt. Everest.)

(c) During 1962, astronaut Alan Shepard took a suborbital path about 100 miles above the earth. How far could he see to the horizon?

Of course, the interesting question is where this formula came from. See Exercise 64 for further discussion.

IV. Extensions

61. How could you find fourth roots on your calculator if you had only the "$\sqrt{\ }$" button?

62. (a) Use a calculator to estimate each of the following to four decimal places: $2^{\frac{1}{2}}, 2^{\frac{1}{3}}, 2^{\frac{1}{4}}, 2^{\frac{1}{5}}, 2^{\frac{1}{10}}, 2^{\frac{1}{100}}, \ldots$

(b) What can you say about $2^{\frac{1}{n}}$, if n is very large?

63. Ramanujan, an Indian mathematician who, in spite of a limited formal education, had a phenomenal talent for discovering profound formulas, gave the following curious approximation to π:

$$\pi = \sqrt{\sqrt{\frac{2{,}143}{22}}}$$

See if you can find another such curious approximation of π that agrees to four decimal places.

64. Derive the formula $V_m = 1.22\sqrt{A}$ of Exercise 60. You will need to find an estimate of the radius of the earth.

7.2 OPERATIONS ON RATIONAL EXPONENTS

Rational Exponents

We now proceed to answer a question raised in Example 1 of Section 7.1, that is, what does $\left(\frac{2}{5}\right)^{\frac{2}{3}}$ mean? Recall that our definition of $x^{\frac{1}{n}}$ was motivated in part by wishing to have properties of integral exponents hold true for rational exponents. For example, one of these properties was $(x^m)^n = x^{mn}$. If it is to hold true for rational exponents, then

$$
\begin{aligned}
8^{\frac{2}{3}} &= 8^{\frac{1}{3}(2)} \\
&= (8^{\frac{1}{3}})^2 \\
&= (2)^2 \\
&= 4
\end{aligned}
\quad \text{or} \quad
\begin{aligned}
8^{\frac{2}{3}} &= 8^{\left(\frac{1}{3}\right)(2)} \\
&= (\sqrt[3]{8})^2 \\
&= (2)^2 \\
&= 4
\end{aligned}
$$

Also

$$16^{\frac{10}{4}} = (16^{\frac{1}{4}})^{10} = (2)^{10} = 1{,}024$$
$$16^{\frac{5}{2}} = (16^{\frac{1}{2}})^5 = (4)^5 = 1{,}024$$

Notice!

Thus, we may as well consider fractional exponents in simplest form.

Now we define rational exponents in general.

Definition

If $\frac{m}{n}$ is a rational number in simplest form, then

$$x^{\frac{m}{n}} = (x^{\frac{1}{n}})^m = (\sqrt[n]{x})^m$$

provided $\sqrt[n]{x}$ is a real number.

NOTE: Even though $x^{\frac{m}{n}}$ is sometimes defined when x is negative, namely, when n is odd, we will continue to assume all variables represent positive numbers. ■

Properties of Rational Exponents

Since both $x^{\frac{1}{n}}$ and $x^{\frac{m}{n}}$ were defined so that the properties of integral exponents would remain valid, it comes as no surprise that all properties of exponents and definitions are valid for rational exponents, both positive and negative.

Properties of Rational Exponents

For any real numbers $x \neq 0$, $y \neq 0$, and any *rational* numbers a and b in simplest form:

1. $x^a \cdot x^b = x^{a+b}$
2. $(x^a)^b = x^{ab}$
3. $(xy)^a = x^a y^a$
4. $\frac{x^a}{x^b} = x^{a-b}$

and $x^{-b} = \frac{1}{x^b}$ (by definition)
whenever all quantities represent real numbers.

EXAMPLE 1

Compute.

(a) $32^{\frac{3}{5}}$ **(b)** $(-32)^{\frac{3}{5}}$ **(c)** $-32^{\frac{3}{5}}$ **(d)** $32^{-\frac{3}{5}}$

(e) $-32^{-\frac{3}{5}}$ **(f)** $16^{\frac{2}{3}}$ **(g)** $16^{\frac{3}{2}}$ **(h)** $16^{-\frac{2}{3}}$

(i) $(1.085)^{\frac{7}{4}}$ **(j)** $\left(\frac{2}{5}\right)^{-\frac{2}{5}}$

Solutions

(a) $32^{\frac{3}{5}} = (32^{\frac{1}{5}})^3$ — Property 2.

$= 2^3$ — Definition of $x^{\frac{1}{n}}$.

$= 8$

(b) $(-32)^{\frac{3}{5}} = [(-32)^{\frac{1}{5}}]^3$ — Property 2. Study very carefully the various uses of the negative sign.

$= (-2)^3$ — Definition of $x^{\frac{1}{n}}$.

$= -8$

(c) $-32^{\frac{3}{5}} = -(32^{\frac{1}{5}})^3$ — Property 2.

$= -(2)^3$ — Definition of $x^{\frac{1}{n}}$.

$= -8$

(d) $32^{-\frac{3}{5}} = \frac{1}{32^{\frac{3}{5}}}$ — Definition of x^{-n}.

$= \frac{1}{8}$ — Part (a)

(e) $-32^{-\frac{3}{5}} = \frac{-1}{32^{\frac{3}{5}}}$ — Definition of x^{-n}. We could also write $-(32^{\frac{1}{5}})^{-3} = -(2)^{-3} = \frac{-1}{2^3}$.

$= -\frac{1}{8}$

(f) $16^{\frac{2}{3}} = (16^{\frac{1}{3}})^2 = (2.52)^2 = 6.35$ — Property 2.

(g) $16^{\frac{3}{2}} = (16^{\frac{1}{2}})^3 = 4^3 = 64$ — Property 2.

(h) $16^{-\frac{2}{3}} = \frac{1}{16^{\frac{2}{3}}} = \frac{1}{6.35} = 0.157$ — Definition of x^{-n} and part (f).

(i) $1.085^{\frac{7}{4}} = 1.085^{1.75} = 1.153$ — Rewrite $\frac{7}{4}$ as 1.75. A calculator helps.

(j) $\left(\frac{2}{5}\right)^{-\frac{2}{5}} = \frac{1}{\left(\frac{2}{5}\right)^{\frac{2}{5}}} = \frac{1}{0.4^{0.4}} = \frac{1}{0.693} = 1.442$ — Definition of x^{-n} and calculator. ■

EXAMPLE 2

Write each of the following as an equivalent expression in radical form. Assume $x > 0$.

(a) $x^{\frac{2}{3}}$ **(b)** $x^{\frac{-2}{3}}$ **(c)** $x^{\frac{2}{-3}}$

Solutions

(a) $x^{\frac{2}{3}} = (x^{\frac{1}{3}})^2$ — Property 2.

$= (\sqrt[3]{x})^2$

(b) $x^{\frac{-2}{3}} = (x^{\frac{1}{3}})^{-2}$

$= (\sqrt[3]{x})^{-2}$

$= \frac{1}{(\sqrt[3]{x})^2}$ — Recall that $x^{-n} = \frac{1}{x^n}$.

(c) $x^{\frac{2}{-3}} = x^{\frac{-2}{3}}$ — $\frac{2}{-3} = \frac{-2}{3}$ or $-\frac{2}{3}$

$= \frac{1}{(\sqrt[3]{x})^2}$ — Same as (b). ■

EXAMPLE 3

Write each as an equivalent expression using a rational exponent: $x > 0$.

(a) $(\sqrt{x})^3$ **(b)** $\frac{1}{(\sqrt[5]{x})^3}$ **(c)** $(\sqrt[3]{x})^6$ **(d)** $\sqrt[4]{\sqrt{x}}$

Solutions

(a) $(\sqrt{x})^3 = (x^{\frac{1}{2}})^3$ Definition of $x^{\frac{1}{2}}$.

$= x^{\frac{3}{2}}$ Property 2.

(b) $\dfrac{1}{(\sqrt[5]{x})^3} = (\sqrt[5]{x})^{-3}$ $\dfrac{1}{x^n} = x^{-n}$

$= (x^{\frac{1}{5}})^{-3}$

$= x^{\frac{-3}{5}}$

(c) $(\sqrt[3]{x})^6 = (x^{\frac{1}{3}})^6$ Definition of $x^{\frac{1}{3}}$.

$= x^{\frac{6}{3}}$ Property 2.

$= x^2$ $\frac{6}{3} = 2$

(d) $\sqrt[4]{\sqrt{x}} = (\sqrt{x})^{\frac{1}{4}} = (x^{\frac{1}{2}})^{\frac{1}{4}} = x^{\frac{1}{8}}$ Property 2. ■

EXAMPLE 4

Perform the operations and write each as an equivalent expression with a positive exponent. Assume x is a positive number.

(a) $x^{\frac{2}{3}} \cdot x^{\frac{1}{4}}$ **(b)** $x^{\frac{2}{3}} \cdot x^{-\frac{1}{4}}$ **(c)** $\dfrac{x^{\frac{2}{3}}}{x^{-\frac{1}{4}}}$ **(d)** $(x^{\frac{2}{3}})^{\frac{1}{4}}$

Solutions

(a) $x^{\frac{2}{3}} \cdot x^{\frac{1}{4}} = x^{\frac{2}{3}+\frac{1}{4}}$ Property 1.

$= x^{\frac{11}{12}}$

(b) $x^{\frac{2}{3}} \cdot x^{-\frac{1}{4}} = x^{\frac{2}{3}-\frac{1}{4}}$ Property 1.

$= x^{\frac{5}{12}}$

(c) $\dfrac{x^{\frac{2}{3}}}{x^{-\frac{1}{4}}} = x^{\frac{2}{3}-\left(-\frac{1}{4}\right)}$ Property 4. Same as (a)

$= x^{\frac{11}{12}}$

(d) $(x^{\frac{2}{3}})^{\frac{1}{4}} = x^{\frac{1}{6}}$ Property 2. ■

EXAMPLE 5

Assume all variables represent positive quantities. Use the properties of rational exponents to write an equivalent expression with positive exponents. Find the roots whenever possible. Evaluate each expression for $x = 3$, $y = 4$, and $z = 2$.

(a) $\left(\dfrac{8x^3}{y^6}\right)^{\frac{2}{3}}$ **(b)** $\left(\dfrac{-8x^2yz^{-2}}{27x^{-4}yz^2}\right)^{\frac{1}{3}}$

Solutions

(a) $\left(\dfrac{8x^3}{y^6}\right)^{\frac{2}{3}} = \dfrac{8^{\frac{2}{3}}x^{3\left(\frac{2}{3}\right)}}{y^{6\left(\frac{2}{3}\right)}}$ Property 3.

$= \dfrac{4x^2}{y^4}$ Property 2.

$= \dfrac{4 \cdot 3^2}{4^4} = \dfrac{36}{256} = 0.140$ When $x = 3$ and $y = 4$.

(b) $\left(\dfrac{-8x^2yz^{-2}}{27x^{-4}yz^2}\right)^{\frac{1}{3}} = \left(-\dfrac{8x^6z^{-4}}{27}\right)^{\frac{1}{3}}$ Express fraction within the parentheses in simplest form first. Use Property 4 to write $\dfrac{x^2}{x^{-4}} = x^6$ and $\dfrac{z^{-2}}{z^2} = z^{-4}$.

$= \dfrac{(-8)^{\frac{1}{3}}(x^6)^{\frac{1}{3}}(z^{-4})^{\frac{1}{3}}}{27^{\frac{1}{3}}}$ Property 3.

$= \dfrac{-2x^2z^{-\frac{4}{3}}}{3}$ Definition of $x^{\frac{1}{n}}$.

$= \dfrac{-2x^2}{3z^{\frac{4}{3}}}$ Definition of x^{-n}.

$= \dfrac{-2 \cdot 3^2}{3 \cdot 2^{\frac{4}{3}}}$ When $x = 3$ and $z = 2$.

$= \dfrac{-18}{3 \cdot 16^{\frac{1}{3}}}$

$= \dfrac{-18}{7.56}$

$= -2.38$ ■

Notice that it is often easier to evaluate these expressions *after* using the properties of exponents.

In many of our examples, we are expressing fractions in simplest form. Therefore, this is a good time to remind ourselves of the importance of distinguishing between factors and terms, as is shown in the following example.

EXAMPLE 6 Write each expression with positive exponents. Express fractions in simplest form.

(a) $\dfrac{x^{-\frac{1}{2}}x^{\frac{1}{2}}}{x}$ **(b)** $\dfrac{x^{-\frac{1}{2}} + x^{\frac{1}{2}}}{x}$

Solutions

(a) $\dfrac{x^{-\frac{1}{2}} \cdot x^{\frac{1}{2}}}{x} = \dfrac{x^{-\frac{1}{2}+\frac{1}{2}}}{x}$ $x^{-\frac{1}{2}}$ and $x^{\frac{1}{2}}$ are *factors*, so Property 1 holds.

$= \dfrac{x^0}{x} = \dfrac{1}{x}$

(b) $\dfrac{x^{-\frac{1}{2}} + x^{\frac{1}{2}}}{x} = \dfrac{\dfrac{1}{x^{\frac{1}{2}}} + x^{\frac{1}{2}}}{x}$ $x^{-\frac{1}{2}}$ and $x^{\frac{1}{2}}$ are *terms*, so we cannot apply any of the properties of exponents directly, and we write $x^{-\frac{1}{2}}$ as $\frac{1}{x^{\frac{1}{2}}}$.

$= \dfrac{\dfrac{1 + x^{\frac{1}{2}} \cdot x^{\frac{1}{2}}}{x^{\frac{1}{2}}}}{x}$ Find an equivalent expression for the numerator: $(x^{\frac{1}{2}})(x^{\frac{1}{2}}) = x$.

$= \dfrac{1 + x}{x^{\frac{1}{2}} \cdot x}$ Multiply the top and bottom by $x^{\frac{1}{2}}$.

$= \dfrac{1 + x}{x^{\frac{3}{2}}}$ Property 2 applied to denominator.

There are many other ways to write (b), but in each case, we get the same answer.

EXAMPLE 7

(a) Multiply $(x^{\frac{1}{2}} - y^{\frac{1}{2}})(x^{\frac{1}{2}} + y^{\frac{1}{2}})$.

(b) Multiply $x^{\frac{2}{3}}(x^{\frac{3}{4}} - x^{-\frac{2}{3}})$.

Solutions

(a) This looks familiar. Notice that if we let $x^{\frac{1}{2}} = a$ and $y^{\frac{1}{2}} = b$, then we have $(a - b)(a + b) = a^2 - b^2$, which is indeed a familiar problem. Therefore

$$(x^{\frac{1}{2}} - y^{\frac{1}{2}})(x^{\frac{1}{2}} + y^{\frac{1}{2}}) = (x^{\frac{1}{2}})^2 - (y^{\frac{1}{2}})^2$$
$$= x - y$$

(b) $x^{\frac{2}{3}}(x^{\frac{3}{4}} - x^{-\frac{2}{3}}) = x^{\frac{2}{3}}(x^{\frac{3}{4}}) - x^{\frac{2}{3}}(x^{-\frac{2}{3}})$ The distributive property.

$= x^{\frac{2}{3}+\frac{3}{4}} - x^{\frac{2}{3}-\frac{2}{3}}$ Property 2.

$= x^{\frac{17}{12}} - x^0$

$= x^{\frac{17}{12}} - 1$ Definition of x^0.

EXAMPLE 8

A bacteria culture starts with 5,000 cells and doubles every hour. How many cells will there be after $3\frac{1}{2}$ hours?

Solution

Using a Table We start by creating a table for the growth of the number of cells B after t hours.

After t hours	*Number of cells B*
0	5,000
1	$2 \cdot 5{,}000 = 10{,}000$
2	$2 \cdot (2 \cdot 5{,}000) = 2^2 \cdot 5{,}000 = 20{,}000$
3	$2 \cdot (2^2 \cdot 5{,}000) = 2^3 \cdot 5{,}000 = 40{,}000$
4	$2^4 \cdot 5{,}000 = 80{,}000$
t	$2^t \cdot 5{,}000$

Table 7.3

Using a Graph

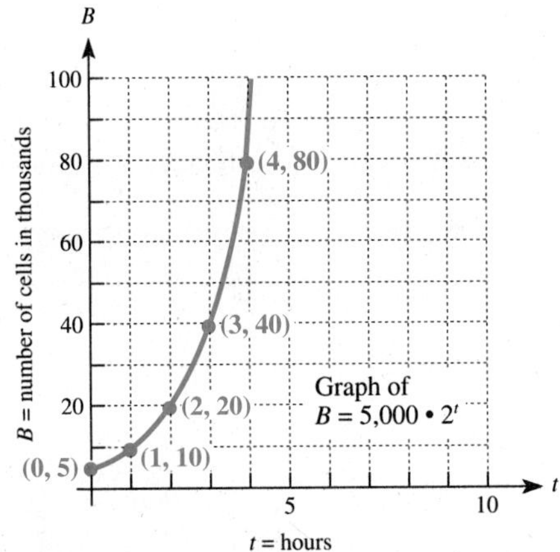

Figure 7.10

We see that the cell population after $3\frac{1}{2}$ hours would be between 40,000 and 80,000. In general, $B = 2^t \cdot 5{,}000$ hours after t hours. So after $3\frac{1}{2}$ hours, we have

$$B = 2^{\frac{7}{2}} \cdot 5{,}000 = \text{about } 56{,}570 \text{ cells.}$$

NOTE: If we had started with B_0 cells instead of 5,000 cells, we would have had

$$B = B_0 \cdot 2^t \text{ cells after } t \text{ hours.}$$

EXAMPLE 9 We now return to the problem of light filtering through water. If the intensity of light is reduced $\frac{3}{5}$ for each meter of water, what is the intensity of light at a depth of $\frac{2}{3}$ meter?

Solution

Refer to Table 7.1 and Figure 7.1 for the table and graph of this problem. Recall that we found that $L(n) = \left(\frac{2}{5}\right)^n$ where $L(n)$ is the amount of light that comes through at n meters of depth. So, intensity of light $= \left(\frac{2}{5}\right)^{\frac{2}{3}} \approx 0.54$. At $\frac{2}{3}$ meter, the intensity of light is only 0.54, or 54% of what it is at the surface.

EXAMPLE 10

In a cider mill, each press of the piston squeezes out 10% of the juice remaining in a batch of apples.

(a) How much juice remains in the apples after ten presses?

(b) How many presses are needed to squeeze out 80% of the juice?

Solution

The amount of juice left $P(n)$ is a function of the number of presses, n.

Using a Table

Number of presses	*$P(n)$ = fraction of original amount left*
0	$1.00 = 100\%$
1	$0.9 = 90\%$ (10% out means 90% left)
2	$0.9 \times 0.9 = 0.9^2 = 0.81 = 81\%$
3	$0.9 \times 0.81 = 0.9^3 = 0.73 = 73\%$
10	$0.9^{10} = 0.35 = 35\%$ left after ten presses
n	0.9^n

Table 7.4

(a) From the table we read that 35% of the juice is left after 10 presses.

(b) We must keep trying values of n until the percent left is at or below 20% in order to remove 80% of the juice from the apples.

Using a Graph

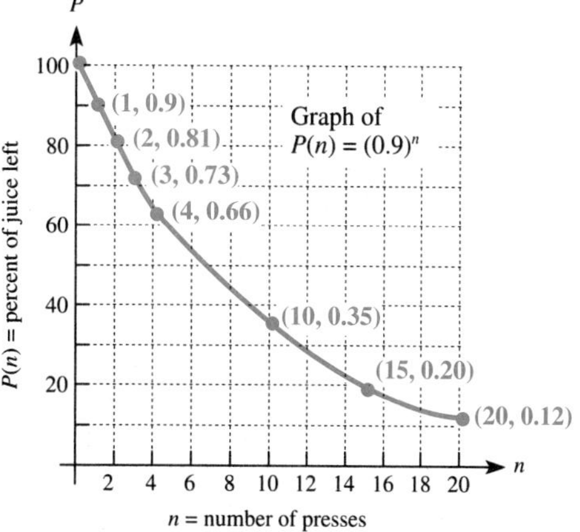

Figure 7.11

From the graph we read: (a) If $n = 10$, then 35% is left. (b) 80% of the juice is squeezed out if $n \approx 15$.

Using an Equation The amount of juice remaining after one stroke is 90%, or 0.9. After two strokes, it is (0.9)(0.9), and the amount of juice after n strokes is $(0.90)^n$, where n is the number of presses.

(a) If $n = 10$, then the amount of juice remaining is $(0.9)^{10} \approx 0.35$ or 35% of the juice remains after 10 presses.

(b) If 80% of the juice is squeezed out, then 20% of the juice remains. Thus $0.20 = (0.9)^n$. We can use our calculators to estimate n. Note from part (a) that $n > 10$. Since $(0.9)^{15} \approx 0.20$, $n \approx 15$ presses. (You may have to try several integers before arriving at 15.) ■

7.2 EXERCISES

We assume any variable is positive throughout this exercise set.

I. Procedures

1–8. Write each of the following expressions in an equivalent radical form.

1. $5^{\frac{1}{5}}$ **2.** $32^{\frac{3}{4}}$ **3.** $-32^{\frac{3}{4}}$

4. $(-32)^{\frac{3}{5}}$ **5.** $(17.5)^{\frac{2}{3}}$ **6.** $x^{\frac{5}{6}}$

7. $(x^2y)^{\frac{2}{3}}$ **8.** $(x^2 - 3)^{\frac{3}{4}}$

9–16. Write each expression in an equivalent exponential form. State the property you used to find the answer.

9. $\sqrt{11}$ **10.** $\sqrt[3]{\frac{x}{y}}$ **11.** $\sqrt[4]{8x^3}$

12. $\sqrt[5]{(a + b)^3}$ **13.** $\sqrt[5]{-y^3}$ **14.** $-\sqrt[5]{y^3}$

15. $-\sqrt[6]{12x^3}$ **16.** $\sqrt{\sqrt[5]{x}}$

17–30. Compute each of the following expressions. Try them first mentally or with pencil and paper.

17. $64^{\frac{2}{3}}$ **18.** $-64^{\frac{2}{3}}$ **19.** $(-64)^{\frac{2}{3}}$

20. $-64^{-\frac{2}{3}}$ **21.** $\left(\frac{16}{25}\right)^{\frac{3}{2}}$ **22.** $\dfrac{81^{\frac{3}{4}}}{8^{\frac{2}{3}}}$

23. $\dfrac{1}{(81)^{-\frac{3}{4}}}$ **24.** $(-125)^{\frac{1}{3}}(25^{-\frac{1}{2}})$

25. $(0.000027)^{\frac{4}{3}}$ **26.** $(8.1 \times 10^{-7})^{\frac{3}{4}}$

27. $\dfrac{16^{\frac{1}{2}} \cdot 16^{\frac{3}{4}}}{4}$ **28.** $\dfrac{36^{\frac{1}{2}} \cdot 25^{\frac{3}{2}}}{150}$ **29.** $\dfrac{\sqrt{64 \times 10^2}}{8^{\frac{2}{3}} \cdot 4^{\frac{3}{2}}}$

30. $(\sqrt[4]{2^8 \cdot 3^4})^2$

31–36. Use your calculator to approximate each of the following expressions to three decimal places.

31. $11^{\frac{3}{2}}$ **32.** $(1.075)^{\frac{8}{3}}$

33. $\left(\frac{-32}{3}\right)^{\frac{2}{3}}$ **34.** $(135)^{\frac{2}{5}}(419)^{\frac{3}{2}}$

35. $(1.16 \times 10^9)^{\frac{3}{2}}$ **36.** $\dfrac{3(1.13)^{\frac{5}{3}}}{7}$

37–58. Use the properties of exponents to write each of the following as equivalent expressions using only positive exponents. Find the nth roots wherever possible. In Exercises 37, 42, 47, 52, and 57 evaluate for $x = 2$, $y = 3.1$.

37. $x^{\frac{3}{4}}x^{\frac{1}{2}}$ **38.** $(x^{\frac{3}{4}})^{\frac{1}{2}}$ **39.** $\dfrac{x^{\frac{3}{4}}}{x^{\frac{1}{2}}}$

40. $(16x^{\frac{1}{2}})^{\frac{3}{4}}$ **41.** $(169x^6)^{-\frac{3}{2}}$ **42.** $(-32x^{10})^{\frac{2}{5}}$

43. $(27x^3y^{12})^{\frac{2}{3}}$ **44.** $\dfrac{x^{\frac{3}{4}}x^{\frac{3}{4}}}{x^{-\frac{1}{2}}}$ **45.** $\left(\dfrac{-32x^{10}}{y^{15}}\right)^{-\frac{3}{5}}$

46. $\left(\dfrac{-32x^{-10}y^{-5}}{x^{-15}y^{15}}\right)^{\frac{3}{5}}$ **47.** $\dfrac{8(x^6y^{-3})^{\frac{2}{3}}}{x^0y^3}$

48. $\left(\dfrac{8x^4y^{-2}}{x^0y^6}\right)^{\frac{3}{4}}$

49. $\sqrt[3]{8x^3y^6}$

50. $\dfrac{\sqrt[5]{x^{10}y^{\frac{2}{3}}}}{(x^3y^5)^{\frac{1}{3}\frac{2}{3}}}$

51. $(\sqrt[4]{x^3y^5})^{\frac{4}{3}}$

52. $\sqrt{x\sqrt{x}}$

53. $x^{\frac{3}{4}}(x^{\frac{1}{2}} + x^{-\frac{1}{2}})$

54. $x^{\frac{3}{4}} \cdot x^{\frac{1}{2}} \cdot x^{-\frac{1}{2}}$

55. $(x^{\frac{1}{2}} - 2^{\frac{1}{2}})(x^{\frac{1}{2}} + 2^{\frac{1}{2}})$

56. $(x^{\frac{1}{2}} + y^{\frac{1}{2}})^2$

57. $\dfrac{x^{-\frac{2}{3}} - x^{\frac{1}{3}}}{x^{\frac{1}{3}}}$

58. $\dfrac{x^{\frac{3}{4}}(x^{\frac{1}{4}} + x^{\frac{3}{4}})}{x^{\frac{1}{4}}}$

II. Concepts

59. (a) Show that $(x + y)^{\frac{2}{3}} \neq (xy)^{\frac{2}{3}}$.

(b) Show that $(x + y)^{\frac{2}{3}} \neq x^{\frac{2}{3}} + y^{\frac{2}{3}}$.

60. (a) Under what conditions will $x^{\frac{3}{5}}$ be negative?

(b) Under what conditions will $x^{\frac{3}{4}}$ be negative?

61–64. State whether each of the following statements is always true (AT), sometimes true (ST), or never true (NT). For any that are sometimes true, give an example to show when they are true and when they are false.

61. $2 \cdot 4^n = 8^n, n > 1$

62. $-x^{-n} = x^n$

63. If $x > 0$, then $x^{\frac{1}{4}} \leq x^{\frac{1}{5}}$.

64. $x^{\frac{1}{2}} + y^{\frac{1}{2}} > (xy)^{\frac{1}{2}}$

65. Graph $y = x^{\frac{2}{3}}$ and $y = x^{-\frac{2}{3}}$ on the same set of axes. What do you notice about these graphs?

66. Graph $y = x^{\frac{3}{2}}$ and $y = x^{-\frac{3}{2}}$ on the same set of axes. What do you notice about these graphs?

67. For the most part, we have been assuming that all variables are positive to help focus on the properties of exponents and radicals. To emphasize this, discuss the difference between $x^{\frac{1}{2}}$ when x is positive and the same expression when x is negative. Use examples and their graphs in your discussion.

III. Applications

68. The number of bacteria in a certain culture doubles every hour. If we start with 100 bacteria, how many would there be after

(a) $5\frac{1}{2}$ hours?

(b) 140 minutes?

(c) t hours?

(d) Graph the relationship in part (c).

69. A certain West Coast city has a population that doubles every 10 years. If the present population is 120,000, then what will the population be in

(a) 10 years? (b) 50 years? (c) n years?

(d) Graph the relationship in part (c).

70. A glass 1 cm thick allows only 80% of the light to filter through. How much light will filter through a glass

(a) 2 cm thick? (b) $\frac{1}{4}$ cm thick? (c) $\frac{7}{3}$ cm thick?

(d) c centimeters thick?

(e) Graph the relationship in part (d).

71. A sheet of plastic 1 millimeter (mm) thick reduces the amount of sunlight by 10%. How much sunlight will filter through a sheet of plastic

(a) 3 mm thick? (b) $\frac{4}{5}$ mm thick?

(c) $\frac{10}{3}$ mm thick?

(d) m millimeters thick?

(e) Graph the relationship in part (d).

72. A vacuum pump operates such that on each stroke, it removes 2% of the gas in a chamber.

(a) After 50 strokes, how much gas remains in the chamber?

(b) Estimate the number of strokes needed to remove 60% of the gas from the chamber.

73. So far in our discussion of compound interest, we have stated that the interest is compounded annually. If we have read the bank advertisements or have actually invested any money, we notice that the money is sometimes compounded semiannually, quarterly, monthly, daily, or even continuously. The formula then becomes

$$A_t = P\left(1 + \frac{I}{n}\right)^{nt}$$

where

I = rate of interest (in decimal notation)

t = number of years

n = number of times per year it is compounded

A_t = amount accumulated after t years

P = principal at the beginning

(a) A savings account pays 5.5% compounded quarterly. How much money do you have at the end of 3 years if you invest $1,000 to start?

(b) How much money will you have at the end of $6\frac{1}{2}$ years?

74. We want to invest $1,000 for 5 years. Which of the following savings plans is our best choice?

(a) $6\frac{3}{4}$% interest compounded daily

(b) 7% interest compounded quarterly

75. For Exercise 74, calculate the interest received at the end of one quarter for each savings plan.

76. The use of nuclear power plants and the possibility of accidents raise the problem of determining how much radioactive contamination is present at the site of the plant and its surroundings at any particular time. The following equation is used to approximate the radioactivity (radiation level) A at any given time t after a nuclear explosion if the radioactivity A_0 at unit time is known:

$$A = A_0 t^{-\frac{1}{2}}$$

(a) If the radiation level is three units six hours after an explosion, find A_0.

(b) Compare the amount of radioactivity after one hour and after seven hours. How much does the level of radioactivity decrease in this time period? Assume A_0 is the amount found in part (a).

77. A certain radioactive isotope of iodine decays according to the following formula: $A = A_0 e^{-0.02t}$, where A_0 is the number of grams of the isotope at the start, A is the number of grams of isotope at the end of t days, and e = 2.71828. If we start with 10 grams of the isotope, how many grams of radioactive isotope of iodine are left after

(a) 100 days?

(b) 360 days?

(c) 600 days?

78. The earth's population was 5.25 billion in 1990. It is growing at a rate that will double the population in about 40 years.

(a) What will be the population in the year 2010?

(b) What was the population in the year 1970? Use an almanac to check the accuracy of your answer.

(c) How does this estimate compare with the prediction used in Example 2 of Section 7.1 for the year 2000?

79. The cooking time T in hours for a turkey of weight W pounds is given (approximately) by the formula

$$T = 3.25\left(\frac{W}{7}\right)^{\frac{2}{3}}.$$

(a) Find the time necessary to cook an 18-pound turkey.

(b) If the weight of the turkey is doubled, what happens to the cooking time?

IV. Extensions

80. Use your calculator to find the sequence

$$\sqrt{2}, \sqrt{\sqrt{2}}, \sqrt{\sqrt{\sqrt{2}}}, \sqrt{\sqrt{\sqrt{\sqrt{2}}}}, \sqrt{\sqrt{\sqrt{\sqrt{\sqrt{2}}}}}, \ldots$$

(a) Continue this for at least 32 times.

(b) What happens? How can this be true?

(c) Write the above sequence using rational exponents.

(d) What is the general form for the nth term of this sequence of numbers?

81. What is wrong with the following argument?

$$-2 = (-8)^{\frac{1}{3}} = (-8)^{\frac{2}{6}} = \sqrt[6]{(-8)^2} = \sqrt[6]{64} = 2$$

Therefore $-2 = 2$. *Hint:* Carefully review the pertinent definitions.

7.3 OPERATIONS ON RADICALS

Recall that the radical $\sqrt[n]{x^m}$ is just another name for $x^{\frac{m}{n}}$. Since we have just finished finding equivalent expressions involving $x^{\frac{m}{n}}$, we could find equivalent radical expressions by first rewriting radical expressions as rational exponents and then proceeding to find equivalent expressions with rational exponents. Indeed, this is true. However, many people prefer to work solely with radicals. In fact, many applications use radical notation rather than rational exponent notation. We want to be proficient in working with expressions involving both radicals and rational exponents.

For example, the following are algebraic models that involve radicals:

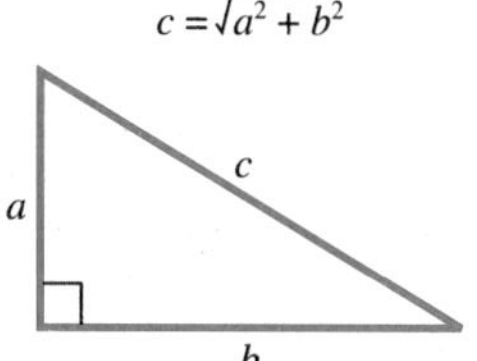

Figure 7.12

Where c is the hypotenuse of a right triangle and a and b are the legs of the right triangle.

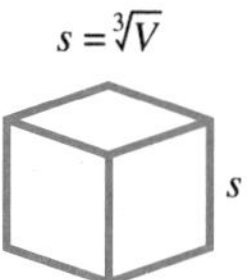

Figure 7.13

Where s is the length of an edge of a cube with volume V.

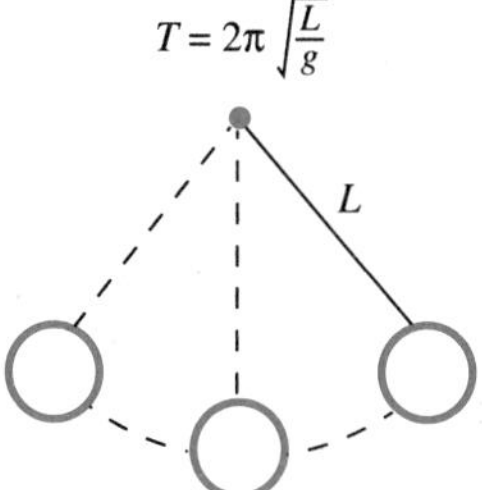

Figure 7.14

Where T is the time in seconds needed for a pendulum with length L to make a complete swing and g is a constant, 980 cm/sec^2, due to gravity.

$T = 0.2\sqrt[4]{N^3}$

Where T is the time in minutes it takes an average adult to memorize a string of N symbols.

Figure 7.15

Writing Equivalent Expressions with Radicals

One way to write an equivalent expression with radicals is to find the indicated root.

EXAMPLES 1–3

Find the indicated root.

1. $\sqrt{10{,}000}$ **2.** $\sqrt[3]{-64}$ **3.** $\sqrt[6]{64}$

Solutions

1. $\sqrt{10{,}000} = \sqrt{10^4}$
$= \sqrt{(10^2)^2}$
$= 10^2$

Definition.

2. $\sqrt[3]{-64} = \sqrt[3]{(-4)^3}$
$= -4$

Recall that $\sqrt[n]{x^n}$ equals x, when n is odd. Note that we have not assumed a positive number. To be sure, check the answer: $(-4)^3 = -64$.

3. $\sqrt[6]{64} = \sqrt[6]{2^6}$
$= 2$ ■

To find the indicated root, it is helpful to ask ourselves the following questions: In Example 1, what number squared is 10,000? In Example 2, what number cubed is -64?

Sometimes it takes some work to rewrite the number under the radical in the form x^n so that computation becomes easy. For example, what happens if the number under the radical cannot be put in the form x^n? Of course, we can also use our calculators to find roots if we are using only numbers, as in Examples 1–3. Consider the following examples:

EXAMPLES 4–5

Write an equivalent expression and find the indicated root.

4. $\sqrt{50}$ **5.** $\sqrt[6]{6{,}400}$

Solutions

4. $\sqrt{50} = \sqrt{5^2 \cdot 2}$
but $50 \stackrel{?}{=} \square^2$

We recall that we can rewrite radicals as rational exponents.

$$\begin{aligned}
\sqrt{5^2 \cdot 2} &= (5^2 \cdot 2)^{\frac{1}{2}} && \text{Definition of } \sqrt[n]{\ }. \\
&= (5^2)^{\frac{1}{2}} 2^{\frac{1}{2}} && \text{Property of exponents.} \\
&= 5 \cdot 2^{\frac{1}{2}} \\
&= 5\sqrt{2} \\
&\approx 7.07 && \text{Calculator approximation.}
\end{aligned}$$

5. Similarly,

$$\begin{aligned}
\sqrt[6]{6{,}400} &= \sqrt[6]{2^6 \cdot 10^2} && 6{,}400 \neq \square^6\text{, so we apply the definition of } \sqrt[n]{\ }. \\
&= (2^6 \cdot 10^2)^{\frac{1}{6}} \\
&= (2^6)^{\frac{1}{6}} \cdot (10^2)^{\frac{1}{6}} \\
&= 2 \cdot 10^{\frac{1}{3}} && \text{Properties of exponents.} \\
&= 2\sqrt[3]{10} && \text{Definition of } x^{\frac{1}{n}}. \\
&\approx 4.31 && \text{Calculator approximation.}
\end{aligned}$$

There is a pattern emerging in Examples 4 and 5. We see that

$$\sqrt{50} = \sqrt{25 \cdot 2} = \sqrt{25}\sqrt{2}$$

and

$$\sqrt[6]{6{,}400} = \sqrt[6]{64 \cdot 100} = \sqrt[6]{64}\sqrt[6]{100}$$

or more generally,

$$\sqrt[n]{ab} = (ab)^{\frac{1}{n}} = a^{\frac{1}{n}} b^{\frac{1}{n}} = \sqrt[n]{a}\sqrt[n]{b}$$

Property 1 for Radicals

$$\sqrt[n]{ab} = \sqrt[n]{a}\sqrt[n]{b} \quad \text{if} \quad \sqrt[n]{a}, \sqrt[n]{b} \text{ exist.}$$

To summarize how to find the indicated root, factor the radicand into factors that are a power of the indicated root, if possible. Then remove each such factor from under the radical sign by applying the definition, $\sqrt[n]{x^n} = x$, if $x > 0$. This strategy also works for radicals containing algebraic expressions.

EXAMPLES 6–9

Assume all variables are positive, and find the indicated root.

6. $\sqrt[3]{16}$ **7.** $\sqrt[5]{-320}$ **8.** $\sqrt{32x^5y^2}$ **9.** $\sqrt[5]{-32x^7y^{10}z^3}$

Solutions

6. $\sqrt[3]{16} = \sqrt[3]{8 \cdot 2}$ — Factor the radicand, writing the factors with exponents equal to the index if possible.

$= \sqrt[3]{(2)^3(2)}$

$= 2\sqrt[3]{2}$ — Property 1 for radicals.

7. $\sqrt[5]{-320} = \sqrt[5]{-32 \cdot 10}$ — Factor the radicand, writing the factors with exponents equal to the index if possible.

$= \sqrt[5]{(-2)^5 \cdot 10}$

$= \sqrt[5]{(-2)^5}\sqrt[5]{10}$ — Property 1 for radicals.

$= -2\sqrt[5]{10}$ — $\sqrt[n]{x^n} = x$ when n is odd.

8. $\sqrt{32x^5y^2} = \sqrt{4^2 \cdot 2 \cdot (x^2)^2 \cdot x \cdot y^2}$ — Factor the radicand, writing each factor with exponents equal to the index, if possible.

$= \sqrt{4^2}\,\sqrt{(x^2)^2}\,\sqrt{(y^2)}\,\sqrt{2x}$ — Property 1 for radicals.

$= 4x^2y\sqrt{2x}$ — $\sqrt[n]{x^n} = x$ when n is even, and $x > 0$.

9. $\sqrt[5]{-32x^7y^{10}z^3} = \sqrt[5]{(-2)^5x^5 \cdot x^2 \cdot (y^2)^5 \cdot z^3}$ — Factor the radicand so that as many factors as possible have exponents equal to the index.

$= \sqrt[5]{(-2)^5x^5(y^2)^5}\,\sqrt[5]{x^2z^3}$

$= -2xy^2\sqrt[5]{x^2z^3}$ — Property 1 for radicals. ■

NOTE: In the resulting equivalent expression (after finding the indicated root), the exponent of each factor in the radicand is less than the index (indicated root). This fact can be used to check whether we have taken out the largest indicated root. ■

The following example contains several problems in which errors are frequently made. Pay attention!

EXAMPLE 10

Assume all variables are positive. Find the indicated root.

(a) $\sqrt{4 + 9}$ **(b)** $\sqrt{4 \cdot 9}$ **(c)** $\sqrt{a^2 + b^2}$

(d) $\sqrt{a^2b^2}$ **(e)** $\sqrt{a^4 + a^2b^2}$ **(f)** $\sqrt[3]{8 + 19}$

Solutions

(a) $\sqrt{4 + 9} = \sqrt{13}$ — 13 has no square *factor.* Since 4 and 9 are *not factors,* we cannot apply Property 1 for radicals. With a calculator, we can approximate $\sqrt{13} \approx 3.61$.

(b) $\sqrt{4 \cdot 9} = \sqrt{36}$

$= 6$ — Since 4 and 9 are factors, we can apply Property 1: $\sqrt{4 \cdot 9} = \sqrt{4}\,\sqrt{9} = 2 \cdot 3 = 6$

(c) $\sqrt{a^2 + b^2}$ — $a^2 + b^2$ has no square factor. Note: $\sqrt{a^2 + b^2} \neq a + b$, since $(a + b)^2 \neq a^2 + b^2$

(d) $\sqrt{a^2b^2} = ab$ Where $a, b \geq 0$.

(e) $\sqrt{a^4 + a^2b^2} = \sqrt{a^2(a^2 + b^2)}$

$= a\sqrt{a^2 + b^2}$ Where $a \geq 0$.

(f) $\sqrt[3]{8 + 19} = \sqrt[3]{27}$

$= 3$ ■

The property $\sqrt[n]{ab} = \sqrt[n]{a}\sqrt[n]{b}$ can also be used to multiply expressions such as $\sqrt{2}\sqrt{18}$ and $\sqrt[3]{2x}\sqrt[3]{4x^2}$.

EXAMPLES 11–12

Multiply and find the indicated root.

11. $\sqrt{2}\sqrt{18}$

12. $\sqrt[3]{2x}\sqrt[3]{4x^2}$

Solutions

11. $\sqrt{2}\sqrt{18} = \sqrt{2 \cdot 18}$

$= \sqrt{36}$

$= 6$

12. $\sqrt[3]{2x}\sqrt[3]{4x^2} = \sqrt[3]{(2x)(4x^2)}$

$= \sqrt[3]{8x^3}$

$= 2x$ ■

NOTE: In Example 11, when we multiply and proceed to find the indicated root, we are creating a family of equivalent expressions. Thus, the expressions

$$\sqrt{2}\sqrt{18} \qquad \sqrt{2 \cdot 18} \qquad \sqrt{36} \qquad 6$$

are all equivalent. Similarly, in Example 12 the expressions

$$\sqrt[3]{2x}\sqrt[3]{4x^2} \qquad \sqrt[3]{(2x)(4x)^2} \qquad \sqrt[3]{8x^3} \qquad 2x$$

are all equivalent. They are different names for the same quantity. Another way to show that these expressions are equivalent is to graph them. (Recall that equivalent expressions have the same graph.) ■

EXAMPLE 13

Show that the following expressions are equivalent by graphing. Assume $x > 0$.

$$\sqrt{16x^3} \qquad 4\sqrt{x^3} \qquad 4x^{\frac{3}{2}} \qquad 4x\sqrt{x}$$

Solution

Let

$$Y1 = \sqrt{16x^3}$$
$$Y2 = 4\sqrt{x^3}$$
$$Y3 = 4x^{\frac{3}{2}}$$
$$Y4 = 4x\sqrt{x}$$

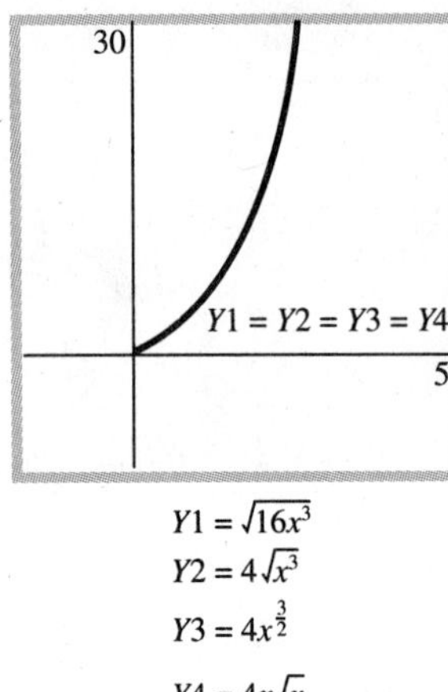

Figure 7.16

NOTE: The graphing strategy is relatively easy to use if the expression contains only one variable. ■

EXAMPLE 14

If a given cube has volume V_1, then the length of one of its sides, s_1, is

$$s_1 = \sqrt[3]{V_1}.$$

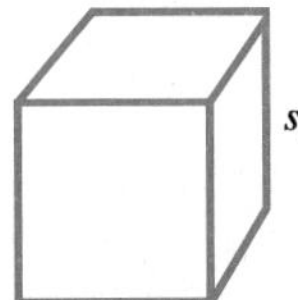

Figure 7.17

If the volume of the cube is doubled, what is the ratio of the length of the side of the original cube, s_1, to the length of the side, s_2, of the new cube?

Solution

To get an idea of what the ratio might be, let's try a specific example. Suppose the volume of the original cube is 3 cubic cm. Then, $s_1 = \sqrt[3]{3}$. The volume of the new cube is 6. The length of one of its sides, s_2, is $\sqrt[3]{6}$. The ratio of the length of the original side to the new side is

$$\frac{s_1}{s_2} = \frac{\sqrt[3]{3}}{\sqrt[3]{6}} = \frac{\sqrt[3]{3}}{\sqrt[3]{3}\sqrt[3]{2}} = \frac{1}{\sqrt[3]{2}}$$

In general,

$$\frac{s_1}{s_2} = \frac{\sqrt[3]{V_1}}{\sqrt[3]{2V_1}} = \frac{\sqrt[3]{V_1}}{\sqrt[3]{2}\,\sqrt[3]{V_1}} = \frac{1}{\sqrt[3]{2}}$$

We could also find the ratio of s_2 to s_1, which is

$$\frac{s_2}{s_1} = \frac{\sqrt[3]{2V_1}}{\sqrt[3]{V_1}} = \frac{\sqrt[3]{2}}{1}$$

Thus, s_2 is $\sqrt[3]{2}$ times as large as s_1. ■

This example suggests a way to divide radicals.

Property 2 for Radicals

If $b \neq 0$, then

$$\sqrt[n]{\frac{a}{b}} = \frac{\sqrt[n]{a}}{\sqrt[n]{b}}, \text{ if } \sqrt[n]{a}, \sqrt[n]{b} \text{ exist.}$$

EXAMPLES 15–17

Express the fractions in simplest form. Recall that this means no common factors in the numerator and denominator. Assume all variables are positive.

15. $\sqrt[3]{\dfrac{16x^5y^{-1}}{2x^{-3}y^8}}$ **16.** $\dfrac{\sqrt[4]{32x^{-1}y^3}}{\sqrt[4]{2x^{11}y}}$ **17.** $\dfrac{\sqrt[3]{64}}{\sqrt{8}}$

Solutions

15. $\sqrt[3]{\dfrac{16x^5y^{-1}}{2x^{-3}y^8}} = \sqrt[3]{\dfrac{8x^8}{y^9}}$ The fraction under the radical is in simplest form.

$= \dfrac{\sqrt[3]{8x^8}}{\sqrt[3]{y^9}}$ Property 2 for radicals.

$= \dfrac{\sqrt[3]{8}\,\sqrt[3]{(x^2)^3x^2}}{\sqrt[3]{(y^3)^3}}$ Property 1 for radicals.

$= \dfrac{2x^2\sqrt[3]{x^2}}{y^3}$ $\sqrt[n]{x^n} = x$, when n is odd.

16. $\dfrac{\sqrt[4]{32x^{-1}y^3}}{\sqrt[4]{2x^{11}y}} = \sqrt[4]{\dfrac{32x^{-1}y^3}{2x^{11}y}}$ Property 2 for radicals.

$= \sqrt[4]{\dfrac{16y^2}{x^{12}}}$ Equivalent fraction under the radical.

$= \dfrac{\sqrt[4]{16y^2}}{\sqrt[4]{x^{12}}}$ Property 2 for radicals.

$= \dfrac{2\sqrt[4]{y^2}}{x^3}$ Property 1 for radicals and $\sqrt[n]{x^n}$.

17. $\dfrac{\sqrt[3]{64}}{\sqrt{8}}$

CAREFUL! Note that the indices are different, so we cannot apply the rule $\dfrac{\sqrt[n]{a}}{\sqrt[n]{b}} = \sqrt[n]{\dfrac{a}{b}}$. We can try to find the indicated root of the radical in the numerator and denominator separately. ■

$= \dfrac{\sqrt[3]{4^3}}{\sqrt{2^2 \cdot 2}}$ Factor the radicands.

$= \dfrac{4}{2\sqrt{2}}$ Apply Property 1 for radicals to both the numerator and denominator.

$= \dfrac{2}{\sqrt{2}}$ ■

By calculator $\frac{2}{\sqrt{2}} \approx 1.414$. The exact value can be found as follows:

$\dfrac{2}{\sqrt{2}} = \dfrac{2}{\sqrt{2}} \cdot \dfrac{\sqrt{2}}{\sqrt{2}}$ Multiply by $1 = \dfrac{\sqrt{2}}{\sqrt{2}}$.

$= \dfrac{2\sqrt{2}}{\sqrt{2^2}}$ Property 1 for radicals.

$= \dfrac{2\sqrt{2}}{2}$ $\sqrt[n]{x^n} = x$

$= \sqrt{2}$ Equivalent fraction.

EXAMPLES 18–19

Find an equivalent expression that does not have radicals in the denominator. Express fractions in simplest form. Assume all variables are positive.

18. $\dfrac{\sqrt{2}}{\sqrt{3}}$

19. $\dfrac{2x}{\sqrt[4]{x^2}}$

Solutions

In each example, we will multiply the numerator and denominator by the same number, which will remove the radical from the denominator. The "multiplier" is the number that will convert the denominator to $\sqrt[n]{x^n}$.

18. $\dfrac{\sqrt{2}}{\sqrt{3}} = \dfrac{\sqrt{2}}{\sqrt{3}} \cdot \dfrac{\sqrt{3}}{\sqrt{3}}$ Multiply by $1 = \dfrac{\sqrt{3}}{\sqrt{3}}$.

$= \dfrac{\sqrt{2 \cdot 3}}{\sqrt{3^2}}$ Multiplication of radicals.

$= \frac{\sqrt{6}}{3}$ $\sqrt[n]{x^n}$. This is now in simplest form. Note: 3 is *not* a factor of the numerator.

19. $\frac{2x}{\sqrt[4]{x^2}} = \frac{2x}{\sqrt[4]{x^2}} \cdot \frac{\sqrt[4]{x^2}}{\sqrt[4]{x^2}}$ Multiply by $1 = \frac{\sqrt[4]{x^2}}{\sqrt[4]{x^2}}$.

$= \frac{2x\sqrt[4]{x^2}}{\sqrt[4]{x^4}}$ $\sqrt[4]{x^4} = x$, when $x > 0$.

$= 2\sqrt[4]{x^2}$ Equivalent fraction.

$= 2\sqrt{x}$ Note: $\sqrt[4]{x^2} = x^{\frac{2}{4}} = x^{\frac{1}{2}} = \sqrt{x}$. ■

These techniques effectively remove radicals from the denominator of a fraction and thereby provide an equivalent fraction whose denominator is a rational number. A word of caution: In Example 18, some people might be tempted to write $\frac{\sqrt{6}}{3}$ as $\sqrt{2}$. Resist the temptation! $\left(\frac{\sqrt{6}}{3} \neq \sqrt{2}\right)$ Using the calculator, $\sqrt{6} \approx 2.4495$ and $\frac{\sqrt{6}}{3} \approx 0.0816 \neq \sqrt{2} \approx 1.414$. Also, in the first step of Example 19, $\left[\frac{2x}{\sqrt[4]{x^2}}\right] \cdot \left[\frac{\sqrt[4]{x^2}}{\sqrt[4]{x^2}}\right]$, some people might be tempted to remove $\sqrt[4]{x^2}$ from the numerator and denominator. However, then we would be right back where we began at $\left(\frac{2x}{\sqrt[4]{x^2}}\right)$, which still has a radical in the denominator. The first step to find an equivalent expression without a radical in the denominator is to multiply by 1. That is, we are applying the fundamental property of fractions: $\left(\frac{a}{b}\right) = \left(\frac{a}{b}\right) \cdot \left(\frac{k}{k}\right) = \frac{(ak)}{(bk)}$. In this case, k is chosen in order to rewrite the denominator in the form $\sqrt[n]{x^n}$.

Operations Involving Radicals

We will now examine algebraic expressions that involve adding and subtracting radicals, and apply properties and techniques we have already learned. For example, the distributive property is very important. Recall that in the statement $2x^2 + 3x^2 = (2 + 3)x^2 = 5x^2$, x^2 is the common factor. In the following problems radicals are used as common factors.

EXAMPLES 20–22

Perform the indicated operations. Assume all variables are positive.

20. $4\sqrt{7} + 2\sqrt{7}$

21. $3\sqrt{24} + 5\sqrt{54}$

22. $\sqrt{16ab^3} - 9\sqrt{a^3b} + \sqrt{25a^3b^3}$

Solutions

20. $4\sqrt{7} + 2\sqrt{7} = (4 + 2)\sqrt{7}$ Distributive property.

$= 6\sqrt{7}$

21. $3\sqrt{24} + 5\sqrt{54} = 3\sqrt{4 \cdot 6} + 5\sqrt{9 \cdot 6}$ — Factor each radicand.

$= 3\sqrt{4}\sqrt{6} + 5\sqrt{9}\sqrt{6}$ — Multiplication of radicals (Property 1).

$= 3 \cdot 2\sqrt{6} + 5 \cdot 3\sqrt{6}$ — $\sqrt[n]{x^n}$

$= 6\sqrt{6} + 15\sqrt{6}$

$= 21\sqrt{6}$ — Distributive property.

22. $\sqrt{16ab^3} - 9\sqrt{a^3b} + \sqrt{25a^3b^3}$

$= 4b\sqrt{ab} - 9a\sqrt{ab} + 5ab\sqrt{ab}$ — Find the indicated roots.

$= (4b - 9a + 5ab)\sqrt{ab}$ — Distributive property. ■

The strategy in each example above is to first find the indicated root of each radical to determine whether each term will have a common factor and then apply the distributive property.

In the following examples, we combine multiplication, division, addition, and subtraction of radicals.

EXAMPLES 23–24

Multiply each of the following, $x > 0$.

23. $\sqrt{3}(\sqrt{2} + \sqrt{3})$ **24.** $(\sqrt{x} - \sqrt{2})(\sqrt{x} + \sqrt{2})$

Solutions

23. $\sqrt{3}(\sqrt{2} + \sqrt{3}) = \sqrt{3}\sqrt{2} + \sqrt{3}\sqrt{3}$ — Distributive property.

$= \sqrt{6} + 3$

24. $(\sqrt{x} - \sqrt{2})(\sqrt{x} + \sqrt{2}) = (\sqrt{x})^2 - (\sqrt{2})^2$

$= \sqrt{x^2} - \sqrt{2^2}$

$= x - 2$ ■

Observe that in Example 24 we have no radical in our answer. We call the binomials $\sqrt{a} + \sqrt{b}$ and $\sqrt{a} - \sqrt{b}$ **conjugates** of each other. Multiplying two conjugates together eliminates the radicals. We use this fact to express the following fraction in simplest form.

EXAMPLE 25

Express the fraction in simplest form without a radical in the denominator.

25. $\dfrac{2}{\sqrt{3} - 1}$

Solution

25. $\dfrac{2}{\sqrt{3} - 1} = \dfrac{2}{\sqrt{3} - 1} \cdot \left(\dfrac{\sqrt{3} + 1}{\sqrt{3} + 1}\right)$ — Multiply by $1 = \left(\frac{\sqrt{3} + 1}{\sqrt{3} + 1}\right)$. Notice that $\sqrt{3} + 1$ is the conjugate of $\sqrt{3} - 1$.

$= \dfrac{2(\sqrt{3} + 1)}{(\sqrt{3})^2 - (1)^2}$ — $(a + b)(a - b) = a^2 - b^2$

$= \dfrac{2(\sqrt{3} + 1)}{3 - 1}$

$$= \frac{2(\sqrt{3} + 1)}{2}$$
$$= \sqrt{3} + 1$$

Strategies for Working with Radicals

For most problems, we can use either radicals or exponents to write equivalent expressions. It is a matter of preference. For example,

$$\frac{\sqrt[3]{8x^7y}}{\sqrt[3]{xy^{-2}}} = \sqrt[3]{\frac{8x^7y}{xy^{-2}}} \quad \text{or} \quad \frac{\sqrt[3]{8x^7y}}{\sqrt[3]{xy^{-2}}} = \frac{(8x^7y)^{\frac{1}{3}}}{(xy^{-2})^{\frac{1}{3}}}$$
$$= \sqrt[3]{8x^6y^3} \qquad\qquad = \left(\frac{8x^7y}{xy^{-2}}\right)^{\frac{1}{3}}$$
$$= 2x^2y \qquad\qquad = (8x^6y^3)^{\frac{1}{3}}$$
$$= 2x^2y$$

Sometimes it is necessary to change radicals to rational exponents. For example, to multiply $\sqrt[3]{x^2}\sqrt{x}$, we cannot apply Property 1 for radicals, since the indices are different. Hence

$$\sqrt[3]{x^2}\sqrt{x} = x^{\frac{2}{3}}x^{\frac{1}{2}} = x^{\frac{4}{6}}x^{\frac{3}{6}} = x^{\frac{7}{6}} = x\sqrt[6]{x}$$

Here is another application of radicals.

EXAMPLE 26

The time, T, in minutes needed to memorize a string of N symbols is

$$T(N) = 0.2\sqrt[4]{N^3}$$

(a) Find the time needed to memorize 20 symbols.

(b) Use a graph to estimate the time needed to memorize 30 symbols.

(c) Use a graph to estimate the number of symbols that can be memorized in five minutes.

Solutions

(a) $T(N) = 0.2\sqrt[4]{20^3} \approx 1.89$ minutes

(b) From the graph, $T \approx 2.56$ minutes

(c) From the graph, $N \approx 73$ symbols

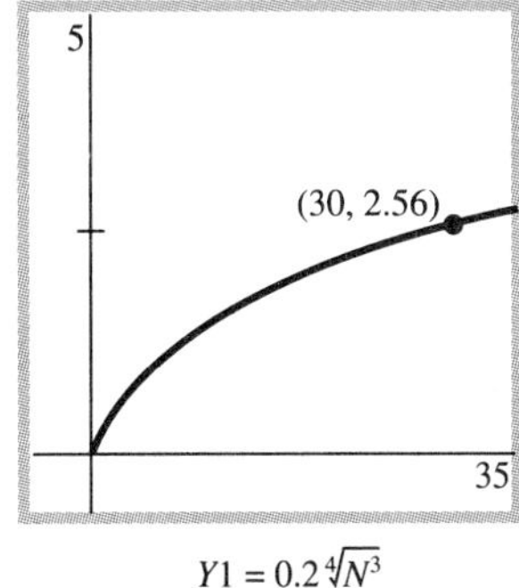

$Y1 = 0.2\sqrt[4]{N^3}$

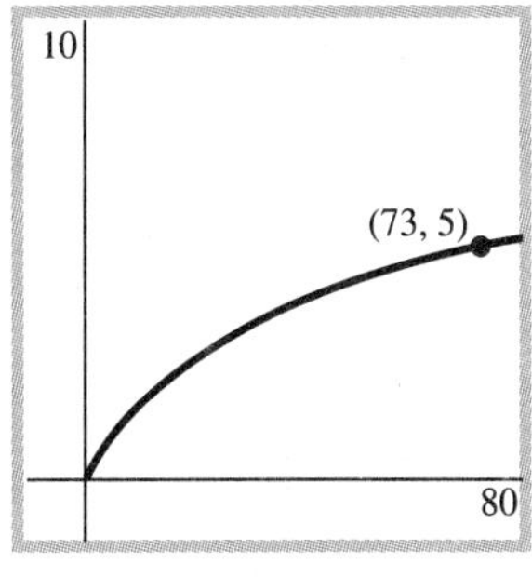

$Y1 = 0.2\sqrt[4]{N^3}$

Figure 7.18

7.3 EXERCISES

Assume all variables are positive for this set of exercises.

I. Procedures

1–12. Find an equivalent expression by extracting the indicated roots.

1. $\sqrt[3]{-250}$

2. $\sqrt{6{,}300}$

3. $\sqrt[5]{-24{,}200}$

4. $\sqrt[4]{1.6 \times 10^{-7}}$

5. $\sqrt{25 + 144}$

6. $-\sqrt{x^3}$

7. $\sqrt{8x^6}$

8. $\sqrt[3]{-27x^4y^{13}}$

9. $ab\sqrt[4]{32a^5b^6}$

10. $2a\sqrt{18ab^5}$

11. $\sqrt{16x^2 - 4x}$

12. $\sqrt{49y^2 + 16y^2}$

13–30. Perform the indicated operations. Express all fractions in simplest form without radicals in the denominator.

13. $\sqrt{2} \cdot \sqrt{5} \cdot \sqrt{10}$

14. $\sqrt[3]{4}\sqrt[3]{2}$

15. $\dfrac{5}{\sqrt{5}}$

16. $\dfrac{12\sqrt{6}}{3\sqrt{2}}$

17. $\dfrac{\sqrt[3]{4}}{4\sqrt[3]{2}}$

18. $\dfrac{(2\sqrt{3})^2}{6}$

19. $2\sqrt{5} - 3\sqrt{5} - 2\sqrt{5}$

20. $4\sqrt{12} - 2\sqrt{3}$

21. $\sqrt{48} + 8\sqrt{18}$

22. $\sqrt{125} + 2\sqrt{27} - \sqrt{20} + 3\sqrt{12}$

23. $(\sqrt{2} - 2)(\sqrt{2} + 2)$

24. $\dfrac{(1 - \sqrt{3})^2}{2}$

25. $(\sqrt{5} - \sqrt{2})(\sqrt{5} + \sqrt{2})$

26. $(2\sqrt{3} + \sqrt{5})(3\sqrt{3} - 2\sqrt{5})$

27. $\dfrac{5}{\sqrt{5} - 1}$

28. $\dfrac{\sqrt{3}}{\sqrt{3} + 1}$

29. $\dfrac{3 + 2\sqrt{3}}{3 - 2\sqrt{3}}$

30. $\dfrac{2 + \sqrt{10}}{\sqrt{2} + \sqrt{5}}$

31–43. Perform the indicated operations. Express all fractions in simplest form without radicals in the denominator.

31. $\sqrt{x^3}\sqrt{x}$

32. $\sqrt{8x^3y^2} \cdot \sqrt{12x^2y^3}$

33. $\sqrt[3]{4x^2y}\sqrt[3]{4x^4y^3}$

34. $\sqrt{\dfrac{x^2}{y}}$

35. $\dfrac{2x}{\sqrt{x}}$

36. $\dfrac{\sqrt[4]{24x^3y}}{\sqrt[4]{8xy^3}}$

37. $\dfrac{\sqrt{x^8y^3}\sqrt{xy}}{\sqrt{2xy}}$

38. $3\sqrt{2x} + \sqrt{8x}$

39. $\sqrt[3]{2xy^2} - \sqrt[3]{16xy^5}$

40. $\sqrt{300x^3y} - \sqrt{27xy^5} + \sqrt{12xy}$

41. $x\sqrt[3]{x^4} - 2\sqrt[3]{xy^6} - \sqrt[3]{-8x}$

42. $\dfrac{a}{1 - \sqrt{a}}, a \neq 1$

43. $\dfrac{a + 1}{\sqrt{a} + 1}$

II. Concepts

44. Is $x = \left(\dfrac{1 - \sqrt{5}}{2}\right)$ a solution of $x^2 - x - 1 = 0$? Why?

45. Is $\dfrac{\sqrt{2x}}{x} = \sqrt{2}, x > 0$? Why?

46. **(a)** Show that $\sqrt{x^2 + y^2} \neq x + y$.

(b) Which side of the equation is usually larger?

(c) When are the two expressions equal?

(d) Explain the difference between $\sqrt{x^2 + y^2}$ and $\sqrt{x^2y^2}$.

47–49. Use your calculator to compute each of the following expressions.

47. $\sqrt{\dfrac{1 - (1.1)^2}{4}}$

48. $\sqrt{2 + \sqrt{2 + \sqrt{2}}}$

49. $\dfrac{\sqrt[3]{4.1 \times 10^9}}{1.2}$

50–53. Use graphs to determine whether the expressions in each set are equivalent; $x > 0$.

50. $\sqrt[3]{x^2}$ $\qquad \dfrac{x}{\sqrt[3]{x}}$ $\qquad x^{\frac{2}{3}}$

51. $\sqrt{x^3}$ $\qquad x^{\frac{3}{2}}$ $\qquad x\sqrt{x}$

52. $\sqrt[16]{x^6}$ $\qquad x^{\frac{3}{8}}$ $\qquad \sqrt[8]{x^3}$

53. $\sqrt{x}$ $\qquad \dfrac{1}{\sqrt{x}}$ $\qquad x^{\frac{1}{2}}$

54 and 55. Find three expressions equivalent to the given expression. Assume all variables are positive.

54. $\sqrt[3]{x} \cdot \sqrt[3]{x^2}$ **55.** $2\sqrt{2}$

56. **(a)** Suppose $x > 1$. Arrange the following expressions from largest to smallest.

$$\sqrt[3]{x^2} \quad \sqrt{x} \quad \sqrt[3]{x} \quad \sqrt{x^2} \quad \sqrt{x^3}$$

Hint: Make a chart to get a sense of each expression and then graph each expression on the same axis.

(b) How do the expressions compare for $0 < x < 1$?

57. Compare the graphs of

$$y = x^2 + 4 \quad y = x^2 \cdot 4 \quad y = x + 2 \quad y = \sqrt{x^2 + 4}$$

58. We have been assuming that all variables are positive when dealing with radicals and fractional exponents. What happens if we allow variables to take on negative values under the radical? Explain, using examples with graphs in your discussion.

III. Applications

59–73. The following exercises are examples of applications where radicals are used. Solve them.

59. The time, T, in minutes, that it takes an average adult to memorize N symbols is given by

$$\text{Time}(N) = T(N) = 0.2\sqrt[4]{N^3}$$

Find the time to memorize

(a) 32 symbols.

(b) 100 symbols.

(c) Use a graph to find the number of symbols that can be memorized in one minute. In 5 minutes.

60. The length of a side, s, of a cube with volume V is given by

$$s = \sqrt[3]{V}$$

Find the length of a side of a cube if its volume is

(a) 10,000,000 cm^3.

(b) 9.852 cm^3.

(c) Use a graph to find the volume if the length of a side is 3.2 cm. If it is 15 cm.

61. What happens to an edge of a cube if its volume is tripled? Show this relationship on a graph.

62. What happens to an edge of a square if its area is tripled? Show this relationship on a graph.

63. If the volume of a cube is tripled, what is the ratio of the lengths of the original side to the new side?

64. The volume, V, of a sphere is related to its surface area, A, by the formula

$$\text{Volume}(A) = 0.094\sqrt{A^3}$$

What is the volume of a balloon whose surface area is

(a) 1,000 cm^2?

(b) 75 cm^2?

(c) Use a graph to find the surface area of the balloon if the volume is 2,000 cm^3.

65. **(a)** If the surface area of a sphere is doubled, what is the ratio of the volume of the original sphere to the volume of the new sphere?

(b) Repeat part (a) if the surface area is tripled.

66. The speed v (in meters per second) necessary for a satellite to stay in orbit around the earth is given by the formula

$$v = \sqrt{\frac{4 \times 10^{14}}{d}}$$

where d is the distance of the satellite from the center of the earth in meters. Calculate the velocity of a satellite that is 5.2×10^7 meters from the center of the earth.

67. The time T (in seconds) needed for a pendulum to make a complete swing is given by the formula

$$T = 2\pi\sqrt{\frac{L}{g}}$$

where L is the length of the pendulum and $\pi \approx 3.142$ and $g = 980$ cm/sec^2. Find the time needed to complete a swing if the length is

(a) 50 cm.

(b) 100 cm.

68. The formula $v = \sqrt{12L}$ *estimates* the speed v of a car from the length of its skid marks on wet pavement (upon braking).

(a) How fast was a car going if the skid marks are 50 feet long?

(b) How fast was a car going if the skid marks are 100 feet long?

69. The formula $v = \sqrt{24L}$ estimates the speed v of a car from the length of its skid marks on *dry* pavement.

(a) How fast was a car going if the skid marks are 50 feet long?

(b) How fast was a car going if the skid marks are 100 feet long?

70. The Pythagorean theorem states that $a^2 + b^2 = c^2$, where a and b are the legs of a right triangle with hypotenuse c. (See Examples 12 and 13 of Section 7.1 for the distance formula.)

(a) Find a if $b = 10$ and $c = 20$.

(b) Find a if $b = 6$ and $c = 26$.

71. A 15-foot ladder is placed against the wall of a house, just reaching the bottom of a window whose sill is 10 feet from the ground. Suddenly, the ladder slips 4 feet down the wall of the house. How far is the foot of the ladder from the base of the house?

72. During World War I, the Central Powers, not anticipating a long, dragged-out war, began to worry about the health of their people. They needed a quick measurement for malnutrition. It was found experimentally that for a healthy person, the cube of a person's sitting height is approximately ten times his or her weight in grams. This ratio is called *peledisi,* and it is computed as follows:

$$\text{peledisi} = \frac{\sqrt[3]{10 \times (\text{weight in grams})}}{\text{sitting height in centimeters}} \times 100\%$$

Sitting height is the distance from the top of the head to the chair. A well-nourished adult has a peledisi very near 100%, an undernourished adult's peledisi is less than 100%, and an obese person's is over 100%.

(a) Describe the health of an adult whose weight is 77,000 grams and whose sitting height is 100 cm.

(b) Can you determine your peledisi?

73. The relationship between the mean distance d of the planets from the sun (taking $d = 1$ for earth) and their periods T (length of one revolution in years) can be approximated by Bode's law, $T = \sqrt{d^3}$ and $d = \sqrt[3]{T^2}$. Use this formula to complete the following table:

Planet	*d*	*T*
Mercury	0.387	_____
Venus	0.723	_____
Earth	1.000	_____
Mars	1.523	_____
Jupiter	_____	11.861
Saturn	_____	29.457
Uranus	_____	84.008
Neptune	_____	164.784
Pluto	_____	248.35

IV. Extensions

74. (a) If a number is doubled, what happens to its square root?

(b) If a number is tripled, what happens to its square root?

(c) If a number is multiplied by itself 100 times, what happens to its square root?

75. Repeat Exercise 74 for cube roots.

76. The golden rectangle revisited: The ratio of the sides of the golden rectangle is equal to the golden ratio. That is,

$$l = \frac{1 + \sqrt{5}}{2} w$$

l

w $\quad l = \frac{2}{\sqrt{5} - 1} w$

(a) Show that

$$\frac{1 + \sqrt{5}}{2} \approx 1.618$$

(b) A method for constructing a golden rectangle starting with a square whose sides are of length x is given below.

1. Let O be the midpoint of the bottom edge. Let P be one of the top corners. Label corners A and B, as directed in the diagram.

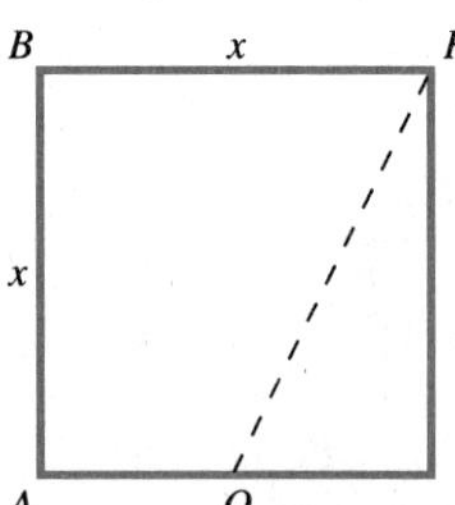

2. Then use the compass to draw an arc with radius equal to OP intersecting the extended bottom side at point C.

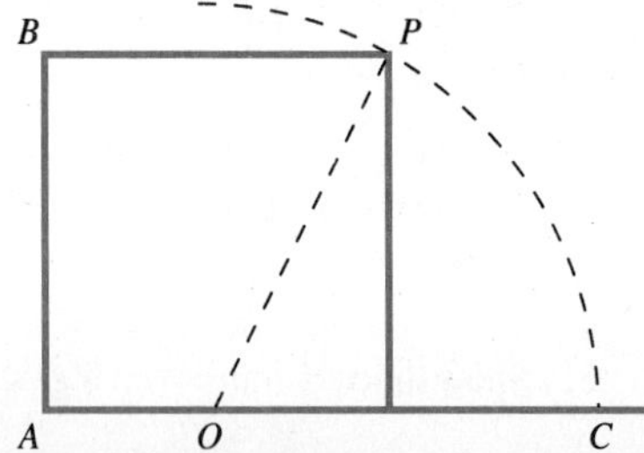

3. Construct a perpendicular to the bottom edge at C.

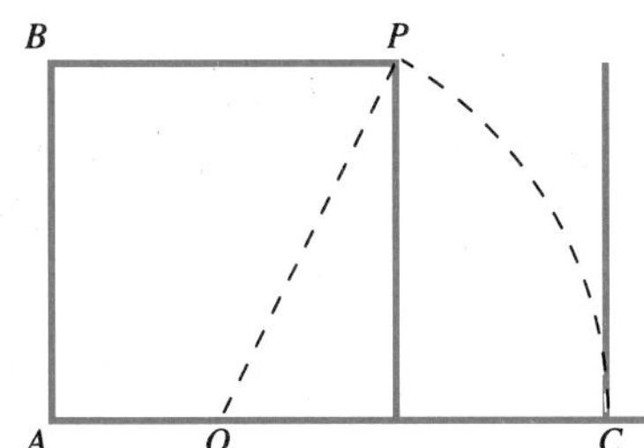

4. Extend the top edge BP to intersect the perpendicular at a point D.

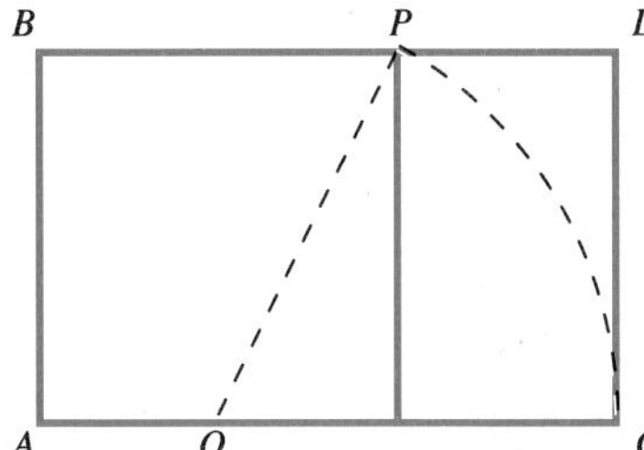

We claim that the rectangle $ABCD$ is a golden rectangle. Prove this. You will need the Pythagorean theorem.

77. Look up the method for constructing a regular pentagon (a high school geometry text will do), and show how the golden ratio is used in this construction.

78. Using the distance formula (in Examples 12 and 13 of Section 7.1), show that the distance between the points $(3a, 3b)$ and $(3c, 3d)$ is three times the distance between the points (a, b) and (c, d).

79. Show how the formula in Exercise 64 relating the volume, V, and the surface area, A, of a sphere is developed. *Hint:* $V = \frac{4}{3}\pi r^3 \quad A = 4\pi r^2$

7.4 EXPONENTIAL AND RADICAL EQUATIONS AND FUNCTIONS

Exponential Equations

We have been working with **radical** and **exponential equations** not only in this chapter but in earlier chapters. For example, if the population of amoebas doubles every hour, and if we start with an initial population of P_0, then the number of amoebas P after t hours is given by $P = P_0 2^t$. This is an exponential equation. Up to now, we were usually given P_0 and t, and then we were asked to find P. Consider the following example from Chapter 2, in which we are given P and now must find t.

EXAMPLE 1

Suppose the population of amoebas doubles every hour. If we start with 20 amoebas,

(a) How many will we have after 10 hours?

(b) How long will it take the population to grow to 1,280 amoebas?

(c) Find the population after 7.5 hours.

(d) Find the time it takes to grow 6,000 amoebas.

Solutions

Throughout the book we have been using three different ways to represent problems: tables, graphs, and equations. We will use and compare all three methods to solve this problem.

Using a Table

Time in hours	*Number of amoebas*
0	20
1	40
2	80
3	160
4	320
5	640
6	1,280
7	2,560
8	5,120
9	10,240
10	20,480

Table 7.5

(a) After 10 hours, the number of amoebas is 20,480.

(b) It takes 6 hours to produce 1,280 amoebas.

(c) In 7.5 hours, the number of amoebas is between 2,560 and 5,120.

(d) It will take between 8 and 9 hours to produce 6,000 amoebas.

Using a Graph The pattern in the chart suggests a function. The population P depends on time, t. We can write Population (time) or P(time) or, briefer yet, $P(t)$. If P is the population after t hours, then we have the equation

$$P(t) = 20(2^t).$$

We can solve this problem by graphing the function and then finding the value of P that corresponds to $t = 10$. To find t when $P = 1{,}280$, we find the value of t that corresponds to $P = 1{,}280$.

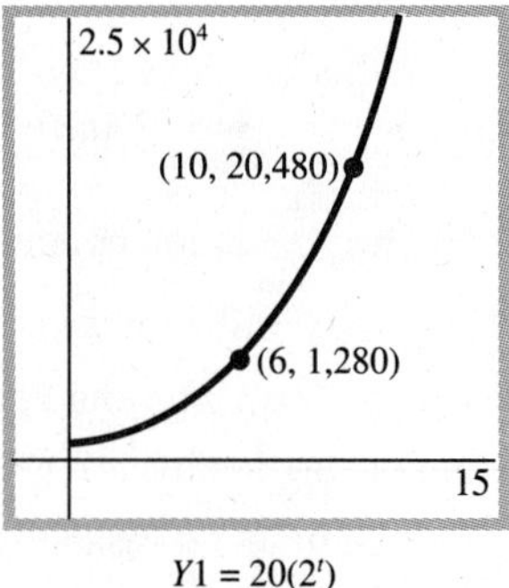

$Y1 = 20(2^t)$

Figure 7.19

From the graph we find the following:

(a) After 10 hours, 20,480 amoebas have been produced.

(b) Six hours are needed to produce 1,280 amoebas.

(c) We find 7.5 hours between 7 and 8 on the graph. So the population is between $20(2^7)$ and $20(2^8)$, or between 2,560 and 5,120. The table is not complete. But from the graph we can estimate that the number of amoebas is 3,600.

(d) From the graph we estimate it will take 8.2 hours to produce 6,000 amoebas.

Using an Equation We can use the function, $P(t) = 20(2^t)$.

(a) We substitute 10 hours for t and compute.

$$P(t) = 20(2)^{10} = 20(1{,}024) = 20{,}480 \text{ amoebas}$$

(b) If we start with 20 amoebas and substitute 1,280 amoebas for P, we have

$$\begin{aligned} 1{,}280 &= 20(2)^t \\ 64 &= 2^t \\ 2^6 &= 2^t \end{aligned}$$

The only time $2^6 = 2^t$ can be true is when $t = 6$. Checking our answer in the original equation, if $t = 6$, then

$$\begin{aligned} 1{,}280 &\stackrel{?}{=} 20(2^6) \\ &\stackrel{?}{=} 20(64) \\ &\stackrel{\checkmark}{=} 1{,}280 \end{aligned}$$

(c) We substitute 7.5 for t in the equation and find that $P \approx 3{,}620$.

(d) Substituting 6,000 for P in the equations, we have

$$\begin{aligned} 6{,}000 &= 20(2^t) \\ \text{or} \quad 300 &= 2^t \end{aligned}$$

But what power of 2 yields 300? We could use the calculator to guess and check, or we could graphically estimate the x-coordinate of the point of intersection of $Y1 = 2^x$ and $Y2 = 300$. ■

For parts (a) and (b) all three methods were relatively easy to use because the values could be read directly from the table, the graph was easy to read, and the equation was relatively easy to solve. Sometimes the values are not easy to generate in a table. In addition, we sometimes come up with equations we do not yet have the techniques to solve. In Example 1, parts (c) and (d), the graph is a valuable tool for solving the problem.

In this section we will be working with simple exponential equations of the form

$$a^x = b$$

Strategies for Solving Exponential Equations in One Variable

1. Use a table to read the values directly. Sometimes the table can also help suggest a pattern that can be generalized as an equation.
2. Graph the equation (function) and then estimate the solution from the graph.
3. Solve an equation. Use algebraic properties to find an equivalent equation in which the base of the quantities on both sides of the equations is the same. Set the exponents equal to each other, and solve the equation.

EXAMPLES 2–3

Solve for x.

2. $2^x = 4^{\frac{3}{2}}$

3. $2^{x-1} = 4^{-3}$

Solutions

2–3. Using a Graph In Example 2, graph $Y1 = 2^x$ and find x when $Y1 = 4^{\frac{3}{2}}$ or $Y1 = 8$ (Figure 7.20). The answer is $x = 3$. In Example 3, graph $Y2 = 2^{x-1}$. Find x when $Y2 = 4^{-3}$ or $Y2 = \frac{1}{64}$ (Figure 7.21). The answer is $x = -5$.

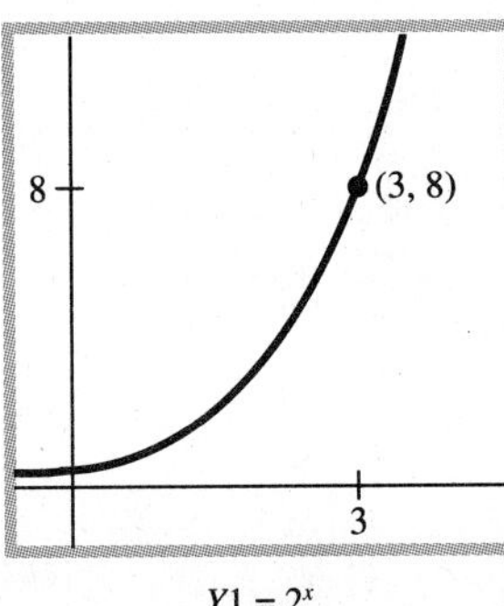

Figure 7.20

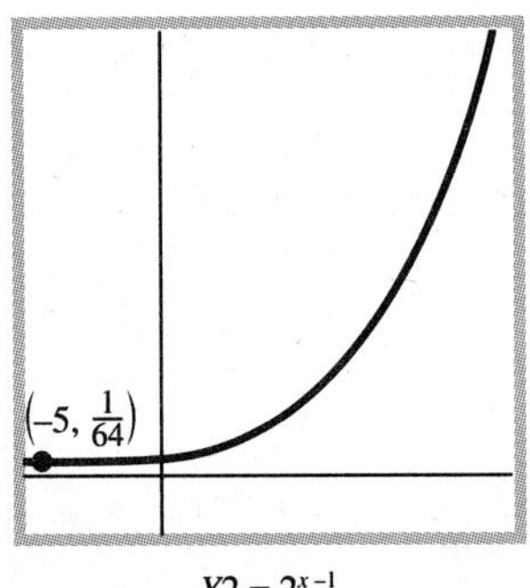

Figure 7.21

Using an Equation

2. $2^x = 4^{\frac{3}{2}}$ The bases are not the same, so we try to express the right side with base 2. Rewrite 4 as 2^2.

$2^x = (2^2)^{\frac{3}{2}}$

$2^x = 2^3$ Property of exponents.

$x = 3$ Bases are the same.

Checking our answer,

$2^3 \stackrel{?}{=} 4^{\frac{3}{2}}$

$2^3 = 8 = (\sqrt{4})^3 \stackrel{\checkmark}{=} 4^{\frac{3}{2}}$ Yes, it checks.

3. $2^{x-1} = 4^{-3}$

$2^{x-1} = (2^2)^{-3}$

$2^{x-1} = 2^{-6}$ Bases are the same.

Thus

$$x - 1 = -6$$

$$x = -5$$

Checking our answer,

$$2^{-5-1} \stackrel{?}{=} 4^{-3}$$

$$2^{-6} = \frac{1}{2^6} = \frac{1}{64} = \frac{1}{4^3} \stackrel{\checkmark}{=} 4^{-3}$$ Yes, it checks.

EXAMPLES 4–5

Solve for x.

4. $3^2 \cdot 3^x = 9^{2x-1}$

5. $3 \cdot 9^{x+1} = \sqrt[3]{27} \cdot 3^x$

Solutions

We could use a table, but when the equation is given it is usually quicker to use a graph or solve an equation.

4. Using a Graph Graph the two equations,

$$Y1 = 3^2 \cdot 3^x \quad \text{and} \quad Y2 = 9^{2x-1}$$

and find the x-coordinate of the point of intersection: $x = 1.33$, or $\frac{4}{3}$ (see Figure 7.22).

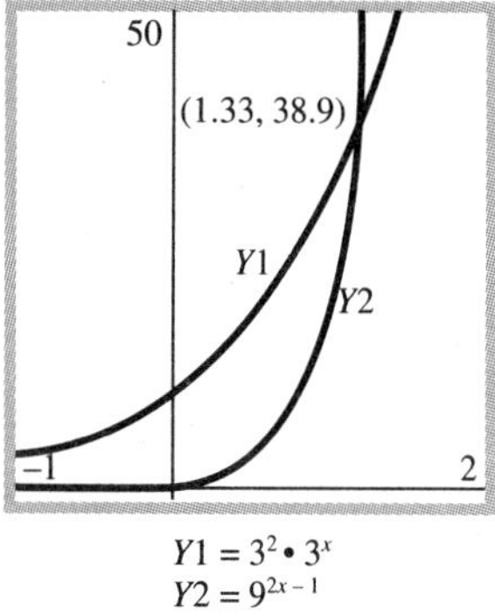

$Y1 = 3^2 \cdot 3^x$
$Y2 = 9^{2x-1}$

Figure 7.22

Using an Equation

$$3^2 \cdot 3^x = 9^{2x-1}$$ Write everything with base 3. Since the bases are the same, the exponents must be equal.

$$3^2 \cdot 3^x = (3^2)^{2x-1}$$

$$3^{2+x} = 3^{4x-2}$$

Therefore

$$2 + x = 4x - 2$$

$$4 = 3x$$

$$\frac{4}{3} = x$$

Check:

$$3^2 \cdot 3^{\frac{4}{3}} \stackrel{?}{=} 9^{2\left(\frac{4}{3}\right)-1}$$
$$3^{2+\frac{4}{3}} \stackrel{?}{=} 9^{\frac{8}{3}-1}$$
$$3^{\frac{10}{3}} \stackrel{?}{=} 9^{\frac{5}{3}}$$
$$3^{\frac{10}{3}} \stackrel{?}{=} (3^2)^{\frac{5}{3}}$$
$$3^{\frac{10}{3}} \stackrel{\checkmark}{=} 3^{\frac{10}{3}}$$

5. Using a Graph From the graph we have $x = -2$.

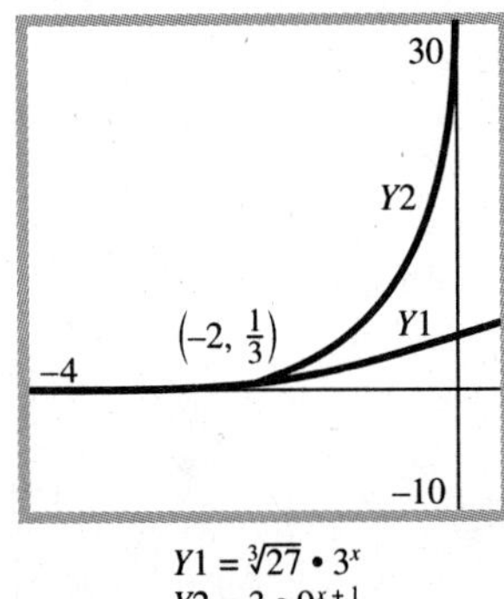

$Y1 = \sqrt[3]{27} \cdot 3^x$
$Y2 = 3 \cdot 9^{x+1}$

Figure 7.23

Using an Equation

$$3 \cdot 9^{x+1} = \sqrt[3]{27} \cdot 3^x$$
$$3 \cdot (3^2)^{x+1} = 3 \cdot 3^x$$
$$3 \cdot 3^{2x+2} = 3^{x+1}$$
$$3^{2x+3} = 3^{x+1}$$

The bases are the same. Thus the exponents must be equal.

Therefore

$$2x + 3 = x + 1$$
$$x = -2$$

Check:

$$3 \cdot 9^{-2+1} \stackrel{?}{=} \sqrt[3]{27} \cdot 3^{-2}$$
$$3 \cdot 9^{-1} \stackrel{?}{=} 3 \cdot 3^{-2}$$
$$\frac{3}{9} \stackrel{?}{=} \frac{3}{3^2}$$
$$\frac{3}{9} \stackrel{\checkmark}{=} \frac{3}{9}$$

We should be able to solve these simpler equations in Examples 2–5 by using algebraic properties as well as by graphing. With many exponential equations, however, we may not be lucky enough to have all quantities involve the same base. Such equations will be considered in Chapter 9.

Radical Equations

In Section 7.3, we saw some interesting applications using radicals. We now investigate some of these problems in more detail.

EXAMPLE 6

The velocity v (in meters per second) of an object reaching a height of d meters is given by $v = 4\sqrt{d}$. If we launch a missile with the velocity of 32 meters per second, what height will the missile reach? (Assume there is no air friction.)

Solution

6. Using a Graph From the graph we read $d = 64$.

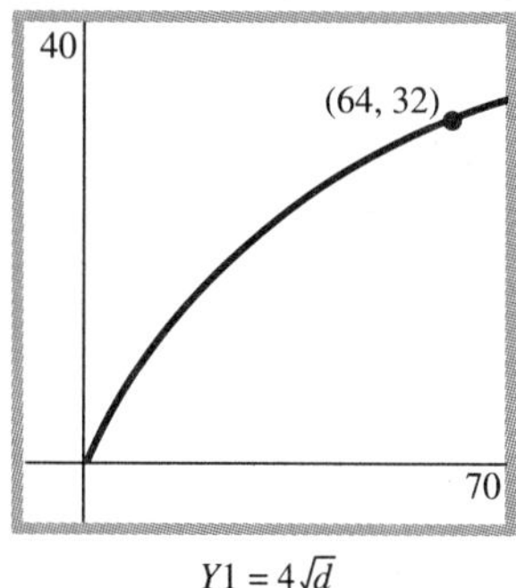

$Y1 = 4\sqrt{d}$

Figure 7.24

Using an Equation

$32 = 4\sqrt{d}$

How do we find d? We can proceed as follows:

$8 = \sqrt{d}$ **Divide both sides by 4.**

$8^2 = (\sqrt{d})^2$ **Square both sides.**

$64 = d$

Checking, we have

$32 \stackrel{?}{=} 4\sqrt{64}$

$32 \stackrel{\checkmark}{=} 4 \cdot 8$ **It checks.** ■

The above equations suggest how we can solve a radical equation algebraically. We can try to transform it into an equivalent linear or quadratic equation and then solve it as we did before. To find an equivalent equation not involving radicals, we can square both sides of the equation if square roots are involved; cube both sides of the equation if cube roots are involved; and so on. Alternatively, we can solve radical equations by graphing.

EXAMPLES 7–9

Solve for x algebraically and by graphing. Check the answer.

7. $\sqrt[3]{3x + 4} = 8$

8. $\sqrt{2x} - \sqrt{x + 1} = 0$

9. $3 + \sqrt{4x + 1} = 0$

Solutions

7. Using an Equation

$$\sqrt[3]{3x + 4} = 8$$

$$(\sqrt[3]{3x + 4})^3 = 8^3 \quad \text{Cube both sides.}$$

$$3x + 4 = 512$$

$$3x = 508$$

$$x = \frac{508}{3}$$

Using a Graph

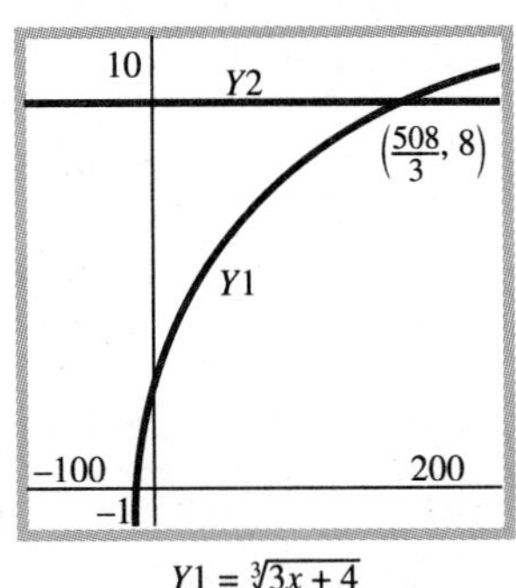

$Y1 = \sqrt[3]{3x+4}$
$Y2 = 8$

Figure 7.25

Check:

$$\sqrt[3]{3\left(\frac{508}{3}\right) + 4} \stackrel{?}{=} 8$$

$$\sqrt[3]{512} \stackrel{\checkmark}{=} 8$$

8. Using an Equation

$$\sqrt{2x} - \sqrt{x + 1} = 0$$

$$\sqrt{2x} = \sqrt{x + 1}$$

$$(\sqrt{2x})^2 = (\sqrt{x + 1})^2$$

$$2x = x + 1$$

$$x = 1$$

We try to get only one radical on each side of the equation. It is easier to square. We could square as is: $(\sqrt{2x} - \sqrt{x + 1})^2$ but it is more work. Try it.

Using a Graph

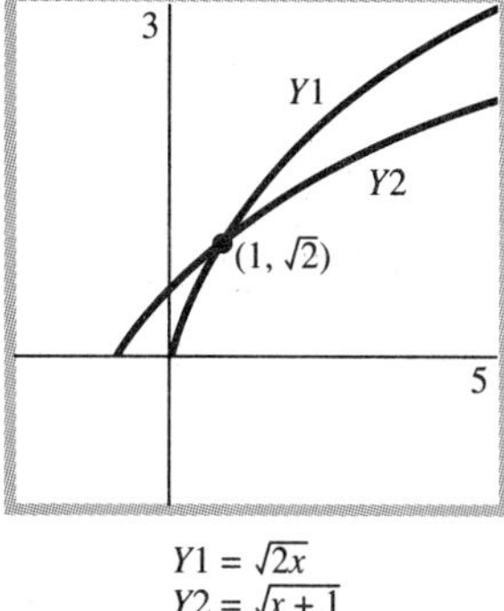

$Y1 = \sqrt{2x}$
$Y2 = \sqrt{x+1}$

Figure 7.26

Check:

$$\sqrt{2(1)} - \sqrt{1 + 1} \stackrel{?}{=} 0$$
$$\sqrt{2} - \sqrt{2} \stackrel{?}{=} 0$$
$$0 \stackrel{\checkmark}{=} 0$$

9. Using an Equation

$$\sqrt{4x + 1} + 3 = 0$$
$$\sqrt{4x + 1} = -3$$
$$(\sqrt{4x + 1})^2 = (-3)^2 \quad \text{Square both sides.}$$
$$4x + 1 = 9$$
$$4x = 8$$
$$x = 2$$

Using a Graph

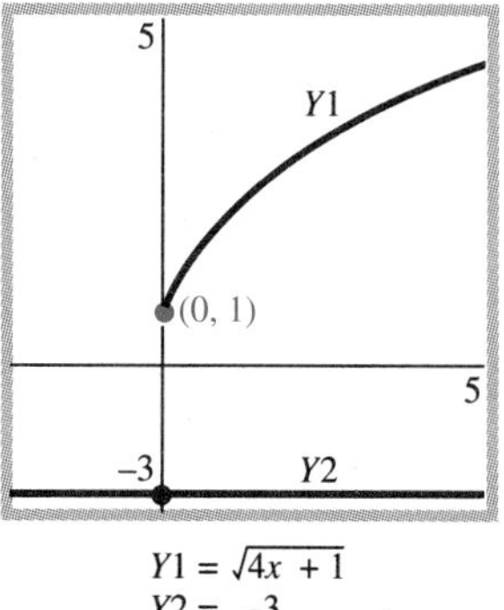

$Y1 = \sqrt{4x + 1}$
$Y2 = -3$

Figure 7.27

Check:

$$\sqrt{4(2)+1}+3 \stackrel{?}{=} 0$$
$$3+3 \stackrel{?}{=} 0$$
$$6 \neq 0$$ **It doesn't check, so there is *no solution*.**

In Example 9, we could see that the equation $\sqrt{4x+1}+3=0$ has no solutions, since $\sqrt{4x+1}=-3$ and the symbol $\sqrt{(\ )}$ always means "take the positive square root of ( )." Clearly, -3 is not positive. On the graph the line $Y2=-3$ does not intersect the graph of $Y1=\sqrt{4x+1}$.

Example 9 illustrates the primary obstacle in solving radical equations—that squaring, cubing, and so on, both sides of an equation may introduce extra or **extraneous roots.** That is, there may be roots to the new equation (obtained by squaring both sides of the original equations) that are not roots to the original equation. One way to determine all the roots to the original equation is to check all roots of the "derived" equation in the original equation and to reject those that do not work. Another way is to check where the graphs cross. To illustrate why extraneous roots may occur, examine the following example. In the equation,

$$x = 3$$

there is only one root, namely, 3. However, if we square both sides of this equation, we obtain

$$x^2 = 3^2$$
$$x^2 = 9$$ **At this step, we could have taken the square root of both sides: $\sqrt{x^2}=\pm\sqrt{9}$.**
$$x^2 - 9 = 0$$
$$(x+3)(x-3) = 0$$

which has roots $x=3$ or $x=-3$. Notice that -3 is an extraneous root, one that does not check in the *original equation*, $x=3$.

EXAMPLES 10–12

Solve algebraically and by graphing. Check the answers by substituting.

10. $x+\sqrt{7-3x}=1$ **11.** $5=\sqrt{3x+13}-x$

12. $\sqrt{2x+2}-\sqrt{x+2}=1$

Solutions

10. Using an Equation

$$x+\sqrt{7-3x} = 1$$
$$\sqrt{7-3x} = 1-x$$ **It is always easiest to isolate the radical on one side of the equation, and put all other terms on the other side.**
$$(\sqrt{7-3x})^2 = (1-x)^2$$
$$7-3x = 1-2x+x^2$$
$$0 = x^2+x-6$$

$$0 = (x + 3)(x - 2)$$
$$x = -3 \quad \text{or} \quad x = 2$$

Using a Graph

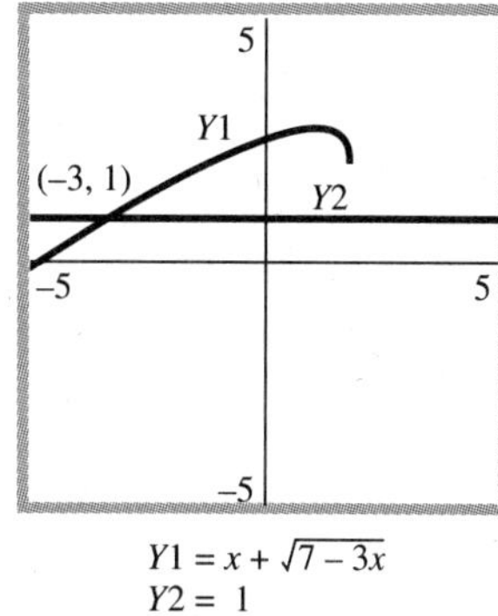

$Y1 = x + \sqrt{7 - 3x}$
$Y2 = 1$

Figure 7.28

Check: When $x = 2$,

$$2 + \sqrt{7 - 3 \cdot 2} = 2 + 1 = 3$$
$$3 \neq 1$$

So 2 is an extraneous root. When $x = -3$,

$$-3 + \sqrt{16} = -3 + 4 \stackrel{?}{=} 1$$
$$-3 + \sqrt{16} = -3 + 4 \stackrel{\checkmark}{=} 1$$

So -3 checks and $x = -3$ is the only solution.

Notice that squaring both sides can lead to a quadratic equation instead of a linear equation. However, the procedure for solving was the same—solve the quadratic, and check its roots in the original equation to see if any of them are extraneous roots.

11. Using an Equation

$$5 = \sqrt{3x + 13} - x$$
$$5 + x = \sqrt{3x + 13}$$
$$(5 + x)^2 = (\sqrt{3x + 13})^2$$
$$25 + 10x + x^2 = 3x + 13$$
$$x^2 + 7x + 12 = 0$$
$$(x + 3)(x + 4) = 0$$
$$x = -3 \quad \text{or} \quad x = -4$$

Using a Graph

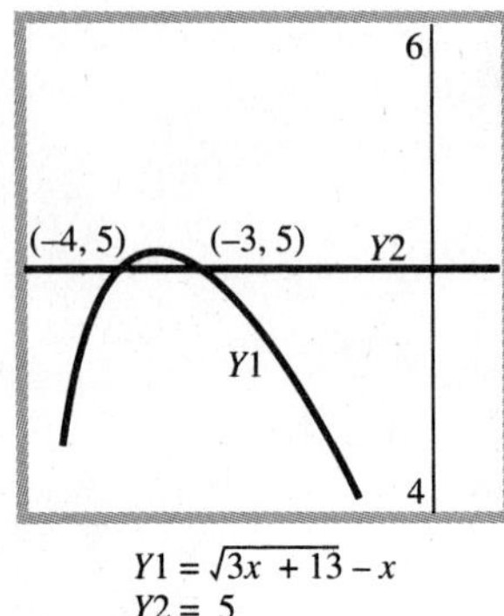

$Y1 = \sqrt{3x + 13} - x$
$Y2 = 5$

Figure 7.29

Check: When $x = -3$,

$$5 \stackrel{?}{=} \sqrt{3(-3) + 13} - (-3)$$
$$5 \stackrel{?}{=} \sqrt{4} + 3$$
$$5 \stackrel{\checkmark}{=} 5$$

When $x = -4$,

$$5 \stackrel{?}{=} \sqrt{3(-4) + 13} - (-4)$$
$$5 \stackrel{?}{=} \sqrt{1} + 4$$
$$5 \stackrel{\checkmark}{=} 5$$

Both $x = -3$ and $x = -4$ are solutions.

12. Using an Equation

$$\sqrt{2x + 2} - \sqrt{x + 2} = 1$$
$$\sqrt{2x + 2} = 1 + \sqrt{x + 2}$$ Put one radical on each side of the equation to simplify squaring.
$$(\sqrt{2x + 2})^2 = (1 + \sqrt{x + 2})^2$$
$$2x + 2 = 1 + 2\sqrt{x + 2} + x + 2$$
$$x - 1 = 2\sqrt{x + 2}$$
$$(x - 1)^2 = (2\sqrt{x + 2})^2$$ Square again.
$$x^2 - 2x + 1 = 4(x + 2)$$
$$x^2 - 6\text{x} - 7 = 0$$
$$(x - 7)(x + 1) = 0$$
$$x = 7 \quad \text{or} \quad x = -1$$

Using a Graph

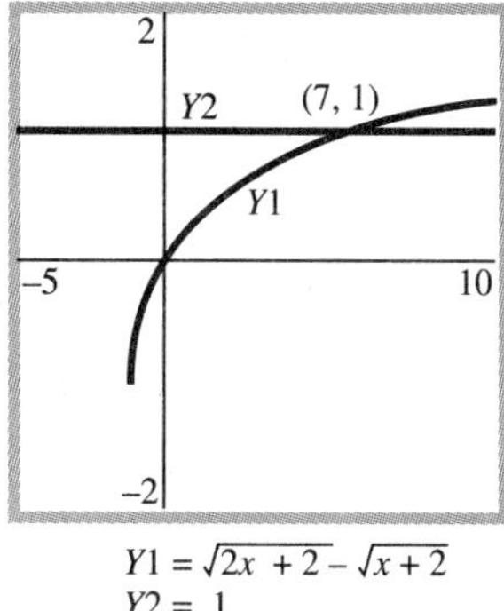

$Y1 = \sqrt{2x+2} - \sqrt{x+2}$
$Y2 = 1$

Figure 7.30

Check: When $x = 7$,

$$\sqrt{16} - \sqrt{9} \stackrel{?}{=} 1$$
$$4 - 3 \stackrel{?}{=} 1$$
$$1 \stackrel{\checkmark}{=} 1$$

When $x = -1$,

$$\sqrt{0} - \sqrt{1} \stackrel{?}{=} 1$$
$$-1 \neq 1$$

So $x = -1$ is extraneous; $x = 7$ is the only solution. ■

In Examples 10–12, we can solve the equations algebraically or by graphing. If we use graphing, then care must be taken to set the range of x-values and y-values so that a solution can be read from the graph.

EXAMPLE 13 Find all points on the x-axis that are 5 units from the point (6, 3).

Solution

There are two solutions, as you can tell from the diagram on the next page. If we let $(x, 0)$ represent the desired point, then from the distance formula, we obtain

$$\sqrt{(x-6)^2 + (0-3)^2} = 5$$
$$(x-6)^2 + 9 = 25$$
$$(x-6)^2 = 16$$
$$x - 6 = \pm 4$$
$$x = 2 \quad \text{or} \quad 10$$

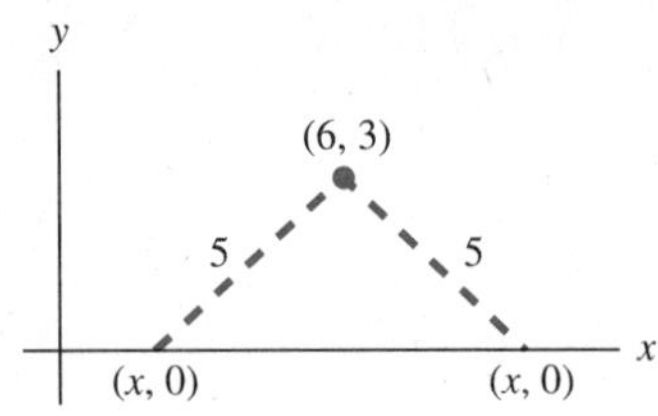

Figure 7.31

Both values check in the original equation, so the desired points are (2, 0) and (10, 0).

Strategies for Solving a Radical Equation

1. Use a table to read the values directly.
2. Graph the equation (often two equations, one for each side of the original equation), and then read the solution from the graph.
3. Use an equation.
   (a) Isolate the radical to one side of the equality, if possible.
   (b) Square both sides, cube both sides, and so on.
   (c) Solve the resulting equation (usually a linear or quadratic equation).
   (d) Check for extraneous roots.

For most radical equations (or any other equation), graphing or solving the equation using algebraic properties are usually the most convenient solution methods. A table is useful to get a sense of the algebraic pattern. The graph is usually relatively easy to read and gives a visual sense of the pattern that the equation represents. We close with a few applications involving radical equations.

EXAMPLE 14

Find the dimensions of a cylinder if the height h is $\frac{2}{3}$ as large as the radius r and the volume v is 1,525 cm^3.

Solution

Using an Equation The volume of a cylinder is

$$v = \pi r^2 h$$

Here the height

$$h = \frac{2}{3}r$$

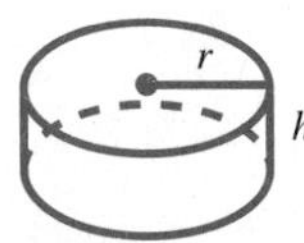

Figure 7.32

Therefore

$$v = \pi r^2\left(\frac{2}{3}r\right)$$

$$v = \frac{2}{3}\pi r^3$$

$$1{,}525 = \frac{2}{3}\pi r^3$$

$$r^3 = 728.10$$

$$\sqrt[3]{r^3} = \sqrt[3]{728.10}$$ We take the cube root of both sides of the equation.

$$r \approx 8.996$$

Check: If $r = 8.996$, then

$$h = \frac{2}{3}(8.996)$$

$$1{,}525 \stackrel{?}{=} \frac{2}{3}(8.996)\ \pi\,(8.996)^2$$

$$1{,}525 \stackrel{\checkmark}{=} 1{,}524.78 \stackrel{\checkmark}{\approx} 1{,}525$$

Using a Graph If we graph $v = \frac{2}{3}\pi r^3$ [Figure 7.33(a)], then we can read the value of r that corresponds to $v = 1{,}525$ cm^3. The graph allows us to read many values for either v or r. This is much quicker than having to stop and solve an equation for each value that we want to find. We can also solve this by graphing $Y1 = \frac{2}{3}\pi r^3$ and $Y2 = 1{,}525$ [Figure 7.33(b)].

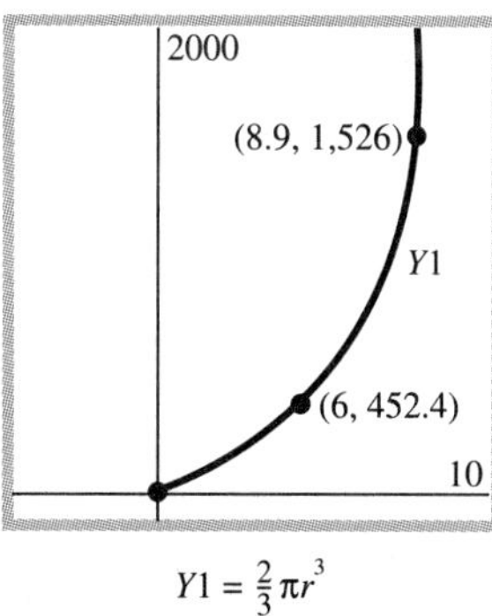

$Y1 = \frac{2}{3}\pi r^3$

Figure 7.33 (a)

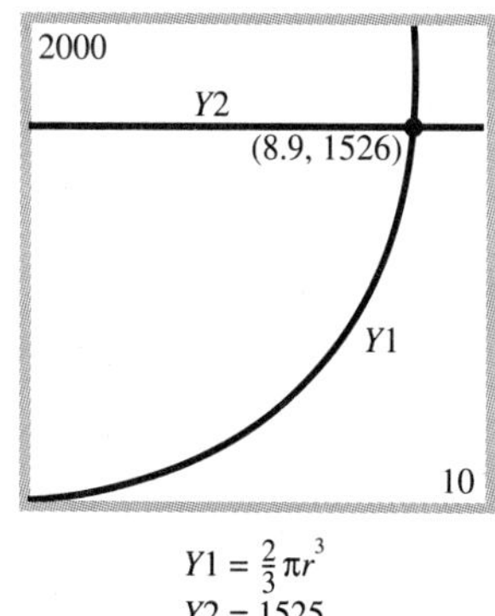

$Y1 = \frac{2}{3}\pi r^3$
$Y2 = 1525$

Figure 7.33 (b)

EXAMPLE 15 The pulse rate p (in beats per minute) of an adult H meters tall is

$$p = \frac{94}{H^{\frac{1}{2}}}$$

(a) Solve for H.

(b) Find H if $p = 70$ beats per minute.

Solutions

Using an Equation

(a)

$$p = \frac{94}{H^{\frac{1}{2}}}$$

$$H^{\frac{1}{2}}p = 94 \qquad \text{Isolate } H.$$

$$H^{\frac{1}{2}} = \frac{94}{p}$$

$$\left(H^{\frac{1}{2}}\right)^2 = \left(\frac{94}{p}\right)^2 \qquad \text{Square both sides.}$$

$$H = \left(\frac{94}{p}\right)^2$$

(b) $H = \left(\dfrac{94}{70}\right)^2$

$H \approx 1.8$ meters

Using a Graph **(b)** A graph can be used to find the value of H. Again the advantage here is that we can graph the equation, $p = \frac{94}{H^{\frac{1}{2}}}$, directly and then read the values directly from the graph. See Figure 7.34(a). We could also solve this by graphing $Y1 = \frac{94}{H^{\frac{1}{2}}}$ and $Y2 = 70$ [Figure 7.34(b)].

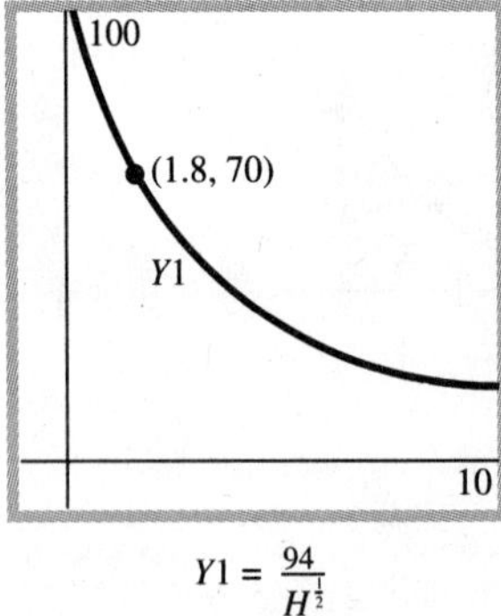

$Y1 = \frac{94}{H^{\frac{1}{2}}}$

Figure 7.34 (a)

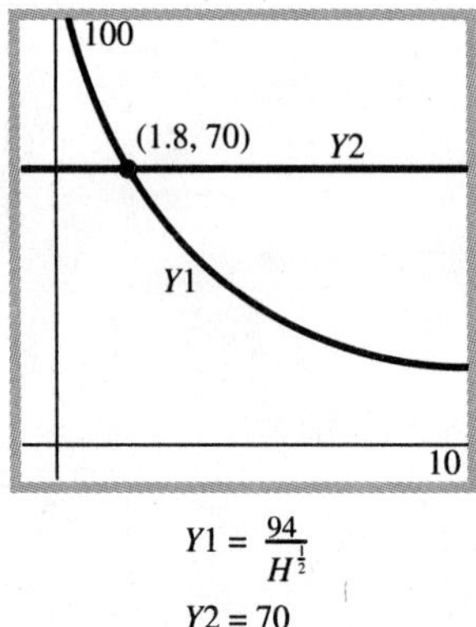

$Y1 = \frac{94}{H^{\frac{1}{2}}}$

$Y2 = 70$

Figure 7.34 (b)

The following problem uses a bit of whimsy to illustrate the use of an exponential function.

EXAMPLE 16

Suppose Leaping Leonard signs a contract with an NBA team for \$1 million a year for 25 years. But a rookie for another NBA team signs for \$1 the first year, \$2 the second year, \$4 the third year, \$8 the fourth year, and so on for 25 years. Considering only money from salaries, who will have the most money at the end of 25 years? (What is the chance an NBA player could last 25 years?)

Solution

Using a Table

Year	*Leonard's salary (total)*	*Rookie's yearly salary*	*Rookie's salary (total)*
1	\$ 1 million	\$ 1	\$ 1
2	2 million	2	3
3	3 million	4	7
4	4 million	8	15
10	10 million	512	1,021
15	15 million	16,384	32,767
20	20 million	524,288	1,048,575
24	24 million	8,388,608	16,777,215
25	\$25 million	\$16,777,216	\$33,554,431

Table 7.6

Even though the rookie's salary is growing very slowly, by the end of the 25th year he has received \$33,554,431, while Leonard has received only \$25 million. If Leonard makes some wise investments, he could end up with more then \$33,000,000.

Using an Equation Leonard's salary, S_L, after t years is $S_L = 10^6 t$. If we observe the pattern in the rookie's total salary, it is a power of 2 (double the amount received that year minus 1). That is, the rookie's salary, S_R, after t years is $S_R = 2^t - 1$. If we substitute 25 for t and solve for S_L or S_R, we get the same values as we did in Table 7.5.

Using a Graph We can graph the equations representing the salaries of Leonard and the rookie. We can read the value of the salary on each graph when $t = 25$. The beauty of the graph is that we can visually see the linear growth of Leonard's salary and the exponential growth of the rookie's salary. In addition, we can read an infinite number of values from the graph. If we want to know when the two equations intersect or have the same value, then the graph is the best way to find this point, $t = 24.68$ and $S = \$24{,}679{,}889$.

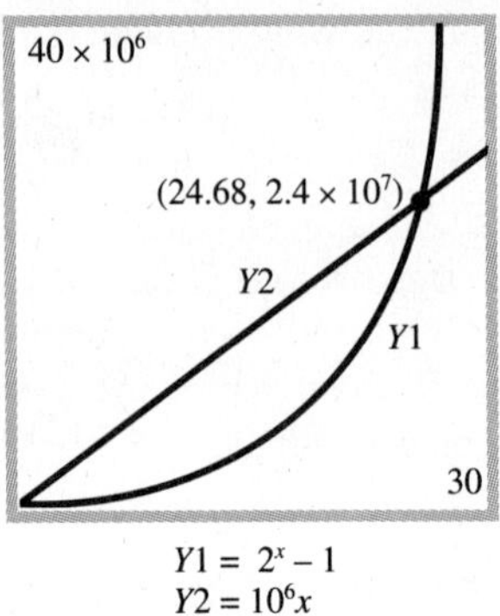

$Y1 = 2^x - 1$
$Y2 = 10^6 x$

Figure 7.35

Many of the exponential equations we have studied thus far are of the form $A = A_0 x^m$. If m is negative, these problems are referred to as decay problems. Examples include rate of decay (half-life) of radioactive isotopes (used, for example, in carbon-14 dating of ancient artifacts) and the forgetting curve of debt payments. If m is positive, these problems are generally called growth problems. Examples include interest accumulated on a savings account, growth of biological cells, population expansion, the learning curve, or inflation. Chapter 9 contains many more examples of growth and decay problems. So far we have been asked to find A or A_0, given x and m. We will now look at a few cases where we need to find the exponent m.

EXAMPLE 17

Approximately how long will it take to double an investment of \$100 if it is invested at 8% annual interest compounded quarterly?

Solution

Using an Equation The formula for compound interest is

$$A(t) = P\left(1 + \frac{I}{n}\right)^{nt}$$

Where
$A(t)$ is the total amount accumulated
P is the principal at the beginning
I is the rate of interest
n is times per year it is compounded
t is the number of years

Therefore

$$\begin{aligned} P &= \$100 \\ I &= 0.08 \\ n &= 4 \text{ times a year} \\ A(t) &= 200 \text{ (double } P) \end{aligned}$$

We have

$$200 = 100\left(1 + \frac{0.08}{4}\right)^{4t}$$

or

$$2 = (1.02)^{4t}$$
$$2 = [(1.02)^4]^t \qquad (1.02)^4 \approx 1.08$$
$$2 = (1.08)^t$$

Now we are left to find t. At this point, we will have to guess for t by raising 1.08 to various powers until we approximate 2:

$$(1.08)^4 \cong 1.36 \qquad \text{Too small.}$$
$$(1.08)^8 \cong 1.85 \qquad \text{Too small, but closer.}$$
$$(1.08)^9 \cong 1.999 \cong 2 \qquad \text{Very close.}$$
$$(1.08)^{10} \cong 2.16 \qquad \text{Too large.}$$

In Chapter 9 on logarithms, we will develop a strategy using logarithms that does not involve guess and check. Therefore, our \$100 will double in approximately nine years if we invest it at 8% annual interest compounded quarterly.

Using a Graph We can also use a graph to find the time needed for the investment to double. Graph $A(t) = 100(1 + 0.02)^{4t}$. See Figure 7.36(a). We could also graph $Y1 = 100(1 + 0.02)^{4t}$ and $Y2 = 2$. See Figure 7.36(b).

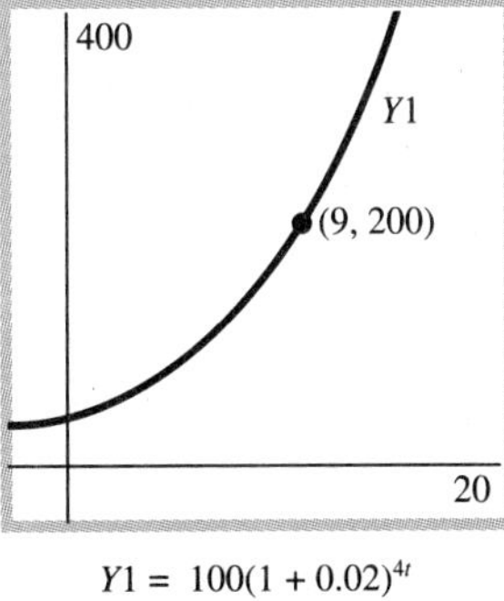

$Y1 = 100(1 + 0.02)^{4t}$

Figure 7.36 (a)

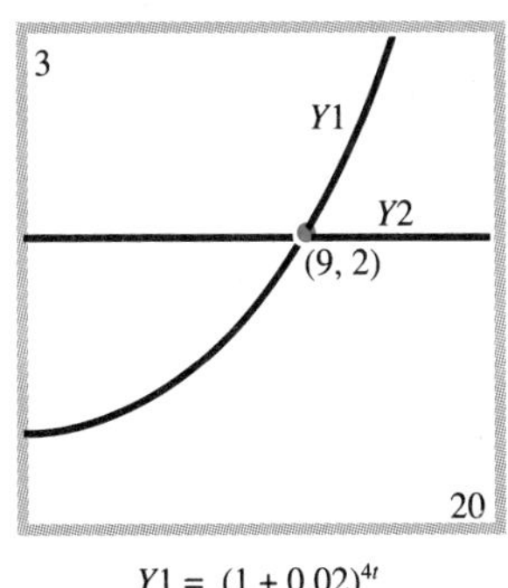

$Y1 = (1 + 0.02)^{4t}$
$Y2 = 2$

Figure 7.36 (b)

Let's take one last look at solving an exponential equation by graphing. We will solve the following equation using graphs in three different ways. For some problems there may be more than three choices. You may use any of these methods to solve equations with graphs.

EXAMPLE 18 Solve the following equation by graphing.

$$2^{x+1} = 100$$

Solution 1

Let $2^{x+1} = y$. Graph the equation and find x when $y = 100$. From the graph (Figure 7.37), $x = 5.63$.

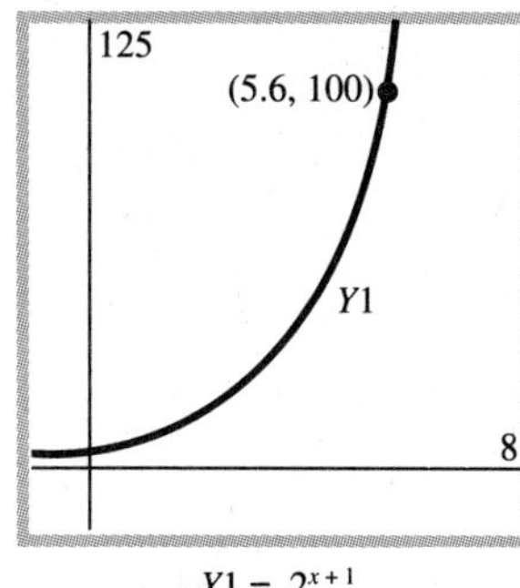

$Y1 = 2^{x+1}$

Figure 7.37

Solution 2

Let $2^{x+1} = Y1$ and $Y2 = 100$. Graph both equations and find the point of intersection: $x = 5.63$ and $y = 100$ (Figure 7.38).

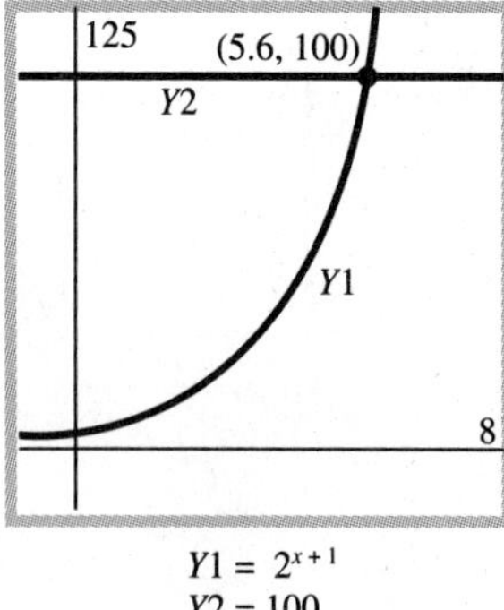

$Y1 = 2^{x+1}$
$Y2 = 100$

Figure 7.38

Solution 3

Rewrite $2^{x+1} = 100$ as $2^{x+1} - 100 = 0$. Let $y = 2^{x+1} - 100$. Find x when $y = 0$: $x = 5.63$ (Figure 7.39).

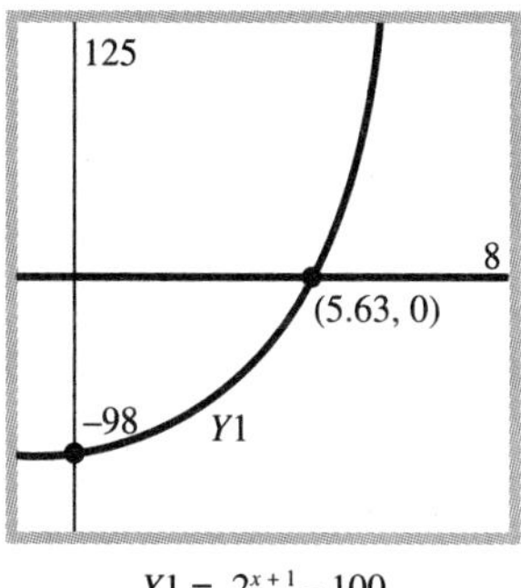

$Y1 = 2^{x+1} - 100$

Figure 7.39

7.4 EXERCISES

I. Procedures

1–34. Solve each of the following equations algebraically or graphically. Check your answers by substitution.

1. $2^x = 4$

2. $2^x = 1$

3. $2^x = 2^{-3}$

4. $2^x = \frac{1}{16}$

5. $2^{x+1} = \frac{1}{64}$

6. $5^{-\frac{1}{x}} = 125$

7. $2 \cdot 2^{x-2} = 64$

8. $4 \cdot 2^x = \frac{1}{64}$

9. $3^{2x} = 3^6$

10. $16^{x+3} = 2^5$

11. $9^{4x} = 27^{x^2-1}$

12. $2^{3x-1} = 32^{\frac{4}{5}}$

13. $5^{2x} = 5^{x-1}$

14. $3^{x+1} \cdot 9^{2x} = 27^{x+1}$

15. $4 \cdot 8^{x+1} = 32^{x-1}$

16. $2^x \cdot 2^{x+2} = 4^{x+1}$

17. $\sqrt{4x} = 12$

18. $\sqrt{6x} - 1 = 1$

19. $\sqrt{2x+1} + 3 = 0$

20. $2\sqrt{x} = \sqrt{3x+4}$

21. $\sqrt{3x-15} = \sqrt{2x}$

22. $\sqrt{6x-5} - x = 0$

23. $3 + x = \sqrt{6x+13}$

24. $x - 1 = \sqrt{7-x}$

25. $x + \sqrt{4x+1} = 5$

26. $2x + \sqrt{3(x+3)+1} = 3x + 4$

27. $4 = \sqrt{3x+10} - x$

28. $x = 2 + \sqrt{x-2}$

29. $\sqrt{x^2-8} = \sqrt{8-x^2}$

30. $\sqrt{x+2} - \sqrt{x} = 2$

31. $\sqrt{x} + \sqrt{x+3} = 3$

32. $\sqrt{x+4} + \sqrt{x-4} = 4$

33. $\sqrt{2x+6} - \sqrt{x+4} = 1$

34. $\sqrt{2x+5} + 2\sqrt{x+6} = 5$

35–40. Find at least one value of x that satisfies the following equations.

35. $x^5 = -32$

36. $x^3 = -125$

37. $x^3 = \frac{1}{8}$

38. $x^2 \cdot x^2 = 256$

39. $x^{\frac{3}{2}} = 8$

40. $x^{\frac{3}{4}} = 27$

II. Concepts

41–44. State whether each of the following statements is always true (AT), sometimes true (ST), or never true (NT). For each statement that is sometimes true, give an example of when it is true and when it is false.

41. $2^x \cdot 4^y = 8^{x+y}, x \neq 0, y \neq 0$

42. $10^n(1{,}000^n - 10^n) = 10^{2n}(10^{2n} - 1)$

43. $\sqrt{(a-b)^2} = a - b$

44. $\sqrt{a^2 - b^2} = a - b$

45. Describe the processes that can be used to solve a radical equation, and then use the following equation as an example to illustrate your strategies:

$$\sqrt{3x+1} - \sqrt{x} = 6$$

46. Describe the processes that can be used to solve an exponential equation, and then use the following equation as an example to illustrate your strategies:

$$1{,}024 - 2^{x+1} = 0$$

47 and 48. Graph each set of equations on the same axes. Describe the behavior of the graphs. What are the similarities and differences among the graphs in each set?

47. $y = 2^x \qquad y = 2^{-x} \qquad y = 2^x + 1$

48. $y = \sqrt{x} \qquad y = \sqrt{x+1} \qquad y = \sqrt{x} + 1$

49 and 50. In the following exercises, the person who solved the problem *has made an error.* Find the error, and solve the problem correctly.

49. Solve for x: $2 \cdot 4^x = 64$

$$2 \cdot 4^x = 8^x$$
$$8^x = 64$$
$$8^x = 8^2$$
$$x = 2 \quad \text{Wrong answer.}$$

50. Solve for x: $\sqrt{x+6} - x = 2$

$$x + 6 - x^2 = 4$$
$$-x^2 + x + 6 = 4$$
$$-x^2 + x + 2 = 0$$
$$x^2 - x - 2 = 0 \quad \text{Solve the quadratic equation.}$$
$$x = 2 \quad \text{or} \quad x = -1 \quad \text{Both are wrong answers.}$$

III. Applications

51–60. Solve each of the following problems.

51. A certain amoeba doubles every hour. If we start with 4 amoebas, approximately how long will it take to acquire 1,024 amoebas?

52. A different strain of amoeba doubles every 30 minutes. How many *hours* will it take to acquire 5,120 amoebas if we start with 10 cells?

53. Approximately how long will it take to double an investment of $1,000 if it is invested at 9.6% annual interest that is

(a) compounded quarterly?

(b) compounded monthly?

54. How long will it take a certain city with a population of 10,000 to triple in size if its population is growing at the rate of 10% a year?

55. The population of a town is 120,000 and is increasing at the rate of 3.5% each year.

(a) What is the population, P, after t years?

(b) What is the population after 20 years?

(c) Approximately how long will it take the population to double?

56. The population of a colony of rats in a certain city is estimated to be 500,000. The town hires an exterminator who promises to decrease the population each year by half of what it was the year before.

(a) Write an equation for the population of rats, P, after t years.

(b) How long will it take to get the population down to 250,000 rats?

(c) How long will it take to get the population down to 10 rats?

57. The price of a car in 1980 was $7,500, and in 1990 it was $17,000. Estimate the average annual rate of inflation.

58. During 1990 a family of four buying groceries paid $50 for the same purchase that cost $35 in 1980. What was the average rate of inflation per year from 1980 to 1990?

59. During 1990 the average rate of inflation was 4%.

(a) Assuming this rate remains constant, what will a pound of hamburger cost in 2000 if it costs $1.89 in 1990?

(b) What will a college education cost in 2000 if the average cost in 1990 is $6,000 a year?

60. If inflation is currently 10% a year and a Volkswagen Jetta costs $9,000, how many years will it be before it will cost $13,000? Assume inflation remains at 10% a year.

61. Suppose a rookie in another league in Example 16 was paid by the following schedule:

- $1 for the first year
- $3 for the second year
- $9 for the third year
- $27 for the fourth year
- $81 for the fifth year, etc.

(a) How much money would he receive in the tenth year? What would be his total salary for the first ten years?

(b) How much money would he receive in the nth year? What would be his total salary for the first n years?

(c) What is the first year in which his total salary will exceed \$50 million?

62. The amount of annual sales, S (in millions of dollars), of the AB Company is related to the number of years, t, since it was formed, as shown by the equation

$$\text{Sales(time)} = S(t) = 2\,(1.25)^t$$

(a) What is the annual sales figure in the tenth year?

(b) In what year will annual sales reach \$10 million?

63. Newton's law of cooling states that the temperature, T, of a heated cup of coffee is related to the time, t (in minutes), by the equation

$$T = 120e^{-0.2t} + 65$$

(e is a constant approximately equal to 2.72).

(a) What is the temperature of the coffee before the cooling process begins?

(b) What is the temperature of the coffee after five minutes? Ten minutes? Twenty minutes?

(c) How long will it take the coffee to cool to room temperature (65 degrees)?

64. Pelidisi is a ratio used to determine the malnutrition in a person based on height (ht) and weight (wt). The ratio is

$$\text{pelidisi} = \frac{\sqrt[3]{10 \times (\text{weight in grams})}}{\text{sitting height in centimeters}} \times 100\%$$

A pelidisi of 98%–102% indicates a healthy person.

(a) What is the pelidisi ratio for an adult with a sitting height of 44 *inches* (from chair to the top of the head) and a weight of 220 pounds? How would you describe the health of this individual?

(b) A certain adult with a pelidisi of 100% has a sitting height of 30 *inches*. How much does this person weigh? *Hint:* Solve for weight first.

65. The formula for approximating the velocity v of a car based on the length of its skid marks l (in feet) is

$$v = \sqrt{12l} \text{ on wet pavement}$$

$$v = \sqrt{24l} \text{ on dry pavement}$$

(a) If a car is traveling 55 mph, what will be the length of the skid marks on wet pavement?

(b) On dry pavement?

66. Find all points on the y-axis that are five units from the point (4, 3).

67. Find all points on the line $y = 2$ that are twice as far from (3, 0) as they are from the origin.

68. The volume of a cube is 64×10^6 mm^3. Find the length of an edge of the cube.

69. The volume of a sphere is 600 m^3. What is the radius of the sphere?

70. Find the radius of a cone whose volume is 225π cm^3 and whose height is $\frac{5}{3}$ as large as the radius.

71. The volume of a cylinder whose height is twice the radius is 315 mm^3. What is the surface area of the cylinder?

72. The volume, V, of a sphere is related to its surface area, A, by the formula

$$V = 0.094A^{\frac{3}{2}}$$

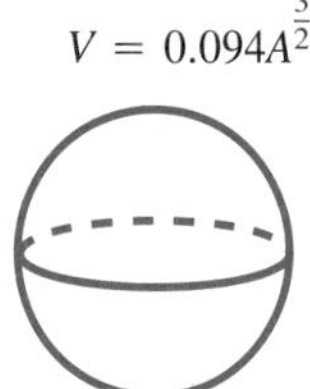

(a) What is the surface area of a soap bubble whose volume is 10 cm^3?

(b) What is the surface area of a hot air balloon whose volume is 899 m^3?

73. In the triangle shown, find x if the area of the triangle is 10.

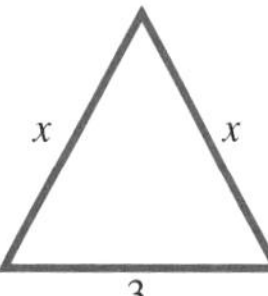

IV. Extensions

74. An oil company would like to ship oil in 10,000-gallon containers. The containers are to be made out of a specified metal and can be shaped like a cone, cylinder, rectangular box, or sphere. The company would like to keep the cost of the container to a minimum. Which shape of container do you think the company should pick if cost is the only consideration? Why?

75. Solve for x: $\sqrt{x} + \sqrt{x-7} = \dfrac{21}{\sqrt{x-7}}$

76. Check on a calculator and then verify algebraically:

(a) $\sqrt{5 + 2\sqrt{6}} = \sqrt{2} + \sqrt{3}$

(b) $\sqrt{9 - 2\sqrt{14}} = \sqrt{7} - \sqrt{2}$

77. The *arithmetic mean* of two real numbers is simply their average. If the two numbers are x and y, then their arithmetic mean is $\frac{x+y}{2}$. The *geometric mean* of these two numbers is $\sqrt{xy}$. For what values of x and y is the arithmetic mean

(a) equal to the geometric mean?

(b) greater than the geometric mean?

78. Much is said about interest being compounded monthly, daily, or even continuously. How much difference is there between the various plans? In the compound interest formula

$$A = P\left(1 + \frac{r}{n}\right)^{nt}$$

let $P = \$1$, let $t = 1$ year, and assume $r = 1\%$. Calculate A for interest compounded annually, quarterly, monthly, weekly, daily, and hourly. What is your conclusion about the various savings plans? Repeat for $P = \$1{,}000$ and $r = 10\%$.

79. (a) In the following pattern with right triangles, find the lengths of the sides p, q, r, and s.

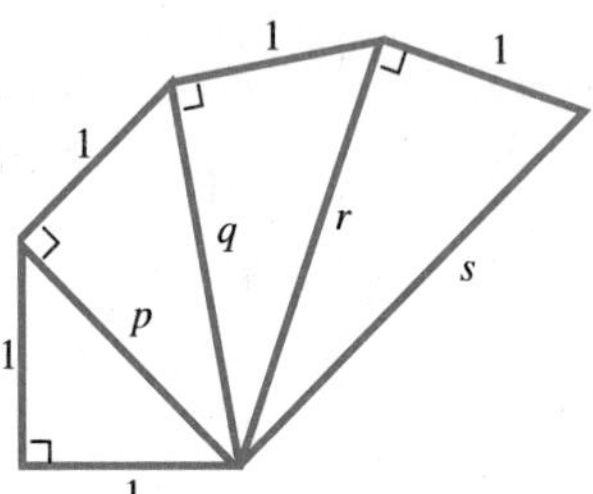

(b) If the pattern continues, what will be the length of the sides of the next two triangles?

CHAPTER 7 SUMMARY

Some of the applications involving exponents from Chapter 2 were reexamined. For example, if the intensity of light filtering through water is reduced by three-fifths for each meter of water, then the amount of sunlight at a depth of n meters is $\left(\frac{2}{5}\right)^n$. By letting $n = \frac{2}{3}$ of a meter, we were motivated to study rational exponents so that we could calculate $\left(\frac{2}{5}\right)^{\frac{2}{3}}$. A similar question motivated a need to define $\left(\frac{2}{5}\right)^{\frac{1}{3}}$, or nth roots.

A discussion of the operations on rational exponents and radicals culminated with the solution of exponential and radical equations. Again, many variations of the applications involving exponents from Chapter 2 were solved, together with several new applications. We looked at three different methods of solution: using a table, using a graph, and using an equation.

Important Words and Phrases

principal nth root	exponential equation
distance formula	radical equation
rational exponent	extraneous root
radical (symbol)	conjugate
radicand	exponential function
index	

Important Properties and Procedures

- Properties of rational exponents:

$$x^a x^b = x^{a+b}$$
$$(x^a)^b = x^{ab}$$
$$(xy)^a = x^a y^a$$
$$\frac{x^a}{x^b} = x^{a-b}$$

where $x \neq 0$, $y \neq 0$, and a and b are rational numbers in simplest form.

- Properties of radicals: $\sqrt[n]{ab} = (\sqrt[n]{a})(\sqrt[n]{b})$;

$$\sqrt[n]{\frac{a}{b}} = \frac{\sqrt[n]{a}}{\sqrt[n]{b}} \text{ if } \sqrt[n]{a}, \sqrt[n]{b} \text{ exist}$$

- Determining the existence of real nth roots
- Determining the distance between any two points in a plane
- Finding equivalent expressions with radicals and rational exponents
- Calculating with radicals and rational exponents
- Solving an exponential equation
- Solving a radical equation

REVIEW EXERCISES

I. Procedures

1–12. Compute each of the following without calculators.

1. $\sqrt{625}$

2. $\sqrt[3]{-216}$

3. $256^{\frac{1}{4}}$

4. $\left(32^{-\frac{1}{5}}\right)^2$

5. $-64^{\frac{2}{3}}$

6. $\sqrt[4]{2}\sqrt[4]{8}$

7. $\dfrac{\sqrt[3]{16}}{\sqrt[3]{2}}$

8. $9^{\frac{3}{2}} \cdot 3$

9. $\dfrac{125^{\frac{2}{3}}}{25^{-\frac{1}{2}}}$

10. $(81^{\frac{3}{4}})(8^{\frac{2}{3}})$

11. $5^0 - \dfrac{1}{\sqrt[3]{-8}} + 64^{\frac{2}{3}} - 4^{-\frac{1}{2}}$

12. $25^{-\frac{1}{2}} + \dfrac{1}{625^{-1}} + 7^0 - \dfrac{1}{\sqrt[3]{125}} - |6 - 31|^2$

13–32. Write an equivalent expression for each expression; express all answers with positive exponents only and without radicals in the denominator. Assume all variables are positive.

13. $\sqrt[3]{-27y^8}$

14. $\sqrt[4]{256x^8y^5}$

15. $\left(-32^{-\frac{1}{5}}x^{\frac{3}{2}}\right)^2$

16. $\left(x^{\frac{3}{5}}y^{-\frac{1}{3}}\right)\left(x^{\frac{4}{5}}y^{\frac{1}{3}}\right)$

17. $\dfrac{-32^{\frac{1}{5}}y^{-1}}{2y^{-\frac{4}{3}}}$

18. $\dfrac{-27x^{\frac{3}{2}}y^{-2}}{3x^{\frac{1}{2}}y^{\frac{1}{3}}}$

19. $\dfrac{4(x^3y^{-2})^{\frac{1}{6}}}{x^{\frac{1}{2}}y^{\frac{2}{3}}}$

20. $\left(\dfrac{16x^{\frac{4}{3}}y^{-2}}{x^4y^2}\right)^{\frac{3}{4}}$

21. $\left(\dfrac{-8x^6y^5}{x^3y^{-1}}\right)^{-\frac{1}{3}}\left(\dfrac{3x^{-2}}{y^{-1}}\right)^0$

22. $\dfrac{\sqrt{7}\sqrt[3]{49}}{(\sqrt[n]{(x^2)^n}}$

23. $\left(\dfrac{-3x^9y^{-10}x^0}{x^{\frac{4}{3}}y^{-\frac{5}{2}}}\right)x^6y^6$

24. $\dfrac{\sqrt[4]{16x^8y^{10}}}{2x^2y^2}$

25. $\dfrac{x}{\sqrt{x}}$

26. $\dfrac{3 - x}{\sqrt{x}}$

27. $\dfrac{9 - x}{3 - \sqrt{x}}$

28. $\dfrac{y^2 - a^2}{\sqrt{a} + \sqrt{y}}$

29. $\sqrt{18} - \sqrt[3]{2} + \sqrt{72}$

30. $\sqrt[3]{5xy^2} - \sqrt[3]{3x^4y^2} + \sqrt[3]{40xy^8}$

31. $\dfrac{2^{-1}}{3^{-1}} - \dfrac{2}{3}$

32. $\dfrac{(x^{-2} - y^{-2})^{-1}}{(xy)^{-2}}$

33–44 Solve for x using two methods: a graph and an equation. Check your answers.

33. $(x-3)^2 = 16$

34. $4 = \sqrt{2x+6}$

35. $\sqrt[3]{2-x} - 3 = 0$

36. $25^x = 125$

37. $3 \cdot 9^x = 27^{-\frac{2}{3}}$

38. $2 \cdot 4^{2x+1} = 32^{\frac{2}{5}}$

39. $9^{2x-1} = 27^{\frac{4}{3}}$

40. $\sqrt{11+2x} - x = 6$

41. $2 - \sqrt{4+x} = x$

42. $\sqrt{x+1} - 1 = \sqrt{x}$

43. $\sqrt{x+1} - 1 = \sqrt{x-2}$

44. $x - 3\sqrt{x} = 4$

II. Concepts

45–52. Are these expressions true or false? If a statement is false, then give an example that shows it is false. If a statement is true, show why it is true.

45. $5 \cdot 25^x = 5^{2x+1}$

46. $\sqrt[5]{x^5 - y^{10}} = x - y^2$

47. $\sqrt[5]{x^5} = x$

48. $\sqrt[4]{32} = 2^{\frac{5}{4}}$

49. $9^x \cdot 3 = 27^x$

50. $(y^2 - a^2)^{\frac{1}{2}} = y - a$

51. $\left(x^{\frac{2}{3}}\right)^{-1} = -x^{\frac{2}{3}}$

52. $\sqrt[4]{x}\sqrt[3]{x} = \sqrt[12]{x^7}$

53 and 54. Find three expressions equivalent to each.

53. $\sqrt{x}$

54. $y^{\frac{2}{3}}$

III. Applications

55–59. Solve each problem two different ways, choosing from among a table, graph, or equation.

55. A city that has a population of 15,000 triples its population every 5 years. If this trend continues, what will the population be

(a) 5 years from now? **(b)** 10 years from now?

(c) 25 years from now? **(d)** 32 years from now?

(e) n years from now?

56. A certain northern industrial city is estimated to be losing 10% of its population every 10 years. If the population is now 890,000 and this trend continues, what will the population be

(a) 5 years from now? **(b)** 10 years from now?

(c) 30 years from now? **(d)** 35 years from now?

(e) n years from now?

57. **(a)** If \$10,000 is to be invested for 10 years, which of the following savings plans is the best choice: 10.5% compounded monthly or 11% compounded annually?

(b) Calculate the interest recorded at the end of $2\frac{1}{2}$ years for each plan.

58. In making apple cider, each press of the piston squeezes out $\frac{2}{5}$ of the juice. How much juice remains in the apples after seven presses?

59. A sheet of plastic 1 mm thick allows 75% of the light to filter through. How much light will filter through a sheet of plastic that is

(a) 2 mm thick? **(b)** $\frac{1}{3}$mm thick?

(c) $\frac{21}{4}$mm thick?

60. If the annual rate of inflation is 10%, what will be the cost of a car in ten years if it costs \$6,000 today and inflation remains at 10% a year?

61. For Exercise 60, approximate the number of years it will take for the price of the car to double.

62. A baseball diamond is a square whose distance (sides) between bases is 90 feet. How far must the second baseman, standing on the base, throw to reach home plate?

63. If you were standing on the top of the Sears Tower in Chicago (height = 442 meters), how far could you see to the horizon on a clear day?

64. The speed necessary for a bicycle to travel around a corner upright is given by $s = 4\sqrt{r}$, where s is the speed in miles per hour and r is the radius in feet of the corner. How fast is a bicycle traveling if it turns the corner at a distance of 2 feet ($r = 2$) from the curb?

65. The time t, in minutes, necessary to learn a set of unknown symbols in a psychology test is

$$t = 0.2n^{\frac{3}{4}}$$

where n is the number of symbols memorized by an average adult. Find the time necessary to memorize a list of 16 symbols.

66. Find all real values of y for which the points $(1, y)$, $(1, -y)$, and $(-2, 0)$ are the vertices of an equilateral triangle.

67. Newton's law of cooling relates the temperature, T, of a heated cup of coffee after the time, t (in minutes), has elapsed. The formula is

$$T = 150e^{-0.02t} + 65;\ e \text{ is a constant} \approx 2.72$$

(a) What is the temperature of the coffee after it has just been poured?

(b) What is the temperature of the coffee after four minutes?

(c) After how many minutes will it reach a room temperature of 70 degrees?

68. A coroner also uses Newton's law of cooling to determine the time of death of a corpse. The temperature, T, of a corpse is related to the time elapsed, t (in hours), by the formula

$$T = 40e^{-0.05t} + 58; \; e \text{ is a constant} \approx 2.72$$

(a) What is the temperature after two hours?

(b) How long has a man been dead if the temperature of his body if 78 degrees?

69. The ABC Company calculates that the percent, P, of the market penetrated by its new product after N years is given by the formula

$$P = 100 - 70e^{-0.75N}; \; e \text{ is a constant} \approx 2.72$$

(a) What percent of the market has been penetrated after five years?

(b) How long will it take to penetrate 30% of the market?

CHAPTER

A Closer Look at Functions

You have been using functions throughout this book. In this chapter, we take a closer look at functions and review many of the concepts studied previously using the language of functions.

Source: B. F. Peterson / West Stock

8.1 DEFINING FUNCTIONS

EXAMPLES 1–9

Throughout the preceding chapters, we studied many situations where one quantity depended on or was associated with another quantity. We studied how

- *Example 1* The daily rental charge for a car depended on the number of miles driven (Chapter 3),
- *Example 2* The height of a golf ball depended on the amount of time elapsed since the ball was hit (Chapter 6),
- *Example 3* The amount of sunlight depended on the depth of the water into which it shines (Chapter 7), and
- *Example 4* A positive number was associated with its positive and negative square roots (several chapters).

In all four examples, we have an association between two sets of numbers. In the second example, the association is between a set of positive numbers representing the number of seconds elapsed since the ball was hit and another set of positive numbers representing the height of the golf ball in feet. In the fourth example, the association is between the set of all positive real numbers and the set of all real numbers representing square roots (both positive and negative).

There are many other examples of associations between two sets of objects (the objects need not be numbers and the associations need not be functions):

- *Example 5* When you look up a name in a telephone directory, you are associating the name of a person with his or her telephone number.
- *Example 6* When you measure a person's height, you are associating that person with a number representing his or her height.
- *Example 7* When you talk to your sister, you are making an association between you and a member of the set of women.
- *Example 8* When you study genetics, you may be interested in the number of mutations associated with a certain amount of radiation.
- *Example 9* When you look at one of the bars in Figure 8.1, you are associating a year from 1990 to 1995 with the amount of profit made during that year by a certain company.

A **function** is a special type of association in which you can always predict or determine exactly one element in the second set that depends on any given element in the first set.

Using a computer analogy, for any input (in the first set), there is only one output (in the second set). Let us take another look at the preceding examples to see when we can make such a prediction or association; that is, to see when a given input has only one output.

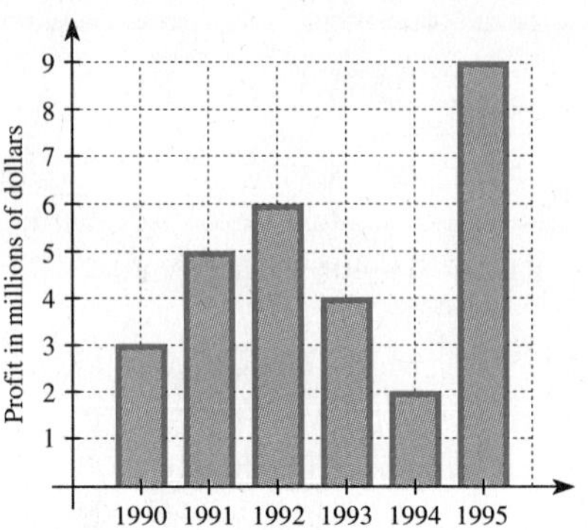

Figure 8.1

- *Example 1* If we are given the number of miles driven (an element in the first set), we *can* determine the car rental charge (an element in the second set) perhaps by using a formula such as $C = 15.95 + 0.23m$ (Chapter 3).

  input: number of miles; output: charge for driving m miles

- *Example 2* If we are given the time elapsed since a golf ball was hit, we *can* determine its height using a formula similar to $H = 80t - 16t^2 + 10$ (see Chapter 6).

  input: time in seconds; output: height in feet

- *Example 3* If we are given the depth of the water, we *can* predict the amount of sunlight at that depth by using the formula $L = \left(\frac{2}{5}\right)^n$ (Chapter 7).

  input: depth; output: amount of sunlight

- *Example 4* If we are given a positive number x, we *cannot* determine exactly one number whose square is x, since there are two (one positive and another negative).

  For each input (number), there are two possible outputs.

- *Example 5* If we are given a person's name to look up in the phone book, we *cannot* always determine exactly one phone number, because the person may have no phone, more than one phone, or an unlisted number.

  For each input (name), there could be no, one, or several possible outputs (telephone numbers).

- *Example 6* For a certain person, we *can* determine his or her (unique) height (for example, by measuring).

  input: person; output: person's height

- *Example 7* Any given person can have no, one, or several sisters.

For each input (person), there could be no, one, or several possible outputs (sisters).

- *Example 8* If we are given a certain amount of radiation, we *cannot* predict the number of mutations, since there may be none, one, or many.

  For each input (amount of radiation), there could be no, one, or several possible outputs (mutations).

- *Example 9* If we are given a particular year from 1990 to 1995, we *can* determine the amount of profit earned that year by looking at the appropriate bar on the bar graph.

  input: year; output: amount of profit in the year

In each case, we have two sets of objects. The first set is called the **domain,** and the second set is called the **range.** We try to associate each element in the domain D with an element in the range R. Several of the associations were "unruly." Some elements in D had no "partner" in R. Others had many "partners" in R. We are interested, though, in an association where *every element in D has exactly one "partner" in R.* Such an association is called a **function.**

Definition

A **function** is a rule of correspondence in which every element of a set D is associated with one and only one element of a second set R. Every element of R is associated with or corresponds to at least one element in D.

The set D is called the **domain** of the function and the set R is called the **range** of the function. Every input in the domain has exactly one output in the range.

In mathematics, the two sets are usually sets of numbers or data that can be represented by a variable. For each value of the variable in the domain, there is only one value of the variable in the range. The variable in the range *depends on* the variable in the domain.

Among our first nine examples, we can see that Examples 1, 2, 3, 6, and 9 are functions.

To denote the rule of correspondence, we usually use letters such as f, g, h, F, G, H, but any symbol can be used. If x is an element of the domain D of a function and f is the rule of correspondence, then $f(x)$ will be the element in the range R associated with x. For example, if $f(x) = 3x + 2$, then $f(0) = 2, f(2) = 8, f(-3) = -7$.

For the five functions in Examples 1, 2, 3, 6, and 9, we list the domain D, the range R, and the rule of correspondence:

- *Example 1* D = {Positive integers representing the number of miles driven}
  R = {Positive real numbers representing the car rental charge in dollars}
  f: $f(m) = 15.95 + 0.23m$ (the charge $f(m)$ depends on the number of miles driven, m)

- *Example 2* D = {all real numbers t satisfying $0 \leq t \leq 5.12$}
  R = {all real numbers h satisfying $0 \leq h \leq 110$}
  h: $h(t) = 80t - 16t^2 + 10$
- *Example 3* D = {Positive real numbers, which represent possible depths in feet}
  R = {Positive real numbers less than 1, representing fractional amounts of sunlight visible at a depth of n feet}
  L: $L(n) = \left(\frac{2}{5}\right)^n$
- *Example 6* D = {adults in the world}
  R = {many real numbers h satisfying $23 \leq h \leq 107$}
  f: For each person, associate his or her height in inches.
- *Example 9* D = {1990, 1991, 1992, 1993, 1994, 1995}
  R = {2, 3, 4, 5, 6, 9 (millions of dollars)}
  f: To each year, associate the amount of profit made during that year.

Notice that in Examples 6 and 9, there is no algebraic formula for the rule of correspondence.

To determine whether an association is a function, then, you must ask two questions:

Q1. Is every element in the domain associated with an element in the range?

Q2. Is every element in the domain associated with exactly *one* element in the range?

In four of the nine examples, we must answer "no" to at least one of these questions:

- *Example 4* The answer to question 2 is "no": Each element in the domain is associated with *two* elements in the range.
- *Example 5* The answer to both questions 1 and 2 is "no." Question 1: Everyone does *not* have a phone. Question 2: A person could have *more* than one phone.
- *Example 7* The answer to both questions is "no." Question 1: Not every person has a sister. Question 2: Some people have more than one sister.
- *Example 8* The answer to question 2 is "no": A certain amount of radiation could produce no, one, or many mutations.

Remember: In a function, *every element* in the domain *must be used once and only once.* If f denotes the rule of correspondence and x is an element in the domain D, then $f(x)$ is the element in the range corresponding to x. This symbol $f(x)$ is called "the *value* of f at x." ■

Describing Functions

Throughout this book, we have seen how we can represent a function as

1. a table
2. a graph, and
3. a rule or formula (equation).

A *table* is simply a listing of the elements in the domain and range; each element in the range and its corresponding element in the domain are written next to each other. For example, the Michigan State University student directory associates each Michigan State University student with his or her student number. Here D = {all Michigan State University students} and R = {all student numbers}. Part of the table might look like this:

Name	*Student number*
Able, William B.	597234
Abner, Lil	612037
Absent, Al Ways	555111

Table 8.1

A sales tax table, an income tax table, and a parcel post rate table are other examples. Another example is

x	$f(x)$
1	2
3	4
−2	−1
4	0
−3	4

Table 8.2

Here $D = \{1, 3, -2, 4, -3\}$ and $R = \{2, 4, -1, 0, 4\}$. Not all tables, however, are functions, as seen in the next example.

EXAMPLE 10 Below are five tables. In each case, determine whether the table represents a function; that is

Q1: Is every element in the domain associated with an element in the range?

Q2: Is every element in the domain associated with exactly one element in the range?

The domain is in the left-hand column (or the top).

(a)

x	$f(x)$
1	0
2	4
3	1
−1	4

Table 8.3

(b)

z	$g(z)$
1	0
2	4
3	1
1	4

Table 8.4

(c)

State	*Capital*
Mich.	Lansing
Ohio	Columbus
Iowa	Des Moines
.	.
.	.
.	.

Table 8.5

(d)

Name	*Type of car owned*
J. Smith	Plymouth
F. Klein	Honda
T. Wong	Volkswagen
.	.
.	.
.	.

Table 8.6

(e)

Number of teams	2	3	4	5	6	7	8	9	10	11	12
Number of games	2	6	12	20	30	42	56	72	90	110	132

Table 8.7 Number of Games on a ''Home-and-Home'' Schedule

Solutions

(a) Yes. This is a function since each x is associated with only one $f(x)$.

(b) No, $g(1) = 0$ and $g(1) = 4$. The answer to Q2 is "no."

(c) Yes, each state has only one capital.

(d) No, a person may own more than one car (the answer to Q2 is "no"). The answer to Q1 is "no" if the first column contains the name of a person who does not own a car.

(e) Yes, since there is only one possibility for the number of games associated with each possibility for the number of teams (in the league). The number of games depends on the number of teams. ■

A federal income tax table is another example of a function.

EXAMPLE 11

The table shows the income tax rates for single returns filed by April 1991. If your income was \$27,000 during 1990, how much federal tax should you pay?

Taxable income	*Tax*
\$0	\$0
\$20,350	\$3,052.50 + 28% of the amount over \$20,350
\$49,300 or over	\$11,158 + 31% of the amount over \$49,300

Table 8.8

Solution

By looking at the table, we see the amount of tax is \$3,052.50 + 0.28(27,000 − 20,350) = \$4,914.50 ■

A *graph* can also be used to describe a function. It can be thought of as the "life history" of the function. Linear functions, quadratic functions, and exponential functions are some we have seen before. The domain is some portion (sometimes all) of the x-axis. The range is some part (or all) of the y-axis. If (x, y) is a point on the graph, then x, in the domain of the function, corresponds to y, in the range of the function. We write $y = f(x)$. See Figure 8.2.

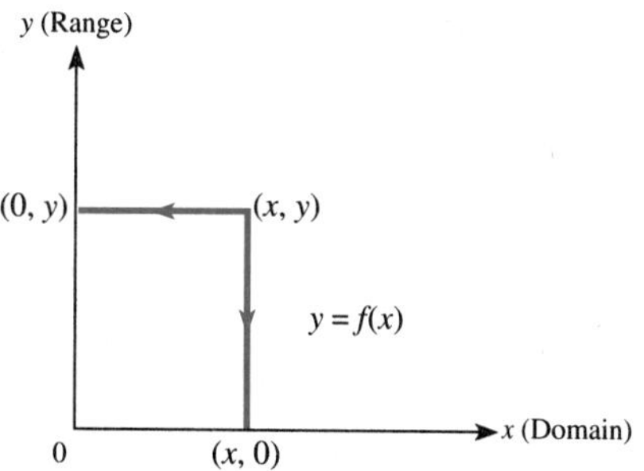

Figure 8.2

When we examine the graph of the linear function $f(x) = x - 1$, we see (Figure 8.3) that $D = \{\text{all real numbers}\}$, since every real number is the x-coordinate of some point on the line, and $R = \{\text{all real numbers}\}$, since every real number is also the y-coordinate of some point on the line. In the language of functions, $f(3) = 2$, $f(4) = 3$, $f(17) = 16$, and in general, $f(x) = x - 1$.

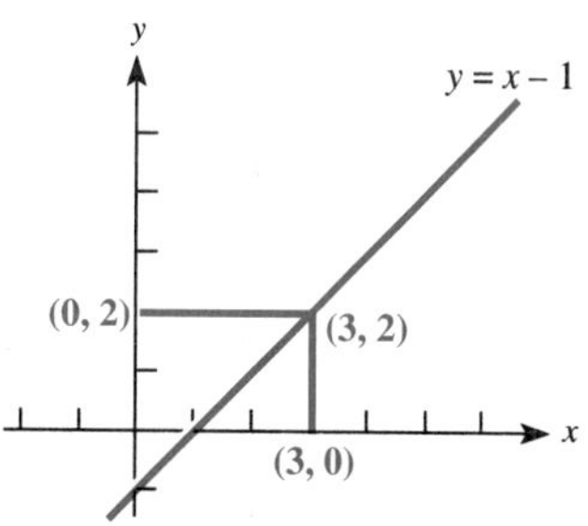

Figure 8.3

Just as in the case with a table, not all graphs represent functions. One test to determine whether a graph represents a function is called the *vertical line test.* Any vertical line cuts the graph of a function at most one time. If a vertical line cuts a graph twice or more, as in Figure 8.4, then that graph would not be of a function, because x, in the domain, would be associated with both y_1 and y_2 in the range. Graphs of functions will be studied in more detail in Section 8.3.

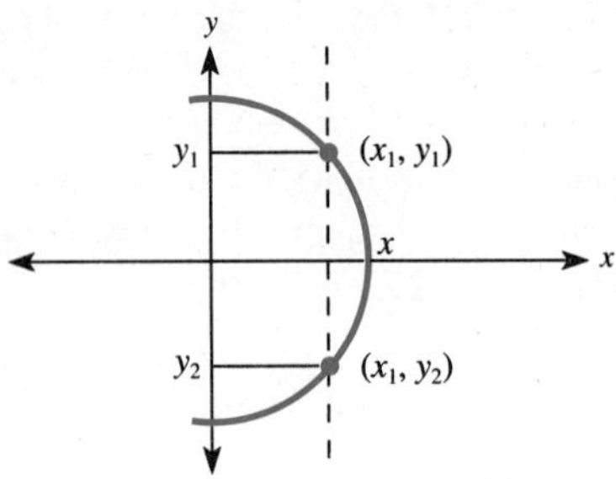

Figure 8.4

EXAMPLES 12–17

Which of the following graphs represent a function $y = f(x)$? If the graph is a function, locate the point $(0, f(a))$ on the y-axis (in the range).

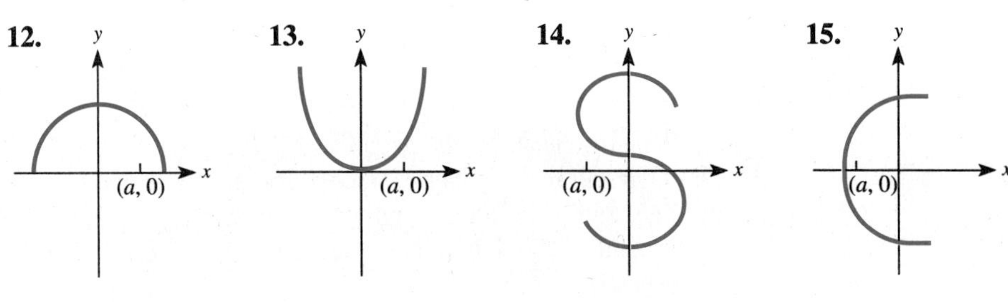

Figure 8.5 Figure 8.6 Figure 8.7 Figure 8.8

16.

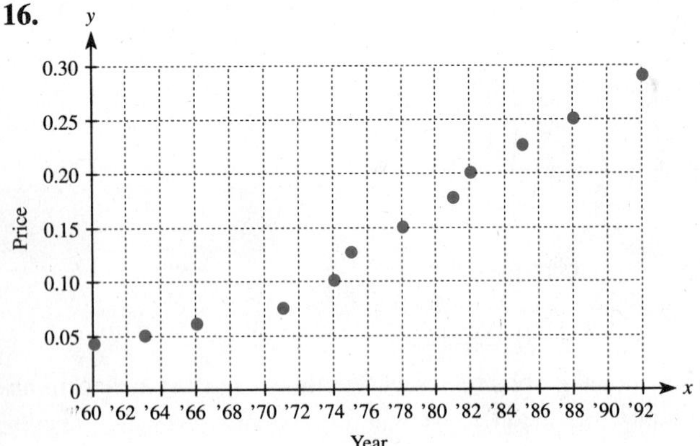

Figure 8.9 Price of First-Class Stamp

17.

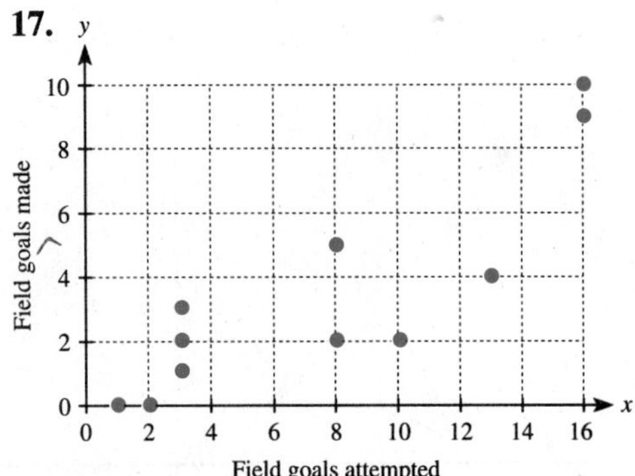

Figure 8.10 Houston Rockets Game Statistics

Solutions

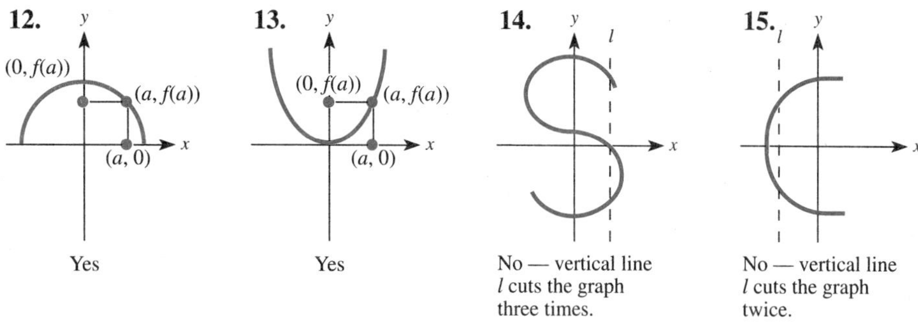

Figure 8.11 Figure 8.12 Figure 8.13 Figure 8.14

16. Since each year is associated with only one price, this graph is a function.

17. This scatterplot is not a function since several players attempted 3, 8, and 16 field goals. The vertical lines $x = 3$, $x = 8$, and $x = 16$ intersect the graph at more than one point. ■

Sequences

A function whose domain is the set of positive integers (or some subset of them) has a special name. First let us consider an example.

EXAMPLE 18

Suppose you tell a rumor to three people in one hour. Then you tell three more people during the next hour, and you keep telling three more people during each succeeding hour. In this case, assume none of these people tells anyone else about the rumor (see Figure 8.15).

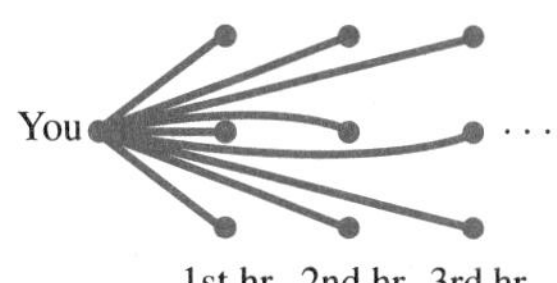

Figure 8.15

(a) How many people learn the rumor during the tenth hour?

(b) How many people, excluding you, know the rumor at the end of ten hours?

Solutions

In each case, we will make a table indicating the hour H, the number of people N learning the rumor during that hour, and the total number of people T, excluding you, who know the rumor at the end of that hour.

H	1	2	3	4	5	10
N	3	3	3	3	3	3

Table 8.9

H	1	2	3	4	5	10
T	3	6	9	12	15	30

Table 8.10

We see $N = 3$ for each hour. Hence when $H = 10$, $T = 3 + 3 + 3 + 3 + 3 + 3 + 3 + 3 + 3 + 3 = 30$. The number of people, N, learning the rumor each hour can be described by the list N: 3, 3, 3, 3, . . . while the number of people, T, who know the rumor at the end of each hour can be T: 3, 6, 9, 12, ... Both lists are examples of sequences.

A sequence is a list of numbers where the order in which the numbers are given is important. For Example 18, the values of N (and T) in each of the tables can be thought of as a list of numbers. Since the position of a number in this list indicates the hour when it occurred, the order of the numbers in the list is important.

Definition

A **finite sequence** is a function $a(n)$ whose domain is the subset of the natural numbers $\{1, 2, 3, 4, \ldots, n\}$. The elements in the range of the function $a(1), a(2), a(3), \ldots, a(n)$ are called the **terms** of the sequence and are denoted by $a_1, a_2, a_3, \ldots, a_n$. Thus $a_1 = a(1)$ is the *first term*, $a_2 = a(2)$ is the *second term*, and so on. The expression $a_n = a(n)$, which defines the sequence, is called the **general term** or the ***n*th term.**

An **infinite sequence** is one whose domain is the set of all natural numbers.

In Example 18, we have the sequences (for N):

$$3, 3, 3, 3, \ldots, 3 \quad \text{or} \quad a_n = 3 \quad \text{for } n = 1, 2, 3, \ldots, 10 \text{ and}$$

$$3, 6, 9, 12, \ldots \quad \text{or} \quad a_n = 3n \quad \text{for } n = 1, 2, 3, \ldots, 10.$$

When describing a sequence, we write only the elements in the range if the domain is understood to be $\{1, 2, 3, \ldots, n\}$ or $\{1, 2, 3, \ldots\}$.

The number of people, including you, who *know* the rumor hour by hour is given by the sequence:

$$1, 4, 7, 10, 13, \ldots$$

Such a sequence in which the terms are equally spaced is called an **arithmetic sequence,** or **arithmetic progression.** An arbitrary arithmetic sequence has the form

$$a, a + d, a + 2d, a + 3d, \ldots$$

and we can describe the general term by

$$a_n = a_1 + (n - 1)d, \text{ for } n \geq 1. \qquad a_1 = a \text{ above.}$$

The quantity d is called the **common difference.** It is the amount by which any two consecutive terms in the sequence always increase or decrease.

For the example above, we have

$$a_1 = 1, d = 3, \text{ so that } a_n = 1 + (n - 1)3 = 3n - 2$$

NOTE: Not every sequence can be described by a formula. An example is the sequence of decimal digits of π:

$$3, 1, 4, 1, 5, 9, 2, \ldots$$

This sequence is a list of numbers given in a well-defined way, but no formula exists for its general term. ■

In mathematics, as well as other areas, the most useful functions are those that can be described by a rule or formula. We have, in fact, studied many functions described by rules in preceding chapters. For example, the formula $F = \frac{9}{5}C + 32$, or in functional notation, $F(C) = \frac{9}{5}C + 32$, is a linear function. The formula $s = 16t^2$, or $s(t) = 16t^2$, is a quadratic function. The formula $L = \left(\frac{2}{5}\right)^n$, or $L(n) = \left(\frac{2}{5}\right)^n$, is an exponential function. We will study all these functions in more detail shortly.

Even though tables are convenient, they are often lengthy and may not be complete. Graphs are complete, but they are only approximate. A function described by a rule or formula is complete and accurate.

8.1 EXERCISES

I. Procedures

1–6. Which of the following tables describe functions? For those that are functions, find $f(1)$.

1.

x	$f(x)$
0	2
1	3
2	4
1	−3
2	4

2.

x	$f(x)$
0	2
1	3
2	4
−1	−3
−2	−4

3.

x	$f(x)$
0	2
1	3
2	4
−1	3
−2	4

4.

x	$f(x)$
0	4
1	4
2	4
3	−7
4	8
5	10

5.

x	$f(x)$
0	4
1	3
2	2
3	1
4	0
5	5

6.

x	$f(x)$
0	4
1	3
2	2
1	4
4	0
5	2

7–16. Which of the following graphs describe functions $y = f(x)$? For those that are functions, estimate $f(1)$.

7.

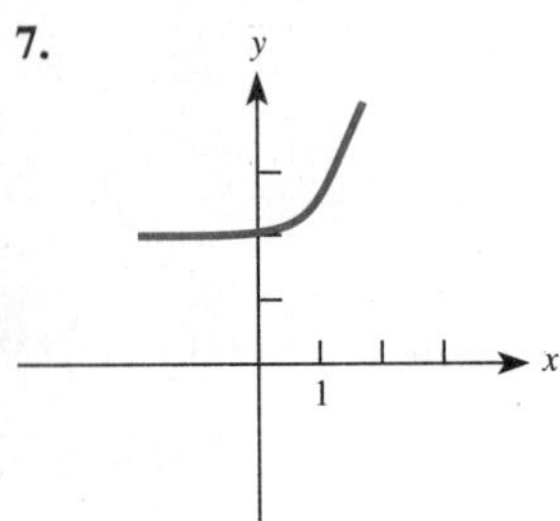

8.

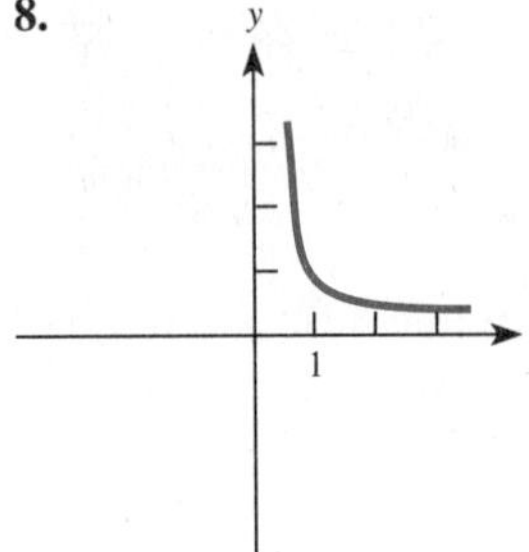

9.

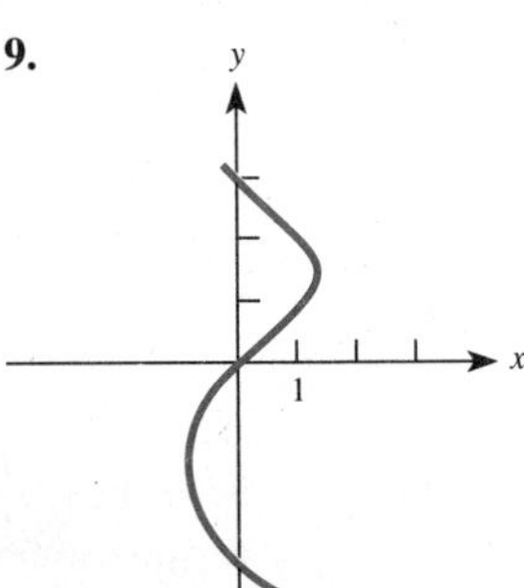

10.

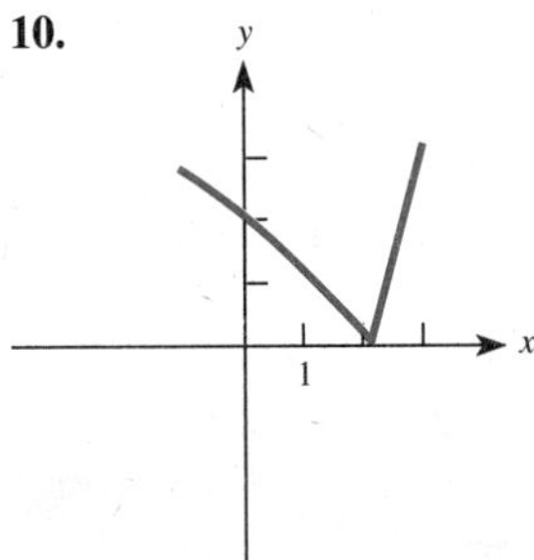

11.

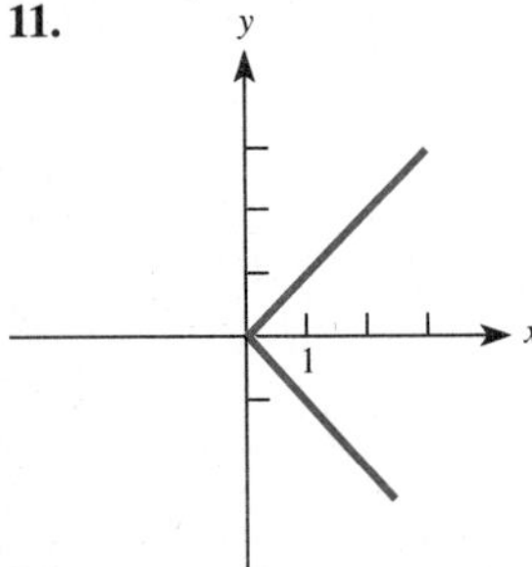

12.

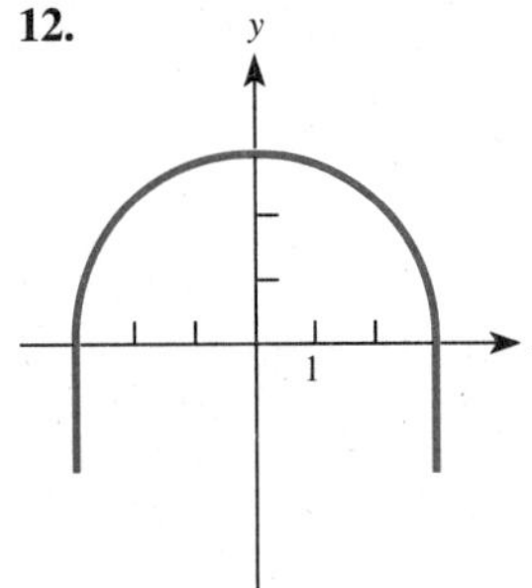

13.

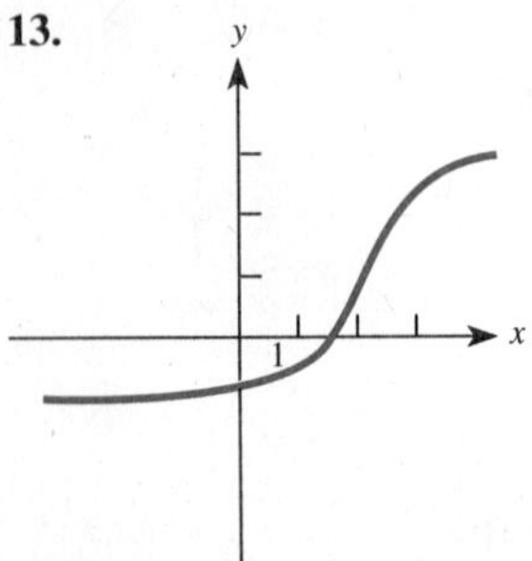

14.

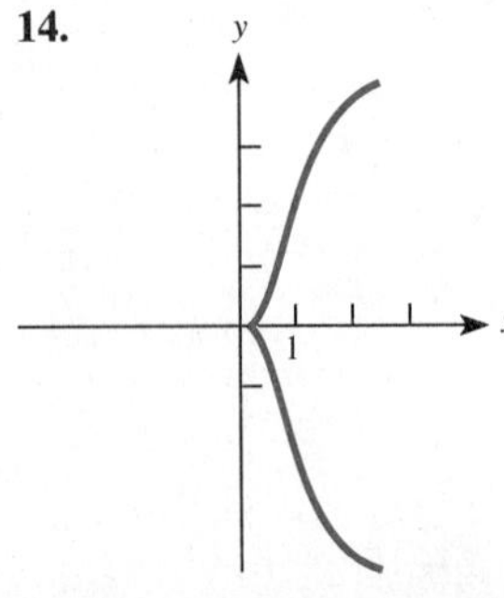

15.

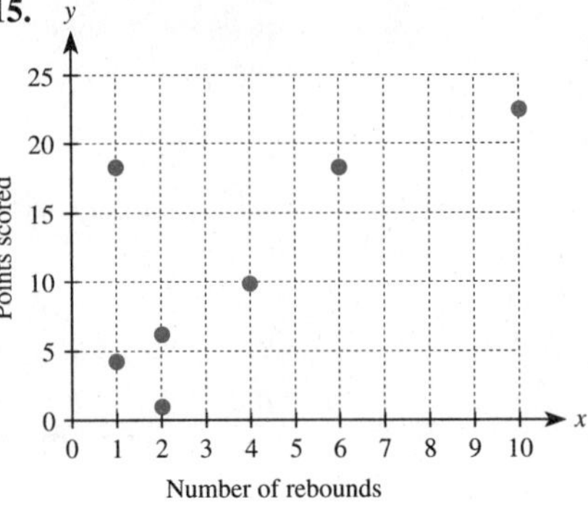

16.

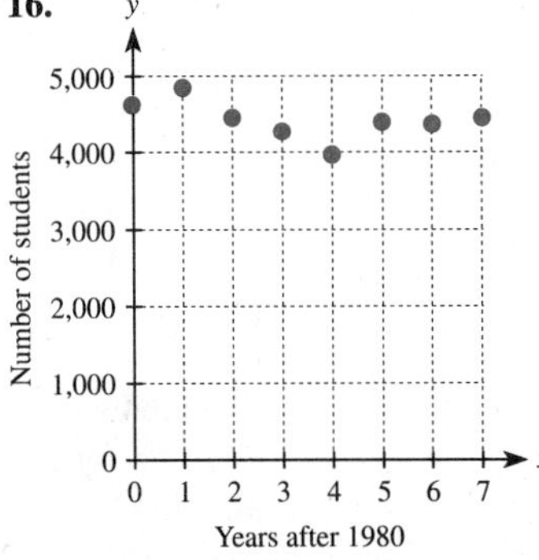

17–24. For each of the following, assume that the domain is $D = \{0, 1, 2, 3, 4, 5\}$. Write down a formula f that describes each function below and find $f(4)$. To each x in D, associate

17. three times x.

18. three more than twice x.

19. the square of x plus 1.

20. four less than three times x.

21. the perimeter of a square whose side is x.

22. the distance covered by a car in four hours traveling $10x$ mph.

23. the square root of x.

24. x divided by 2.

II. Concepts

25. Give three examples of tables describing functions used in everyday life.

26. Give three examples of graphs describing functions used in everyday life.

27. Give three examples of functions discussed in an example in a previous chapter.

28. Give two examples of sequences whose general term cannot be given by a formula.

29 and 30. Find the next element in each of the following sequences. *Note:* These two exercises are word puzzles.

29. O, T, T, F, F, S, S, E, N, ?

30. T, T, T, F, F, S, S, E, N, ?

31–36. For each of the following physical objects, describe a domain D, a range R, and a function f that the object could represent.

31. a gasoline pump

32. a "take a number" at a bakery

33. a meter stick

34. a dictionary

35. a calculator

36. a teacher's grade book

37. Write a paragraph defining a function. Include an example of a table, a graph, a rule, and a sequence.

38. Give several examples of relationships between two sets of data that are *not* functions. Explain why.

III. Applications

39. Refer to the graph in Example 16.

(a) What was the cost of a first-class stamp in 1986?

(b) In what year did the cost of a first-class stamp first exceed 20¢?

(c) What was the increase in the price of a first-class stamp from 1970 to 1991?

(d) Between what two years was there the greatest single percent increase in the price of a first-class stamp?

40. See Example 11 of this section, and answer the following questions.

(a) What was the tax on a taxable income of $32,000?

(b) If your tax was $4,750, what was your taxable income?

41. Study the table below and answer the following questions.

(a) If your weight is 170 pounds, how many calories do you need to walk fast for two hours?

(b) If your weight is 160 pounds, how many calories do you need to sleep normally for eight hours?

(c) What are some things you could do to burn up at least 500 calories in one hour if your weight is 200 pounds?

Activity done	*Calories needed per pound, per hour*
Sleeping (soundly)	0.4
Sleeping (fitfully)	0.5
Sitting (no movement)	0.6
Standing (relaxed)	0.7
Sewing (by hand)	0.75
Dressing (or undressing)	0.8
Singing	0.85
Typewriting (average speed)	0.9
Washing dishes, ironing, or dusting	0.95
Sweeping floors (broom or vacuum)	1.0
Exercising (lightly)	1.25
Walking (approximately 2.8 mph)	1.5
Trade work (carpentry or plumbing)	1.75
Exercising (active)	1.9
Walking fast (4 mph)	2.0
Going down steps	2.25
Loading (or unloading) heavy objects	2.5
Exercising (heavy)	2.75
Active sports (tennis or swimming)	3.25
Running (approximately 5.5 mph)	3.75
Exercising (very heavy)	4.0
Going up steps	7.0

Caloric Energy Requirement Chart

42. The figure on the next page illustrates graphically the motion of two cyclists traveling to meet each other. Using the graph, determine:

(a) the time of departure and arrival for each cyclist

(b) when and where the cyclists met

(c) where and when the cyclists had their rest

(d) how much time it took each cyclist to meet the other

(e) the average speed of each cyclist during her trip.

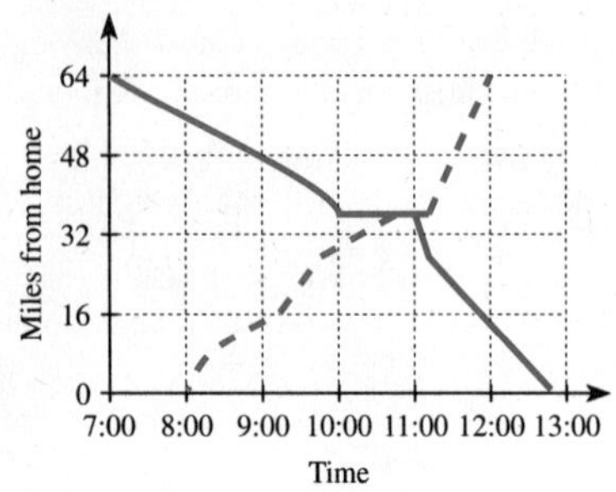

IV. Extensions

43–46. Each of the functions is given by a table. Express the function using a rule.

43.

x	$f(x)$
1	4
2	7
3	10
4	13
5	16
6	19

44.

x	$f(x)$
1	1
2	10
3	25
4	46
5	73
6	106

45.

x	$f(x)$
1	120
2	60
3	40
4	30
5	24
6	20

46.

x	$f(x)$
1	7
2	9
3	13
4	21
5	37
6	69

47–50. Find the function $f(x)$ that is represented by each sequence of calculator keys (using algebraic logic). Enter a number x, then press the keys in sequence.

47. [×] 4 [+] 3 [=]

48. [x^2] [×] 2 [+] 3 [=]

49. [×] 2 [+] 1 [=] [√]

50. [y^x] 3 [×] 2 [+] 3 [=]

51. Let $D = \{1, 2, 3\}$ and $R = \{4, 5, 6\}$.

(a) How many functions are there whose domain is D and whose range is a subset of R? For example, two such functions are

x	1	2	3
$f(x)$	4	4	4

and

x	1	2	3
$g(x)$	6	5	4

(b) In how many of the functions in part (a) does the range of the function equal R, as in $g(x)$?

52. Answer the questions in Exercise 51 if $D = \{1, 2, 3, 4\}$ and $R = \{5, 6, 7, 8\}$.

53. Answer the questions in Exercise 51 if

$$D = \{1, 2, 3, \ldots, n\} \text{ and } R = \{n + 1, n + 2, \ldots, 2n\}$$

Your answer should be a formula (function), with n as the variable.

8.2 EVALUATING FUNCTIONS: A CLOSER LOOK

In this section, we investigate functions given by a formula or an equation. Such a function assigns to every element in the domain one and only one element in the range. In this section we revisit many concepts studied previously in the context of functions.

EXAMPLE 1

(a) Give a function, $A(s)$, for the area of a square in terms of its side, s.

(b) Find the domain and range of $A(s)$.

(c) Sketch the graph of $A(s)$.

(d) If the length of the side of a square is 3.5 cm, what is its area?

(e) If the area of a square is 20 cm^2, what is the length of its side?

(f) If the length of the side of the square is increased by 20%, by what factor is the area increased? See the figure below.

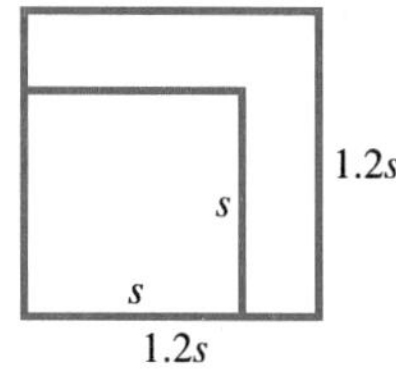

Figure 8.16

Solutions

(a) $A(s) = s^2$

(b) $D = \{\text{positive real numbers}\}$ Length is positive.
$R = \{\text{positive real numbers}\}$ Squares are positive.

(c)

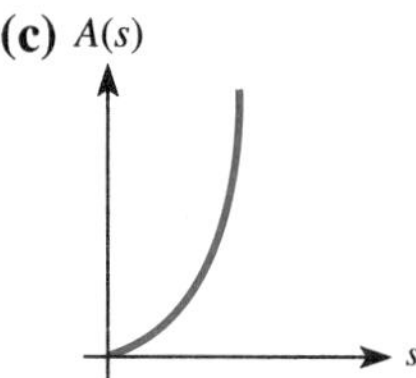

Figure 8.17

(d) $A(3.5) = (3.5)^2 = 12.25 \text{ cm}^2$ Evaluate the function.

(e) If $A(s) = 20$, then $s^2 = 20$ or $s = \sqrt{20} = 2\sqrt{5}$ cm. Solve the equation.

(f) $\dfrac{A(1.2s)}{A(s)} = \dfrac{1.44s^2}{s^2} = 1.44$ Compute the ratio of the areas to determine the factor by which the area is increased.

The area of the "new" square is 1.44 times as large as that of the original square. ■

In part (d) of the preceding example, we **evaluated a function.** Given a function f and a value x in the domain of f, we found $f(x)$, the value in the range of f associated with x. To compute $f(a)$, then, we simply "plug in" a for x and perform the indicated operations.

In part (e) of the example, we **solved an equation.** Given a function f and a value $f(x)$ in the range of f, we found the value(s) of x in the domain associated with $f(x)$. We have solved many types of equations in this book: linear, quadratic, exponential, and radical, to mention a few.

In Chapter 6, you studied an example similar to the following one.

EXAMPLE 2 The height above ground level of a golf ball hit from an elevated tee 10 feet above ground level is approximated by the function

$$h(t) = 80t - 16t^2 + 10$$

where t is in seconds and $h(t)$ is in feet.

(a) Sketch the graph of $h(t)$.

(b) How high is the ball 3 seconds after it is hit?

(c) How long will it take the ball to hit the ground?

(d) Find the maximum height reached by the ball.

(e) Find the domain and range of $h(t)$.

Solutions

Using a Graph

(a)

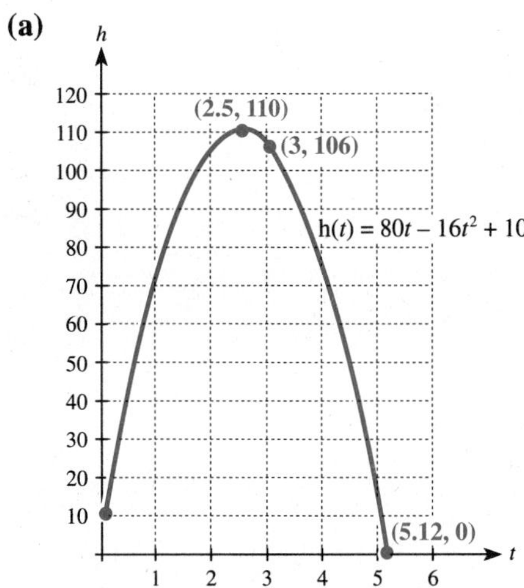

Figure 8.18

Using an Equation

(b) The height at the end of 3 seconds is

$$\begin{aligned} h(3) &= 80(3) - 16(3)^2 + 10 \\ &= 106 \text{ feet} \end{aligned}$$

Evaluate $h(t)$ when $t = 3$.

(c) The ball hits the ground when $h(t) = 0$:

$$-16t^2 + 80t + 10 = 0$$

Solve the equation $h(t) = 0$.

$$8t^2 - 40t - 5 = 0$$

$$t = \frac{40 + \sqrt{1{,}760}}{16} \approx 5.12 \text{ seconds}$$

Only the positive root makes sense.

(d) The vertex is (2.5, 110) so the maximum height of the ball is 110 feet.

(e) $D = \{t | 0 \leq t \leq 5.12\}$
$R = \{h | 0 \leq h \leq 110\}$ ■

EXAMPLE 3

Let $g(x) = 80x - 16x^2 + 10$.

(a) Find the domain and range of $g(x)$.

(b) Sketch the graph of $g(x)$.

Solutions

(a) $D = \{\text{all real numbers}\}$ — $g(x)$ does *not* represent a physical situation.
$R = \{y | y \leq 110\}$ — Maximum value of $g(x)$ is 110.

(b)

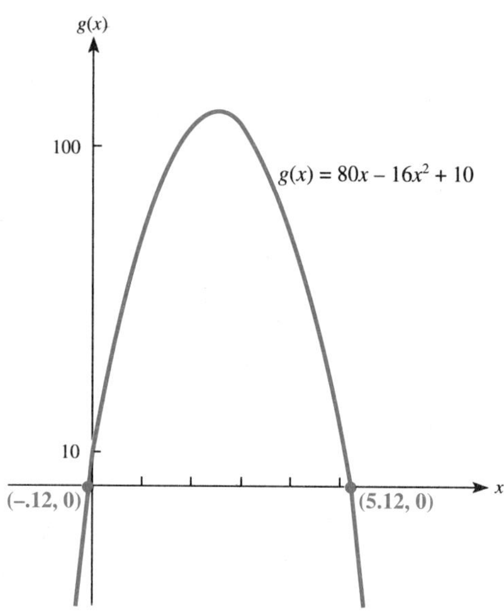

Figure 8.19

NOTE: Although Examples 2 and 3 have the same equation, they have different graphs. Example 2 represents a physical situation in which the graph is a piece of the complete graph of $g(x) = 80x - 16x^2 + 10$ in Example 3. ■

Remember that a sequence is a function whose domain is the set of positive integers.

EXAMPLE 4

(a) Write out the first six terms of the sequence whose general term a_n is

(i) $4n + 3$ **(ii)** $\dfrac{n(n + 1)}{2}$

(b) Sketch the graph of each sequence.

Solutions

(a) (i) 7, 11, 15, 19, 23, 27 — Substitute $n = 1, 2, 3, 4, 5, 6$ in each case. In part (i) $a_1 = 4 \cdot 1 + 3 = 7, a_2 = 4 \cdot 2 + 3 = 11$, and so on.

(ii) 1, 3, 6, 10, 15, 21 — In (ii), $\frac{1(1 + 1)}{2} = 1, \frac{2(2 + 1)}{2} = 3$; and so on.

(b) (i)

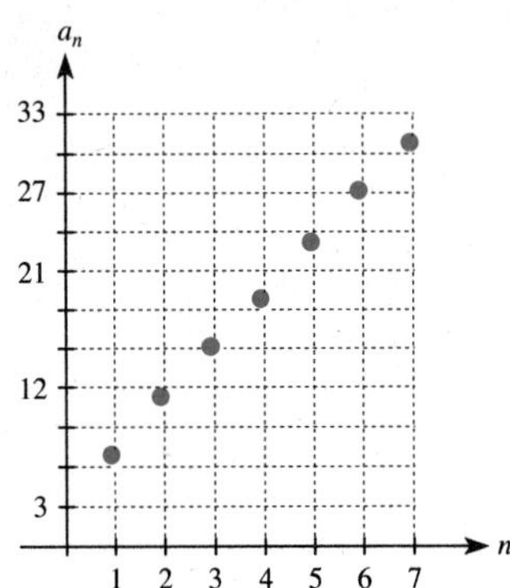

Figure 8.20

Graph is set of dots.

(ii)

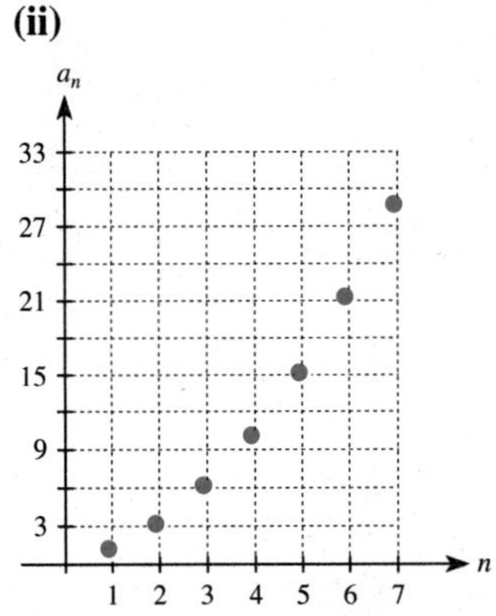

Figure 8.21

Graph is set of dots.

EXAMPLE 5 Find the twentieth term in the arithmetic sequence 54, 50, 46, 42, . . .

Solution

$$\begin{aligned} a_{20} &= a_1 + 19d \\ &= 54 + 19(-4) \\ &= -22 \end{aligned}$$

Terms are "equally spaced," so the sequence is arithmetic. $a_1 = 54, d = -4$

In the next few examples, we will explore further the ideas of evaluating a function and solving an equation. At the same time, we will review a number of concepts studied earlier.

When it does not represent a physical situation, the domain of a function is taken to be the largest subset of the real numbers for which the rule is defined (yields a real number). The domain, then, consists of all real numbers except some "bad" ones that cause the rule to be undefined. Such bad values are usually those for which a denominator of a fractional expression equals 0, or for which the radicand of a square root expression (or any even root) is negative. In addition, if the function models a physical situation, real numbers for which the rule makes no physical sense are also excluded.

EXAMPLE 6

Let $h(x) = 5x^2 - 3\sqrt{x+1} + x^{\frac{2}{3}} + \frac{2}{x}$.

(a) Find $h(8)$, $h(\pi)$, $h(-0.374)$.

(b) Find the domain of $h(x)$.

(c) Sketch the graph of $h(x)$.

Solutions

(a) We can think: $h(\ ) = 5(\ )^2 - 3\sqrt{(\ ) + 1} + (\ )^{\frac{2}{3}} + \frac{2}{(\ )}$. Then we replace the blank with the given number.

$$h(8) = 5(8)^2 - 3\sqrt{8+1} + 8^{\frac{2}{3}} + \frac{2}{8}$$

$$= 320 - 9 + 4 + \frac{1}{4}$$

$$= 315.25$$

$$h(\pi) = 5\pi^2 - 3\sqrt{\pi + 1} + \pi^{\frac{2}{3}} + \frac{2}{\pi}$$

$$\approx 49.35 - 6.11 + 2.15 + 0.64$$

$$\approx 46.03$$

$$h(-0.374) = 5(-0.374)^2 + 3\sqrt{-0.374 + 1} + (-0.374)^{\frac{2}{3}} + \frac{2}{-0.374}$$

$$\approx 0.699 - 2.374 + 0.519 - 5.348$$

$$\approx -6.504$$

Some calculators will not calculate $\sqrt[3]{-0.374}$. Calculate $\sqrt[3]{0.374}$ and then multiply by -1.

(b) We analyze when each term in the function is defined (x is assumed to be a real number).

$5x^2$ is defined for all x.

$\sqrt{x+1}$ is defined for $x \geq -1$. The square root is defined only for nonnegative quantities.

$x^{\frac{2}{3}}$ is defined for all x.

$\frac{2}{x}$ is defined for $x \neq 0$. $x = 0$ makes the denominator of $\frac{2}{x}$ equal to 0.

Putting these together, we see that

$D = \{x | x \geq -1 \text{ and } x \neq 0\}$

(c)

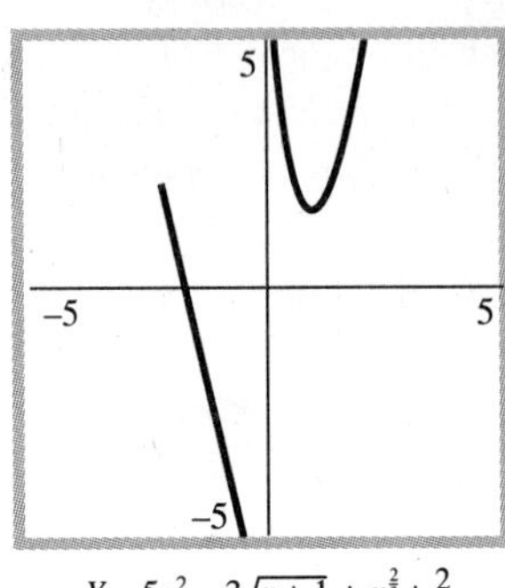

$Y = 5x^2 - 3\sqrt{x+1} + x^{\frac{2}{3}} + \frac{2}{x}$

Figure 8.22

On some graphing calculators you must enter $x^{\frac{2}{3}}$ as $\sqrt[3]{x^2}$ or the left portion of the graph is omitted. No graph for $x \leq -1$ as expected.

Functions, just like numbers, can be added, subtracted, multiplied, and divided.

EXAMPLE 7

Let $f(x) = 3x - 4$ and $g(x) = 5x + 7$.

(a) Find $[g(x) - f(x)]$.

(b) Find $[g(x) - f(x)]^2$.

(c) Find $\dfrac{g(a+h) - g(a)}{h}$.

(d) Find all real x for which $3f(x) = g(x)$.

(e) Find all real x for which $4f(x) - 5g(x) < 0$.

(f) Find all real x for which $|f(x)| = 7$.

Solutions

For all parts, remember $f(x) = 3x - 4$ and $g(x) = 5x + 7$.

(a) $(5x + 7) - (3x - 4) = 2x + 11$ Operations on polynomials (see Chapter 2).

(b) $(2x + 11)^2 = 4x^2 + 44x + 121$ Use result of part (a).

(c)
$$\frac{g(a+h) - g(a)}{h} = \frac{5(a+h) + 7 - (5a + 7)}{h}$$
$$= \frac{5a + 5h + 7 - 5a - 7}{h}$$
$$= \frac{5h}{h} = 5 \text{ if } h \neq 0.$$

Evaluating $g(a + h)$ is similar to evaluating $g(3)$; substitute $(a + h)$ for x.

(d) Using an Equation

$$\begin{aligned} 3(3x - 4) &= 5x + 7 \\ 9x - 12 &= 5x + 7 \\ 4x &= 19 \\ x &= \frac{19}{4} = 4.75 \end{aligned}$$

Solve a linear equation (Chapter 3).

Using a Graph

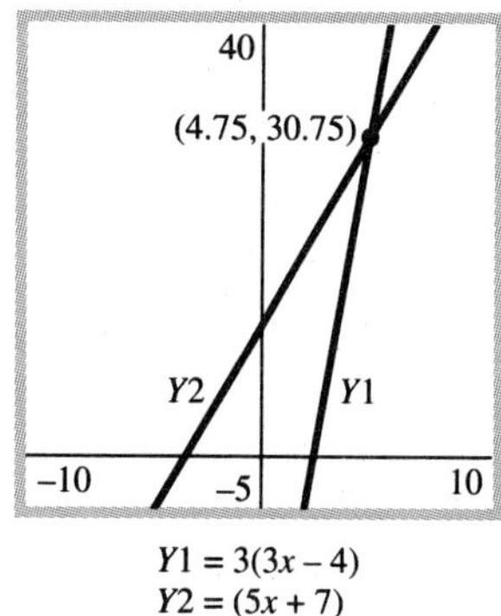

$Y1 = 3(3x - 4)$
$Y2 = (5x + 7)$

Figure 8.23

(e) Using an Equation

$$\begin{aligned} 4(3x - 4) - 5(5x + 7) &< 0 \\ 12x - 16 - 25x - 35 &< 0 \\ -13x - 51 &< 0 \\ x &> -\frac{51}{13} \approx -3.92 \end{aligned}$$

Solve a linear inequality (see Section 3.4).

Using a Graph

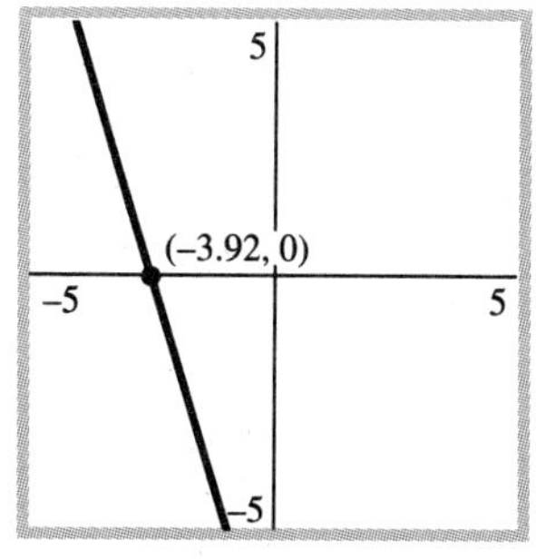

$Y1 = 4(3x - 4) - 5(5x + 7)$

Figure 8.24

(f) Using an Equation

$|3x - 4| = 7$

$3x - 4 = 7 \quad \text{or} \quad 3x - 4 = -7$ — Solve an absolute-value equation (see Section 3.4).

$3x = 11 \quad \text{or} \quad 3x = -3$

$x = \dfrac{11}{3} \quad \text{or} \quad x = -1$

Using a Graph

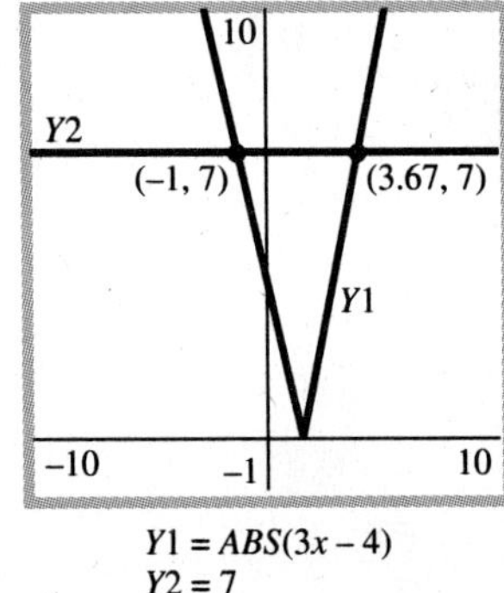

$Y1 = ABS(3x - 4)$
$Y2 = 7$

Figure 8.25

EXAMPLE 8

Let $f(x) = x - 2$, $g(x) = \sqrt{x}$, and $h(x) = 3x - 8$. Find all real x for which

(a) $[f(x)]^2 = h(x)$

(b) $[h(x)]^2 = [3g(x)]^2$

(c) $f(x) = g(x)$

Solutions

(a) Using an Equation For all parts, remember $f(x) = x - 2$, $g(x) = \sqrt{x}$, and $h(x) = 3x - 8$.

$$(x - 2)^2 = 3x - 8$$
$$x^2 - 4x + 4 = 3x - 8$$
$$x^2 - 7x + 12 = 0$$
$$(x - 4)(x - 3) = 0$$ Solve a quadratic equation by factoring.
$$x = 3 \quad \text{or} \quad x = 4$$

Using a Graph

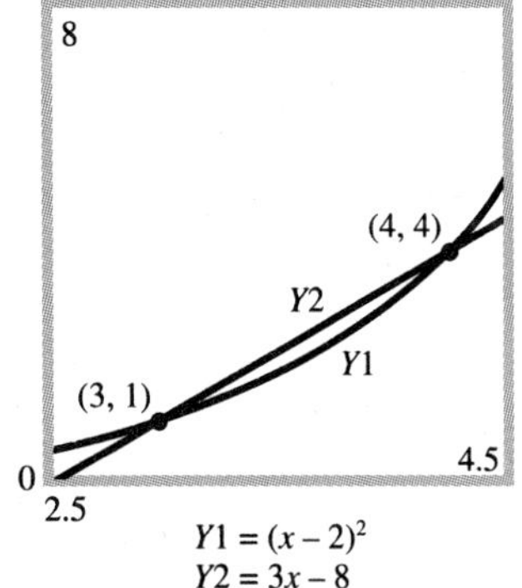

Figure 8.26

Find an appropriate viewing window. Make several attempts.

(b) Using an Equation

$$(3x - 8)^2 = (3\sqrt{x})^2$$

$$9x^2 - 48x + 64 = 9x$$

$$9x^2 - 57x + 64 = 0$$

$$x = \frac{57 \pm \sqrt{(-57)^2 - 4 \cdot 9 \cdot 64}}{18}$$ Quadratic formula.

$x \approx 1.46$ or 4.87

Using a Graph

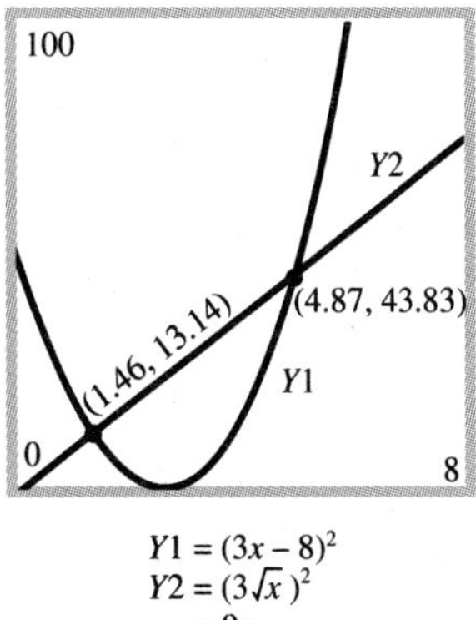

Figure 8.27

(c) Using an Equation

$$x - 2 = \sqrt{x}$$ Solve a radical equation.

$$(x - 2)^2 = (\sqrt{x})^2$$

$$x^2 - 4x + 4 = x$$

$$x^2 - 5x + 4 = 0$$

$$(x - 4)(x - 1) = 0$$

$$x = 4$$ $x = 1$ is extraneous.

Using a Graph

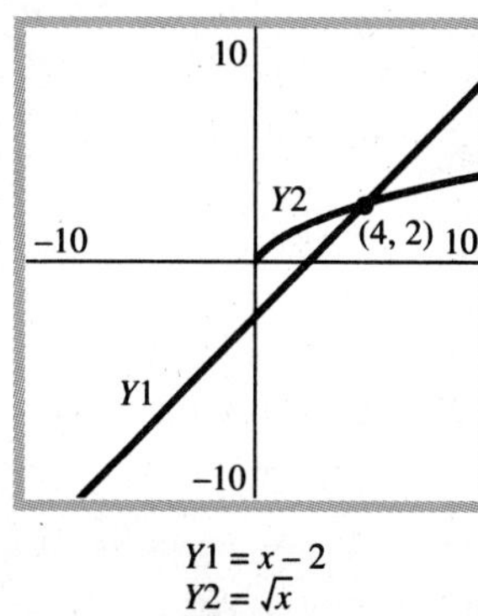

$Y1 = x - 2$
$Y2 = \sqrt{x}$

Figure 8.28

8.2 EXERCISES

I. Procedures

1–8. For each of the following functions $f(x)$,
(a) compute $f(0), f(-1), f(2.3)$, and $f(a + 1)$.
(b) Suppose $D = \{-1, 0, 2, 3\}$. Find the range of f for this set D.

1. $f(x) = 3x + 2$ **2.** $f(x) = 4 - 3x$

3. $f(x) = -x^2$ **4.** $f(x) = x^2 + 2$

5. $f(x) = x^2 + 2x - 1$ **6.** $f(x) = 2x^2 - 3x + 4$

7. $f(x) = \dfrac{x + 3}{x - 1} - \dfrac{x}{x + 4}$

8. $f(x) = \dfrac{x^2}{x - 1} - \dfrac{1}{x + 2}$

9 and 10. For each sequence, find the first six terms and sketch its graph.

9. $a_n = 2n - 5$ **10.** $a_n = \dfrac{n - 1}{n + 1}$

11–16. Find the fifteenth term in each of the following arithmetic sequences in which either a_n or the first four terms are given.

11. $a_n = 3 + 7n$ **12.** $a_n = 54 - 11n$

13. 100, 91, 82, 73 **14.** 3, 9, 15, 21

15. 12, 6, 0, -6

16. 0.123, 0.246, 0.369, 0.492

17–20. Find x so that each of the following is an arithmetic sequence.

17. $x, 5, 7$

18. $-0.35, 0.75, x$

19. $-8, x, 4$

20. $-\frac{4}{3}, x, \frac{11}{12}$

21–36. Let $f(x) = x^2 + 3x - 4$ and $g(x) = -x + 3$.

21–30. Compute.

21. $f(0) - g(0)$

22. $\frac{f(0)}{g(0)}$

23. $f(-1) \cdot g(3) - f(4) \cdot g(0)$

24. $f(1) \cdot g(2) + f(3) \cdot g(4)$

25. $f(a + 1) - g(a + 1)$

26. $f(a) + ag(a)$

27. $f(g(1))$

28. $g(f(1))$

29. $[f(x) - xg(x)]^2$

30. $[f(x) + xg(x)]^2$

31–36. Find all real x satisfying the given equation. Solve both algebraically and graphically.

31. $g(x) = 0$

32. $f(x) = 0$

33. $f(x) = g(x)$

34. $f(x) = -xg(x)$

35. $3g(x) > 2$

36. $-3g(x) < 4$

37 and 38. Let $f(x) = \sqrt{x - 3}$.

37. Compute

(a) $f(12)$ **(b)** $f(2)$ **(c)** $\frac{f(15)}{f(6)}$ **(d)** $f(f(12))$

38. Find all real x satisfying $f(x) = x - 3$. Solve both algebraically and graphically.

39 and 40. Let $g(x) = x^{\frac{2}{3}} - x^{-\frac{2}{3}}$.

39. Compute

(a) $g(8)$ **(b)** $g(-27)$ **(c)** $g(g(8))$

40. Compute

(a) $g(a^3)$ **(b)** $g(a) + g(-a)$

41 and 42. Let $h(x) = |2x - 3|$.

41. Compute

(a) $h(3)$ **(b)** $h(3) + h(-3)$ **(c)** $h\left(x + \frac{3}{2}\right)$

42. Find all real x satisfying the given statement. Solve both algebraically and graphically.

(a) $h(x) = 4$ **(b)** $h(x) < 5$ **(c)** $h(x) > 5$

43–52. Find the largest subset of the real numbers that could be the domain of the indicated function.

43. $f(x) = 3x - 2$

44. $g(x) = 4 - 3x$

45. $f(x) = -3x^2 + x + 1$

46. $g(x) = x^2 + 4$

47. $f(x) = \frac{4}{3x - 2} + 1$

48. $g(x) = 1 - \frac{3}{2x + 3}$

49. $f(x) = \sqrt{2x - 3}$

50. $g(x) = \sqrt{8 - 5x} + 1$

51. $f(x) = \sqrt{2 - x} + \frac{1}{x + 3}$

52. $g(x) = \sqrt{3x - 2} + \frac{3}{x - 4}$

II. Concepts

53–57. Find three examples of a nonconstant function $f(x)$ satisfying the given condition. Sketch their graphs.

53. $f(0) = 2$ and $f(2) = 3$

54. $f(x) > 10$ for all real x

55. $f(x) \leq 4$ for all real x

56. $f(x) = f(-x)$ for all real x

57. $f(x) = -f(-x)$ for all real x

58. Write a paragraph explaining how to find the domain of a function. Include a function describing a physical situation as one of your examples.

III. Applications

59–67. Solve each of the following problems.

59. Suppose the length of the radius of a circle is

(a) doubled **(b)** reduced by 25%.

What happens to the perimeter? To the area?

60. Suppose the length of the side of an equilateral triangle is

(a) tripled **(b)** halved.

What happens to the perimeter? To the area?

61. If the length of the edge of a cube is

(a) doubled **(b)** reduced by one third,

what happens to the surface area? To the volume?

62. The surface area of a sphere of radius r is $S(r) = 4\pi r^2$. The volume $V(r) = \frac{4}{3}\pi r^3$. If the radius of the sphere is

(a) increased by 50% **(b)** halved,

what happens to the surface area? To the volume?

63. If the length and the width of a rectangle are both

(a) doubled **(b)** reduced by 10%,

what happens to the perimeter? To the area?

64. If the length of a rectangle is doubled and the width is tripled, what happens to the area?

65. If the length, the width, and the height of a rectangular box are tripled, what happens to the

(a) volume? **(b)** surface area of the box?

66. If the radius of a cylinder is doubled and the height is halved, what happens to the

(a) volume? **(b)** total surface area of the cylinder?

67. The "Great Shape Up" exercise program tells you to do five push-ups in each workout the first week and then add three per week.

(a) How many push-ups per workout are you doing in the twenty-seventh week?

(b) If you continued this program for one year, working out four times per week, how many push-ups would you have done?

IV. Extensions

68. Consider these properties that a function might have. Let $f(x) = x^2$, $g(x) = 3x$, and $h(x) = 2\sqrt{x}$. Which of these properties are true for f? For g? For h?

$H(a + b) = H(a) + H(b)$ **Property 1**

$H(3a) = 3H(a)$ **Property 2**

$H(ab) = H(a)H(b)$ **Property 3**

In other words, let $H = f$ and then check to see if the properties are true, and so on.

69. If $f(x)$ is a function for which $f(a + b) = f(a) + f(b)$ for all real numbers a and b, prove that $f(0) = 0$. Can you give an example of such a function?

70. If $f(x)$ is a function for which $f(a + b) = f(a)f(b)$ for all real numbers a and b, prove that $f(0) = 0$ or $f(0) = 1$. Can you give an example of such a function?

71. As n gets large, what happens to $a_n = \sqrt[n]{n}$?

72. Let a_n be the nth digit to the right of the decimal point in the decimal expansion of $\frac{1}{7}$. Find a_{9999}. (For example, $a_1 = 1, a_2 = 4, a_3 = 2$.)

73. Functions such as $y = f(x) = 2x^2 + 3$ are functions of *one variable* since the value y in the range depends on only one value, x, in the domain. There are many situations where more than one variable is involved. Two examples are $P(l, w) = 2l + 2w$ (perimeter of a rectangle) and $V(l, w, h) = lwh$ (volume of a rectangular box). We can write $P(8, 6) = 2(8) + 2(6) = 28$.

(a) Give several more examples of functions of more than one variable.

(b) Heron's formula

$$A(a, b, c) = \sqrt{s(s - a)(s - b)(s - c)},$$

where

$$s = \frac{a + b + c}{2}$$

represents the area of a triangle whose sides are a, b, and c. What does $A(3, 5, 7)$ represent? Find its value.

74. Let $f(x, y) = 3x - 4y + 7$, $g(x, y) = 4x + 3y - 13$, $h(x, y) = x^2 - 4y + 3$. Find all pairs of real numbers (x, y) satisfying the system of equations

(a) $f(x, y) + g(x, y) = 0$; $f(x, y) - g(x, y) = 0$ **(b)** $f(x, y) = 0$; $h(x, y) = 0$

75. If $3g(x) - 2g\left(\frac{1}{x}\right) = x^2$ for all $x \neq 0$, find $g(2)$.

8.3 GRAPHING FUNCTIONS

In Section 8.1, we showed that a graph can be used to describe a function. We begin a detailed study of graphing by examining the graphs of three different types of functions.

Some Types of Graphs

EXAMPLE 1

Sketch the graph of the function that can be represented by the bar graph below or by the table.

x	$f(x)$
1990	3
1991	5
1992	6
1993	4
1994	2
1995	9

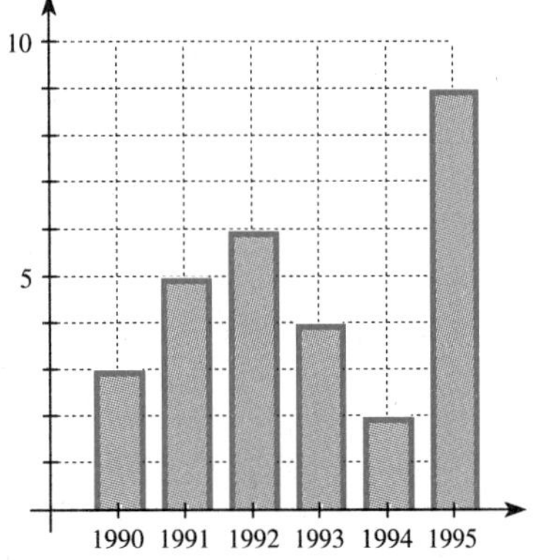

Figure 8.29

The function represents the profit (in millions of dollars) earned by a company from 1990 to 1995.

Solution

The graph consists of the six points shown in Figure 8.30.

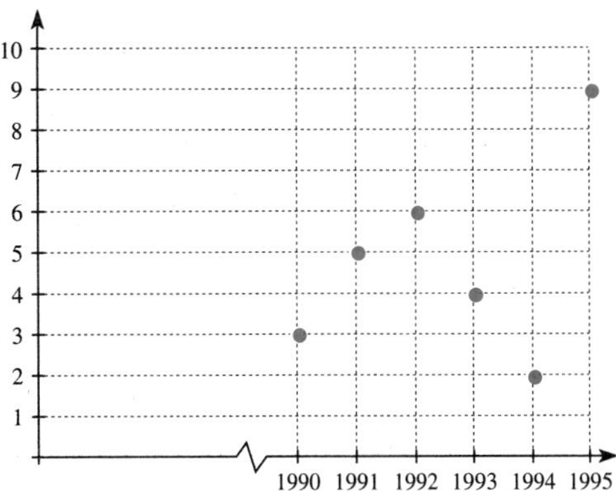

Figure 8.30

We could rename the years 1990–1995 by 0, 1, 2, 3, 4, and 5, and the graph becomes the one shown in Figure 8.31.

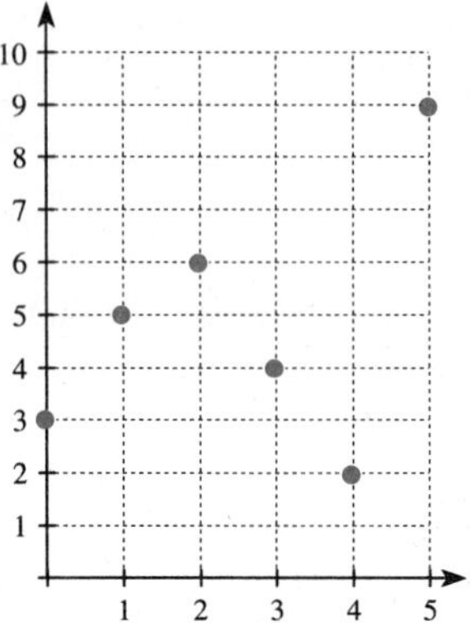

Figure 8.31

A function whose graph consists of a set of isolated points is called a **discrete function.** A sequence is another example of a discrete function.

EXAMPLE 2

Sketch a graph of the function that describes the amount you would pay for parking in a visitor lot: 60¢ for the first hour, 30¢ each additional hour, $2.50 maximum charge.

Solution

The amount you pay for any fraction of an hour is shown in the graph in Figure 8.32; parking eight hours or more costs $2.50.

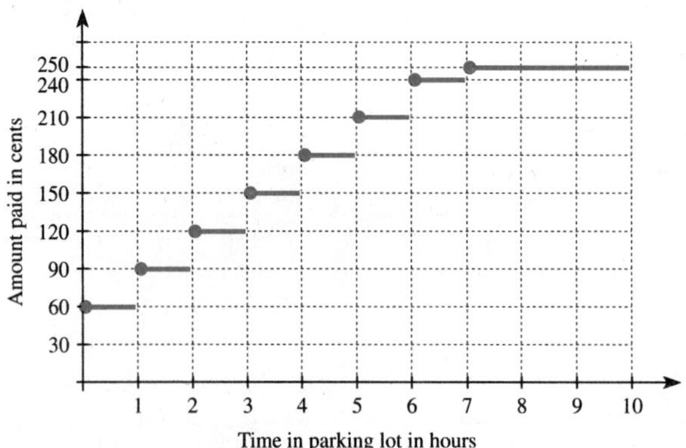

Figure 8.32

A function with a graph similar to the one shown in Example 2 is often called a **step function.**

EXAMPLE 3

Suppose you drop a rubber ball. Sketch the graph of the function that shows how the height of the ball depends on the time elapsed since you dropped it.

Solution

You would need a stopwatch and a ruler to draw an accurate graph, but it would probably resemble the one below.

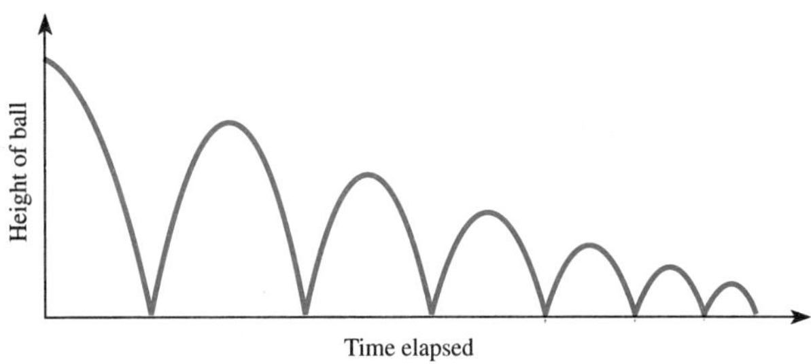

Figure 8.33

Graphing Some Basic Polynomial Functions

In this section, we concentrate on graphs of polynomial functions. Recall that a polynomial function is a function of the form

$$p(x) = a_n x^n + a_{n-1}x^{n-1} + \cdots + a_1 x + a_0$$

where n is a positive integer and $a_n \neq 0$. In earlier chapters, we studied many applications of linear functions, $p(x) = a_1 x + a_0$, and quadratic functions, $p(x) = a_2 x^2 + a_1 x + a_0$.

By this time, we know well the graphs of linear functions (lines) and quadratic functions (parabolas). We now examine the graphs of some other basic polynomial functions. For simplicity, we use y instead of $p(x)$. That is, we denote the elements in the range by y instead of $p(x)$. Let's recall the graphs of some basic polynomials.

EXAMPLE 4

Graph $y = x^3$.

Solution

Using a Table

x	0	1	2	3	4	5	10
x^3	0	1	8	27	64	125	1,000

Table 8.11

Using a Graph

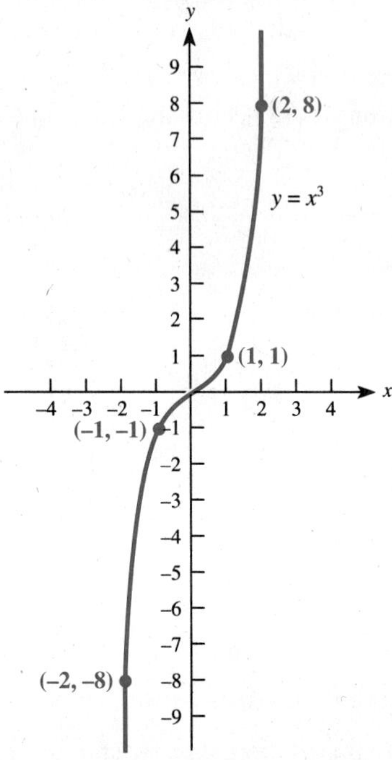

Figure 8.34

We observe that the graph rises rapidly as x gets larger. It rises faster than $y = x^2$. When $x < 0$, the values of x^3 are the negatives of those for the corresponding values of x. Symbolically, $(-x)^3 = -x^3$. Therefore the point $(-x, -x^3)$ is as far below the x-axis as (x, x^3) is above the x-axis. "Zooming in," we can check some values of x between 0 and 1. The graph "flattens out" near the origin:

Using a Table

x	0.1	0.2	0.4	0.5	0.8
x^3	0.001	0.008	0.064	0.125	0.512

Table 8.12

Using a Graph

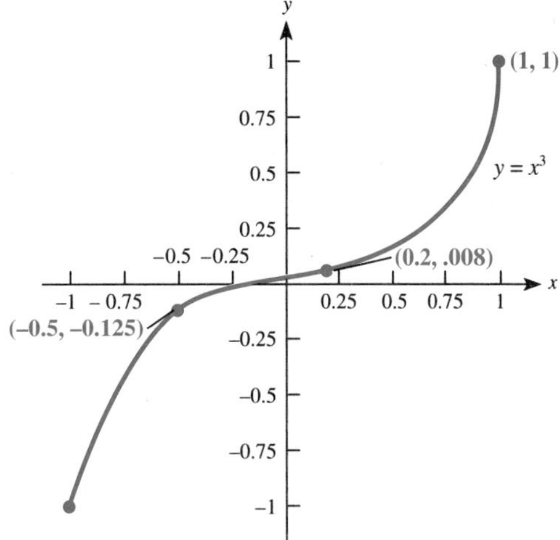

Figure 8.35

From the graph, we also observe that the domain and the range of the function $y = x^3$ are all real numbers. ■

EXAMPLE 5 Graph $y = x^4$.

Solution

Again we make a table of values.

x	x^4
0	0
1	1
2	16
3	81
4	256
5	625
10	10,000

Table 8.13

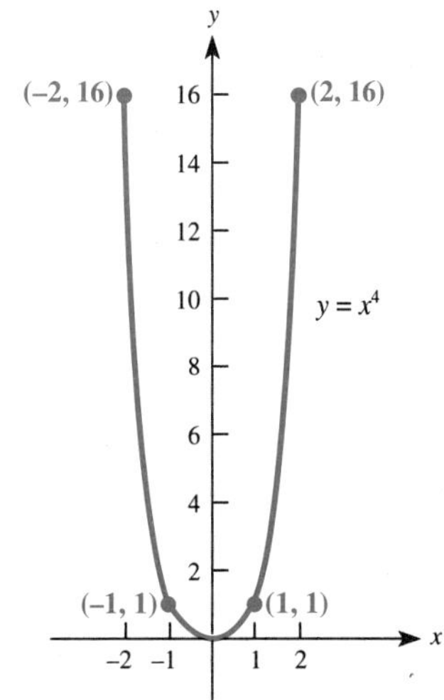

Figure 8.36

For $x \geq 0$, the shape is much like that of $y = x^3$ (Figure 8.34), but it rises faster as x gets larger. However for $x < 0$, $(-x)^4 = x^4$, so the portion of the graph for $x < 0$ is the mirror image in the y-axis of the portion for $x \geq 0$ (Figure 8.36). As in the case of the parabola $y = x^2$, this graph is symmetric with respect to the y-axis. In Figure 8.37 we see that the graph near 0 is "flatter" than that of $y = x^3$. The domain is all real numbers, but the range consists of only nonnegative real numbers.

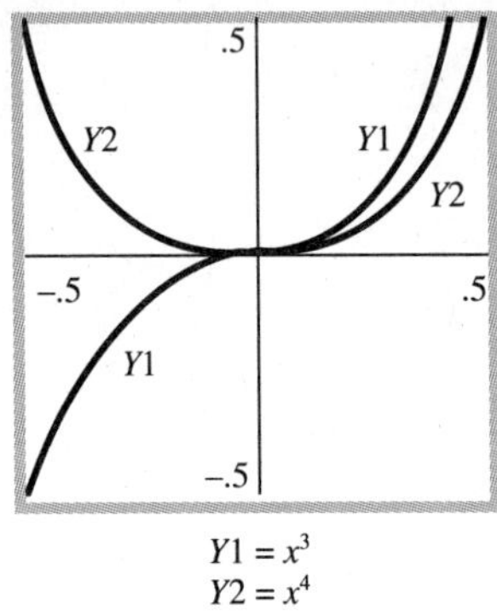

$Y1 = x^3$
$Y2 = x^4$

Figure 8.37

Graphing Some Basic Nonpolynomial Functions

Several of the examples in previous sections and in Chapter 7 involved square roots and cube roots. We now examine the graphs of these two nonpolynomial functions.

EXAMPLE 6 Graph

(a) $y = \sqrt{x}$ **(b)** $y = \sqrt[3]{x}$

Solutions

(a) The domain is $x \geq 0$. Recall $\sqrt{x}$ is the principal square root.

x	0	0.5	1	2	3	5	10	100
$\sqrt{x}$	0	0.707	1	1.414	1.732	2.236	3.162	10

Table 8.14

Since $\sqrt{0.5} \approx 0.707$, the graph lies above the line $y = x$ for $0 \leq x \leq 1$.

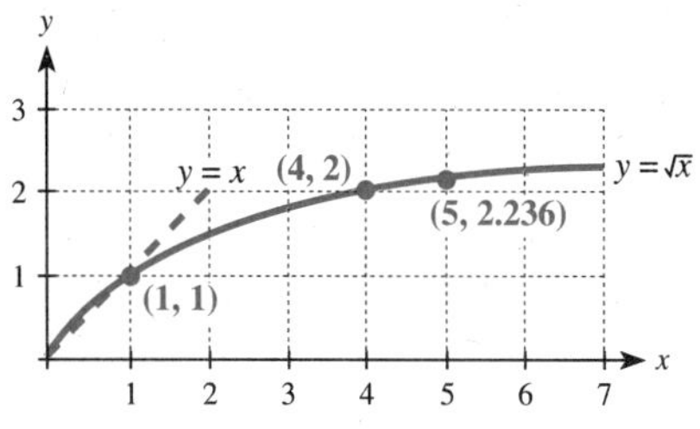

Figure 8.38

(b) The domain is all real numbers, since the cube root of a negative number is defined.

x	0	0.5	1	2	3	5	10	100
$\sqrt[3]{x}$	0	0.793	1	1.260	1.442	1.710	2.154	4.64

Table 8.15

The graph lies above $y = x$ for $0 \leq x \leq 1$, and below $y = x$ for $-1 < x < 0$. Since $\sqrt[3]{-x} = -\sqrt[3]{x}$, the point $(-x, -\sqrt[3]{x})$ is as far below the x-axis as $(x, \sqrt[3]{x})$ lies above the x-axis. The graph is shown below.

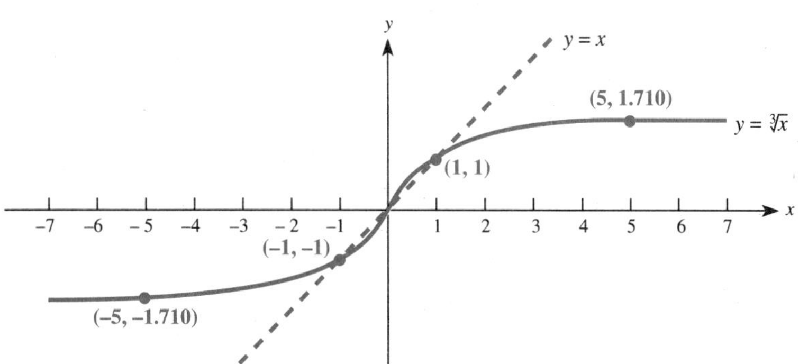

Figure 8.39

For $x \geq 0$ the graph $y = \sqrt[3]{x}$ rises more slowly than that of $y = \sqrt{x}$ (Figure 8.40).

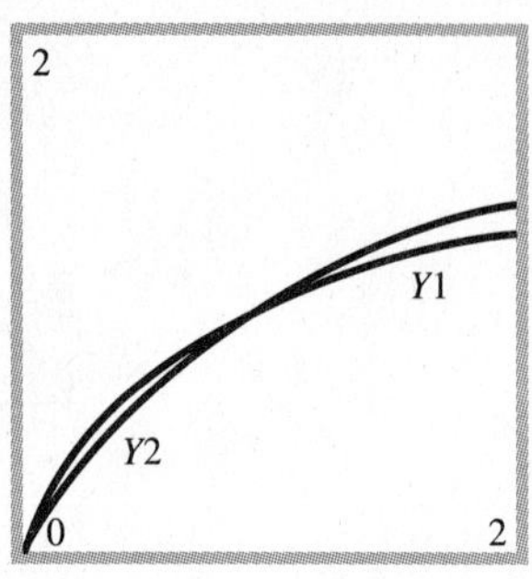

Figure 8.40

For completeness we include the graphs of three nonpolynomial functions considered in earlier chapters.

EXAMPLE 7

(a) $y = |x|$ **(b)** $y = \dfrac{1}{x}$ **(c)** $y = \dfrac{1}{x^2}$

Solution

(a)

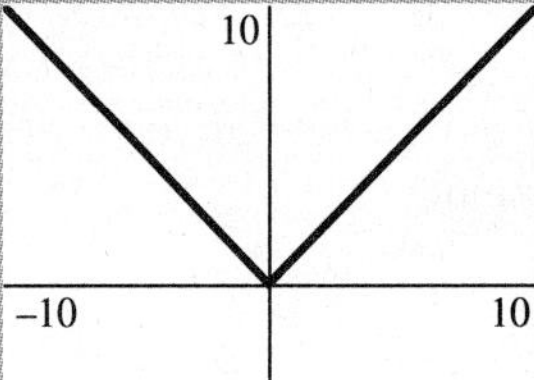

$Y = |x|$

Figure 8.41

(b)

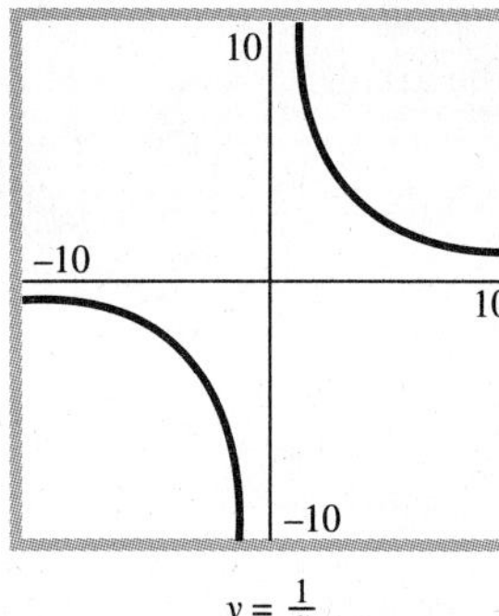

$y = \frac{1}{x}$

Figure 8.42

(c)

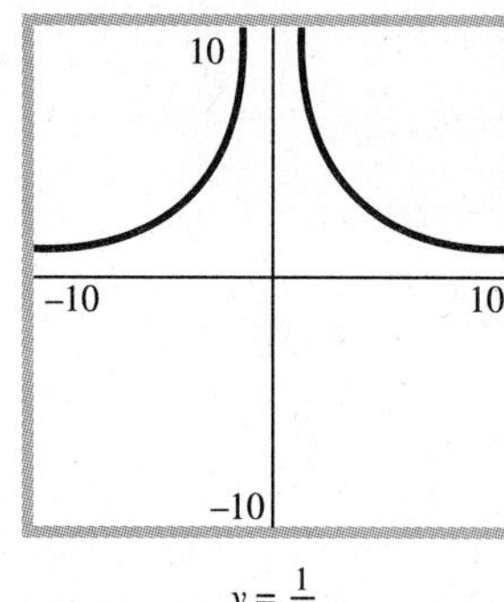

$y = \frac{1}{x^2}$

Figure 8.43

Five Graphing Principles

Functions in real life are rarely as simple as those we have just graphed. For example, the speed of sound is not $V = V(t) = \sqrt{t}$, but

$$V = V(t) = \frac{1{,}087\sqrt{273 + t}}{16.32}$$

The graphs of these more complicated functions are, however, closely related to those of the simpler functions. Some of these relationships can be seen by examining several general graphing principles. We recall an example from Chapter 6.

EXAMPLE 8

Graph $y = x^2 - 2$ and $y = x^2 + 2$ on the same set of axes.

Solution

The graphs are related to that of $y = x^2$. Thus the graph of $y = x^2 + 2$ is two units higher than that of $y = x^2$, while the graph of $y = x^2 - 2$ is two units lower (see Figure 8.44).

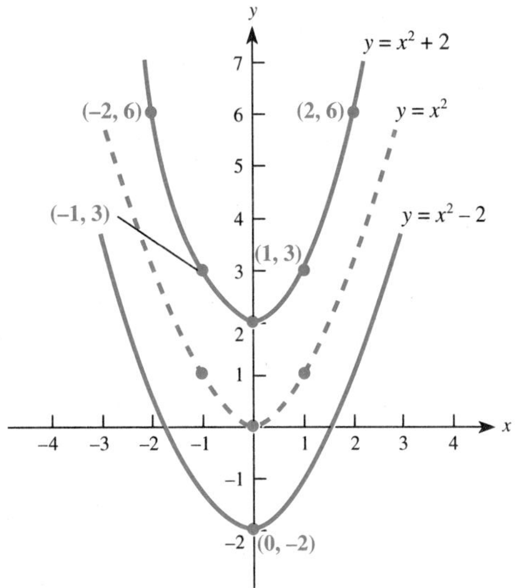

Figure 8.44

In general, for all *positive* numbers c, the graph of $y = f(x) + c$ is the graph of $y = f(x)$ shifted c units upward, while the graph of $y = f(x) - c$ is the graph of $y = f(x)$ shifted c units downward.

EXAMPLE 9

Graph $y = (x - 2)^2$ and $y = (x + 2)^2$ on the same set of axes.

Solution

Again these graphs are related to that of $y = x^2$. We see that the graph of $y = (x - 2)^2$ is the graph of $y = x^2$ shifted two units to the right, while the graph of $y = (x + 2)^2$ is the graph of $y = x^2$ shifted two units to the left (see Figure 8.45).

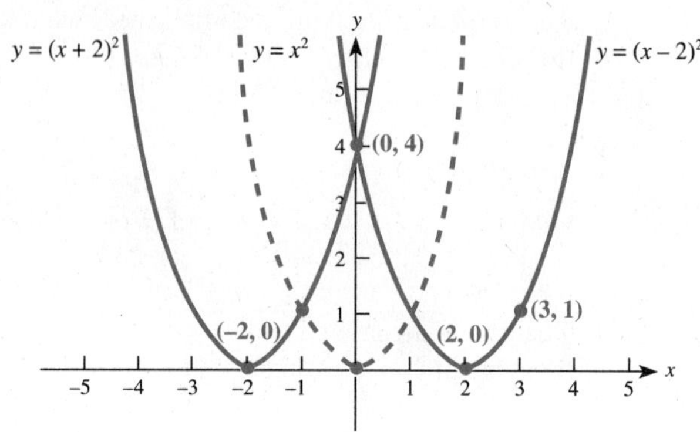

Figure 8.45

In general, for all *positive* numbers c, the graph of $y = f[x + (-c)] = f(x - c)$ is the graph of $y = f(x)$ shifted c units to the right, while the graph of $y = f[x - (-c)] = f(x + c)$ is the graph of $y = f(x)$ shifted c units to the left.

Another graphing principle is contained in the next example.

EXAMPLE 10 Graph $y = 3x^2$ and $y = -3x^2$ on the same set of axes.

Solution

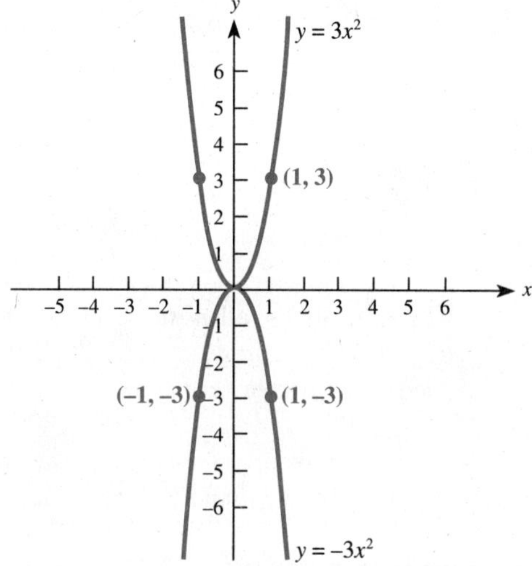

Figure 8.46

In general, the graph of $y = -f(x)$ is the mirror image in the x-axis of the graph of $y = f(x)$.

We now summarize the five graphing principles.

Five Graphing Principles

Suppose $c > 0$. The graph

1. $y = f(x) + c$
2. $y = f(x) - c$
3. $y = f(x - c)$
4. $y = f(x + c)$

is the graph of $y = f(x)$ shifted c units upward / downward / to the right / to the left (respectively).

5. The graph of $y = -f(x)$ is the mirror image in the x-axis of the graph of $y = f(x)$.

We close this section with several examples illustrating these graphing principles. To test your understanding of the graphing principles, first make a rough sketch using paper and pencil, then use a graphing calculator.

EXAMPLES 11–15

For each of the functions, **(a)** sketch its graph, making use of the graphing principles; **(b)** find the x- and y-intercepts; and **(c)** find the domain, D, and range, R.

11. $y = |x| + 2$ **12.** $y = \dfrac{1}{x^2} - 3$ **13.** $y = \dfrac{1}{x - 1}$

14. $y = \sqrt[3]{x + 4}$ **15.** $y = -\sqrt{x}$

Solutions

11. **(a)** The graph of $y = |x| + 2$ is the graph of $y = |x|$ shifted 2 units upward.

(b) x-intercept: none
y-intercept: $(0, 2)$

(c) $D = \{\text{all real numbers}\}$
$R = \{y | y \geq 2\}$

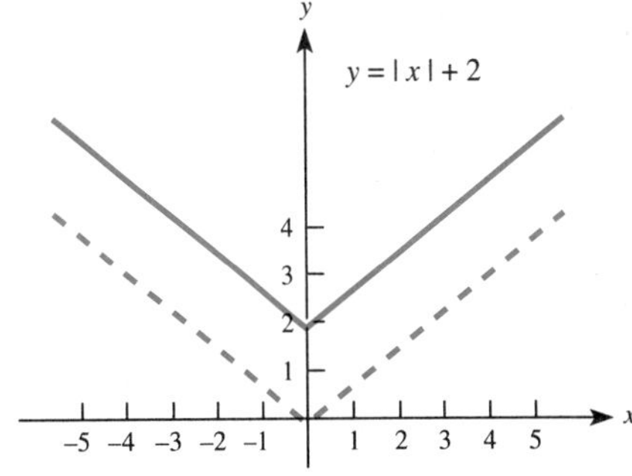

Figure 8.47

12. **(a)** The graph of $y = \dfrac{1}{x^2} - 3$ is the graph of $y = \dfrac{1}{x^2}$ shifted 3 units downward.

(b) x-intercepts: $\left(\dfrac{\sqrt{3}}{3}, 0\right)\left(-\dfrac{\sqrt{3}}{3}, 0\right)$

y-intercepts: none

(c) $D = \{\text{all real numbers} \neq 0\}$
$R = \{y \mid y > -3\}$

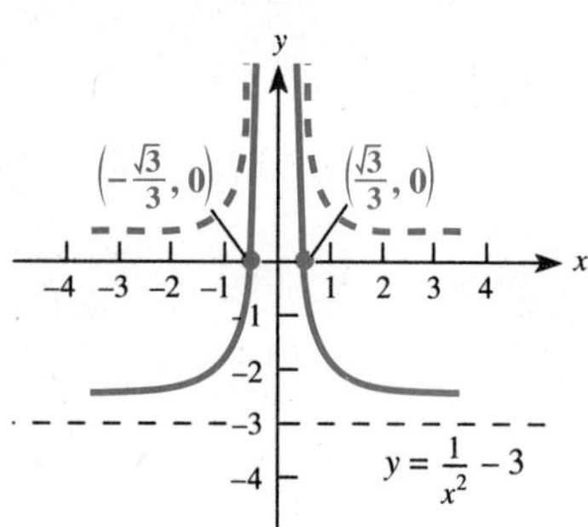

Figure 8.48

13. (a) The graph of $y = \dfrac{1}{x-1}$ is the graph of $y = \dfrac{1}{x}$ shifted 1 unit to the right.

(b) x-intercept: none
y-intercept: $(0, -1)$

(c) $D = \{\text{all real numbers} \neq 1\}$
$R = \{\text{all real numbers} \neq 0\}$

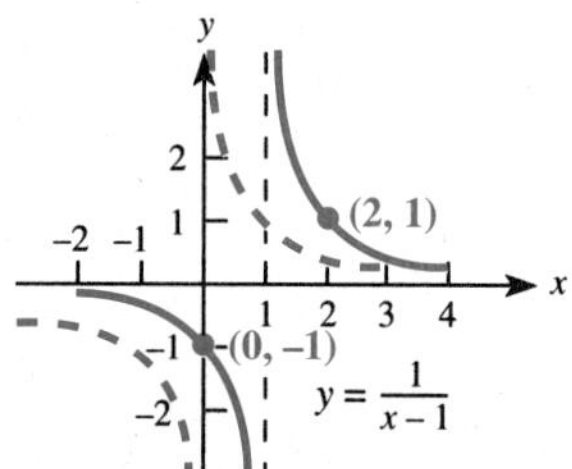

Figure 8.49

14. (a) The graph of $y = \sqrt[3]{x+4}$ is the graph of $y = \sqrt[3]{x}$ shifted 4 units to the left.

(b) x-intercept: $(-4, 0)$
y-intercept: $(0, \sqrt[3]{4})$

(c) $D = R = \{\text{all real numbers}\}$

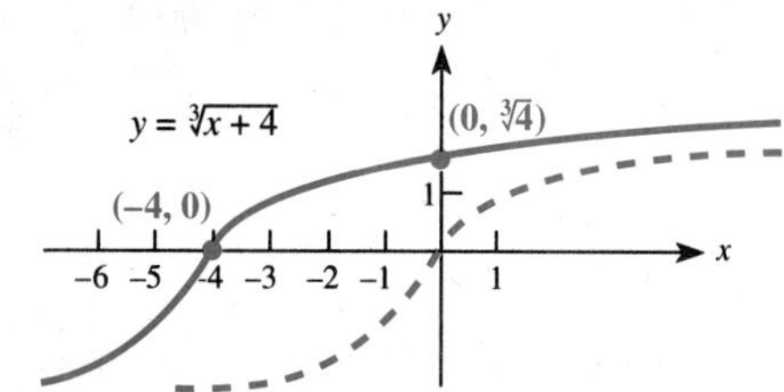

Figure 8.50

15. (a) The graph of $y = -\sqrt{x}$ is the mirror image of $y = \sqrt{x}$ in the x-axis.

(b) x-intercept $=$ y-intercept $= (0, 0)$

(c) $D = \{x \mid x \geq 0\}$; $R = \{y \mid y \leq 0\}$.

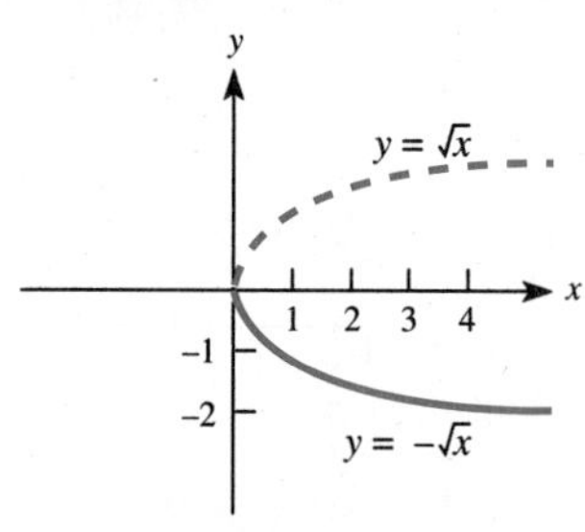

Figure 8.51

In these last examples, we study some graphs in which two of the graphing principles are used in constructing the same graph.

EXAMPLES 16–19

For each of the functions, **(a)** sketch its graph, making use of the graphing principles; **(b)** find the x- and y-intercepts; and **(c)** find the domain, D, and range, R.

16. $y = \dfrac{1}{x - 1} + 2$

17. $y = \sqrt[3]{x + 4} - 3$

18. $y = \dfrac{1{,}087\sqrt{273 + t}}{16.32}$

19. $y = |x + 2| - 3$

Solutions

16. (a) The graph of $y = \dfrac{1}{x - 1} + 2$ is the graph of $y = \dfrac{1}{x}$ shifted 1 unit to the right and *then* 2 units upward.

(b) x-intercept: (0.5, 0)
y-intercept: (0, 1)

(c) $D = \{x | x \neq 1\}$
$R = \{y | y \neq 2\}$

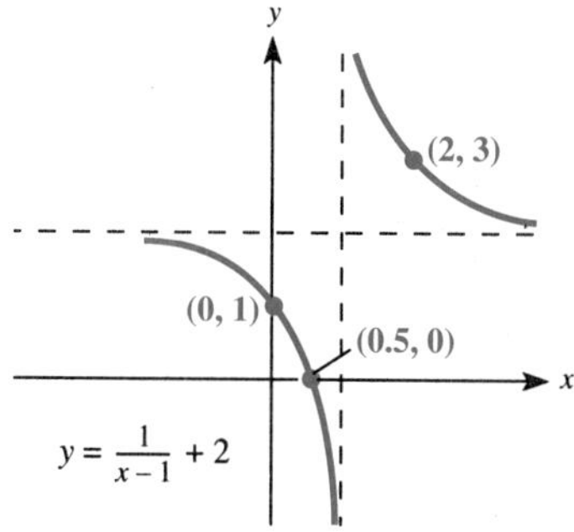

Figure 8.52

17. (a) The graph of $y = \sqrt[3]{x + 4} - 3$ is the graph of $y = \sqrt[3]{x}$ shifted 4 units to the left and *then* 3 units downward.

(b) x-intercept: (23, 0)
y-intercept: $(0, \sqrt[3]{4} - 3)$

(c) $D = R = \{\text{all real numbers}\}$

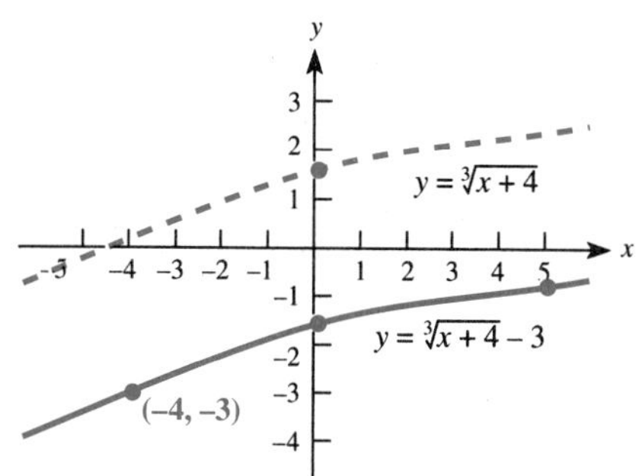

Figure 8.53

18. (a) The graph of $y = \dfrac{1{,}087\sqrt{273 + t}}{16.32}$ is the graph of $y = \sqrt{x}$ shifted 273 units to the left, then changed so it rises faster than $y = \sqrt{x}$ since $\dfrac{1{,}087}{16.32} > 0$.

(b) x-intercept: $(-273, 0)$

y-intercept: $\left(0, \dfrac{1{,}087\sqrt{273}}{16.32}\right) \approx (0, 1{,}100)$

(c) $D = \{x | x \geq -273\}$
$R = \{y | y \geq 0\}$

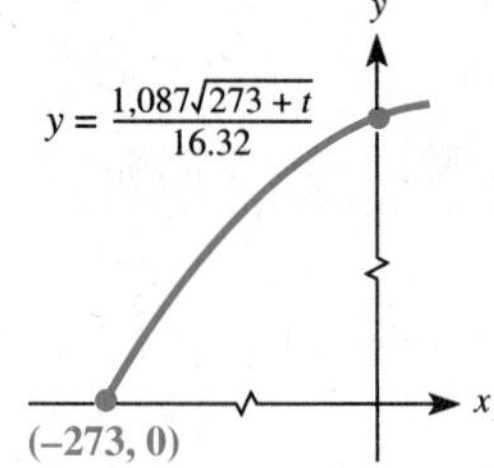

Figure 8.54

19. **(a)** The graph of $y = |x + 2| - 3$ is the graph of $y = |x|$ shifted 2 units to the left and *then* 3 units downward.

(b) x-intercepts: $(1, 0), (-5, 0)$
y-intercept: $(0, -1)$

(c) $D = \{\text{all real numbers}\}$
$R = \{y | y \geq -3\}$

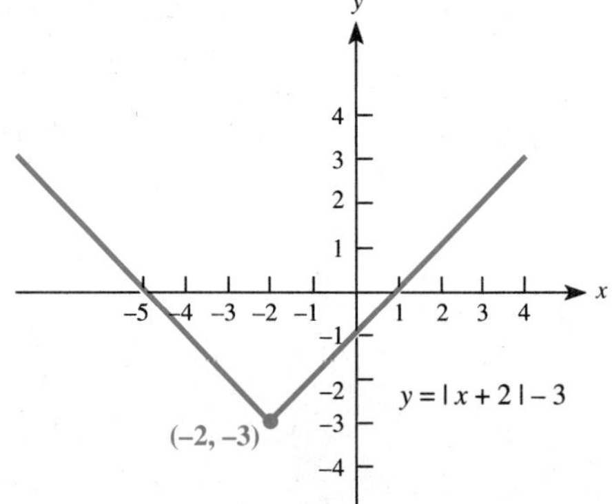

Figure 8.55

Not all functions, however, have such well-behaved graphs. If you have ever seen the output of a lie detector test, an electrocardiogram, or a sound wave, you know graphs can behave erratically. (See Figure 8.56.)

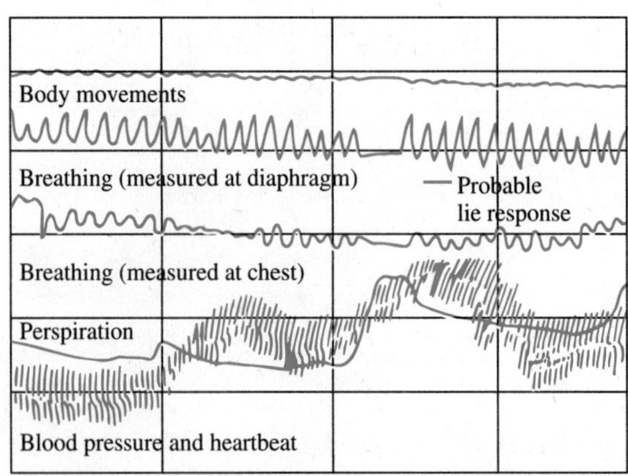

Figure 8.56

8.3 EXERCISES

I. Procedures

1–6. Which of the following graphs represent functions? Find the domain and range of those that are functions.

1. **2.**

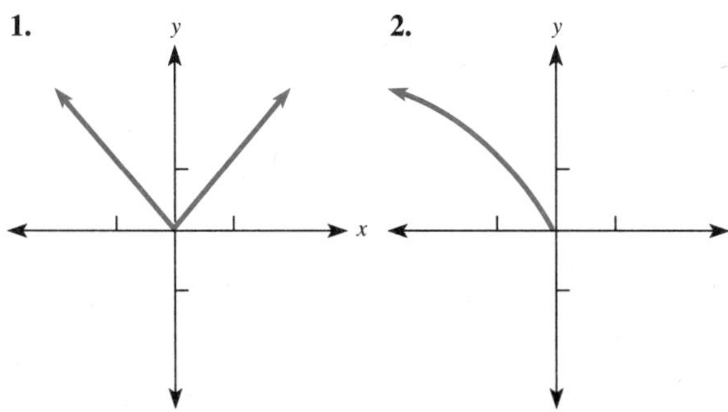

3. **4.**

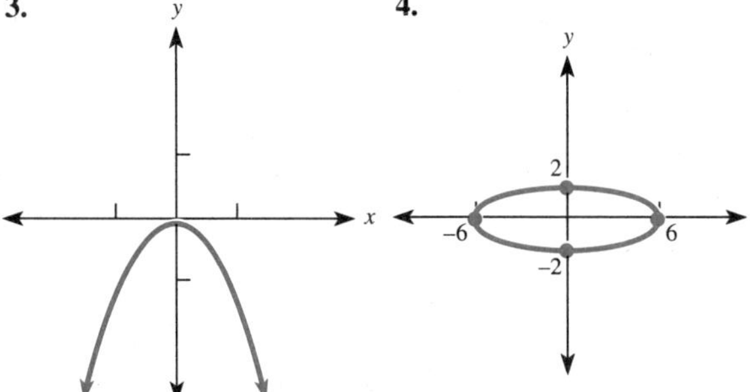

5. **6.**

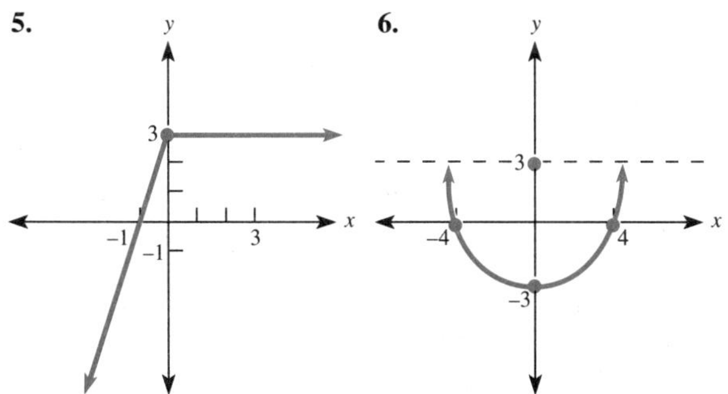

7–16. Roughly sketch the graphs of the following functions and then use a graphing calculator.

7. $f(x) = x^5$

8. $g(x) = x^6$

9. $f(x) = 2x^4$

10. $g(x) = -2x^3$

11. $f(x) = \sqrt[4]{x}$

12. $g(x) = \sqrt[5]{x}$

13. $f(x) = \dfrac{1}{x^3}$

14. $g(x) = \dfrac{1}{x^4}$

15. $f(x) = \dfrac{1}{\sqrt{x}}$

16. $g(x) = \dfrac{1}{\sqrt[4]{x}}$

17–42. For each of the functions, **(a)** roughly sketch its graph, making use of the graphing principles; **(b)** find the x- and y-intercepts; and **(c)** find the domain and range. Then use a graphing calculator.

17. $y = x^2 + 3$

18. $y = -2x^2$

19. $y = 2(x - 2)^2$

20. $y = (x + 3)^2$

21. $y = 2 - x^2$

22. $y = (x + 3)^2 - 1$

23. $y = x^3 + 1$

24. $y = x^4 - 1$

25. $y = \sqrt{x - 3}$

26. $y = \sqrt[3]{x + 1}$

27. $y = \dfrac{1}{x} - 2$

28. $y = \dfrac{1}{x^2} + 3$

29. $y = |x - 3|$

30. $y = |3 - x| + 1$

31. $y = -\sqrt{x + 2}$

32. $y = -\dfrac{1}{x^2}$

33. $y = (x - 4)^3 + 3$

34. $y = (x - 4)^2 - 4$

35. $y = (x + 2)^4 - 1$

36. $y = (x + 1)^5 + 8$

37. $y = 1 + \sqrt{x + 3}$

38. $y = \sqrt[3]{x - 3} - 1$

39. $y = \dfrac{1}{(x + 2)^2} + 2$

40. $y = \dfrac{1}{x - 7} + 4$

41. $y = -2x^3 + 4$

42. $y = 4 - 3x^4$

II. Concepts

43. The graph of $y = x^2$ is *symmetric* with respect to the y-axis, since the y-axis acts like a mirror for this graph. Find three other functions whose graphs are symmetric with respect to the y-axis.

44. Can you find any functions whose graphs are symmetric with respect to the x-axis?

45. Can you find any functions whose graphs are symmetric with respect to the line $y = -x$?

46. Do you observe any relationship between the graph of $y = x^3$ and that of $y = \sqrt[3]{x}$? Find three other "pairs" of functions whose graphs have this relationship.

47. Explain the use of the word "range" in describing a function and in sketching a graph on a graphing calculator.

48. In sketching a graph on a graphing calculator, a key step is determining an appropriate range. Write a paragraph on how you determine a suitable range for a given graph. Give some examples.

49. Carefully sketch the graphs of $y = x^2$, $y = x^3$, $y = x^4$ for $-1 \le x \le 1$, $-1 \le y \le 1$ on the same coordinate system. Describe your observations.

50. Carefully sketch the graphs of $y = \sqrt{x}$ and $y = \sqrt[3]{x}$ for $-1 \le x \le 1$, $-1 \le y \le 1$ on the same coordinate system. Describe your observations.

51. Carefully sketch the graphs of $y = \frac{1}{x}$ and $y = \frac{1}{x^2}$ for $-1 \le x \le 1$, $-1 \le y \le 1$ on the same coordinate system. Describe your observations.

52. When you "complete the square" on a quadratic polynomial, you rewrite it in the form $y = a(x - h)^2 + k$. How does this form help in sketching its graph?

53. Give at least three examples of functions whose graphs contain the points $(1, 1)$ and $(2, 4)$.

54. Suppose you plan to drive at a *constant* speed from Chicago to St. Louis via Springfield. Here are four variables that depend on the speed you choose to drive and four graphs that show the relationships of these variables to the speed you choose.

- *CD* = your distance from Chicago after three hours
- *HD* = the time it takes you to drive 200 miles
- *SLD* = your distance from St. Louis after three hours
- *SD* = your distance from Springfield after three hours

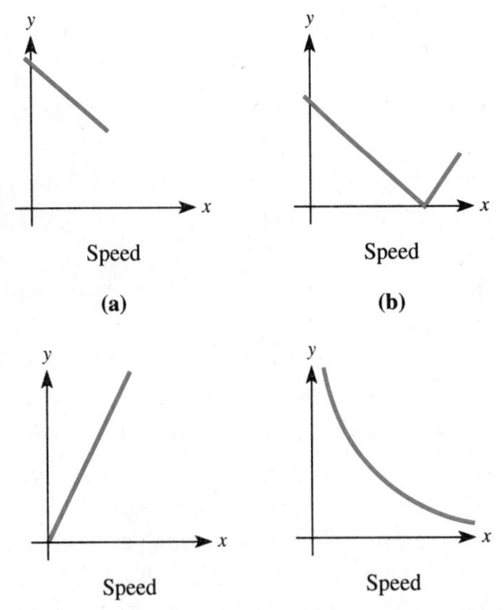

Label the vertical axis of each graph with one of the four variables. Ask yourself: "As your speed increases, what happens to each of the four variables?" Finally, give reasons for your choices.

55. Are these two functions identical? Give a reason for your answer.

$$f(x) = \frac{\sqrt{x+3}}{\sqrt{x-3}} \qquad g(x) = \sqrt{\frac{x+3}{x-3}}$$

III. Applications

56. Sketch the graph of $f(x) = \frac{1}{x + 13}$ in each of the following viewing rectangles:

(a) x-interval $(-5, 5)$; y-interval $(-2, 2)$

(b) x-interval $(-10, 10)$; y-interval $(-5, 5)$

(c) x-interval $(-15, 15)$; y-interval $(-15, 15)$

(d) x-interval $(-20, 20)$; y-interval $(-20, 20)$

(e) What do you notice?

57. **(a)** Graph the function $f(x) = x^3 - 3x + 1$.

(b) Locate all roots and turning points accurate to two decimal places.

(c) Based on your graph in part (a), sketch the graph of $g(x) = x^3 - 3x$ without using your calculator.

(d) Based on your graph in part (a), sketch the graph of $g(x) = -x^3 + 3x - 1$ without using your calculator.

58–61. Sketch the graphs of each of the following functions. Decide what you think would be a reasonable domain for the function.

58. $f(r) = \frac{240}{r}$; the time it takes to make a 240-mile trip at an average speed of r mph.

59. $g(t) = 442 - 5t^2$; the approximate height m (in meters) *above* the ground of a ball t seconds after it is dropped from the top of the Sears Tower in Chicago.

60. $p(H) = \frac{94}{\sqrt{H}}$; the approximate pulse rate (beats per minute) of an adult who is H meters tall.

61. $V(L) = \sqrt{12L}$; the approximate velocity in feet per second of a car that left skid marks of length L feet on wet pavement.

62–74. Sketch, as best you can, the graph of a function that describes each of the following situations. You will have to supply most of the data yourself, and the graphs will be approximate.

62. The cost of mailing a first-class letter as a function of the number of ounces it weighs.

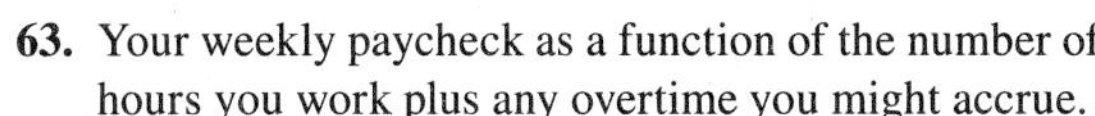

63. Your weekly paycheck as a function of the number of hours you work plus any overtime you might accrue.

64. The cost of a taxi ride as a function of the number of miles traveled.

65. The cost of a taxi ride, including a tip, as a function of the number of miles traveled.

66. The sales tax on a washing machine as a function of its cost.

67. The cost of renting a car for one day as a function of the number of miles driven.

68. The height of your head from the ground when riding on a ferris wheel as a function of the number of seconds since you were seated. Assume

(a) you are the only rider.

(b) every chair on the ferris wheel is full.

69. The amount of light you receive from a lamp as a function of the distance you are sitting from the lamp.

70. The altitude of a football as a function of the number of seconds since it was punted.

71. The distance you have traveled as a function of how long you have been traveling (at a constant speed).

72. Your height above the water in a swimming pool as a function of the number of seconds since you jumped off the (high) diving board.

73. The amount of air in your lungs as a function of time.

74. The amount of water in the bathtub as a function of the amount of time since you removed the stopper.

75–79. Select five of the graphs used in Exercises 66–74. Describe another situation that could be depicted by each graph.

IV. Extensions

80. Given the graph of $g(x)$ (which shows its domain and range), sketch the graph of $h(x)$ and give its domain and range.

(a) $h(x) = g(x) - 3$

(b) $h(x) = g(x + 1)$

(c) $h(x) = -2g(x)$

(d) $h(x) = 1 - g(x - 3)$

(e) $h(x) = 2g(x + 1)$

(f) $h(x) = |g(x) - 1|$

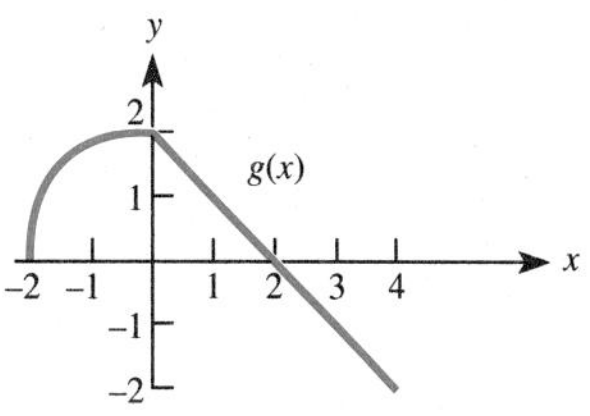

81–84. If the graph of a function is continuous (has no "hole" or "breaks") for $a \leq x \leq b$, and $f(a)$ and $f(b)$ have opposite signs, then the equation $f(x) = 0$ has a solution between a and b.

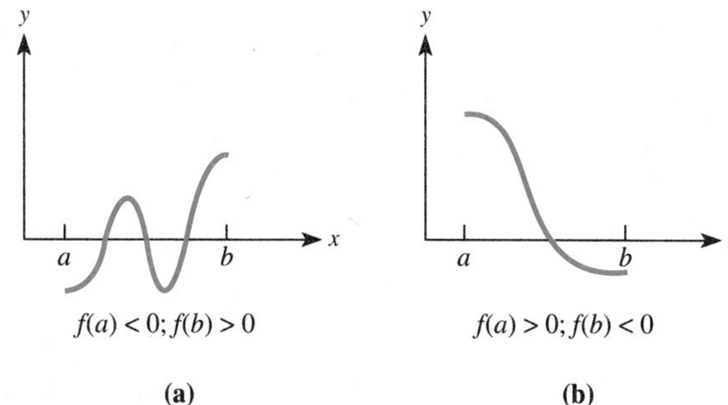

$f(a) < 0; f(b) > 0$

(a)

$f(a) > 0; f(b) < 0$

(b)

For example, if $f(x) = 3x^2 - 4$, then $f(1) = -1$ and $f(2) = 8$, so the graph of $f(x)$ crosses the x-axis between 1 and 2, and the equation $f(x) = 0$ has a solution between 1 and 2. From Chapter 7, we know this solution is $x = \sqrt{\frac{4}{3}} \approx 1.55$.

For each of the following functions, locate (approximately—between consecutive integers) the solutions to the equation $f(x) = 0$. Use a graphing calculator.

81. $f(x) = 3x^3 - 7x + 2$

82. $f(x) = x^4 - 7x + 1$

83. $f(x) = \sqrt{x + 2} - \dfrac{7}{x}$

84. $f(x) = \dfrac{1}{x + 1} - \dfrac{2}{x^2 + 1}$

85. Find the domain and range of the given function.

(a) $f(x) = \sqrt{x^2 - 5x - 14} - 3$

(b) $g(x) = \dfrac{1}{\sqrt{x^2 - 5x - 14}} + 2$

(c) $h(x) = \sqrt{\dfrac{1}{x^2 - x - 2} - \dfrac{1}{4}}$

8.4 FINDING AND OPTIMIZING FUNCTIONS

In the previous sections of this chapter, we were given a function $f(x)$ and asked to do such things as compute $f(3)$, find all real x for which $f(x) = 0$, or sketch the graph of $f(x)$. In short, we defined the notion of a function and then showed how some of the skills and concepts learned in earlier chapters could be viewed in terms of functions. In this section, we take another look at some of the applications discussed in previous chapters and show how they can be interpreted as problems involving functions. The concept of a function, then, can truly be thought of as a universal algebraic concept.

Some of our first applications involved proportions and other linear equations. These applications can be viewed as problems involving linear functions:

EXAMPLE 1

See Example 3 in Section 3.2. Nine acres of grazing land are needed to graze 2 cows. How many acres are needed to graze 2,000 cows?

Solution

Since 9 acres are required for 2 cows, it follows that 4.5 acres are required for 1 cow. This situation can be described by the function $g(c) = 4.5c$, where c is the number of cows and $g(c)$ is the number of acres of grazing land required. For 2,000 cows, then $g(2{,}000) = 9{,}000$, so 9,000 acres are required. ■

EXAMPLE 2

See Example 1 in Section 4.4. The We Try Harder car rental company charges \$19.95 plus 19¢ a mile per day, while the U Drive It Company charges \$15.95 plus 23¢ per mile.

(a) How far can you travel in one day for \$100.00?

(b) Which company should you choose if you wish to rent a car for one day?

Solutions

The daily charges for the We Try Harder Company can be given by the linear function $f(x) = 19.95 + 0.19x$; those for the U Drive It Company by $g(x) = 15.95 + 0.23x$. In

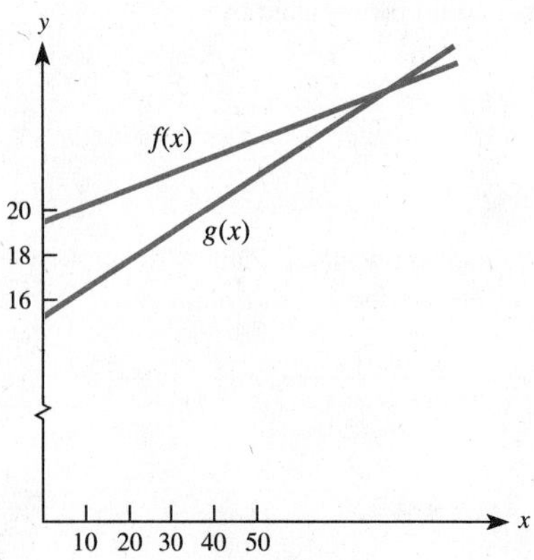

Figure 8.57

each case x is the number of miles driven. These functions are graphed in Figure 8.57. For part (a), we seek the value of x for which $f(x) = 100$. For part (b), we choose the We Try Harder Company whenever $f(x) < g(x)$; we choose the U Drive It Company whenever $g(x) < f(x)$.

EXAMPLE 3

Find the equation of the linear function whose graph contains the points $(-1, 1)$ and $(2, 7)$.

Solution

We seek the linear function $f(x) = ax + b$ satisfying $f(-1) = 1$ and $f(2) = 7$. Thus

$$-a + b = 1 \quad f(-1) = 1$$
$$2a + b = 7 \quad f(2) = 7$$

Therefore

$$-3a = -6 \quad \text{Solve the system.}$$
$$a = 2$$
$$b = 3$$

and the desired function is $f(x) = 2x + 3$.

The key to solving each of these problems is to *find a function that describes or models the situation*. In the first three examples, the function is a linear polynomial, but it need not be. We usually *assume* (because of the nature of the problem) that we are looking for a function of a certain type, and then we find such a function. We hope, then, that this function closely approximates the real-life situation.

Recall that a sequence is a type of discrete function.

EXAMPLES 4–6

The first five terms of a sequence are given. Find a formula for the general term a_n in each case.

4. $1, 3, 5, 7, 9$

5. $\frac{1}{3^3}, \frac{1}{4^3}, \frac{1}{5^3}, \frac{1}{6^3}, \frac{1}{7^3}$

6. $\left(1 + \frac{1}{1}\right)^1, \left(1 + \frac{1}{2}\right)^2, \left(1 + \frac{1}{3}\right)^3, \left(1 + \frac{1}{4}\right)^4, \left(1 + \frac{1}{5}\right)^5$

Solutions

Finding a pattern for a general term a_n can often be tricky, if not impossible. In fact, many different patterns can be suggested by the first few terms in a sequence. The task, then, is to find some link between n, the number of the term, and a_n, the value of that term. For Examples 4–6, we give one possible general term, but we stress that other answers are possible.

4. $a_n = 2n - 1$ The terms are each one less than $2, 4, 6, 8, \ldots, 2n$.

5. $a_n = \frac{1}{(n + 2)^3}$ The number cubed (in the denominator) is 2 more than the number of the term.

6. $a_n = \left(1 + \frac{1}{n}\right)^n$ ■

In the preceding examples, we gave a "natural" answer for a_n, but as we remarked, many different correct answers are possible. In each case, we are asked to determine a general pattern based on looking at the first five cases. Many different patterns, however, can all begin in the same way. Thus, in Example 4,

$$a_n = (n - 1)(n - 2)(n - 3)(n - 4)(n - 5) + (2n - 1)$$

or

$$a_n = \frac{n^2 - (n - 1)^2}{(n - 1)(n - 2)(n - 3)(n - 4)(n - 5) + 1}$$

are also reasonable answers. The inclusion of such questions on intelligence tests is often criticized, because such a question could have more than one "right answer."

Direct Variation

There are many instances in which our experience tells us the type of function we are seeking.

In Example 1, we had two quantities—the number of cows and the amount of grazing land. As one quantity (the number of cows) gets larger (or smaller), the other quantity (the amount of grazing land needed) also gets larger (or smaller). This relationship is also present between the radius and the circumference of a circle. As the radius of a circle increases, both the circumference and the area of the circle increase. We say the circumference $C = 2\pi r$ is **directly proportional to** (or varies directly as) the radius r of the circle. The area $A = \pi r^2$ is directly proportional to (or varies directly as) the square of the radius r^2.

A situation in which the growth of one object is directly proportional to the growth of another object can be described by a function of the form

$$y = f(x) = kx^p$$

where p is a positive rational number and k is a real number.

Definition

The quantity y is **directly proportional to** (or **varies directly as**) the pth power of the quantity x if $y = f(x) = kx^p$, where p is a positive rational number and k is a real number called the **constant of variation.**

In Example 1, $g(c) = 4.5c$, the constant of variation k is 4.5, and the exponent p is 1. For $C(r) = 2\pi r$, $p = 1$ and $k = 2\pi$. For $A(r) = \pi r^2$, $p = 2$, and $k = \pi$.

EXAMPLE 7

Suppose y is directly proportional to x^2, and $y = 18$ when $x = 3$. Express y as a function of x by finding the constant of variation.

Solution

$$y = f(x) = kx^2 \qquad y \text{ is directly proportional to } x^2.$$

$$18 = 9k \qquad f(3) = 18\ (y = 18 \text{ when } x = 3).$$

$$2 = k$$

Therefore

$$y = f(x) = 2x^2$$

EXAMPLE 8

The faster a car goes, the longer will be the skid marks if the car is forced to stop suddenly. In fact, the length L of the skid marks of a car's tires (when the brakes are applied) varies directly as the square of the velocity v of the car. If skid marks of 37.5 feet are produced by a car going 30 mph on a dry concrete road,

(a) express the length of the skid marks L as a function of the speed v of the car.

(b) find the length of the marks produced by a car going 55 mph.

Solutions

(a) $L = kv^2$ $\qquad$ *L varies directly as v^2.*

$37.5 = k \cdot 30^2$ $\qquad$ $L(30) = 37.5$

Therefore

$$k = \frac{37.5}{900} = \frac{1}{24} \quad \text{and} \quad L = \frac{v^2}{24}$$

(b) $L(55) = \dfrac{55^2}{24} \approx 126$ feet $\qquad$ We seek $L(55)$.

Inverse Variation

If you are taking a car trip to Grandma's house, the faster you drive, the less time it takes to get there. In this situation, there are two quantities that can vary—the average rate and the time. As the average rate gets larger (or smaller), the time required gets smaller (or larger).

A situation similar to the drive to Grandma's house, in which as one quantity gets larger (or smaller), the other gets smaller (or larger), can be described by a function of the form

$$y = f(x) = \frac{k}{x^p}$$

where p is a positive rational number and k is a real number.

Definition

The quantity y is **inversely proportional to (varies inversely as)** the pth power of the quantity x if

$$y = f(x) = \frac{k}{x^p}$$

where p is a positive rational number and k is a real number called the **constant of variation.**

On a 120-mile trip to Grandma's house, the time t and the average rate r are related by the function $t = f(r) = \frac{120}{r}$; t is inversely proportional to r, $p = 1$, and $k = 120$.

EXAMPLE 9

Suppose z is inversely proportional to $\sqrt{w}$, and $z = 3$ when $w = 18$. Express z as a function of w by finding the constant of variation.

Solution

$$z = f(w) = \frac{k}{\sqrt{w}} \qquad z \text{ is inversely proportional to } \sqrt{w}.$$

$$3 = \frac{k}{\sqrt{18}} \qquad f(18) = 3$$

$$k = 3\sqrt{18} = 9\sqrt{2}$$

Therefore

$$z = f(w) = \frac{9\sqrt{2}}{\sqrt{w}}$$

EXAMPLE 10

Many situations obey an "inverse square law." For example, the loudness of the sound of a stereo speaker is inversely proportional to the square of the distance of the listener from the speaker.

(a) Express the loudness L in decibels as a function of the distance d from the speaker.

(b) When you are sitting 10 feet from the speaker, the loudness is 40 decibels. What is the loudness when you are 6 feet from the speaker?

(c) If you are sitting in one location and then move to one that is twice as far from the speaker, how does the loudness change?

Solutions

(a) $L = L(d) = \dfrac{k}{d^2}$ Loudness L is inversely proportional to the square of the distance d.

$$40 = \frac{k}{10^2} \qquad L(10) = 40$$
$$k = 4{,}000$$

Therefore

$$L = \frac{4{,}000}{d^2}$$

(b) $L(6) = \dfrac{4{,}000}{36} \approx 111.1$ decibels We seek $L(6)$.

This is very loud indeed.

(c) $$\frac{L(2d)}{L(d)} = \frac{\frac{k}{(2d)^2}}{\frac{k}{d^2}} = \frac{1}{4} \qquad \text{We seek the ratio } \frac{L(2d)}{L(d)}.$$

Therefore the sound is only one-fourth as loud. ■

Here is a problem involving both direct and inverse variation.

EXAMPLE 11

The pitch (frequency) of a violin string is directly proportional to the tension on the string and inversely proportional to its length. If the length is increased by 10% and the tension by 20%, by what percent is the pitch increased?

Solution

$$p = p(t, l) = \frac{kt}{l}$$

$p(t, l)$ is a function of two variables. Pitch p is directly proportional to the tension t and inversely proportional to the length l.

$$\frac{p(1.2t, 1.1l)}{p(t, l)} = \frac{\frac{1.2kt}{1.1l}}{\frac{kt}{l}} = \frac{1.2}{1.1} \approx 1.09$$

We seek the ratio $\dfrac{p(1.2t, 1.1l)}{p(t, l)}$.

Therefore the pitch is increased by approximately 9%. ■

It is instructive to compare rough graphs of functions of the form $f(x) = kx^p$ and $g(x) = \dfrac{k}{x^p}$ for various values of p. (See Figure 8.58.)

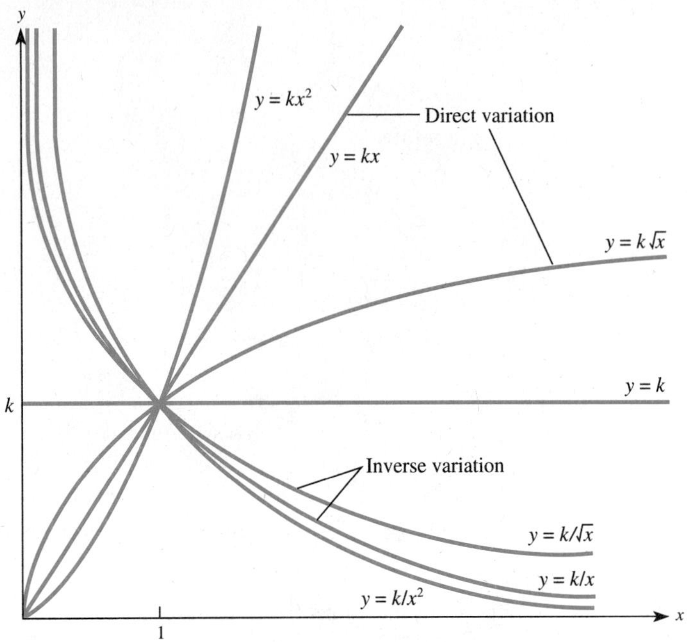

Figure 8.58

Joint Variation

Just as the volume V of a box changes when its length l, its width w, or its height h changes, there are many situations where more than two quantities can vary at the same time. As l, w, or h increase (or decrease), V increases (or decreases). This situation is an example of **joint variation.**

EXAMPLE 12

The safe load for a beam of given length varies jointly as the breadth b and the square of the depth d. (See Figure 8.59.) An old house has floor beams a full 3 inches $\times$ 10 inches (that is, $b = 3$ inches, $d = 10$ inches). These beams are 2 feet apart. The owner wishes to replace these beams with new ones that are 2 inches $\times$ 8 inches (actually, $b = 1.5$ inches, $d = 7.5$ inches). If the two types of wood are essentially the same, how far apart must the new beams be placed in order for the new floor to have the same carrying capacity as the old floor?

Solution

The general formula for the safe load of any beam is

$$S = kbd^2$$

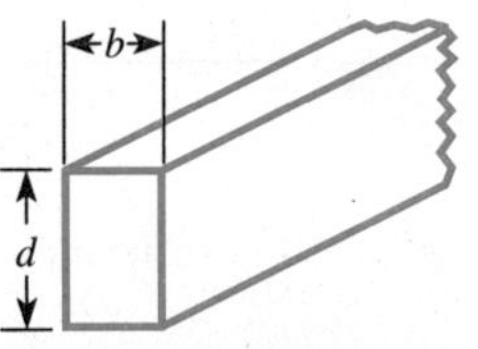

Figure 8.59

Therefore

$$S_0 = k_0 \cdot 3 \cdot 10^2 = 300k_0 \qquad S_0 = \text{safe load on one old beam.}$$

and

$$S_n = k_n \cdot 1.5(7.5^2) = 84.375k_n \qquad S_n = \text{safe load on one new beam.}$$

Now

$$k_0 = k_n \qquad \text{Type of wood is the same in both beams.}$$

Therefore

$$\frac{S_0}{S_n} = \frac{300k_0}{84.375k_n} \approx 3.55 \qquad \text{Ratio represents how much stronger the old beam is than the new one.}$$

An old beam is 3.55 times as strong as a new beam (plausible, since the old beam is larger), so we would adjust the space between beams accordingly. Every old beam must be replaced by 3.55 new beams, so the space between two new beams is

$$\frac{1}{3.55} \times 2 \text{ feet} \approx 0.56 \text{ feet} \approx 6.72 \text{ inches}$$

This example shows that some variation problems can be solved without explicitly calculating the constant of variation. ■

Since a linear model can be written as $f(x) = mx + b$, we need two pieces of data to determine m and b (see Example 3). If a situation can be represented by a quadratic model, we need three pieces of information to determine a, b, and c in the model $g(x) = ax^2 + bx + c$.

EXAMPLE 13 Assume that a quadratic function is a reasonable model for a punt in football. Suppose a ball is punted from 5 feet above the ground. Its height is 55 feet after 1 second and 59 feet after 3 seconds.

(a) Find a model for this situation.

(b) Sketch its graph.

(c) How high was the ball after 2 seconds?

(d) What was the "hang time" for this punt—the number of seconds until it hit the ground (assuming the returner elected not to catch it)?

(e) What was the maximum height reached by this ball?

Solutions

(a) If $g(x) = ax^2 + bx + c$, then we know $g(0) = 5$, $g(1) = 55$, and $g(3) = 59$. Since $g(0) = 5$, $c = 5$. To find a and b, then, we must solve the linear system

$$\begin{array}{lllll} \text{Equation 1} & 55 = a + b + 5 & \text{or} & 50 = a + b & \text{For } x = 1 \text{ sec.} \\ \text{Equation 2} & 59 = 9a + 3b + 5 & \text{or} & 54 = 9a + 3b & \text{For } x = 2 \text{ sec.} \end{array}$$

Thus $6b = 396$, or $b = 66$, and $a = -16$ **Multiply Equation 1 by 9. Then subtract Equation 2.**

$$g(x) = -16x^2 + 66x + 5.$$

(b)

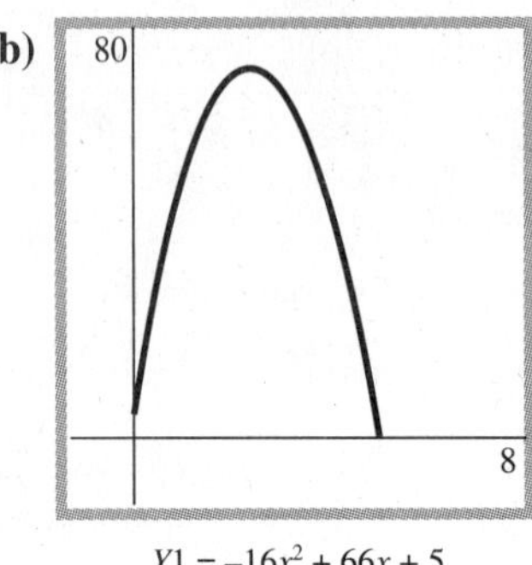

$Y1 = -16x^2 + 66x + 5$

Figure 8.60

(c) $g(2) = 73$. The ball was 73 feet high after 2 seconds.

(d) Find x so $g(x) = 0$. Using the quadratic formula,

$$x = \frac{-66 \pm \sqrt{(-66)^2 - 4(-16)(5)}}{-32} = 2.0625 \pm 2.1370 = -0.0745 \text{ or } 4.2 \text{ sec.}$$

The hang time was approximately 4.2 seconds.

(e) The maximum height, reached at $\frac{-66}{-32} = 2.0625$ seconds, was $g(2.0625) = 73.06$ feet.

We could use the matrix capabilities of a graphing calculator to solve this system of three equations in three variables a, b, and c. For parts (c), (d), and (e), we could first graph the function $g(x)$ using an appropriate viewing rectangle. Then we could

(c) estimate $g(2)$ by finding the point on the graph whose x-coordinate is closest to 2

(d) estimate the roots by finding the points on the graph whose y-coordinates are closest to 0

(e) estimate the maximum height by visually estimating the highest point of the graph.

Notice that there could be some small differences in the answers obtained by the two methods. ■

Optimizing Functions

Some of the most interesting applications involve finding the maximum or minimum value of a certain function. To find the maximum height reached by a golf ball, we had to find the maximum value of a quadratic function, which we did by finding the vertex of the parabolic graph of the function.

One approach here is simply to use guess and check and common sense to find the maximum or minimum value of a function. A calculator is invaluable.

EXAMPLE 14

The safe load for a beam of given length varies jointly as the breadth b and the square of the depth d (see Example 12). What is the approximate breadth and depth of the strongest beam that can be cut from a cylindrical log of diameter 25 cm?

Solution

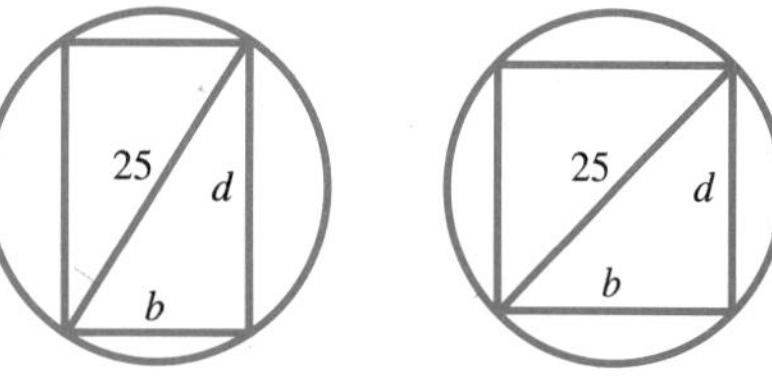

Figure 8.61 **Figure 8.62**

Since s varies jointly as b and d, we know $s = kbd^2$. The problem, now, is to estimate the values of b and d for which s is the largest. We make a table and proceed by guess and check.

The Pythagorean theorem tells us

$$d^2 = 25^2 - b^2 = 625 - b^2$$

b	d^2	$s = kbd^2$
10	525	5,250k
11	504	5,544k
12	481	5,772k
13	456	5,928k
14	429	6,006k
15	400	6,000k
16	369	5,904k
17	336	5,712k

Table 8.16

Looking at the table, we see that the largest value of s occurs around $b = 14$, since s is decreasing to the left and right of $b = 14$. To get a more accurate estimate, we could make another table for $13 < b < 15$. We show, here, only the portion for $14 < b < 15$, since s is increasing for the other values of b.

b	d^2	$s = kbd^2$
14.0	429.00	6,006.0k
14.1	426.19	6,009.3k
14.2	423.36	6,011.7k
14.3	420.51	6,013.3k
14.4	417.64	6,014.0k
14.5	414.75	6,013.9k
14.6	411.84	6,012.7k
14.7	408.91	6,011.0k
14.8	405.96	6,008.2k
14.9	402.99	6,004.6k
15.0	400.00	6,000.0k

Table 8.17

By examining this table, we see that the maximum value for s occurs when $b \approx 14.4$ and $d^2 \approx 417.64$, so that $d \approx 20.4$. ■

EXAMPLE 15 You can make a "popcorn box" from a rectangular piece of heavy paper by following the sequence of steps in Figures 8.63–8.66. Suppose $l = 24$ units and $w = 12$ units.

(a) Find the dimensions of all boxes whose volume is 200 cubic units.

(b) Find the dimensions of the box of maximum volume.

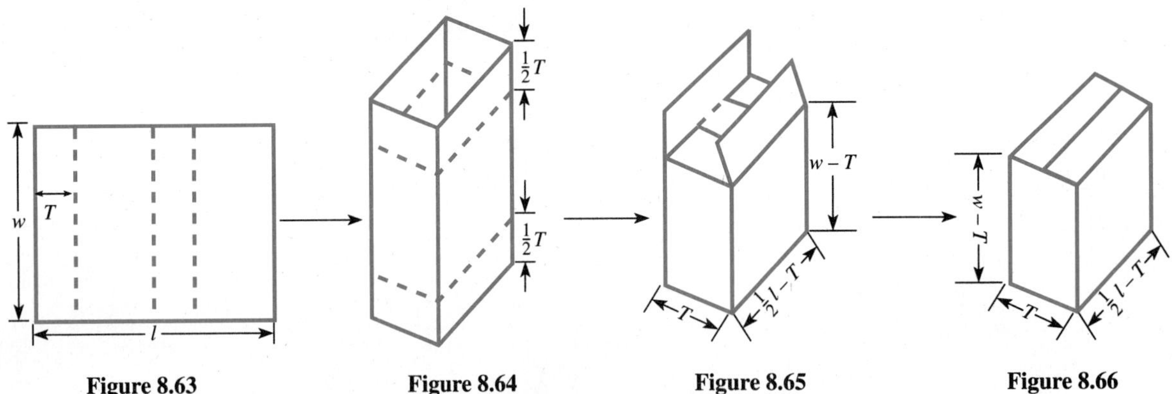

Figure 8.63 Figure 8.64 Figure 8.65 Figure 8.66

Solutions

Let $T = x$ in the figures. Express the volume, V, in terms of the height, x, by $V = lwh = (12 - x)(12 - x)x = x(12 - x)^2$. Graph $Y1 = x(12 - x)^2$ using an appropriate viewing rectangle. In this case, using the standard one where the x-interval: $(-10, 10)$ and y-interval: $(-10, 10)$ yields a very misleading figure. To help find an appropriate viewing rectangle, it is helpful to realize that a suitable domain is $0 \leq x \leq 12$ (or something slightly larger to better view the screen). You need to do some computation of functional values to find a suitable range. (See Figure 8.67)

(a) Graph $Y2 = 200$ to show that the x-coordinates of the points of intersection of the two graphs are $x = 2$ and $x \approx 6.42$.

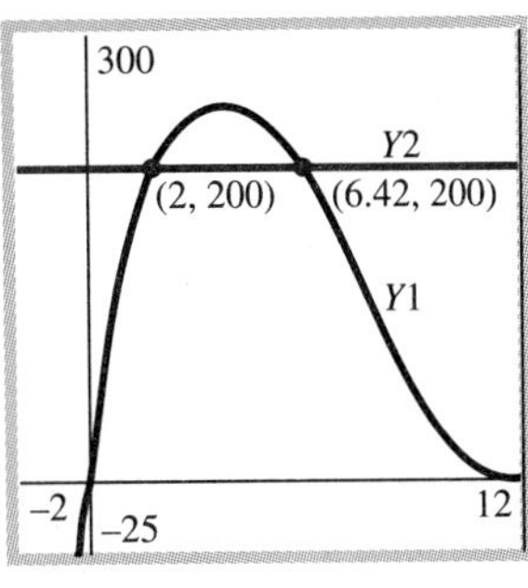

Figure 8.67

(b) The maximum value of V is 256 for a box with dimensions $8 \times 8 \times 4$.

8.4 EXERCISES

I. Procedures

1–6. Write an appropriate function containing a constant of variation k that describes each of the following situations.

1. F varies directly as the cube of y.
2. u is inversely proportional to the square root of w.
3. W varies inversely as the cube root of s.
4. A is directly proportional to the square of r.
5. S varies jointly as a, b^2, and c^3.
6. S varies jointly as the square of u and the square root of v.

7–12. Express y as a function of the other variables.

7. y is directly proportional to $\sqrt{x}$ and $y = 20$ when $x = 100$.
8. y varies directly as x^3 and $y = 11.2$ when $x = 2$.
9. y varies inversely as x^2 and $y = 0.25$ when $x = 4$.
10. y is inversely proportional to $\sqrt{x}$ and $y = 0.25$ when $x = 4$.
11. y varies jointly as r^2 and h, and $y = 4\pi$ when $r = 2$ and $h = 3$ (y is the volume of a cone, the radius of whose base is r and whose height is h).
12. y varies jointly as u, $\frac{1}{\sqrt{V}}$, and w^2; and $y = 10$ when $u = 1$, $V = 4$, and $w = 9$.
13. Find a linear function whose graph contains the points $(2, -1)$ and $(3, 7)$.
14. Find a quadratic function of the form $g(x) = ax^2 + b$ whose graph contains the points $(-1, 5)$ and $(3, 21)$.
15. Find the quadratic function $h(x) = ax^2 + bx + c$ whose graph contains the points $(-1, 2)$, $(0, 3)$, and $(2, 11)$.
16. Find the quadratic function $h(x) = ax^2 + bx + c$ whose graph contains the points $(-1, 5)$, $(1, 5)$, and $(2, 11)$.

17–24. Find a possible expression for the general term a_n. (Remember: more than one correct answer is possible.)

17. $1, 5, 9, 13, 17, \ldots$
18. $34, 24, 14, 4, -6, \ldots$
19. $\frac{1}{2}, \frac{2}{3}, \frac{3}{4}, \frac{4}{5}, \frac{5}{6}, \ldots$
20. $3, 2, \frac{5}{3}, \frac{3}{2}, \frac{7}{5}, \ldots$
21. $\frac{2}{5}, \frac{3}{10}, \frac{4}{17}, \frac{5}{26}, \frac{6}{37}, \ldots$
22. $3, 5, 9, 17, 33, \ldots$
23. $1, -2, 3, -4, 5, \ldots$
24. $\frac{-1}{2}, \frac{2}{9}, \frac{-3}{28}, \frac{4}{65}, \frac{-5}{126}, \ldots$

25–32. Find the next two terms in each of the following sequences. Explain the general term rule you used. Once again, more than one answer is possible.

25. 3, 5, 8, 12, 17, . . .

26. 2, 12, 21, 29, 36, . . .

27. 1, 3, 4, 7, 11, . . .

28. 1, 2, 6, 24, 120, . . .

29. 1, 1, 2, 2, 3, 4, 4, 8, 5, 16, 6, 32, . . .

30. 100, 100, 75, 50, 50, 25, 25, . . .

31. 2, 3, 5, 7, 11, 13, . . .

32. 6, 8, 10, 14, 15, 21, 22, . . .

II. Concepts

33. Robert Ringer, in his book *Winning Through Intimidation,* claims the results a person obtains are inversely proportional to the degree to which he is intimidated. Do you agree?

34. One of Murphy's Laws states that the chance that a piece of bread will fall to the dining room carpet with the buttered side down is directly proportional to the cost of the carpet. Formulate a "Murphy's Law" of your own.

35–40. Do you think each of the following pairs of quantities are directly proportional, inversely proportional, or neither? Give a reason for your answer.

35. A person's age and his or her attention span.

36. The length of a person's shadow and the time of day.

37. The number of dogs pulling a dog sled and the speed of the sled.

38. The size of a diamond and its value.

39. The price of a candy bar and its size.

40. The number of coils in a mattress and its price.

41–44. True or false. Give a reason.

41. If x varies directly as y, then y varies directly as x.

42. If x varies inversely as y, then y varies inversely as x.

43. If z varies jointly as x and y, then x varies jointly as y and z.

44. If x varies inversely as y^2, then y varies inversely as $\sqrt{x}$.

III. Applications

Solve each of the following problems.

45. The time it takes for a pendulum (on a clock) to make one swing (back and forth) is directly proportional to the square root of its length. A 1-meter pendulum takes approximately 8 seconds to complete one swing.

(a) Express the time T as a function of the length l of the pendulum.

(b) Sketch its graph.

(c) How long does it take a 2-meter pendulum to complete one swing? Approximately how long is the pendulum if it takes 3 seconds to make one swing?

46. The maximum distance you can see from the top of a tall building varies directly as the square root of the height of the building (measured in kilometers). You can see 65.7 km from the top of the Canadian National Tower in Toronto (1,135 feet high).

(a) Express the distance D as a function of the height h of the building.

(b) Sketch its graph.

(c) How far can you see from the top of the Sears Tower in Chicago (1,454 feet)? *Hint:* Watch the *units*—feet and kilometers.

47. The intensity of illumination is inversely proportional to the square of the distance from the light source. If the distance between your book and the light source is doubled, how is the illumination changed? Sketch a rough graph to show how the illumination depends on the distance from the light source.

48. The pressure at any point on a submarine varies directly as its depth. If the pressure on a submarine 50 feet below the surface is 322 pounds per square foot, what is the pressure if the submarine descends to 125 feet below the surface?

49. The weight of an object varies directly as its distance from the center of the earth if it is within the earth, but inversely as the square of the distance from the center if above the earth.

(a) Sketch a rough graph of the function that describes how the weight of the object depends on its distance from the center of the earth supposing that the weight is zero at the center of the earth and 100 at the earth's surface. (Assume the radius of the earth is 4,000 miles.)

(b) If you weigh 150 pounds on the surface of the earth, how much would you weigh 1,000 miles above the earth's surface? On the moon (240,000 miles from the center of the earth)?

50. The speed of a skidding car is directly proportional to the square root of the length of a skid. On a dry concrete road, a car going 40 mph would skid about 67 feet. If the skid marks were 120 feet long, what was the approximate speed of the car?

51. The collision impact (kinetic energy) of an automobile varies jointly as its mass and the square of its speed. If the speed is doubled but the mass is reduced by 10%, what happens to the collision impact of the car?

52. The cost of insulating an attic floor varies jointly with the area of the floor and the thickness of the insulation. If the area and the thickness are both increased by 20%, by what percent is the cost increased?

53. The volume of a sphere is directly proportional to the cube of its radius. A recipe for apple pie calls for eight apples, each 4 inches in diameter. How many apples, each 3 inches in diameter, would be needed to make this pie?

54. An ox is tied by a rope 20 yards long in the center of a field and eats all the grass within its reach in $2\frac{1}{2}$ days. How many days would it have taken the ox to eat all the grass within its reach if the rope had been 10 yards longer (from an 1897 textbook on algebra!)?

55. The "threshold weight" (the crucial weight above which the mortality risk increases dramatically) of a man 40 to 49 years old is directly proportional to the cube of the man's height. If the threshold weight of a man 5 feet 10 inches is 184 pounds, what is the threshold weight of a man 6 feet 2 inches tall?

56. Newton's Law of Gravitation says the gravitational attraction between two bodies is directly proportional to the product of their masses and inversely proportional to the square of the distance between them. If the masses of the two objects are doubled and the distance between them is halved, what happens to the gravitational attraction?

57. Boyle's Law in chemistry states that the volume of a fixed amount of gas (at a constant temperature) is inversely proportional to the pressure applied to the gas. Suppose a pressure of 46 pounds per square inch (psi) compresses the gas to a volume of 350 cubic feet.

(a) Express the volume V as a function of the pressure p.

(b) What pressure is necessary to compress the gas to a volume of 250 cubic feet?

(c) Could you compress to 0 volume?

58. Another example of an inverse-square law is the fact that the strength of a radio signal varies inversely as the square of the distance to the transmitter. How much stronger is the signal if your home is 5 miles from the transmitter rather than 15 miles from the transmitter?

59. Suppose an ice cream seller at a summer fair assumes that the amount, A, of ice cream he will sell is directly proportional to the number of people who come to the fair, directly proportional to the temperature in excess of 65° F, and inversely proportional to the selling price.

(a) What is a reasonable model for this situation?

(b) What is a reasonable domain for the values of the temperature, T?

(c) Make up and solve several problems you based on this model.

60. The selling price for a hand tool, P, consists of fixed amount A plus an amount, B, proportional to the square of the normal size, s, so that $P = A + Bs^2$. If a 20-mm tool sells for \$7.50 and a 50-mm tool sells for \$15.00, what is the cost of a 60-mm tool?

61. Suppose the price of a television set varies quadratically with the diagonal of its screen. A 9-inch set costs \$430, a 13-inch set costs \$400, and a 19-inch set costs \$570.

(a) Find a model for this situation.

(b) Sketch its graph.

(c) According to this model, what would be the cost of a 25-inch model?

(d) Check the prices of television sets of a given brand in a newspaper ad or consumer magazine. Is a quadratic model reasonable? Why do you suppose the price increases much more for sets with very small (for example, 3-inch) or very large (for example, 44-inch) screens?

62. As a meteor enters the earth's atmosphere, it rapidly becomes hot so it looks like a "falling star." The degree to which it is heated depends on how fast it is traveling. We have the data where S = speed in kilometers per second and t = highest temperature in degrees Celsius.

S	5	6	7	8	9
t	11,250	16,200	22,050	28,800	36,450

Express S as a quadratic function of t.

63. The number of games played in a basketball league season depends on the number of teams in the league. Suppose each team plays every other team in the league twice (home and home). Make a table and then find a relationship between the number of games played in a season and the number of teams in the league.

64. Rotollo's Pizza charges the following prices for a cheese pizza:

- Small (10-inch): $3.15
- Medium (12-inch): $4.00
- Large (14-inch): $4.95
- Jumbo (16-inch): $6.00

(a) Express the cost of the pizza as a quadratic function of its diameter (in inches).

(b) If adding sausage adds 60¢ to the price of a small pizza, how much should be added to the price of the other three sizes when sausage is added?

65–68. Solve each optimization problem using tables, graphs, and/or a graphing calculator.

65. Using the same 24-by-12 sheet of heavy paper as in Example 15, build a "pizza box" as indicated in the accompanying figure. Find the dimensions of the box(es) whose volume

(a) equals 105

(b) is maximized.

(c) Can you generalize your solution to part (b)?

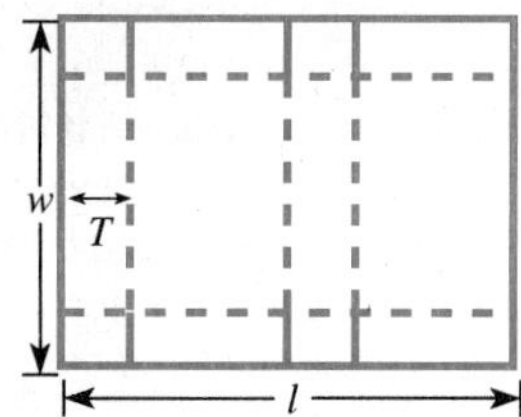

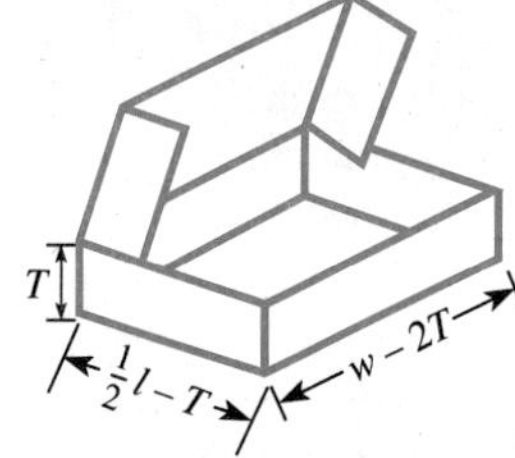

66. A (right circular) cylindrical can is to hold 100 cm^3. Find the radius and height of all cans whose surface area

(a) equals 200 cm^2

(b) is minimized.

(c) Can you generalize your solution to part (b)?

67. A topless box with a square base has a volume of 8 cubic meters. The material for the base costs $8 per square meter, and the material for the sides costs $6 per square meter. Find the dimensions of the box of least cost.

68. The power, P, generated by a waterwheel is given by $P = 0.5v(V - v)^2$, where V is the velocity of the stream and v is the velocity of the waterwheel. Compare the speed of the waterwheel when the power is at a maximum for $V = 4$ feet/second and $V = 20$ feet/second. (Is the maximum speed of the second five times that of the first?)

IV. Extensions

69. **(a)** Find the equation of the quadratic polynomial, $p(x)$, with x-intercepts of 0 and 6 and a maximum at (3, 12), and graph it. [*Hint:* Start with $p(x) = x(x - 6)$.]

(b) Sketch each of the following graphs, giving the x-intercepts and the coordinates of the maximum point. Then sketch the graphs using a graphing calculator and compare them to your hand-drawn sketches.

1. $p(2x)$ **2.** $2p(x)$ **3.** $p(x - 2)$
4. $p(x + 2)$ **5.** $p(x - 1) + 3$

70. **(a)** If $y = f(x)$ varies directly as x, show that

$$\frac{f(x_1)}{f(x_2)} = \frac{x_1}{x_2}$$

(b) What is the analogous equation if y varies inversely as x? Justify your answer.

71. Suppose y is directly proportional to x, and let $y = f(x)$.

(a) Does $f\left(\frac{1}{a}\right) = \frac{1}{f(a)}$?

(b) Does $f(ab) = f(a)f(b)$?

(c) Does $f(a + 1) = f(a) + 1$?

(d) Show $\frac{f(a + h) - f(a)}{h} = f(1)$.

72. Suppose y is inversely proportional to x, and let $y = f(x)$.

(a) Does $f\left(\frac{1}{a}\right) = \frac{1}{f(a)}$?

(b) Does $f(a + b) = f(a) + f(b)$?

(c) Does $f(ab) = af(b)$?

CHAPTER 8 SUMMARY

The main purpose of this chapter was to take a closer look at the function concept—a universal concept for the rest of the book and subsequent mathematics courses. Functions can be represented by a rule of correspondence between two sets in which every element in the first set, the domain, can be associated with exactly one element in the second set, the range. Functions can be represented by tables, graphs, or formulas (equations). Many of the skills and concepts learned earlier were reviewed using functional notation. Both polynomial and nonpolynomial functions were included. Many of the applications involved finding a function that models a physical situation using direct, inverse, and joint variation. Many of the concepts were explored with a graphing calculator.

Important Words and Phrases

function
domain
range
function of one variable
constant function
linear function
quadratic function
polynomial function
evaluate a function
discrete function
step function
continuous function
directly proportional
inversely proportional
joint variation
constant of variation

Important Properties and Procedures

- Determining whether an equation, a table, or a graph represents a function
- Determining the domain and range of a function
- Evaluating a function
- Graphing functions using five graphing principles
- Finding functions using direct, inverse, or joint variation
- Optimizing a function

REVIEW EXERCISES

I. Procedures

1–4. Which of the following tables or graphs represent functions? If the table or graph is a function, find $f(1)$.

1.

x	0	1	2	3	4	5
$f(x)$	3	3	4	5	6	6

2.

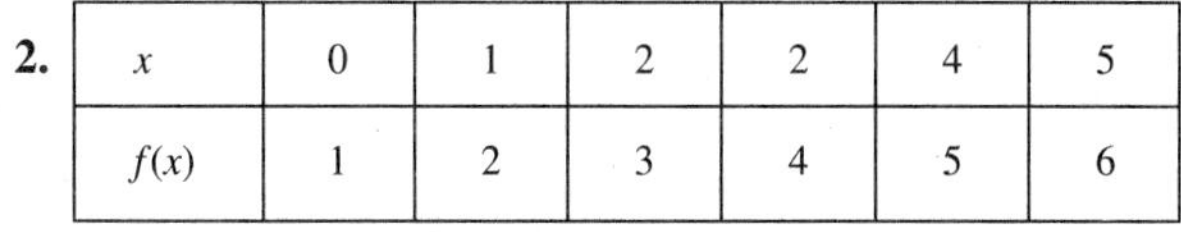

x	0	1	2	2	4	5
$f(x)$	1	2	3	4	5	6

3.

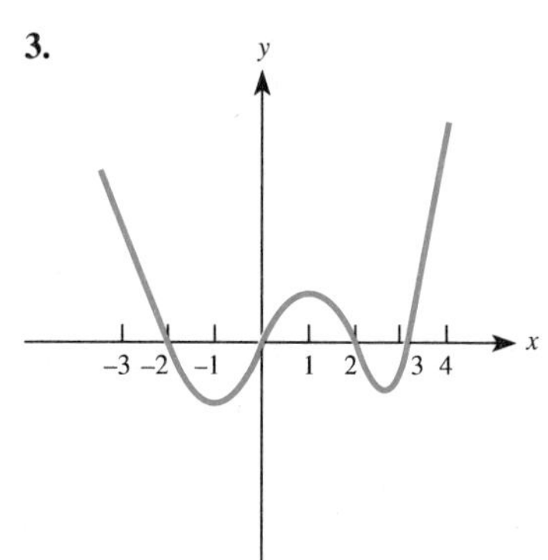

4.

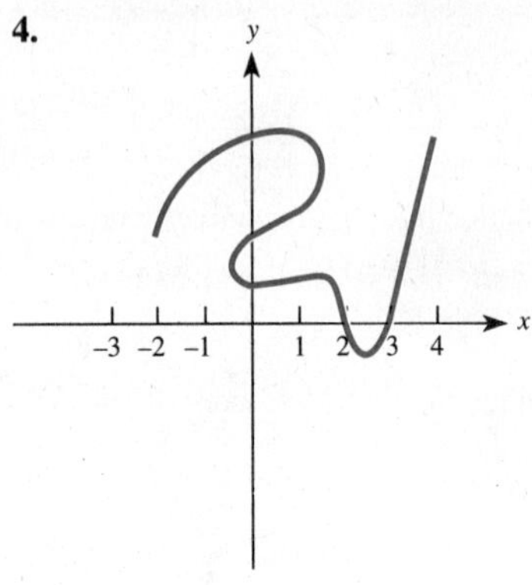

5–8. Evaluate.

5. $f(x) = x^3 + 2|x| - 3$, if $x = -2$

6. $g(x) = \frac{4}{x} + 2\sqrt{x-1} - x^{\frac{2}{5}} + \pi$, if $x = 3$

7. $h(x) = 3x^5 - 2x^4 + 13x^3 - 10x^2 + x - 7$, if $x = 5$

8. $w(x) = x^4 - 19x^3 + 15x^2 - 23$, if $x = 3.2$

9–14. Let $f(x) = 3x - 4$. Solve each of the following equations or inequalities.

9. $f(x) = 3f(x) + 2$ **10.** $f(x) = f(x^2)$

11. $\sqrt{f(x)} = 2$ **12.** $(f(x))^2 = f(x^2)$

13. $f(x) < 5f(x) + 1$ **14.** $|f(x)| < 1$

15–20. For each of the functions, **(a)** roughly sketch its graph, making use of the graphing principles; **(b)** find the x- and y-intercepts; and **(c)** find the domain and range. Then use a graphing calculator.

15. $y = x^3 - 1$ **16.** $y = \frac{1}{(x+2)^2}$

17. $y = \sqrt{x-2} + 1$ **18.** $y = (x+1)^4 - 3$

19. $y = \frac{2}{x-2} + 1$ **20.** $y = |3x - 2| + 1$

21. If w is directly proportional to y^2, and $w = 16$ when $y = 2$, express w as a function of y.

22. If z is inversely proportional to $\sqrt{x}$, and $z = 2$ when $x = 24$, express z as a function of x.

23. Find the largest subset of the real numbers that could be the domain of the function $f(x) = \sqrt{x-3} + |x+3| + \frac{2}{x-4}$.

24. Find a linear function whose graph contains the points $(2, -1)$ and $(4, 9)$.

II. Concepts

25 and 26. For each of the following physical objects, describe a domain D, a range R, and a function f that the object could represent.

25. A scale at a produce counter in a supermarket.

26. A class list.

27. Let $p(x) = \sqrt{x}$; $q(x) = \frac{1}{x}$, and $r(x) = x^3$. Express each of the following in terms of p, q, or r. For example, $y = \sqrt{x+1} - 3$ is $y = p(x+1) - 3$.

(a) $y = 1 + \frac{2}{x}$ **(b)** $y = \sqrt{7-x}$

(c) $y = (2x)^3$ **(d)** $y = \frac{1}{\sqrt{x+2}}$

28 and 29. Give an example of a function $f(x)$ satisfying

28. $f(0) = 2$ and $f(x) > 3$ for $x > 1$

29. $2f(x) = f(2x)$ for all real x

30. If x is directly proportional to y, and y is directly proportional to z, is x directly proportional to z?

31. If x is inversely proportional to y, and y is inversely proportional to z, is x inversely proportional to z?

32–35. The graph of $f(x)$ is shown. Sketch the graph of each of the following functions.

32. $-f(x)$ **33.** $|f(x)|$ **34.** $f(x) + 2$ **35.** $f(x - 2)$

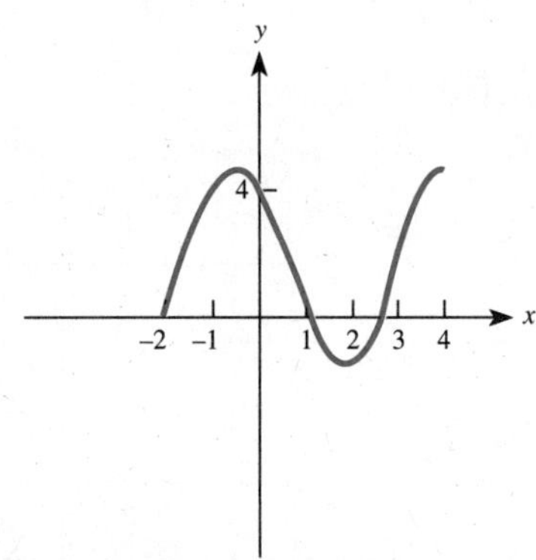

III. Applications

36.

	Kwh	BTUs
Human being	0.1	341
100-watt light bulb	0.1	341
Color TV	0.2	682
Toaster	1.1	3,750
Room air conditioner	3.0	10,240
Electric clothes dryer	4.0	13,600
Oven (baking a turkey)	4.5	15,400
Small car at 45 mph	60.0	205,000
Large car at 70 mph	270.0	921,500

Energy Required by Typical Consumers During One Hour of Normal Operation or Activity

(a) Name two functions that are described by this table. Give the domain and range of each.

(b) How many Kwh are required to run a color television and an air conditioner for six hours?

(c) List several ways you could consume 6 Kwh in one hour.

37.

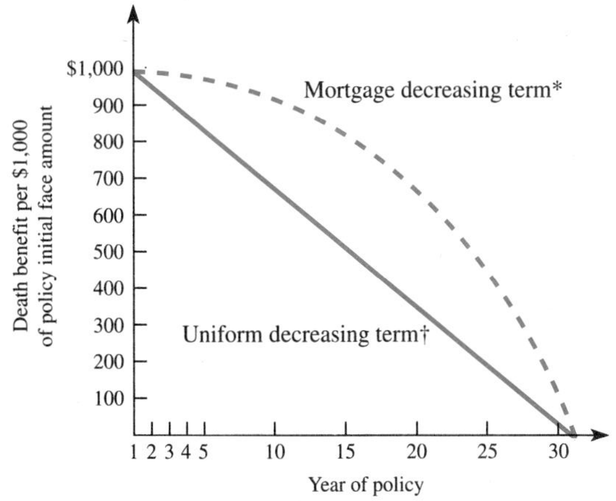

*Sample policy: Allstate Life
†Sample policy: Metropolitan Life
(Source: *Consumer Reports*)

(a) There are two graphs in this figure. Describe the function associated with each one.

(b) How much is each decreasing term policy worth after 10 years?

(c) In approximately what year is the difference between the two policies the greatest?

(d) Find the linear function that represents the uniformly decreasing term policy.

38. If the length of the edge of a cube is

(a) tripled, (b) reduced by 40%,

what happens to the surface area? To the volume?

39. The initial velocity v (in meters per second) required for an object to reach a height of h meters is given by

$$v = v(h) = 64h$$

(a) Sketch the graph of this function. What is a reasonable domain for the function?

(b) What velocity is necessary for the object to reach a height of 5,000 meters?

40 and 41. Sketch, as best you can, the graph of a function that describes each of the following situations.

40. The amount you pay in a downtown parking lot as a function of the number of hours you park.

41. The speed of a car that started from rest and then was "floored" as a function of the time elapsed after "flooring" it.

42. The amount of force you need to exert on a wrench to loosen a rusted nut is inversely proportional to the length of the wrench.

(a) If a 6-inch wrench requires 200 pounds of force, express the length of the wrench as a function of the required force.

(b) If you could exert a force of only 75 pounds, how long a wrench would you need?

(c) If you could exert 300 pounds of force, how long a wrench would you need?

43. The amount of water that can flow through a pipe is directly proportional to the square of the diameter of the pipe (since the flow depends on the cross-sectional area).

(a) If a pipe of diameter 10 cm can serve 60 people, express the number of people served as a function of the diameter of pipe.

(b) How many people could be served by a pipe of diameter 30 cm?

(c) What size pipe is required for a new subdivision housing 1,200 people?

44. The time required for an elevator to lift a weight varies jointly as the weight and distance through which it is lifted *and* inversely as the power of the motor. If an elevator with a 400-horsepower motor takes 4.5 seconds to raise 850 kg a distance of 10 meters, how long will it take that same elevator to lift 2,500 kg a distance of 20 meters?

45. An advertisement for swimming-pool fencing gives the following prices:

Pool size	*Cost*
15 feet	\$159
18 feet	\$193
21 feet	\$237
24 feet	\$291

(a) Find a quadratic function that gives the price as a function of the pool size.

(b) How much would fencing for a 30-foot pool cost?

46. An advertisement for end-of-season clearance prices on solar covers for circular swimming pools gives the following prices:

Pool size, feet	15	18	21	24	27
Price, dollars	442	457	485	528	586

(a) Find a quadratic function giving the price in terms of the pool size.

(b) How much would a cover for a 30-foot pool cost?

47. Find the minimum value of the function

$$f(x) = 3x^2 + \frac{1}{\sqrt{x}}$$

48. A window with a perimeter of 16 feet has the shape of a rectangle and a semicircle. What is the greatest area of such a window? *Hint:* $A = \frac{\pi r^2}{2} + 2rx$; $2r + 2x + \pi r = 16$, so $x = \frac{16 - 2r - \pi r}{2}$.

Thus $A = \frac{\pi r^2}{2} + 2r\left(\frac{16 - 2r - \pi r}{2}\right) = 16r - 2r^2 - \frac{\pi}{2}r^2$.

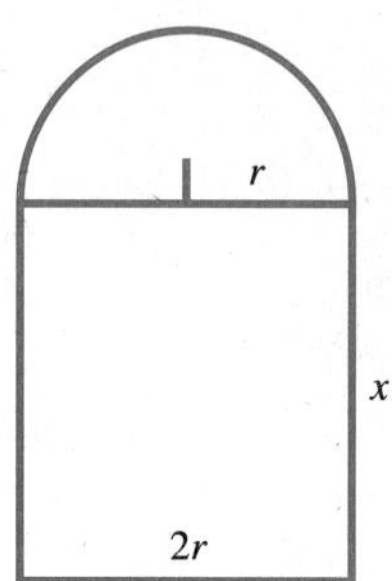

49. Postal regulations require that the length plus the girth of any package sent by parcel post be at most 108 inches. Find the length and height of the largest box with a square base that could be sent by parcel post.

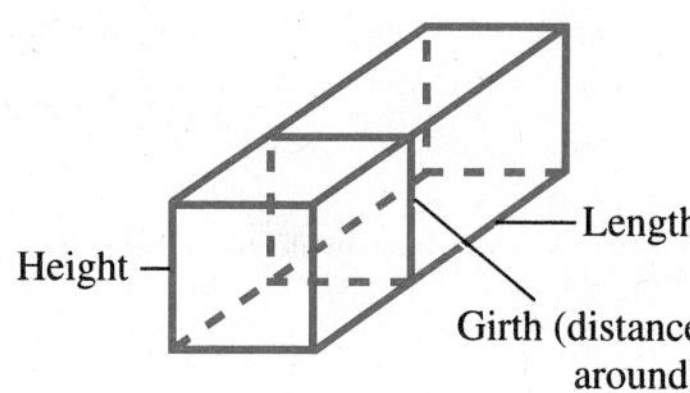

50. Focusing a camera involves moving the lens so it is an appropriate distance from the film. For an object far away, the lens is at the infinity setting. The distance, D, the lens must move from the infinity setting depends on the focal length, F, of the lens and the distance, x, from the object to the lens and is given by

$$D(x) = \frac{F^2}{x - F}.$$

(a) Compare the way D varies for close objects (with small x-values) with the way it varies for distant objects. Sketch the graphs for several values of F.

(b) Why is it difficult to focus on an object that is very close?

CHAPTER

9 Exponential and Logarithmic Functions

We have studied some polynomial functions of the form $p(x) = a_n x^n + \ldots + a_1 x + a_0$, and some "root" functions of the form $g(x) = \sqrt[n]{x}$ for positive numbers n. A polynomial function grows rapidly as x increases; root functions, on the other hand, grow slowly. In this chapter we return to the exponential function, which we already studied in Chapters 2 and 7. The exponential function grows faster than any polynomial function. We also study its inverse, the logarithmic function, which grows more slowly than any root function. Both functions have many interesting applications.

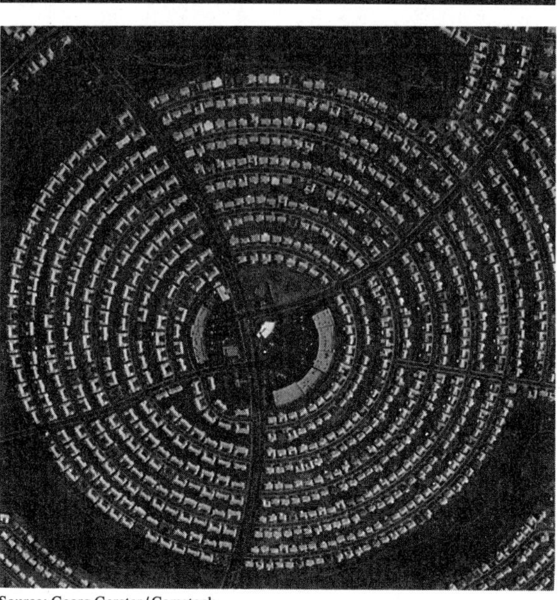

Source: Georg Gerster / Comstock

9.1 EXPONENTIAL FUNCTIONS

We begin by returning to an example of population growth.

EXAMPLE 1

A bacterial population doubles every hour. Each cell grows to a certain size and divides sometime during a one-hour time period. Beginning with one cell, how many cells will there be at the end of

(a) 12 hours? **(b)** 4 hours and 20 minutes?

How long will it take to produce

(c) 512 cells? **(d)** 750 cells?

Solutions

Using a Table Let t denote the number of hours elapsed and N denote the number of cells produced:

t	0	1	2	3	4	5	6	7	8	9	10	11	12
N	1	2	4	8	16	32	64	128	256	512	1,024	2,048	4,096

Table 9.1

(a) We want the value of N corresponding to $t = 12$, namely, $2^{12} = 4{,}096$.

(b) We want the value of N corresponding to $t = \frac{13}{3}$ (4 hours, 20 minutes), namely,

$$2^{\frac{13}{3}} = \sqrt[3]{2^{13}} = \sqrt[3]{8{,}192} \approx 20 \text{ (actually, 20.16)}$$

Using a Graph

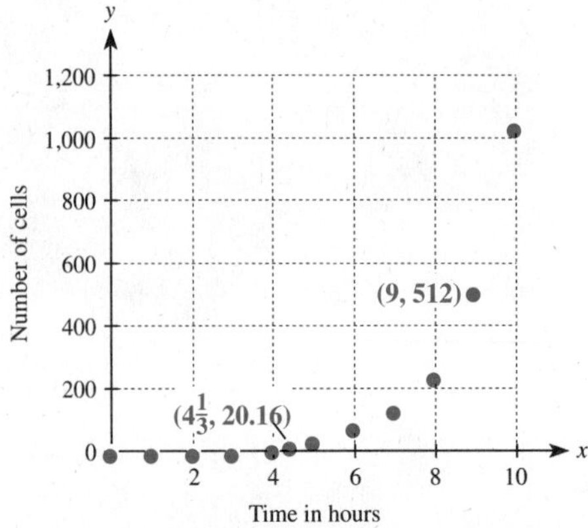

Figure 9.1

(c) We want the "reverse" of what we wanted in the first two parts. That is, we want the value of t corresponding to $N = 512$. By looking at the table and graph, we see $t = 9$.

(d) We want the value of t corresponding to $N = 750$. By looking at the table and graph, we observe $9 < t < 10$, but we as yet have no direct way to calculate t. Using a calculator, we see that $2^{9.5} \approx 724$ and $2^{9.6} \approx 776$. Finally, $2^{9.55} \approx 750$. Thus it takes approximately 9.55 hours to produce 750 cells. ■

In parts (a) and (b), we were seeking certain values for the number of cells. We wanted to know how the quantity N depended on the quantity t. This relationship can be described by the *exponential function* $N(t) = 2^t$. In parts (c) and (d), we were seeking certain values of the time t. We wanted to know how the quantity t depended on the quantity N. Though it is certainly related to the exponential function, this relationship is not yet clear. We will investigate it in the next section.

Definition and Graph of the Exponential Function

In Chapter 7, we defined exponential expressions b^r, where r was a rational number or an integer. For example, $(-2)^5 = -32$, $2^3 = 8$, $(32)^{-\frac{3}{5}} = \frac{1}{8}$, and $2^{\frac{1}{2}} = \sqrt{2} \approx 1.414$. Such expressions obeyed the four properties of exponents: For any real numbers $x > 0$, $y > 0$, and any rational numbers a and b,

$$\text{Property 1: } x^a x^b = x^{a+b} \qquad \text{Property 2: } (x^a)^b = x^{ab}$$

$$\text{Property 3: } (xy)^a = x^a y^a \qquad \text{Property 4: } \frac{x^a}{x^b} = x^{a-b}$$

Exponential expressions lead to exponential functions—functions where the variable is an exponent.

Definition

An **exponential function** is a function of the form $f(x) = b^x$, where the *base* $b > 0$, $b \neq 1$, the domain is *all real numbers*, and the range is *all positive real numbers.*

EXAMPLE 2

Sketch the graph of

(a) $y = 3^x$ **(b)** $y = \left(\frac{1}{3}\right)^x$

Solutions

(a)

x	-3	-2	-1	0	1	2	3	4	5
3^x	0.037	0.111	0.333	1	3	9	27	81	243

Table 9.2

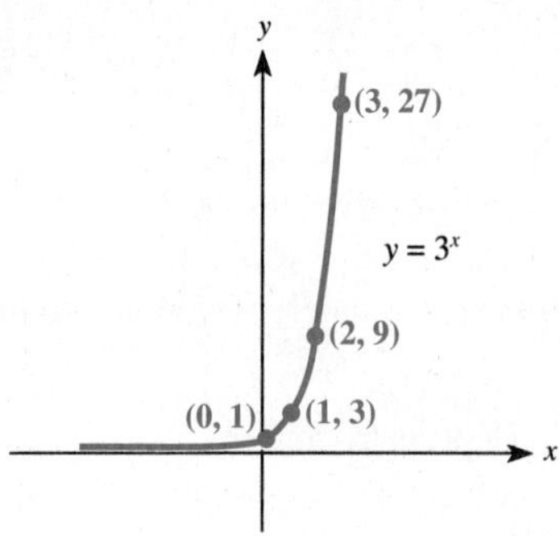

Figure 9.2

When we draw the graph of $y = 3^x$ as a smooth curve with no gaps (that is, continuous), we have *assumed* that expressions like $3^{\sqrt{2}}$ are defined. By calculator, $3^{\sqrt{2}} \approx 4.73$. We also insist that the base b is positive to ensure that the graph of $y = b^x$ is continuous. If $b = -2$, for example, $(-2)^{\frac{1}{2}}$ and $(-2)^{-\frac{5}{4}}$ are not real numbers, so the graph of $y = (-2)^x$ would have many holes in it. Note that the base, 3, is greater than 1, and the graph increases as x gets larger.

(b)

x	-3	-2	-1	0	1	2	3
$\left(\frac{1}{3}\right)^x$	27	9	3	1	0.333	0.111	0.037

Table 9.3

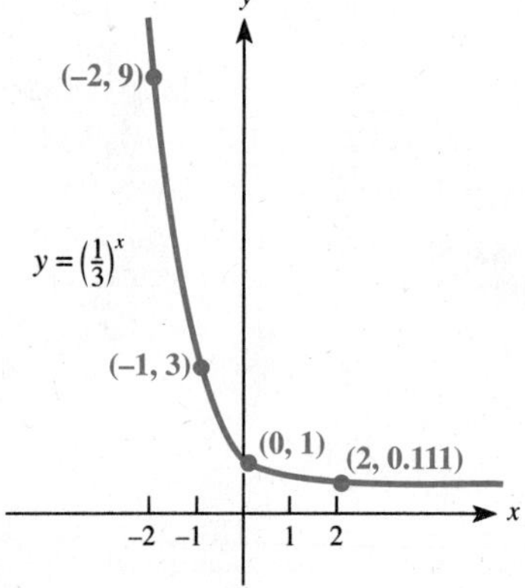

Figure 9.3

Note that the base, $\frac{1}{3}$, is less than 1, and the graph decreases. ■

We now summarize the properties of the exponential function.

Properties of the Exponential Function

For each $b > 0$, the function $f(x) = b^x$ is called the **exponential function to the base *b*.** The function has the following properties:

1. The domain is all real numbers.
2. The range is all positive real numbers.
3. If $b > 1$, as x increases, $f(x)$ increases (see Figure 9.2 in Example 2).
4. If $0 < b < 1$, as x increases, $f(x)$ decreases (see Figure 9.3 in Example 2).

It is useful to be able to roughly sketch graphs of some simple exponential functions.

EXAMPLES 3–7

For each of the functions,

(a) roughly sketch its graph, making use of the graphing principles (see Section 8.4);

(b) find the x- and y-intercepts; and

(c) find the domain, D, and range, R. Then use a graphing calculator.

3. $y = 2^x$ **4.** $y = 2^{-x}$ **5.** $y = -2^x$ **6.** $y = 2^x - 3$

7. $y = 2^{x-2}$

Solutions

3. (a)

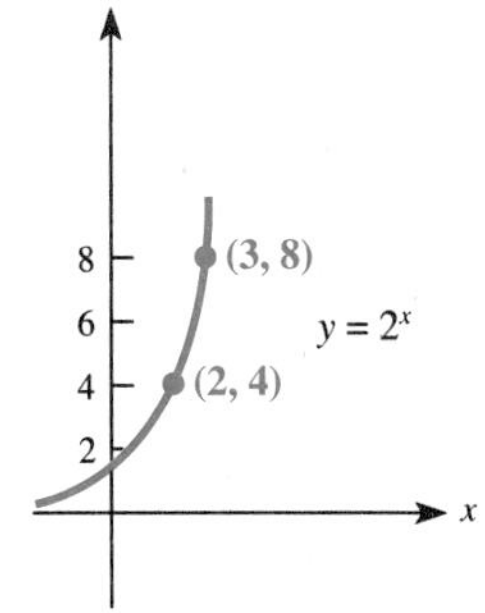

This graph has the same general shape as $y = 3^x$.

Figure 9.4

(b) x-intercept: none; y-intercept: $(0, 1)$

(c) $D = \{\text{all real numbers}\}$; $R = \{y \mid y > 0\}$

4. (a)

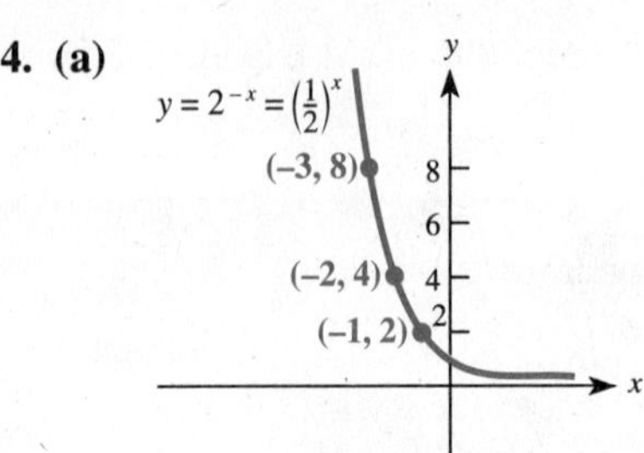

This graph is the mirror image of $y = 2^x$ in the y-axis. Notice that $y = 2^{-x}$ is the same as $y = \left(\frac{1}{2}\right)^x$, an exponential function whose base b satisfies $0 < b < 1$.

Figure 9.5

(b) x-intercept: none; y-intercept: (0, 1)

(c) $D = \{\text{all real numbers}\}$; $R = \{y|y > 0\}$

5. (a)

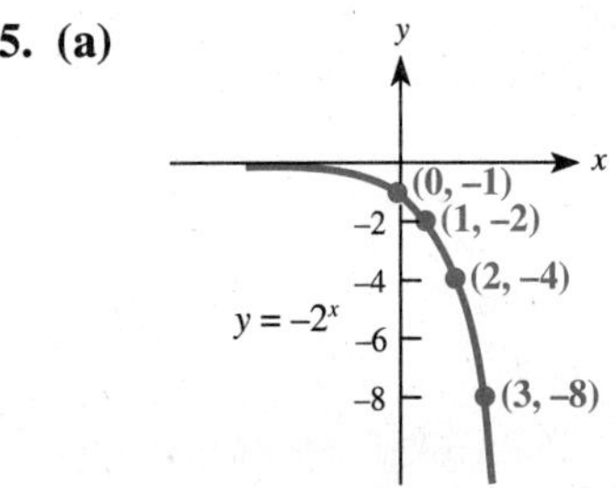

This graph is the mirror image of $y = 2^x$ in the x-axis (graphing principle 5).

Figure 9.6

(b) x-intercept: none; y-intercept: $(0, -1)$

(c) $D = \{\text{all real numbers}\}$; $R = \{y|y < 0\}$

6. (a)

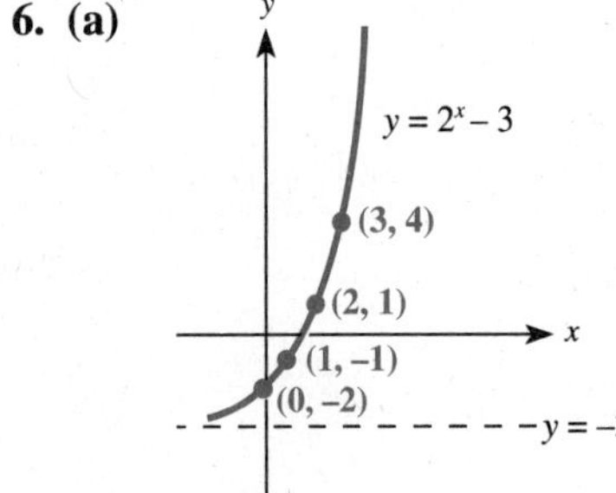

This graph is the graph of $y = 2^x$ moved 3 units downward (graphing principle 2).

Figure 9.7

(b) x-intercept: ?; y-intercept: $(0, -2)$

(c) $D = \{\text{all real numbers}\}$; $R = \{y|y > -3\}$

The x-intercept cannot be found at this time. It can be approximated with a graphing calculator.

7. (a)

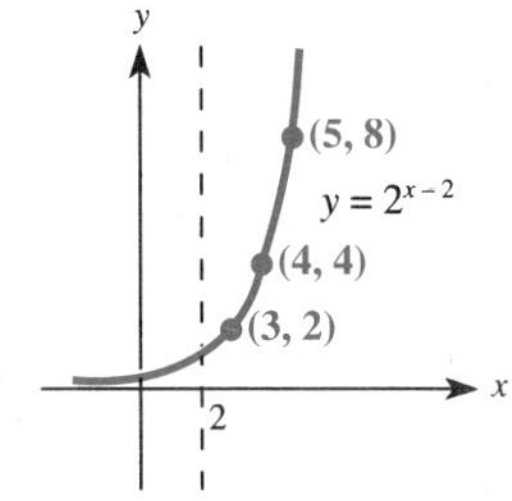

This graph is the graph of $y = 2^x$ moved 2 units to the right (graphing principle 3).

Figure 9.8

(b) x-intercept: none; y-intercept: $\left(0, \frac{1}{4}\right)$

(c) $D = \{\text{all real numbers}\}$; $R = \{y|y > 0\}$ ■

Evaluating Exponential Functions

Evaluating an exponential function answers the question, "What is the xth power of b?" As we have seen in Chapter 7, except for a few small values of x, it is best to evaluate $f(x) = b^x$ using a calculator. Values of $f(x)$ containing more digits than your calculator displays are given in scientific notation.

EXAMPLE 8

Approximate with a calculator.

(a) 3^7 **(b)** $3^{4.37}$ **(c)** 3^{101} **(d)** 3^{-4}

(e) $3^{-13.274}$ **(f)** $3^{\sqrt{2}}$ **(g)** 3^{1001}

Solutions

You could use the $\boxed{y^x}$ or $\boxed{\wedge}$ key.

(a) 2,187 **(b)** 121.624 **(c)** 1.5461×10^{48}

(d) 0.0123 **(e)** 0.00000046 **(f)** 4.729

(g) Your calculator possibly indicated an error, because the number is too large. Most calculators hold numbers between 10^{-99} and 10^{99}, but $3^{1001} > 10^{99}$. Such a number can be approximated by using logarithms, as we shall see in Section 9.3. ■

Exponential Growth and Decay

One classical use of the exponential function, first discussed in Chapter 7, is the study of **exponential growth and decay.** We give only a brief introduction to this important topic in this book.

You have probably heard the phrase "the population is growing exponentially." Mathematically, this means that the situation can be described by an increasing exponential function of the form

$$B = B(t) = B_0 b^{kt}$$

where

$$b > 1$$
$$B_0 = \text{amount present at time } t = 0$$
$$k = \text{a positive constant that depends on the rate of growth}$$

When t is increased by 1, $B(t)$ is multiplied by b^k.

EXAMPLE 9

Here are several examples of exponential growth from previous chapters:

(a) The growth of a bacterial population is given by

$B(t) = 2^t$ — $B_0 = 1$, since one cell was present at the beginning, $k = 1$, since the number of bacteria doubles every hour, and t = number of hours.

The exponential growth of this culture is given by the graph (Figure 9.9).

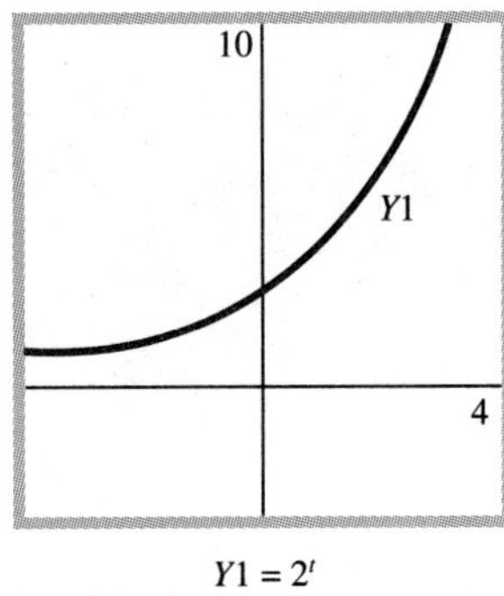

$Y1 = 2^t$

Figure 9.9

(b) Assuming the present growth rate continues, the world's population (in billions) will be

$$P(t) = (5.25)2^{\frac{t}{40}}$$

where t is the number of years after 1990.

$P_0 = 5.25$, the population in 1990. $k = \frac{1}{40}$, since the population doubles every 40 years and t = number of years.

(c) If an annual inflation rate of 12% continues, an item costing \$100 today will cost $C(t) = 100(1.12)^t$ in t years.

$C_0 = 100$; today's cost. $k = 1$, since cost increases 12% (is multiplied by 1.12) every year, and t = number of years.

(d) If the inflation rate increases 6% every six months, then an item costing \$100 today will cost $C(t) = 100(1.06)^{2t}$ in t years.

$C_0 = 100$ and $k = 2$, since cost increases 6% twice a year and t = number of years.

EXAMPLE 10

Find the cost in eight years of an item costing \$100 today assuming the model in

(a) Example 9(c). **(b)** Example 9(d).

Solutions

(a) $P = 100(1.12)^8 = \$247.60$ Evaluate $P(t) = 100(1.12)^t$ when $t = 8$.

(b) $P = 100(1.06)^{16} = \$254.04$ Evaluate $P(t) = 100(1.06)^{2t}$ when $t = 8$.

Notice that the two prices are not the same.

EXAMPLE 11

A savings account pays 5.5% interest compounded quarterly. What is the effective annual percentage rate (APR) on such an account?

Solution

Recall from Chapters 2 and 7: $A = P_0\left(1 + \frac{r}{n}\right)^{nt}$.

P_0 is the initial investment.
r is the annual interest rate.
n is the number of times per year the interest is compounded.
A is the new principal.
t = number of years.

Suppose we put \$100 in such a savings account and calculate the amount of interest at the end of one year ($t = 1$). The APR can, then, be determined.

$$A = 100\left(1 + \frac{0.055}{4}\right)^4$$

$$\approx 105.61$$

Therefore the interest earned is \$5.61, and the APR = 5.61%.

If $k > 0$, the graph of $B(t) = B_0b^{kt}$ is increasing. If $k < 0$, then the graph of $B(t)$ decreases. It represents decline or decay, as we see in the next example.

EXAMPLE 12

In living matter, the amount of carbon-14, an isotope of normal carbon-12, is constant. When the organism dies, the carbon-14 atoms decay into nitrogen-14 with an emission of radiation such that there will only be half as many carbon-14 atoms every 5,570 years (approximately). Thus the *half-life* of carbon-14 is 5,570 years.

By measuring the emission of radiation (a delicate job, especially for very small and very old samples), scientists can infer the population of carbon-14 atoms, compare this to the population of carbon-14 atoms in similar living matter, and estimate when the matter died.

If A_0 is the original amount of carbon-14 present, then the amount A of carbon-14 remaining after t years is given by the function

$$A = A(t) = A_0 2^{-\frac{t}{5{,}570}}$$

(a) Let $A_0 = 100$ and roughly sketch the graph of the decay of carbon-14 by plotting $A(t)$.

(b) How much carbon-14 is present in an organism that originally contained 100 mg and died 3,000 years ago?

Solutions

(a)

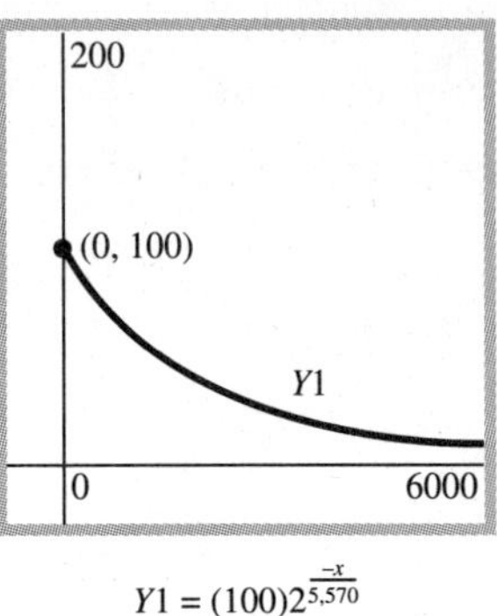

$Y1 = (100)2^{\frac{-x}{5{,}570}}$

Figure 9.10

(b) $A(3{,}000) = (100)2^{-\frac{3000}{5570}} \approx 68.84$ mg ∎

The Natural Base

The base of the exponential function that occurs most frequently in applications is, surprisingly, an irrational number. To see how such a number could arise, let us recall from Chapter 7 the formula for the amount of money accumulated in an account earning compounded interest:

$$A_t = P_0\left(1 + \frac{I}{n}\right)^{nt}$$

where

I = rate of interest
t = number of years
n = number of times per year the interest is compounded
P_0 = initial principal
A_t = amount accumulated after t years.

To simplify the problem, suppose we invest \$1 for one year at an interest rate of 100%, that is, $P_0 = I = t = 1$. Then

$$A_1 = \left(1 + \frac{1}{n}\right)^{n}$$

What happens as n increases? That is, if we invest \$1 for one year at 100%, how much difference does it make how often the interest is compounded? Let's find the amount earned if interest is compounded annually, semiannually, quarterly, monthly, weekly, daily, hourly, and so on, by letting $n = 1, 2, 4, 12, 52, 360, 8{,}640, \ldots$

n	$1 + \frac{1}{n}$	$\left(1 + \frac{1}{n}\right)^n$
1	2	2
2	1.5	2.25
4	1.25	2.44
12	1.083	2.61
52	1.019	2.69
360	1.0027	2.715
8,640	1.0001157	2.7181
518,400	1.000001929	2.71826
31,104,000	1.0000000321	2.7182818

Table 9.4

Notice that you receive substantial increases until the weekly calculation, but beyond that, the increases are insignificant. In fact, no matter how often the interest is compounded, the amount accumulated does not exceed a number that is approximately 2.7182818 and is denoted by the symbol e. The **natural base *e*** has many applications since it describes continuous growth. If interest were compounded continuously, we could rewrite the compound-interest formula as the function

$$A_t = Pe^{It} \qquad P = \text{principal},\ I = \text{interest rate},\ t = \text{number of years.}$$

We see that compound interest is another example of exponential growth, since it has the form

$$B = B_0 e^{kt},\ k > 0$$

EXAMPLE 13

Smallville had a population of 15,000 during 1990 and was growing according to the growth function $y = y_0 e^{0.04t}$, where t is given in years. What will the population be in the year 2010?

Solution

$$y = 15{,}000e^{0.04(20)} \approx 33{,}383 \text{ people} \qquad t = 20 \text{ for the year 2010.}$$

EXAMPLE 14

The decay of radioactive argon-39 is described by the decay function

$$y = y_0 e^{-0.173t}$$

where y milligrams (mg) of argon remain after t minutes. If we begin with 200 mg of argon-39, then how much is left after ten minutes?

Solution

$$y = 200e^{-0.173(10)} \approx 35.5 \text{ mg}$$

Geometric Sequences

A geometric sequence is a discrete exponential function. Let us first consider an example.

Your annual salary, year by year, consisting of an initial salary of \$15,000 and annual raises of 9%, is given by (in thousands of dollars):

$$15,\ 15(1.09),\ 15(1.09)^2,\ 15(1.09)^3,\ \ldots.$$

Each new term in this sequence is found by multiplying the previous term by 1.09, or, the quotient of any two consecutive terms is

$$\frac{15(1.09)}{15} = \frac{15(1.09)^2}{15(1.09)} = \frac{15(1.09)^3}{15(1.09)^2} = 1.09.$$

Any sequence with this property is a **geometric sequence** or **geometric progression.** An arbitrary geometric sequence has the form

$$a_1,\ a_1r,\ a_1r^2,\ a_1r^3,\ \ldots$$

and we can describe the general term by $a_n = a_1r^{n-1}$. The quantity r is called the **common ratio.** In this case, $a_1 = 15$, $r = 1.09$, $a_n = 15(1.09)^{n-1}$.

A geometric sequence is an exponential function of the form $f(n) = ab^n$, where the domain is a subset of the natural numbers.

EXAMPLE 15

(a) Find the twentieth term in the geometric sequence

$$5,\ 10,\ 20,\ 40,\ \ldots$$

(b) Sketch the graph of the sequence.

Solutions

(a) $a_{20} = a_1r^{19}$ $\qquad \frac{10}{5} = \frac{20}{10} = \frac{40}{20} = 2$, so the sequence is geometric, with $a_1 = 5, d = 2.$

$= 5(2)^{19}$

$= 2{,}621{,}440$

(b)

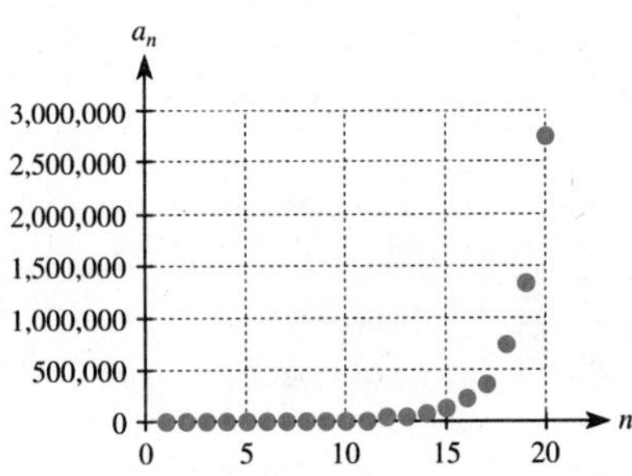

Figure 9.11

EXAMPLES 16–17

Find the general term for each of the following sequences.

16. A geometric sequence with first term 4 and common ratio 5.

17. $\frac{3}{2}, \frac{3}{4}, \frac{3}{8}, \frac{3}{16}, \ldots$

Solutions

16. $a_n = 4(5)^{n-1}$ $\quad a_n = a_1 r^{n-1}; a_1 = 4, r = 5$

17. The quotient of any two consecutive terms is

$$\frac{\frac{3}{4}}{\frac{3}{2}} = \frac{\frac{3}{8}}{\frac{3}{4}} = \cdots = \frac{1}{2}$$

so this is a geometric sequence. The first term $a_1 = \frac{3}{2}$, and the common ratio $r = \frac{1}{2}$. Thus

$$a_n = \frac{3}{2}\left(\frac{1}{2}\right)^{n-1}$$

Note: This sequence is decreasing. ■

9.1 EXERCISES

I. Procedures

1–16. Approximate, using a calculator.

1. $5^{4.37}$

2. $6^{2.02}$

3. $8^{-2.17}$

4. $7^{-0.014}$

5. $(3.25)^{101}$

6. $(-3.25)^{101}$

7. $3^{4.56} + 7^{1.23}$

8. $6^{2.37} - 4^{3.26}$

9. $2^{1.02} \cdot 5^{3.79}$

10. $7^{3.16} \cdot 8^{-2.14}$

11. $\frac{2^{1.02}}{5^{3.79}}$

12. $\frac{8^{2.37}}{5^{3.18}}$

13. $9^{\sqrt{2}}$

14. $(\sqrt{2})^9$

15. $4^{1,001}$

16. $(-4)^{1,001}$

17–22. Find the fifteenth term in each of the following geometric sequences in which either a_n or the first four terms are given.

17. $a_n = 3 \cdot \left(\frac{3}{2}\right)^{n-1}$

18. $a_n = 5\left(-\frac{1}{2}\right)^{n-1}$

19. $-\frac{3}{4}, \frac{9}{16}, -\frac{27}{64}, \frac{81}{256}$

20. $-2, 4, -8, 16$

21. $12, 6, 3, \frac{3}{2}$

22. $1, 1.05, 1.1025, 1.157625$

23–26. Find x so that each of the following is a geometric sequence.

23. $x, 10, 50$

24. $8, 2, x$

25. $1.2, x, 1.728$

26. $-\frac{3}{2}, x, -\frac{8}{27}$

27–46. For each of the functions, **(a)** sketch its graph, making use of the graphing principles as much as possible; **(b)** if possible, find the x- and y-intercepts; and **(c)** find the domain, D, and range, R.

27. $y = 5^x$

28. $y = 4^x$

29. $y = 3^{-x}$

30. $y = 4^{-x}$

31. $y = 3^{x+1}$

32. $y = 4^{x-1}$

33. $y = 3^{x+2}$

34. $y = 4^x - 3$

35. $y = 2 - 3^x$

36. $y = 4 - 4^x$

37. $y = 2 - 3^{-x}$

38. $y = 1 - 4^{-x}$

39. $y = 2 - 3^{x+1}$

40. $y = 4 - 4^{1-x}$

41. $y = (1.06)^x$

42. $y = (2.02)^{x+1}$

43. $y = e^{-x}$

44. $y = 1 + e^{-x}$

45. $y = \frac{1}{\sqrt{2\pi}} e^{\frac{-x^2}{2}}$ (the normal curve)

46. $y = 100(1 - 3^{-\frac{x}{10}})$
(a Heines curve used in mental testing)

47 and 48. How many real solutions do each of the following equations have? Estimate each root with a graphing calculator.

47. **(a)** $3^x = 4$ **(b)** $3^x = -4$ **(c)** $3^{-x} = 4$
(d) $3^{-x} = -4$

48. **(a)** $e^x = 10$ **(b)** $e^{-x} = 10$ **(c)** $e^x = -10$
(d) $e^{-x} = -10$

II. Concepts

49. A piece of paper is 0.003 inches thick. Folding it once gives it a thickness of two pieces. How thick would the "stack" of paper be if you folded the original piece of paper 50 times? (Physically, seven or eight folds is the maximum for a piece of paper of any size, so you should tear and stack.)

50. Suppose you tried to plot $y = 2^x$ on graph paper with a scale of one unit equals $\frac{1}{16}$ inch. Find the length and width of the piece of graph paper needed to plot $y = 2^x$ for $-12 \leq x \leq 12$. What are the range values on a graphing calculator?

51–53. Mark each of the following statements AT (always true), ST (sometimes true), or NT (never true). If a statement is ST, give an example where it is true and one where it is false.

51. The graph of $y = b^x$ contains the point (0, 1).

52. If $f(x) = b^x$, then $f(c + d) = f(c) \cdot f(d)$ for real numbers c and d.

53. If $f(x) = b^x$, then $f(2c) = 2f(c)$ for real numbers c.

54. Each of the three accompanying graphs is an exponential growth model of a function of the form $f(x) = ab^x$.

(a) For which of the three functions, A, B, or C, is the value of a the greatest?

(b) For which of the three functions, A, B, or C, is the value of b the greatest?

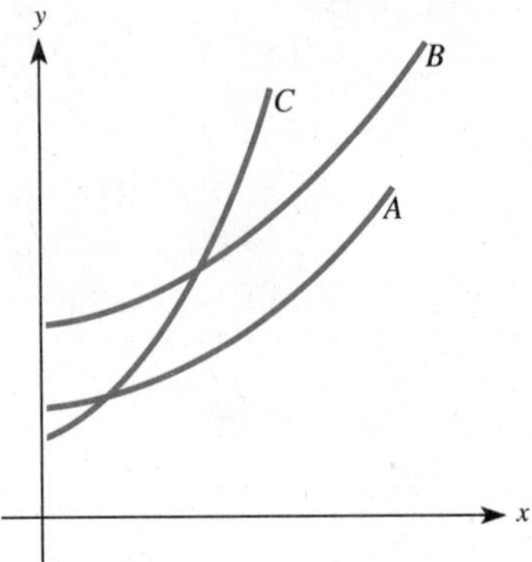

55. Write a paragraph comparing the functions $f(x) = x^2$ and $g(x) = 2^x$.

III. Applications

56–67. Solve each of the following problems.

56. The world's population during 1975 was approximately 4 billion.

(a) If the population doubles every 35 years, then the world's population t years after 1975 is given by $P(t) = 4 \cdot 2^{\frac{t}{35}}$. Estimate the world's population in the year 2001.

(b) If the world's population increases by 1.6% per year, estimate the world's population in 2001.

(c) If the world's population t years after 1975 is given by $E(t) = 4e^{0.016t}$, estimate the world's population in 2001.

57. You deposit \$1,500 in a savings account today. How much will you have at the end of five years from today if the interest is

(a) 5.25%, compounded quarterly?

(b) 5%, compounded continuously?

58. The world's population in 1990 was 5.25 billion. If the population doubles every 40 years, estimate the world's population in 2010.

59. Suppose you drop a Super Ball from a window 15 meters above the ground. The ball bounces to 80% of its previous height with each bounce.

(a) How far does the ball travel, up and down, between the second and the third bounce?

(b) How far does the ball travel between the first bounce and the eighth bounce?

60. Suppose you start a chain letter (illegal in the United States) by sending a copy of it to five of your friends. In the letter, you ask each recipient to send a copy of the letter to five friends.

(a) Assuming no one breaks the chain, how many people will have received the letter after eight mailings?

(b) After how many mailings will every person in the United States have received a copy of the letter (assuming no person gets the letter more than once)?

61. You bought a diamond ring for \$2,000. How much will this ring be worth in ten years if

(a) its value increases 12% each year?

(b) its value after t years is given by $V(t) = 2{,}000e^{0.12t}$?

62. Your bank advertises money-market certificates that pay 8.35% interest and mature in three months. What is the effective annual interest rate on such a certificate?

63. The atmospheric pressure m miles above sea level is given by $P(m) = 14.2e^{-0.021m}$. Find the atmospheric pressure

(a) in Denver, the "mile-high" city.

(b) at the top of Mount McKinley (20,320 *feet* above sea level).

(c) in Death Valley (282 *feet* below sea level).
Find a reasonable domain for the function $P(m)$, and then sketch its graph.

64. The decomposition of the radioactive isotope ^{11}C is given by $F(t) = F_0 2^{\frac{-t}{20}}$, where F_0 is the original amount and t is in minutes. If there were 20 grams of ^{11}C initially, how many grams were present after

(a) 10 minutes? **(b)** 20 minutes? **(c)** 1 hour?

65. Suppose one of your ancestors bought one acre of the Louisiana Purchase in 1803. The price of the Louisiana Purchase was \$27,670 for 828,000 square miles. Assuming an annual inflation rate of 5%, what is the value of this acre of land today (640 acres = 1 square mile)?

66. If you borrow \$60,000 for a home loan at an annual percentage rate of 12%, you can calculate your monthly payment using the formula:

$$P(n) = \frac{60{,}000(0.01)(1.01)^n}{(1.01)^n - 1}$$

where n is the number of months of the loan. (12% annual rate = 1% per month)

(a) Find the monthly payment on a 30-year loan.

(b) Find the domain and range of this function.

(c) Verify that 600 is not in the range of the function. Give a reason why this monthly payment is not possible. (What would happen if it were possible?)

67. The luminous intensity I at a depth of x meters in clean sea water is given by $I(x) = I_0 e^{-2x}$, where I_0 is the intensity at sea level. Relative to I_0, by what percent does the intensity decrease going from

(a) sea level to a depth of $\frac{1}{3}$ meter?

(b) $\frac{1}{3}$ meter to 1 meter?

IV. Extensions

68. By experimenting on your calculator, find the largest power of the following numbers that your calculator will display.

(a) 3 **(b)** e **(c)** 9 **(d)** 12

69. By experimenting on your calculator, find the largest value of x for which the following sequence does not result in "error."

x [e^x] [e^x] [e^x]

70–73. Find the *number* of solutions and then approximate the solutions using a graphing calculator for each of the following equations.

70. $2^x = x + 3$. *Hint:* Sketch the graphs of $y = 2^x$ and $y = x + 3$ on the same coordinate system.

71. $e^{-x} = x^2$

72. $(1.5)^x = 1.5x$

73. $2e^x = x$

74. For what values of x is

(a) $3^x = x^3$ **(b)** $3^x < x^3$

75. It seems e^x and $1 + x$ are very close when x is a small positive number. Investigate this phenomenon by making a table of values of the form

x				
e^x				
$1 + x$				

(a) By carefully drawing a graph, show that $e^x > 1 + x$ for $x > 0$.

(b) Find the largest value of x for which $e^x - (1 + x) < 0.005$.

76. It is clear that $2^3 < 3^2$, but $4^5 > 5^4$. By experimenting with your calculator, see if you can find the range of values of a and b for which $a^b < b^a$ if $a < b$.

77. You probably feel that quantities that grow exponentially increase very fast—faster than polynomials. Find a real number N so that $b^x > x^n$ whenever $x > N$ for the given values of b and n.

(a) $b = 2, n = 2$

(b) $b = 2, n = 4$

(c) $b = 1.01, n = 1$

(d) $b = 1.01, n = 2$

78. Why must the base, b, of an exponential function be greater than 0? Try to graph $y = b^x$ for $b = -2$ on your graphing calculator. What happens? Make a table of values and try to sketch the resulting graph. Try a few more such examples, and then answer the question.

9.2 LOGARITHMIC FUNCTIONS

Definition of the Logarithmic Function

In the last section, we considered the following example:

EXAMPLE 1

A bacterial population doubles every hour. Beginning with 1 cell, how long does it take to produce 750 cells?

Solution

We seek the value of t for which $2^t = 750$ (≈ 9.55 by trial and error). ■

In Chapter 7 we considered an example similar to this one.

EXAMPLE 2

Approximately how long will it take to double an investment of $100 at an annual interest rate of 8% compounded quarterly?

Solution

$$A = P\left(1 + \frac{I}{n}\right)^{nt} \qquad \text{Compound interest formula.}$$

$$200 = 100\left(1 + \frac{.08}{4}\right)^{4t} \qquad A = 200, P = 100, n = 4, I = 0.08$$

$$2 = (1.02)^{4t}$$

$$2 = (1.08)^t \qquad (1.02)^{4t} = [(1.02)^4]^t \approx (1.08)^t$$

So we seek the value of t for which $(1.08)^t = 2$ (≈ 9 by trial and error). ■

We observe that the solutions to both problems are essentially the same. In each case, we wanted to find the value of a certain exponent. Such an exponent is called a *logarithm*. Formally, we have the following definition.

Definition

For each $b > 1$, there is a function $L(x) = \log_b x$ called the **logarithmic function with respect to base b.** The statement $y = \log_b x$ is equivalent to the statement $b^y = x$. Note that in b^y the exponent is y, where $y = \log_b x$.

The word *logarithm* comes from the two Greek words "logos," meaning reckoning or reason, and "arithmos," meaning number.

We stress the fact that a logarithm, $\log_b x$, is an exponent. It is the power (exponent) to which the base b must be raised to obtain x, that is, $b^{\log_b x} = x$. Observe:

Logarithmic statement	*Exponential statement*
$\log_2 8 = 3$	$2^3 = 8$
$\log_5 15{,}625 = 6$	$5^6 = 15{,}625$
$\log_7 \sqrt{7} = \frac{1}{2}$	$7^{\frac{1}{2}} = \sqrt{7}$
$\log_4\left(\frac{1}{16}\right) = -2$	$4^{-2} = \frac{1}{16}$
$\log_{10} 0.00001 = -5$	$10^{-5} = 0.00001$
$\log_2 750 \approx 9.55$	$2^{9.55} \approx 750$
$\log_{1.08} 2 \approx 9$	$1.08^9 \approx 2$

Table 9.5

To help understand that a logarithm is an exponent, we will solve the following logarithmic equations by using the exponential equation associated with it.

EXAMPLES 3–10

Solve for x.

3. $\log_4 x = \frac{3}{2}$ **4.** $\log_4 x = -\frac{3}{2}$ **5.** $\log_4 64 = x$

6. $\log_4 \left(\frac{1}{64}\right) = x$ **7.** $\log_4 (-64) = x$ **8.** $\log_4 (3x + 1) = 2$

9. $\log_4 64 = x + 2$ **10.** $\log_x 16 = 4$

Solutions

3. $\log_4 x = \frac{3}{2}$ $\log_4 x = \frac{3}{2}$ is equivalent to $4^{\frac{3}{2}} = x$.

$$4^{\frac{3}{2}} = x$$

$$8 = x$$

4. $\log_4 x = -\frac{3}{2}$

$$x = 4^{-\frac{3}{2}} = \frac{1}{8}$$

5. $\log_4 64 = x$

$$4^x = 64 = 4^3$$

$$x = 3$$ See Section 7.4.

6. $\log_4 \left(\frac{1}{64}\right) = x$

$$4^x = \frac{1}{64} = 4^{-3}$$

$$x = -3$$

7. $\log_4 (-64) = x$

$$4^x = -64$$

No solution; $4^x > 0$ for all x.

8. $\log_4 (3x + 1) = 2$

$$4^2 = 3x + 1$$

$$16 = 3x + 1$$

$$15 = 3x$$

$$5 = x$$

9. $\log_4 64 = x + 2$

$$4^{x+2} = 64$$

$$4^{x+2} = 4^3$$

$$x + 2 = 3$$

$$x = 1$$

10. $\log_x 16 = 4$

$$x^4 = 16$$

$$x = 2 \quad \text{or} \quad x = -2$$

$$x = 2$$ Reject $x = -2$, since the base x must be positive.

Some additional properties of $L(x) = \log_b x$ that you may have observed in the examples are as follows.

Properties of the Logarithmic Function

1. $L(1) = 0$ for any $b > 1$.
2. Domain of $L(x)$ is all positive real numbers.
3. Range of $L(x)$ is all real numbers.
4. As x increases, $L(x)$ increases.
5. $\log_b x = \log_b y$ if and only if $x = y$.

The relationship between the exponential function and the logarithmic function is the same as that between the cubing function and the cube root function. When we cube a number c, we obtain c^3. If we then take the cube root of c^3, we return to c. Alternatively, if we take the cube root of c to obtain $\sqrt[3]{c}$ and then cube it, we again return to c.

The functions $f(x) = x^3$ and $g(x) = \sqrt[3]{x}$ are said to be **inverse functions** because they "undo" each other. Cubing "undoes" taking the cube root, and vice versa (see Figure 9.12). Formally,

$$\sqrt[3]{x^3} = x \qquad \text{Cube, then take the cube root.}$$
$$(\sqrt[3]{x})^3 = x \qquad \text{Take the cube root, then cube.}$$

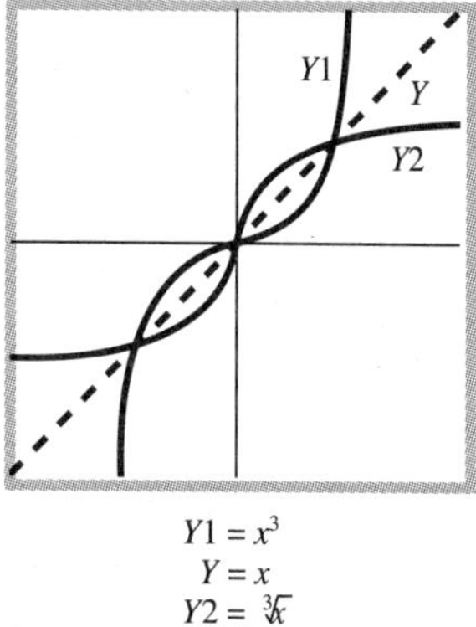

$Y1 = x^3$
$Y = x$
$Y2 = \sqrt[3]{x}$

Figure 9.12

In the same way, $E(x) = b^x$ and $L(x) = \log_b x$ are inverse functions. Raising to a power "undoes" taking the logarithm, and vice versa. For example, take $b = 3$ and $x = 9$. If we raise 3 to the ninth power, we obtain 19,683. Then taking the logarithm of 19,683 to the base 3, we return to 9. Alternatively, if we take the logarithm of 9 to the base 3, we obtain 2. If we then raise 3 to the second power, we again return to 9. Symbolically,

$$3^9 = 19{,}683 \qquad \text{and} \qquad \log_3 19{,}683 = 9$$
$$\log_3 9 = 2 \qquad \text{and} \qquad 3^2 = 9$$

Take another look at Table 9.5, preceding Example 3. In general, we have the relationships

$$b^{\log_b x} = x \quad \text{and} \quad \log_b b^x = x$$

We now see why INV log or INV ln can be used to calculate exponents on some calculators.

On a graph the exponential and logarithmic functions are symmetric about the line $Y = x$, just like $Y1 = x^3$ and $Y2 = \sqrt[3]{x}$ in Figure 9.12.

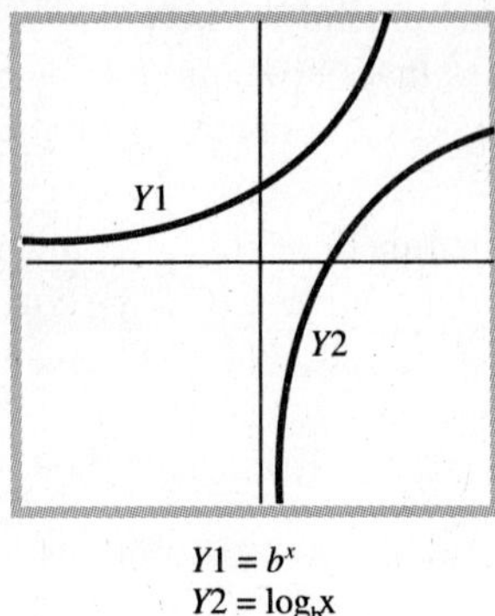

$Y1 = b^x$
$Y2 = \log_b x$

Figure 9.13

Graphing the Logarithmic Function

EXAMPLE 11 Sketch the graph of $y = \log_3 x$.

Solution

Using a Table Note: $y = \log_3 x$ if and only if $3^y = x$. First make a table to get a look at the behavior of the function for some specific values of x.

x	$\frac{1}{27} = 3^{-3}$	$\frac{1}{9} = 3^{-2}$	$\frac{1}{3} = 3^{-1}$	$1 = 3^0$	$3 = 3^1$	3^2	3^4
$\log_3 x$	-3	-2	-1	0	1	2	4

Table 9.6

Using a Graph

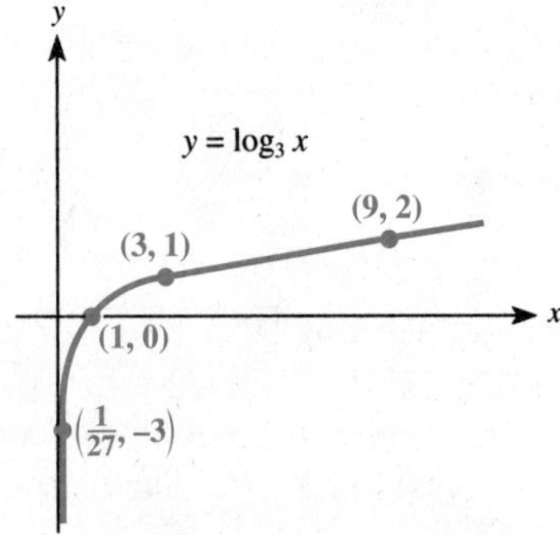

Figure 9.14

The graphs of $L(x) = \log_b x$ have the same general shape for every $b > 1$ as the graph in Example 11. Notice there are no points to the left of the y-axis where $x < 0$. ■

EXAMPLES 12–16

For each of the functions,

(a) roughly sketch its graph, making use of the graphing principles;

(b) find the x- and y-intercepts; and

(c) find the domain, D, and range, R.

Then use a graphing calculator.

12. $y = \log_{10} x$ **13.** $y = 3 + \log_{10} x$ **14.** $y = -\log_{10} x$

15. $y = \log_{10} (x - 2)$ **16.** $y = \log_{10} (-x)$

Solutions

12. (a)

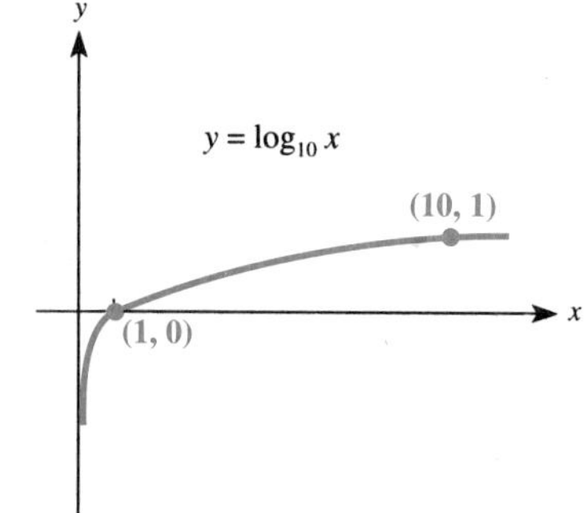

This graph has a shape similar to $y = \log_3 x$.

Figure 9.15

(b) x-intercept: $(1, 0)$; y-intercept: none

(c) $D = \{x|x > 0\}$; $R = \{\text{all real numbers}\}$

13. (a)

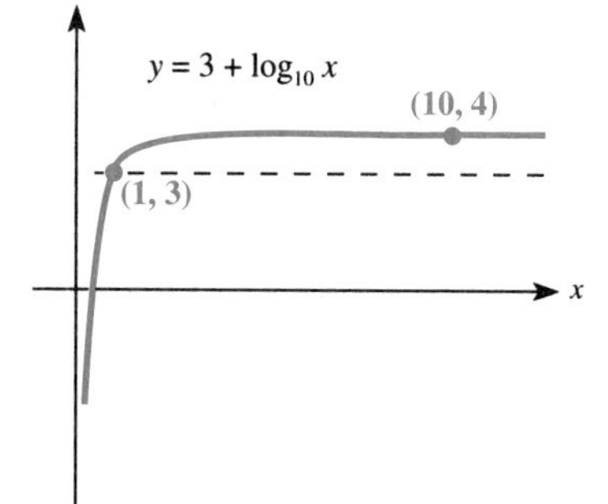

This is the graph of $y = \log_{10} x$ shifted 3 units upward (graphing principle 1).

Figure 9.16

(b) x-intercept: $\left(\frac{1}{1000}, 0\right)$; y-intercept: none

(c) $D = \{x|x > 0\}$; $R = \{\text{all real numbers}\}$

14.  **(a)**

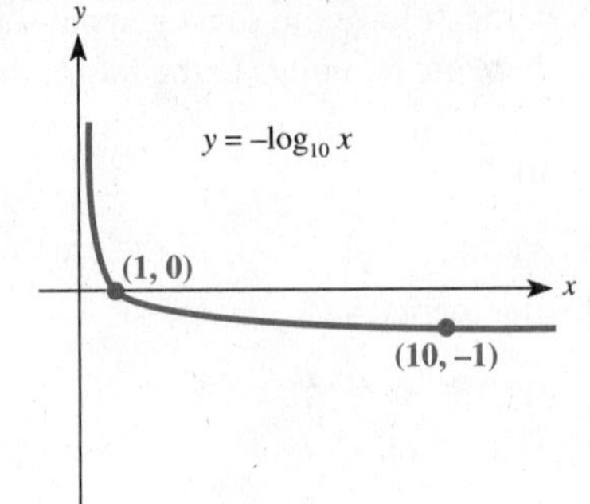

This graph is the mirror image of $y = \log_{10} x$ in the x-axis (graphing principle 5).

Figure 9.17

(b) x-intercept: (1, 0); y-intercept: none

(c) $D = \{x|x > 0\}$; $R = \{\text{all real numbers}\}$

15. (a)

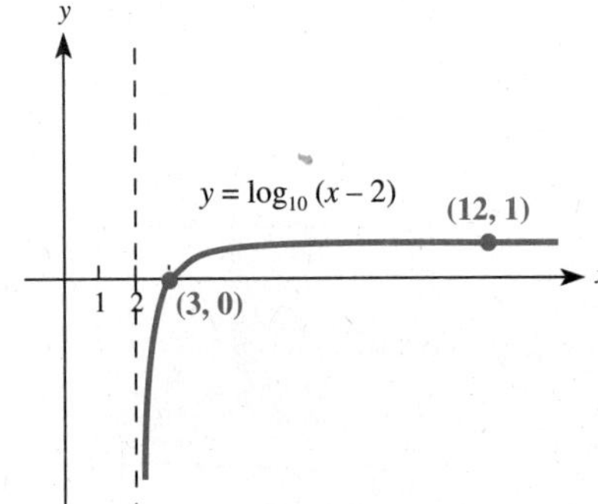

This is the graph of $y = \log_{10} x$ shifted 2 units to the right (graphing principle 3).

Figure 9.18

(b) x-intercept: (3, 0); y-intercept: none

(c) $D = \{x|x > 2\}$; $R = \{\text{all real numbers}\}$

16. (a)

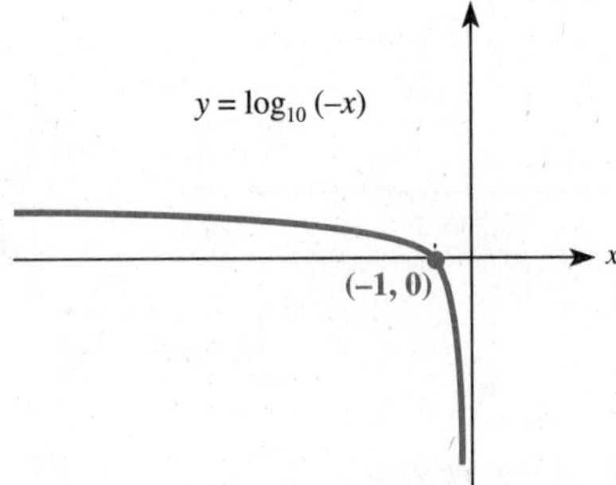

This graph is the mirror image of $y = \log_{10} x$ in the y-axis.

Figure 9.19

(b) x-intercept: $(-1, 0)$; y-intercept: none

(c) $D = \{x|x < 0\}$; $R = \{\text{all real numbers}\}$ ■

Evaluating Logarithmic Functions: Common Logs and Natural Logs

While the logarithmic function $L(x) = \log_b x$ is defined for any base $b > 1$, the most useful bases are $b = 10$ and the natural base $b = e$. The function $L(x) = \log_{10} x$ is called the **common logarithm** and is usually denoted simply by *log x*. The function $L(x) = \log_e x$ is called the **natural logarithm** and is usually denoted by *ln x*. Both log x and ln x can be evaluated on a calculator using the keys LOG and LN, respectively. Since the logarithm is an exponent, evaluating the logarithmic function $L(x) = \log_b x$ answers the question: "What power of b is x?"

EXAMPLE 17

Using a calculator, find

(a) log 324.7 **(b)** log 0.0023 **(c)** ln 324.7 **(d)** ln 0.0023
(e) log (3.24×10^{37}) **(f)** ln (3.24×10^{-17}) **(g)** log (-37.47)

Solutions

(a) 2.5115 **(b)** -2.638 **(c)** 5.783 **(d)** -6.075
(e) 37.511 **(f)** -37.968
(g) "Error," since the domain of the logarithmic function is the positive real numbers. In other words, you can't take the log of a negative number.

Computational Properties of Logarithms

To compute the logs of large and small numbers expressed in scientific notation, we can make use of several algebraic properties of logarithms that they inherit from exponentials.

Properties of Logarithms

For any base $b > 1$ and any positive real numbers r and s,

1. $\log_b (rs) = \log_b r + \log_b s$
2. $\log_b \left(\frac{r}{s}\right) = \log_b r - \log_b s$
3. $\log_b r^s = s \log_b r$

We have been using these properties of exponents since Chapter 2. In Property 1, for example, if

$$r = b^x \quad \text{or} \quad \log_b r = x$$

and if

$$s = b^y \quad \text{or} \quad \log_b s = y$$

then

$$rs = b^{x+y} \quad \text{or} \quad \log_b rs = x + y$$

Thus $\log_b r + \log_b s = \log_b rs$, since $b^x b^y = b^{x+y}$. The other two properties can be verified in a similar way.

These properties, especially the third one, will be used extensively in the next section, which is concerned with solving equations involving logarithmic and exponential functions.

Logarithmic Growth

Though less common than exponential growth, many quantities are said to grow logarithmically. A classic example of such growth is given by the Weber-Fechner law of learning, which says, in the case of sound, that the intensity of sound is directly proportional to the logarithm of the power of the sound. That is,

$$N = N(I) = 10 \log \left(\frac{I}{I_0}\right)$$

where

N = number of decibels (intensity of the sound)

I = power of the sound in watts per square centimeter

I_0 = constant representing the power of the sound just below the threshold of hearing (approximately 10^{-16} watts per square centimeter)

The graph of $N = 10 \log \left(\frac{I}{I_0}\right)$ is given in Figure 9.20.

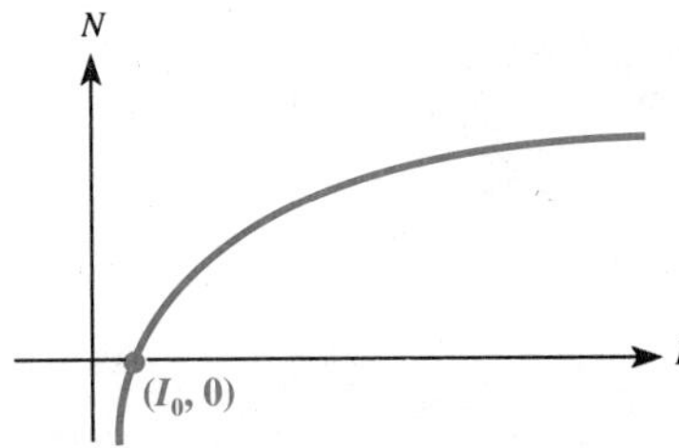

Figure 9.20

A person's perception of other sensations, such as taste, smell, and visual brightness, can be described by similar logarithmic functions.

EXAMPLE 18

(a) Find the number of decibels for normal conversation; power $= 3.16 \times 10^{-10}$ watts.

(b) If the power of a sound is doubled (for example, by turning up the sound), what is the difference in the sound levels?

Solutions

(a) $N(3.16 \cdot 10^{-10}) = 10 \log \dfrac{3.16 \cdot 10^{-10}}{10^{-16}}$ $\qquad I_0 = 10^{-16}$ watt/cm^2.

$$= 10 \log (3.16 \cdot 10^6)$$
$$\approx 10\,(6.50)$$
$$\approx 65 \text{ decibels}$$

(b) $N(2I) - N(I) = 10 \log \left(\dfrac{2I}{I_0}\right) - 10 \log \left(\dfrac{I}{I_0}\right)$

$$= 10 \log \left(\frac{2\frac{I}{I_0}}{\frac{I}{I_0}}\right)$$ Property of logarithms.

$$= 10 \log 2 \approx 3 \text{ decibels}$$

EXAMPLE 19

The *Richter scale* is a well-known method of measuring the magnitude M of an earthquake in terms of the amplitude A of its shock waves:

$$M = M(A) = \log \left(\frac{A}{A_0}\right)$$

where A_0 is a constant representing the amplitude of a norm quake. Serious damage occurs when $M = 5.5$. The magnitude of the great San Francisco quake of 1906 was 8.3 on the Richter scale. How many times stronger was the San Francisco quake than one that causes serious damage ($M = 5.5$)?

Solution

We seek the ratio of the amplitude A_s for the San Francisco quake to A_D, the amplitude for a serious-damage quake. Now

$$8.3 = \log \frac{A_s}{A_0} \qquad 5.5 = \log \frac{A_D}{A_0}$$

$$\frac{A_s}{A_0} = 10^{8.3} \qquad \frac{A_D}{A_0} = 10^{5.5}$$ Definition of logarithm.

$$\frac{A_s}{A_D} = \frac{10^{8.3}}{10^{5.5}} \approx \frac{1.99 \times 10^8}{3.16 \times 10^5} \approx 630 \text{ times as strong}$$

The brightness of stars can be measured using a similar logarithmic function.

9.2 EXERCISES

I. Procedures

1–20. Solve for x using properties of logarithms and exponents.

1. $\log_8 x = 2$

2. $\log_8 x = -2$

3. $\log_8 x^2 = 2$

4. $\log_8 x = \frac{2}{3}$

5. $\log_8 x = -\frac{2}{3}$

6. $\log_8 (3x + 1) = 2$

7. $\log_8 64 = x$

8. $\log_8 \left(\frac{1}{256}\right) = x$

9. $\log_8 (1{,}024) = x + 3$

10. $\log_8 \left(\frac{1}{512}\right) = 3x + 1$

11. $\log_8 \sqrt[4]{8} = x$

12. $\log_8 \sqrt[5]{64} = x$

13. $\log_x 8 = 3$

14. $\log_x 8 = \frac{3}{2}$

15. $\log_x 3 = 0.2$

16. $\log_x 3 = -0.2$

17. $\log_8 (-8) = x$

18. $\log_8 (8) = x$

19. $\log_8 (x) = 2$

20. $\log_8 (-x) = 2$

21–32. Solve for x using a calculator.

21. $\log 329.24 = x$

22. $\log 0.032924 = x$

23. $\log (3.29 \times 10^{24}) = x$

24. $\log (3.29 \times 10^{-24}) = x$

25. $\ln 329{,}240 = x$

26. $\ln (-0.003294) = x$

27. $\ln (3.29 \times 10^{24}) = x$

28. $\ln (3.29 \times 10^{-24}) = x$

29. $\log (3.29 \times 10^{241}) = x$

30. $\ln (3.29 \times 10^{240}) = x$

31. $\ln (3.29 \times 10^{-101}) = x$

32. $\log (3.29 \times 10^{-101}) = x$

33–50. For each of the functions, **(a)** roughly sketch its graph, making use of the graphing principles; **(b)** find, if possible, the x- and y-intercepts; and **(c)** find the domain, D, and range, R. Then use a graphing calculator.

33. $y = \log x$

34. $y = \ln x$

35. $y = 4 + \log x$

36. $y = \ln x + e$

37. $y = 2 - \log x$

38. $y = 3 - \ln x$

39. $y = \log (x - 1)$

40. $y = \ln (x + 2)$

41. $y = \log (3 - x)$

42. $y = \ln (4 - x)$

43. $y = 1 - 2 \log x$

44. $y = 2 \ln x - 5$

45. $y = \log x^2$

46. $y = \ln x^3$

47. $y = \log \sqrt{x}$

48. $y = \ln \sqrt[3]{x}$

49. $y = \log |x|$

50. $y = \ln |x - 2|$

51–54. Find all real x satisfying the given inequality.

51. $-2 < \log x < -1$

52. $e < \ln x < 10$

53. $3 \le \log x \le 4$

54. $-2 \le \ln x \le -1$

II. Concepts

55 and 56. The definition of $L(x) = \log_b x$ required that $b > 1$.

55. Why can't $b = 1$ be a base?

56. Why can't a negative number be a base?

57–59. Mark each of the following statements AT (always true), ST (sometimes true), or NT (never true). If a statement is ST, give an example of when it is true and of when it is false.

57. The graph of $L(x) = \log_b (x)$ contains the point $(1, 0)$.

58. $\log (c + d) = \log c + \log d$ for real numbers c and d.

59. $\log (cd) = (\log c)(\log d)$ for real numbers c and d.

60. **(a)** Can the logarithm of an irrational number ever be rational?

(b) Can the logarithm of a rational number ever be irrational?

61. Find the flaw in the following "proof":

$$2 > 1$$

$$2 \log \frac{1}{2} > \log \frac{1}{2}$$

$$\log \left(\frac{1}{2}\right)^2 > \log \left(\frac{1}{2}\right)$$

$$\left(\frac{1}{2}\right)^2 > \frac{1}{2}$$

$$\frac{1}{4} > \frac{1}{2}$$

62. Solve for x algebraically:

$$(\log x)^2 - 3 \log x + 2 = 0.$$

Then solve by graphing for base 10 and base e.

63. Let $a_1, a_2, \ldots, a_n$ be a geometric sequence. What type of sequence is $\log a_1, \log a_2, \ldots, \log a_n$? Why?

64. Using a graphing calculator, graph $y = \log x$ and $y = \ln x$ on the same set of axes. For what values of x is $\log x > \ln x$?

65. Explain in your own words why $E(x) = b^x$ and $L(x) = \log_b x$ are "inverses" of each other. Include graphs in your explanation.

III. Applications

Solve each of the following problems.

66. The acidity or alkalinity of any solution is determined by the concentration of hydrogen ions $[H^+]$ in the substance. It is measured on a pH scale, using the formula $\text{pH} = -\log [H^+]$. A neutral solution, such as distilled water, has a pH of 7. Acids have pH < 7, bases have pH > 7.

(a) Find the pH of rain ($[H^+] = 6.31 \cdot 10^{-7}$).

(b) Find the pH of vinegar ($[H^+] = 1.26 \cdot 10^{-3}$).

(c) Tomatoes have $[H^+] = 6.3 \cdot 10^{-5}$. Are they acidic or alkaline?

67. Find the number of decibels (see Example 18) of

(a) normal conversation ($I = 3.16 \times 10^{-10}$ watts/cm^2).

(b) a rock concert ($I = 5.23 \times 10^{-6}$ watts/cm^2).

68. If the power of a sound is

(a) tripled,

(b) increased by 30%,

then what is the difference in the sound levels?

69. Two tones of the same frequency register 45 decibels and 60 decibels, respectively, on a meter. What is the ratio of their actual physical intensities?

70. An earthquake in India during 1897 registered 8.7 on the Richter scale (see Example 19). How many times stronger than

(a) the San Francisco quake in 1906

(b) an earthquake that begins to cause serious damage

was the Indian earthquake?

71. Stars have been classified into magnitudes according to their brightness. Stars in the first six magnitudes are visible with the naked eye, but those of higher magnitudes (which are fainter) require telescopes. The function

$$M = M(d) = 8.8 + 5.1 \log d$$

gives the relationship between the magnitude M of the faintest star visible with a telescope whose lens diameter is d inches. What is the highest magnitude of a star visible with a

(a) 6-inch home telescope?

(b) 200-inch Hale telescope at Mount Palomar, California?

72. An earthquake in San Francisco in 1989 registered 7.1 on the Richter scale.

(a) How many times stronger is this quake than one that causes serious damage?

(b) If an earthquake were twice as intense as this one, what would it measure on the Richter scale?

73. The basic unit of brightness of a star is its magnitude. The magnitudes of two stars are related to their intensities by the equation

$$m_1 - m_2 = 2.5 \log \frac{I_2}{I_1}.$$

The magnitude of a full moon is -13, while the magnitude of Venus is -4.6.

(a) Which object is brighter?

(b) What is the ratio of the intensities (brightness) of the two objects?

IV. Extensions

74. Prove the following properties of logarithms:

(a) Property 2 **(b)** Property 3

(*Hint:* Use properties of exponents.)

75. By experimenting with your calculator, find the smallest x for which the sequence x [ln] [ln] [ln] does not result in "error."

76–79. Find the *number* of solutions to the following equations. Then estimate the solutions using a graphing calculator.

76. $2^{-x} = \ln x$

77. $e^{-x} = \ln x$

78. $x + \log x = 0$

79. $\log x = x^2 - 2$

80. Find all values of a and b for which $\log (a + b) = \log a + \log b$. Justify your answers.

81. Find all possible values for the product $\log_b a \cdot \log_a b$, where $a > 1$, $b > 1$. Justify your answers.

82. Since decibels are logarithmic, they cannot be added directly. In fact, to "add two sounds," we use the formula

$$IL(\text{total}) = 10 \log \left(10^{\frac{IL_1}{10}} + 10^{\frac{IL_2}{10}}\right)$$

where IL_j = intensity of sound j.

(a) Find the combined intensity, in decibels, of two office machines, one generating 70 dB and the other generating 76 dB.

(b) For most calculations, however, engineers use a rule of thumb that says if the difference in intensities of two sounds is 6 dB, then we add 1 to the larger intensity to determine the total intensity of the two sounds. Prove this rule of thumb.

83. It is possible to define the logarithmic function for $0 < b < 1$. Compare the graphs of $y = \log_2 x$ and $y = \log_{\frac{1}{2}} x$ on the same coordinate system. What happens? Try a few more such examples and then answer the question.

9.3 EXPONENTIAL AND LOGARITHMIC EQUATIONS

There are many applications that require the solution of an equation involving a logarithmic or exponential function.

Antilogs: Inverse Logarithms

EXAMPLE 1 Estimate $3^{1,001}$.

Solution

One way to estimate the size of a large number, say $3^{1,001}$, is to express it in scientific notation. The powers of 10 are used as "benchmarks" in some sense. Suppose

$$3^{1,001} = m \cdot 10^c, \text{ where } 1 < m < 10 \qquad m \cdot 10^c \text{ is scientific notation.}$$

Then

$$\begin{aligned} \log 3^{1,001} &= \log (m \cdot 10^c) \\ 1{,}001 \log 3 &= \log m + c \qquad \text{Property 3, Property 1.} \\ 477.598 &= \log m + c \end{aligned}$$

Now c is a positive integer, and $0 < \log m < 1$, since $1 < m < 10$. Consequently, we must have

$$\log m = 0.598 \quad \text{and} \quad c = 477$$

Thus

$$3^{1,001} = m \cdot 10^{477}$$

We now need to solve the equation

$$\log m = 0.598$$

to finish the problem. The solution m to this equation is called the **antilog** or inverse log of 0.598.

You can compute the antilog m on a calculator using the sequence [INV] [log] or the [10^x] key. Thus antilog $0.598 \approx 3.963$ and $3^{1,001} \approx 3.963 \cdot 10^{477}$, a number with 478 digits to the left of the decimal point. The [10^x] key is used since, by definition, $\log m = 0.598$ is equivalent to $m = 10^{0.598}$. ■

EXAMPLE 2 Solve for x using a calculator.

(a) $\log x = 0.6789$ **(b)** $\log x = 2.6789$ **(c)** $\log x = -3.6789$

(d) $\ln x = 0.6789$ **(e)** $\ln x = 2.6789$ **(f)** $\ln x = -3.6789$

(g) $\log (3x + 1) = 4.6789$ **(h)** $\ln x^2 = -2.6789$

Solutions

(a) $x = 4.7742$ (a)–(f): In each case, we compute antilog x.

(b) $x = 477.42$

(c) $x = 0.0002$

(d) $x = 1.9717$

(e) $x = 14.5690$

(f) $x = 0.0253$

(g)
$$\begin{aligned} 3x + 1 &\approx 47{,}741.93 \\ 3x &\approx 47{,}740.93 \\ x &\approx 15{,}913.64 \end{aligned}$$

(h)
$$\begin{aligned} x^2 &= 0.0686 \\ x &\approx \pm 0.2620 \end{aligned}$$

or $\ln x^2 = 2 \ln x$ Property 3.

$$\begin{aligned} 2 \ln x &= -2.6789 \\ \ln x &= -1.33945 \\ x &\approx 0.2620 \end{aligned}$$

Since Property 3 is valid only when $x > 0$, we must realize -0.2620 is also a solution. ■

NOTE: Each equation in Example 2 could be solved by locating the point on the graph of the appropriate logarithmic function. For example, in part (h) we could use the graph $y = \ln x^2$ (see Figure 9.21). ■

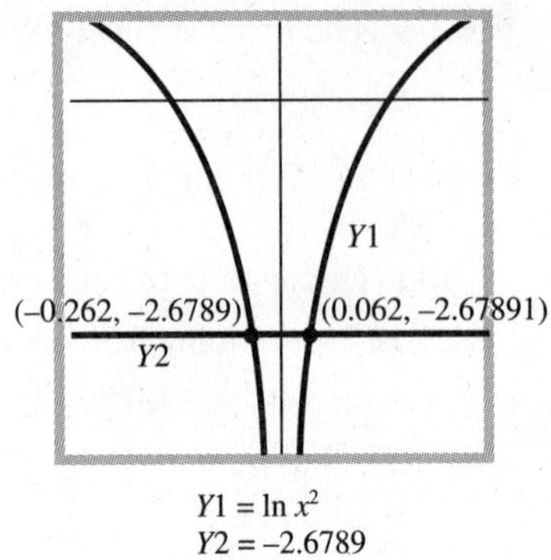

$Y1 = \ln x^2$
$Y2 = -2.6789$

Figure 9.21

Other Exponential and Logarithmic Equations

The next example should seem familiar by now.

EXAMPLE 3

The world's population in 1975 was about 4 billion and doubling every 35 years. Assuming this rate of increase continues, estimate the year in which the world's population will reach

(a) 32 billion. **(b)** 37 billion.

Solutions

If $P(x)$ denotes the population in the year x, then $P(x) = 4 \cdot 2^{\frac{(x-1975)}{35}}$.

(a)

$$32 = 4 \cdot 2^{\frac{(x-1975)}{35}}$$

We seek the value of x for which $P(x) = 32$ (see Section 7.4).

$$8 = 2^{\frac{(x-1975)}{35}}$$

$$2^3 = 2^{\frac{(x-1975)}{35}}$$

$$3 = \frac{x - 1975}{35}$$

$$2{,}080 = x$$

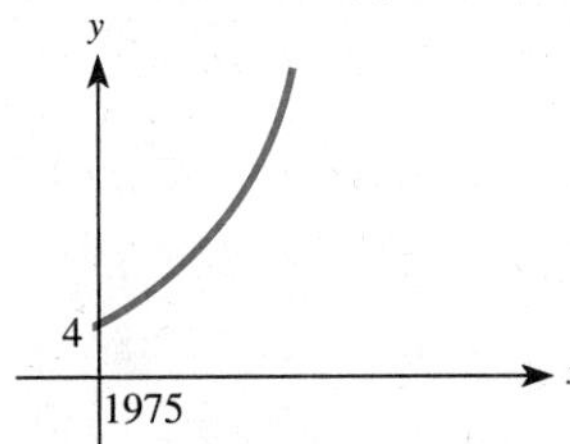

Figure 9.22

(b)

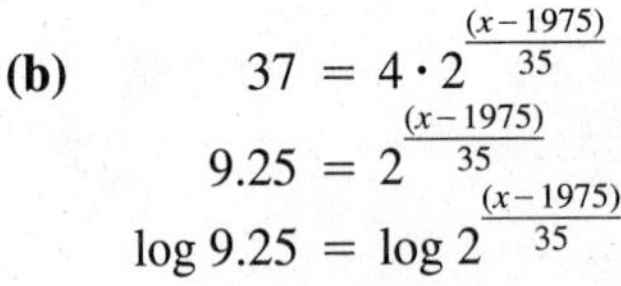

$$37 = 4 \cdot 2^{\frac{(x-1975)}{35}}$$

$$9.25 = 2^{\frac{(x-1975)}{35}}$$

First divide both sides by 4.

$$\log 9.25 = \log 2^{\frac{(x-1975)}{35}}$$

9.25 is not an integral power of 2, so we take the log of both sides of the equation, that is if $a = b$, then $\log a = \log b$.

$$\log 9.25 = \left(\frac{x - 1975}{35}\right) \log 2$$

Property 3 of logs (see Section 9.2).

$$\frac{x - 1975}{35} = \frac{\log 9.25}{\log 2} \approx 3.21$$

$$x \approx 2{,}087.33$$

Therefore, the population will reach 37 billion around April 1, 2087. ■

As the solution to Example 3(b) indicates, the basic strategy in solving an exponential equation is to "take the log" of both sides of the equation. It makes no difference if we use common logs or natural logs.

EXAMPLES 4–5

Solve for x.

4. $2^x = 750$ **5.** $2^{3x+1} = 5^{2x-7}$

Solutions

4. Using an Equation

$$\begin{aligned} 2^x &= 750 \\ \log 2^x &= \log 750 \\ x \log 2 &= \log 750 \\ x &= \frac{\log 750}{\log 2} \approx \frac{2.875}{0.301} \\ x &\approx 9.55 \end{aligned}$$

See Example 1, Section 9.1. Take the (common) log of both sides of the equation. Property 3 for logs.

or

$$\begin{aligned} 2^x &= 750 \\ \ln 2^x &= \ln 750 \\ x \ln 2 &= \ln 750 \\ x &= \frac{\ln 750}{\ln 2} \approx \frac{6.62}{0.693} \\ x &\approx 9.55 \end{aligned}$$

Take the (natural) log of both sides of the equation. Property 3 for logs.

Using a Graph

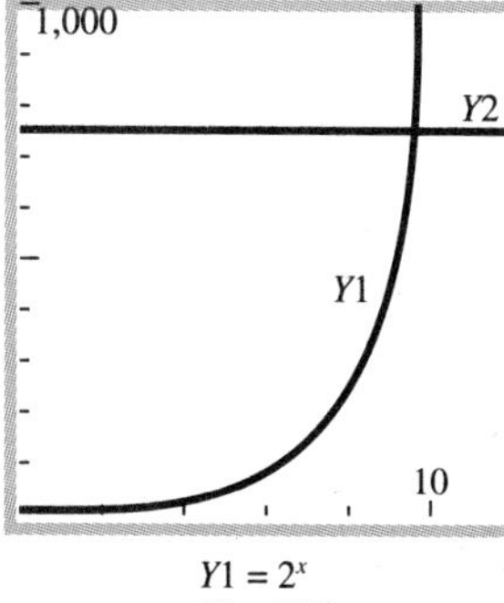

$Y1 = 2^x$
$Y2 = 750$

Figure 9.23

Find the x-coordinate of the point of intersection of the graphs.

5. Using an Equation

$$\begin{aligned} 2^{3x+1} &= 5^{2x-7} \\ \log 2^{3x+1} &= \log 5^{2x-7} \end{aligned}$$

Take the log of both sides. Property 3.

$$(3x + 1) \log 2 = (2x - 7) \log 5$$

$$x(3 \log 2 - 2 \log 5) = -7 \log 5 - \log 2$$ Distributive law and simplify.

$$x = \frac{-7 \log 5 - \log 2}{3 \log 2 - 2 \log 5}$$

$$x = \frac{7 \log 5 + \log 2}{2 \log 5 - 3 \log 2}$$ We prefer to use as many plus signs as possible.

$$x \approx 10.496$$

Using a Graph

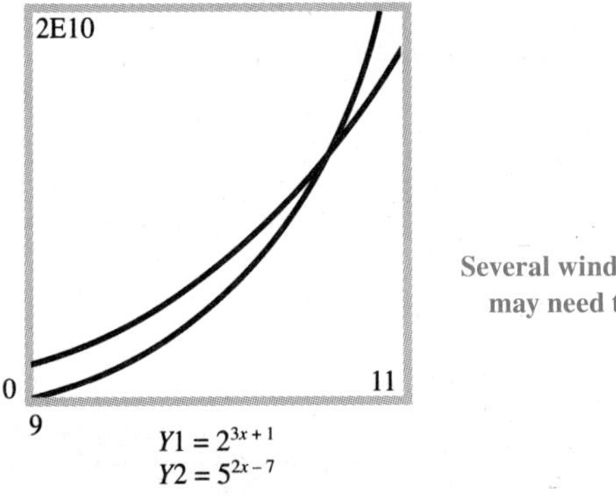

Several windows may need to be tried.

Figure 9.24

Now we turn to some applications of exponential and logarithmic equations.

EXAMPLE 6

Translucent materials have the property of reducing the intensity of light passing through them. A 1-mm-thick sheet of a certain translucent plastic reduces the intensity of light by 8%. How many such sheets are necessary to reduce the intensity of a beam of light to 25% of its original value?

Solution

Using an Equation Each sheet of plastic reduces the intensity of the light to 92% of its original value. Thus

$$(0.92)^n = 0.25$$

$$\log (0.92)^n = \log (0.25)$$

$$n \log (0.92) = \log (0.25)$$

$$n = \frac{\log (0.25)}{\log (0.92)} \approx 16.62$$

Therefore 17 sheets will do.

Using a Graph

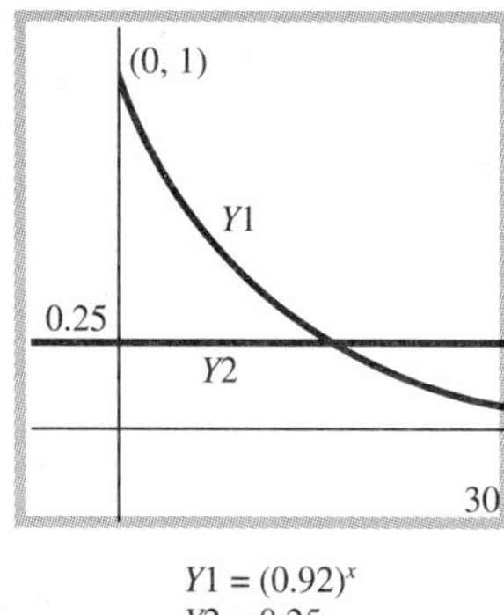

$Y1 = (0.92)^x$
$Y2 = 0.25$

Figure 9.25

EXAMPLE 7

How long will it take to double your money if it is invested at 8% interest compounded continuously?

Solution

$$A_t = Pe^{It}$$

Therefore

$$2P = Pe^{0.08t} \qquad \text{Doubling your money means } A_t = 2P.$$

$$2 = e^{0.08t} \qquad \text{First divide by } P.$$

$$\ln 2 = \ln e^{0.08t} \qquad \text{Take the (natural) log of both sides.}$$

$$\ln 2 = 0.08t \qquad \text{Property 3 of logs.}$$

$$t = \frac{\ln 2}{0.08} \approx 8.66 \text{ years}$$

In many long-term accounts, you may withdraw money only at six-month intervals without an interest penalty. Thus you could withdraw your money in 8.5 years and receive slightly less than twice your principal, or wait until 9 years have passed and receive slightly more.

Finding a Growth or Decay Function

We now consider the problem of how to find a growth or decay function that describes a certain situation. We have considered this problem in earlier chapters for other types of functions:

Chapter 4: Finding a line (linear function); finding a "fitted line" whose graph "comes close" to containing several points.

Chapter 6: Finding a parabola (quadratic function) whose graph contains three points.

Chapter 8: Finding a function to describe a situation wherein two quantities are directly proportional or inversely proportional; finding a quadratic model.

Finding a growth (or decay) function $B = B(t) = B_0 b^{kt}$ essentially means finding the value of the constant k. The base b we use is really immaterial, as we shall see in the next example.

EXAMPLE 8 The number of bacteria in a culture grows from 100 to 400 in 12 hours.

(a) Find a growth function describing this situation.

(b) How many bacteria will be present in 18 hours?

Solutions

We know $B = B(t) = B_0 b^{kt}$, $B_0 = 100$, and $B(12) = 400$. Here are three different approaches to the problem.

1. **Base 2:** Suppose $b = 2$. Then

$$\begin{aligned} 400 &= 100 \cdot 2^{12k} \\ 4 &= 2^{12k} \\ 2^2 &= 2^{12k} \\ 2 &= 12^k \\ k &= \frac{1}{6} \end{aligned}$$

Therefore

$$B = B(t) = 100 \cdot 2^{\frac{t}{6}}$$

For part (b), $B(18) = 100 \cdot 2^{\frac{18}{6}} = 800$ bacteria

2. **Solve for the quantity b^k**

$$\begin{aligned} 400 &= 100(b^k)^{12} \\ 4 &= (b^k)^{12} \\ 4^{\frac{1}{12}} &= \sqrt[12]{4} = b^k \\ b^k &= 1.122462 \end{aligned}$$

Thus

$$B = B(t) = 100 \cdot (1.122462)^t$$

Therefore

$$B(18) = 100 \cdot (1.122462)^{18} = 800$$

3. **Base e:** Suppose $b = e$. Then

$$\begin{aligned} 400 &= 100e^{12k} \\ 4 &= e^{12k} \\ \ln 4 &= 12k \\ k &= \frac{\ln 4}{12} \approx 0.11552453 \end{aligned}$$

Therefore

$$B = B(t) = 100 \cdot e^{0.11552453t}$$

For part (b),

$$B(18) = 100 \cdot e^{0.11552453(18)} \approx 800 \quad ■$$

The functions $B(t) = 100 \cdot 2^{\frac{t}{6}}$, $B(t) = 100 \cdot (1.122462)^t$ and $B(t) = 100e^{0.11552453t}$ are, in fact, *three names for the same function,* because

$$2^{\frac{1}{6}} \approx e^{0.11552453} \approx 1.122462.$$

You can also sketch a graph of the three functions. Thus the base we select in trying to determine a growth or decay function does not matter. We will usually choose the natural base e.

EXAMPLE 9

A radioactive substance has a half-life of 138 days. How long will it take for 90% of a sample of this substance to decay?

Solution

$B = B(t) = B_0e^{-kt}$ — Decay function; choose base e. $B(138) = 0.5B_0$ (half-life); find k.

$0.5B_0 = B_0e^{-138k}$

$0.5 = e^{-138k}$ — First divide by B_0.

$\ln 0.5 = -138k$

$k = \dfrac{\ln 0.5}{-138} \approx 0.005$

$B = B(t) = B_0e^{-0.005t}$

$0.1B_0 = B_0e^{-0.005t}$ — We seek the value of t for which $B(t) = 0.1B_0$ (90% decayed means 10% left). Solve for t.

$0.1 = e^{-0.005t}$

$\ln (0.1) = -0.005t$

$t = \dfrac{\ln (0.1)}{-0.005} \approx 460.5$ days ■

EXAMPLE 10

The population of Dallas, Texas, is given by the following table

Year	*Population*
1970	844,000
1980	905,000
1990	1,007,000

(a) Find a growth function that describes the increase from 1970 to 1980.

(b) Was this function an accurate prediction of the population in 1990?

Solutions

(a)

$$P = P(t) = P_0 e^{kt}$$

$$905{,}000 = 844{,}000e^{10k}$$ **$t = 0$ represents the year 1970, so $P_0 = 844{,}000$.**

$$\frac{905}{844} = e^{10k}$$ **$t = 10$ represents the year 1980.**

$$\ln\left(\frac{905}{844}\right) = 10k$$

$$k = \frac{\ln\left(\frac{905}{844}\right)}{10} \approx 0.007$$

Therefore

$$P = P(t) = 844{,}000e^{0.007t}$$

(b) $P(20) = 844{,}000e^{0.007(20)}$ **$P(20)$ = predicted population for 1990.**

$\approx 971{,}000$

Therefore the population of Dallas increased at a faster rate from 1980 to 1990 than from 1970 to 1980.

9.3 EXERCISES

I. Procedures

1–44. Solve for x. Use a calculator when appropriate.

1. $\log x = 0.3741$

2. $\log x = 2.3741$

3. $\log x = -1.3471$

4. $\ln x = 0.3741$

5. $\ln x = 2.3471$

6. $\ln x = -1.3471$

7. $\log(3x + 1) = 2.3471$

8. $\log(1 - 2x) = 0.3741$

9. $\ln(1 + 3x) = -1.3471$

10. $\ln x = -0.0374$

11. $\log x^2 = 3$

12. $\ln 2^{x^2} = 4$

13. $\log(x^2 + 2x) = 1$

14. $\ln(x^2 + x) = 1$

15. $\log(2x - x^2) = 1$

16. $\ln(x - x^2) = 1$

17. $3^x = 50$

18. $3^{-x} = 50$

19. $3^x = -50$

20. $3^{-x} = -50$

21. $7^x = 100$

22. $7^{-x} = 100$

23. $3^{2x} = 4$

24. $5^{3x} = 2$

25. $(1.12)^x = 20$

26. $(2.04)^x = 37$

27. $(0.97)^x = 0.5$

28. $(0.97)^x = 1.5$

29. $e^x = 4$

30. $e^{-x} = 4$

31. $e^{x^2} = 5$

32. $e^{-x^2} = 5$

33. $2^x = 3^{x-1}$

34. $5^{2x-3} = 7^{3x-4}$

35. $2^{x+2} = 7^{x-2}$

36. $e^{3x-1} = 2^{4x}$

37. $2^x \cdot 5 = 10^x$

38. $3^x \cdot 9 = 27^{x-1}$

39. $2^{x^2} \cdot 4 = 8^{3x}$

40. $e^x \cdot e^{-x^2} = 4e^{3x}$

41. $e^x = e^{-x}$

42. $e^{x^2} = e^{-x^2}$

43. $2^x = e^{0.6931x}$

44. $10^x = 3^{2.096x}$

45 and 46. Solve for t.

45. $I = \frac{E}{R}(1 - e^{-\frac{Rt}{L}})$

46. $S = a\left(\frac{1 - r^t}{1 - r}\right)$

II. Concepts

47 and 48. Use graphs as part of your explanation.

47. Explain why the growth functions $B(t) = B_0 \cdot 3^{\frac{t}{5}}$ and $E(t) = E_0 e^{0.2197t}$ are essentially the same.

48. Find k so that the decay functions $B(t) = B_0 2^{-\frac{t}{10}}$ and $E(t) = E_0 e^{-kt}$ are essentially the same.

49–51. Which number is larger?

49. 300^{301} or 301^{300}?

50. $e^{1,000}$ or 10^{434}?

51. n^{n+1} or $(n + 1)^n$?

52. Write a paragraph comparing the exponential function and the logarithmic function. Use graphs as part of your explanation.

53. Use the properties of logarithms to explain why $d = 100c$ if $\log c = 0.6789$ and $\log d = 2.6789$.

III. Applications

54–76. Solve each of the following problems.

54. How long will it take you to double your money if it is invested at 9% interest compounded continuously?

55. If the cost of an item increases at the same rate as the inflation rate, then how long will it take the cost of the item to triple if the annual inflation rate is

(a) 5%? **(b)** 8%? **(c)** 12%? **(d)** 15%?

56. The cost of some items has decreased over the years. One such item is a calculator.

(a) Suppose the price of a certain model is \$109.95. If the price decreases by 23% a year, how long will it take before the price reaches \$39.95?

(b) The price of a basic scientific calculator was \$125.00 during 1974, and the price of (essentially) the same model was \$14.95 during 1979 and later. By what percentage did the price decrease, on the average, per year from 1974 to 1979?

57. To oversimplify, suppose a car depreciates at an annual rate of 18%. How long does it take for a car to lose one-half of its value?

58. The population of Smallville has increased by 8% per year for the past few years. Its population is currently 35,560. If the population continues to increase at the same rate, when will it reach 60,000?

59. The price of milk is \$1.95 per half gallon. If the price of milk goes up at an annual rate of 5%, how long will it take the price to reach \$2.75 per half-gallon?

60. The world's population in 1990 was approximately 5.25 billion and doubling every 40 years. Assuming this rate continues, estimate the year in which the population will reach

(a) 32 billion **(b)** 37 billion

61. One-quarter of an acre of land is required to provide food for one person. The world contains 10 billion acres of arable land. If we assume the population continues to increase at 1.6% per year, the population t years after 1975 is given by $P(t) = 4 \cdot e^{0.016t}$. In what year will the world reach its practical maximum of 40 billion people? Rework this problem using the data in Exercise 60.

62. How long would it take \$1 invested at 1% interest compounded continuously to earn \$1,000,000?

63. Another way to express the decay function for carbon-14 used in archeological dating is

$$P(t) = P_0 e^{-\frac{t \ln 2}{5,570}}$$

In charcoal found in Java, the amount of carbon-14 was $\frac{1}{16}$ that of a contemporary living sample. Estimate the age of the charcoal.

64. Cesium-137, a dangerous radionuclide produced by the Chernobyl disaster of 1986, has a half-life of 30.3 years. In approximately what year will the amount of radiation be reduced to 10% of that present at the time of the disaster?

65. If the temperature T at time t of a steel ingot cooling from 350°C to room temperature of 15°C can be modeled by $T(t) = 15 + 350e^{-kt}$, where $k = 0.31$,

(a) Sketch its graph.

(b) Find the temperature of the ingot at $t = 3.0$ hours.

(c) Find the temperature of the ingot at $t = 10.0$ hours.

(d) When does the ingot reach 20°C?

(e) When does the ingot reach room temperature?

66. Many people feel the trade-in value of a car decreases 30% each year.

(a) Suppose you now own a car with a trade-in value of \$8,000. How much will it be worth three years from now?

(b) In how many years will the trade-in value of this car be \$1,000?

(c) If the car is now two years old, what was its trade-in value when it was new?

(d) The car cost $18,000 when it was new. Why is there a difference between this number and your answer to part (c)?

67. The distances between markings on an AM radio dial are not uniform. Assume the frequency varies exponentially with its distance from the left end of the dial, and the frequencies are between 53 and 162 kilohertz.

(a) Find an appropriate model for this situation, assuming the dial is 12 cm long.

(b) How can you use your model to calculate the locations of various frequencies on the dial?

(c) Calculate the location of several of your favorite stations on the dial.

(d) Check the accuracy of this model by checking your answers to part (c) on an actual radio dial.

68. What is the largest power of the following quantities that can be displayed on your calculator?

(a) 3 (b) e (c) 12 (d) 237

69. How many digits are in the large prime $2^{756,839} - 1$?

70. A growth law is given by $y = y_0e^{kt}$, where t is in years. Find k if this quantity

(a) doubles every ten years.

(b) increases 4% per year.

(c) increases 23% per year.

71. A decay law is given by $y = y_0e^{-kt}$, where t is in years. Find k if this quantity

(a) decreases 3% per year.

(b) decreases 18% per year.

(c) is half gone after seven years.

72. The populations of Cleveland, Ohio, and Phoenix, Arizona, are given in the following table:

	1970	*1980*	*1990*
Cleveland	751,000	574,000	506,000
Phoenix	584,000	790,000	983,000

(a) In each case, find a function that describes the population from 1970 to 1980.

(b) Was this function an accurate prediction of the population during 1990?

73. In a new wildlife preserve, it is estimated that fuzzy rabbits will triple every 3.7 years and pink-eared rabbits will quadruple every 4.1 years. If initially there were 46 fuzzy rabbits and 24 pink-eared rabbits, how long would it take for there to be the same number of each of these two types of rabbits?

74. The number of bacteria in a culture grew from 100 to 300 in 16 hours.

(a) Find a growth function describing this situation.

(b) How many bacteria will be present in 30 hours?

75. Radium has a half-life of 1,620 years. How long will it take for 80% of a radium sample to decay?

76. Phosphate compounds act as a fertilizer for algae, and these compounds are responsible for the exponential growth of algae in lakes and streams. Suppose the number of algae increases from 10 to 20 in 2 hours. Estimate the number of algae in the lake in 24 hours.

IV. Extensions

77. According to a rule of thumb, money invested at an annual percentage rate of r doubles in $\frac{72}{r}$ years (compounded annually). For what values of r is this rule approximately correct?

78. Some bankers feel that the Rule of 72 (see Exercise 77) is not accurate enough. For what other values of N near 72 do you think a Rule of N would be more accurate? Give a convincing argument for your answer.

79. If $a > 0$ and $b > 0$, show there exists a positive constant k for which $b^x = a^{kx}$.

80. Estimate all solutions to the following equations.

(a) $2^x + 3^x = 15$ (b) $x^x = 100$

CHAPTER 9 SUMMARY

New questions such as "How long will it take a certain colony of bacteria to double?" motivated a need for an efficient method to calculate exponents. The properties, the graphs, and the equations of two inverse functions—the exponential function and the logarithmic function—were studied. Exponential growth and decay was the principal example. Many of the concepts were explored with a graphing calculator.

Important Words and Phrases

exponential function	common logarithm
logarithmic function	natural logarithm
exponential growth	antilog
exponential decay	inverse function
the natural base	geometric sequence

Important Properties and Procedures

- Graphing exponential and logarithmic functions
- Evaluating exponential and logarithmic functions
- Properties of logarithms: $\log_b (rs) = \log_b r + \log_b s$

  $$\log_b \left(\frac{r}{s}\right) = \log_b r - \log_b s$$

  $$\log_b (r^s) = s \log_b r$$

  where $b > 1$ and r and s are positive real numbers.
- Calculating with logarithms
- Solving exponential and logarithmic equations
- Finding a growth or decay function

REVIEW EXERCISES

I. Procedures

1–8. Solve for x without using a calculator.

1. $\log_2 128 = x$ **2.** $\log_8 16 = x$

3. $\log_4 x = 3$ **4.** $\log_4 x = 2.5$

5. $\log_x 64 = 3$ **6.** $\log_x \left(\frac{4}{9}\right) = -2$

7. $\log_{17} 17^{10} = x$ **8.** $10^{\log_{10} 5} = x$

9–14. Calculate each of the following expressions with a calculator.

9. $4^{3.6}$ **10.** $e^{-1.7}$ **11.** $\log 432.67$

12. $\log 0.043267$ **13.** $\ln 0.5736$ **14.** $\ln 573.6$

15–20. For each of the functions, **(a)** sketch its graph, making use of the graphing principles; **(b)** find, if possible, the x- and y-intercepts; and **(c)** find the domain, D, and range, R.

15. $y = 3^x - 2$ **16.** $y = e^{-x} + 1$

17. $y = \log x - 1$ **18.** $y = \ln (2x)$

19. $y = \ln (x - 3)$ **20.** $y = 2^{x+3} - 1$

21–30. Solve for x. Use a calculator when appropriate.

21. $\log x = 0.1257$ **22.** $\ln x = -0.3458$

23. $\ln x = 2.3584$ **24.** $\log x = -3.7612$

25. $4^x = 100$ **26.** $e^{2x+3} = 100$

27. $3(1.09)^{4x} = 19$ **28.** $2^{3x} = e^{x+1}$

29. $5^{3x} = 7^{2x+1}$ **30.** $3^x 9^{2x} = 27^{3x}$

31–34. How many solutions do each of the following equations have? Estimate the solutions using a graphing calculator.

31. $3^x = 6x$ **32.** $e^{-x} = 1 - x^2$

33. $\log x = 3^{-x}$ **34.** $e^x = x^4$

II. Concepts

35. Suppose you wish to plot $y = \log x$ on graph paper, with a scale of 1 unit $= \frac{1}{16}$ inch. Find the approximate length and width of the graph needed to plot this graph for $0.5 < x < 100$.

36. Find k so that the growth functions $B(t) = 2^{\frac{t}{7}}$ and $E(t) = e^{kt}$ are the same.

37–40. Give, if possible, a function that satisfies the given condition.

37. $f(x + y) = f(x)f(y)$ for all positive real numbers x and y.

38. $f(xy) = f(x) + f(y)$ for all positive real numbers x and y.

39. $f(x + y) = f(x) + f(y)$ for all positive real numbers x and y.

40. $f(xy) = f(x)f(y)$ for all positive real numbers x and y.

41. Does the difference between the logarithms of two consecutive natural numbers decrease as the numbers decrease? Justify your answer.

III. Applications

Solve each of the following problems.

42. The pH of a substance, a measure of its acidity or alkalinity, is given by $\text{pH} = -\log [H^+]$, where $[H^+]$ is the concentration of hydrogen ions in the substance. Find the hydrogen ion concentration $[H^+]$ of a skin cleanser with a pH of 5.5.

43. Levels of sound above 90 decibels are considered dangerous. A jet airliner takeoff has a power of $10^{12}I_0$ watts/cm^2. Is this level dangerous? (See Section 9.2.)

44. The cooking time for a turkey of weight w pounds is approximately $T = 3.25\left(\frac{w}{7}\right)^{\frac{2}{3}}$, where T is the time in hours. What is the weight of the largest turkey that could be cooked in four hours?

45. Suppose the population of Niceville was 50,000 during 1990, and it is increasing at a rate of 11% per year.

(a) Estimate the population in 2000.

(b) When will the population triple?

46. The population of Grandeville was 250,000 in 1990, and it decreases at a rate of 3% per year.

(a) Estimate the population in 2000.

(b) When will the population reach 150,000?

47. Growth and decay laws both have the form $y = y_0e^{kt}$, where k is positive for a growth law and negative for a decay law. Find k if a substance

(a) increases by 50% in five years.

(b) decreases by 30% in seven years.

48. A large prime number is $2^{19,937} - 1$.

(a) Find the number of digits in this large number.

(b) Find the number of digits in the product $2^{19,936}(2^{19,937} - 1)$. This is a perfect number.

49. The population of a bacteria sample grows by 5% in ten minutes.

(a) Find a growth law to describe this situation, where t is in minutes.

(b) How many bacteria will be present in two hours if there were 100 bacteria initially?

(c) How long will it take for the population of any sample of bacteria to double?

50. The air pressure of the earth's atmosphere decreases exponentially (obeys a decay law) with the altitude above the earth's surface. The air pressure is 14.7 pounds per square inch (psi) at sea level and 13.5 psi at 2,000 feet.

(a) Find the decay law for this situation.

(b) Sketch its graph.

(c) Predict the air pressure at the cruising altitude of army airplanes—35,000 feet.

(d) How far above the earth will the air pressure become negligible?

51. It is said that the island of Manhattan was purchased for \$24 in 1626.

(a) Suppose the \$24 had been invested in an account paying 6% APR. What would it be worth today?

(b) Investigate the worth today of some of the famous buildings in Manhattan.

52. How many digits are in the largest known (as of 1993) prime, $2^{756,839} - 1$?

53. You have two parents and four grandparents.

(a) How many great-grandparents do you have?

(b) Assume a new generation is born every 25 years. How many years ago were 1,000,000 of your ancestors alive?

(c) Assume a new generation is born every 25 years. How many years ago were 10,000,000 of your ancestors alive?

(d) The population of Britain did not reach 10,000,000 until well into the nineteenth century. If your ancestors came from Britain, how is this possible?

54. In a radioactive decay process, it takes 70 days for one-third of the original mass of a substance to disintegrate. What is the half-life of this substance?

55. When sound passes through a glass window, its intensity is reduced according to the formula

$$I_o - I_i = I_o(1 - 10^{-k}),$$

where I_o and I_i are the inside and outside intensities, respectively, and k is a constant. If the intensity of street traffic is 6.5×10^{-5} and of ordinary conversation is 5.0×10^{-6}, find the value of k that this window should have so the traffic noise is reduced inside the room to "conversation level."

56. The population of the United States was (approximately) 205 million in 1970, 225 million in 1980, and 249 million in 1990.

(a) Using the data for 1970 and 1980, find a growth model that expresses the population in terms of the number of years after 1970.

(b) Was the population for 1990 accurately predicted by the growth model in part (a)? If not, what was the percent error?

(c) According to the growth model in part (a), when would the U.S. population have reached 249 million?

CUMULATIVE REVIEW: CHAPTERS 1–9

I. Procedures

1–12. Perform any indicated operations and express all fractions that occur as answers in simplest form.

1. $6 - 2(4 - 2x) - [-(x - 2) + 7]$

2. $(x - 3)^2 - (x - 3)(x - 1)$

3. $(3x^{\frac{3}{2}} y^{\frac{5}{2}})(2x^{\frac{1}{2}} y^{-\frac{3}{4}})$

4. $\sqrt[3]{(8x^0 y^6)^3}\,(x^2 y)$

5. $\dfrac{16a^2 (b^3)^3 c}{2^3 a^{-3} b^7 c^0}$

6. $\dfrac{-96(x^2 y^{-1} z)^{-2}}{6^2 x^4 y^{-1} z}$

7. $\dfrac{\sqrt[5]{32x^5 y^6}}{2xy}$

8. $1 - \dfrac{x + 1}{x} - \dfrac{1}{x^2 - x}$

9. $\dfrac{2x^2 + 5x - 3}{x^2 - 9} \cdot \dfrac{(x - 3)^2}{4x^2 - 4x + 1}$

10. $\left(1 - \dfrac{1}{x}\right) \div \dfrac{x^2 - 2x + 1}{1 - x}$

11. $\dfrac{1}{x - 2} - \dfrac{1}{x} + \dfrac{x}{2 - x}$

12. $\left[\dfrac{x^2 - 7x + 6}{x^2 + 6x + 8} \cdot \dfrac{(x^2 + 3x - 4)}{x^2 - 36}\right] \div \dfrac{(x - 1)^2}{x + 2}$

13–21. Solve each equation, inequality, or system.

13. $\dfrac{x}{x - 1} - \dfrac{2}{x + 1} = 1$

14. $3 - \dfrac{2(x - 1)}{5} = \dfrac{x}{2}$

15. $x^2 - 5x = 10$

16. $(x - 3)(x - 2) = 6$

17. $\sqrt{8x + 1} - x = 2$

18. $\sqrt{2x - 6} + 3 = x$

19. $5 - 2(x + 1) \geq x + 1$

20. Solve for (x, y):

$$x + y = 10$$
$$3x - 2y = 5$$

21. Solve for (x, y):

$$y^2 + 2x = 3$$
$$x + 3y = 6$$

22–27. For each function, state the domain and range and sketch the graph.

22. $f(x) = \dfrac{1}{x - 1}$

23. $f(x) = x^2 - 2x + 5$

24. $f(x) = x^{\frac{2}{3}}$

25. $f(x) = (x + 1)^4 - 2$

26. $f(x) = 4^{x+1} - 2$

27. $y = \log(x - 2) - 3$

II. Concepts

28. **(a)** Solve the following problem in two ways:

   1. Change the problem to one that has only rational exponents, and then use the properties of rational exponents to express the answer in simplest form.

   2. Use the properties of radicals to express the answer in simplest form.

(b) Which method do you prefer? Why?

$$\frac{\sqrt[3]{x^3y^2z}}{\sqrt[3]{8xy^{-1}z^4}}$$

29. Suppose you are given the coordinates for three points in the plane. Explain how you could check whether the three points are vertices of

(a) an equilateral triangle.

(b) a right triangle.

30. A student claims that the following two expressions are equivalent. Is she correct? Why? How can you use a graph to help explain your answer?

$$\frac{x^2-4}{x-2}, \qquad \sqrt{(x+2)^2}$$

31. Give an example of a function $f(x)$ satisfying

(a) $f(0) = 1$ and $f(x) > 2$ for $x > 1$

(b) $2f(x) = f(2x)$ for all x.

32–36. The graph of $f(x)$ is given. Sketch the graph of each of the following functions.

32. $-f(x)$ **33.** $|f(x)|$ **34.** $f(x) + 3$

35. $f(x - 3)$ **36.** $3f(x)$

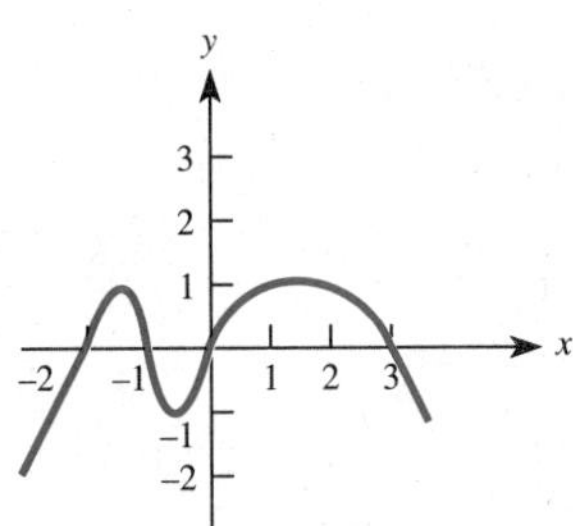

37. Write a short description of the important concepts and processes in the course. Give examples to support your discussion. Include a short discussion of how you perceive the value of the course for reaching your future goals.

38–46. Multiple choice. For each problem, select the best answer.

38. If a line has slope $-\frac{1}{2}$ and passes through the point (0, 2), then the equation of the line is

(a) $2y = 4x - 1$ **(b)** $2x = -y + 2$

(c) $x = 2y - 2$ **(d)** $2y = -x + 4$

(e) none of these

39. $16^{-\frac{3}{4}} =$

(a) -12 **(b)** 12 **(c)** -8 **(d)** $-\frac{1}{8}$ **(e)** $\frac{1}{8}$

40. The domain of the function $f(x) = \sqrt{x^2 + 1}$ is

(a) all real numbers $x \geq 0$

(b) all real numbers $x \leq 1$

(c) all real numbers $x \geq -1$

(d) all real numbers except $x = -1$

(e) all real numbers

41. $x^{-1} + x^{-2} =$

(a) $\dfrac{x^2}{x+1}$ **(b)** $\dfrac{x+1}{x^2}$ **(c)** x^{-3} **(d)** x^2

(e) none of these

42. If $x \geq 0$, then $\sqrt{x^3}$ is equivalent to all of the following *except*

(a) $x\sqrt{x}$ **(b)** $x \cdot x^{\frac{1}{2}}$ **(c)** $x^{\frac{3}{2}}$

(d) $x^{\frac{2}{3}}$ **(e)** $\sqrt{x \cdot x^2}$

43. The lines represented by the equations below

$$y = 2x + 3$$
$$2y = 4x - 1$$

(a) are perpendicular **(b)** are parallel

(c) coincide

(d) intersect at one point, but are not perpendicular

(e) none of these

44. If $m \neq 0$, then $\dfrac{m^{10} - m^2}{m^2} =$

(a) $m^{10} - 1$ **(b)** m^{10} **(c)** $m^5 - 1$

(d) $m^8 - 1$ **(e)** none of these

45. If $-2(2 + x) < x + 2$, then

(a) $x > -2$ **(b)** $x < 6$ **(c)** $x > 2$

(d) $x < -2$ **(e)** $x > 0$

46. If $x \neq 1$, then $\dfrac{x-1}{\sqrt{x}-1} =$

(a) $\sqrt{x}$ **(b)** $\sqrt{x+1}$ **(c)** $\sqrt{x-1}$ **(d)** $\dfrac{x-1}{x}$

(e) none of these

47–54. True or false: Determine whether each statement is true or false. Explain your answers.

47. If $f(x) = e^x + \log x$, then the domain of f is the set of all real numbers.

48. If $A \neq 0$, then $(A + 3)^2 = A^2 + 9$.

49. The line $2y - 3 = x$ is perpendicular to the line $2x - 3 = y$.

50. $\dfrac{6.26 \times 10^{-15}}{3.59 \times 10^{-12}} > 1$.

51. $27^{\frac{2}{3}} = 9$.

52. If $m \neq 0$, then $\dfrac{m}{\sqrt{m}} = \sqrt{m}$.

53. If $f(x) = x^2 - 3x - 1$, then $f(-1) = 3$.

54. If the graph of $y = 2x^2 - b$ passes through the origin, then $b = 2$.

55–57. Short answers.

55. For what values of c does the following equation have exactly one solution?

$$x^2 - 6x + c = 0$$

56. For what values of c is the following equation an identity?

$$2 - 3(5 - 3x) = c + 4x + 5(x + 1)$$

57. If $f(x) = \sqrt{x+1}$ and $g(x) = x^3$, then,

$$f(3) + g(2) = \underline{\hspace{3cm}}?$$

III. Applications

58. Ms. Smith invests \$100,000. She invests part of the money in bonds paying an average of $5\frac{1}{2}\%$ interest per year and the rest in mutual funds that pay an average of 8% interest per year. How much should she invest in each if she wants her total interest per year to be \$7,000?

59. The State College football stadium holds 80,000 people. A regular ticket costs \$15, while a student ticket costs \$7.50. If the total expenses for the game are \$375,000, what is the maximum number of student tickets that could be allotted to make a \$600,000 profit with a full house?

60. A balloon is being filled with air such that the volume of air in the balloon doubles every 4 seconds.

(a) If there are 10 cubic inches of air in the balloon now, how many cubic inches will there be 12 seconds from now?

(b) If the balloon bursts when there are 2,560 cubic inches of air in it, how long will it be before the balloon bursts? (Start with 10 cubic inches.)

61. Suppose it takes Sam twice as long to drive the 600 miles from Lansing to Calumet as it does for Pat to drive 250 miles from Lansing to Chicago. What is the average rate of speed for each person if Sam drives 10 miles per hour faster than Pat?

62. The ABU pep club wishes to sell Prune Bowl sweatshirts. It costs \$2,500 to print the shirts, including permission to use the ABU and Prune Bowl logos, advertising, etc. If the shirts cost \$8 and sell for \$15, how many must the pep club sell to make a profit equal to 30% of total costs?

63. A missile is launched from the ground. The height $h(t)$ in feet of the missile from the ground after t seconds is given by the equation

$$h(t) = 84t - 12t^2$$

(a) How high is the missile after one second?

(b) How many seconds after launching will it take the missile to hit the ground?

(c) At what time does the missile hit its maximum height? What is the maximum height?

64. The UDI car rental agency charges \$26 per day plus 21¢ a mile. The WTH car rental agency charges \$19 per day plus 27¢ a mile.

(a) What is charged by each agency and which car rental agency is the least expensive if you plan on driving 500 miles in one day?

(b) How many miles must be driven in one day for the two agencies' charges to be equal?

65. The length of a rectangle is 5 feet more than its width. If the area of the rectangle is 750 square feet, what are the dimensions of the rectangle?

66. Suppose you make \$20,000 a year and are offered the following choice for a pay raise:

- Plan 1: \$500 at the end of each six months, or
- Plan 2: \$1,000 at the end of each year.

Which is the better plan? Why? (*Hint:* how much money do you have at the start of the sixth year under each plan?)

67. The time $T(N)$, in minutes, necessary to learn a set of unknown symbols in a psychology test is $T(N) = 0.2N^{\frac{3}{4}}$, where N is the number of symbols memorized by an average adult.

(a) Find the time necessary to memorize a list of 16 symbols.

(b) How many symbols can be memorized in two minutes?

(c) Describe the graph for this function. What are a realistic domain and range for this problem?

68. A certain southern city is estimated to be tripling its population every 50 years. If the population is 100,000 and this trend continues,

(a) what will the population be in 25 years? In 100 years?

(b) In how many years will the population be 750,000?

69. A farmer wants to fence in a rectangular pasture with 3,600 meters of fencing. What is the largest area she can enclose with the fencing, and what are the dimensions of the rectangular pasture?

70. An open-top box (rectangular prism) is to be made from a rectangular sheet of paper by cutting the same size square from each corner and folding. If the dimensions of the rectangle are 15 cm and 20 cm, what is the maximum volume of box that can be made?

71. Examine the pattern below.

(a) Compute the product for each row.

If the pattern continues,

(b) write out the next two rows with their products.

(c) What is the product of the tenth row? Fiftieth row? Nth row?

(d) In what row is the product equal to $\frac{1}{100}$?

(e) Graph the pattern: Put the row number on the horizontal axis and the corresponding product on the vertical axis. Describe the graph.

(f) What is the first row in which the product is less than 0.005?

$$\left(1 - \tfrac{1}{2}\right) =$$

$$\left(1 - \tfrac{1}{2}\right)\left(1 - \tfrac{1}{3}\right) =$$

$$\left(1 - \tfrac{1}{2}\right)\left(1 - \tfrac{1}{3}\right)\left(1 - \tfrac{1}{4}\right) =$$

$$\left(1 - \tfrac{1}{2}\right)\left(1 - \tfrac{1}{3}\right)\left(1 - \tfrac{1}{4}\right)\left(1 - \tfrac{1}{5}\right) =$$

$$\left(1 - \tfrac{1}{2}\right)\left(1 - \tfrac{1}{3}\right)\left(1 - \tfrac{1}{4}\right)\left(1 - \tfrac{1}{5}\right)\left(1 - \tfrac{1}{6}\right) =$$

72. A table giving the elapsed time of play of a tape recorder, T, in terms of the counter reading, r, follows:

r	100	200	300	400	500	600	700	800
T (sec.)	205	430	676	943	1,231	1,540	1,870	2,221

(a) Find the quadratic polynomial that fits this data.

(b) Use this polynomial to give the elapsed time when the counter reads 455.

ANSWERS TO SELECTED EXERCISES

To the Student

If you need further help with intermediate algebra, you may want to obtain a copy of the *Student's Solutions Manual* that goes with this book. It contains solutions to all the odd-numbered exercises and all the cumulative review exercises. Your college bookstore either has this book or can order it for you.

In this section we provide the answers that we think most students will obtain when they work the exercises using the methods explained in the text. If your answer does not look exactly like the one given here, it is not necessarily wrong. In many cases there are equivalent forms of the answer that are correct. For example, if the answer section shows $\frac{3}{4}$ and your answer is 0.75, you have obtained the right answer but written it in a different (yet equivalent) form. Unless the directions specify otherwise, 0.75 is just as valid an answer as $\frac{3}{4}$.

In general, if your answer does not agree with the one given in the text, see whether it can be transformed into the other form. If it can, then it is the correct answer. If you still have doubts, talk with your instructor.

CHAPTER 1

1.1 (page 14)

I. 1.(a) $4n$; n an integer **3.(a)** $10n - 1$; n an integer **5.(a)** n^3; n an integer **II. 7.** $ab = ba$; it is true for all numbers **9.** $a^2 - b^2 = (a + b)(a - b)$; it is true for all numbers **III. 11.(a)**

Length	*Width*	*Perimeter*	*Area*
1	9	20	9
2	8	20	16
3	7	20	21
4	6	20	24
5	5	20	25
6	4	20	24
7	3	20	21
8	2	20	16
9	1	20	9

$L = -W + 10$

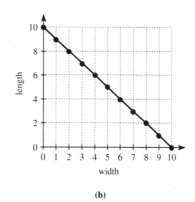

(b)

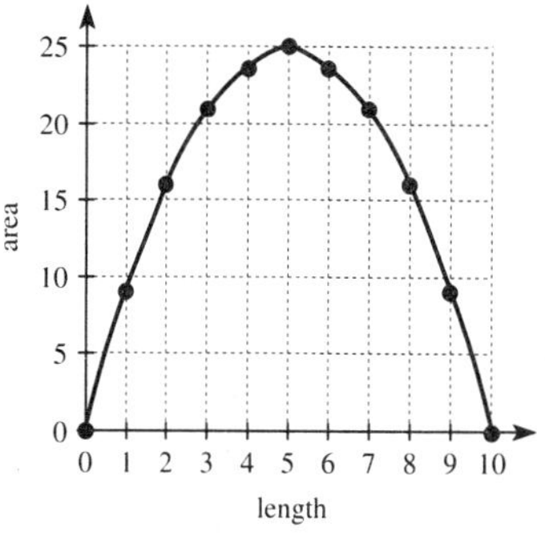

(c)

13. $P = 4s$; s is the side of a square **15.** $A = \frac{1}{2}ab$; a, b are the lengths of each side of a triangle **17.** $C = 2\pi r$; r is the radius of a circle **19.** $S = 6a^2$; a is the edge of a cube **21.** $A = (2r)^2 - \pi r^2$; r is the radius of a circle **23.** $t = \frac{350}{r}$; t is the time, r is the average rate **25.** $A = 25q$; q is the number of quarters **27.** $I = PT(0.0725)$; P is the principal, T is the number of years, I is the interest rate **29.** $c = 28 + (0.12)m$; m is the number of miles, c is the cost **31.** tax $= 2{,}550 + (x - 17{,}000)0.28$; $17{,}000 < x \leq 41{,}000$ **IV. 33.** $x + (x + 1) + (x + 2) + (x + 3) = 4x + 6$

35.(a) those of the form $\frac{n(n+1)}{2}$

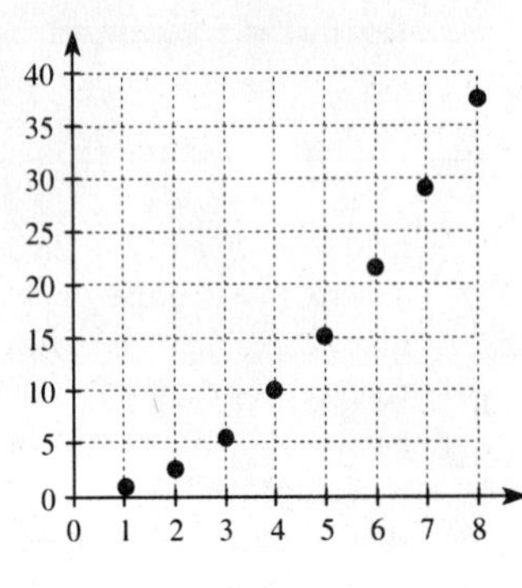

(b)

37.(a) 15 **(b)** 45, $\frac{n(n-1)}{2}$

1.2 (page 32)

I. 1. length = 5 **3.** length = 4.1231056 **5.** $A = 4(2 + 5)$ or $A = 4(2) + 4(5)$ **7.** 3 [7 | 2]

9. 5 [6 | 10] **11.** always 2; $\frac{3x+6}{3} - x = 2$ **13.** $x + 2$; x is the original number; $2(x + 4) - 6 - x = x + 2$

15. largest = 3.5, smallest = $-\sqrt{8}$ **17.** true **19.** true **21.** false **23.** true **25.** true **27.(a)** $\{a \mid a \geq -8\}$ **(b)** −8 0

29.(a) $\{m \mid m < -2 \text{ or } m > 1\}$ **(b)** −2 1

31.(a) $\{w \mid w < 3 \text{ and } w > 0\}$ or $\{w \mid 0 < w < 3\}$ **(b)** 0 3

33.(a) 0 4 **(b)** x is less than or equal to 4.

35.(a) −3 3 **(b)** x is greater than 3 or x is less than −3

37.(a) entire number line

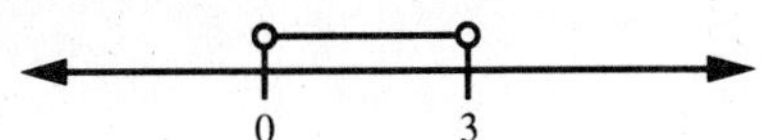

(b) x is less than or equal to 5 or x is greater than 2

39. 6 4 2 0 −2 −4 −6 −8; 1 2; D F G B A C E

41. $A(0, 0)$, $B(3, 0)$, $C(2, 1)$, $D(0, 3)$, $E(-3, 5.75)$, $F(-3, -3)$, $G(0, -3)$, $H(1.5, -4)$

II. 43. 10 **45.** 3 **47.** $3(2 + 4 + 1)$ **49.** Answers vary: 3, 4, 5 **51.** H **53.** D **55.** A **57.** Answers vary:

The first number of the point corresponds to the position on the x-axis (horizontal axis), the second number corresponds to the position on the y-axis (vertical axis). So, locate and go to the position of the first number on the x-axis, then from that position go up or down depending on the second number. **III. 59.** $P \geq 2{,}500{,}000 - 1{,}006{,}877 = 1{,}493{,}123$ **61.** $78 \leq S \leq 100$ **63.** $6.95t - (25 + 5.25t) > 0$
65.(a) $7 + 6 > 5, 6 + 5 > 7, 7 + 5 > 6$, yes **(b)** $3 + 4 > 8$, no **(c)** $8.3 + 7.62 > 15.905, 7.62 + 15.905 > 8.3, 8.3 + 15.905 > 7.62$, yes **67.** $187.2 \leq v \leq 242$ **69.** $\$4.25 \leq w \leq \$10.75$ **71.(a)** \$36 **(b)** 390 miles
(c) the cost is \$12 plus 20¢ per mile **73.(a)** 4,096 flies **(b)** 11 hours

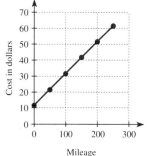

(c) The number of flies doubles every unit of time.

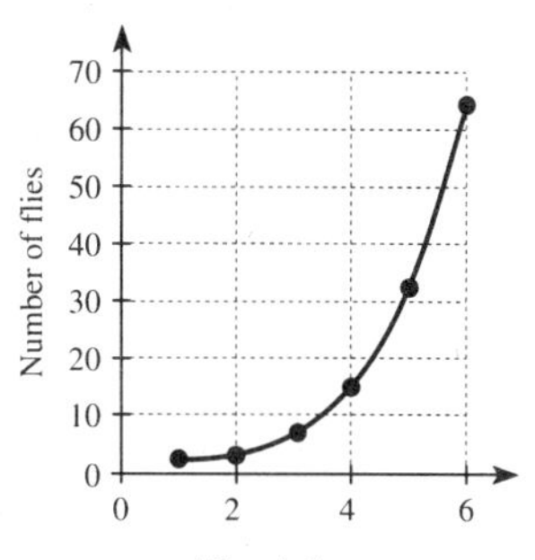

IV. 75.(a) 13×16 **(b)** 60

77. a, because you are adding less to a to equal b **79.** -4 **81.** Answers vary: **(a)** (0, 2), (2, 0), (0, −2), (−2, 0)
(b) (0, 0), (0.5, 0), (0, 0.5), (−0.5, 0) **(c)** (1, 1), (−1, 1), (1, −1), (−1, −1) **83.** A little less than half the length of the straw.

1.3 (page 48)

I. 1. $\frac{13}{6}$ **3.** $\frac{7}{12}$ **5.** $\frac{7}{25}$ **7.** $\frac{32}{9}$ **9.** $\frac{2}{3}$ **11.** 1 **13.** $\frac{2}{9}$ **15.** $\frac{4}{3}$ **17.** $\frac{25}{16}$ **19.** $\frac{17}{12}$ **21.(a)** 3 **(b)** 3.096
23.(a) 3 **(b)** 2.583 **25.(a)** 0 **(b)** 0.183 **27.(a)** 1,000 **(b)** 1,167.296 **29.** (a), (b), (d), (e), (g) **II. 31.(a)** $\frac{15}{28}$
(b) $\frac{28}{15}$ **(c)** $\frac{12}{35}$ **(d)** $\frac{12}{35}$ **33.** $\frac{5}{8}$ **35.** Answers vary: $\frac{24}{10} = \frac{12}{5} = \frac{48}{20}$

37.

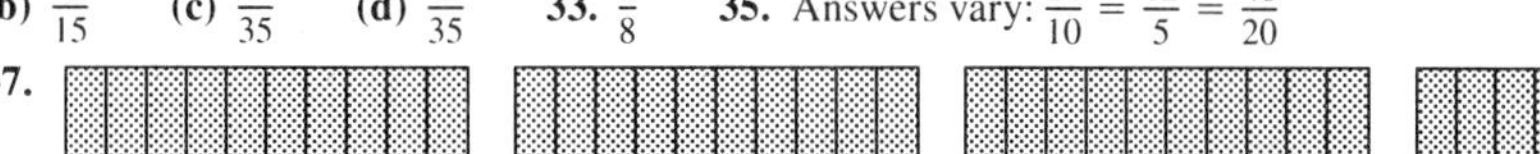

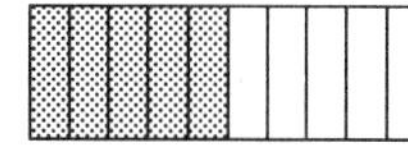

$3\frac{5}{10} = \frac{35}{10}$

39.

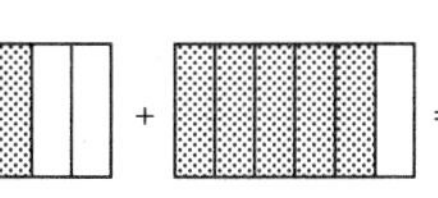

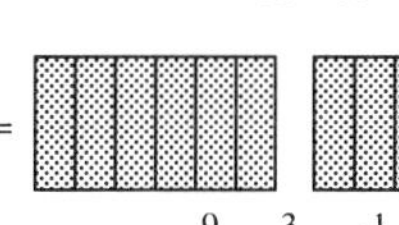

$\frac{2}{3}$ + $\frac{5}{6}$ = $\frac{9}{6}$ or $\frac{3}{2}$ or $1\frac{1}{2}$

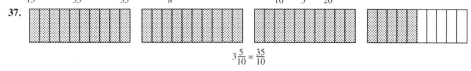

41. Answers vary: $\frac{10}{185} = \frac{2}{37}, \frac{10}{186} = \frac{5}{93}, \frac{10}{187}$ **43.** 3

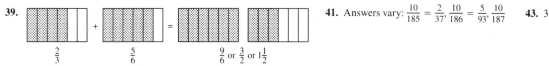

45. $\frac{1}{2}$ or 0.5 **47.** 6 **49.** $\frac{3}{2}$ **51.** Answers vary **53.** Answers vary **55.** Answers vary **III. 57.(a)** $\frac{781}{3{,}125}, \frac{2441406}{9765625}$

(b) Add the previous sum's numerator and denominator to get the next sum's numerator; the denominator is 5^n, where n is the nth row

(c) 0.25 **(d)**

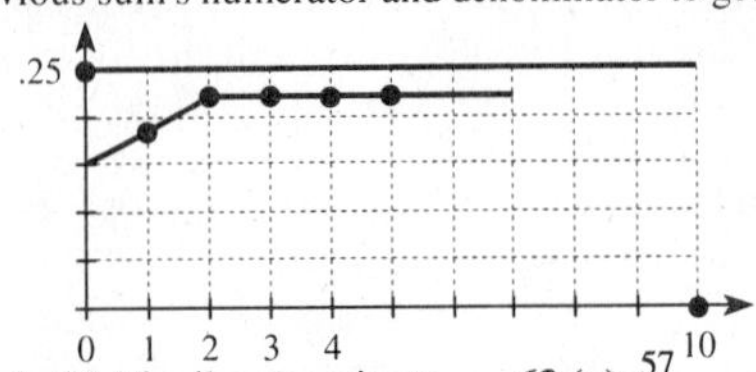

59.(a) $\frac{127}{128}, \frac{1{,}023}{1{,}024}$ **(b)** no **(c)** Answers vary

61. 2,289.38 mph, 38.16 miles per minute **63.(a)** $\frac{57}{20}$ **(b)** $\frac{20}{57}$

65.(a) Hong Kong **(b)** New York
67. 32% shaded, 68% unshaded **69.** Mary
71. 28.8% increase **73.** 36.8% decrease **75.** 0.5%

77.(a)

rent and utilities:	28%
car:	17%
insurance:	7%
food:	19%
misc.:	20%

(b) between 9% and 10%
79.(a) 1.865 inches **(b)** 0.62 inches **81.** 368 times **83.(a)** 17.97 mpg **(b)** 31.45 mph **85.** Answers will vary, for example: **A.** Speed is increasing; skier may be going down hill. **B.** Speed is decreasing; skier may be climbing or has just slowed down. **C.** Speed is increasing. **D.** Speed is decreasing faster than at point *B.* **E.** Speed is zero; skier has stopped. **F.** Speed is about to decrease. **G.** Speed is decreasing; skier may be going up hill or may be coming to a stop.

IV. 87. Pattern repeats after five numbers **89.(a)** Answers vary **(b)** $\frac{1}{1\cdot 2} + \frac{1}{2\cdot 3} + \frac{1}{3\cdot 4} + \cdots + \frac{1}{n(n+1)} = \frac{n}{n+1}$ **(c)** yes

1.4 (page 62)

I. 1. 5 **3.** 7 **5.** 36 **7.** 0 **9.** 9 **11.** −48 **13.** −8 **15.** −4 **17.** −4 **19.** 0 **21.** $\frac{1}{4}$ **23.** $\frac{5}{3}$ **25.** 24 **27.** $-16\frac{5}{6}$ **29.** −30 **31.** −29 **33.** $\frac{2}{9}$ **35.** 0 **37.** $-\frac{7}{5}$ **39.** 3 **41.** 5 **43.** 18 **45.** 2.542 **47.** −0.091

49. 12.520 **51.** mean = 10.5, median = 6.5 **53.** mean = 1.604, median = $\frac{1}{2}$ **II. 55.** Only 2 is multiplied by (6 − 2); −3 **57.** added the top numbers instead of multiplying: $\frac{(-3)(-5)}{6-2} = \frac{15}{4}$ **59.** −6 **61.** −3 **63.** −13 **65.** > **67.** >

69.(a)

$x - y$	$y - x$	$\lvert x - y\rvert$	$\lvert y - x\rvert$
3	−3	3	3
4	−4	4	4
5	−5	5	5
6	−6	6	6
7	−7	7	7
8	−8	8	8
9	−9	9	9

(b) $\lvert x - y\rvert = \lvert y - x\rvert$ **71.(a)** $\lvert x\rvert = 4$ **(b)** 4, −4 **73.(a)** $\lvert x - 5\rvert = 4$

(b) 1, 9 **75.(a)** $\lvert x + 4\rvert = 3$ **(b)** −1, −7 **77.(a)** x is 5 units from origin **(b)** −5, 5 **79.(a)** x is 5 units from the point 8 **(b)** 3, 13 **81.(a)** x is 3 units from the point −1 **(b)** −4, 2 **83.(a)** Answers vary: −10, −2, −1, −3, −4, 5, 6, 7, 8, −46 **(b)** an infinite number **85.** AT **87.** NT **89.** ST, 0 is the only exception **III. 91.** 1,175 years **93.** 10 fewer patients **95.(a)** 30-yard line, **(b)** 25-yard line **97.** loss of $1.45 **99.** −1,493 feet **101.(a)** 309 meters **(b)** 9,243 meters **(c)** 6,280 meters **103.(a)** 6.4 yards gained **(b)** 5 yards gained **(c)** Answers vary **105.(a)** Answers vary: 10,000; 11,000; 15,000; 16,000; 17,000; 20,000; 30,000; 41,000 **(b)** Answers vary **IV. 107.(a)** 8 **(b)** −1000 **(c)** 1001 **d.** even #: $\frac{-n}{2}$; odd #: $\frac{n+1}{2}$ **(e)** 199 **(f)** yes **109.** Answers vary: (4 + 10 − 2)(3) **111.** Answers vary: (8 − 6)(4) ÷ 2 **113.** Answers vary: (8 − 6) ÷ (4 − 2) **115.(a)** 0 **(b)** 996 + 997 + 998 + 999 + 1000 = 4990

CHAPTER 1 SUMMARY REVIEW

I. 1. $\frac{43}{21}$ **3.** $\frac{21}{4}$ **5.** $-\frac{1}{3}$ **7.** -11 **9.** $\frac{1}{18}$ **11.** -10 **13.** -1 **15.** 0 **17.** 10 **19.** $\frac{6}{7}$ **21.** -4
23. -3 **25.(a)** $4(2 + 6)$ **(b)** $(4 \times 2) + (4 \times 6)$ **27.**

29.

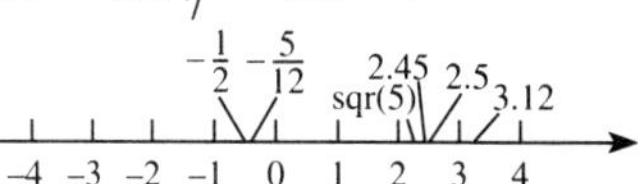

31.(a) 22 **(b)** 20.756 **33.(a)** $|x| = 3$ **(b)** $x = 3, -3$ **35.(a)** $|x - 3| = 2$ **(b)** $x = 1, 5$ **37.** $<$ **39.** $>$
41. $>$ **43.** $>$ **45.** $>$ **47.** $x \geq -5$ **49.** $-3 < x < 6$ **51.** $-3 \leq x \leq 3$
53.(a)

(b) x is less than 1 **55.(a)**

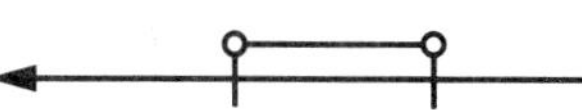

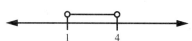

(b) x is greater than 1 and less than 4 **II. 57.** false **59.** false **61.** false **63.** false **65.** 2 **67.** Answers vary
69. -25 **71.** -7 **73.** 2 **75.** 9 **77.** Answers vary: $\frac{21}{100}, \frac{11}{50}, \frac{23}{100}, \frac{6}{25}$ **79.** Answers vary **III. 81.** $\frac{15}{8}$
83.(a) $\frac{8{,}000}{9}$ **(b)** 10,666.67 miles **85.** 18.4° C **87.** 8.333 . . . yards **89.** 24.308 cm **91.(a)** no **(b)** yes **93.(a)** \$238
(b) $\$354 \leq x \leq \$404$ **95.(a)**

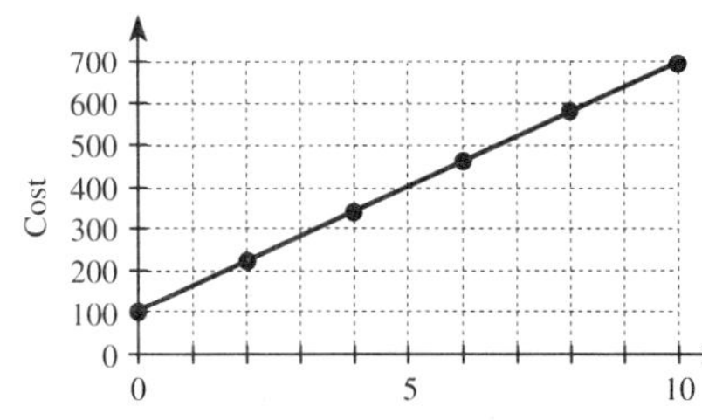

(b) \$1,120, 15 credits

(c) tuition is \$100 plus \$60 per credit **97.(a)** 45.3% decrease **(b)** 13.3% increase **99.** Answers will vary; the peak time for the number of soft drink cans sold is at 12 P.M. which coincides with the lunch period; the least demand is 0 cans at 6 P.M. when the company is closed. Other peak times may coincide with breaks in the workers' day.

CHAPTER 2

2.1 (page 83)

I. 1. 64 **3.** -1 **5.** 16 **7.** 16 **9.** $\frac{64}{125}$ **11.** $\frac{243}{4}$ **13.** x^{13} **15.** a^6b^3 **17.** $-8a^6b^3$ **19.** $0.0002x^5$ **21.** $-x^4y^3$
23. $\frac{2}{x^6}$ **25.** $\frac{9y}{b}$ **27.** $-\frac{0.9b^2}{a^3}$ **29.** $2^{2m+n}x^{m+n}$ **31.** $(2x + 3y)^5$ **33.** 108 **35.** 1; 1 **II. 37.** x^4 **39.** 3^{10}
41. $2(a^2b)^6$ **43.** $(-2ab)^2$ **45.** $(2a)^2b^3c^4$ **47.** Answers vary: $x^{18}, (x^9)^2, (x^3)^6$ **49.** Answers vary: $\frac{9 \cdot 32}{4^3 \cdot 3^3}, \frac{288}{12^3}, \frac{1}{6}$ **51.** $2x$
53. $3(x + 5)$ **55.** $6x$ **57.** $\frac{500}{x}$ **59.** $4x$ **61.** $x^2 + x$ **63.** $10x + 50$ **65.** $\frac{x}{4}$ **67.** $4x$

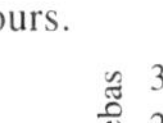

69. No, only $2^{20} \approx 1$ million people learned. **III. 71.(a)** $\approx 4{,}186{,}666.7\ \text{cm}^3$ **(b)** $10^6\ \text{cm}^3$ **(c)** $1.0612\ \text{cm}^3$
(d) $0.011775\ \text{cm}^3$ **(e)** $2.533\ \text{cm}^3$ **73.(a)** 2 **(b)** 4 **(c)** 16 **(d)** 256
(e) The number doubles every 3 hours.

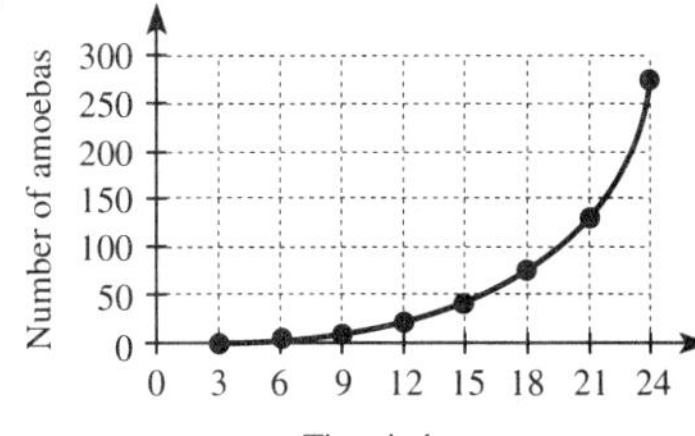

75.(a) Estimate A **(b)** Estimate B; Estimate A

77.(a) $2^3 = 8$ times as much **(b)** $2^3 = 8$ times as much **(c)** $2^3 = 8$ times as much **79.** \$3,125
IV. 81.(a) 6 **(b)** 3 **83.** $2^2 - 1$; $2^3 - 1$; $2^5 - 1$, $2^7 - 1$; $2^n - 1$ when n is a composite number
85. With n toppings, there are 2^n possible pizzas, so $n \geq 20$ since $2^{20} = 1{,}048{,}576$.

2.2 (page 97)

I. 1. $\frac{1}{27}$ **3.** $-\frac{1}{4}$ **5.** $\frac{64}{27}$ **7.** 10^6 **9.** 100.1 **11.** $\frac{1}{0.0004} = 2{,}500$ **13.** 0.08 **15.** $\frac{1}{2}$ **17.** $-\frac{80}{9}$ **19.** 1 **21.** $\frac{1}{x^9}$
23. x^5 **25.** $\frac{1}{x^3}$ **27.** x^5 **29.** $\frac{y^4}{2x}$ **31.** $-2x^5y^5$ **33.**

x	y	$x^{-2}y^3$	$\frac{x^{-2}}{y^3}$	$x^{-2} + y^3$
2	2	2	$\frac{1}{32}$	$\frac{33}{4}$
-2	2	2	$\frac{1}{32}$	$\frac{33}{4}$
2	-2	-2	$-\frac{1}{32}$	$-\frac{31}{4}$
-2	2	-2	$-\frac{1}{32}$	$-\frac{31}{4}$
-10	10	10	$\frac{1}{10^5}$	$\frac{100{,}001}{100}$
100	-10	$-\frac{1}{10}$	$-\frac{1}{10^7}$	$-\frac{9{,}999{,}999}{10{,}000}$

35.(a) 0.000693 **(b)** 0.000000074694 **(c)** 423,400,000 **(d)** 300,200 **37.** 420 **39.** 0.1 **41.** 6×10^{-7}
43.(a) 1,000 **(b)** 980.392 **45.(a)** 0 **(b)** .0000006 **II. 47.** 266 **49.** 79.42
51.

53.

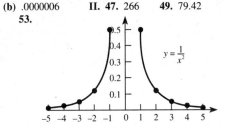

55. Answers vary **57.(a)** $n \approx 170$ **(b)** $n \approx 60$ **59.** NT **61.** ST [true if n is odd] **63.** ST [true if $x < 0$]
III. 65. 5.84×10^8 miles, or 2π A.U. **67.(a)** 1,000 nanoseconds **(b)** 1,000 micromicroseconds; 1,000,000 micromicroseconds
69.(a) $\approx 1.116 \times 10^7$ miles **(b)** $\approx 6.696 \times 10^8$ miles **(c)** $\approx 1.607 \times 10^{10}$ miles **(d)** $\approx 5.866 \times 10^{12}$ miles
71. 72 hamburgers **73.** 1,833 times as heavy **75.** 6.75×10^{-12} cm^3 **IV. 77.(a)** 10 **(b)** 31 **(c)** 90
79. Yes: assuming 10 ft^3 per person, a cubic mile would hold about 14.7 billion people.

2.3 (page 108)

I. 1. a product; $x + 2$ is not a factor. **3.** a product; $x + 2$ is a factor **5.** a product; $x + 2$ is a factor **7.** no **9.** $6x + 2xy$
11. $1.8x - 13.5y$ **13.** $-5xy - 10y$ **15.** $3x^3 + 2x^2 - 2x$ **17.** $5x(a - 5)$ **19.** $3a(1 + 2b + 5) = 3a(6 + 2b) = 6a(3 + b)$
21. $(6y + 1)(b + 3)$ **23.** $(3 + a)(11)$ **25.** $11 - 10a$ **27.** $-3x + 4y$ **29.** $5x^2 + 2x + 1$ **31.** $y^2 + 2x - 1$
33. $2x^{10} - 5x^5 + 7x^2$ **35.(a)** 1 **(b)** binomial **37.(a)** 2 **(b)** trinomial **39.(a)** 4 **(b)** none **II. 41.** $4 - t$
43. $\frac{4}{3}$ **45.** $(3 + x)y = 3y + xy$ **47.** $3(x + 2)$ **49.** $6a + 6(a + 1)$ **51.(a)** Answers vary: $12 = 3 \cdot 4$; $12 = 2 \cdot 2 \cdot 3$
(b) Answers vary: $12 = 3 + 9$; $12 = 3 + 4 + 5$ **(c)** Answers vary: $12 = 2(2 + 4)$ **53.** 6 **55.** 2 **57.** Answers vary

59. Answers vary **61.**

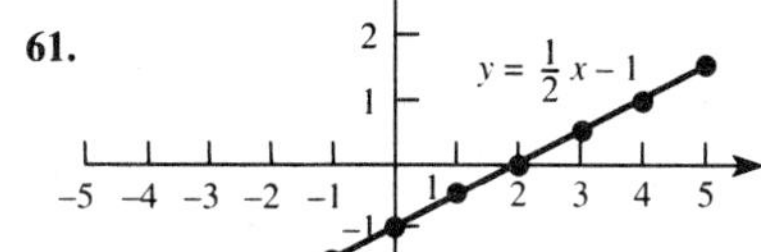

(a)

$y = -\frac{1}{2}x - 1$

(b)

63.

$y = x^2 + x$

(a)

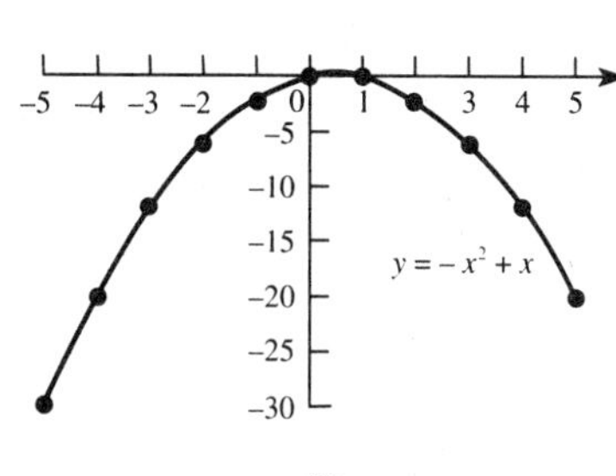

(b)

65. They are the same.

III. 67. sum $= n + (6n + 4) = 7n + 4$ **69.** profit $= 6x - 4x = 2x$ **71.** cost $= 10 + 0.24x$ for $x \leq 50$; $22 + 0.15(x - 50)$ for $x > 50$ **73.** distance $= 60t + 50t = 110t$ **75.** $3x + 4y$ **77.** $1.60x + 1.85y$ **79.** $4x + 3y$ **81.** $2.50x + 2.50y + 2.50z = 2.50(x + y + z)$ **83.** $2 \cdot 20 + 2x = 2(20 + x)$ **85.(a)** $(2q^2 - \pi q^2) \approx 0.86q^2$ **(b)** $\pi q^2 - (1.414q)^2 \approx 1.14q^2$ **(c)** The second figure has a larger area. **87.(a)** $\pi r^2 h - \frac{4}{3}\pi r^3 = \pi r^2\, h - \frac{4}{3}r$ **(b)** ≈ 134 cm^3

IV. 89.(d) half the sum of the four border numbers **(e)** yes **(f)**

	p	q	
r	pr	qr	$pr + qr$
s	ps	qs	$ps + qs$
	$pr + ps$	$qr + qs$	equal

(g) Answers vary **91.** \$1 the first day; \$2 the second day; etc.

2.4 (page 121)

I. 1. $-6x^3y^3$ **3.** $-6x^3y - 2xy^2 + 2xy$ **5.** $x^2 - 1$ **7.** $25x^2 - 4$ **9.** $x^2 + 2x + 1$ **11.** $25x^2 + 20x + 4$ **13.** $x^2 + 5x + 6$ **15.** $3x^2 + 7x + 2$ **17.** $6x^2 + 13x + 6$ **19.** $40x^2 - 32x + 6$ **21.** $4 - x^2$ **23.** $10x^2 - 17x + 6$ **25.** $a^2x^2 - b^2$ **27.** $x^3 - 2x + 1$ **29.** $x^5 + x^2 - x + 1$ **31.** $\frac{6y^4}{x}$ **33.** $a + 2b$ **35.** $\frac{3}{2}x - y$ **37.** $9x^2 - x + 1$ **39.** $2(1 - a^2b)$ **41.** $\frac{4x + y}{4xy}$ **43.** $\frac{2x - 1}{x^2}$ **45.** $-(n - 1) = 1 - n$ **II. 47.**

	x	3
x	x^2	$3x$
2	$2x$	6

$(x + 3)(x + 2) = x^2 + 5x + 6$

49.

	x	3
x	x^2	$3x$
3	$3x$	9

$(x + 3)(x + 3) = x^2 + 6x + 9$ **51.(a)** $(x + a)$ by $(x + 2a)$ **(b)** $2(2x + 3a)$ **53.** Answers vary

55. $(x + 3), (x + 2)$ **57.** $(x + 5), (x - 5)$ **59.** $(x + 1), (x^2 - x + 1)$ **III. 61.(a)** $\frac{3}{r}$ **(b)** $\frac{2(wh + lh + hw)}{lwh}$ **(c)** $\frac{6}{e}$ **(d)** $\frac{2(r + h)}{rh}$ **(e)** $\frac{3(s + r)}{rh}$ **63.(a)** $\pi A^2 - \pi B^2 = \pi(A - B)(A + B)$ **(b)** 215.875

65. $2a(3a + 2) - a(2a + 1) = 4a^2 + 3a = 2\left(\frac{a}{2}\right)(3a + 2) + 2a\left[\frac{3a + 2 - (2a + 1)}{2}\right]$ **67.** 0; has no value after 20 years **69.** -0.078; no, expansion occurs only when temperature is high. **71.** \$5,675.94 **73.** \$5,223

IV. 75. 21, 624, 216, 4,209; $(10t + 5 + d)(10t + 5 - d) = (10t + 5)^2 - d^2$. Square the average as in #74 and then subtract d^2—e.g. $43 \times 47 = 45^2 - 2^2 = 2025 - 4 = 2021$. **77.(a)** 5,050 **(b)** 250,500 **(c)** 100,500 **(d)** 669,001
79. Answers vary; $a^2 + 6b^2 = 5ab$; no **81.** $[n(2n + 1)]^2 + [n(2n + 1) + 1]^2 + \cdots + [n(2n + 1) + n]^2 = [n(2n + 1) + (n + 1)]^2 + \cdots + [n(2n + 1) + 2n]^2$; yes **83.** middle product is 2 more than first and last product: $(n + 1)(n + 2) - n(n + 3) = 2$

CHAPTER 2 SUMMARY REVIEW

I. 1. 3^4 **3.** 3^7 **5.** $\frac{244}{27}$ **7.** 3^9 **9.** $\frac{2}{3x}$ **11.** $\frac{3y^5}{x^5}$ **13.** $-3x^4y^4$ **15.** $a - 1$ **17.** $9 + x - 6x^2$
19. $2m - m^2n + n^2 - 1$ **21.** $-18x - 1$ **23.** $3x^2 + 4x - 11$ **25.** $\frac{5x^4y^2 - y^3}{2}$ **27.** -4 **29.(a)** 24,000 **(b)** 19,421.6835
II. 31. $(7 + 5)a - a = 11a$ **33.** $(7 + 5)(a - a) = 0$ **35.** True **37.** False: $(1 + 2)^2 \neq 1^2 + 2^2$ **39.** False: $x = 0$
41. False: it is always positive. **43.(a)** (a), (c), (d), (h) **(b)** (f), (h) **45.** the sum and difference of the two expressions
47.(a)

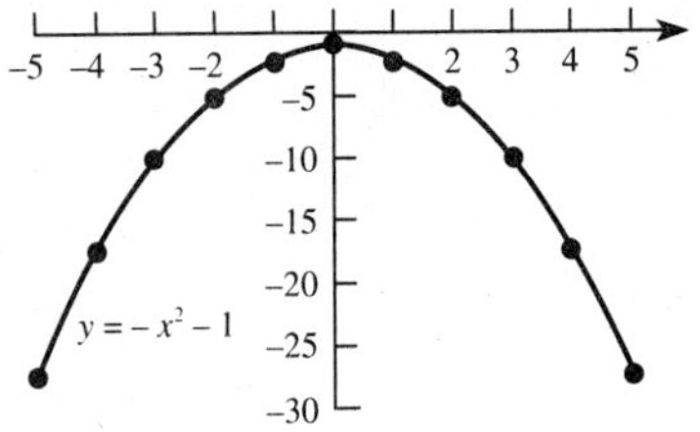

(b)

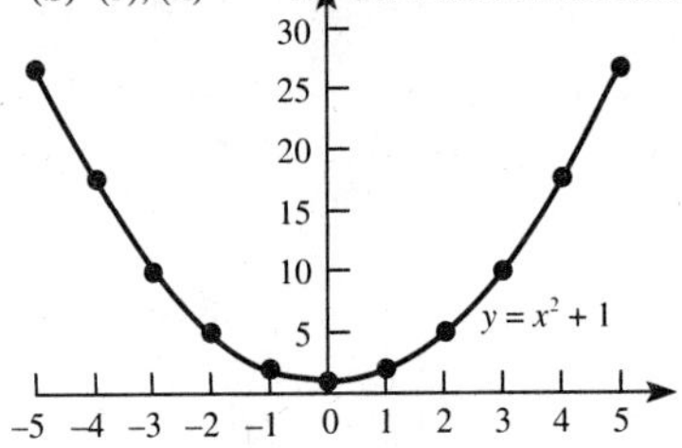

III. 49.(a) \$16,699.32 **(b)** yes **51.(a)** $120t$ **(b)** 5 hours **53.(a)** $\frac{1}{36}$ **(b)** $\frac{1}{6^{10}}$ **(c)** $\frac{1}{6^n}$ **55.** 19.8
57. $0.10x + 0.25y$ **59.(a)** $2x^2$ **(b)** $(x + 3)$ by $3x$ **(c)** $8x + 6$ **61.** 2, 4, 8, 32, 1,024, 1.2677×10^{30}; around 7; not much

CUMULATIVE REVIEW: CHAPTERS 1 AND 2

I. 1. $-\frac{67}{30}$ **3.** $-\frac{5}{64}$ **5.** 0 **7.** 26 **9.** $5a + 15$ **11.** $3m^{18}n^{10} - 2$ **13.** $x < -1$ or $x > 3$ **15.** $0 \leq x < 10$
17.

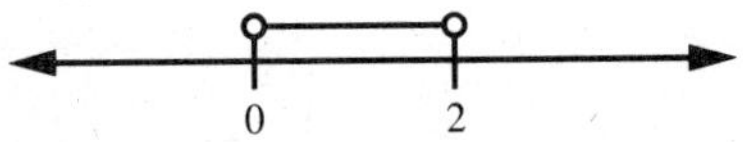

II. 19. NT **21.** AT **23.** AT **25.** (d) **27.** (e) **29.** Answers vary
III. 31.(a) 120 ml **(b)** 24% **33.(a)** \$52.50 **(b)** $\$113.50 \leq c \leq \$134.75$ **35.(a)** $-1.1°$ C **(b)** $-1.5°$ C
(c) $-20°\text{ C} \leq R \leq 15°\text{ C}$ **37.(a)** \$10,000 **(b)** (10, 15) **(c)** around \$2000 **39.(a)** 40 miles **(b)** $13t - 25$
41.(a) \$1,200 **(b)** \$3,000 **(c)** $100N$ **(d)** The one in Exercise 40. It gives more money over time.

CHAPTER 3

3.1 (page 141)

I. 1. $x = 13$ **3.** $x = -30$ **5.** $x = 5$ **7.** $x = -2$ **9.** $x = 48{,}543.69$ **11.** $x = 6$ **13.** $x \approx 256.41$
15. $x \approx 23{,}292.21$ **17.** $x \approx 0.0590$ **19.** $x \approx 5.89$ **21.** $x = 60$ **23.** $x = 2{,}000$ **25.** $m = \frac{F}{a}$ **27.** $d = \frac{C}{\pi}$
29. $t = \frac{I}{pr}$ **31.** $h = \frac{A}{b}$ **33.** $b = \frac{F}{a + c}$ **35.** $a = L - nd$ **II. 37.(a)** $x = \frac{c - b}{a}$ **(b)** $x = \frac{c - by}{a}$ **39.** reciprocals of each other **41.** Answers vary; all have $x = 3$ as a solution **43.** Answers vary; (0, 3000) by (0, 60,000)

III. 45.(a)

m	C
50	30
75	30
100	30
125	33.75
150	37.5
175	41.25
200	45
225	48.75
250	52.5

(b)

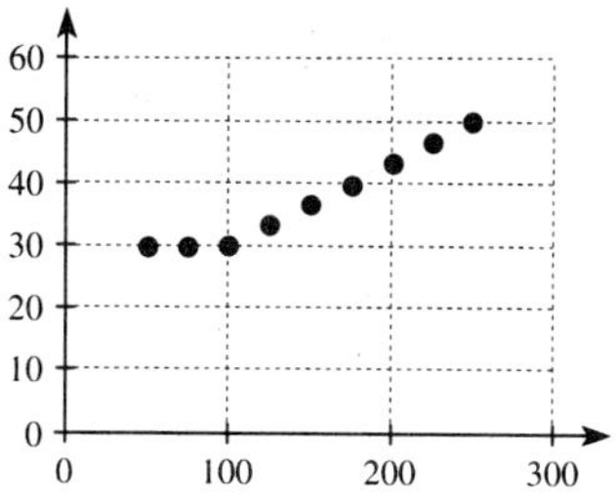

(c) $C = \begin{cases} 30 & \text{if } M \leq 100 \\ 30 + 0.15(M - 100) & \text{if } M > 100 \end{cases}$ **(d)** \$63

(e) 400 miles **47.(a)** 3,000 miles **(b)** 2,903 miles **49.** 42 hours, 36 minutes **51.** 12.3 days **53.** \$48.20 **55.(a)** yes **(b)** no **(c)** yes **(d)** yes **(e)** yes **57.(a)** 8,000 widgets **(b)** 12,000 widgets **59.** 17.5 years **61.** \$39.98 **63.(a)** 10 sets **(b)** 9 sets **65.(a)** 4.15 miles **(b)** 3.695 miles **67.** 92 or 93 **IV. 69.(a)** multiples of 3 **(b)** of the form $4k + 2$ **(c)** multiples of 5 **71.** 16.78%; you did not borrow all the money for the whole year

3.2 (page 153)

I. 1. $x = 6.25$ **3.** $x = 21.6$ **5.** $Z \approx 22.62$ **7.** $Y = 1\frac{11}{13}$ **9.** $W \approx 6.22 \times 10^6$ **11.** $z = 51$ **13.** $a = 28.5$ **15.** $z = 14.5$ **II. 17.** \$24.40x **19.** $\frac{13y}{168}$ **21.** Answers vary: probably not proportional **23.** $\frac{4}{15} = \frac{32}{120}$; maybe—I probably couldn't eat 32 pancakes **25.** $\frac{8}{4} = \frac{12}{6}$; no—pizza cost related to area of pizza **27.** $\frac{15}{3} = \frac{20}{4}$; no—three pieces require two cuts, and each cut takes 7.5 minutes **29.** (a) **31.** Answers vary; $\frac{7}{6} = \frac{55}{x}$ and $\frac{6}{7} = \frac{x}{55}$ **33.** Answers vary **III. 35.** 6 cups **37.** 99 games; 99 − 63; 61% **39.** 24,000 miles **41.** 1,980 bulbs **43.** 1,170 pounds **45.(a)** 11.25 miles **(b)** 3,144 calories **47.(a)** 30 grams **(b)** 34 tablets **49.(a)** $\approx 5.87 \times 10^{12}$ miles **(b)** $\approx 9.45 \times 10^{15}$ meters **51.** Not unless the other person is the same height. **53.(a)** $46\frac{2}{3}$ feet **(b)** $\approx$ 178.57 meters **55.(a)** 6,250 trout **(b)** 1,667 **57.** $\approx$17.73 N/cm^2 **59.(a)** 4 times; 4 times **(b)** $\frac{1}{6}$ **(c)** 20 times **61.** 4:1 **63.** \$250, 960 **65.** 21.875 cm by 13.125 cm **67.** 14.29 cm by 10.71 cm **69.(a)** 36°, 54°, 90° **(b)** no **71.** 1,150 stocks **73.(a)** $\approx$ 13.3 million **(b)** $\approx$ 51.4 million **75.** 1,600; between 1,333 and 2,000 **IV. 77.(a)** 22.4 cm **(b)** 7.2 cm^2; 627.2 cm^2 **(c)** 87.1:1; 9.3; no **(d)** ratio of areas is square of ratio of corresponding sides. **(e)** r^2 **79.(a)** 173.28 in^2 **(b)** true **81.** Lockett—assuming 200 hits in 600 at bats (or other sets of plausible values); Lockett's final average is higher; $\frac{209}{612} > \frac{207}{608}$

3.3 (page 171)

I. 1. $x = -3$ **3.** $x = -\frac{3}{4}$ **5.** $x = \frac{5}{11}$ **7.** $x = -\frac{13}{14}$ **9.** $x = 3.3$ **11.** $x = -1$ **13.** $x = 15$ **15.** $x = \frac{b+d}{a-c}$; $a \neq c$ **17.** $x = ab$; $a, b \neq 0$ **19.** $y = -\frac{9}{4}$ **21.** $x = 14$ **23.** $x = 13$ **25.** $x = -5$; conditional **27.** $x = 0$; conditional **29.** x = all real numbers; identity **31.** $x = 5$; conditional **33.** $x = 1$; conditional **35.** $x = 1$; conditional **37.** x = all real numbers; identity **39.** $x = 1$; conditional **41.** $x \approx 0.7261$ **43.** $x \approx -3.2762$ **45.** $b_1 = \frac{2A - b_2h}{h}$ **47.** $y = 1$ **II. 49.** no solution **51.** $x = 4$ **53.** no solution **55.** $c = -1$ **57.(a)** $c = 15$ **(b)** $c \neq 15$ **59.** No; no solution if $a = c$ and $b \neq d$ **61.** Answers vary **63.** Answers vary

III. 65. Rent Acme if you drive 66 miles or less.

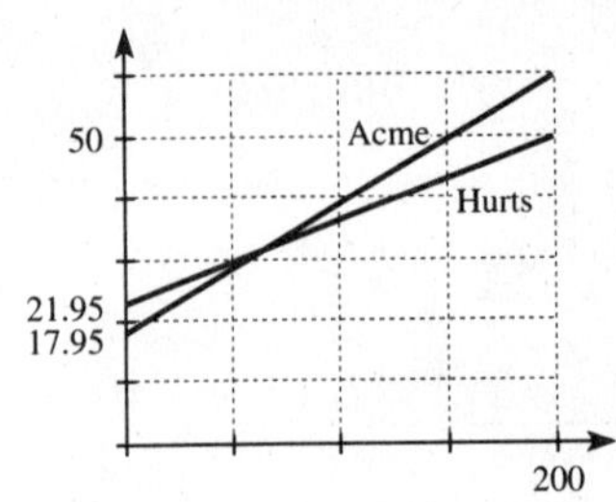

67. 20,000 million board feet is break-even point **69.(a)** 1.25 hours **(b)** 5 hours after they start walking **71.(a)** $x \approx 2.385$ cm **(b)** no values of x **(c)** $x = 17.04$ cm **73.(a)** 1,875 seats **(b)** 6,875 seats **75.** 4.54 miles **77.** Two days **79.** 250,000 widgets; the first one for fewer than 250,000 widgets **81.** 12 miles **83.(a)** length $\approx$ 366.35 meters; width $\approx$ 244.24 meters **(b)** lane 2: 2π meters ahead
lane 3: 4π meters ahead
lane 4: 6π meters ahead
lane 5: 8π meters ahead
lane 6: 10π meters ahead

IV. 85. $13\frac{1}{3}$ mph **87.** 5.73 feet **89.** all acute angles **91.** 60%

3.4 (page 189)

I. 1. $x < 6$

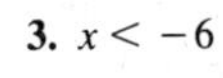

3. $x < -6$

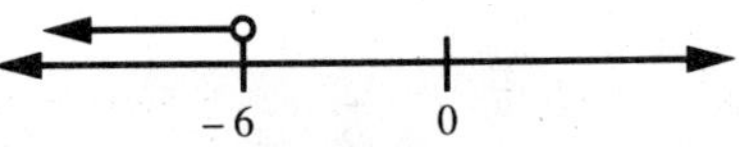

5. $-8 < x < 8$

7. $-6 \leq x \leq 10$

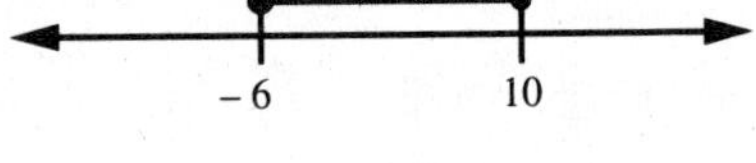

9. $-10 \leq x \leq 6$

11. $x \geq 3$ or $x \leq -3$

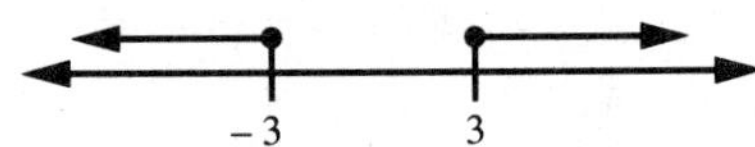

13. $-\frac{8}{3} \leq x \leq 2$

15. $5 < x < 9$

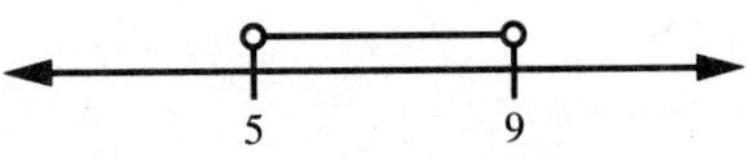

17. $x \geq -\frac{8}{3}$

19. $x \leq 4$

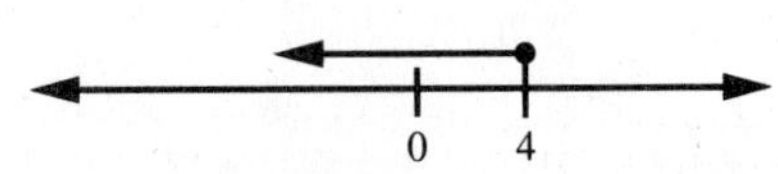

21. $x < 3$

23. $x \geq -\frac{5}{2}$

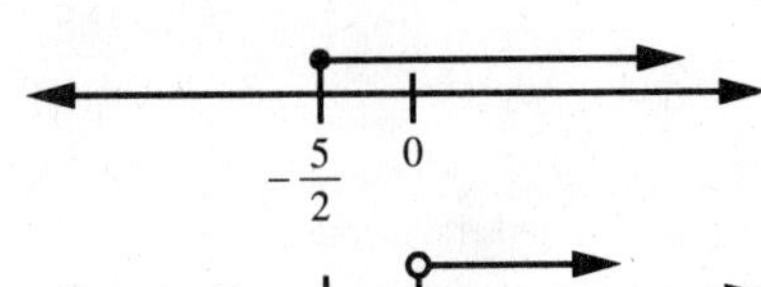

25. $x > \frac{1}{4}$

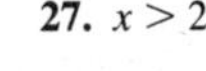

27. $x > 2$

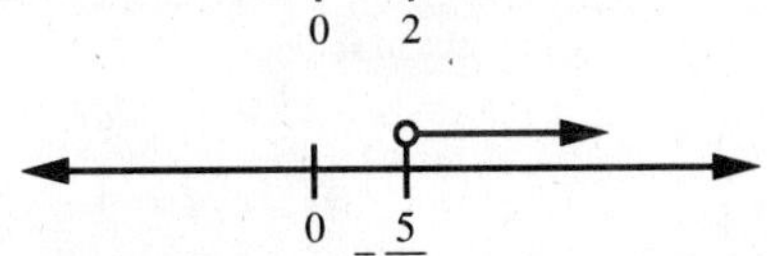

29. no solution

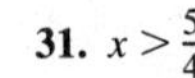

31. $x > \frac{5}{4}$

33. $6 < x < 10$

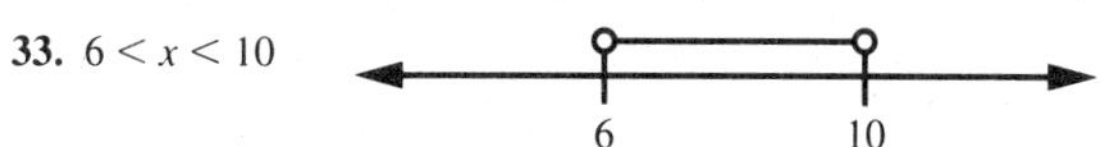

35. $x \geq 3$ or $x \leq 0$

37. $-2 < x < 12$

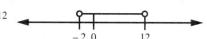

39. $-\frac{1}{2} \leq x \leq \frac{7}{2}$

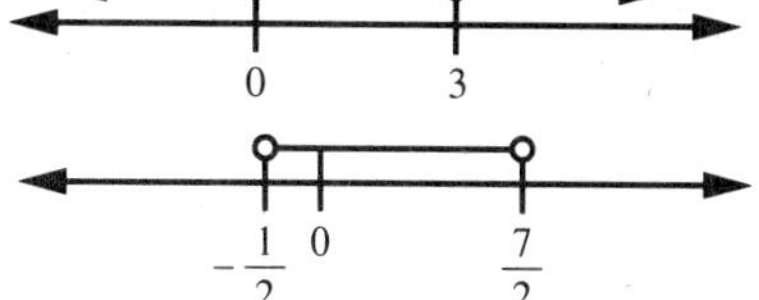

II. **41.** AT **43.** ST; true only if $a = b$ **45.** ST; true if $b > 0$ **47.(a)** F, J **(b)** C, H **(c)** A, E, G **(d)** B, D
49. Both $|x - 3|$ and $|3 - x|$ represent the distance between x and 3 **51.** Answers vary **53.(a)** $c \leq 3$ **(b)** $c > 3$
55.(a) $p < y - x$ **(b)** $p < y - x$ **III.** **57.** $28.29 \leq g \leq 31.89$, where g = gas mileage
59. when less than 50 checks are written **61.(a)** 4 miles **(b)** 3.3 miles **63.** 7 or more vacuums
65.(a) $\$58.50 \leq$ mark up $\leq \$82.50$ **(b)** cannot be done **67.** $16 \leq n \leq 25$, where n is the first integer **69.(a)** $x > 11$
(b) $2 < x < 11$ **71.** between 15 and 35 minutes **73.** 420,000 **75.(a)** 3.6 pounds or more **(b)** no **77.(a)** $C > 33.3°$
(b) $F < 77°$ **(c)** $0° < C < 20°$ **79.** $7.674 < A < 7.687$ **81.(a)** $M \leq 200$ **(b)** $200 \leq M \leq 233$
(c) $M \geq 234$

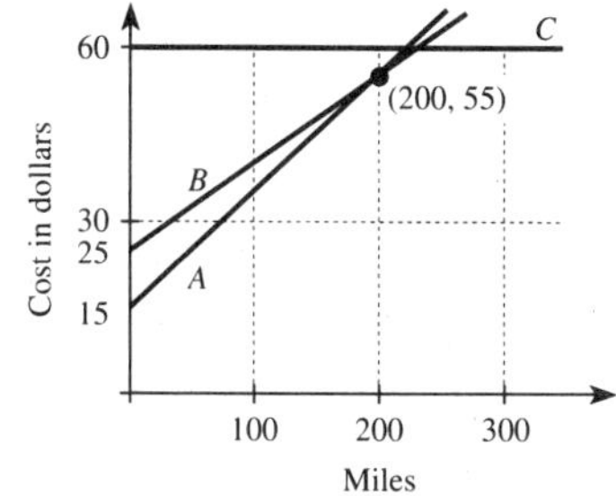

IV. **83.** 0.14863 oz.

CHAPTER 3 SUMMARY REVIEW

I. **1.** $x = 75$ **3.** $x = 5$ **5.** $x = 3.4$ **7.** $x = -\frac{4}{3}$ **9.** $x = -\frac{9}{13} \approx -0.6923$ **11.** $x = 2.4$ **13.** $x = 1$ **15.** $x = 3$
17. $x = -\frac{1}{2}$ **19.** $x = \frac{11}{4}$ **21.** $-\frac{8}{9}$ **23.** $x = \frac{5}{4}$ **25.** no solution **27.** $x \approx 0.1013$ **29.** $x \approx -9.6366$
31. $g = \frac{2s}{t^2}; t \neq 0$ **33.** $h = \frac{A - 2\pi r^2}{2\pi r}; r \neq 0$ **35.** $x > \frac{1}{4}$ **37.** no solution
39. no solution **41.** $x \leq -34.5$

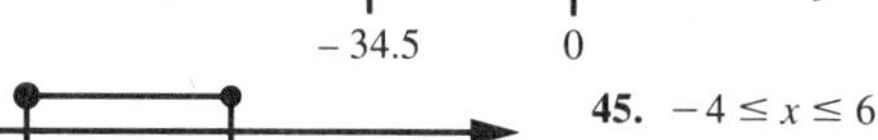

43. $-3 \leq x \leq 9$ **45.** $-4 \leq x \leq 6$

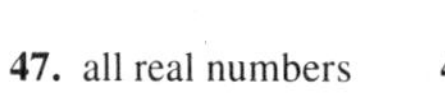

47. all real numbers **49.** $-1 < x < 3$

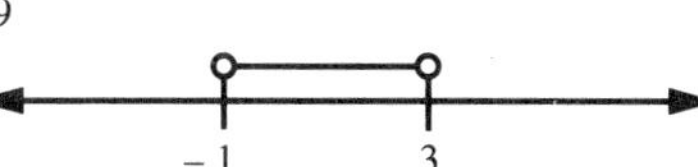

II. **51.** False; it has no solution

53. True; $\left(3 = -\frac{5}{5} + \frac{2(6)}{3}\right)$ **55.** False; $3 < 4$ but $\frac{1}{3} \nless \frac{1}{4}$ **57.** True; $ar = b$; $cs = d \Rightarrow ab(rs) = cd$ **59.** Answers vary
III. **61.** Linda: 3.6 miles; Mark: 2.4 miles **63** 10,000 men **65.** 30°, 60°, 90° **67.** 68 hits

69.(a)

Weekly sales	A	B
0	0	75
100	11	82
500	55	110
1000	110	145
1500	165	180
2000	220	215

(b)

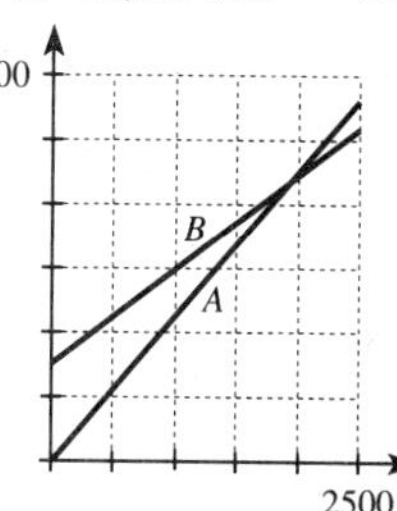

(c) \$1,875 **(d)** more than \$1,875 **(e)** Choose *B*

71.(a) 5.56 feet **(b)** $w = \dfrac{2V}{d(a + b)}$ **(c)** $b = \dfrac{2V}{dw} - a$ **73.** $x = 12.4$ **75.** $\approx 9{,}583{,}333$
77. end of the seventh year **79.** 11.5% **81.** $85.75

CHAPTER 4

4.1 (page 208)

I. 1. none **3.** (a), (b), (c) **5.** Checkpoints: Answers vary

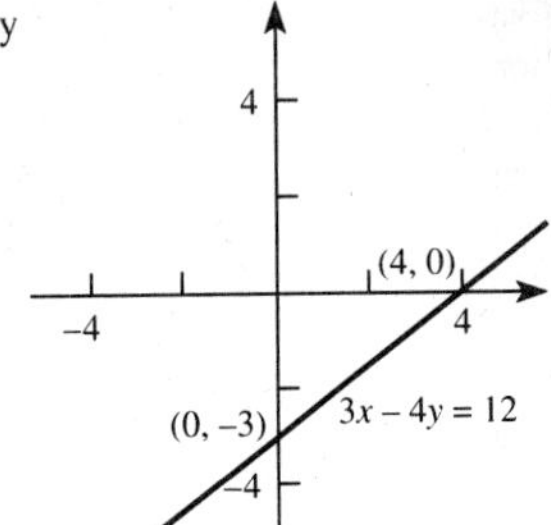

7.

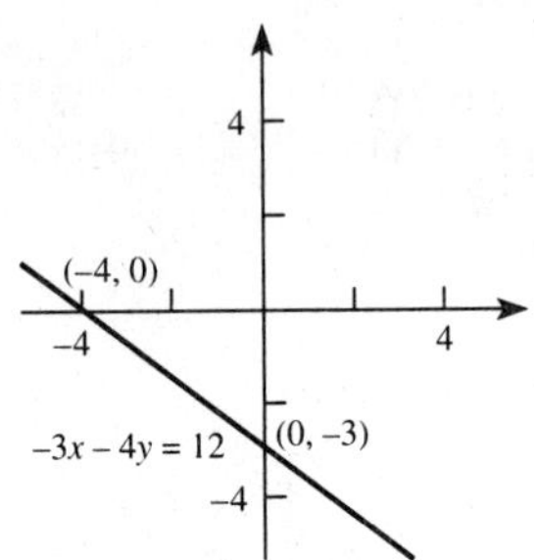

9.

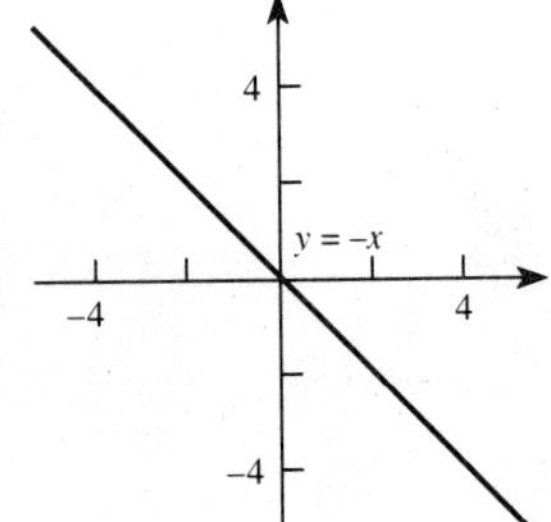

11.

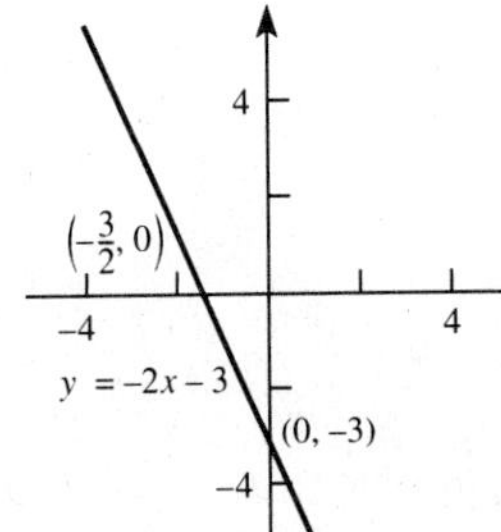

13.

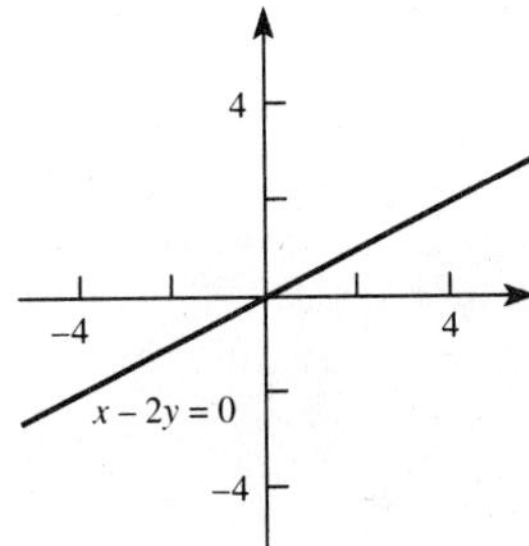

15.

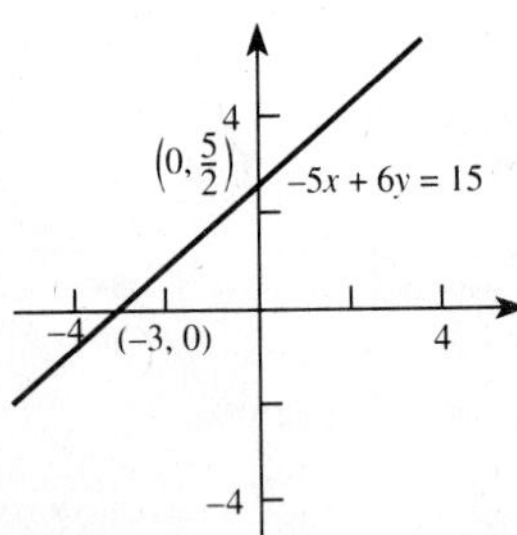

17.

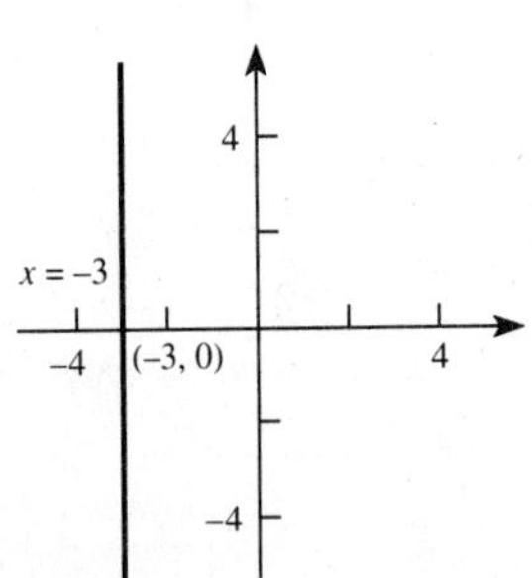

19.

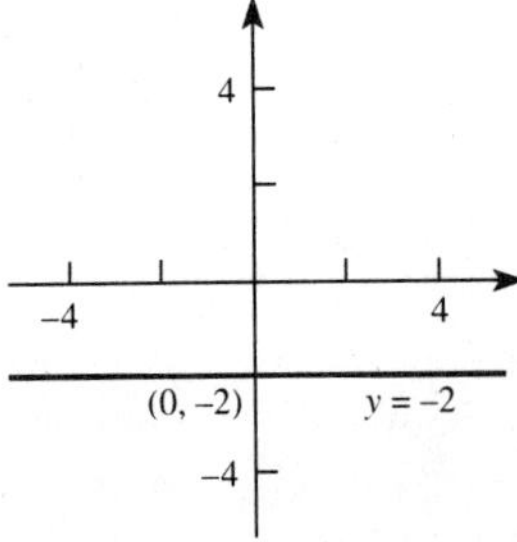

21.

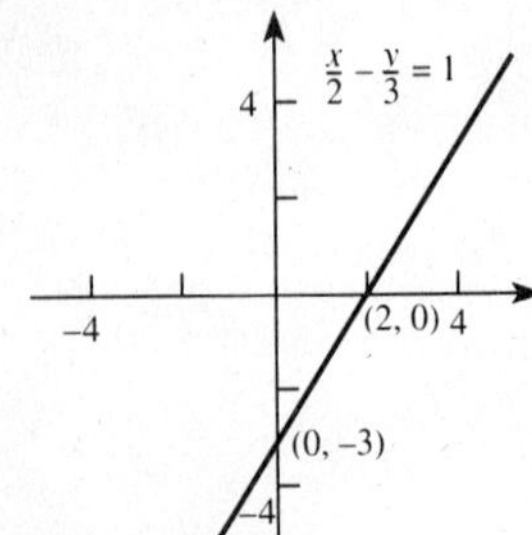

23.

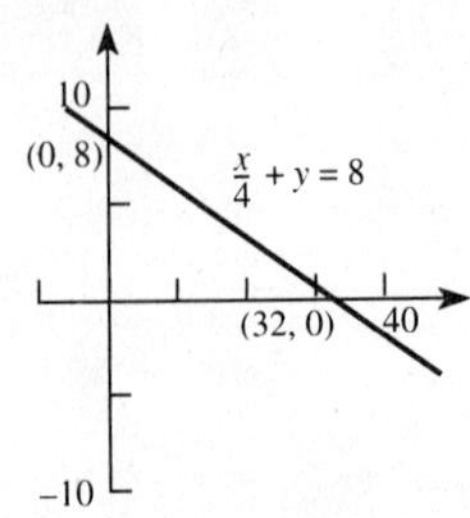

II. 25. Origin is common point

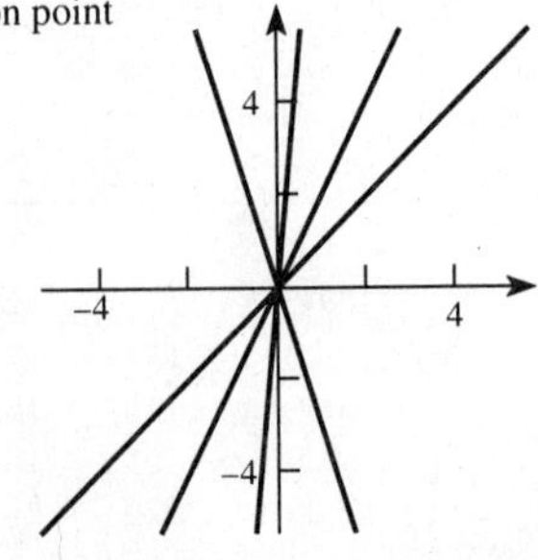

27. y-intercept is 1

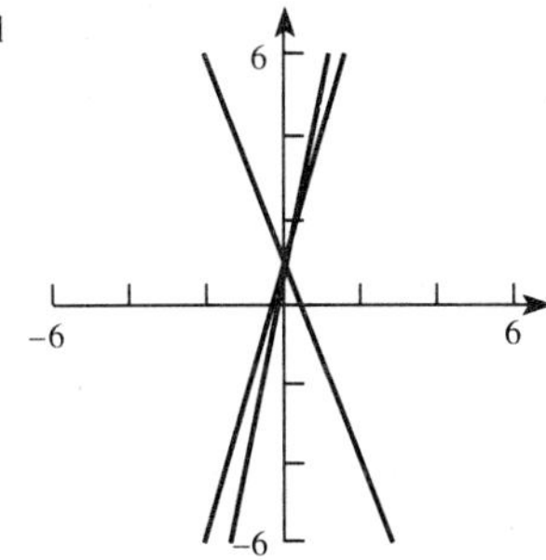

29. Lines perpendicular

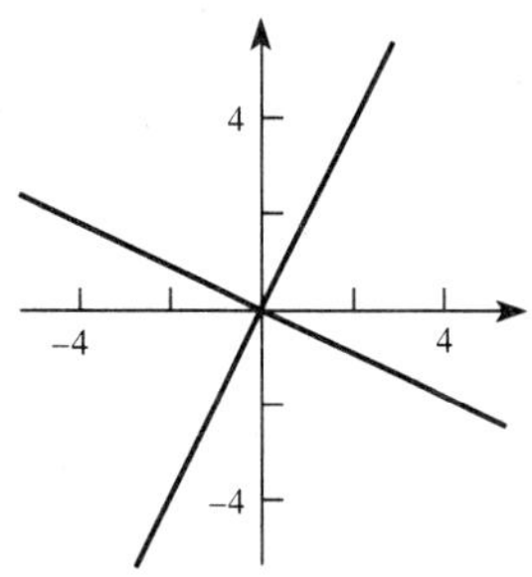

31. ST; true when $a = 1, b = 1$; false when $a = 0, b = 1$

33. AT **35.** Answers vary; probably not since population growth is usually exponential. **37.** [0, 100] by [0, 600] will work

39. Answers vary

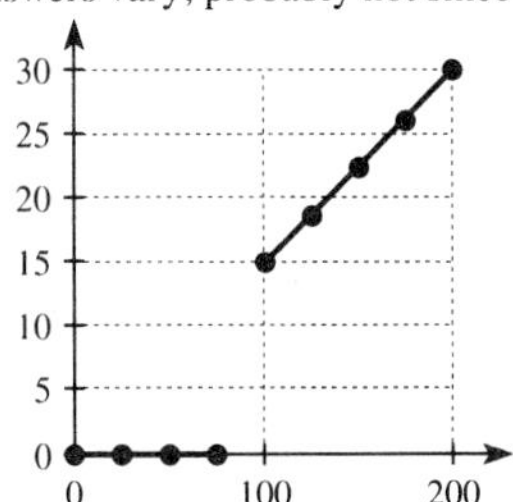

III. 41. $y = 0.20x + 23$

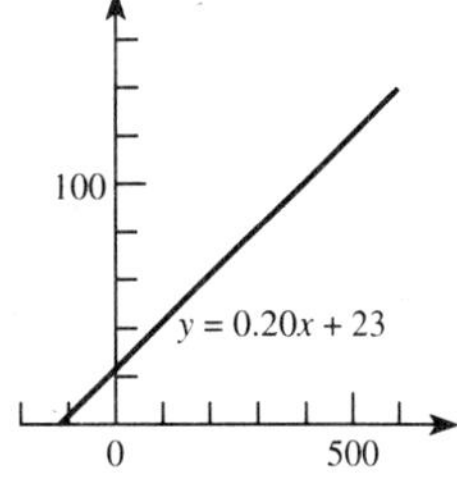

(a) x-intercept: -115, no y-intercept: 23, yes, fixed cost per day **(b)** \$93 **(c)** 635 miles **43.**

$y = 2{,}284 + 0.15(x - 15{,}200)$; $x \geq 15{,}200$ **(a)** x-intercept: -26.67, no y-intercept: 4 **(b)** \$2,704, \$2,104 **(c)** \$17,307 **(d)** no; accurate for $15{,}200 \leq x \leq 19{,}200$ **45.** **(a)** $C = 0.2M + 22$ **(b)** 40 miles

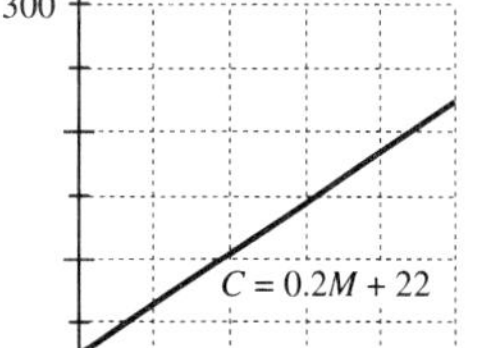

(c) \$62 **(d)** \$244 **47.** **(a)** $50y + 80x = 800$ **(b)** 8.75 hours

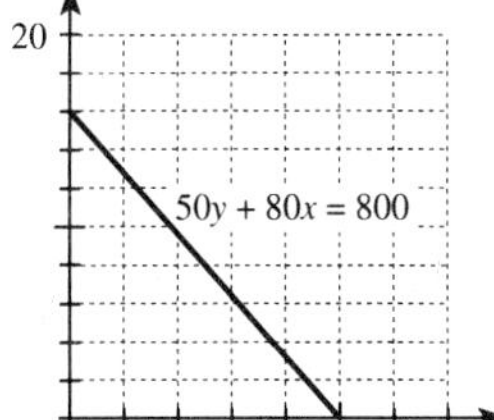

49.

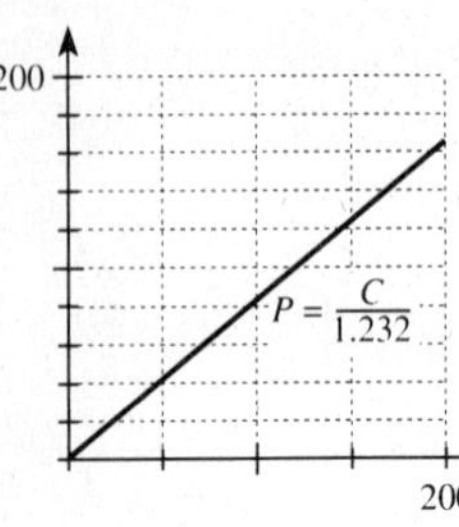

(a) $P = \frac{C}{1.232}$ **(b)** \$81.17 **(c)** \$19.40 **51.**

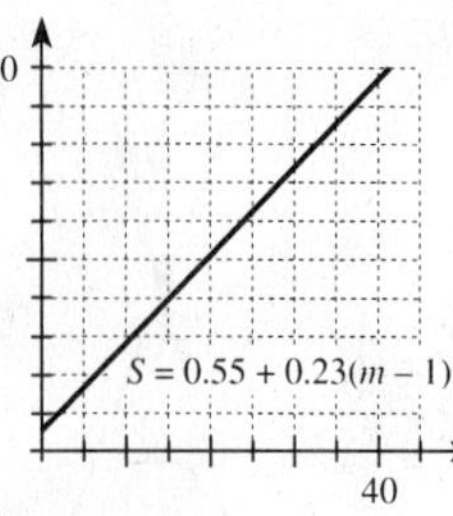

(a) $S = 0.55 + 0.23(m - 1)$ **(b)** \$6.76 **(c)** 29 minutes **53.(a)** $F = 2C + 30$ **(b)**

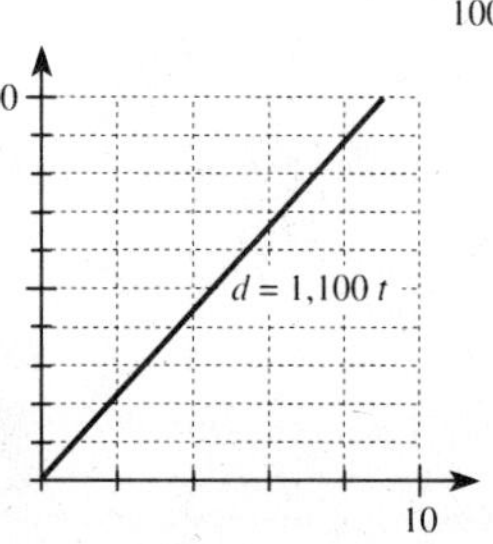

(c) (10, 50) **(d)** 10°C corresponds to 50°F **(e)** 2°F **(f)** when $C < 10$ **55.**

(a) $d = 1100t$ **(b)** 4.8 seconds **(c)** 7150 feet **57.(a)** 25 meters **(b)** Betty: 50 meters Ann: 58 meters **(c)** Ann, Betty **(d)** 7 units of time **(e)** 70 meters **59.(a)** \$5,000 **(b)** 120 widgets **(c)** \$6000 **(d)** \$4000 **(e)** ≈ 4500 (extend graph) **(f)** ≈ 200 (extend graph) **IV. 61.(a)** Score = $7t + 3f$ **(b)** Score 46 if $t = 1$ and $f = 13$; $t = 4, f = 6$; score 42 for $(t, f) = (6, 0); (3, 7); (0, 14)$ **(c)** 1, 2, 4, 5, 8, 11

63.

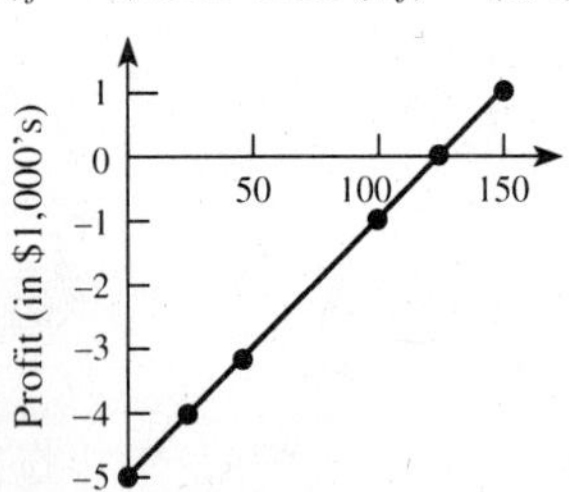

4.2 (page 224)

I. 1.(a) 4 **(b)** (0, –1) **(c)**

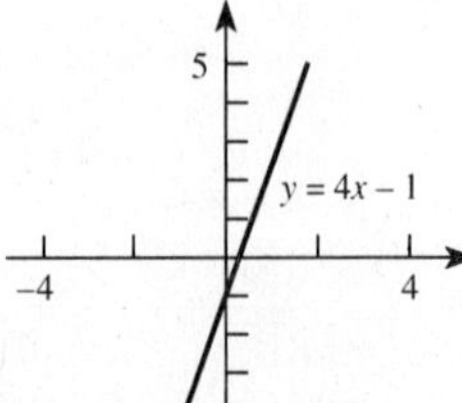

3.(a) $\frac{10}{3}$ **(b)** (0, –10) **(c)**

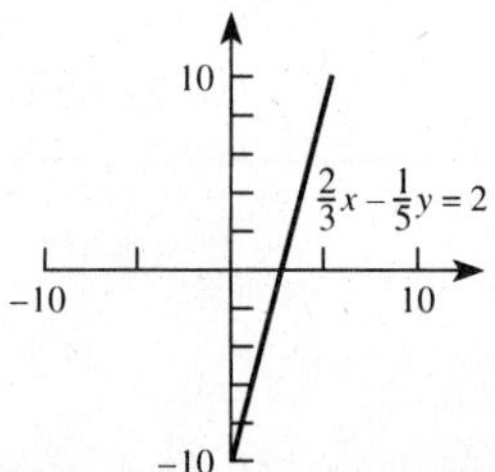

5.(a) undefined **(b)** none **(c)**

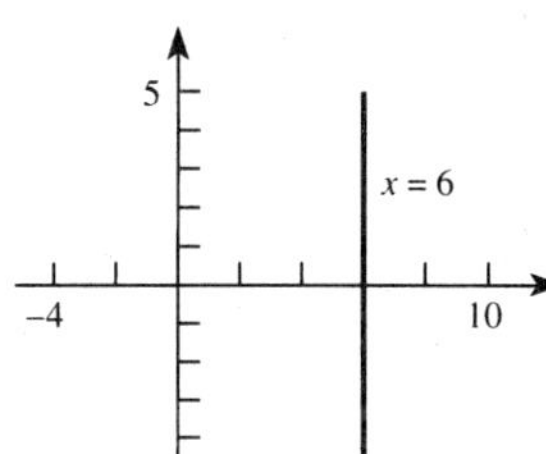

7.(a) -1 **(b)** $(0, 0)$ **(c)** $y = -x$

9.(a) $\frac{5}{6}$ **(b)** $\left(0, \frac{5}{2}\right)$ **(c)** $y = \frac{5}{6}x + \frac{5}{2}$ **11.(a)** 0 **(b)** $(0, -4)$ **(c)** $y = -4$ **13.** $y = 2x - 3$

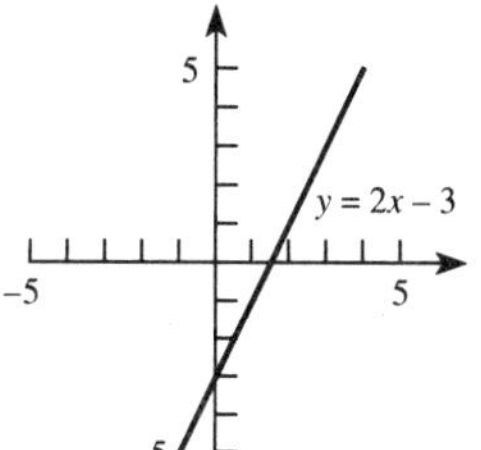

15. $y = -\frac{8}{5}x + \frac{7}{10}$

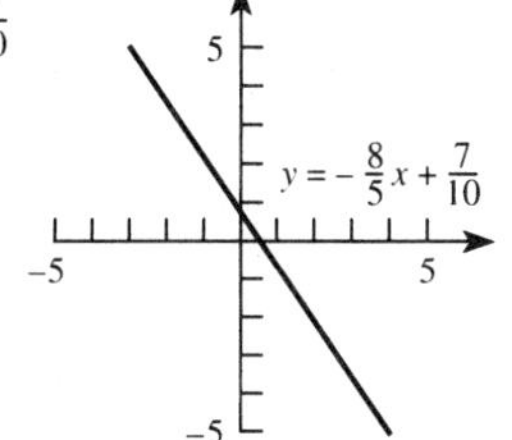

17. $x = 2$

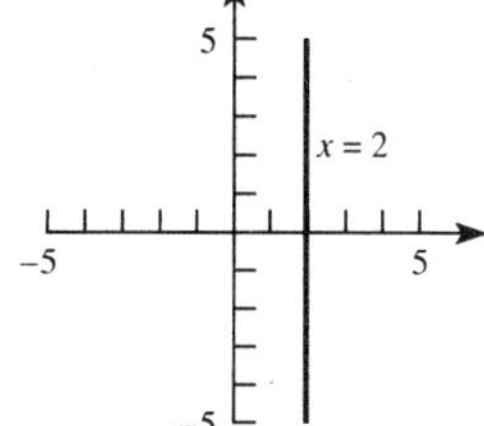

19. $y = -\frac{1}{2}x - \frac{5}{8}$

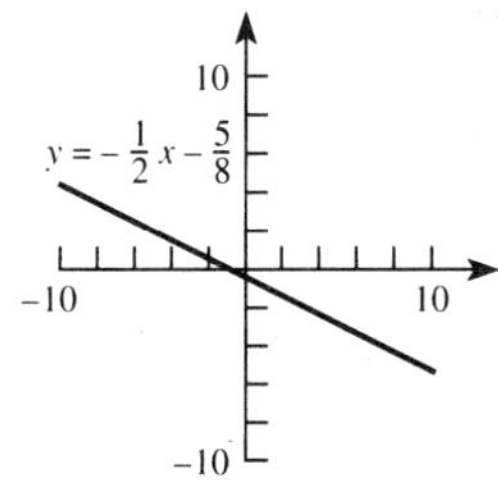

21. $y = -\frac{3}{2}x - 1$

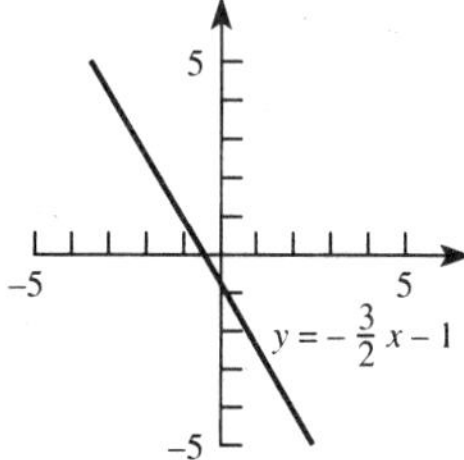

23. $y = -x - 1$

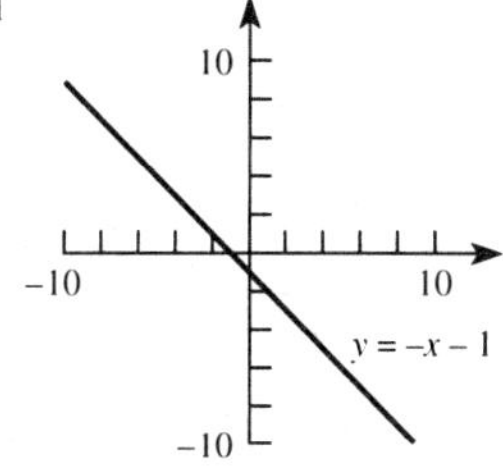

25. $y = 0$

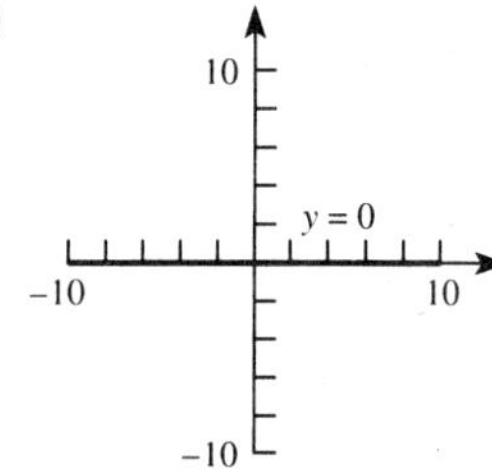

27. $y = -\frac{1}{7}x + \frac{11}{7}$

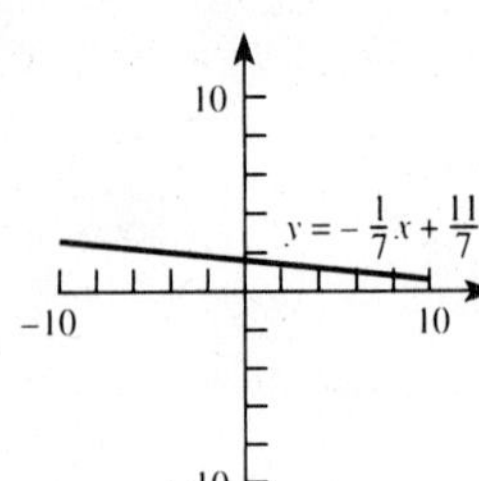

29. $y = 0.234x - 4.586$

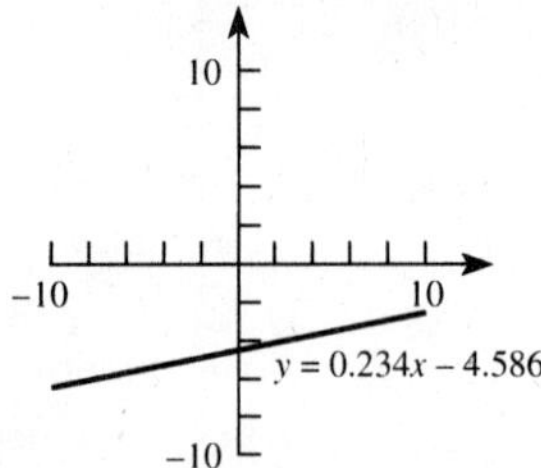

31. $y = 4x - 20$

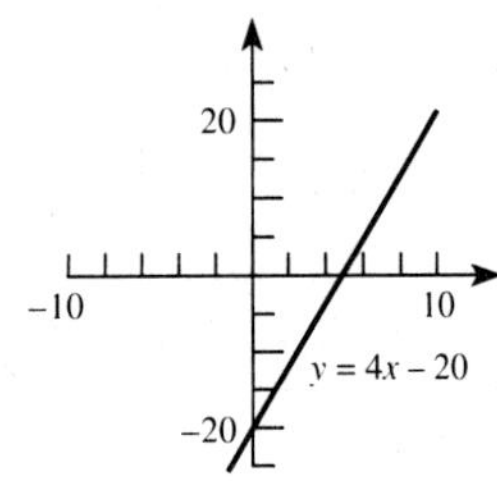

II. 33. Answers vary; of the form $y = mx$

35. Answers vary; $x = -1$, $y = 2x + 2$, $y = x + 1$ **37.** Answers vary; $y = 2x$, $y = 2x + 5$, $y = 2x - 1$; slope must be 2
39. Answers vary; $x = 1$, $x = 2$, $x = 3$ **41.** Answers vary; $x = -1$, $y = x + 1$, $y = 2x + 2$
43. Answers vary; $y = -x + 2$, $y = -0.75x + 2$, $y = -x + 2.5$ **45.** yes
47. The lines are always perpendicular. As a changes, the two graphs rotate about the origin.
III. (some graphs omitted) 49. $y = 25 + 0.17(x - 100)$; $x \geq 100$ The slope is the cost per mile for each mile more than 100; y-intercept is cost for driving 100 miles. **(a)** \$122.92 **(b)** 688 miles

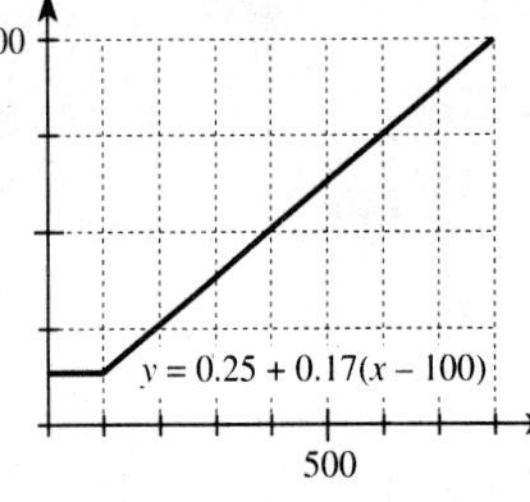

51. $y = 2.88x - 5{,}698$; x = year Slope is approximate number of new cable channels per year; the y-intercept is irrelevant because it predicts how many channels there were in the year 0.

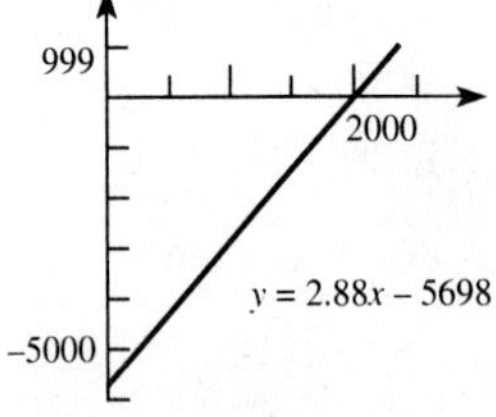

In 2001, the average subscriber will receive 65 channels. This is probably low, so the model is valid for a small range of values for x.

53. $y = \frac{20}{3}x - 301.7$; x = height Slope is increase in weight for each 1-inch increase in height; y-intercept is meaningless.

(a) 198.3 pounds **(b)** 72.3 inches; valid for reasonable heights: $58 \leq x \leq 75$

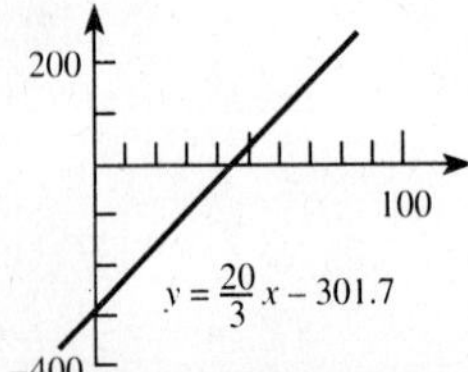

55. $y = 0.21x + 0.20$; x = number of minutes Slope is price increase per 1-minute increase in time of call: y-intercept is the additional charge for the first minute. **(a)** \$5.45 **(b)** about 23 minutes; the equation is valid when $x > 0$

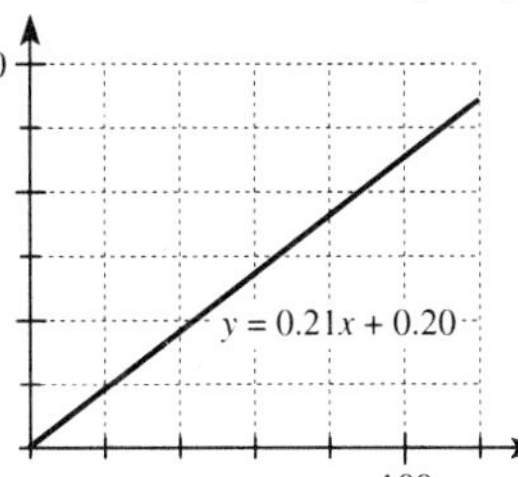

57. $y = 25x - 15$; x = price per case Slope is increase in supply for each \$1 increase in price.

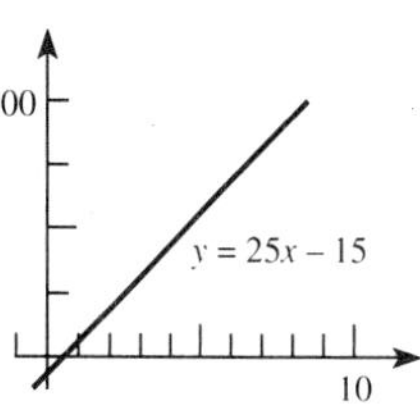

(a) 15 cases **(b)** 60 cases; the equation is valid only when $x \geq \$0.60$ **59.(a)** $y = -0.0053x + 4.28$; x = time in minutes. **(b)** The equation predicts that the mile will be run in 1942 in 4.03 minutes, which is .07 minutes faster than the actual record. The equation predicts that the mile will be run in 1966 in 3.90 minutes, which is .04 minutes slower than the actual record. **(c)** 3.72 minutes

61. $y = \frac{5}{3}x + 455$ **(a)** -273°C **(b)** 538.3 cc **63.(a)** $y = -112.50x + 7600$ **(b)**

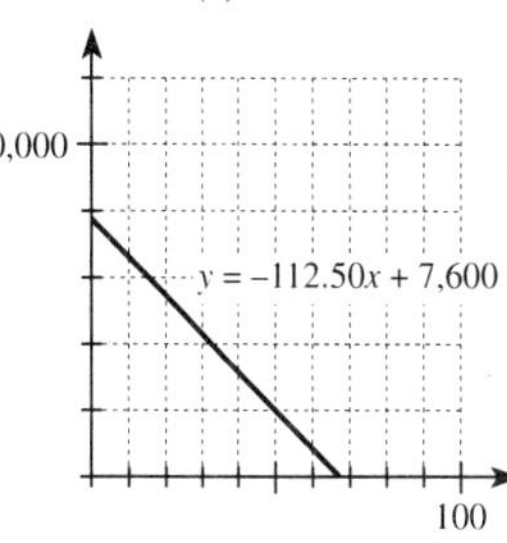

(c) 49.8 months **(d)** slope = amount of money lost per month that the car is kept **(e)** 67.6 months **(f)** \$7,600 **(f)** The car loses a lot of value over the first two years. **65.(a)** $y = 0.23x + 56.2$ **(b)** A male's life expectancy increases 0.23 years for each year after 1920 that he was born. **(c)** 71.15—approximately the same as value in table **(d)** 2002 **67.(b)** \$25,000 **(c)** \$1.25 **(d)** SB **(e)** $22{,}000 \leq x \leq 55{,}000$ **IV. 69.** Compare with $y = mx + b$, $m = \frac{y_2 - y_1}{x_2 - x_1}$

71. A is the origin. $y = 0.7x$ (AB) $y = 0.1x + 12$ (BC) $y = -0.1x + 18$ (CD) $y = -0.7x + 42$ (DE)

4.3 (page 238)

I. (some graphs omitted) 1.(a)

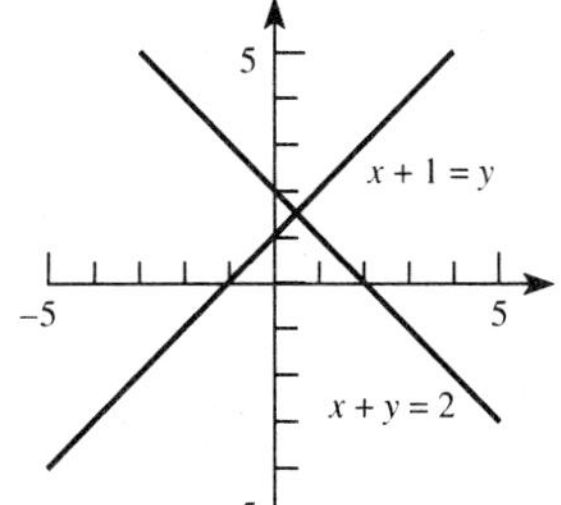

(b) perpendicular **(c)** slopes: 1, -1

3.(a)

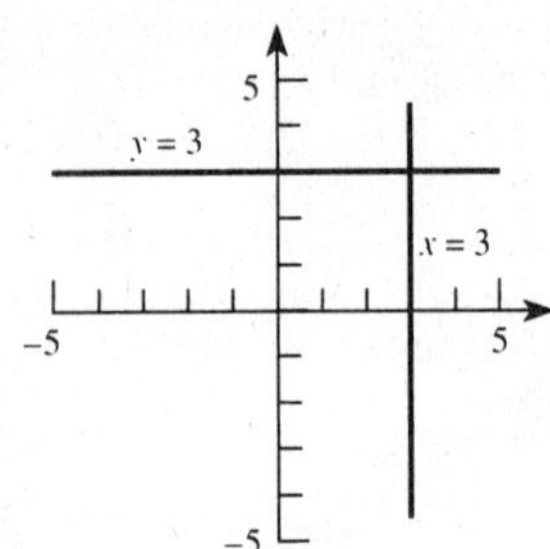

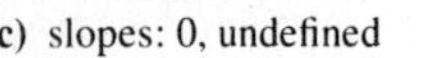

(b) perpendicular **(c)** slopes: 0, undefined **5.(a)**

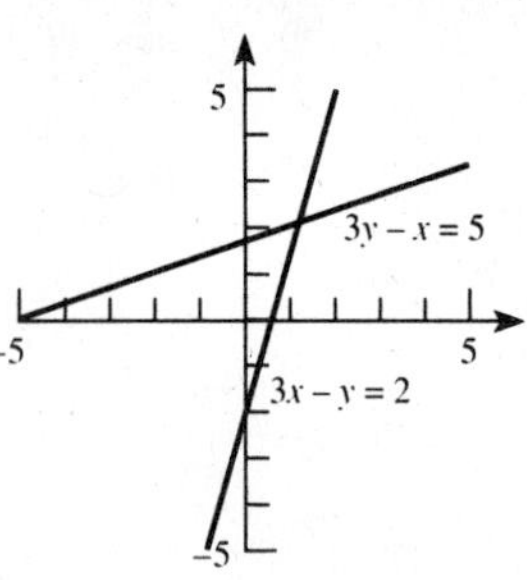

(b) neither **(c)** slopes: 3, $\frac{1}{3}$ **7.(b)** same line: $x = -\frac{1}{4}$ **(c)** slopes: undefined, undefined **9.** $y = 3x + 5$ **11.** $2x - 3y = -8$ **13.** $x = -4$ **15.** $y = -\frac{1}{3}x - 4$ **17.** $y = -\frac{3}{2}x + \frac{7}{2}$ **19.** $y = 4$ **II. 21.** Answers vary; $y = 2x - 1, y = 2x + 3, y = 2x - 5$ **23.** Answers vary; $y = -\frac{1}{2}x, y = -\frac{1}{2}x + 1, y = -\frac{1}{2}x - 3$ **25.** (0, 1) **27.** Answers vary; (0, 4), (3, 0) **29.** Answers vary **31.** Answers vary **33.** $y = \frac{x}{2} - \frac{1}{2}$ for (a) through (d). **III. 35.** Equation of line: answers vary **(a)** ≈ 2,900 **(b)** ≈ 5,900 **(c)** ≈ \$9.50 **37.** Equation of line: answers vary **(a)** ≈ 640 **(b)** ≈ 800 **39.(a)** $y = 0.342x + 57.4$; x = number of years after 1920 **(b)** 79.63 years **(c)** 2000 **41.** Answers vary **(a)** ≈ 3.6 minutes **(b)** ≈ 2000 **43.** $y = -0.0062x + 3.96$; x = number of years after 1954, 3.96 minutes **(a)** 3.67 minutes **(b)** 2000 **45.** $y = 0.0082x + 0.0226$; x = number of years after 1960 **(a)** 35¢ **(b)** 2018 **IV. 47.** $y = 3x - 1$

4.4 (page 257)

I. 1. (7, 3)

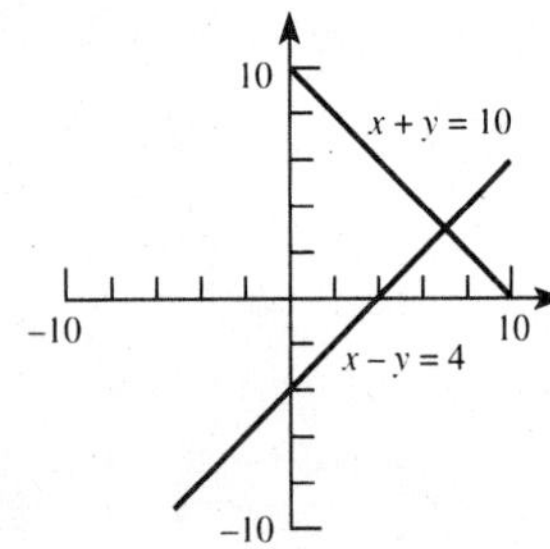

3. (4, 1)

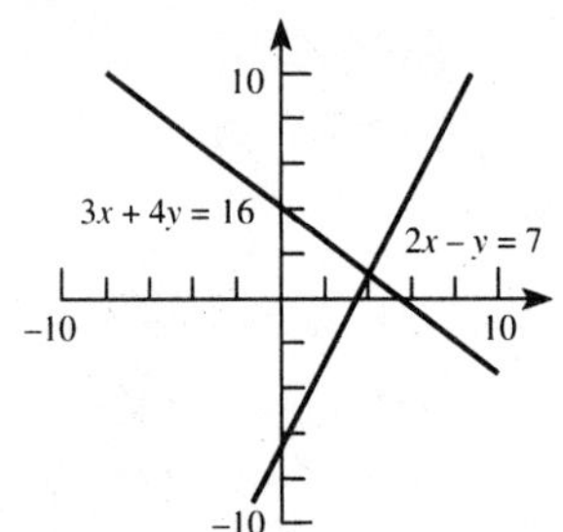

5. (17,3)

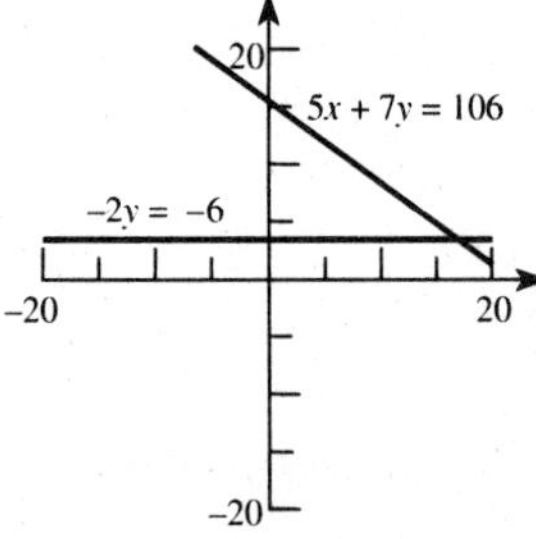

7. (1, 1)

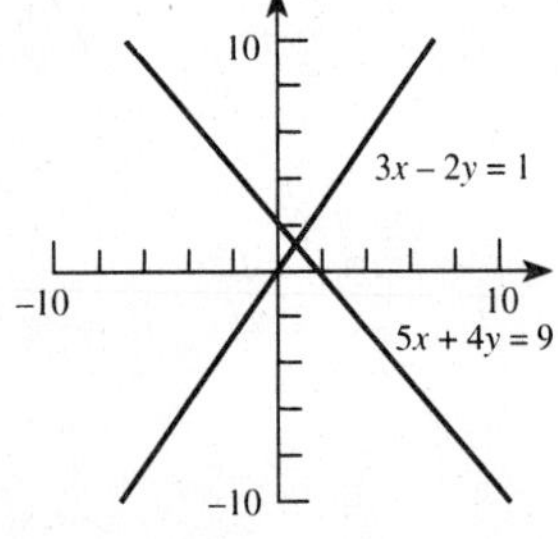

9. (5, −3)

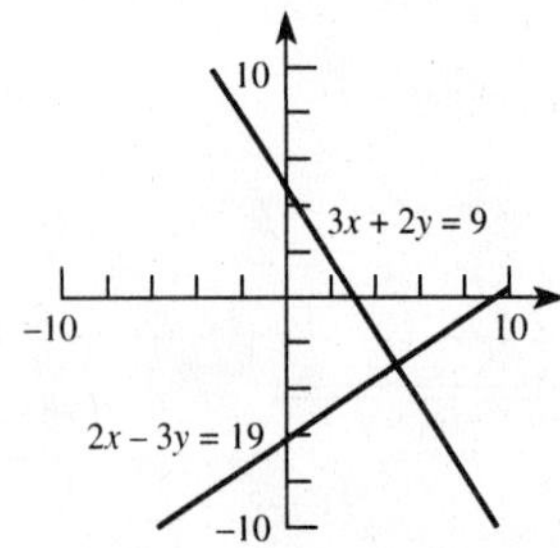

11. $\left(\frac{119}{46}, \frac{1}{46}\right)$

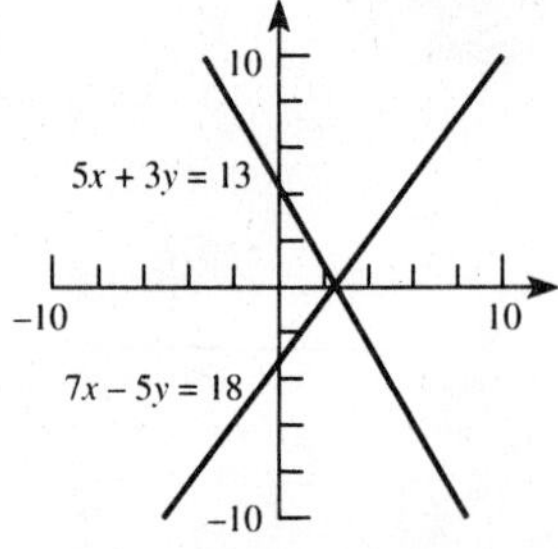

13. no solution

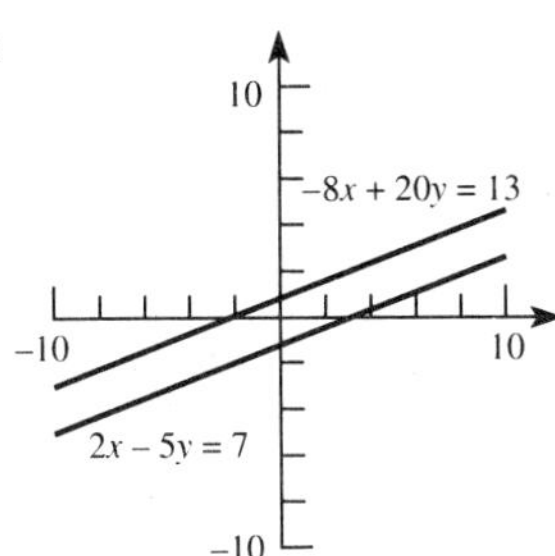

15. $\left(\frac{a+b}{2}, \frac{a-b}{2}\right)$ **17.** $\left(\frac{md+nb}{ad+bc}, \frac{an-cm}{ad+bc}\right)$ $ad \neq -bc$ **19.** (1.3, 2.4)

21. (1.298, 2.393) **23.** $\left(\frac{3}{2}, \frac{1}{2}\right)$ **25.** (2, 3, 4) **27.** (3, −2, 0) **29.** (10, −1, 3) **31.** (1, 2, 3) **33.** $c \neq -2$

II. 35. Answers vary $x + 2y = 0$ $2x + 3y = 0$ **37.** Answers vary $x + y = 0$; $x + y = 1$ lines have same slope and are parallel.
39. Answers vary $x + y + z = 0$ $2x + y + z = 0$ $3x + y + z = 0$
41. Answers vary $x + y + z = 1$ $x + y + z = 2$ $x + y + z = 3$ **43.** All contain the point (−1, 2); any line of the form $nx + (n + 1)y = n + 2$ contains the point (−1, 2) **45.(a)** infinitely many solutions **(b)** one solution (if $x \neq 2$) **(c)** no solutions

III. 47. 25, 21

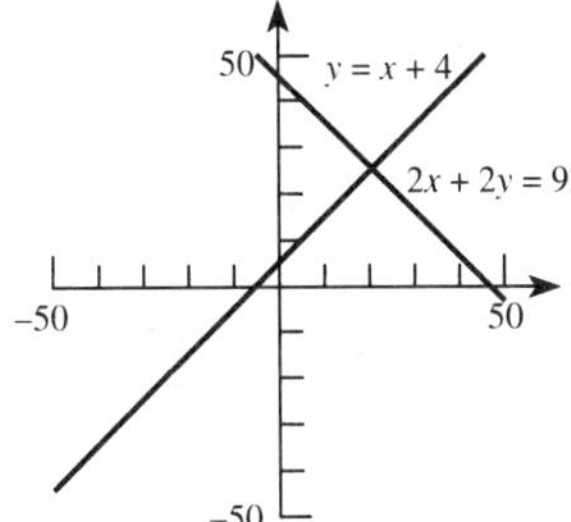

49. \$3 and \$4

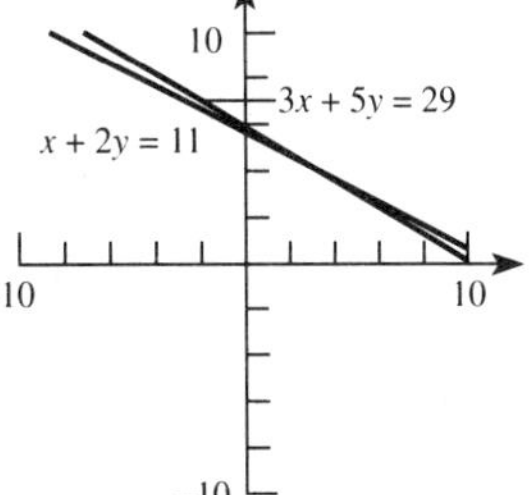

51. 1575 nickels and 1325 dimes

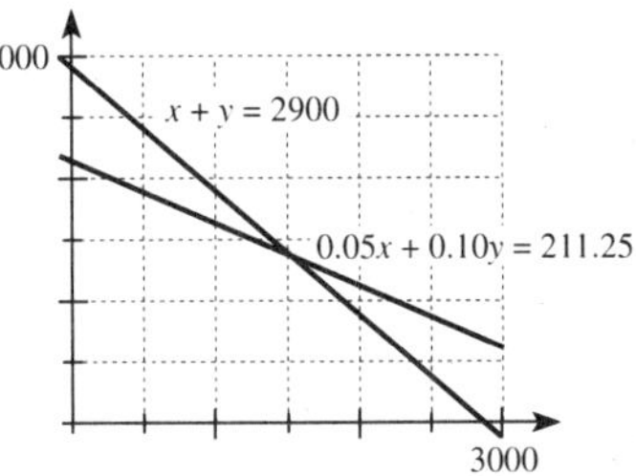

53.(a) 800 miles **(b)** 782 miles

55. 60 ounces and 80 ounces **57.** 357.14 pounds and 642.86 pounds **59.** 64,000 **61.** 54,000 Democrats, 40,000 Republicans, 6,000 Independents **63.** 2.449×10^7 units of coal and 4.422×10^7 units of electricity **IV. 65.** (3, 2)
67.(a) {18, 13, 20, 19, 17, 15, 14, 21, 16} {15, 10, 17, 16, 14, 12, 11, 18, 13} {12, 1, 8, 3, 7, 6, 13, 2}
(b) common sum = 3 × middle entry; sum of four corner entries = 4 × middle entry

CHAPTER 4 SUMMARY REVIEW

I. 1.(a)

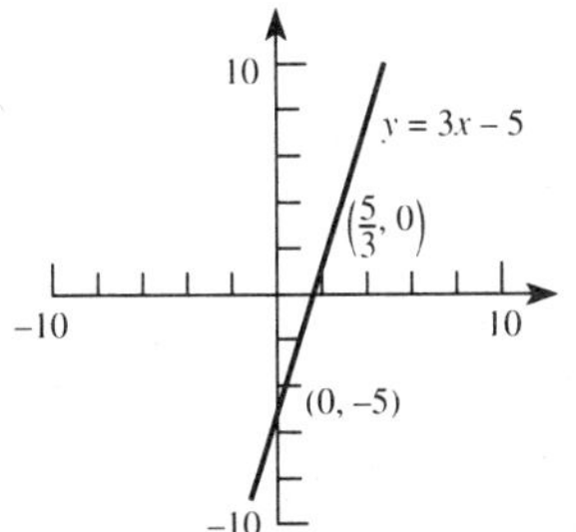

(b) y-intercept = (0, −5); x-intercept = $\left(\frac{5}{3}, 0\right)$ **(c)** slope = 3

3.(a)

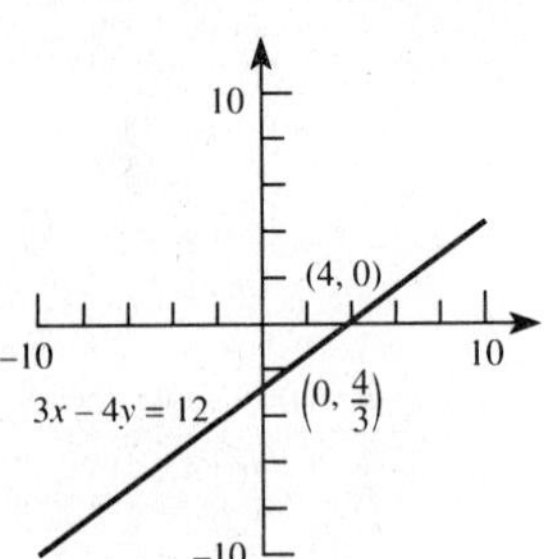

(b) y-intercept $= (0, -3)$; x-intercept $= (4, 0)$ **(c)** slope $= \frac{3}{4}$

5.(a)

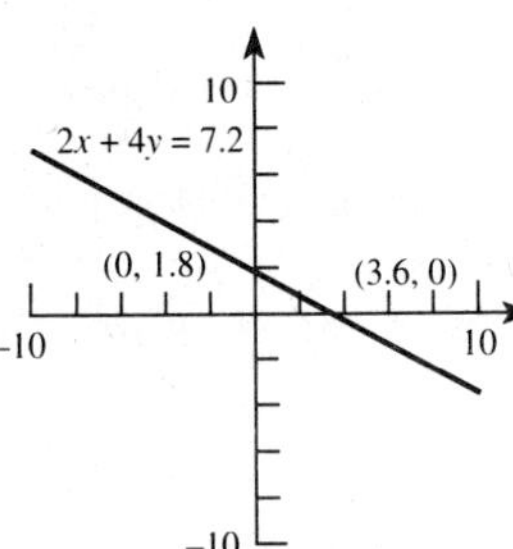

(b) y-intercept $= (0, 1.8)$; x-intercept $= (3.6, 0)$ **(c)** slope $= -0.5$

7.(a)

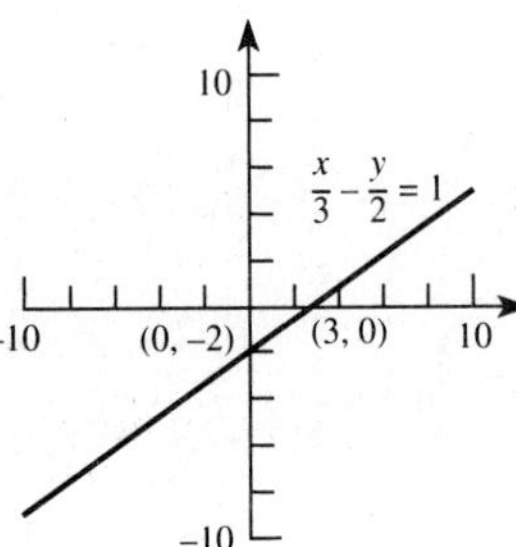

(b) y-intercept $= (0, -2)$; x-intercept $= (3, 0)$ **(c)** slope $= \frac{2}{3}$

9. $y = -2x + 3$

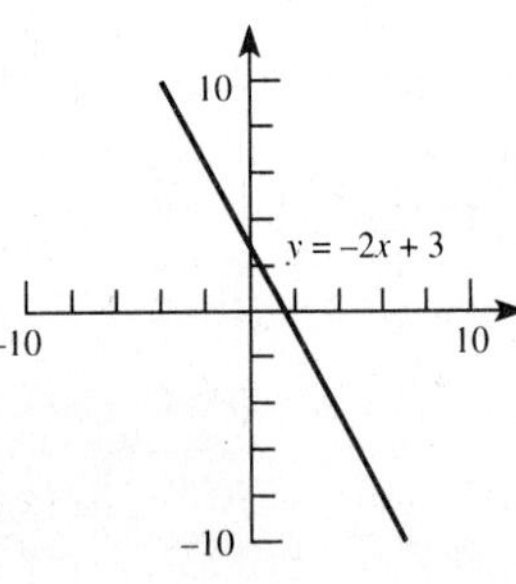

11. $y = \frac{1}{2}x + 2$

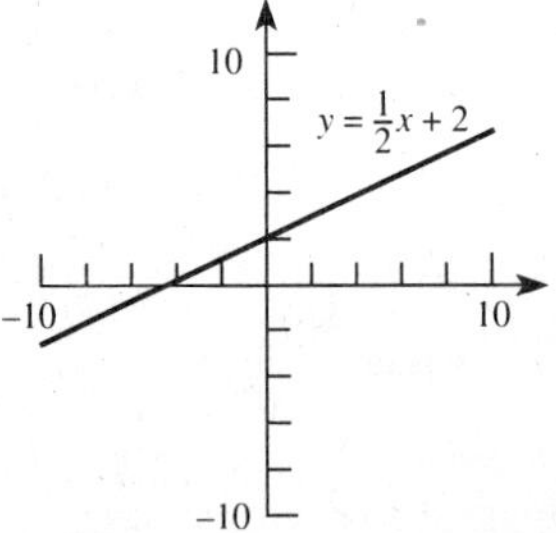

13. $y = -x - 5$

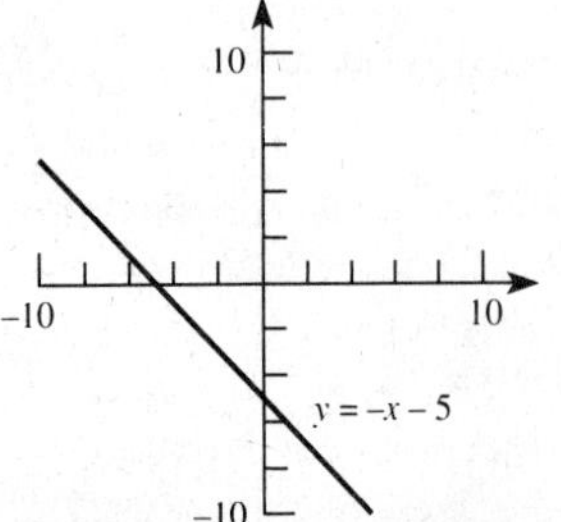

15. $y = -2.471x + 1.962$

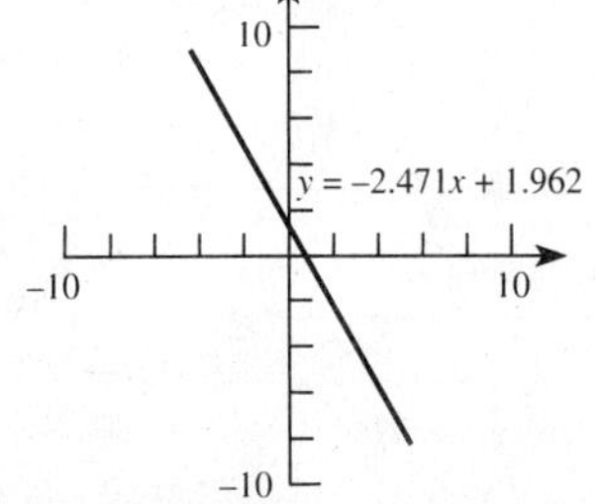

17. $y = 2x + 6$

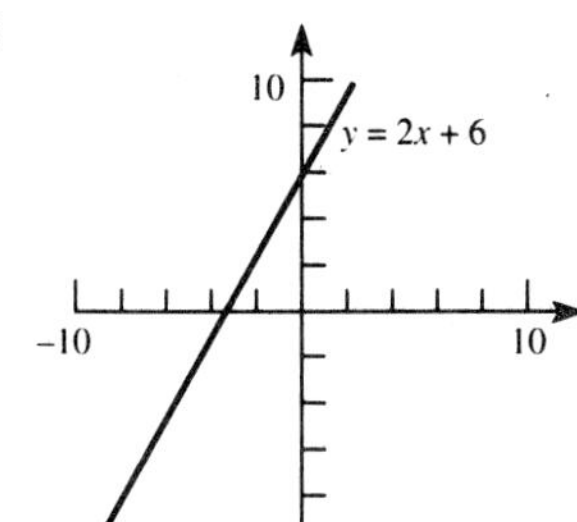

19. $y = -\frac{1}{5}x + 1$

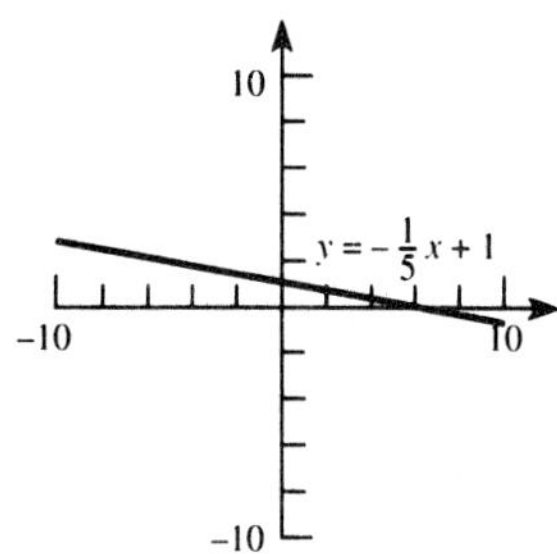

21. $(-1, 2)$

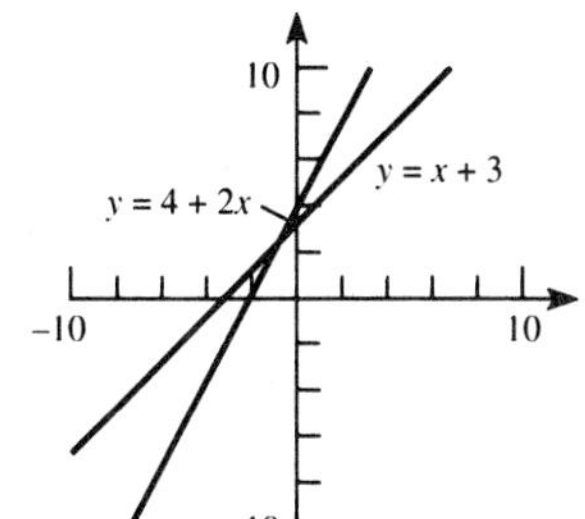

23. no solution

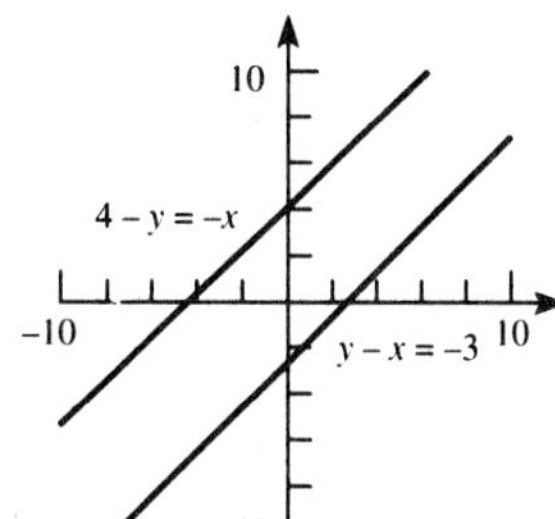

25. $x = \frac{3}{2}, y = \frac{1}{2}$

27. $x = -1, y = -3$ **29.** $x = 24, y = 8$ **31.** $x = -1, y = -1, z = 3$ **33.** $\left(0, \frac{3}{2}\right), \left(1, \frac{5}{4}\right), \left(\frac{7}{4}, \frac{1}{2}\right)$, (2, 0), and (0, 0)

II. 35. $c = -3$ **37.** $c = 0$ **39.** $c = \frac{1}{3}$ **41.** Answers vary **43.** $c = 16$ **III. 45.(a)** cost = $0.27m + 19$

(b) 116.7 miles **(c)**

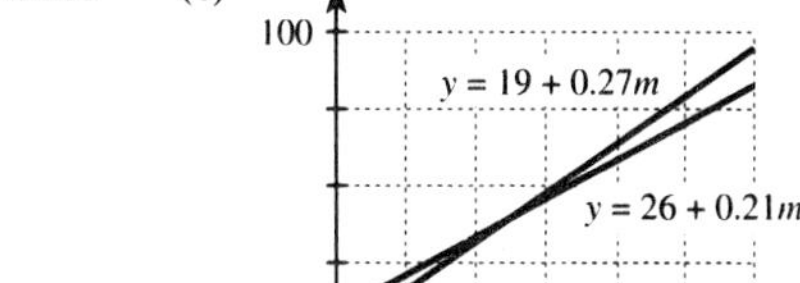

(d) cheaper at 100 miles driving WTH; cheaper at 500 miles driving UDI

47.(a) 12.875 mm **(b)** 392 grams **49.(a)** $2x + y = 46$ **(b)**

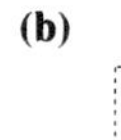

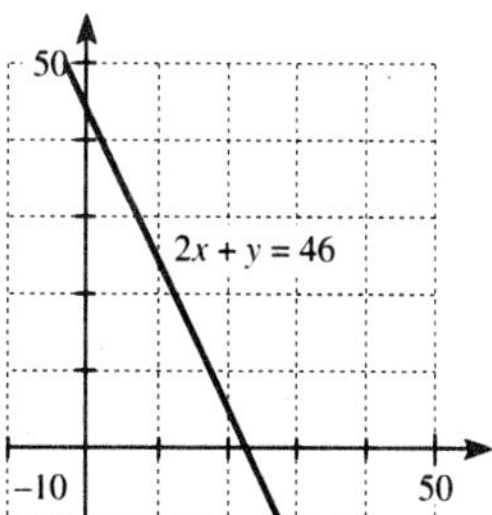

(c) Answers vary: (2, 42); (4, 38); (6, 34) **51.** 12 feet wide and 15 feet long

53. average speed of freight train = 45 mph; average speed of passenger train = 75 mph

55.(a) $2w + t = 60$ **(b)** 2 wins, 6 ties; 3 wins, 5 ties; 4 wins, 4 ties; 5 wins, 3 ties; 6 wins, 2 ties; 7 wins, 1 tie; 8 wins, 0 ties

57.(a) \$700 **(b)** 50 figgles: cost = \$1,200; revenue = \$800
100 figgles: cost = \$1,800; revenue = \$1,800
300 figgles: cost = \$4,000; revenue = \$5,000

(c) break-even point = 100 figgles, \$1,800 for cost

(d) 50 figgles: profit at \$400
100 figgles: profit = \$0
300 figgles: profit at \$1,000

(e) about 170 figgles

59.(a) $\frac{400}{7}x - \frac{2600}{7} = -100x + 1800$; supply = demand $\times$ = price **(b)** price = 0 supply = -371 nonsense; price = 0, demand = 1800 **(c)** supply = 0, price = \$6.50 possible; demand = 0, price = 18 possible **(d)** for each \$1 increase in price, demand decrease by 100, supply increases by 57 **(e)** Answers vary **61.** Answers vary **63.** $y = 48.07x + 104.5$; x = number of years after 1977 **(a)** \$1,210,110 **(b)** 2006

CUMULATIVE REVIEW: CHAPTERS 1–4

I. 1.(a) $\frac{5}{12}$ **(b)** $-\frac{67}{30}$ **3.** 2 **5.** $-\frac{8}{9}$ **7.** $\frac{243}{5}$ **9.** $-3 + 10x - x^2$ **11.** $-16a^3bc^{-1}$ **13.** $-6x - 2$ **15.** $x = \frac{7}{3}$ **17.** $x = -12$ **19.** $x = -\frac{7}{4}, y = -\frac{9}{4}$ **21.** no solutions **23.** $\left\{x \middle| x \geq \frac{1}{5}\right\}$

0 $\frac{1}{5}$

25. $x = 13$ or $x = -5$ **27.** $\{x | 0 < x < 2\}$ **29.** $|x - 1| = 3$ **II. 31.** (c) **33.** (f) **35.** (d) **37.** true **39.** false, let $x = -1$ **41.** true, x^2 is always positive, so $\frac{1}{x^2}$ is too **43.** (c) **45.** (b) **47.** (e) **49.** (c) **51.** (c) **53.** (d) **55.** -16, -26, $34 - 10(n - 1)$, assuming $n \geq 1$ **57.** $\frac{1}{243}, \frac{1}{729}, \frac{1}{3^n}$ **59.(a)** lines with slope of 2 **(b)** lines with slope of $-\frac{1}{2}$
III. 61.(a) $\frac{32}{3}$ **(b)** $\frac{16}{9} < x < \frac{5}{9}$ **63.(a)** 27,000 people **(b)** $3^n \cdot 1{,}000$ **65.** midnight **67.** $-1.1°$C **69.** 680 elk **71.** \$7750 **73.(a)** 40.96% **(b)** $(0.8)^n \cdot 100\%$ **75.(a)** 16 years **(b)** \$35,000

CHAPTER 5

5.1 (page 287)

I. 1. $(y + 2)(y + 1)$ **3.** $(w - 7)(w + 5)$ **5.** $x^2 + x + 1$ **7.** $(x - 5)(x + 5)$ **9.** $(9x - 7)(9x + 7)$ **11.** $(x + 7)^2$ **13.** $(0.4q + 1)^2$ **15.** $(2x - 3)(x + 5)$ **17.** $(4x + 3)(3x - 5)$ **19.** $(3y + 1)(y + 3)$ **21.** $4x^2 + x - 21$ **23.** $xy(3x - 2y)$ **25.** $y^2(y^2 + 3y + 1)$ **27.** $3(4x - 5)(2x + 3)$ **29.** $0.3(x - 5)(x + 5)$ **31.** $7y^2(y + 1)(2y - 3)$ **33.** $50x^2y^2 - 50xy + 1$ **35.** $y^2(4y - 3)(3y + 5)$ **37.** $(x - 1)(x + y)$ **39.** $-3(x + 1)$ **41.** $(x - y)(2 + z)$ **43.** $(x - 2y)(x + 5)$ **45.** $(x - 1)(a - b)^2$ **47.** $(2x^2 - y)(2x^2 + y)(4x^4 + y^2)$ **49.** $(x^2 - 4)^2 = (x - 2)^2(x + 2)^2$ **51.** $\frac{5ac^3}{9b}$ **53.** $3x - 2$ **55.** $6x^4z^3 - 3x^3y + 5y^2z^7$ **57.** $-x$ **59.** $\frac{x - 1}{x + 1}$ **61.** $\frac{x + 3}{x + 5}$ **63.** $\frac{3x + 1}{3x}$ **65.** $3x - y + 4$ **67.** $\frac{3x - y}{x + 1}$ **II. 69.** $y + 3$ **71.** $x + 4$ **73.(b)** 10; $x^2 + 7x + 10 = (x + 2)(x + 5)$ **75.** $x^2 + 2x + \boxed{1} = (x + 1)^2$ **77.** $x^2 - \boxed{6}x + 9 = (x - 3)^2$ **79.** $\boxed{4x^2} - 20x + 25 = (2x - 5)^2$ **81.** $\frac{x}{x + 1}$ **83.** $-(x + 1)$ **85.** (b), (d), (e) **87.** $\frac{x^4 - x^2}{x^3 - x^2} = \frac{x^2(x^2 - 1)}{x^2(x - 1)} = \frac{x^2(x + 1)(x - 1)}{x^2(x - 1)} = x + 1$ **89.** Answers vary: $\frac{3x^2 + 3x}{2x + 2}, \frac{3x^2 - 3x}{2x - 2}, \frac{3x^3}{2x^2}$ **91.(a)** Answers vary **(b)** Answers vary: $\frac{2(x^2 - 9)}{x + 3}, \frac{4x^2 - 36}{2x + 6}$ **III. 93.(a)** $\frac{10}{3}$ hours **(b)** $\frac{50}{21}$ hours **(c)** $\frac{50}{18 - x}$ hours **(d)** $\frac{50}{18 + x}$ hours **(e)** Answers vary; $\frac{25}{7}$ hours; 6.25 hours **95.(a)**

999
Time
100
Percent

(b) 1.11 hours **(c)** It approaches infinity **(d)** Answers vary; 42 hours, 1.67 hours, 6.67 hours **(e)** Answers vary; 37.5%, 85.7% **97.(a)** $10^{14} - 16 = 99{,}999{,}999{,}999{,}984$ **(b)** .9999999975 **99.** 11 **IV. 101.(b)** $(x - 3)(x^2 + 3x + 9)$ **(c)** $(2x + 5)(4x^2 - 10x + 25)$ **(d)** $(4x + y)(16x^2 - 4xy + y^2)$ **(e)** $(0.2x^2 - y^2)(0.04x^4 + 0.2x^2y^2 + y^4)$ **(f)** $2(3xy - 2z^2)(9x^2y^2 + 6xyz^2 + 4_z^4)$ **(g)** $0.000003(x + 30y)(x^2 - 30xy + 900y^2)$

5.2 (page 298)

I. 1. $\frac{8}{21}$ **3.** $\frac{3}{y}$ **5.** $\frac{3y^2}{x^5}$ **7.** $\frac{(x+1)(x-3)}{15x^2}$ **9.** $-\frac{x^2}{x+3}$ **11.** $\frac{2}{x-y}$ **13.** 1 **15.** $\frac{y(y-2)}{(y+1)(y-1)}$ **17.** $\frac{x(x+y)}{a+b}$ **19.** x **21.** $\frac{x+2}{2}$ **23.** $3x(x+1)$ **25.** $4(m-1)$ **27.** $\frac{m(m+3)^2}{2}$ **29.** $\frac{(x^2-1)^2}{x^3(x+2)} = \frac{(x-1)^2(x+1)^2}{x^3(x+2)}$ **31.** $\frac{2y(x-y)^2}{3(25x^2-y^2)(x-2y)} = \frac{2y(x-y)^2}{3(5x+y)(5x-y)(x-2y)}$ **33.** $\frac{(x+2)(x-7)^2}{x+3}$ **35.** $\frac{(x-1)^2(2x+7)}{12(5x-3)}$ **37.** $2x-1$ **39.** $\frac{2(x-y)}{x(x+y)}$ **II. 41.** $\frac{(x+2)^2(x+3)}{2(x^2+4)}$ **43.** Because for all values of x not equal to 2 and -3, the equation gives a fraction with a nonzero numerator. **45.** $\frac{8}{9}$ **47.** $\frac{14}{2x+5}$ **49.** $\frac{10}{9}$ **51.** $\frac{x-2}{x}$ **53.** $\frac{x}{2(x+1)} \cdot \frac{3x^2+3x}{x}$; not necessarily **55.** (a), (b), (d)

III. 57. $\frac{1}{1{,}024}$ **59.** approaches 0; approaches 1 **IV. 61.** In step 5, you cannot divide by $(x-y)$ because $x-y=0$ (since $x=y$).

5.3 (page 309)

I. 1. $\frac{61}{56}$ **3.** $\frac{11}{21}$ **5.** $\frac{3y+2x}{xy}$ **7.** $\frac{20-3y}{4y^2}$ **9.** $\frac{y^2-xy+3z}{xyz}$ **11.** $\frac{3ab+a-b}{b(a-b)}$ **13.** $\frac{(3x-y)(x+2y)}{(x-y)(x+y)}$ **15.** $\frac{5-x^2-3x}{(x-4)(x+3)}$ **17.** $\frac{x(x+6)}{(x+2)(x-2)}$ **19.** $\frac{-4x}{(x+1)(x-1)}$ **21.** $\frac{23ab-12a^2-7b^2}{(2a-b)(2b-a)}$ **23.** $\frac{4x+3-x^2}{(x-1)(x+1)}$ **25.** $\frac{5}{2(x+2)}$ **27.** $-\frac{x^2+6x+45}{2(x-3)(x+3)^2}$ **29.** $\frac{4}{a-2}$ **31.** $\frac{x+2}{x+1}$ **33.** $\frac{a^2-2a+8}{a(a-2)}$ **35.** $\frac{2(x^2+7x-3)}{(x+2)(x+7)(x+8)}$ **37.** $\frac{3a^3+a+2}{3(a+1)(a-1)}$ **39.** $\frac{x-6}{x(x+2)}$ **41.** $\frac{-1}{(x-1)(x+1)}$ **43.** $\frac{6x^3+x^2-28x+7}{(x+2)(x-2)}$ **45.** $x(2x+3)$ **47.** $x(y-x)$ **49.** 1 **51.** $4+x$ **53.** $-\frac{1}{y}$ **55.** $\frac{(y-x)(x-y)}{xy}$ **II. 57.** $\frac{1-x^3}{x}$ **59.** $\frac{x(x+6)}{(x-2)(x+2)}$ **61.** Answers vary: $\frac{1}{3}+\frac{1}{3}$ **63.** Answers vary: $x+\frac{-x}{x+1}$ **65.** Answers vary: $\frac{2}{3}-\frac{1}{3}$ **67.** Answers vary: $1-\frac{x-1}{x}$ **69.(a)** Answers vary: $\frac{1}{a}-\frac{1}{a+1}=\frac{1}{a(a+1)}$ **(b)** always true (for a > 1) **71.(a)** $\left(\frac{a}{b}\right)^2+\frac{b-a}{b}=\frac{a}{b}+\left(\frac{b-a}{b}\right)^2$ or $x^2+(1-x)=x+(1-x)^2$ **(b)** always true

III. 73. $x+\frac{1}{x}$ **75.(a)** 110 mph **(b)** 1,100 mph **(c)** 11,000 mph **(d)** 110,000 mph; certainly when the speeds meet or exceed 7×10^8 mph. **77.(a)** $\frac{1}{3}+\frac{1}{9}+\frac{1}{27}+\frac{1}{81}=\frac{40}{81}$ $\frac{1}{3}+\frac{1}{9}+\frac{1}{27}+\frac{1}{81}+\frac{1}{243}=\frac{121}{243}$ $\frac{1}{3}+\frac{1}{9}+\frac{1}{27}+\frac{1}{81}+\frac{1}{243}+\frac{1}{729}=\frac{364}{729}$ **(b)** $\frac{3^{10}-1}{2\cdot3^{10}}; \frac{3^{50}-1}{2\cdot3^{50}}; \frac{3^N-1}{2\cdot3^N}$ **(c)** It approaches $\frac{1}{2}$. **79.(a)** $\frac{156}{625}; \frac{781}{3{,}125}; \frac{3{,}906}{15{,}625}$ **(b)** $\frac{5^{10}-1}{4\cdot5^{10}}; \frac{5^{50}-1}{4\cdot5^{50}}; \frac{5^N-1}{4\cdot5^N}$ **(c)** It approaches $\frac{1}{4}$. **IV. 81.** $\frac{1}{2}-\frac{1}{n+1}=\frac{n-1}{2(n+1)}$ **83.(a)** Yes ($x=y=24$; $x=28$, $y=21$ are two solutions) **(b)** Yes ($x=y=96$; $x=72$, $y=144$ are two solutions)

5.4 (page 325)

I. 1. 30 **3.** -8 **5.** $-\frac{5}{12}$ **7.** $-\frac{17}{14}$ **9.** 6 **11.** contradiction **13.** $-\frac{3}{2}$ **15.** $\frac{1}{12}$ **17.** contradiction **19.** 3 **21.** 2 **23.** 4 **25.** 5 **27.** 0 **29.** $l=\frac{V}{w^2}$ **31.** $h=\frac{V}{\pi r^2}$ **33.** $R=\frac{W}{I^2}$ **35.** $g=\frac{4\pi^2L}{p^2}$ **37.** $x=\frac{my}{2y-m}$ **39.** $M=\frac{c-1}{w}$ **II. 41.** $x=-\frac{3}{14}$ **43.** $t=\frac{6m}{a-b}$ **45.** contradiction **47.** Answers vary: $\frac{1}{x}-1=0$ **49.** Answers vary: $\frac{1}{x-1}+x=0$ **III. 51.(a)** $t=\frac{K-Km-1}{Km}$ **(b)** $t=0.4815$, $t=0.5110$ **(c)** $m=\frac{K-1}{K+Kt}$ **53.** 32.1430 feet **55.(a)** 0.328 **(b)** 13 hits **57.** $\frac{12}{7}$ hours **59.** 56.52 mph **61.(a)** 58.6 mph **(b)** 5.7 hours (including the 4 hours for the meeting) **63.** 2.71 meters **65.** $x=-\frac{1}{67}$ **67.** Edith's speed $=0.62\cdot$(Bill's speed) **69.** 5 km/h **71.(a)** $V=\frac{Y+L}{P-S-F-T}$ **(b)** 1,600,000 cubic meters **73.(a)** $C=\frac{P-NT-D}{N}$ **(b)** $N=\frac{P-D}{C+T}$ **(c)** \$0.14 per ball **IV. 75.** sides of length 4 and width 4 or length 6 and width 3 **77.** 4, 4 and 3, 6 **79.** $n=3, 4, 6$ **81.** $ad^2+b^2c=0$; $b=d$, $a=-c$ is one case.

CHAPTER 5 SUMMARY REVIEW

I. 1. $3xy(x - 2y^2 + 5xy^2)$ **3.** $2x(xy - 1)(xy + 1)$ **5.** $(x - 3)(x + 5)$ **7.** $10x^2 - 3xy + y^2$ **9.** $(a - 3)(x - y)$ **11.** $\frac{a - 3b}{a + b}$ **13.** $\frac{a(a - 1)}{a + 1}$ **15.** $\frac{6x^2z + 36xy - 5yz}{12x^3y}$ **17.** 1 **19.** $\frac{(x - 6)(x + 1)(a - 6)}{3x(a - 2)}$ **21.** $\frac{-2(x + 2)(x - 2)^2}{3x(x - 4)}$ **23.** $\frac{(11x + 6)}{3}$ **25.** $-x$ **27.** $x = 6$ **29.** $b = -1$ **31.** contradiction **33.** $a = 2$ **35.** $x = \frac{1}{2}$ **37.** $b = \frac{ad}{c}$ **39.** $M = \frac{c - 1}{W}$ **41.** 5,000.05 **43.** 2.002 **II. 45.** (a), (b), (d), (e), (f) **47.** Answers vary: $\frac{28}{50}$ **49.** Answers vary: $\frac{x^3 - 4x}{x^3}$ **51.** Answers vary: **(a)** $1 + \frac{x - x^2 - 1}{x^2 + 1}$ **(b)** $\frac{x^2 + x + 1}{x^2 + 1} - 1$ **53.** Answers vary: (a) $\frac{2x^2}{x + y} \cdot \frac{y}{x}$ **(b)** $\frac{2x^2}{x + y} \div \frac{x}{y}$ **55.** Answers vary **57.(a)** 4, 8 **(b)** $(x + 8)(x + 4)$ **(c)** Answers vary **III. 59.(a)** $\frac{259}{1,296}$; $\frac{1,555}{7,776}$; $\frac{9,331}{46,656}$ **(b)** $\frac{6^{10} - 1}{5 \cdot 6^{10}}$, $\frac{6^{30} - 1}{5 \cdot 6^{30}}$, $\frac{6^N - 1}{5 \cdot 6^N}$ **(c)** It approaches $\frac{1}{5}$. **61.** 3 hours and 20 minutes **63.** 9 hours. Not very realistic because torrential rainstorms rarely last 9 hours. **65.** 20 km/h, 60 km/h **67.** 48 mph **69.** $-\frac{1}{6}$ **71.** $\frac{-6}{55}$ **73.(a)** length $= 2x + 3$, width $= x + 5$ **(b)** $2x + 3$ (top), $x + 5$ (side) **75.** 40 mph **77.** 15 mph **79.** 24 inches **81.** 6

CHAPTER 6

6.1 (page 343)

I. 1. $A = 43$ square meters **3.** $B = 19.87$ board feet **5.** $A = 803.84$ square inches **7.** for $t = 5$, $R = 36.6$; for $t = 10$, $R = -16.8$ **9.** $y = -16x^2$

x	y
0	0
1	−16
−1	−16
2	−64
−2	−64

11. $y = 40x - 10x^2$

x	y
0	0
1	30
2	40
3	30
−1	−50
−2	−120

13. $A = 4\pi r^2$

x	y
0	0
1	4π
2	16π
3	36π
4	64π

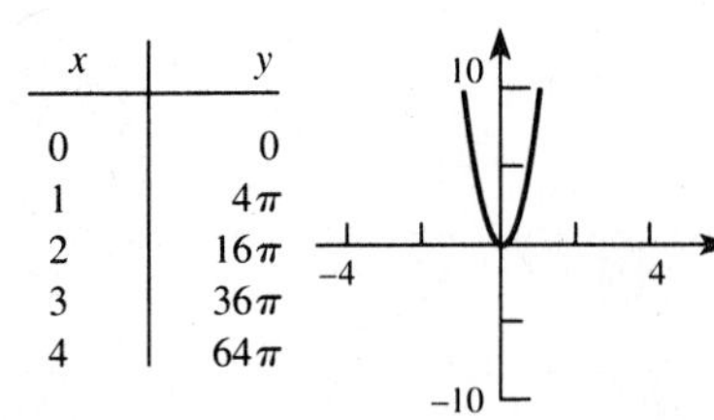

15. $b = 0.22h^2 - 0.71h$

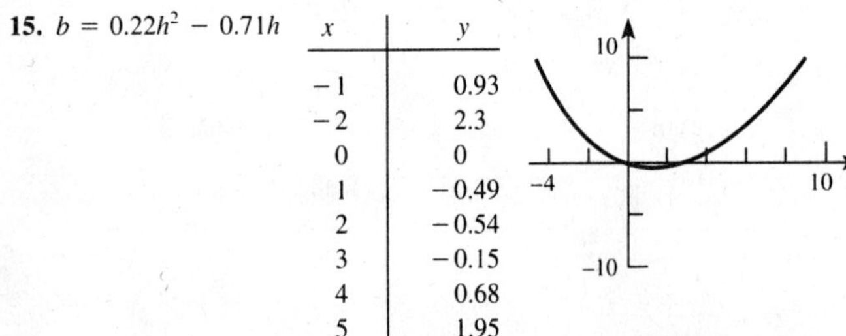

x	y
−1	0.93
−2	2.3
0	0
1	−0.49
2	−0.54
3	−0.15
4	0.68
5	1.95

II. 17. $x \approx 5.66$ **19.** $x \approx -3.62$

21.(a) 15 sec **(b)** 2250 ft **(c)** 45 sec **(d)** 8 sec and 23 sec

23. Answers vary: Estimate points of intersection of $y = 1750$ and the given graph. **25.** Answers vary: visualizing may be easier than calculating; answer is approximate. **III. 27.** $x + \frac{1}{x} = 9, x = 8.9, .1$ **29.** $14^2 + \left(\frac{1}{2^x}\right)^2 = x^2$ or $\frac{3}{4}x^2 = 196, x = 16.2$
31.(a) $628{,}000 = x^2 - 6x$, where x is the number of tubes **(b)** Answers vary: $x \approx 800$ **33.(a)** $15 = \pi(w + 1)^2 - \pi(1)^2$ or $15 = \pi w^2 + 2\pi w$ **(b)** Answers vary: $w \approx 1.4$ **35.(a)**

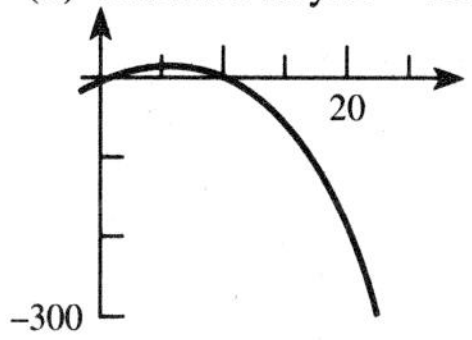

20
−300

(b) for $x = 0, R = -10$, so there is a \$10 million loss
for $x = 8, R =$ \$6 million
for $x = 20, R = -$\$210 million, so a loss
(c) $x = 5$ **(d)** two or eight franchises

37.(a) $A = l\left(\frac{2400 - l}{2}\right)$

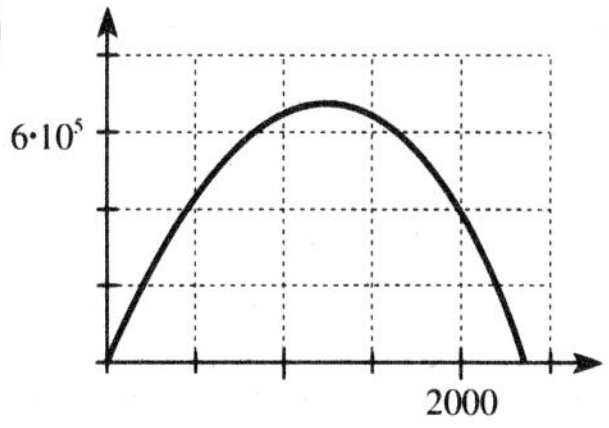

$6 \cdot 10^5$
2000

(b) parabola **(c)** $l = 1200, w = 600$

39.(a) $h = -2t^2 + 10t$

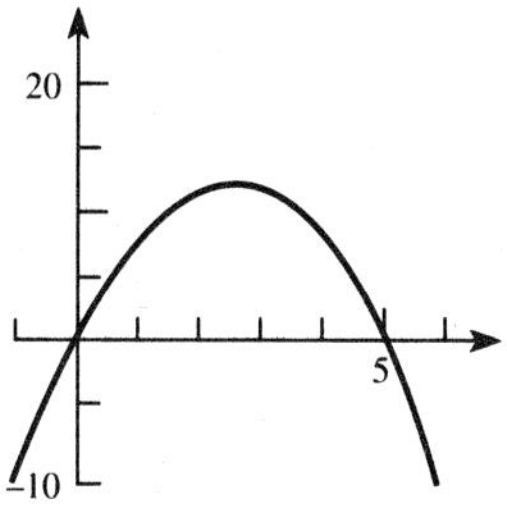

20
5
−10

(b) $t = 2.5$ tenths of a second **(c)** h maximum $= 12.5$ inches

(d) $t \approx 0.7$ tenths of a second
$t \approx 4.3$ tenths of a second (0.07 seconds, 0.43 seconds)

IV. 41.(a) 20 mph: 46 ft; 40 mph: 140 ft; 50 mph: 205 ft; 75 mph: 420 ft

(b)

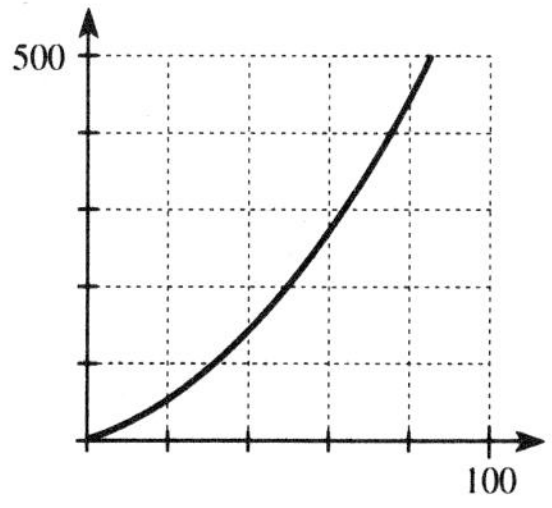

500
100

(c) $V = 33.16$ mph, $d = 100$ ft; $V = 56$ mph, $d = 250$ ft; $V = 67.8$ mph, $d = 350$ ft

6.2 (page 354)

I. (In Exercises 1–15, the graph intersects the x-axis at points whose x-coordinates are the solutions.) **1.** 3, −3 **3.** $0, \frac{7}{3}$ **5.** 6, −1 **7.** 2.5, −3 **9.** 3, −3 **11.** $-\frac{1}{2}$ (graph touches x-axis at $x = -\frac{1}{2}$) **13.** $5, \frac{-2}{3}$ **15.** $\frac{1}{2}, \frac{1}{3}$ **17.** 3, −4 **19.** 1, 4 **21.** −1 **23.** −2 **25.** $\frac{4}{3}$ **27.** no solution **29.** no solution **31.** no solution **II. 33.** can't divide by x; $x = 0$ or $x = 1$ **35.** $x \neq 1$; $x = -2$ only answer **37.** Answers vary **39.** Answers vary: $x^2 + 4x - 21 = 0$ **41.** Answers vary:

$6x^2 + x - 2 = 0$ **43.** Other solution is $\frac{7}{4}$. **III. 45.** $w(2w + 10) = 2{,}100$ $w = 30, 2w + 10 = 70$ **47.** $x + \frac{1}{x} = \frac{41}{20}$ $x = \frac{4}{5}$, or $x = \frac{5}{4}$ **49.** 45 mph **51.** 80 members **53.** after 1 second and after 3 seconds; after 4 seconds

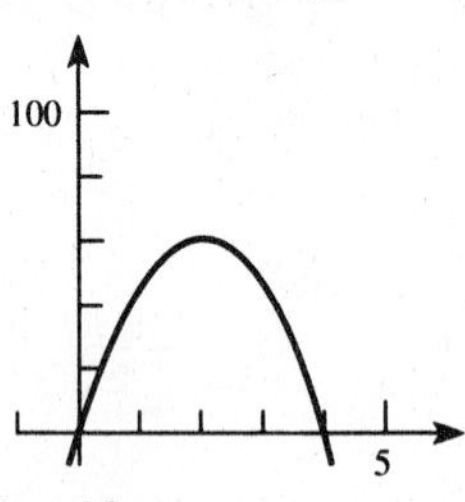

IV. 55. $c = 0, 1, -3, -8, -15, -24, -35, -48, \ldots$; numbers of the form $-n(n + 2)$ **57.** $a = -2, -10, -16$ are three examples **59.** $x = 0, \frac{1}{2}, \frac{-1}{3}, \frac{1}{4}, \frac{-1}{5}, \frac{1}{6}, \frac{-1}{7}, \ldots \frac{(-1)^n}{n}$; $n = 2$ to 100

6.3 (page 377)

I. 1. $16x^2 - 3x - 9 = 0, a = 16, b = -3, c = -9$ **3.** $x^2 + 2x - 6 = 0, a = 1, b = 2, c = -6$ **5.** $18x^2 - 149x - 72 = 0, a = 18, b = -149, c = -72$ **7.** $x^2 - 14 = 0, a = 1, b = 0, c = -14$ **9.** $5(\sqrt{2})$ **11.** $2(\sqrt{2})$ **13.** $5(\sqrt{10})$ **15.** $7(\sqrt{2})$ **17.** $20(\sqrt{3})$ **19.** $3, -3$ **21.** $\sqrt{3}, -\sqrt{3}$ **23.** $8, -6$ **25.** $-0.755, -13.245$ or $\sqrt{39} - 7$, $-\sqrt{39} - 7$ **27.** -4 **29.** 2, 0.5 **31.** $1 \pm \frac{\sqrt{21}}{2}$ **33.** no solution (graphs not shown for Exercises 35–65) **35.** 4.82, -5.82 **37.** $3.08, -1.08$ **39.** 2 **41.** $-2, 3$ **43.** $\frac{3}{4}, -\frac{5}{2}$ **45.** $5, -5$ **47.** $-\frac{1}{5}$ **49.** $0, \frac{4}{7}$ **51.** $0.81, -0.53$ **53.** $2.21, -2.71$ **55.** 1, 6 **57.** $6, -\frac{4}{3}$ **59.** $2.89, -5.89$ **61.** $-0.01, -9.99$ **63.** contradiction **65.** contradiction

II. 67. Answers vary **69.** factored incorrectly: $(3x + 1)(2x - 1)$; $x = -\frac{1}{3}$, or $\frac{1}{2}$ **71.** $x = \frac{6 \pm \sqrt{36 + 84}}{6}$, $x = 2.83$ or -0.83 **73.** $x^2 + 4x - 21 = 0$ **75.** $x^2 - 2 = 0$ **III. 77.** $x^2 - 24 = 0$ $x = 4.9$ **79.** $x = 16$ **81.** $k = 0, 4$

83.(a) **(b)** at $\frac{3}{4}$ seconds and at 2 seconds **(c)** after $2\frac{3}{4}$ seconds

85. $\frac{4\pi}{3} \cdot \left(\frac{3}{2}\right)^3 = 4\pi x^2$ (x = radius of the cylindrical tank) $\rightarrow x = \frac{3\sqrt{2}}{4}$ meters **87.** $l = 560$ m, $w = 40$ m **89.** ≈ 28 mph and ≈ 43 mph

91.(a) **(b)** 1.34 sec or 4.66 sec

93.(a) ≈ 17.75 miles **(b)** ≈ 130 miles **(c)** ≈ 220 miles **95.** $t = \frac{V_0 \pm \sqrt{V_0 - 2gh}}{g}$ **97.(a)** $x = \frac{n \pm \sqrt{n^2 - 4}}{2}$ **b.** 7.87, 0.127, 0,07, -14.93 error, error, 6.85, 0.15, 99.99, 0.01 **(c)** no real solution **99.(a)** cubic equation, $1000(1 + i)^3 = 1225$ **(b)** $\approx 7\%$ **(c)** $\approx 8.8\%$

6.4 (page 389)

I. 1. two complex, 0 times **3.** two real, 2 times **5.** one real, 1 time **7.** two complex, 0 times (graphs omitted from Exercises 9–27) **9.** $x = \frac{4}{3}i$ or $-\frac{4}{3}i$ **11.** 16.5 or 0.48 **13.** $1.5 + 1.66i$ or $1.5 - 1.66i$ **15.** $4 \pm 7i$ **17.** 1.16 or -1.73

19. 0.92 or -1.52 **21.** $\pm\sqrt{2}i$ **23.** 0, $\pm\frac{i}{2}$ **25.** $\pm\sqrt{2}i$ or $\pm\sqrt{3}i$ **27.** $\pm 1.48i$ or ± 0.83 **29.** $13 - i$ **31.** $6\frac{1}{2} - \frac{13}{3}i$ **33.** $38 - 5i$ **II. 35.** $c \leq 4$ **37.** $b \geq 2(\sqrt{10})$ or $b \leq -2(\sqrt{10})$ **39.** $a \leq 1.8$ **41.** $n \geq 2$ or $n \leq -2$ **43.** Answers vary **III. 45(a)** $t = 1.16$ sec or 7.17 sec **(b)** no solution **(c)** 50 ft is below the maximum, while 150 ft is above the maximum **(d)** yes for $t = 3\frac{1}{3}$ or 5 sec **47.** For any real-number value of N **49.(a)** Each number is equal to the sum of the previous two numbers: $(N_i = N_{i-1} + N_{i-2})$ **(b)** The ratio is converging to ≈ 1.618. **(c)** $q = \frac{1 + \sqrt{5}}{2}$; same value as the golden ratio **IV. 51.** Player A: Choose b and then choose a or c so $b^2 - 4ac > 0$.

6.5 (page 408)

I. 1. (0, 0) **3.** (0, -5) **5.** $\left(\frac{5}{12}, \frac{167}{24}\right)$ **7.** $\left(\frac{3}{4}, \frac{25}{32}\right)$ **9.** (0.35, 0.15) **11.** $y = x^2 - 9$

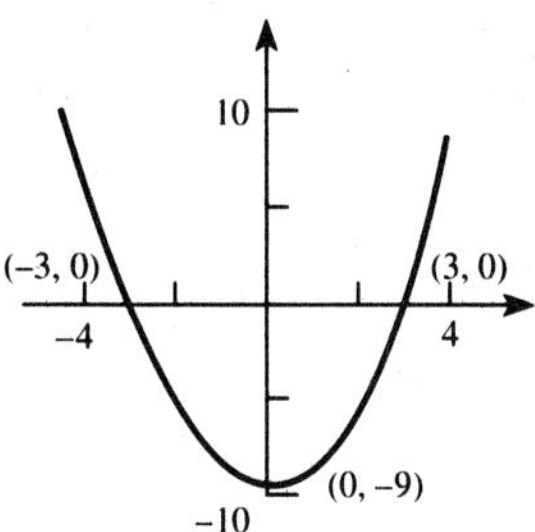

13. $y = x^2 + 6x + 5$

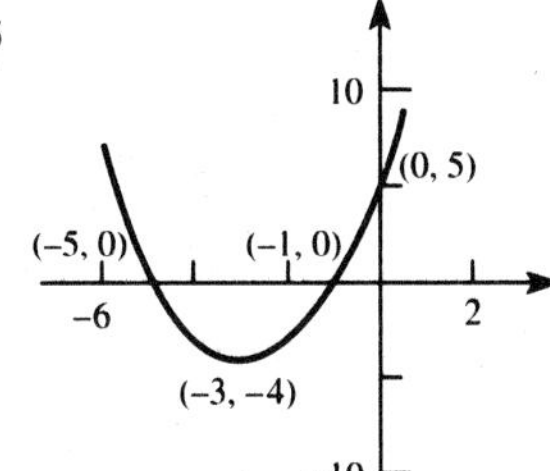

15. $y = -x^2 + 5x - 8$

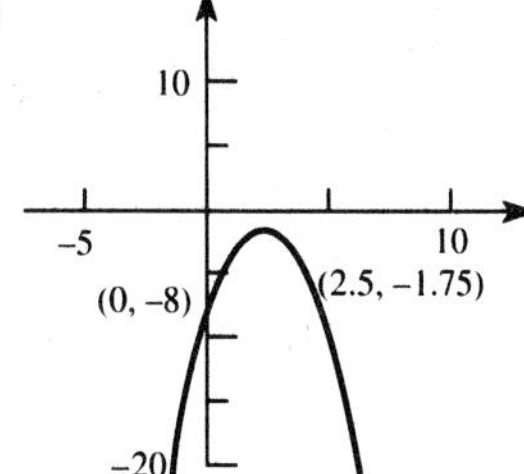

17. $y = 5x + 6$

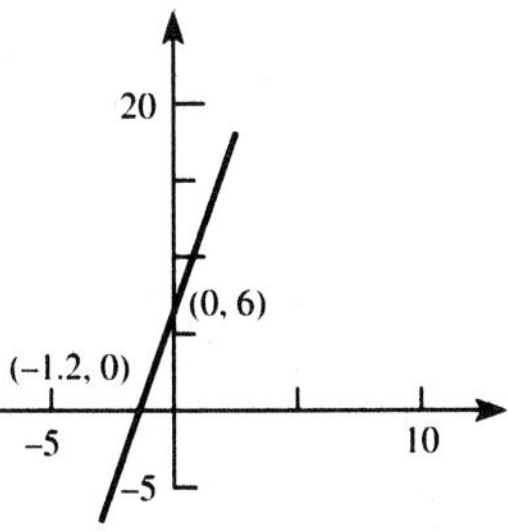

19. $y = 0.8x^2 - 0.16x + 4$

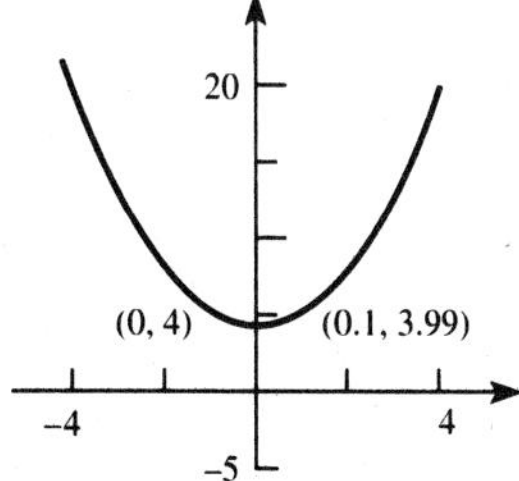

21. $y = -2x + 2$

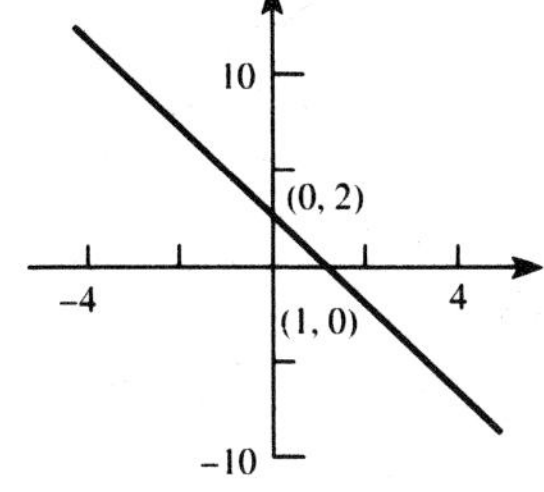

23. $y = 0.85x - 0.25x^2 + 1$

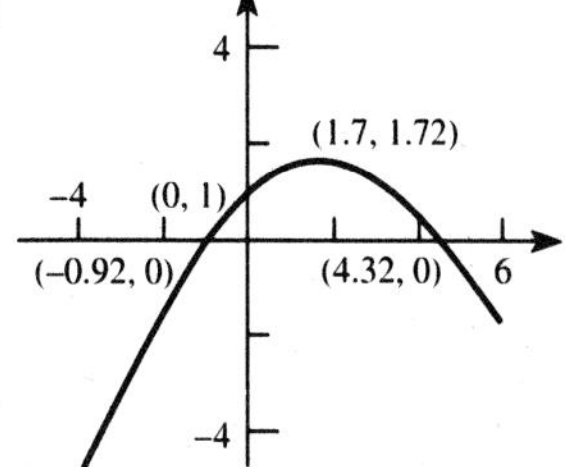

25. $y = x^2 + x + 1$

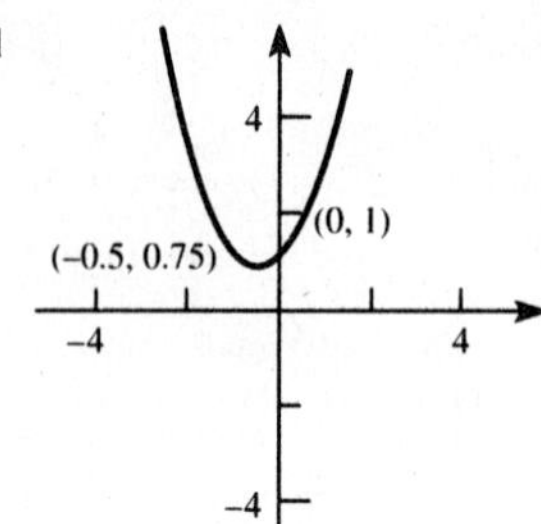

27. $y = x^2 + 5x + 10$

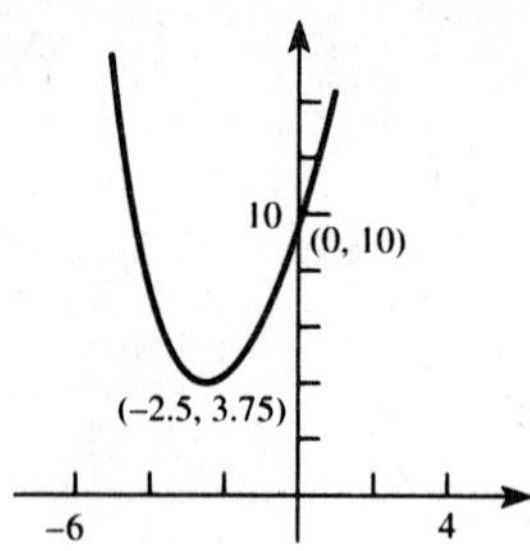

29. $y = 2x^2 + 4x + 2$

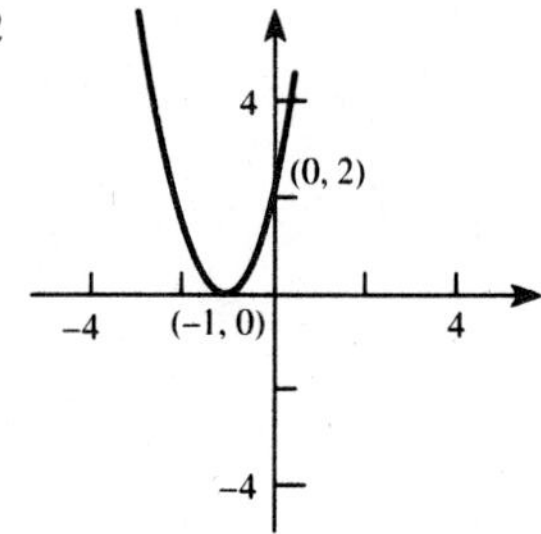

II. 31. (graphs omitted) $a > 1$: narrows; $0 < a < 1$: widens; change sign: flips in x-axis **33.** (graphs omitted) moves graph right or left **35.** Answers vary **37.(a)** $y = ax^2$ **39.(a)** $y = a(x - 2)^2 + 3$ **41.(a)** $y = ax^2 + bx + 5$ **43.** $C < 1$ **45.** $a > -\frac{9}{20}$ **III. 47.(a)** $x \approx 637.9$, so $y \approx 89{,}066$ board feet **(b)** average girth = 22 feet, so average diameter is $\frac{22}{\pi} \approx 7 = 84$ inches, $\approx 34{,}163$ board feet

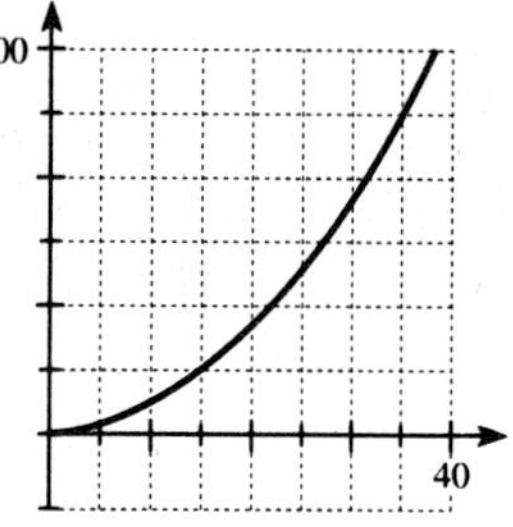

49.(a) $y = 4 + 100x - 16x^2$

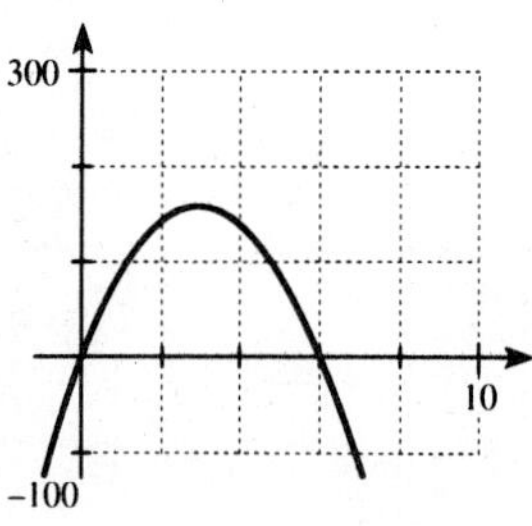

(b) vertex (3.125, 160.25) ft **(c)** 6.29 sec

(d) 4 ft (y-intercept) **51.(a)** $B = 0.8(D - 1)^2 - \frac{D}{2}$

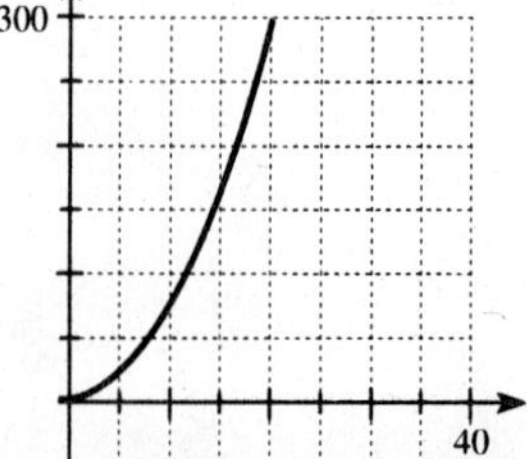

(b) more than 1 inch (more than ≈ 3 inches for a board foot of lumber); 0.8 feet **(c)** 3997 board feet **(d)** ≈ 17 inches
(e) 324,194 board feet **(f)** 125,171 board feet **(g)** No, valid for smaller trees **53.(a)** yes **(b)** 2.75 seconds
(c) $h = 33t - 12t^2$

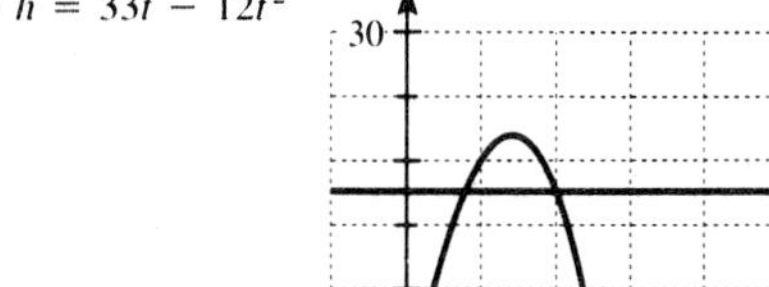

55.(a) $y = 0.5g + 0.008g^2$ **(b)**

1.9 • 10^4
9999

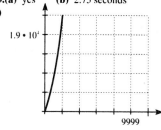

(c) 805,000 fingerlings **(d)** 15,780 trout **57.(a)** $A = 1{,}200l - l^2$ **(b)**

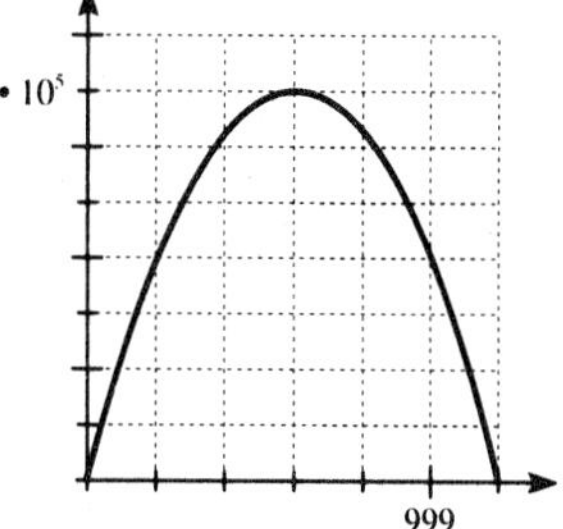

(c) 600 by 6000

IV. 59. Answers vary **61.** all real values of b **63.(a)** $x = \dfrac{-b \pm \sqrt{b^2 - 4ac}}{2a}$ **(b)** obtain x-coordinate of the vertex

65.(a) $y = \dfrac{-b}{2a}$ **(b)** at $y = \dfrac{-b \pm \sqrt{b^2 - 4ac}}{2a}$ if $b^2 - 4ac \geq 0$ **(c)** $a > 0$ opens right; $a < 0$ opens left

67.(a) $y = 0.17x^2 - 0.17x$ **(b)** $y = -2.7x^2 - 0.33x + 6$ **(c)** $y = -0.36x^2 + 2.36x + 1.7$

6.6 (page 420)

I. Graphs omitted for Exercises 1–21 **1.** (0, 0), (4, 16) **3.** (0.76, 1.76), (5.24, 6.24) **5.** (5, 16), (−2, −5) **7.** (−2.54, 3.46), (3.54, 9.54) **9.** (−1.37, 0.51), (1.70, 4.60) **11.** no solutions **13.** (0, 4) **15.** (0.927, 0.336), (21.57, 9.985) **17.** (0.15, 3.9775), (−1.65, 1.2775) **19.** (−4.12, −0.29), (4.43, 0.14) **21.** (7.25, 4.55), (−1.65, 6.33)
II. 23.(a) ≈ 14,000 board feet **(b)** ≈ 7,500 board feet **(c)** loses money **25.** Answers vary **27.** Answers vary
29. Answers vary **31.** Answers vary **33.** Answers vary **35.** $\left(\frac{1}{2}, 2\right)$, (1, 1) **37.** (3.41, 3.41), (−1.91, −1.91)
III. 39. $r = 31.56$ mph, $r + 15 = 46.56$ mph. No, your sister was not speeding. The farthest distance (within the speed limit) is 272 miles.
41.(a) $C = 5x^2 - 15x + 8$; $R = 2x + 2$ **(b)** (3, 8), (0.4, 2.8) **(c)** 0.4 tons (and 3 tons)

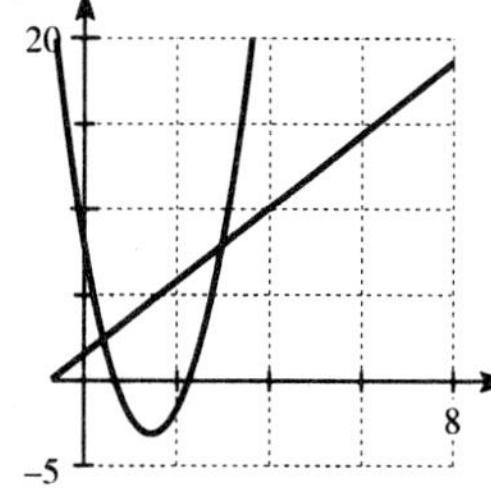

(d) loss of \$18 **43.(a)** $q_s = 6p^2 - 5p - 5, q_d = 20 + 5p - p^2$

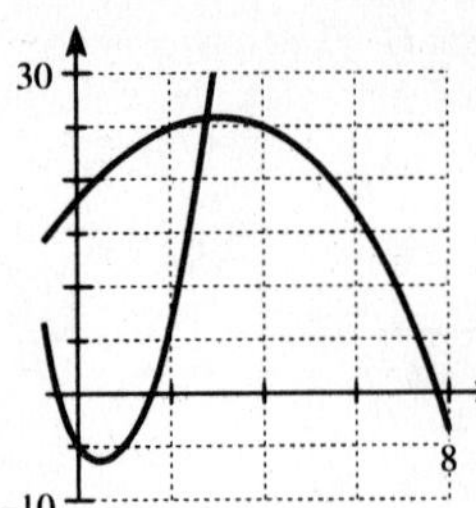

(b) \$2.73 **45.** no **47.** 10 yds

49. 55 apartments, 60 apartments **IV. 51.** (1.46, 0.68) **53.** (2.13, −4.52), (−2.13, −4.52) **55.(a)** $|m| > \sqrt{12}$ **(b)** $|m| = \sqrt{12}$ **(c)** $|m| < \sqrt{12}$ **57.(a)** $c \leq -\frac{23}{8}$ **(b)** $c \geq -0.25$

CHAPTER 6 SUMMARY REVIEW

I. 1. $0, \frac{1}{9}$ **3.** $\frac{1}{3}, -\frac{1}{4}$ **5.** $\frac{-3 \pm \sqrt{21}}{2}$ **7.** $2, \frac{3}{2}$ **9.** complex **11.** real and unequal **13.** 1, 3 **15.** $1 \pm \sqrt{3}i$ **17.** $\frac{29}{36}$ **19.** $-\frac{2}{3}$ **21.** $\pm i$ **23.** $\pm\sqrt{\frac{5}{3}}$ **25.** $\frac{1 \pm \sqrt{241}}{6}$ **27.** $\frac{7 \pm \sqrt{19}}{6}$ **29.** $y = 3x^2 - 27$

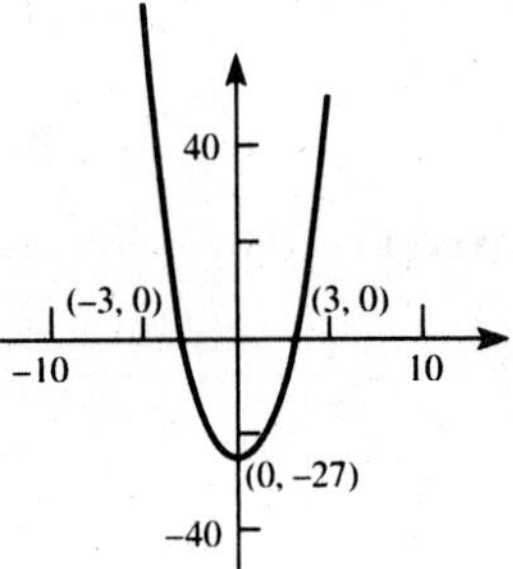

31. $y = 0.75x - 0.125x^2$

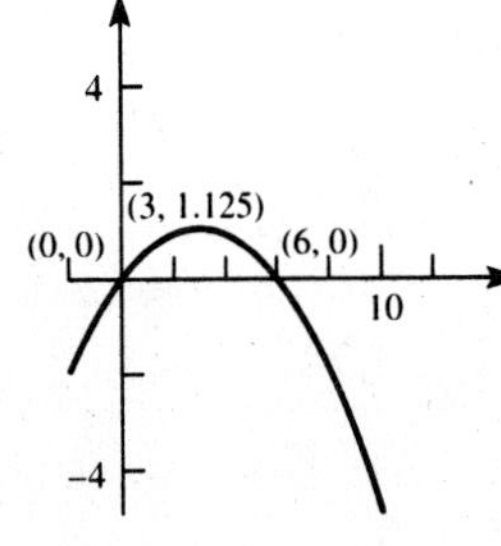

33. $3y = -7x + 6$

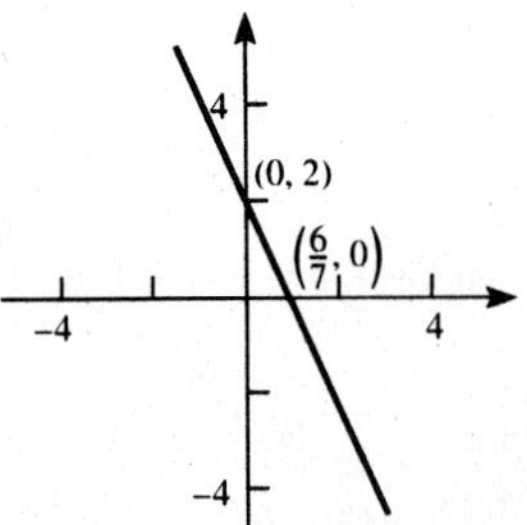

35. $y = -5x^2 - 6$

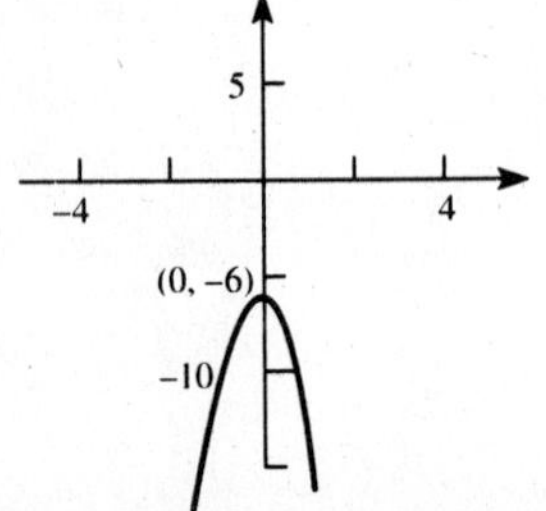

37. $y = -3x^2 - 15x - 18$

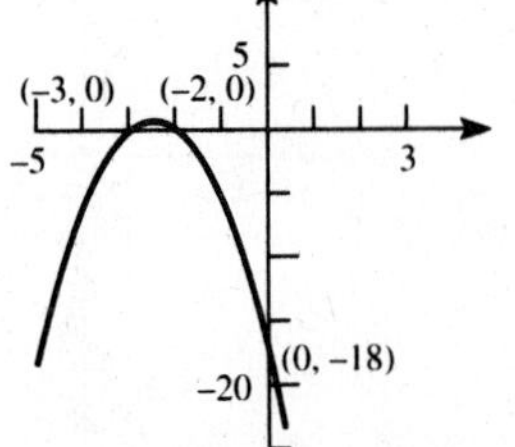

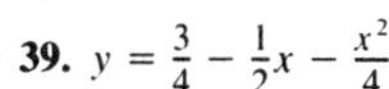

39. $y = \frac{3}{4} - \frac{1}{2}x - \frac{x^2}{4}$

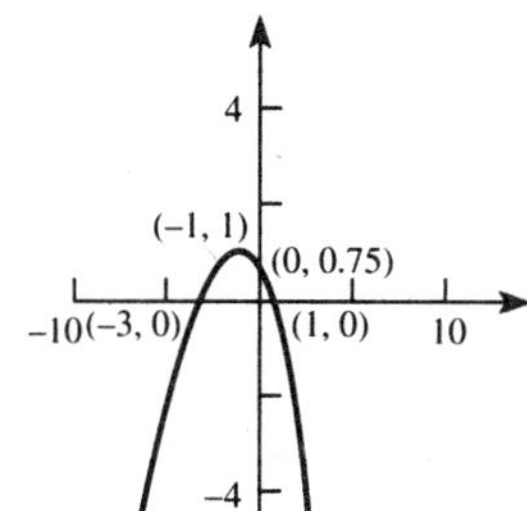

(graphs omitted from Exercises 41–47) **41.** no real solution **43.** (3, 6) (4, 10)

45. no real solution **47.** $(3 \pm \sqrt{5}, 4 \pm \sqrt{5}) \approx (.76, 1.76); (5.24, 6.24)$

II. 49. Answers vary: $x^2 - 5x - 14 = 0$ **51.** $x = \frac{3}{2}$ **53.(a)** 10.29 sec **(b)** 28 feet at 5 sec **55.** Answers vary: **(a)** $y = a(x - 1)^2$ **(b)** $y = a(x - 1)^2 + 1$ **57.** Answers vary: **(a)** $y = ax^2 + bx + 7$ **(b)** $y = ax^2 + bx - 5$ **59.** $a > -\frac{9}{7}$ **61.** graph; narrows or widens **III. 63.** 9% **65.** 53.67 feet (63.67 feet if wire is attached 10 feet below top of tree) **67.(a)** 125 pounds ($w = 0.25$) **(b)** 2 inches; 3.08 inches **69.** 4 cm **71.(a)** at ≈ 0.83 seconds and again ≈ 3.17 seconds **(b)** ≈ 4.12 seconds **73.(a)** $15 at 5 pairs **(b)** same $24 **(c)** 228 pairs **75.** 1724.3 board feet **77.(a)** $A = l\left(1{,}900 - \frac{l}{2}\right)$ **(b)** 1,900 m × 950 m **79.** 30 mph, 40 mph

CUMULATIVE REVIEW: CHAPTERS 1–6

I. 1. 39 **3.** $-10x$ **5.** 6 **7.** 0 **9.** $\frac{x - 3}{2x - 1}$ **11.** $x(x + 5) = x^2 + 5x$ **13.** 3 **15.** $2 < x < 6$ **17.** 0, 5 **19.** $\frac{7 \pm \sqrt{33}}{4}$ **II. 21.** B **23.** E **25.** AT **27.** NT **29.** (e) **31.** (e) **33.** $c = 16$ **35.** Answers vary **37.(a)** (5750, 6800) **(b)** cost $6000, revenue $5000 **(c)** loss of $1000 **(d)** Answers vary **39.(a)** $x + 3, 2x + 8$ **(b)** 6.59 by 15.18 **41.** $1875 per week **43.** 40 mph **45.(a)** $\frac{1}{6}, \frac{1}{20}, \frac{1}{30}, \frac{1}{42}$ **(b)** $\frac{1}{n} - \frac{1}{n + 1} = \frac{1}{n(n + 1)}$ **(c)** graph is set of isolated points including $\left(1, \frac{1}{6}\right), \left(2, \frac{1}{20}\right), \left(3, \frac{1}{30}\right), \left(4, \frac{1}{42}\right)$ **47.(a)** even integer **(b)** even integer **(c)** odd integer **(d)** even **(e)** odd **(f)** even **49.** 4.12 hours **51.** 12 km/h **53.** $x = 1, 2, 3, 4, 5$

CHAPTER 7

7.1 (page 445)

I. 1. 11 **3.** 3 **5.** does not exist **7.** −20 **9.** 4 **10.** $-\frac{3}{2}$ **11.** 1.2 **13.** 100 **15.** −1 **17.** 6.245 **19.** 5.155 **21.** −3.761 **23.** $\sqrt{49}, 7$ **25.** $\sqrt{25x^2}, 5x$ **27.** $\sqrt[3]{27x^3y^{12}}, 3xy^4$ **29.** $-32\sqrt[5]{x^{10}}, -32x^2$ **31.** $\sqrt[3]{64y^{12}}, 4y^4$ **33.** $\sqrt{(x + 2)^2} = x + 2$ **35.** $\sqrt{128} \approx 11.31$ **37.** 8.25 **39.** 22.62 **II. 41.** no sides are $\sqrt{20}, 10, \sqrt{160}$ **43.(a)** −8 **(b)** 625 **(c)** 100,000 **(d)** −1 **45.** Principal square roots are defined to be positive.

47. Answers vary

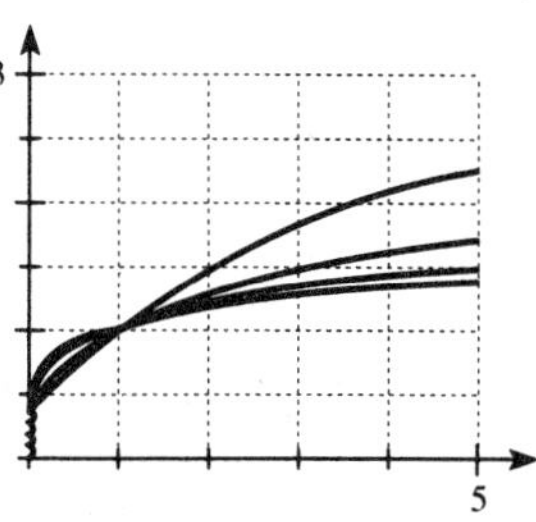

49. True for all $n \neq 0$ or -1

III. 51. 0.754 of the original amount **53.(a)** 21.6% **(b)** 46.5%, 88.0%, 93.8% **55.(a)** 31 mph **(b)** 54.8 mph

57.(a) $v = \sqrt[3]{\dfrac{P}{0.015}}$ **(b)** 20 mph **(c)** 24.48 mph **59.(a)** 11.79 grams **(b)** 11.49 grams

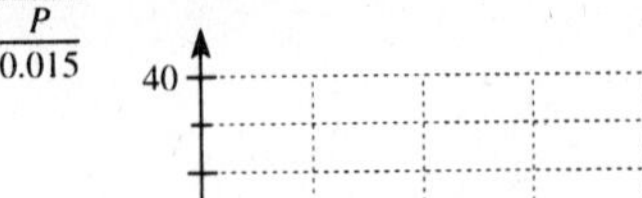

(c) 8.49 grams **(d)** 6 grams **(e)** 3 grams **IV. 61.** Press the square root button twice. **63.** $\sqrt{\sqrt{97.41}}$

7.2 (page 456)

I. 1. $\sqrt[5]{5}$ **3.** $-\sqrt[4]{32^3}$ **5.** $\sqrt[3]{(17.5)^2}$ **7.** $\sqrt[3]{(x^2y)^2}$ **9.** $11^{\frac{1}{2}}$ **11.** $(8x^3)^{\frac{1}{4}}$ **13.** $(-y^3)^{\frac{1}{5}}$ **15.** $-(12x^3)^{\frac{1}{6}}$ **17.** 16 **19.** 16 **21.** $\frac{64}{125}$ **23.** 27 **25.** 0.00000081 **27.** 8 **29.** 2.5 **31.** 36.483 **33.** 4.846 **35.** 3.9508×10^{13} **37.** $x^{\frac{5}{4}}$; 2.378 **39.** $x^{\frac{1}{4}}$ **41.** $\frac{1}{13^3x^9}$ **43.** $9x^2y^8$ **45.** $\frac{y^9}{-8x^6}$ **47.** $\frac{8x^4}{y^5}$; 0.447 **49.** $2xy^2$ **51.** $xy^{\frac{5}{3}}$ **53.** $x^{\frac{5}{4}} + x^{\frac{1}{4}}$ **55.** $x - 2$ **57.** $\frac{1}{x} - 1$; -0.5 **II. 59.** $x = y = 1$ is one counterexample **61.** NT **63.** ST; true if $0 < x < 1$; false if $x > 1$

65. Answers vary $y = x^{\frac{2}{3}}$; $y = x^{-\frac{2}{3}}$

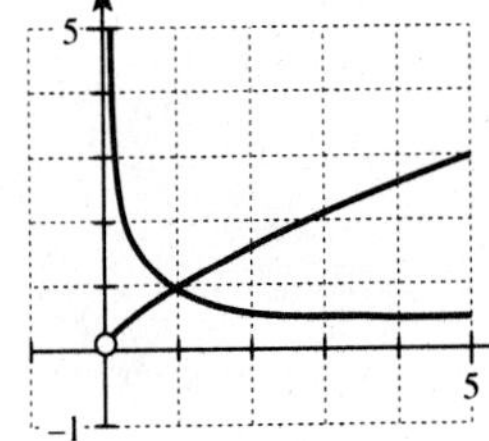

67. Answers vary

III. 69.(a) 240,000 **(b)** 3,840,000 **(c)** $(120{,}000)2^{\frac{n}{10}}$ **(d)**

71.(a) 72.9% **(b)** 91.9% **(c)** 70.4% **(d)** $(0.9)^m \cdot 100\%$ **(e)**

73.(a) \$1,178.07 **(b)** \$1,426.27 **75.(a)** \$1,017.02 **(b)** \$1,017.50 **77.(a)** 1.35 grams **(b)** 0.0075 grams **(c)** 0.0000614 grams **79.(a)** 6.10 hours **(b)** about 1.6 times as long **IV. 81.** You are not allowed to take the sixth root of -8; i.e., $(-8)^{\frac{2}{6}}$ is not defined.

7.3 (page 470)

I. 1. $-5\sqrt[3]{2}$ **3.** $-\sqrt[5]{24{,}200}$ **5.** 13 **7.** $2x^3(\sqrt{2})$ **9.** $2a^2b^2\sqrt[4]{2ab^2}$ **11.** $2\sqrt{4x^2 - x}$ **13.** 10 **15.** $\sqrt{5}$ **17.** $\frac{\sqrt[3]{2}}{4}$
19. $-3\sqrt{5}$ **21.** $4\sqrt{3} + 24\sqrt{2}$ **23.** -2 **25.** 3 **27.** $\frac{5}{4}(\sqrt{5} + 1)$ **29.** $-(7 + 4\sqrt{3})$ **31.** x^2 **33.** $2x^2y\sqrt[3]{2y}$
35. $2\sqrt{x}$ **37.** $\frac{x^4y}{2}\sqrt{2y}$ **39.** $\sqrt[3]{2xy^2}(1 - 2y)$ **41.** $\sqrt[3]{x}(x^2 - 2y^2 + 2)$ **43.** $\frac{(a + 1)(\sqrt{a} - 1)}{a - 1}$ **II. 45.** No, because $\sqrt{2x}$ is not equal to $\sqrt{2}x$ **47.** Cannot be done **49.** 1333.7 **51.** All three are equivalent. **53.** $\sqrt{x}$ and $x^{\frac{1}{2}}$ are equivalent.
55. Answers vary. **57.** Answers vary

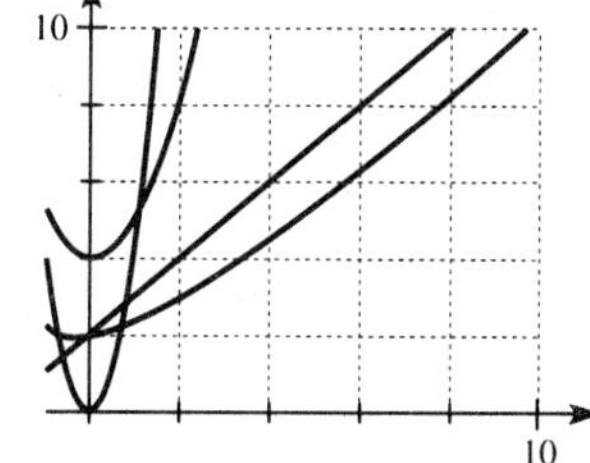

III. 59.(a) 2.69 minutes **(b)** 6.32 minutes
(c) Answers vary: 10, 70

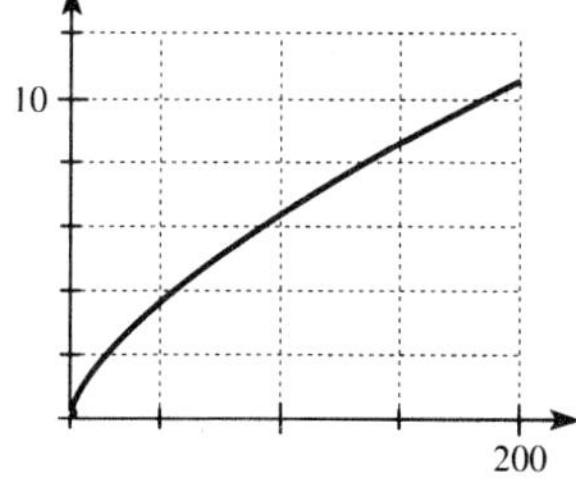

61. It is multiplied by $\sqrt[3]{3}$ **63.** $1{:}\ \sqrt[3]{3}$ **65.(a)** $1{:}2\sqrt{2}$ **(b)** $1{:}3\sqrt{3}$

67.(a) 1.419 seconds **(b)** 2.007 seconds **69.(a)** 34.64 mph **(b)** 48.99 mph **71.** 13.75 feet **73.**

Mercury	.241
Venus	.615
Earth	1.000
Mars	1.880
Jupiter	5.201
Saturn	9.538
Uranus	19.181
Neptune	30.057
Pluto	39.510

IV. 75.(a) It is multiplied by $\sqrt[3]{2}$ **(b)** It is multiplied by $\sqrt[3]{3}$ **(c)** It is multiplied by $\sqrt[3]{100}$ **77.** Answers vary
79. Consider V^2 and A^3

7.4 (page 493)

I. 1. 2 **3.** -3 **5.** -7 **7.** 7 **9.** 3 **11.** $3, -\frac{1}{3}$ **13.** -1 **15.** 5 **17.** 36 **19.** no solution **21.** 15
23. $2, -2$ **25.** 2 **27.** $-2, -3$ **29.** $\sqrt{8}, -\sqrt{8}$ **31.** 1 **33.** 5 **35.** -2 **37.** $\frac{1}{2}$ **39.** 4
II. 41. ST (True when $y = -2x$) **43.** ST; true if $a \geq b$ **45.** Answers vary **47.** Answers vary **49.** $\frac{5}{2}$
III. 51. 8 hours **53.(a)** and **(b):** approximately 7 years **55.(a)** $P = 120{,}000(1.035)^t$ **(b)** 238,775 **(c)** approximately 20 years **57.** approximately 8.5% **59.(a)** \$2.80 **(b)** about \$8880 per year **61.(a)** \$19,683, \$29,524 **(b)** $3^{n-1}, \frac{3^n - 1}{2}$
(c) the eighteenth year **63.(a)** 185° **(b)** 109.15°, 81.24°, 67.20° **(c)** It will never cool to precisely 65 degrees, but it will come very close. **65.(a)** 252.08 feet **(b)** 126.04 feet **67.** $(-1, 2)$ **69.** 5.23 meters **71.** 170.86 m^2 **73.** 6.83
IV. 75. 16 **77.(a)** $x = y$ **(b)** $x \neq y$ **79.(a)** $p = \sqrt{2}, q = \sqrt{3}, r = \sqrt{4} = 2, s = \sqrt{5}$ **(b)** $\sqrt{6}, \sqrt{7}$

CHAPTER 7 SUMMARY REVIEW

I. 1. 25 **3.** 4 **5.** -16 **7.** 2 **9.** 125 **11.** 17 **13.** $-3y^2\sqrt[3]{y^2}$ **15.** $\frac{x^3}{4}$ **17.** $-y^{\frac{1}{3}}$ **19.** $\frac{4}{y}$ **21.** $\frac{1}{-2xy^2}$
23. $-3x^{\frac{41}{3}}y^{\frac{15}{2}}$ **25.** $\sqrt{x}$ **27.** $\sqrt{3}+\sqrt{x}$ **29.** $9\sqrt{2}-\sqrt[3]{2}$ **31.** $\frac{5}{6}$ **33.** $-1, 7$ **35.** -25 **37.** $-\frac{3}{2}$ **39.** $\frac{3}{2}$ **41.** 0
43. 3 **II. 45.** True **47.** True **49.** False, $x = 2$ **51.** False, $x = 8$ **53.** Answers vary. **III. 55.(a)** 45,000
(b) 135,000 **(c)** 3,645,000 **(d)** 16,969,431 **(e)** $(15{,}000)3^{n/5}$ **57.(a)** 10.5% compounded monthly
(b) \$12,987; \$12,981 **59.(a)** 0.5625 **(b)** 0.91 **(c)** 0.221 **61.** between 7 and 8 years **63.** 46.46 miles
65. 1.6 minutes **67.(a)** 210° **(b)** 198.47° **(c)** 135.40 minutes **69.(a)** 98.35% **(b)** happens immediately

CHAPTER 8

8.1 (page 511)

I. 1. no **3.** yes; $f(1) = 3$ **5.** yes; $f(1) = 3$ **7.** yes; $f(1) = 2.5$ **9.** no **11.** no **13.** yes; $f(1) = -.5$ **15.** no
17. $f(x) = 3x; f(4) = 12$ **19.** $f(x) = x^2 + 1; f(4) = 17$ **21.** $f(x) = 4x; f(4) = 16$ **23.** $f(x) = \sqrt{x}; f(4) = 2$
25. Answers vary **27.** Answers vary **29.** *T* ten (one, two, three, . . .) **31.** Answers vary: *D*: number of gallons of gas; *R*: price of gas; *f*: amount you pay for gas **33.** Answers vary: *D*: objects; *R*: real numbers between 0 and 100 (centimeters); *f*(object) = length of object **35.** Answers vary: *D*: numbers; *R*: numbers **37.** Answers vary **III. 39.(a)** 22¢ **(b)** 1985
(c) 23¢ **(d)** 1988 and 1991 **41.(a)** 680 cal **(b)** 512 cal **IV. 43.** $f(x) = 3x + 1$ **45.** $f(x) = \frac{120}{x}$
47. $f(x) = 4x + 3$ **49.** $g(x) = \sqrt{2x+1}$ **51.(a)** 27 **(b)** 6 **53.(a)** n^n **(b)** $n! = n(n-1)(n-2)\ldots 3\cdot 2\cdot 1$

8.2 (page 524)

I. 1.(a) $2, -1, 8.9, 3a + 5$ **(b)** $\{-1, 2, 8, 11\}$ **3.(a)** $0, -1, -5.29, -a^2 - 2a - 1$ **(b)** $\{0, -1, -4, -9\}$
5.(a) $-1, -2, 8.89, a^2 + 4a + 2$ **(b)** $\{-2, -1, 7, 14\}$ **7.(a)** $-3, -\frac{2}{3}, 3.7, \frac{8a+20}{a(a+5)}$ **(b)** $\left\{-\frac{2}{3}, -3, \frac{14}{3}, \frac{18}{7}\right\}$
9. $-3, -1, 1, 3, 5, 7$ **11.** 108 **13.** -26 **15.** -72 **17.** 3 **19.** -2 **21.** -7 **23.** -72 **25.** $a^2 + 6a - 2$
27. 6 **29.** $4x^4 - 16x^2 + 16$ 31–42 Graphs omitted. **31.** 3 **33.** $-2 \pm \sqrt{11}$ **35.** $x < \frac{7}{3}$ **37.(a)** 3
(b) undefined or i **(c)** 2 **(d)** 0 **39.(a)** $\frac{15}{4}$ **(b)** $\frac{80}{9}$ **(c)** ≈ 2.00 **41.(a)** 3 **(b)** 12 **(c)** $|2x|$
43. all real numbers **45.** all real numbers **47.** all real numbers except $\frac{2}{3}$ **49.** all real numbers $\geq \frac{3}{2}$ **51.** $x \leq 2$ and $x \neq -3$
II. 53. Answers vary: $f(x) = .5x + 2$ **55.** Answers vary: $f(x) = 4 - x^2$ **57.** Answers vary: $f(x) = x^3$
III. 59.(a) perimeter doubled, area quadrupled **(b)** perimeter reduced by 25%, area reduced by 43.75%
61.(a) surface area quadrupled, volume multiplied by 8 **(b)** surface area reduced by $\frac{5}{9}$, volume reduced by $\frac{19}{27}$
63.(a) perimeter doubled, area quadrupled **(b)** perimeter reduced by 10%, area reduced by 19% **65.(a)** multiplied by 27
(b) multiplied by 9 **67.(a)** 83 push-ups **(b)** 16,952 push-ups **IV. 69.** Let $a = b = 0$. $f(0) = f(0) + f(0) \rightarrow f(0) = 0; f(x) = kx$ **71.** approaches 1 **73.(a)** Answers vary: $P(l, w) = 2l + 2w$; $V(r, h) = \pi r^2 h$
(b) Area of a triangle whose sides are 3, 5, 7; $A = \frac{15\sqrt{3}}{4}$ **75.** $g(2) = 2.5$

8.3 (page 541)

1. yes; *D*: all real numbers, *R*: $\{y \mid y \geq 0\}$ **3.** yes; *D*: all real numbers, *R*: $\{y \mid y \leq 0\}$ **5.** yes; *D*: all real numbers, *R*: $\{y \mid y \leq 3\}$

7. $f(x) = x^5$

9. $f(x) = 2x^4$

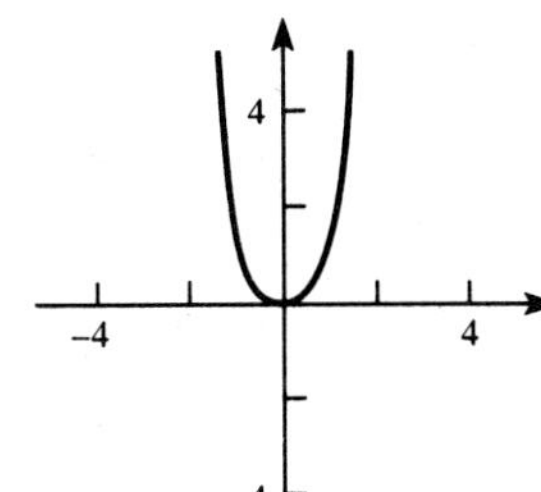

11. $f(x) = \sqrt[4]{x}$

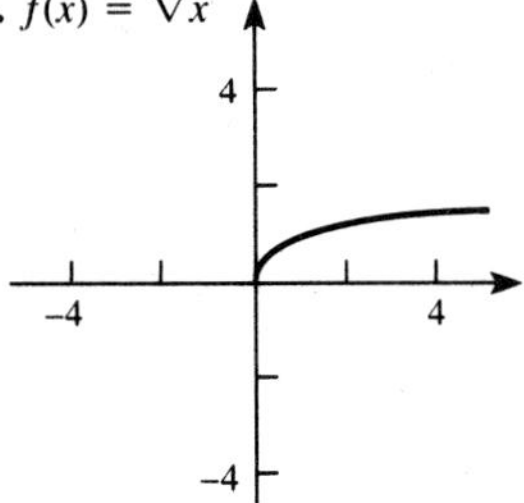

13. $f(x) = \frac{1}{x^3}$

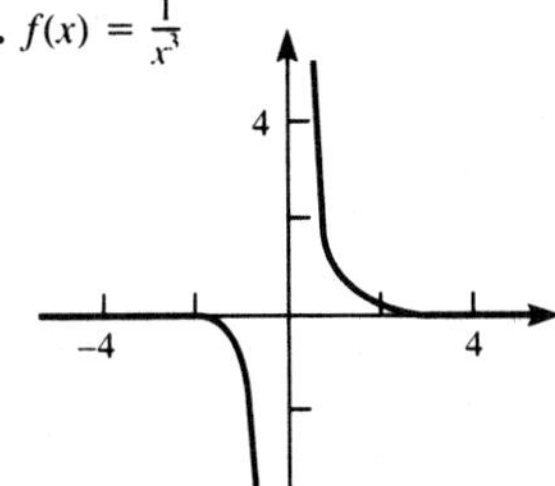

15. $f(x) = \frac{1}{\sqrt{x}}$

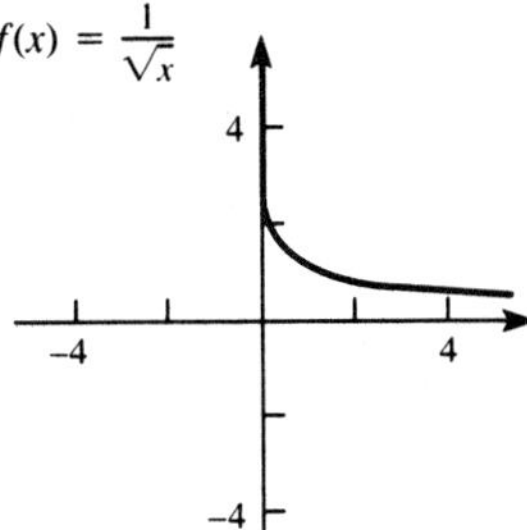

17.(a) $y = x^2 + 3$

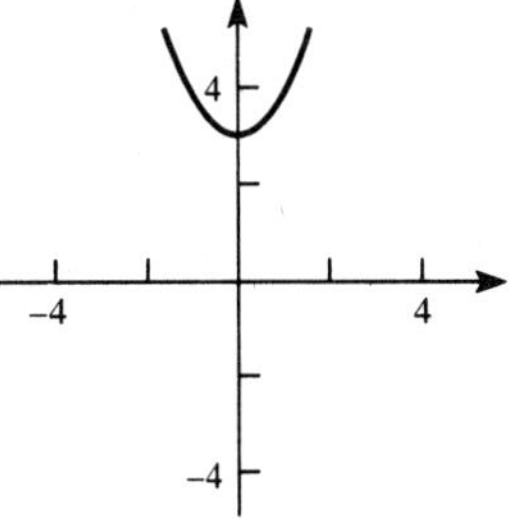

(b) (0,3) **(c)** $D = \{\text{all real numbers}\}$; $R = \{y \mid y \geq 3\}$ **19.(a)** $y = 2(x - 2)^2$

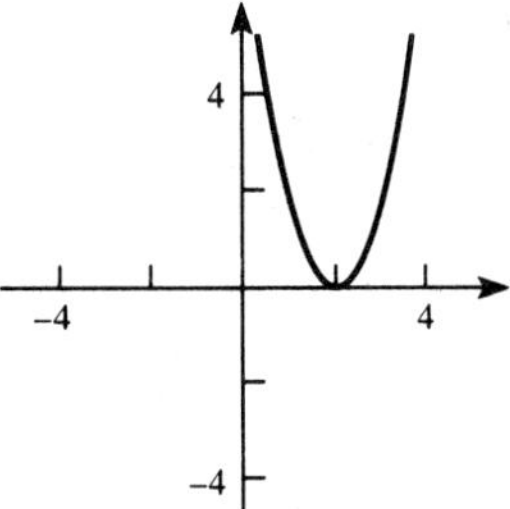

(b) (0,8) (2,0) **(c)** $D = \{\text{all real numbers}\}$; $R = \{y \mid y \geq 0\}$ **21.(a)** $y = 2 - x^2$

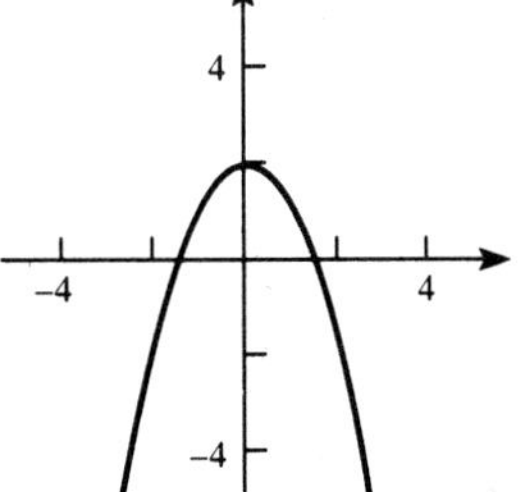

(b) (0,2) $(\sqrt{2},0)$, $(-\sqrt{2},0)$ **(c)** $D = \{\text{all real numbers}\}$; $R = \{y \mid y \leq 2\}$ **23.(a)** $y = x^3 + 1$

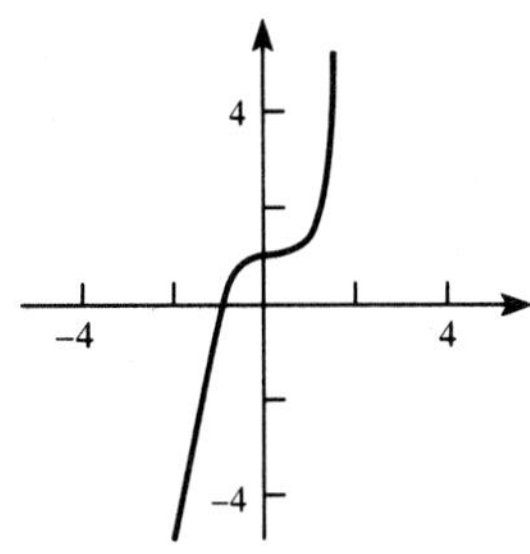

(b) (0,1) (−1,0) **(c)** $D = R = \{\text{all real numbers}\}$ **25.(a)** $y = \sqrt{x - 3}$

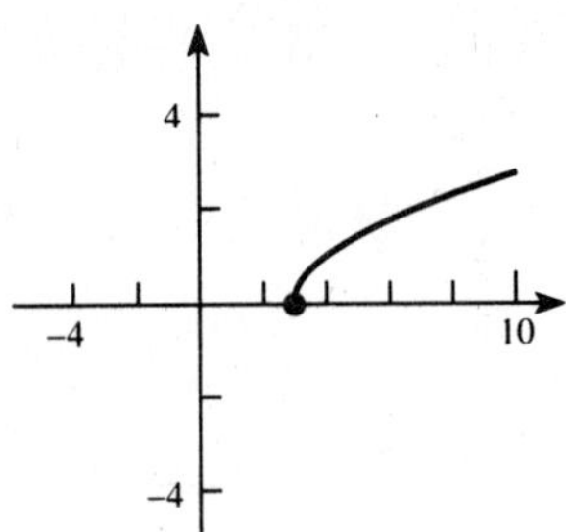

(b) (3,0) **(c)** $D = \{x|x \geq 3\}$; $R = \{y|y \geq 0\}$ **27.(a)** $y = \frac{1}{x} - 2$

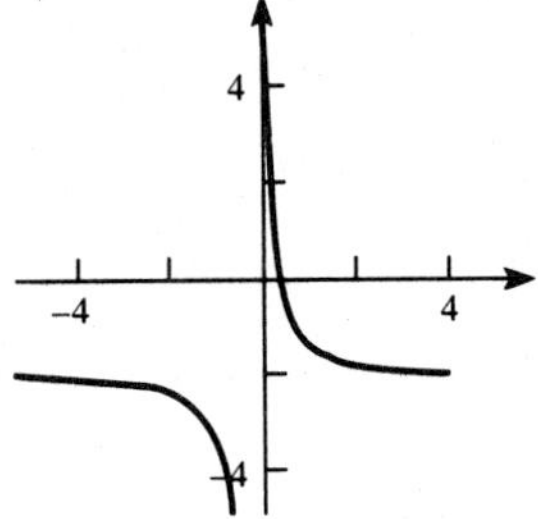

(b) $\left(\frac{1}{2}, 0\right)$ **(c)** $D = \{x|x \neq 0\}$; $R = \{y\,|\,y \neq -2\}$ **29.(a)** $y = |x - 3|$

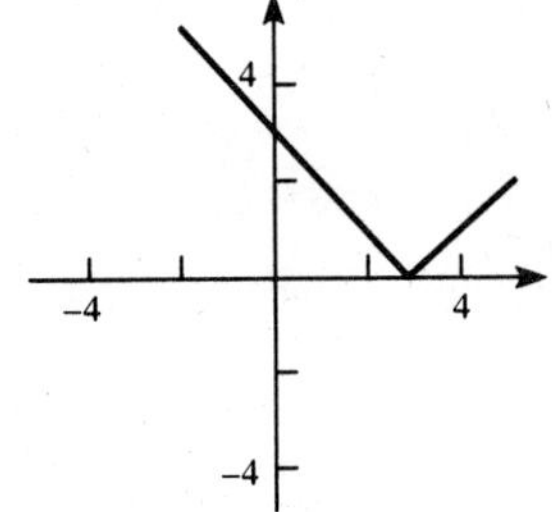

(b) (0,3), (3,0) **(c)** $D = \{\text{all real numbers}\}$; $R = \{y|y \geq 0\}$ **31.(a)** $y = -\sqrt{x + 2}$

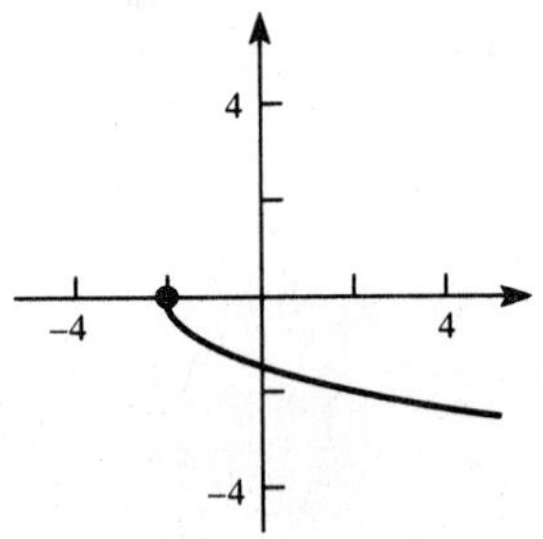

(b) $(0, -\sqrt{2})$, $(-2, 0)$ **(c)** $D = \{x|x \geq -2\}$; $R = \{y|y \leq 0\}$ **33.(a)** $y = (x - 4)^3 + 3$

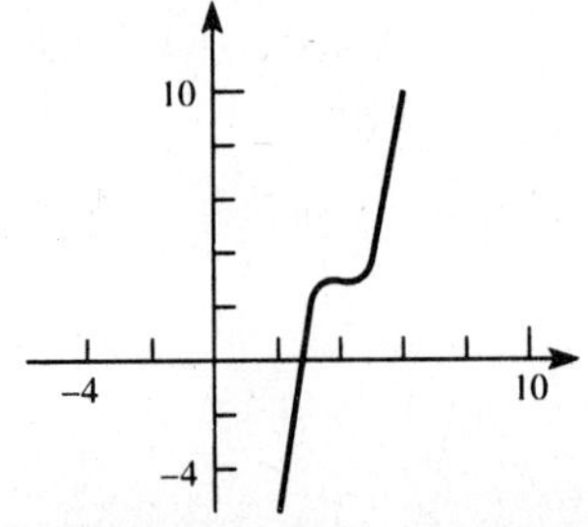

(b) (0, −61) (2.56, 0) **(c)** $D = R =$ {all real numbers} **35.(a)** $y = (x + 2)^4 - 1$

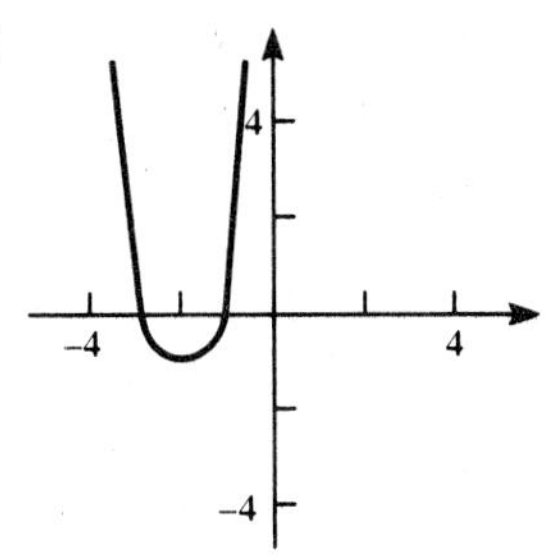

(b) (0, 15); (−1, 0); (−3, 0) **(c)** $D =$ {all real numbers}; $R = \{y \mid y \geq -1\}$ **37.(a)** $y = 1 + \sqrt{x + 3}$

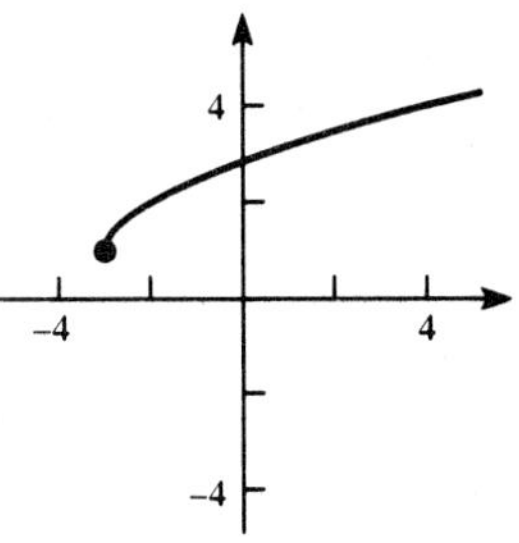

(b) (0, 2.73) **(c)** $D = \{x \mid x \geq -3\}$; $R = \{y \mid y \geq 1\}$ **39.(a)** $y = \dfrac{1}{(x + 2)^2} + 2$

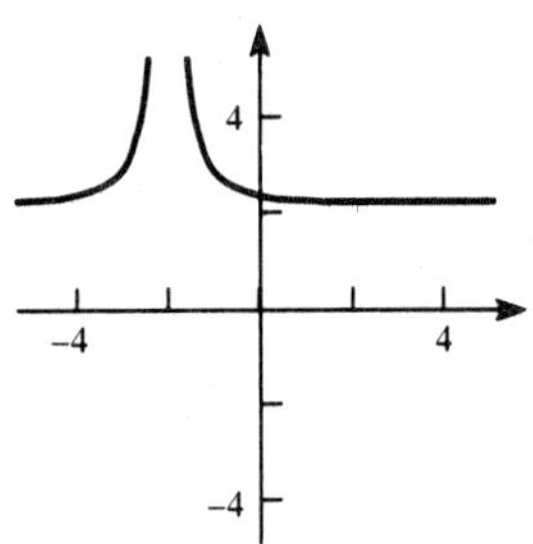

(b) (0, 2.25) **(c)** $D = \{x \mid x \neq -2\}$; $R = \{y \mid y > 2\}$ **41.(a)** $y = -2x^3 + 4$

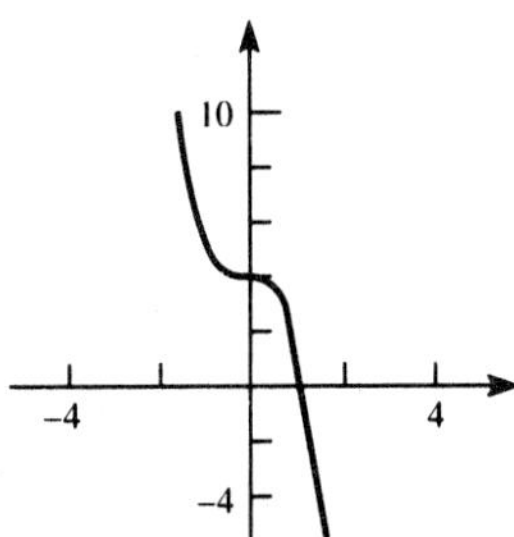

(b) (0, 4); $(\sqrt[3]{2}, 0)$ **(c)** $D = R =$ {all real numbers} **43.** Answers vary: $y = ax^{2k} + b$
45. No (for continuous functions) **47.** Answers vary: range on a graphing calculator includes both the domain and range of the function **49.** Answers vary: $x^4 > x^3 > x^2$ for $x > 1$; $x^4 < x^3 < x^2$ for $0 < x < 1$ **51.** Answers vary **53.** Answers vary

55. No, graphs are not the same. **III. 57.(a)** $y = x^3 - 3x + 1$

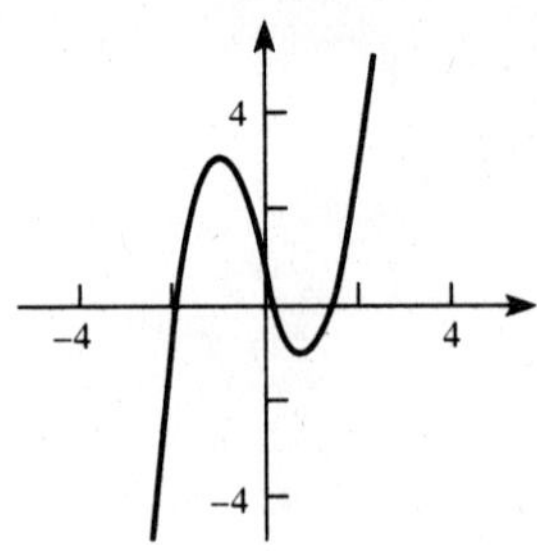

(b) $-1.88, .35, 1.53; (1, -1); (-1, 3)$ **(c)** Translate graph in part (a) down 1 unit. **(d)** Reflect graph in part (a) in the x-axis.

59. $g(t) = 442 - 5t^2$ Answers vary: $D: 0 < t < 10$

61. $V = \sqrt{12L}$ Answers vary: $D: 0 < L < 400$ **63.** Answers vary **65.** Answers vary

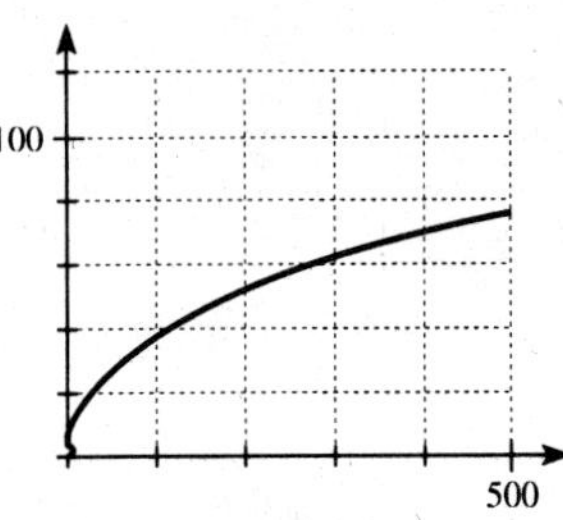

67. Answers vary **69.** Answers vary **71.** Answers vary **73.** Answers vary **75.** Answers vary **77.** Answers vary
79. Answers vary **IV. 81.** $-2 < x < -1, 0 < x < 1, 1 < x < 2$ **83.** $3 < x < 4$ **85.(a)** $D: \{x|x \geq 7 \text{ or } x \leq -2\}$; $R: \{y|y \geq -3\}$
(b) $D: \{x|x > 7 \text{ or } x < -2\}$; $R: \{y|y > 2\}$ **(c)** $D: \{x|-2 < x < -1 \text{ or } 2 < x < 3\}$; $R: \{y|y \geq 0\}$

8.4 (page 555)

I. 1. $F(y) = ky^3$ **3.** $W = \frac{k}{\sqrt[3]{s}}$ **5.** $S = kab^2c^3$ **7.** $y = 2\sqrt{x}$ **9.** $y = \frac{4}{x^2}$ **11.** $y = \frac{1}{3}\pi r^2 h$ **13.** $y = 8x - 17$
15. $y = x^2 + 2x + 3$ **17.** $4n - 3$ **19.** $\frac{n}{n+1}$ **21.** $\frac{n+1}{[(n+1)^2 + 1]}$ **23.** $(-1)^{n+1}n$ **25.** 23, 30 **27.** 18, 29
29. 7, 64, 8, 128 (1, 2, 3, 4, 5, . . . alternating with powers of 2) **31.** 17, 19 (primes) **II. 33.** Answers vary **35.** directly
37. directly **39.** neither (depends on the ingredients) **41.** true **43.** false **III. 45.(a)** $T(l) = 8\sqrt{l}$

(b)

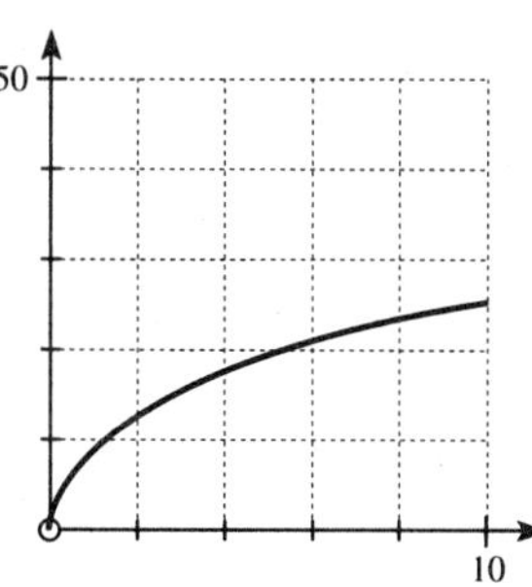

(c) $8\sqrt{2} = 11.3$ seconds, 0.14 meters **47.** $\frac{1}{4}$ as much

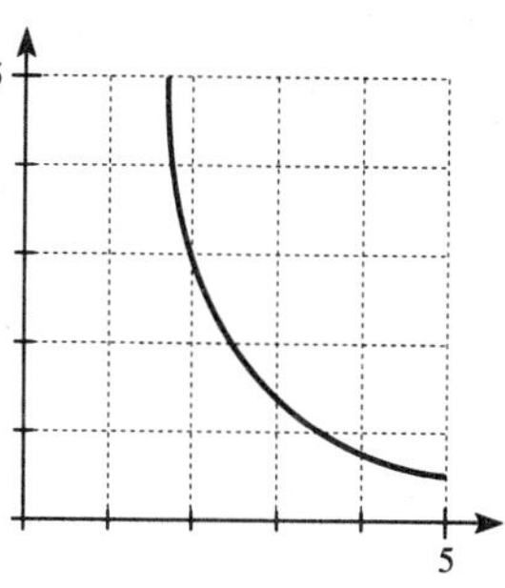

49.(a)

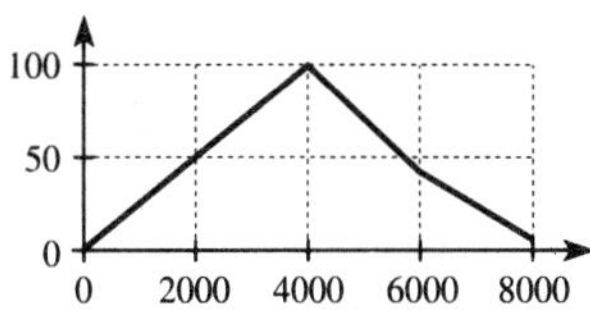

(b) 96 pounds; mathematical model $w = \frac{k}{d^2}$, $k = (4{,}000)^2 150$ yields 0.04 pounds; model may not be valid on moon **51.** 3.6 times as much **53.** 19 apples **55.** 217 pounds **57.(a)** $v = \frac{16{,}100}{P}$ **(b)** 64.4 psi

59. Answers vary: $A = \frac{kn(T - 65)}{p}$, $T \geq 65$ **61.(a)** $p(x) = 3.58x^2 - 86.3x + 916.75$ **(b)**

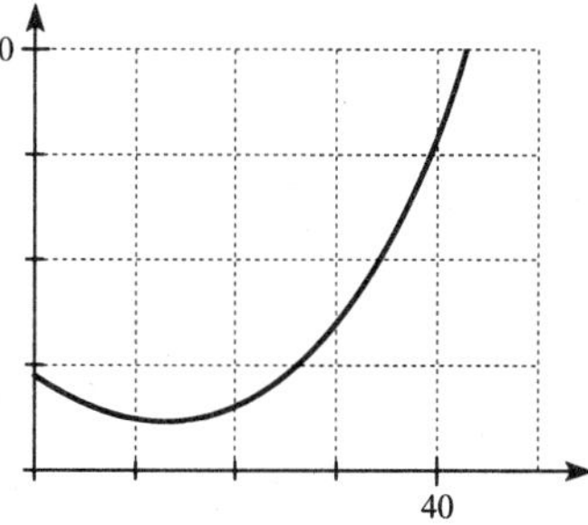

(c) \$996.75 **(d)** Answers vary **63.** $G(n) = n(n - 1)$ **65.(a)** $x = 0.94, 4.43$ **(b)** $x = 2.54$ **(c)** Answers vary
67. $2.29 \times 2.29 \times 1.53$ **IV. 69.** $p(x) = -\frac{4}{3}x(x - 6)$ **71.(a)** no **(b)** no **(c)** no **(d)** $\frac{ka + kh - ka}{h} = k = f(1)$

CHAPTER 8 SUMMARY REVIEW

I. 1. yes; $f(1) = 3$ **3.** yes; $f(1) \approx 1$ **5.** -7 **7.** 9,498 **9.** 1 **11.** $\frac{8}{3}$ **13.** $x > \frac{5}{4}$

15. $y = x^3 - 1$

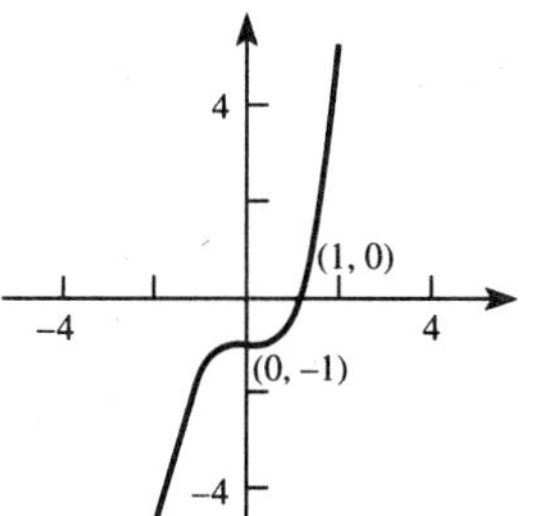

D: {all real numbers}
R: {all real numbers}

17. $y = \sqrt{x - 2} + 1$

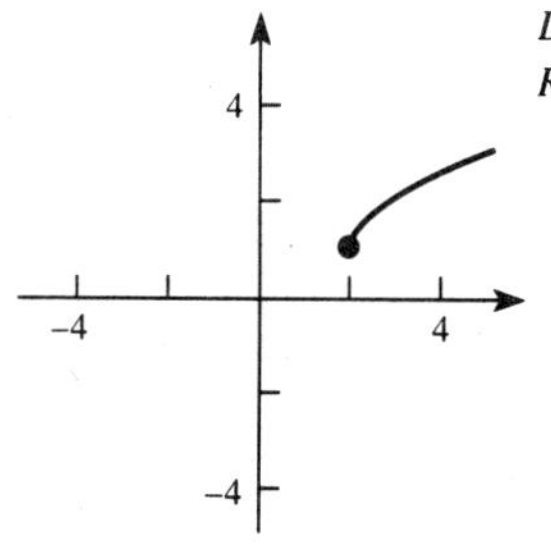

D: $\{x \mid x \geq 2\}$
R: $\{y \mid y \geq 1\}$

19. $y = \frac{2}{(x-2)} + 1$

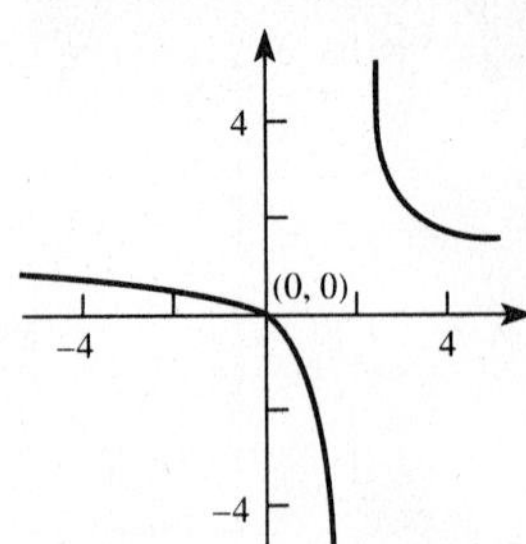

D: {real numbers $\neq 2$}
R: {real numbers $\neq 1$}

21. $W(y) = 4y^2$ **23.** $\{x \mid x \geq 3 \text{ and } x \neq 4\}$

II. 25. D: fruit R: weight f: fruit → weight **27.(a)** $y = 1 + 2q(x)$ **(b)** $y = p(7 - x)$ **(c)** $y = r(2x)$ **(d)** $y = q(p(x + 2))$ **29.** Answers vary **31.** no **33.** $|f(x)|$

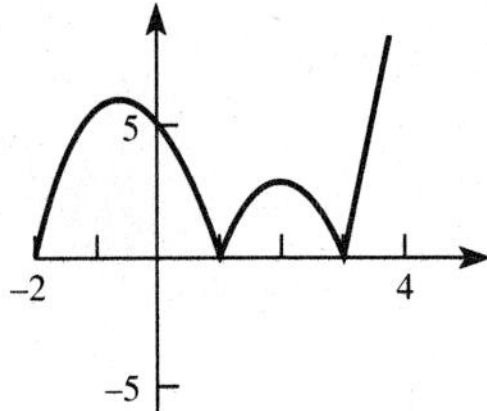

35. $f(x - 2)$

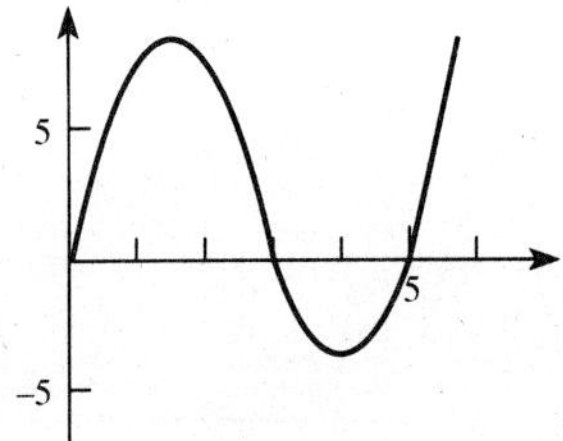

III. 37.(a) Answers vary **(b)** 725 − 925 **(c)** twentieth year **(d)** $f(x) = -32.26x + 1000$ **39.(a)** $v = 64h$

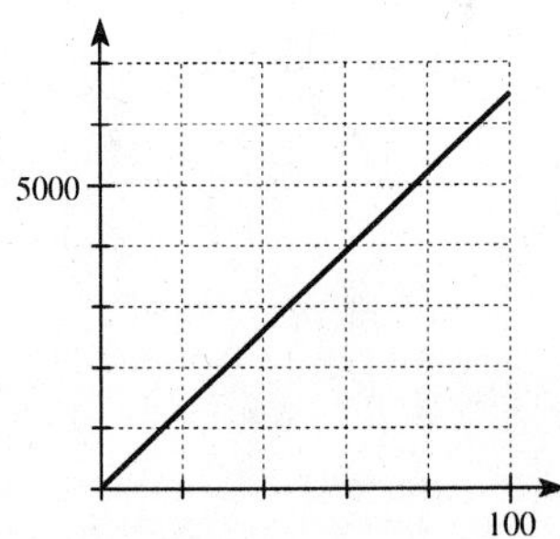

D: Answers vary **(b)** 320,000 m/sec

41. Answers vary **43.(a)** $A = 0.6d^2$ **(b)** 540 people **(c)** 44.8 cm **45.(a)** $C(x) = 0.55x^2 - 7x + 139$ **(b)** \$424 **47.** 2.05 **49.** length = 36, height = 18

CHAPTER 9

9.1 (page 575)

I. 1. 1,133.7 **3.** 0.011 **5.** $5.0144 \cdot 10^{51}$ **7.** 160.8 **9.** 903.96 **11.** 0.0045 **13.** 22.36 **15.** too large for most calculators **17.** $\frac{3^{15}}{2^{14}}$ **19.** $-\left(\frac{3}{4}\right)^{15}$ **21.** $\frac{12}{2^{14}}$ **23.** 2 **25.** 1.44

27. $y = 5^x$

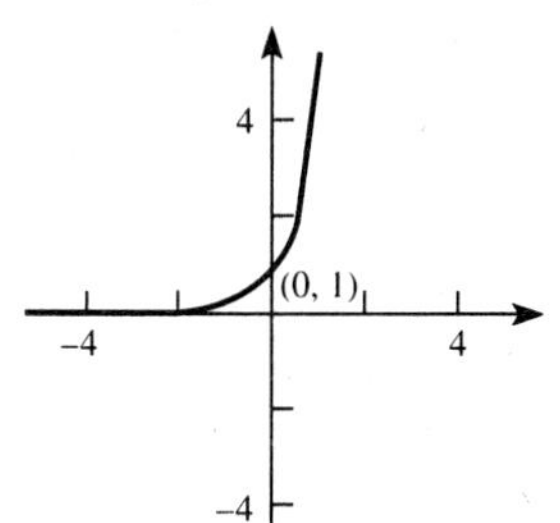

D: {all real numbers}
R: $\{y \mid y > 0\}$

29. $y = 3^{-x}$

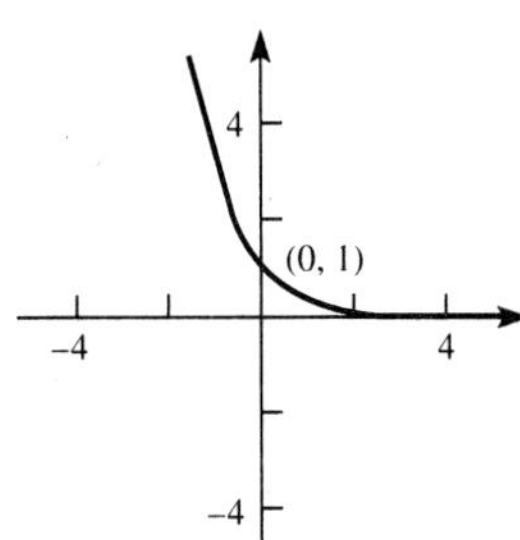

D: {all real numbers}
R: $\{y \mid y > 0\}$

31. $y = 3^{x+1}$

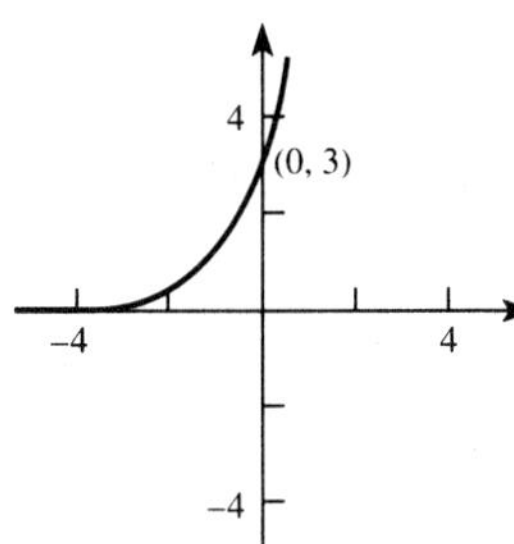

D: {all real numbers}
R: $\{y \mid y > 0\}$

33. $y = 3^{x+2}$

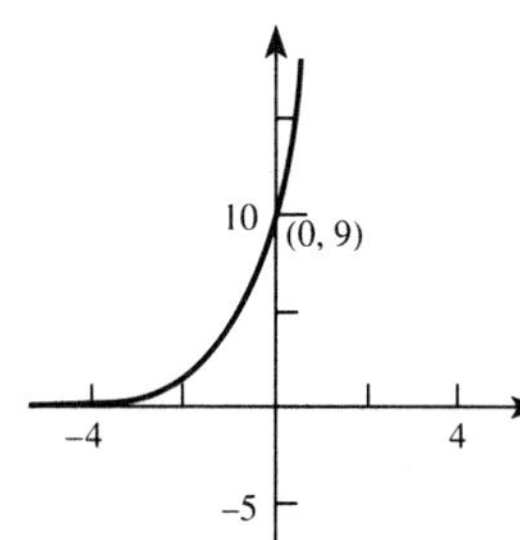

D: {all real numbers}
R: $\{y \mid y > 0\}$

35. $y = 2 - 3^x$

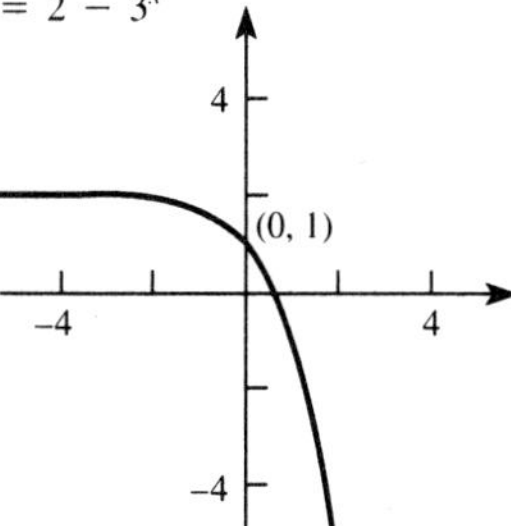

D: {all real numbers}
R: $\{y \mid y < 2\}$

37. $y = 2 - 3^{-x}$

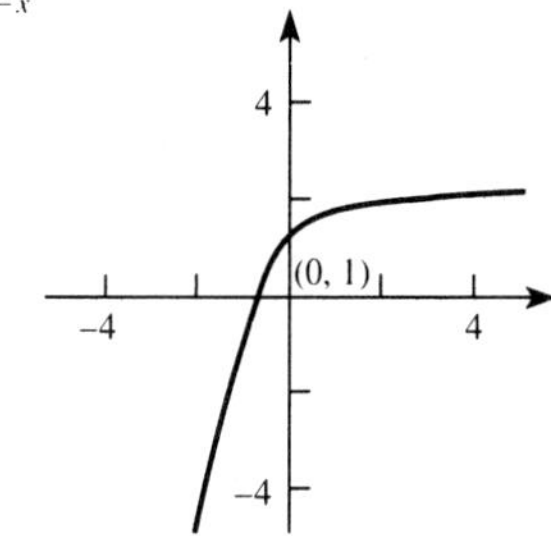

D: {all real numbers}
R: $\{y \mid y < 2\}$

39. $y = 2 - 3^{x+1}$

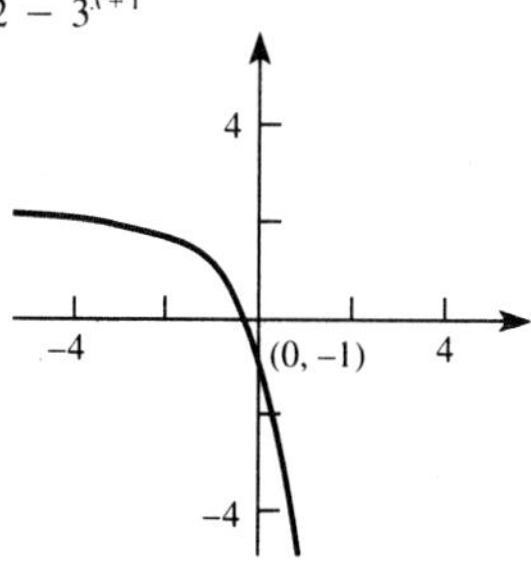

D: {all real numbers}
R: $\{y \mid y < 2\}$

41. $y = (1.06)^x$

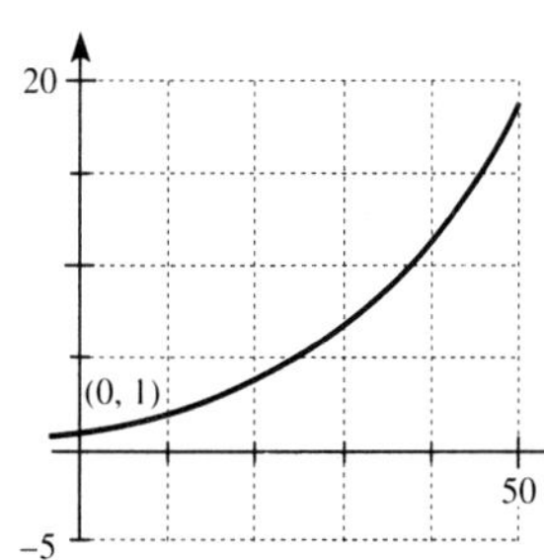

D: {all real numbers}
R: $\{y \mid y > 0\}$

43. $y = e^{-x}$

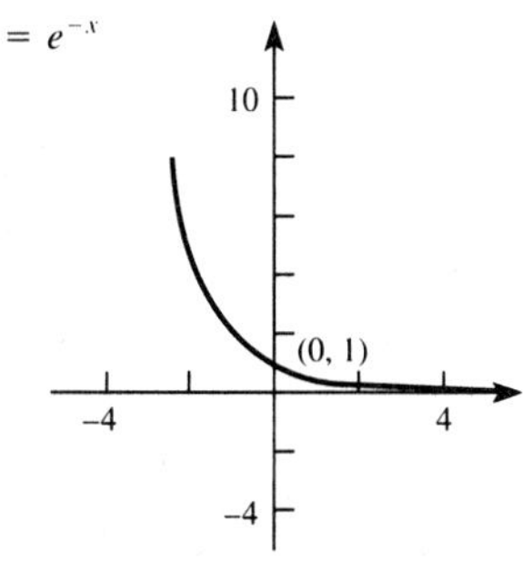

D: {all real numbers}
R: $\{y \mid y > 0\}$

45. $y = \frac{1}{\sqrt{2\pi}}e^{-\frac{x^2}{2}}$

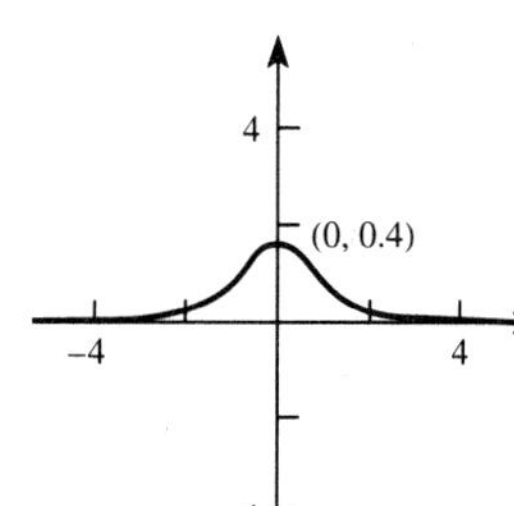

D: {all real numbers}
R: $\left\{y \mid 0 < y < \frac{1}{\sqrt{2\pi}}\right\}$

47.(a) 1; 1.25 **(b)** 0 **(c)** 1; −1.25 **(d)** 0 **II. 49.** 53,309,656 miles **51.** AT **53.** ST (true only when $b = \sqrt{2}$) **55.** Answers vary. **III. 57.(a)** \$1,946.94 **(b)** \$1,926.04 **59.(a)** 19.2 meters **(b)** 94.83 meters **61.(a)** \$6,212 **(b)** \$6,640 **63.(a)** 13.90 **(b)** 13.10 **(c)** 14.22 **65.** \$0.55 in 1993 **67.(a)** 48.66% **(b)** 37.81% **IV. 69.** Answers vary **71.** 1 **73.** 0 **75.(a)**

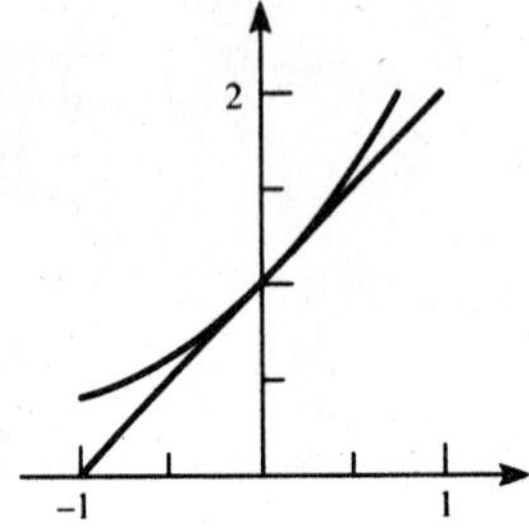

(b) 0.098

77. Answers vary: **(a)** $x > 4$ **(b)** $x > 16$ **(c)** $x > 650$ **(d)** $x > 1{,}475$

9.2 (page 588)

I. 1. 64 **3.** ±8 **5.** $\frac{1}{4}$ **7.** 2 **9.** $\frac{1}{3}$ **11.** $\frac{1}{4}$ **13.** 2 **15.** 243 **17.** no solution **19.** 64 **21.** 2.518 **23.** 24.518 **25.** 12.705 **27.** 56.453 **29.** 241.518 **31.** −231.370 **33.** $y = \log x$,

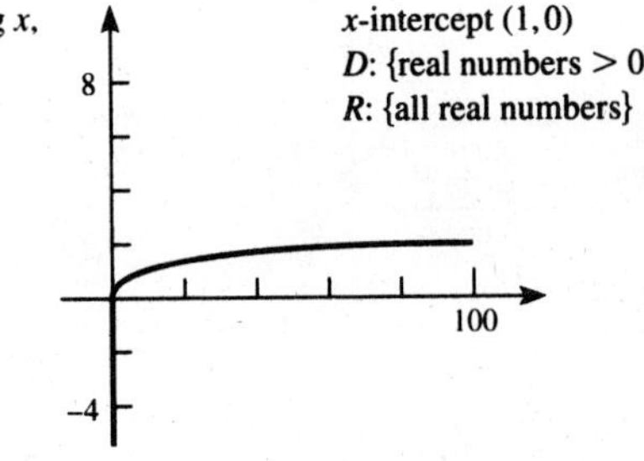

x-intercept (1, 0)
D: {real numbers > 0}
R: {all real numbers}

35. $y = 4 + \log x$,

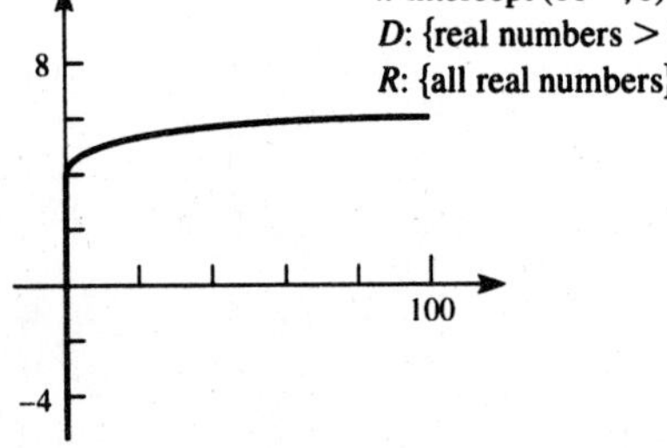

x-intercept $(10^{-4}, 0)$
D: {real numbers > 0}
R: {all real numbers}

37. $y = 2 - \log x$

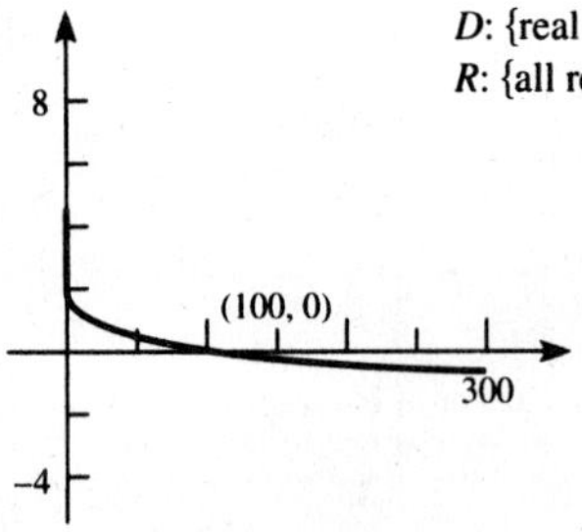

D: {real numbers > 0}
R: {all real numbers}

39. $y = \log(x - 1)$

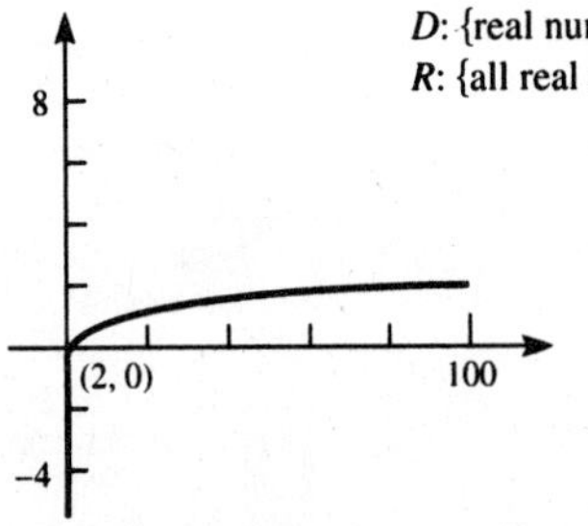

D: {real numbers > 1}
R: {all real numbers}

41. $y = \log(3 - x)$

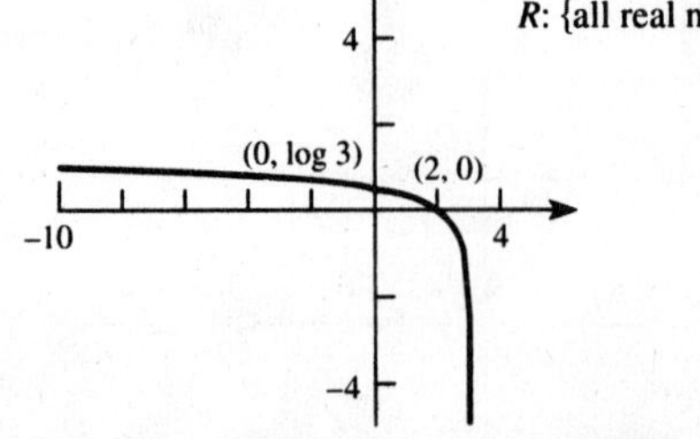

D: {real numbers < 3}
R: {all real numbers}

43. $y = 1 - 2\log x$

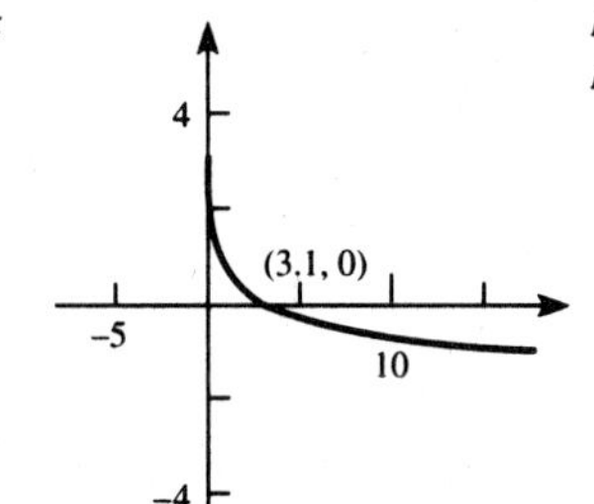

D: {real numbers > 0}
R: {all real numbers}

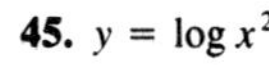

45. $y = \log x^2$

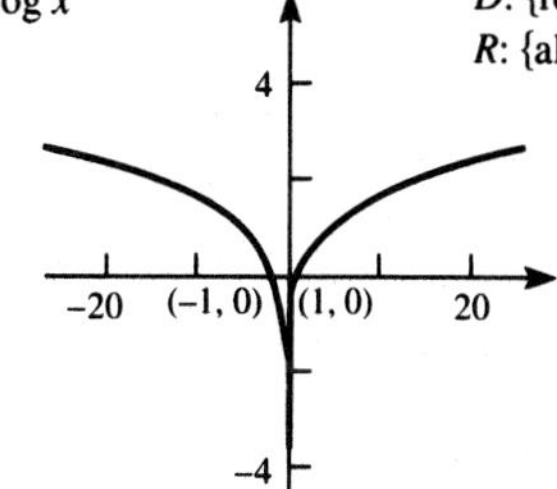

D: {real numbers ≠ 0}
R: {all real numbers}

47. $y = \log \sqrt{x}$

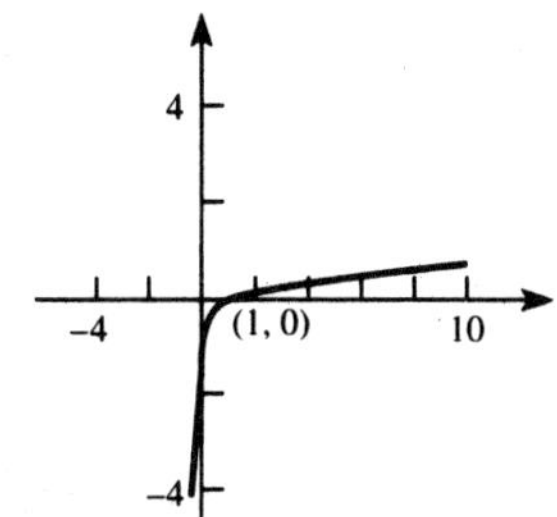

D: {real numbers > 0}
R: {all real numbers}

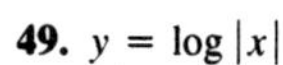

49. $y = \log |x|$

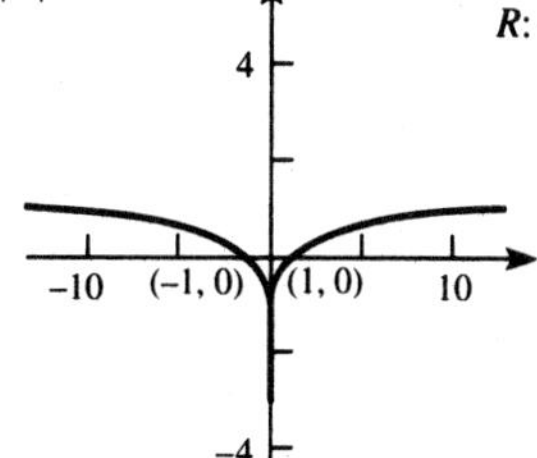

D: {real numbers ≠ 0}
R: {all real numbers}

51. $0.01 < x < 0.1$ **53.** $1{,}000 \le x \le 10{,}000$ **55.** 1 to any power is 1 **57.** AT **59.** ST **61.** When you begin with $2 > 1$ and multiply both sides by $\log \frac{1}{2}$, you are multiplying by a negative number, and so ">" becomes "<." **63.** arithmetic **65.** Answers vary.
III. 67.(a) 65 decibels **(b)** 107 decibels **69.** The second is 31.6 times as strong as the first. **71.(a)** 12.77 magnitudes **(b)** 20.54 magnitudes **73.(a)** full moon **(b)** The moon is 2,291 times as bright as Venus. **IV. 75.** Answers vary. $(x > e)$
77. 1 **79.** 2 **81.** 1 **83.** Answers vary

9.3 (page 598)

I. 1. 2.3665 **3.** 0.0450 **5.** 10.4552 **7.** 73.7941 **9.** -0.2467 **11.** ± 31.6228 **13.** $-1 \pm \sqrt{11}$ **15.** no solution **17.** 3.5609 **19.** no solution **21.** 2.3666 **23.** 0.6309 **25.** 26.4340 **27.** 22.7566 **29.** 1.3863 **31.** ± 1.2686 **33.** 2.7095 **35.** 4.2132 **37.** 1 **39.** 8.7720, 0.2280 **41.** 0 **43.** any real number **45.** $t = \frac{L}{R}\ln\left(1 - \frac{IR}{E}\right)$

II. 47. $3^{\frac{1}{5}} = e^{0.2197}$ **49.** 300^{301} **51.** n^{n+1} **53.** $d = 100c$ so $\log d = \log 100c = \log 100 + \log c = 2 + \log c$
III. 55.(a) 22.5 years **(b)** 14.3 years **(c)** 9.7 years **(d)** 7.9 years **57.** 3.5 years **59.** 7 years **61.** 2,119; 2,107
63. 22,280 years **65.(a)** $T(t) = 15 + 350e^{0.31t}$

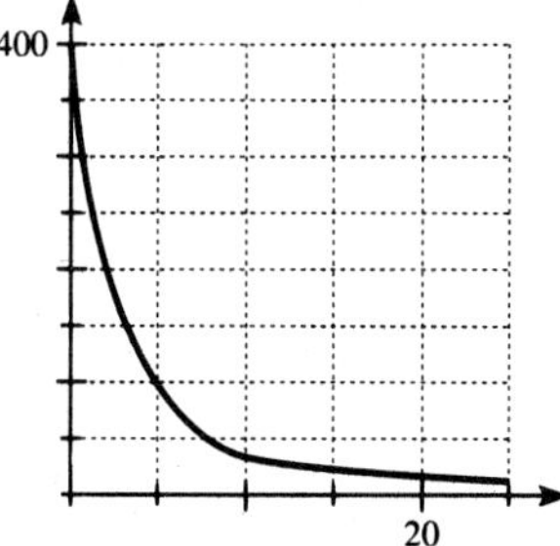

(b) 153°C **(c)** 30.77°C

(d) after 13.70 hours **(e)** never **67.(a)** $F(d) = 53e^{0.0931d}$
(b) Given a frequency, F, note $d = \dfrac{\ln \frac{F}{53}}{0.0931}$ **(c)** Answers vary **(d)** Answers vary **69.** 227,832 digits **71.(a)** 0.0305
(b) .1985 **(c)** 0.0990 **73.** 15.8 years **75.** 3,762 years **IV. 77.** $r = 5\%$ to 20% is accurate to within approximately 0.2 years **79.** If $b^x = a^{kx}$ then $x \ln b = kx \ln a$ or $k = \dfrac{\ln b}{\ln d}$

CHAPTER 9 SUMMARY REVIEW

I. 1. 7 **3.** 64 **5.** 4 **7.** 10 **9.** 147.03 **11.** 2.6362 **13.** -0.5558 **15.** $y = 3^x - 2$

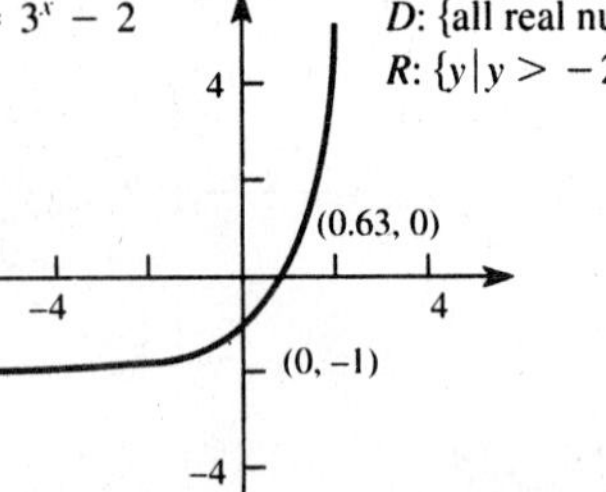

D: {all real numbers}
R: $\{y \mid y > -2\}$

17. $y = \log x - 1$

D: {real numbers > 0}
R: {all real numbers}

19. $y = \ln(x - 3)$

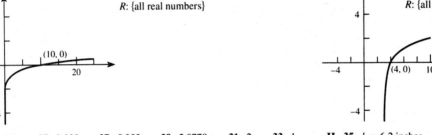

D: {real numbers > 3}
R: {all real numbers}

21. 1.3357 **23.** 10.5740 **25.** 3.322 **27.** 5.355 **29.** 2.0779 **31.** 2 **33.** 1 **II. 35.** $l = 6.2$ inches, $w = 0.144$ inches **37.** $f(x) = 2^x$ **39.** $f(x) = kx$ **41.** yes **III. 43.** yes **45.(a)** 142,000 **(b)** 2,001 **47.(a)** 0.0811 **(b)** -0.0510 **49.(a)** $P = P_0e^{0.0049t}$ **(b)** 180 bacteria **(c)** 141 minutes **51.(a)** ≈43,869,000,000 (in 1992) **(b)** Answers vary **53.(a)** 8 **(b)** 500 years ago **(c)** 581 years ago **(d)** intermarriage **55.** $k = 1.114$

CUMULATIVE REVIEW: CHAPTERS 1–9

I. 1. $5x - 11$ **3.** $6x^2y^{\frac{7}{4}}$ **5.** $2a^5b^2c$ **7.** $\sqrt[5]{y}$ **9.** $\frac{x-3}{2x-1}$ **11.** $\frac{2-x^2}{x^2-2x}$ **13.** $x = 3$ **15.** $x = 6.531$ or -1.531 **17.** $x = 3, 1$ **19.** $x \le \frac{2}{3}$ **21.** $(-3, 3)$ **23.** D: {all real numbers} R: $\{y \mid y \ge 4\}$ **25.** D: {all real numbers} R: $\{y \mid y \ge -2\}$ **27.** D: {real numbers > 2} R: {all real numbers} **II. 29.(a)** Check that the distances between all three points are the same. **(b)** Check that the slopes of two of the lines between the points are negative reciprocals of each other. **31.(a)** Answers vary: $f(x) = 1 + 3x^2$ **(b)** Answers vary: $f(x) = kx$ **33.** $y = |f(x)|$

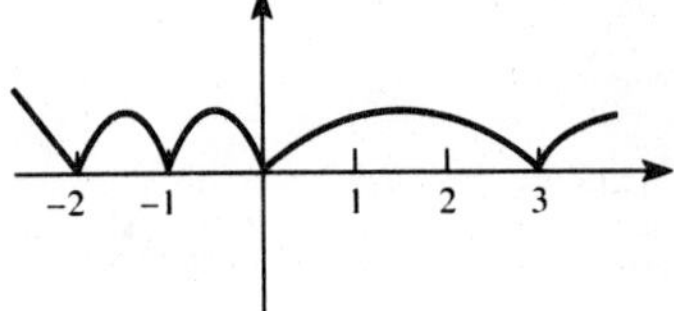

35.

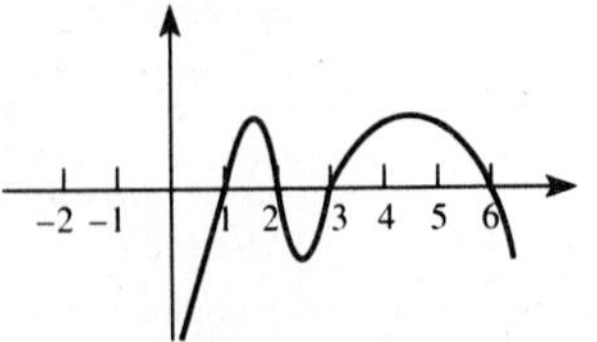

37. Answers vary. **39.** (e) **41.** (b) **43.** (b) **45.** (a) **47.** false **49.** false

51. true **53.** true **55.** $c = 9$ **57.** 10 **III. 59.** 30,000 student tickets **61.** Sam drives 60 mph while Pat drives 50 mph

63.(a) 72 feet **(b)** 7 seconds **(c)** 3.5 seconds, 147 feet **65.** The width is 25 feet and the length is 30 feet. **67.(a)** 1.6 minutes **(b)** 21 symbols **(c)** $T(N) = 0.2\,N^{\frac{3}{4}}$

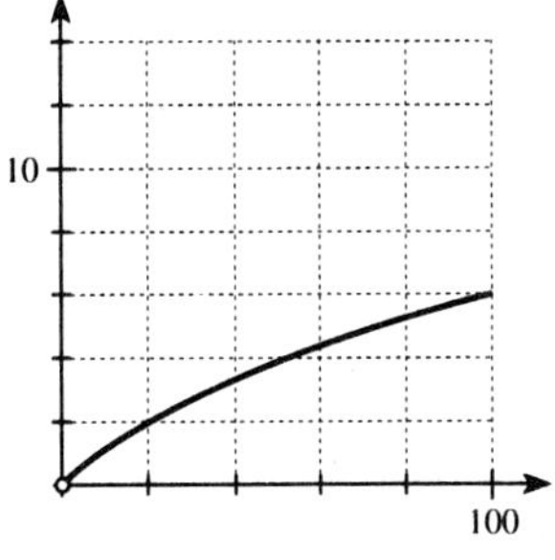

Answers vary.

69. The largest area is 810,000 m^2, and the length and width are both 900 meters. **71.(a)** $\frac{1}{2}, \frac{1}{3}, \frac{1}{4}, \frac{1}{5}, \frac{1}{6}$ **(b)** $\frac{1}{7}$ and $\frac{1}{8}$ **(c)** $\frac{1}{11}, \frac{1}{51}, \frac{1}{N+1}$ **(d)** row 99 **(e)** $f(n) = \frac{1}{n+1}$ **(f)** row 200

Index

GEOMETRIC FORMULAS

Right triangle	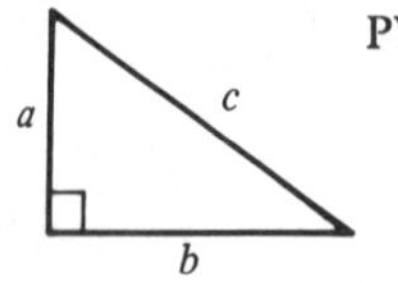 PYTHAGOREAN THEOREM $a^2 + b^2 = c^2$	
	PERIMETER	AREA
Triangle	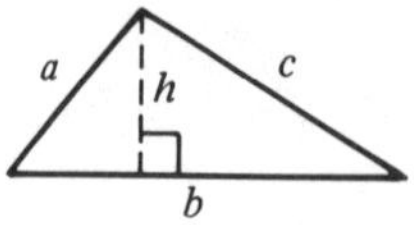 $a + b + c$	$\frac{1}{2}bh$
Rectangle	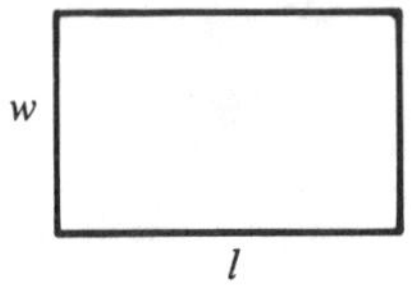 $2w + 2l$	lw
Square	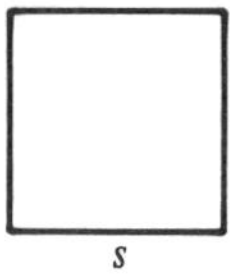 $4s$	s^2
	CIRCUMFERENCE	AREA
Circle	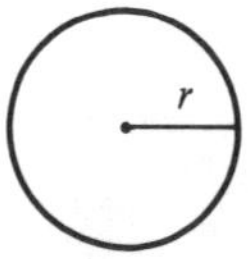 $2\pi r$	πr^2
	SURFACE AREA	VOLUME
Cube	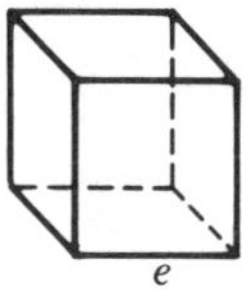 $6e^2$	e^3
Rectangular prism	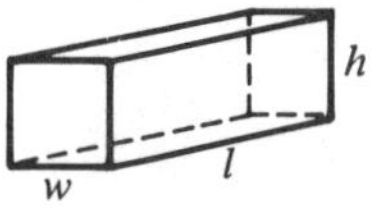 $2wh + 2lh + 2lw$	hlw
Cylinder	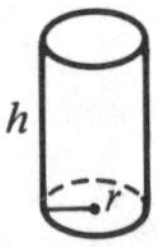 $2\pi r^2 + 2\pi rh$	$\pi r^2 h$
Cone	 $\pi rs + \pi r^2$	$\frac{1}{3}\pi r^2 h$
Sphere	 $4\pi r^2$	$\frac{4}{3}\pi r^3$